Dependence Modeling with Copulas

MONOGRAPHS ON STATISTICS AND APPLIED PROBABILITY

General Editors

F. Bunea, V. Isham, N. Keiding, T. Louis, R. L. Smith, and H. Tong

Monographs on Statistics and Applied Probability 134

Dependence Modeling with Copulas

Harry Joe

University of British Columbia

Vancouver, Canada

CRC Press
Taylor & Francis Group
Boca Raton London New York

CRC Press is an imprint of the
Taylor & Francis Group, an **Informa** business

A CHAPMAN & HALL BOOK

CRC Press
Taylor & Francis Group
6000 Broken Sound Parkway NW, Suite 300
Boca Raton, FL 33487-2742

First issued in paperback 2022

Version Date: 20140523

ISBN 13: 978-1-03-247737-4 (pbk)
ISBN 13: 978-1-4665-8322-1 (hbk)

DOI: 10.1201/b17116

Library of Congress Cataloging-in-Publication Data

Joe, Harry.
  Dependence modelling with copulas / Harry Joe.
      pages cm. -- (Monographs on statistics and applied probability ; 133)
  "A CRC title."
  Includes bibliographical references and index.
  ISBN 978-1-4665-8322-1 (hardcover : alk. paper) 1. Copulas (Mathematical statistics) 2. Dependence (Statistics) 3. Probabilities. I. Title.

  QA273.6.J57 2014
  519.5'35--dc23   2014018932

Visit the Taylor & Francis Web site at
http://www.taylorandfrancis.com

and the CRC Press Web site at
http://www.crcpress.com

Contents

Preface

This book is devoted to dependence modeling with copulas. It is not a revision of *Multivariate Models and Dependence Concepts* (Joe, 1997); the few chapters in the previous book on copula construction, copula families and applications have been substantially expanded to account for much research on copula theory, inference and applications since 1997. While the previous book used tail dependence as a way to distinguish parametric copula families, this book compares copula families and constructions with more kinds of dependence structures and tail properties, and indicates how these properties affect tail inferences such as joint tail probabilities, tail conditional expectations, etc.

Between 1997 and 2012, there have been a number of books that are devoted to copula theory and related topics.

1. Nelsen (1999) and its second edition Nelsen (2006): these are introductory books to copula theory and the material is mainly bivariate; statistical inference and multivariate copula families useful for statistical modeling are not covered.
2. Balakrishnan and Lai (2009): this consists of only bivariate distributions; it has a limited selection of parametric copula families, but a fairly complete presentation of continuous bivariate distributions with commonly used univariate margins.
3. McNeil et al. (2005): this has variations of elliptical and t copula families, as well as some new multivariate constructions, and applications to quantitative finance and insurance.
4. Kurowicka and Cooke (2006): this has graphical models including vines and belief nets.
5. Kurowicka and Joe (2011): this is devoted to the vine copula or pair-copula construction, with properties, inference and applications.
6. Mai and Scherer (2012a): this concentrates on stochastic models and simulation for some copula familes with stochastic representations.
7. Others with copulas and applications to finance, insurance, or environment are Cherubini et al. (2004), Denuit et al. (2005), Salvadori et al. (2007), and Cherubini et al. (2012).

This book contains much recent research on dependence modeling with copulas that is not covered in any of the above-listed books. The vine copula construction has been a big advance for copula modeling with high-dimensional data. The class of vine copula models have the feature of construction from a sequence of bivariate copulas: some applied to pairs of univariate margins and others applied to pairs of univariate conditional distributions. Generalizations of vine copula models are covered, including versions of common and structured factor models that extend from the Gaussian assumption to copulas, by making use of parametrization via partial correlations for Gaussian correlation matrices. Also included are other multivariate constructions and parametric bivariate copula families that have different tail properties, and much more on dependence and tail properties to assist in copula choice for statistical modeling.

For high-dimensional copula applications, numerical methods and algorithms for inference and simulation are important. The increased speed of computing since 1997 has much impact on the kinds of copula models that can now be considered for applications. Almost all of the computations in Joe (1997) were done in the C programming language. The availability of the open source software R (http://www.r-project.org), that links a simpler user interface and code that can be written in C/C++ or Fortran90, provides an easier way to develop implementations of software for high-dimensional copula models.

This book has ordered chapters but there is no natural ordering for the study and use of copulas. Hence there is cross-referencing of later chapters in the earlier chapters. Some examples and content of this book were influenced by much correspondence that I received from *Multivariate Models and Dependence Concepts*. A brief summary of the contents and purposes of the chapters is mentioned here.

1. An introduction to some history of non-Gaussian multivariate distributions and some construction methods of multivariate distributions leading to copulas. There are some data examples that (i) show via plots when classical multivariate statistics can be inappropriate and (ii) show some steps in likelihood inference with copulas.
2. Basic material on dependence, tail behavior and asymmetries for multivariate statistics beyond Gaussian.
3. Copula construction methods covering many dependence structures.
4. A listing of parametric copula families together with dependence and tail properties. This chapter is intended as a reference in the style of the books by Johnson and Kotz.
5. Inference, diagnostics and model selection for copula applications, mostly covering parametric models with likelihood inference.
6. Numerical methods and algorithms for copula applications. The algorithms are presented as pseudocode for those who may want to implement them for learning and applications.
7. A variety of applications of dependence modeling with copulas, covering initial data analysis, model comparisons and inference.
8. Theorems for properties of copulas, with more theoretical details and advanced examples. This chapter is intended as a reference for results to determine dependence and tail properties of multivariate distributions for future constructions of copula models.

Due to time and space constraints, this book does not cover all research topics concerning copulas. The viewpoint of copulas in this book is statistical and largely motivated from applications; some topics that weren't needed for the applications are not included.

Acknowledgements

- Technische Universität München for a John von Neumann Visiting Professorship in 2011 and the opportunity to teach a copula course that formed the beginnings of this book.
- The organizers of vine copula workshops: Roger Cooke, Dorota Kurowicka, Kjersti Aas, Claudia Czado, Haijun Li.
- My Ph.D. students and research collaborators in the past few years, who have shared in the development of new theory and high-dimensional copula models.
- BIRS (Banff International Research Station for Mathematical Innovation and Discovery) for the opportunity of holding two stimulating workshops for which I was a co-organizer: "Composite Likelihood Methods" in April 2012, "Non-Gaussian Multivariate Statistical Models and their Applications" in May 2013.
- The University of British Columbia and its Department of Statistics for a sabbatical leave in the 2012–2013 academic year during which time most of this book was written.
- The Natural Sciences and Engineering Research Council of Canada for research support with a Discovery Grant.
- Three anonymous referees of a 90%-complete book draft before the proofreading stage.
- GNU-Linux and Open Source software; the computer implementations for this book made ample use of open source software tools: GNU/Unix utlities, GNU C, GNU Fortran, R, Perl, LaTeX, and the VIM editor with graphical interface and customized menus.

Information on where to download software and code for the computations in this book will be available at the author's web page of publications.

Harry Joe, Vancouver, BC, Canada.

Notation and abbreviations

Notation

Overloading of notation cannot be avoided with the wide range of topics in this book. The list below indicates the most common usage of notation. Any letter not mentioned has multiple uses.

d for multivariate dimension or number of variables.

n for sample size.

F for cumulative distribution function

G, H, K sometimes for cumulative distribution functions.

$\overline{F}$ for survival function,

$\overline{G}$ sometimes for survival function.

f for density function or probability mass function.

F^{-1} for the (left-continuous) functional inverse of univariate cumulative distribution function, that is, $F^{-1}(p) = \inf\{x : F(x) \geq p\}$ for $0 < p < 1$.

g for density sometimes, more often it is a generic function.

C for copula or multivariate distribution with Uniform$(0,1)$ margins.

C^+ for comonotonicity copula.

$C^{\perp}$ for independence copula.

C^- for bivariate countermonotonicity copula.

c for copula density.

$\widehat{C}$ for the survival/reflected copula.

$\widehat{C}_n$ for the empirical copula with sample size n.

F_S, C_S for marginal distributions when S is a non-empty subset of indices; if S is a sequence $j_1, \ldots, j_2$, the marginal distribution might be written as $F_{j_1:j_2}$.

$F_{S_1|S_2}, C_{S_1|S_2}$ for conditional distributions when S_1, S_2 are subsets of indices.

$f_{S_1|S_2}$ for conditional density.

$C_{jk;S}$ for a bivariate copula applied to conditional distributions $F_{j|S}, F_{k|S}$.

$c_{jk;S}$ for a bivariate copula density of $C_{jk;S}$.

$C_{j|k;S}$ and $C_{k|j;S}$ for the conditional distributions of $C_{jk;S}$.

$\mathbb{P}$ for probability (of a set/event).

$\mathbb{E}$ for expectation.

Var for variance, Cov for covariance and Cor for correlation.

$\mathbb{R}$ for reals, $\mathbb{R}^+$ for non-negative reals.

$\emptyset$ for the *empty set*.

$\perp\!\!\!\perp$ for *independent* random variables or vectors.

ϕ, ϕ_d for standard Gaussian/normal and multivariate Gaussian density.

Φ, Φ_d for standard Gaussian/normal and multivariate Gaussian cumulative distribution function.

$N(\mu, \sigma^2)$ and $N_d(\boldsymbol{\mu}, \boldsymbol{\Sigma})$ for univariate and d-variate Gaussian/normal.

t_ν for the univariate or multivariate t distribution.

$t_{d,\nu}(\cdot; \boldsymbol{\Sigma})$ and $T_{d,\nu}(\cdot; \boldsymbol{\Sigma})$ for the d-variate t density and cumulative distribution function respectively.

t_ν and T_ν sometimes for the univariate t density and cumulative distribution function respectively.

$U(0, 1)$ for Uniform on interval $(0, 1)$.

$\phi_j, \phi_{j\ell}$ for regression coefficents in stochastic representation of multivariate Gaussian distributions parametrized as partial correlation vines; special examples are autoregressive models.

ψ, φ for Laplace(-Stieltjes) transforms.

ψ^{-1} for inverse of Laplace transform.

ψ_j^2 for variance of the uniqueness component in factor analysis and partial correlation vines.

θ, δ most commonly for dependence parameters of copula families or parameters of Laplace transform families.

$\rho_{jk;S}$ for the partial correlation of variables j, k given those indexed in set S.

ρ_e for correlation or partial correlation on an edge of a partial correlation vine.

ρ_S for Spearman's rank correlation.

ρ_N for correlation of normal scores.

ρ_N^-, ρ_N^+ for the lower and upper semi-correlations.

τ for Kendall's tau.

β for Blomqvist's beta, and sometimes parameter or regression coefficient.

λ for tail dependence coefficient.

α for regular variation index, sometimes a parameter and sometimes one minus the confidence level.

ξ for extreme value tail index, sometimes another parameter.

κ for tail order or reciprocal of residual dependence index.

ν most often for positive-valued (degree of freedom) parameter of univariate or multivariate t distribution; also for a measure in integrals to combine the discrete case (ν=counting measure, where an integral can be converted to a sum) and the continuous case (ν=Lebesgue measure).

σ for scale parameter.

π sometimes for a parameter in $[0, 1]$.

$\mathcal{F}$ for the Fréchet class of multivariate distributions with given lower-dimensional margins, for example, $\mathcal{F}(F_1, \ldots, F_d)$ and $\mathcal{F}(F_1, F_{12}, F_{13})$.

$\mathcal{L}_\infty$ for the class of Laplace transforms of positive random variables.

$\mathcal{L}_d$ for the class of Williamson d-monotone functions that map from $[0, \infty)$ onto $[0, 1]$.

$\mathcal{W}_d$ for Williamson d-transform.

$\mathcal{L}_\infty^*$ for the class of infinite differentiable monotone increasing functions ω such that $\omega(0) = 0$, $\omega(\infty) = \infty$ and derivatives alternate in sign. More generally, for $n = 1, 2, \ldots, \infty$.

$$\mathcal{L}_n^* = \{\omega : [0, \infty) \to [0, \infty) \mid \omega(0) = 0, \ \omega(\infty) = \infty, \ (-1)^{j-1}\omega^{(j)} \geq 0, \ j = 1, \ldots, n\},$$

$\mathcal{V}$ for vine.

$\mathcal{T}$ for trees of a vine.

$\mathcal{E}$ for edges of a vine.

$\prec_{\mathrm{c}}$ for concordance ordering.

$\prec_{\mathrm{SI}}$ for stochastic increasing dependence ordering.

$|S|$ for cardinality of a finite set S.

A^c for complement of a set or event A.

$\top$ as a superscript indicates the *transpose* of a vector or matrix, for example, $\boldsymbol{A}^\top$.

$\mathrm{D}_j h$ is a differential operator notation for the first order partial derivative with respect to the jth argument of h.

$\mathrm{D}_{jk}^2 h$ is a differential operator notation for a second order partial derivative of h.

$:=$ or $=:$ is the symbol to define a variable or function on left or right.

$\sim$ is the symbol for *distributed as.*

$\stackrel{d}{=}$ is the symbol for *equality in distribution* or *stochastic equality.*

$\to_d$ is the symbol for *convergence in distribution or law.*

$\stackrel{.}{\sim}$ is the symbol for "asymptotically distribution as."

$\sim$ is also the symbol for asymptotically equivalent.

$:$ is the symbol used for a sequence of integers, e.g., $\boldsymbol{u}_{1:d} = (u_1, \ldots, u_d)$.

Conventions and abbreviations

Below are some abbreviations, as well as explanation of some terminology and fonts.

Vectors when used for vector/matrix operations are assumed to be column vectors. Whether a vector is a row or column vector should be clear from the context. Vectors are boldfaced and the components are subscripted. For example, $\boldsymbol{a}$ is a vector with elements a_j.

The words "non-increasing" and "non-decreasing" are not used; instead "increasing," "strictly increasing", "decreasing" and "strictly decreasing" are used.

i.i.d.: *independent and identically distributed.*

cdf: *cumulative distribution function.*

pdf: *probability density function.*

pmf: *probability mass function.*

MDA: *maximum domain of attraction.*

GEV: *generalized extreme value* (univariate) distribution.

LT: *Laplace transform.*

RV: regular variation (applied to functions at 0 or ∞).

ML and MLE: *maximum likelihood*, and *maximum likelihood estimate* or *estimation.*

IFM: *inference functions for margins.*

SE: *standard error.*

AIC: *Akaike information criterion.*

BIC: *Bayesian information criterion.*

VaR: *Value at Risk.*

CTE: *conditional tail expectation.*

As a carryover from Joe (1997), there is some use of the abbreviated names of parametric families of Laplace transforms and copulas, such as LT families A–M and copula families B1–B8, BB1-BB10, MM1–MM3, MM8.

Chapter 1

Introduction

This chapter introduces some terminology for multivariate non-Gaussian models. Examples are developed to show the basic steps of copula construction, and data analyis and inference for dependence modeling with copulas. These form a preview of the theory, inference and data analysis in subsequent chapters.

The main purposes are the following:

- introduce some terminology for dependence modeling (Section 1.1);
- show some history of the development of multivariate distributions with given margins and how univariate properties are used to construct multivariate counterparts;
- introduce mixtures and conditional independence as construction methods for multivariate distributions;
- show the probability integral transforms to go from a multivariate distribution to a copula;
- show the extreme value limit as another construction method (Section 1.2 for this and the three preceding items);
- introduce the basic theorem on copulas (Section 1.3);
- present some data examples where classical multivariate normal analysis is not appropriate;
- illustrate tail dependence and tail asymmetry through bivariate normal score plots (Section 1.4 for this and the preceding item);
- introduce the typical steps of likelihood inference and model comparisons for parametric copula models (Section 1.5);
- mention alternative approaches for dependence modeling (Section 1.6);
- indicate how copulas have increased in popularity (Section 1.7).

1.1 Dependence modeling

Suppose there are d variables $y_1, \ldots, y_d$, and the data set consists of $\boldsymbol{y}_i = (y_{i1}, \ldots, y_{id})$ for $i = 1, \ldots, n$, considered as a random sample of size n; that is, the $\boldsymbol{y}_i$ are independent and identically distributed (i.i.d.) realizations of a random vector $(Y_1, \ldots, Y_d)$.

For dependence modeling with copulas, the steps are:

(i) univariate models for each of the variables $y_1, \ldots, y_d$;
(ii) copula models for the dependence of the d variables.

More generally, (i) could consist of d univariate regression models with covariates, or d innovation variables of dependent time series.

Examples are two loss variables in insurance, d asset returns in finance, daily maximum wind speeds at d locations, d quantitative traits, d ordinal item responses in a psychological instrument, tumor counts in family clusters, and longitudinal ordinal response.

Choices for univariate parametric families in step (i) with two or more parameters include the following depending on modality, tailweight, and asymmetry:

- unimodal and symmetric, support $(-\infty, \infty)$: Gaussian, t;
- unimodal and asymmetric, support $(-\infty, \infty)$: skew-normal, skew-t;
- unimodal and support on $(0, \infty)$, exponential tail: gamma, Weibull with shape parameter ≥ 1;
- unimodal and support on $(0, \infty)$, subexponential tail: lognormal, Weibull with shape parameter < 1;
- unimodal and support on $(0, \infty)$, heavy tail: Pareto, Burr, generalized beta of the second kind (four parameters, with Burr as special case);
- maxima or minima: 3-parameter generalized extreme value (GEV);
- unimodal and support on non-negative integers: negative binomial, generalized Poisson (with Poisson as a boundary case);
- bimodal and multimodal: finite mixture.

See the Appendix of Klugman et al. (2010) for parametric univariate families useful in insurance. In the above, F is *subexponential* if for X_1, X_2 i.i.d. F with survival function $\overline{F} = 1 - F$, then $\overline{F}^{(2*)}(x) := \mathbb{P}(X_1 + X_2 > x) \sim 2\overline{F}(x)$ as $x \to \infty$. Some examples are the lognormal density with $f(x) = (2\pi)^{-1/2} x^{-1} \sigma^{-1} \exp\{-\frac{1}{2}(\log x - \mu)/\sigma^2\}$ and the Weibull density with $f(x) = \sigma^{-1}(x/\sigma)^{\zeta-1} \exp\{-(x/\sigma)^\zeta\}$ and $0 < \zeta < 1$. The density decreases exponentially as $x \to \infty$, but more slowly than e^{-ax} $(a > 0)$. The tailweight is related to the rate of decrease of of the density as $x \to \pm\infty$; for example, the rate is $|x|^{-\nu-1}$ for the t_ν density and this inverse polynomial rate is slower than subexponential.

After univariate models are chosen, copula models can be considered for the dependence. Parametric copula families can be constructed to satisfy

(a) different types of dependence structures (e.g., exchangeable, conditional independence, flexible positive dependence, flexible with a range of positive and negative dependence) and

(b) different types of tail behavior.

The difficult step in copula construction is the extension from bivariate to multivariate to get flexible dependence. In risk analysis for insurance and finance, the tail properties of univariate and dependence models have an influence on the probability of losses being simultaneously large.

The choice of parametric copula families is similar to the choice of parameter univariate families, in that one considers the shape and tail behavior of the density. For data, it is easier to assess the shape and tail of the density from pairwise normal scores plots (where each variable has been transformed to normality). Simple copula families have unimodal densities with standard Gaussian margins (but not Uniform $U(0, 1)$ margins). There can be different symmetries and asymmetries from copula families with one to three parameters. One-parameter copula families that include independence and perfect dependence mostly are symmetric with respect to $(u_1, u_2) \to (u_1', u_2') = (u_2, u_1)$ in the density (that is, permutation symmetry), and some of these one-parameter copula families are also symmetric with respect to $(u_1, u_2) \to (u_1', u_2') = (1 - u_1, 1 - u_2)$ in the density (this property will be referred to as reflection symmetry).

Examples of copula families with and without reflection symmetry are given in Section 1.2 and some contour plots of the copula densities combined with standard normal $N(0, 1)$ margins are shown in Section 1.4.

1.2   Early research for multivariate non-Gaussian

The earliest books for parametric classes of multivariate distributions are the following.

- Mardia (1970), *Families of Bivariate Distributions*: includes translation method (Johnson's transformation); bivariate binomial, hypergeometric, Poisson, negative binomial, logarithmic, Gaussian, Cauchy, beta, gamma, Pareto, t, F. It also has arbitrary margins for convex combination of independence, comonotonicity and countermonotonicity; Plackett; Morgenstern.

- Johnson and Kotz (1972), *Continuous Multivariate Distributions*: includes multivariate Gaussian, t, beta, gamma, F, extreme value, exponential, Pareto, Burr, logistic.
- Hutchinson and Lai (1990), *Continuous Bivariate Distributions, Emphasising Applications*: unlike the previous two, this includes copula families (but bivariate only).

A standard approach to go from univariate to multivariate parametric family is to generalize a property of the univariate distribution to get the multivariate. This is shown through some examples.

Example 1.1 (Gaussian). Linear combinations of independent univariate Gaussian random variables are Gaussian, and this property can be used to define multivariate Gaussian. Let $Z_1, \ldots, Z_d$ be independent Gaussian (with arbitrary means and variances), then $X_j = a_{j1}Z_1 + \cdots + a_{jd}Z_d$ is univariate Gaussian for $j = 1, 2, \ldots$ when $a_{j1}, \ldots, a_{jd}$ are real constants, and $(X_1, X_2, \ldots, X_d, \ldots)$ is multivariate Gaussian.

Let $\boldsymbol{X} = \begin{pmatrix} X_1 \\ \vdots \\ X_d \end{pmatrix} = \boldsymbol{A} \begin{pmatrix} Z_1 \\ \vdots \\ Z_d \end{pmatrix}$ be constructed based on the $d \times d$ matrix $\boldsymbol{A}$ where $Z_1, \ldots, Z_d$ are i.i.d. N$(0, 1)$ random variables. Then $\boldsymbol{X} + \boldsymbol{\mu}$ is multivariate Gaussian with mean vector $\boldsymbol{\mu}$ and covariance matrix $\boldsymbol{\Sigma} = \boldsymbol{A}\boldsymbol{A}^\top$.

From the Central Limit Theorem, the Gaussian distribution arises as a limit of a scaled sum of weakly dependent random variables with no variable dominating. Quantitative traits are examples of variables which might be justified as approximate Gaussian based on the polygenic model. In general, multivariate Gaussian is a mathematically convenient assumption.

Example 1.2 (Extension to multivariate t with parameter $\nu > 0$).

A stochastic representation is $X = Z/\sqrt{W/\nu} \sim t_\nu$ if $Z \sim$ N$(0, 1)$ and $W \sim \chi_\nu^2$ with $Z \perp\!\!\!\perp W$. A multivariate extension is $(X_1, \ldots, X_d) = (Z_1, \ldots, Z_d)/\sqrt{W/\nu}$, where $(Z_1, \ldots, Z_d)$ is multivariate Gaussian with mean $\boldsymbol{0}$ and covariance/correlation matrix $\boldsymbol{\Sigma}$, and $W \sim \chi_\nu^2$ with $W \perp\!\!\!\perp (Z_1, \ldots, Z_d)$.

Let $R = 1/\sqrt{W/\nu}$, so that $\boldsymbol{X} = R\boldsymbol{Z}$ can be interpreted as a scaled mixture of multivariate Gaussian; each $X_j = RZ_j$ has a univariate t_ν distribution. Note that $\nu > 0$ can be a non-integer value by treating χ_ν^2 as Gamma$(\frac{1}{2}\nu, \text{scale}=2)$. The context of a scale mixture to get heavier tails is relevant to applications where multivariate t is used as a model. The derivation of density $f_{\boldsymbol{X}}$ is given in Section 2.7.

Example 1.3 (Pareto survival function as Gamma mixture of Exponential).

Let $[X|Q = q] \sim$ Exponential(rate $= q/\sigma$), and $Q \sim$ Gamma$(\alpha, \text{scale} = 1)$, where $\alpha > 0$, $\sigma > 0$ or $X = Z/Q$ where $Z \sim$ Exponential(mean $= \sigma$) independent of Q. Then for $x > 0$,

$$\mathbb{P}(X > x) = \overline{F}(x; \alpha, \sigma) = \int_0^\infty \mathbb{P}(X > x|Q = q) f_Q(q) \, \mathrm{d}q$$

$$= \int_0^\infty e^{-qx/\sigma} \cdot \frac{q^{\alpha-1}}{\Gamma(\alpha)} e^{-q} \mathrm{d}q = (1 + x/\sigma)^{-\alpha}.$$

This is the Pareto(α, σ) distribution; α is the shape or tail parameter and σ is the scale parameter. As α increases, the tail becomes lighter. The cdf is

$$F_X(x; \alpha, \sigma) = 1 - (1 + x/\sigma)^{-\alpha}, \quad x \geq 0,$$

and the density is

$$f_X(x; \alpha, \sigma) = F'_X(x; \alpha, \sigma) = \sigma^{-1}\alpha(1 + x/\sigma)^{-\alpha-1}, \quad x > 0.$$

In applications where the Pareto distribution is used as a model, the exponential distribution might be reasonable with a homogeneous subpopulation but the rate parameter varies over the population.

An extension to multivariate Pareto is given below; the distribution is given in Mardia (1962), but not derived as gamma mixture.

Let $Z_1, \ldots, Z_d$ independent Exponential random variables with respective means $\sigma_1, \ldots, \sigma_d$, and let $Q \sim \text{Gamma}(\alpha, 1)$ independent of the Z_j's. The stochastic representation for the multivariate extension is:

$$(X_1, \ldots, X_d) = (Z_1, \ldots, Z_d)/Q = (Z_1/Q, \ldots, Z_d/Q).$$

The dependence comes from a common latent variable Q; this is also an example of a conditional independence model (given a latent variable). From the above, $X_j \sim \text{Pareto}(\alpha, \sigma_j)$. For $x_1, \ldots, x_d > 0$, the joint survival function is:

$$\overline{F}_{X_1,\ldots,X_d}(\boldsymbol{x}) = \mathbb{P}(X_1 > x_1, \ldots, X_d > x_d) = \int_0^\infty \mathbb{P}(Z_1 > x_1 q, \ldots, Z_d > x_d q) f_Q(q)\, \mathrm{d}q$$

$$= \int_0^\infty \left\{\prod_{j=1}^d \mathbb{P}(Z_j > x_j q)\right\} f_Q(q)\, \mathrm{d}q$$

$$= \int_0^\infty \exp\left\{-q \sum_{j=1}^d x_j/\sigma_j\right\} \cdot \frac{q^{\alpha-1}}{\Gamma(\alpha)} e^{-q} \mathrm{d}q = \left[1 + \sum_{j=1}^d x_j/\sigma_j\right]^{-\alpha}. \qquad (1.1)$$

For $d = 2$, $\overline{F}_{X_1,X_2}(x_1, x_2) = [1 + x_1/\sigma_1 + x_2/\sigma_2]^{-\alpha}$ with margins $\overline{F}_{X_1}(x_1) = \overline{F}_{X_1,X_2}(x_1, 0) = (1 + x_1/\sigma_1)^{-\alpha}$ for $x_1 > 0$, and $\overline{F}_{X_2}(x_2) = \overline{F}_{X_1,X_2}(0, x_1) = (1 + x_2/\sigma_2)^{-\alpha}$ for $x_2 > 0$.

Example 1.4 (Burr distribution). The extension of the Burr distribution is similar to the preceding example, and is given in Takahasi (1965) as a gamma mixture of Weibull survival functions.

Let $\overline{F}_Z(z; \zeta, \sigma) = \exp\{-(z/\sigma)^\zeta\}$ be a Weibull(ζ, σ) survival function. Let $[X|Q = q] \sim$ Weibull$(\zeta, \sigma/q^{1/\zeta})$, and $Q \sim \text{Gamma}(\alpha, \text{scale} = 1)$, where $\alpha > 0$, $\zeta > 0$, $\sigma > 0$ or $X = Z/Q^{1/\zeta}$ where $Z \sim \text{Weibull}(\zeta, \sigma)$ independent of Q. Then for $x > 0$,

$$\mathbb{P}(X > x) = \overline{F}(x; \alpha, \sigma) = \int_0^\infty \mathbb{P}(Z > q^{1/\zeta} x) f_Q(q)\, \mathrm{d}q$$

$$= \int_0^\infty e^{-q(x/\sigma)^\zeta} \cdot \frac{q^{\alpha-1}}{\Gamma(\alpha)} e^{-q} \mathrm{d}q = [1 + (x/\sigma)^\zeta]^{-\alpha}.$$

The above is the Burr(α, ζ, σ) survival function. A Burr random variable is also derived as a power of a Pareto random variable. Similarly, a Weibull random variable can be derived as a power of an exponential random variable. For the multivariate Burr distribution based on conditionally independent Weibull$(\zeta_j, \sigma_j q^{-1/\zeta_j})$ random variables, the joint survival function is:

$$\mathbb{P}(X_1 > x_1, \ldots, X_d > x_d) = \int_0^\infty \prod_{j=1}^d e^{-q(x_j/\sigma_j)^\zeta_j} \cdot \frac{q^{\alpha-1}}{\Gamma(\alpha)} e^{-q} \mathrm{d}q$$

$$= \left[1 + \sum_{j=1}^d (x_j/\sigma_j)^{\zeta_j}\right]^{-\alpha}, \quad \boldsymbol{x} \in \mathbb{R}_+^d. \qquad (1.2)$$

$\square$

Now that we have some multivariate distributions, we can use the probability integral transform to transform to $U(0,1)$ random variables. Let $X \sim F_X$ be a continuous random variable. Then $F_X(X) \sim U(0,1)$. Also $\overline{F}_X(X) = 1 - F_X(X) \sim U(0,1)$. The proof is as follows: For $0 < u < 1$, $\mathbb{P}(F_X(X) \leq u) = \mathbb{P}(X \leq F_X^{-1}(u)) = F_X \circ F_X^{-1}(u) = u$ (this is obvious if F_X strictly increasing; some technicalities are needed if F_X is increasing but not strictly increasing). Here, $F_X^{-1}(p) = \inf\{x : F_X(x) \geq p\}$ for $0 < p < 1$; F_X^{-1} is the left-continuous functional inverse of F_X.

Example 1.5 (MTCJ copula). As an example of converting a multivariate survival function to a copula, we use Examples 1.3 and 1.4.

Let $(X_1, \ldots, X_d)$ be d-variate Pareto, as in (1.1); let $F_j = F_{X_j}$. Let $U_j = F_j(X_j)$ and $V_j = \overline{F}_j(X_j)$ for $j = 1, \ldots, d$. The jth univariate survival function is $\overline{F}_j(x) = (1+x/\sigma_j)^{-\alpha} = p$, $F_j^{-1}(p) = \sigma_j[(1-p)^{-1/\alpha} - 1]$ for $0 < p < 1$. Then for $0 \leq v_j \leq 1$ $(j = 1, \ldots, d)$,

$$\mathbb{P}(V_1 \leq v_1, \ldots, V_d \leq v_d) = \mathbb{P}(F_j(X_j) \geq 1 - v_j, j = 1, \ldots, d)$$

$$= \mathbb{P}(X_j \geq F_j^{-1}(1 - v_j), j = 1, \ldots, d) = \left[1 + \sum_{j=1}^{d} \sigma_j^{-1} F_j^{-1}(1 - v_j)\right]^{-\alpha}$$

$$= \left[1 + \sum_{j=1}^{d}(v_j^{-1/\alpha} - 1)\right]^{-\alpha} = \left[v_1^{-1/\alpha} + \cdots + v_d^{-1/\alpha} - (d-1)\right]^{-\alpha}.$$

In order that dependence increases as the parameter increases, this is more commonly written as (Section 4.6):

$$C_{\text{MTCJ}}(v; \theta) = [v_1^{-\theta} + \cdots + v_d^{-\theta} - (d-1)]^{-1/\theta}, \quad 0 \leq v_j \leq 1 \ (j = 1, \ldots, d), \ \theta > 0. \quad (1.3)$$

With a similar derivation from (1.2) with F_j that are univariate Burr, the above obtains. (1.3) is an example of a copula function (which we name as MTCJ for Mardia-Takahasi-Cook-Johnson), and the derivation from Pareto or Burr margins shows the invariance of the copula to monotone increasing transforms of random variables. Kimeldorf and Sampson (1975) obtained the bivariate version and Cook and Johnson (1981) obtained the multivariate copula with the probability integral transformation. Clayton (1978) derived the bivariate distribution through a differential equation.

For deriving the bivariate copula from the cdf F_{12}, we use the relation of the bivariate cdf and survival function:

$$\overline{F}_{12}(x_1, x_2) = 1 - F_1(x_1) - F_2(x_2) + F_{12}(x_1, x_2) = [1 + x_1/\sigma_1 + x_2/\sigma_2]^{-1/\alpha},$$
$$F_{12}(x_1, x_2) = -1 + F_1(x_1) + F_2(x_2) + \overline{F}_{12}(x_1, x_2) = 1 - \overline{F}_1(x_1) - \overline{F}_2(x_2) + \overline{F}_{12}(x_1, x_2),$$
$$\overline{F}_1(x_1) = 1 - F_1(x_1), \quad \overline{F}_2(x_2) = 1 - F_2(x_2), \quad x_1, x_2 \geq 0$$

Then from the bivariate Pareto distribution, for $0 \leq u_1, u_2 \leq 1$,

$$\mathbb{P}(U_1 \leq u_1, U_2 \leq u_2) = \mathbb{P}(X_1 \leq F_1^{-1}(u_1), X_2 \leq F_2^{-1}(u_2))$$
$$= 1 - \mathbb{P}(X_1 > F_1^{-1}(u_1)) - \mathbb{P}(X_2 > F_2^{-1}(u_2)) + \mathbb{P}(X_1 > F_1^{-1}(u_1), X_2 > F_2^{-1}(u_2)$$
$$= 1 - (1 - u_1) - (1 - u_2) + [1 + \{(1 - u_1)^{-1/\alpha} - 1\} + \{(1 - u_2)^{-1/\alpha} - 1\}]^{-\alpha}.$$

The copula from the bivariate cdf (in contrast to that based on the multivariate survival function) is, with $\theta = 1/\alpha$,

$$C_{\text{rMTCJ}}(u_1, u_2; \theta) = u_1 + u_2 - 1 + [(1 - u_1)^{-\theta} + (1 - u_2)^{-\theta} - 1]^{-1/\theta}, \ 0 \leq u_1, u_2 \leq 1, \ \theta > 0. \quad (1.4)$$

The subscript r in rMTCJ is for the copula from the reflection $(V_j = 1 - U_j)$ so the copula is like (1.3), with extra terms from the inclusion-exclusion rule of probability of union of events. For the d-variate extension, there are 2^d terms from the inclusion-exclusion rule. $\square$

For the preceding example, note that a copula is a multivariate cdf with U$(0, 1)$ margins, so if one particular copula is based on a stochastic representation for $(U_1, \ldots, U_d)$, then there is another copula based on $(V_1, \ldots, V_d) = (1 - U_1, \ldots, 1 - U_d)$. Also there are copulas based on $(V_1, \ldots, V_d)$ where V_j is one of U_j or $1 - U_j$, that is, there are 2^d copulas from a single d-variate random vector of dependent U$(0, 1)$ random variables.

The next example shows the extreme value limit as another method to derive a copula.

Example 1.6 (Extreme value limit). We start with the univariate case before the multivariate extension. Consider the maxima of i.i.d. Pareto random variables; without loss of generality we take the scale parameter as 1. Let $Y_1, Y_2, \ldots$ be i.i.d. Pareto$(\alpha, 1)$. Let $M_n = \max\{Y_1, \ldots, Y_n\}$ for $n = 1, 2, \ldots$; M_n is increasing stochastically with n. The extreme value limit is obtained by finding sequences of constants $\{a_n\}, \{b_n\}$ such that $(M_n - a_n)/b_n$ converges to a non-degenerate distribution as $n \to \infty$. Since $\mathbb{P}(M_n \leq s) = [1 - (1 + s)^{-\alpha}]^n$, then as $n \to \infty$,

$$\mathbb{P}(M_n \leq n^{1/\alpha} z - 1) = [1 - n^{-1} z^{-\alpha}]^n \to \exp\{-z^{-\alpha}\}, \quad z > 0,$$

with $a_n = -1$, $b_n = n^{1/\alpha}$.

This is a Fréchet(α) distribution (with scale parameter 1). A Fréchet random variable is the reciprocal of a Weibull random variable.

Next we take a bivariate extreme value limit, as given in Galambos (1975). Let (X_{i1}, X_{i2}), $i = 1, 2 \ldots$, be a sequence of i.i.d. bivariate Pareto$(\alpha, \sigma_1 = 1, \sigma_2 = 1)$ random vectors with distribution in (1.1). Let $(M_{1n}, M_{2n}) = (\max_{1 \leq i \leq n} X_{i1}, \max_{1 \leq i \leq n} X_{i2})$ be the vector of componentwise maxima. Then, for $s_1, s_2 \geq 0$,

$$\mathbb{P}(M_{1n} \leq s_1, M_{2n} \leq s_2) = F_{12}^n(s_1, s_2) = \left[1 - (1 + s_1)^{-\alpha} - (1 + s_2)^{-\alpha} + (1 + s_1 + s_2)^{-\alpha}\right]^n.$$

For $z_1, z_2 > 0$, as $n \to \infty$,

$$\begin{aligned}
\mathbb{P}(M_{1n} &\leq n^{1/\alpha} z_1 - 1, M_{2n} \leq n^{1/\alpha} z_2 - 1) \\
&= \left[1 - n^{-1} z_1^{-\alpha} - n^{-1} z_2^{-\alpha} + (n^{1/\alpha} z_1 + n^{1/\alpha} z_2 - 1)^{-\alpha}\right]^n \\
&\approx \left[1 - n^{-1}\{z_1^{-\alpha} + z_2^{-\alpha} - (z_1 + z_2)^{-\alpha}\}\right]^n \\
&\to \exp\{-[z_1^{-\alpha} + z_2^{-\alpha} - (z_1 + z_2)^{-\alpha}]\} =: H(z_1, z_2).
\end{aligned} \quad (1.5)$$

For $j = 1, 2$, the univariate margin is $H_j(z_j) = \exp\{-z_j^{-\alpha}\}$, $z_j > 0$, $\alpha > 0$, and the inverse cdf is $z_j = H_j^{-1}(u_j) = (-\log u_j)^{-1/\alpha}$ for $0 < u_j < 1$, so that $z_j^{-\alpha} = -\log u_j$. The copula of (1.5) is:

$$C_{\text{Galambos}}(u_1, u_2; \theta) = u_1 u_2 \exp\{([-\log u_1]^{-\theta} + [-\log u_2]^{-\theta})^{-1/\theta}\}, \quad 0 \leq u_1, u_2 \leq 1, \ \theta > 0,$$
$$(1.6)$$

with $\theta = \alpha^{-1} > 0$.

For the trivariate case, in following similar steps, for $0 \leq u_j \leq 1$ $(j = 1, 2, 3)$,

$$\begin{aligned}
C_{\text{Galambos}}(u_1, u_2, u_3; \theta) = \exp\big\{&-(-\log u_1) - (-\log u_2) - (-\log u_3) \\
&+ ([-\log u_1]^{-\theta} + [-\log u_2]^{-\theta})^{-1/\theta} + ([-\log u_1]^{-\theta} + [-\log u_3]^{-\theta})^{-1/\theta} \\
&+ ([-\log u_2]^{-\theta} + [-\log u_3]^{-\theta})^{-1/\theta} \\
&- ([-\log u_1]^{-\theta} + [-\log u_2]^{-\theta} + [-\log u_3]^{-\theta})^{-1/\theta}\big\}.
\end{aligned} \quad (1.7)$$

The pattern extends to higher dimensions with alternating signs depending on the cardinality of the terms in the exponential. $\square$

Below we mention some properties of extreme value copulas, which can be derived as extreme value limits similar to the above example. The copula, with $d = 2$, in Example 1.6 is:

$$C(u_1, u_2; \theta) = u_1 u_2 \exp\{([-\log u_1]^{-\theta} + [-\log u_2]^{-\theta})^{-1/\theta}\}, \quad 0 \le u_1, u_2 \le 1, \ \theta > 0.$$

Transforming to Exponential(1) survival margins with $u_j = e^{-x_j}$ or $-\log u_j = x_j$, $j = 1, 2$, leads to the bivariate survival function:

$$\overline{G}(x_1, x_2; \theta) = \exp\{-x_1 - x_2 + (x_1^{-\theta} + x_2^{-\theta})^{-1/\theta}\}, \quad x_1, x_2 \ge 0.$$

Transforming to Fréchet(1) margins with $u_j = e^{-1/y_j}$ or $-\log u_j = y_j^{-1}$, $j = 1, 2$, leads to the bivariate cdf:

$$H(y_1, y_2; \theta) = \exp\{-y_1^{-1} - y_2^{-1} + (y_1^\theta + y_2^\theta)^{-1/\theta}\}, \quad y_1, y_2 \ge 0.$$

With θ omitted from C, the following homogeneity properties hold more generally for bivariate extreme value copulas: for all $r > 0$,

$$C(u_1^r, u_2^r) = C^r(u_1, u_2), \quad \text{or} \quad C^{1/r}(u_1^r, u_2^r) = C(u_1, u_2), \quad 0 < u_1, u_2 < 1,$$
$$-\log \overline{G}(rx_1, rx_2) = -r \log \overline{G}(x_1, x_2), \quad x_1, x_2 \ge 0,$$
$$-\log H(ry_1, ry_2) = -r^{-1} \log H(y_1, y_2), \quad y_1, y_2 \ge 0.$$

These properties extend to multivariate extreme value copulas.

This section has consisted of an introduction to show how some parametric copula families are derived. More theory and construction methods are given in Chapter 3. The basic theorem on multivariate copulas is given in the next section, followed by some data examples in Section 1.4.

1.3 Copula representation for a multivariate distribution

Concrete examples of multivariate distributions and their conversion to copulas were given in Section 1.2. In this section, we state the general representation of a multivariate distribution as a composition of a copula and its univariate margins; this result is due to Sklar (1959) and is now called Sklar's theorem.

Sklar's theorem and its derivation are based on (1.8) and (1.9) below.

The relevance for dependence modeling with copulas is that for continuous multivariate distributions, the modeling of the univariate marginals and the multivariate or dependence structure can be separated, and the multivariate structure can be represented by a copula.

The *copula* is a multivariate distribution with all univariate margins being U(0, 1). Hence if C is a copula, then it is the distribution of a vector of dependent U(0, 1) random variables.

Theorem 1.1 *For a d-variate distribution $F \in \mathcal{F}(F_1, \ldots, F_d)$, with jth univariate margin F_j, the copula associated with F is a distribution function $C : [0, 1]^d \to [0, 1]$ with U(0, 1) margins that satisfies*

$$F(\boldsymbol{y}) = C\big(F_1(y_1), \ldots, F_d(y_d)\big), \ \boldsymbol{y} \in \mathbb{R}^d. \tag{1.8}$$

(a) *If F is a continuous d-variate distribution function with univariate margins $F_1, \ldots, F_d$, and quantile functions $F_1^{-1}, \ldots, F_d^{-1}$, then*

$$C(\boldsymbol{u}) = F\big(F_1^{-1}(u_1), \ldots, F_d^{-1}(u_d)\big), \quad \boldsymbol{u} \in [0, 1]^d, \tag{1.9}$$

is the unique choice.

(b) If F is a d-variate distribution of discrete random variables (more generally, partly continuous and partly discrete), then the copula is unique only on the set

$$Range(F_1) \times \cdots \times Range(F_d).$$

Proof: This result essentially follows from two properties: (i) if H is a univariate cdf with inverse cdf H^{-1} and $U \sim U(0,1)$, then $H^{-1}(U) \sim H$; (ii) if H is a *continuous* univariate cdf and $Y \sim H$, then $H(Y) \sim U(0,1)$. Hence, if $\boldsymbol{Y} \sim F$ and F is continuous, then $(F_1(Y_1), \ldots, F_d(Y_d)) \sim C$, and if $\boldsymbol{U} \sim C$, then $(F_1^{-1}(U_1), \ldots, F_d^{-1}(U_d)) \sim F$.

If F is a d-variate distribution of some or all discrete random variables, then the copula associated with F is not unique. If H is a *non-continuous* or discrete univariate cdf and $Y \sim H$, then $H(Y)$ does not have a $U(0,1)$ distribution. The copula satisfying (1.8) is non-unique. It would be unique only on $Range(F_1) \times \cdots \times Range(F_d)$, since C in (1.8) would only be required to be defined on this set. Such a C must satisfy $C(1, \ldots, 1, u_j, 1, \ldots, 1) = u_j$ for $u_j \in Range(F_j)$ for $j = 1, \ldots, d$, and then it can be extended to a multivariate distribution with $U(0,1)$ margins. $\qquad\square$

Similarly, if $\overline{F}$ is a continuous d-variate survival function with univariate survival functions $\overline{F}_1, \ldots, \overline{F}_d$, and left-continuous inverse functions $\overline{F}_1^{-1}, \ldots, \overline{F}_d^{-1}$, then the distribution of $(V_1, \ldots, V_d) = (\overline{F}_1(Y_1), \ldots, \overline{F}_d(Y_d))$ is

$$\begin{aligned}
\widehat{C}(\boldsymbol{v}) &= \mathbb{P}\big(\overline{F}_1(Y_1) \leq v_1, \ldots, \overline{F}_d(Y_d) \leq v_d\big) \\
&= \mathbb{P}\big(Y_1 \geq \overline{F}_1(v_1), \ldots, Y_d \geq \overline{F}_d(v_d)\big) \\
&= \overline{F}\big(\overline{F}_1^{-1}(v_1), \ldots, \overline{F}_d^{-1}(v_d)\big), \quad \boldsymbol{v} \in [0,1]^d. \quad (1.10)
\end{aligned}$$

The copula derived in this way from a multivariate survival function is sometimes called the *survival copula*. That is, the survival copula is the copula of $\overline{F}_1(Y_1), \ldots, \overline{F}_d(Y_d)$ when Y_j has survival function $\overline{F}_j$ for $j = 1, \ldots, d$.

The copula can be considered independent of the univariate margins, since if C is a copula, then

$$G(\boldsymbol{y}) = C\big(G_1(y_1), \ldots, G_d(y_d)\big), \quad \boldsymbol{y} \in \mathbb{R}^d,$$

is a distribution (survival) function if $G_1, \ldots, G_d$ are all univariate distribution (survival) functions. If C is parametrized by a (vector) parameter $\boldsymbol{\theta}$, then we call $\boldsymbol{\theta}$ a *multivariate or dependence parameter*.

For an example of non-uniqueness in the discrete case, if F_j is Bernoulli(p_j) for $j = 1, 2$, $F_j(y) = 0$ for $y < 0$, $F_j(y) = 1 - p_j$ for $0 \leq y < 1$, $F_j(y) = 1$ for $y \geq 1$, $Range(F_1) = \{0, 1 - p_1, 1\}$, $Range(F_2) = \{0, 1 - p_2, 1\}$, and the copula is unique only for $C(1 - p_1, 1 - p_2)$. Note that $C(0, 1 - p_2) = 0$ and $C(1, 1 - p_2) = 1 - p_2$, etc., also follow from (1.8)).

The multivariate Gaussian distribution is converted to its copula in the next example.

Example 1.7 (Multivariate Gaussian copula). From the multivariate Gaussian distribution with zero means, unit variances and $d \times d$ correlation matrix $\boldsymbol{\Sigma}$, one gets:

$$C(\boldsymbol{u}; \boldsymbol{\Sigma}) = \Phi_d\big(\Phi^{-1}(u_1), \ldots, \Phi^{-1}(u_d); \boldsymbol{\Sigma}\big), \quad \boldsymbol{u} \in [0,1]^d,$$

where $\Phi_d(\cdot; \boldsymbol{\Sigma})$ is the d-variate Gaussian cdf, Φ is the univariate Gaussian cdf, and Φ^{-1} is the univariate Gaussian inverse cdf or quantile function.

Note that copula families are usually given as cdfs, and then (for likelihood inference), the copula density is obtained by differentiation. If $C(\boldsymbol{u})$ is an absolutely continuous copula cdf, then its density function is

$$c(\boldsymbol{u}) = c(u_1, \ldots, u_d) = \frac{\partial^d C(\boldsymbol{u})}{\partial u_1 \cdots \partial u_d}, \quad \boldsymbol{u} \in (0,1)^d.$$

Next, let $F(\boldsymbol{y}) = C(F_1(y_1), \ldots, F_d(y_d))$ be in $\mathcal{F}(F_1, \ldots, F_d)$, the Fréchet class of multivariate distributions with univariate margins $F_1, \ldots, F_d$. If $F_1, \ldots, F_d$ are absolutely continuous with respective densities $f_j = F_j'$, and C has mixed derivative of order d, then the density of F is

$$f(\boldsymbol{y}) = c\big(F_1(y_1), \ldots, F_d(y_d)\big) \times \prod_{j=1}^{d} f_j(y_j), \quad \boldsymbol{y} \in \mathbb{R}^d.$$

In the case of discrete random variables with cdfs $F_1, \ldots, F_d$, and $F = C(F_1, \ldots, F_d)$, the multivariate probability mass function (pmf) f comes from rectangle probabilities. For the bivariate case where F_1, F_2 have support on integers, the pmf for integers a_1, a_2 is:

$$
\begin{aligned}
f(a_1, a_2) &= \mathbb{P}(a_1 - 1 < Y_1 \leq a_1, \; a_2 - 1 < Y_2 \leq a_2) \\
&= F(a_1, a_2) - F(a_1 - 1, a_2) - F(a_1, a_2 - 1) + F(a_1 - 1, a_2 - 1) \\
&= C\big(F_1(a_1), F_2(a_2)\big) - C\big(F_1(a_1 - 1), F_2(a_2)\big) \\
&\quad - C\big(F_1(a_1), F_2(a_2 - 1)\big) + C\big(F_1(a_1 - 1), F_2(a_2 - 1)\big).
\end{aligned}
$$

For the trivariate pmf, rectangle probabilities involve 8 terms, 4 with + signs, 4 with − signs. For the d-variate pmf, rectangle probabilities involve 2^d terms.

Chapter 3 has methods of construction of copula families with many types of dependence structures from simple to more flexible. Almost all methods are based on continuous multivariate distributions derived through stochastic representations or limits, followed by applying (1.9) to get the copulas. In some cases, a general functional form is obtained from the construction, and then extended by checking for weaker conditions for the mixed derivative to be non-negative.

1.4 Data examples: scatterplots and semi-correlations

In Section 1.2, some copula families were derived, and the preceding section explained how they can be used as multivariate models. To determine whether they are suitable for modeling multivariate data, we need to know the shape of their densities and compare them with scatterplots.

In this section, we indicate some methods to identify and distinguish copula families based on scatterplots of pairs of variables, and show why sometimes classical multivariate statistics based on multivariate Gaussian (copula) is inadequate.

Although copula theory use transforms to $U(0, 1)$ margins, for diagnostics, there are several reasons to convert to standard normal $N(0, 1)$ margins; some reasons are given in this section and others are given in Section 2.16.1.

Let Φ be the standard normal cdf with functional inverse Φ^{-1}. If $(U_1, \ldots, U_d) \sim C$ for a copula C, let $Z_j = \Phi^{-1}(U_j)$ for $j = 1, \ldots, d$. Then $(Z_1, \ldots, Z_d)$ has a multivariate distribution in $\mathcal{F}(\Phi, \ldots, \Phi)$, the Fréchet class of multivariate distributions with $N(0, 1)$ margins.

The empirical transform, with a random sample $(y_{i1}, \ldots, y_{id})$, $i = 1, \ldots, n$, is the following. Order the jth variable as $y_{1:n,j} \leq \cdots \leq y_{i:n,j} \leq \cdots \leq y_{n:n,j}$; the ith ordered value $y_{i:n,j}$ is transformed from i to $\Phi^{-1}((i + a)/(n + 1 + 2a)) = z_{i:n,j}$, where $a \approx -0.5$. The resulting $z_{i:n,j}$ are called the empirical *normal scores*. The empirical parametric transform after fitting $F_j(\cdot; \hat{\theta}_j)$ to $y_{1j}, \ldots, y_{nj}$ is the following: $y_{ij} \to \Phi^{-1}[F(y_{ij}; \hat{\theta}_j)] = z_{ij}$.

For multivariate distributions, after transforms to $N(0, 1)$ margins, the contour plots of the resulting bivariate marginal density functions can be compared with the elliptical contours of the bivariate normal density.

Sharper corners (relative to ellipse) indicate *tail dependence* (Section 2.13). Also tail or reflection asymmetries can be seen. Reflection symmetry on the $N(0, 1)$ scale means that the bivariate density contour plot is symmetric to the $(z_1, z_2) \to (-z_1, -z_2)$ reflection. Examples of contour plots are shown in Figure 1.2 for bivariate distributions mentioned earlier.

Figure 1.1 has normal scores scatterplots for some simulated data: bivariate Pareto, maxima from bivariate Pareto, bivariate normal and bivariate t_5. For the simulated bivariate maxima, note that there is more dependence in the upper corner.

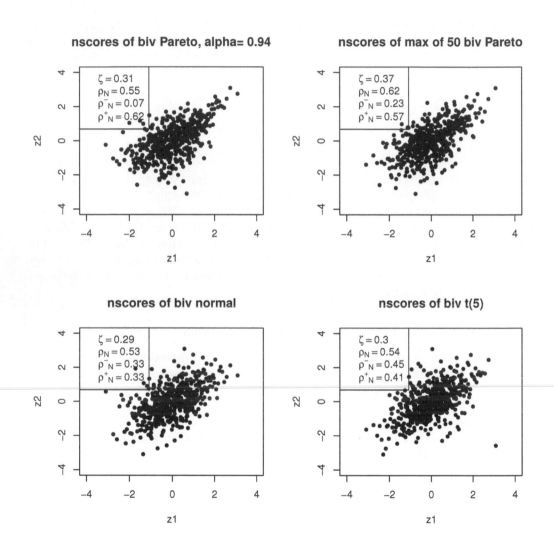

Figure 1.1 *Normal scores plots of simulated data: bivariate Pareto, maxima of bivariate Pareto, bivariate normal, bivariate t_5; ρ_N is the correlation of the normal scores, ρ_N^- is the lower semi-correlation, ρ_N^+ is the lower semi-correlation, and ζ=bvnsemic is the semi-correlation for bivariate normal with correlation ρ_N.*

Contour density plots are shown in Figures 1.2 and 1.3. The latter consists of contour plots of copula densities combined with $N(0, 1)$ margins for some other copulas in Chapter 4. Note the variety of shapes and tail asymmetry that are possible from different copula families. Spearman's rank correlation ρ_S, as defined in (2.45), is fixed at 0.5 for each of the plots.

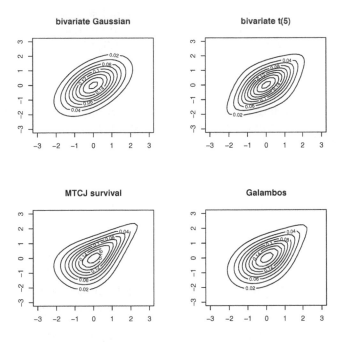

Figure 1.2 *Copulas combined with* $N(0,1)$ *margins and* $\rho_S = 0.5$, *contour plots of density* $c(\Phi(z_1,,\Phi(z_2))\,\phi(z_1)\,\phi(z_2)$.

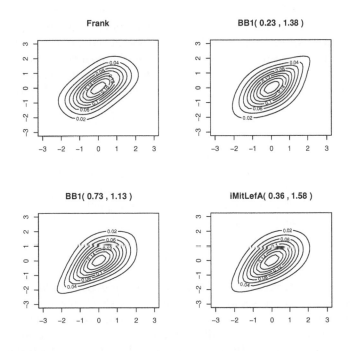

Figure 1.3 *More copulas combined with* $N(0,1)$ *margins and* $\rho_S = 0.5$, *contour plots of density* $c(\Phi(z_1,,\Phi(z_2))\,\phi(z_1)\,\phi(z_2)$.

Next for some real data, Figures 1.4–1.5 have normal scores scatterplots for some data examples, where each variable has been transformed to normal scores. These can be compared with density contour shapes of Figures 1.2 and 1.3.

The plots or corresponding tables show the values of ρ_N (the correlation of the normal scores in the plots), the lower and upper semi-correlations ρ_N^-, ρ_N^+ (which are the correlations of the points in the lower and upper quadrants respectively), and bvnsemic(ρ_N) (the semi-correlation of the bivariate normal distribution with correlation equal to ρ_N, given in (2.59)). When there is tail asymmetry, then the two semi-correlations are quite different. When there is stronger upper (lower) tail dependence than with the bivariate normal, then the upper (lower) semi-correlation is larger than bvnsemic(ρ_N). More details about the semi-correlations are given in Section 2.17.

Because only part of the data are used in computing sample semi-correlations, their standard errors are larger and interval estimates are longer than for dependence measures like ρ_S and ρ_N. However, if there are many variables and a consistent direction to the tail asymmetry based on semi-correlations, this is useful information for choosing potential copula models.

Figure 1.4 shows pairwise plots of maximum daily wind speeds (maxima of 24 hourly measurements) at several stations in Holland. A data source with hourly measurements is http://www.knmi.nl/klimatologie/onderzoeksgegevens/potentiele_wind/. Univariate distributions can be modeled by the generalized extreme value (GEV) family. For data that are maxima, one can expect to see more dependence in the joint upper tail compared with bivariate normal. This is indicated with the summaries of lower and upper semi-correlations.

Figure 1.5 shows pairwise plots of log daily returns of European market indexes, 2003–2006. A data source is: http://quote.yahoo.com. The log return is $y_t = \log(P_t/P_{t-1})$ where P_t is the index value at time t. Univariate distributions are roughly symmetric and heavier-tailed than Gaussian. Both the joint upper and joint lower tails have more dependence than bivariate normal.

Figure 1.6 shows a subset of the insurance loss versus ALAE (allocated loss adjustment expense) data, discussed in Frees and Valdez (1998). For this insurance loss data set, there is tail asymmetry skewed to the upper corner with lower semi-correlation 0.15 and upper semi-correlation 0.37. Also, the joint upper tail has more dependence than would be expected from a bivariate normal distribution with correlation 0.46. Using a bootstrap sample size of 2000, the standard errors for the lower semi-correlation, upper semi-correlation and difference are respectively 0.08, 0.07 and 0.11; the 95% bootstrap confidence interval is $(-0.42, 0.01)$ for the difference of the lower and upper semi-correlations.

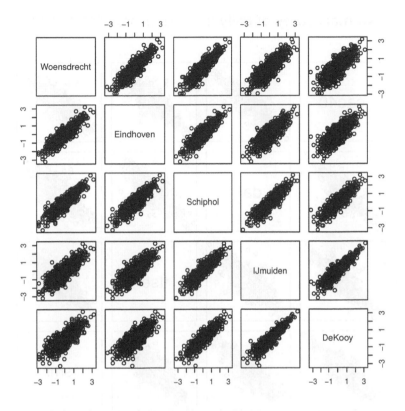

Figure 1.4 *Normal scores plots of maximum daily wind speeds at several stations in Holland; 2-year period from June 2007 to May 2009*

semi-correlations of normal scores: maximum wind speeds					
var.1	var.2	$\hat{\rho}_N$	$\hat{\rho}_N^-$	$\hat{\rho}_N^+$	ζ
Woensdrecht	Eindhoven	0.889	0.725	0.773	0.747
Woensdrecht	Schiphol	0.889	0.686	0.782	0.747
Woensdrecht	IJmuiden	0.806	0.494	0.668	0.605
Woensdrecht	De Kooy	0.793	0.502	0.630	0.586
Eindhoven	Schiphol	0.874	0.658	0.786	0.719
Eindhoven	IJmuiden	0.791	0.488	0.668	0.583
Eindhoven	Dekooy	0.782	0.425	0.654	0.570
Schiphol	IJmuiden	0.880	0.655	0.775	0.731
Schiphol	De Kooy	0.883	0.692	0.765	0.736
IJmuiden	De Kooy	0.907	0.693	0.846	0.782

Table 1.1 *Maximum daily wind speeds at five stations in Holland with $\hat{\rho}_N$ =correlation of normal scores, $\hat{\rho}_N^-$ =lower semi-correlation, $\hat{\rho}_N^+$ =upper semi-correlation, and ζ is the semi-correlation for bivariate Gaussian with correlation $\hat{\rho}_N$. There are 10 pairs for each measure. Based on a stationary bootstrap sample (Section 5.5.1) of size 2000, the approximate 95% confidence intervals for (i) the average of the 10 ρ_N is $(0.82, 0.87)$, (ii) the average of the 10 ρ_N^- is $(0.53, 0.66)$, (iii) the average of the 10 ρ_N^+ is $(0.67, 0.77)$, and (iv) the average of the 10 different $(\rho_N^- - \rho_N^+)$ is $(-0.20, -0.06)$. This is an indication of more dependence in the joint upper tails than the lower tails.*

Normal scores of daily index returns: 2003 to 2006

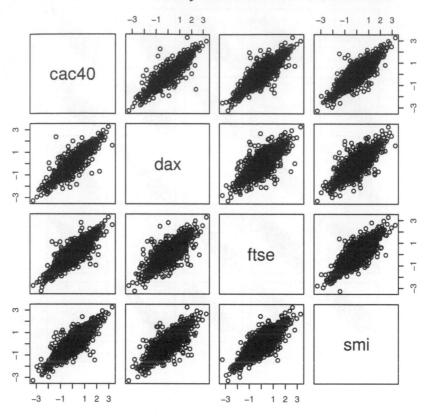

Figure 1.5: *Normal scores plots of log returns of European market indexes 2003–2006.*

semi-correlations of normal scores: Europe 2003-2006					
var.1	var.2	$\hat{\rho}_N$	$\hat{\rho}_N^-$	$\hat{\rho}_N^+$	ζ
CAC40	DAX	0.888	0.821	0.802	0.746
CAC40	FTSE	0.848	0.776	0.687	0.674
CAC40	SMI	0.843	0.760	0.710	0.664
DAX	FTSE	0.771	0.684	0.611	0.554
DAX	SMI	0.786	0.652	0.641	0.576
FTSE	SMI	0.799	0.709	0.617	0.595

Table 1.2 *Log returns of European market indexes 2003–2006, with $\hat{\rho}_N$ =correlation of normal scores, $\hat{\rho}_N^-$ =lower semi-correlation, $\hat{\rho}_N^+$ =upper semi-correlation, and ζ is the semi-correlation for bivariate Gaussian with correlation $\hat{\rho}_N$. As a summary, based on a stationary bootstrap (see Section 5.5.1) sample of size 2000, the approximate 95% confidence intervals for (i) the average of 6 ρ_N is (0.80, 0.84), (ii) the average of 6 ρ_N^- is (0.69, 0.78), (iii) the average of 6 ρ_N^+ is (0.61, 0.72), (iv) the average of 6 different $(\rho_N^- - \rho_N^+)$ is (0.00, 0.12). This is some indication of more dependence in the joint lower tails than in the upper tails, but the difference in the tails is not as much as for the daily maximum wind data with summaries in Table 1.1.*

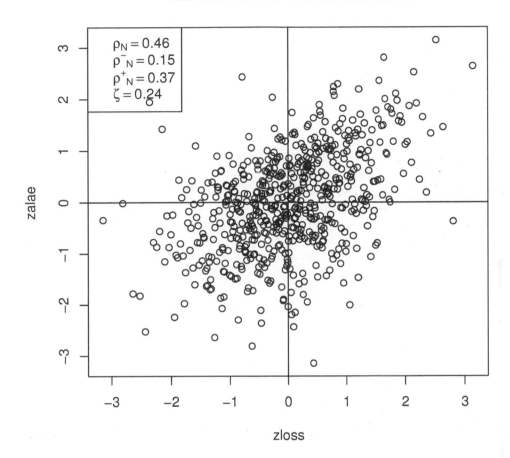

Figure 1.6 *Loss versus ALAE (allocated loss adjustment expense data): normal scores plot of a subset of the Frees-Valdez data, with $\hat{\rho}_N$ =correlation of normal scores, $\hat{\rho}_N^-$ =lower semi-correlation and $\hat{\rho}_N^+$ =upper semi-correlation, ζ is the semi-correlation for bivariate Gaussian with correlation $\hat{\rho}_N$. These summaries give an indication of more dependence in the joint upper tail.*

1.5   Likelihood analysis and model comparisons

Making use of the constructions in Section 1.2 and some preliminary analyses in Section 1.4, this section illustrates some typical steps in the use of likelihood analysis and model comparisons with parametric copula families, for continuous and discrete response data.

Likelihood inference based on multivariate parametric families is the most common approach for model comparison and inferences, especially when the number of observations is of the order of 10^2 to 10^3; semi-parametric and non-parametric methods based on ranks can be applied only in low dimensions and might not work when response variables are discrete or when there are covariates.

For parametric univariate and copula models, background knowledge can be used; alternatively, initial data analysis, diagnostics and plots can suggest suitable candidates.

1.5.1   A brief summary of maximum likelihood

Consider a statistical model $f(\cdot; \boldsymbol{\theta})$, where $\boldsymbol{\theta}$ is parameter vector. For a univariate model, $\boldsymbol{\theta}$ consists of one of more parameters. For a copula model, $\boldsymbol{\theta}$ consists of univariate and dependence parameters.

Suppose the sample size is n, and the data values are $y_1, \ldots, y_n$, considered as a random sample from $f(\cdot; \boldsymbol{\theta})$. For likelihood inference, the maximum likelihood estimator (MLE) $\hat{\boldsymbol{\theta}} = \hat{\boldsymbol{\theta}}_{\text{obs}}$ maximizes the log-likelihood $L(\boldsymbol{\theta}; y_1, \ldots, y_n) = \sum_{i=1}^{n} \log f(y_i; \boldsymbol{\theta})$ and the inverse observed information $(n\hat{\mathcal{I}})^{-1}$, of order n^{-1}, is

$$(n\hat{\mathcal{I}})^{-1} = \left[-\sum_{i=1}^{n} \frac{\partial^2 \log f(y_i; \boldsymbol{\theta})}{\partial \boldsymbol{\theta} \partial \boldsymbol{\theta}^{\top}} \Big|_{\hat{\boldsymbol{\theta}}_{\text{obs}}} \right]^{-1}$$

If $\boldsymbol{\theta}_0$ is the true value, then asymptotically $n^{1/2}(\hat{\boldsymbol{\theta}} - \boldsymbol{\theta}_0) \sim N(\mathbf{0}, \mathcal{I}^{-1}(\boldsymbol{\theta}_0))$ or approximately $\hat{\boldsymbol{\theta}} \stackrel{.}{\sim} N(\boldsymbol{\theta}_0, n^{-1}\mathcal{I}^{-1}(\boldsymbol{\theta}_0))$, where $\mathcal{I}(\boldsymbol{\theta}_0) = \mathbb{E}[-\partial^2 \log f(Y; \boldsymbol{\theta}_0)/\partial \boldsymbol{\theta} \partial \boldsymbol{\theta}^{\top}]$ is the expected Fisher information. Note that $\hat{\mathcal{I}}$ converges to $\mathcal{I}(\boldsymbol{\theta}_0)$ in probability as $n \to \infty$ (weak law of large numbers).

For further inference after maximum likelihood, with the plug-in method, one can use the approximation

$$N\big(\hat{\boldsymbol{\theta}}_{\text{obs}}, n^{-1}\hat{\mathcal{I}}^{-1}\big)$$

for confidence intervals and the delta method, etc. With expected information, one can use

$$N\big(\hat{\boldsymbol{\theta}}_{\text{obs}}, n^{-1}\mathcal{I}^{-1}(\hat{\boldsymbol{\theta}}_{\text{obs}})\big).$$

For the delta method, the normal approximation is:

$$s(\hat{\boldsymbol{\theta}}) \stackrel{.}{\sim} N\big(s(\boldsymbol{\theta}_0), n^{-1}(\nabla s)^{\top} \mathcal{I}^{-1}(\boldsymbol{\theta}_0) \nabla s\big),$$

where s is one-to-one. Then for confidence intervals and regions with the plug-in method, one can use

$$N\big(s(\hat{\boldsymbol{\theta}}_{\text{obs}}), n^{-1}(\nabla s)^{\top}(\hat{\boldsymbol{\theta}}_{\text{obs}}) \hat{\mathcal{I}}^{-1} \nabla s(\hat{\boldsymbol{\theta}}_{\text{obs}})\big),$$

where ∇s is the gradient vector of s. In practice, the *rate of convergence to asymptotic normality* depends on the parametrization, i.e., on s.

Typically, for a well-behaved likelihood, optimization of the log-likelihood is performed numerically using a quasi-Newton minimization function in statistical software, combined with the BFGS method for updating the Hessian matrix.

For comparing different models, say $f^{(1)}, \ldots, f^{(M)}$, model comparisons are made with the Akaike information criterion AIC and Bayesian information criterion BIC:

$$\begin{aligned} \text{AIC} &= -2 \log \text{lik}(\hat{\boldsymbol{\theta}}_{\text{obs}}) + 2(\#\text{parameters}), \\ \text{BIC} &= -2 \log \text{lik}(\hat{\boldsymbol{\theta}}_{\text{obs}}) + (\log n)(\#\text{parameters}), \end{aligned} \tag{1.11}$$

where loglik and #parameters vary over the models $f^{(m)}$. In addition, other context-dependent assessments of model prediction adequacy can be performed.

The negatives of AIC and BIC can be interpreted as penalized log-likelihood functions; BIC has bigger penalty rate as n increases; smaller AIC or BIC values are considered better.

1.5.2   Two-stage estimation for copula models

For copula models, because the dependence modeled through the copula is "separated" from the univariate margins, a two-stage estimation method can be used to assess the adequacy of the univariate and copula models. The two-stage estimation is important from a practical point of view because numerical optimization depends on a good starting point if the dimension of the parameter vector is large.

Consider a parametric copula model:

$$F(y_1, \ldots, y_d; \boldsymbol{\theta}) = C\big(F_1(y_1; \boldsymbol{\eta}_1), \ldots, F_d(y_d; \boldsymbol{\eta}_d); \boldsymbol{\delta}\big)$$

where $\boldsymbol{\eta}_1, \ldots, \boldsymbol{\eta}_d$ are univariate parameters, $\boldsymbol{\delta}$ is a dependence parameter. More generally, there could be covariates, in which case univariate margins are (generalized) regression models.

Suppose data are $(y_{i1}, \ldots, y_{id})$, $i = 1, \ldots, n$, and variables are considered as realizations of continuous random variables. Steps for the two-stage estimation method are the following.

- Univariate margins: find good fitting parametric models, get $\tilde{\boldsymbol{\eta}}_j$ via a univariate maximum likelihood for $j = 1, \ldots, d$.
- Copula likelihood: fix the univariate parameters and estimate $\boldsymbol{\delta}$ by maximizing the likelihood

$$L_C = \sum_{i=1}^{n} \log c\big(F_1(y_{i1}; \tilde{\boldsymbol{\eta}}_1), \ldots, F_d(y_{id}; \tilde{\boldsymbol{\eta}}_d); \boldsymbol{\delta}\big)$$

to get $\tilde{\boldsymbol{\delta}}$; different copula models could be compared via AIC/BIC (equations in (1.11)) with univariate parameters held fixed.

- Full log-likelihood: use two-stage estimates to start iterative numerical optimization and maximize

$$L_F = \sum_{i=1}^{n} \log\big\{c\big(F_1(y_{i1}; \boldsymbol{\eta}_1), \ldots, F_d(y_{id}; \boldsymbol{\eta}_d), \boldsymbol{\delta}\big) f_1(y_{i1}; \boldsymbol{\eta}_1) \cdots f_d(y_{id}; \boldsymbol{\eta}_d)\big\}$$

over $(\boldsymbol{\eta}_1, \ldots, \boldsymbol{\eta}_d, \boldsymbol{\delta})$ to get MLE $(\hat{\boldsymbol{\eta}}_1, \ldots, \hat{\boldsymbol{\eta}}_d, \hat{\boldsymbol{\delta}})$ and asymptotic covariance matrix $\boldsymbol{V}$.

If data are discrete, joint probabilities come from rectangle probabilities computed via the copula cdf (see Section 5.3), but otherwise the above procedure can be used.

Estimation consistency checks of the model are the following.

- If $(\hat{\boldsymbol{\eta}}_1, \ldots, \hat{\boldsymbol{\eta}}_d, \hat{\boldsymbol{\delta}})$ and $(\tilde{\boldsymbol{\eta}}_1, \ldots, \tilde{\boldsymbol{\eta}}_d, \tilde{\boldsymbol{\delta}})$ differ a lot relative to the standard errors from $\boldsymbol{V}$, then either the copula model or the univariate models may be inadequate.
- For the case of continuous variables with no covariates, the maximum pseudo likelihood approach of Genest et al. (1995a) uses empirical distribution functions $\hat{F}_1, \ldots, \hat{F}_d$ and maximizes

$$L_{\text{pseudo}}(\boldsymbol{\delta}) = \sum_{i=1}^{n} \log c\big(\hat{F}_1(y_{i1}), \ldots, \hat{F}_d(y_{id}); \boldsymbol{\delta}\big)$$

in $\boldsymbol{\delta}$. This can be compared with the two-stage estimation procedure as another consistency check. For empirical distributions, one can use

$$\hat{F}_j(y) = n^{-1}\Big[\sum_{i=1}^{n} I(y_{ij} \leq y) - \tfrac{1}{2}\Big];$$

the additive factor of $-\tfrac{1}{2}$ is to ensure that $\hat{F}$ is strictly in the interval $(0, 1)$ so that the density c has arguments which are not at the boundary.

1.5.3 *Likelihood analysis for continuous data: insurance loss*

The steps in the preceding subsection are illustrated for a subsample of $n = 600$ uncensored observations in the loss-ALAE data set, shown in Figure 1.6.

For the loss-ALAE data set, both variables are right-skewed and heavy-tailed, and typical parametric families in this case are the 2-parameter Pareto and 3-parameter Burr distributions. The bivariate normal scores plots and semi-correlations show some skewness to the upper corner but otherwise the plots suggest unimodality, so the MTCJ copula in (1.4) and Galambos in (1.6) may be suitable 1-parameter copula families. Other families with two or more parameters are given in Chapter 4.

Summaries of the data and likelihood analysis are shown in Tables 1.3–1.5. From the univariate fits, the AIC suggests Burr as the better model for ALAE and Pareto as the better model for loss. The two-stage estimation procedure uses these as the margins of the copula model and compares different copulas (Galambos, MTCJ and bivariate Gaussian), and then the full log-likelihoods (with 6 parameters) can be optimized. The fit of the bivariate Gaussian copula can be considered as a baseline to see how much improvement there is in the log-likelihood when a copula family with upper tail dependence but not lower tail dependence is used.

summary	loss	ALAE
Minimum	10	15
Q1	3500	2502
Median	12000	5702
Mean	38135	11582
Q3	30000	13354
Maximum	2173595	160365

Table 1.3: *Loss-ALAE data subset of $n = 600$: univariate summaries.*

parameter	Pareto estimate (SE)	Burr estimate (SE)
	univariate for ALAE	
α	2.92 (0.46)	2.12 (0.48)
ζ		1.10 (0.06)
σ	2.23 (0.45)	1.42 (0.43)
	univariate for loss	
α	1.32 (0.13)	1.42 (0.27)
ζ		0.97 (0.06)
σ	1.67 (0.26)	1.87 (0.55)
summary	Pareto	Burr
	model comparison for ALAE	
-loglik	644.7	643.5
AIC	1293.4	1292.9
BIC	1302.2	1306.1
	model comparison for loss	
-loglik	1194.0	1193.9
AIC	2392.0	2393.8
BIC	2400.8	2407.0

Table 1.4 *Loss-ALAE data subset of $n = 600$: MLEs and standard errors (SEs) for univariate models.*

	Parameter estimates and SEs			
parameter	univariate	Galambos	rMTCJ	Gaussian
α_{ALAE}	2.12	2.04 (0.40)	1.87 (0.37)	2.61 (0.63)
ζ_{ALAE}	1.10	1.10 (0.06)	1.11 (0.06)	1.06 (0.06)
σ_{ALAE}	1.42	1.35 (0.36)	1.21 (0.31)	1.87 (0.61)
α_{loss}	1.32	1.31 (0.12)	1.26 (0.12)	1.39 (0.14)
σ_{loss}	1.67	1.63 (0.24)	1.52 (0.23)	1.82 (0.28)
δ_{full}		0.71 (0.05)	0.79 (0.09)	0.46 (0.03)
δ_{2stage}		0.70	0.76	0.46
δ_{pseudo}		0.70	0.75	0.46
-loglik		1756.3	1758.8	1765.6

Table 1.5 *Loss-ALAE data subset of $n = 600$: comparison of copula models, with univariate margins of Burr for ALAE/10000 and Pareto for loss/10000; variables are scaled so that σ's is not large.*

The estimates of the dependence parameter δ for the Galambos and MTCJ copulas with the two-stage estimation method and pseudo-likelihood method are similar. Also the estimates of the univariate parameters are similar in the two-stage and full likelihood estimation methods. The Galambos copula is the best fit based on log-likelihood (equivalently AIC/BIC because the number of parameters is 6 for all three copula models); and its MLEs from the full likelihood are closest to those from the univariate likelihoods. For this example, one can match the bivariate plots of normal scores of the two variables to contour plots of the densities of the Galambos and MTCJ copulas. The MTCJ copula is more skewed than the Galambos copula towards the upper corner; this is made more precise with the theory of Section 2.16. It is shown in Section 5.7 through Kullback-Leibler divergences that likelihood inference can differentiate among different tail behaviors.

An analysis of the entire loss-ALAE data set is given in Section 7.4.

1.5.4 Likelihood analysis for discrete data: ordinal response

In this subsection, a multivariate longitudinal ordinal data set is used to illustrate the steps of Section 1.5.2 and show how copula models can be used with discrete variables. The data set comes from a study on the psychological effects of the accident at the Three Mile Island nuclear power plant in 1979. An original source is Fienberg et al. (1985) and the data set was used in Section 11.2 of Joe (1997) for comparing a few copula-based models. Here, we fit some simpler copula models based on the multivariate distributions introduced earlier in this chapter.

The study focuses on the changes in levels of stress of mothers of young children living within 10 miles of the plant. Four waves of interviews were conducted in 1979, 1980, 1981 and 1982, and one variable measured at each time point is the level of stress (categorized as low, medium, or high from a composite score of a 90-item checklist). Hence stress is treated as an ordinal response variable with three categories, coded with 0,1,2 for low, medium and high respectively. We do some simple analysis and modeling for the subset of $n = 115$ mothers in subgroup living within 5 miles of the plant.

The data are shown in Table 1.9, with frequencies of 4-vectors in $\{0, 1, 2\}^4$. Univariate and bivariate marginal frequency tables are shown in Tables 1.6 and 1.7 respectively. The univariate distributions are unimodal, and the two-way tables are unimodal with positive stochastic increasing dependence. Hence it is reasonable to summarize each bivariate margin with Spearman rank correlations adjusted for ties and polychoric correlations; the latter are the latent correlations based on fitting the discretized bivariate Gaussian model with the univariate margins estimated from the sample univariate probability mass functions.

See Section 2.12.7 for the mathematical forms of these dependence measures for ordinal data. Table 1.7 also has bivariate expected counts based on the fitted discretized bivariate Gaussian model; they show that this model might be underestimating a little in some (2,2) [or High-High] cells and overestimating in some (0,0) [or Low-Low] cells. This comparison of bivariate observed frequencies to those expected under a Gaussian model is the discrete analogue of the bivariate normal scores plot.

Let $\boldsymbol{y}_i = (y_{i1}, \ldots, y_{id})$ for $i = 1, \ldots, n$ be an ordinal vector response for the ith subject with each $y_{ij} \in \{0, 1, \ldots, K-1\}$. The discretized multivariate Gaussian model assumes that the $\boldsymbol{y}_i$ is the realization of a random vector $\boldsymbol{Y}_i = (Y_{i1}, \ldots, Y_{id})$ such that

$$Y_{ij} = k \quad \text{if} \quad \zeta_{j,k} < Z_{ij} \leq \zeta_{j,k+1},$$

where $(Z_{i1}, \ldots, Z_{id}) \sim \mathrm{N}_d(\boldsymbol{0}, \boldsymbol{R})$ and $-\infty = \zeta_{j,0} < \zeta_{j,1} < \cdots < \zeta_{j,K-1} < \zeta_{j,K} = \infty$ are the cutpoints of the normal random variable Z_j. This is a common model for multivariate ordinal response; it is also called the multivariate probit model.

Let $U_{ij} = \Phi(Z_{ij})$ for all i, j so that $\boldsymbol{U}_i = (U_{i1}, \ldots, U_{id})$ is a vector of $\mathrm{U}(0,1)$ random variables, and let $a_{j,k} = \Phi(\zeta_{j,k})$ for all j, k. Then the $a_{j,k}$ are cutpoints on the $\mathrm{U}(0,1)$ scale. The joint distribution of $\boldsymbol{Y}_i$ is

$$\begin{aligned}
\mathbb{P}(Y_{ij} = k_j, \, j = 1, \ldots, d) &= \mathbb{P}(\zeta_{j,k_j} < Z_{ij} \leq \zeta_{j,k_j+1}, \, j = 1, \ldots, d) \\
&= \mathbb{P}(a_{j,k_j} < U_{ij} \leq a_{j,k_j+1}, \, j = 1, \ldots, d).
\end{aligned}$$

This rectangle probability can be evaluated as a d-dimensional integral. For $d = 2$, there are many efficient numerical methods such as Donnelly (1973). For $d = 3$ and 4, the numerical integration method of Schervish (1984) is one method that can be used, and for moderate $d \geq 3$, the randomized quasi-Monte Carlo method of Genz and Bretz (2009) is reasonably accurate and quick. In the special case where $\boldsymbol{R}$ is an exchangeable correlation matrix with a common positive correlation for any pair (or $\boldsymbol{R}$ has the 1-factor structure), the multivariate Gaussian rectangle probability can be converted to a 1-dimensional integral for fast computation for any dimension $d \geq 3$.

As an alternative to the discretized multivariate Gaussian model, a copula model would specify a parametric distribution for $\boldsymbol{U}_i$. Note that the discretized multivariate Gaussian model is also a copula model with the Gaussian copula given in Example 1.7. One idea in early developments of copula models for multivariate discrete response was to find parametric copula families with closed form cdfs to avoid numerical integration.

Because the bivariate analysis suggests that the dependence is not far from exchangeable dependence and overall there is slightly more probability in the joint upper tail than expected with discretized bivariate Gaussian, we illustrate copula modeling with the 4-variate extension of the Galambos copula in (1.7).

For the two-stage estimation method, the univariate cutpoints $a_{j,k}$ are estimated from the univariate empirical cdfs and converted to $\zeta_{j,k}$ by transforming with Φ^{-1}. Then with these held fixed, the second stage is to maximize the log-likelihood in the dependence parameters. In terms of the 4-variate copula $C = C(\cdot; \boldsymbol{\delta})$, this is:

$$\sum_{i=1}^{n} \log \mathbb{P}(Y_{ij} = y_{ij}, \, j = 1, \ldots, 4),$$

where

$$\mathbb{P}(Y_{ij} = k_j,\ j = 1,\ldots,d) = C(a_{1,k_1+1}, a_{2,k_2+1}, a_{3,k_3+1}, a_{4,k_4+1}; \boldsymbol{\delta})$$
$$-C(a_{1,k_1}, a_{2,k_2+1}, a_{3,k_3+1}, a_{4,k_4+1}; \boldsymbol{\delta}) - C(a_{1,k_1+1}, a_{2,k_2}, a_{3,k_3+1}, a_{4,k_4+1}; \boldsymbol{\delta})$$
$$-C(a_{1,k_1+1}, a_{2,k_2+1}, a_{3,k_3}, a_{4,k_4+1}; \boldsymbol{\delta}) - C(a_{1,k_1+1}, a_{2,k_2+1}, a_{3,k_3+1}, a_{4,k_4}; \boldsymbol{\delta})$$
$$+C(a_{1,k_1}, a_{2,k_2}, a_{3,k_3+1}, a_{4,k_4+1}; \boldsymbol{\delta}) + C(a_{1,k_1}, a_{2,k_2+1}, a_{3,k_3}, a_{4,k_4+1}; \boldsymbol{\delta})$$
$$+C(a_{1,k_1}, a_{2,k_2+1}, a_{3,k_3+1}, a_{4,k_4}; \boldsymbol{\delta}) + C(a_{1,k_1+1}, a_{2,k_2}, a_{3,k_3}, a_{4,k_4+1}; \boldsymbol{\delta})$$
$$+C(a_{1,k_1+1}, a_{2,k_2}, a_{3,k_3+1}, a_{4,k_4}; \boldsymbol{\delta}) + C(a_{1,k_1+1}, a_{2,k_2+1}, a_{3,k_3}, a_{4,k_4}; \boldsymbol{\delta})$$
$$-C(a_{1,k_1}, a_{2,k_2}, a_{3,k_3}, a_{4,k_4+1}; \boldsymbol{\delta}) - C(a_{1,k_1}, a_{2,k_2}, a_{3,k_3+1}, a_{4,k_4}; \boldsymbol{\delta})$$
$$-C(a_{1,k_1}, a_{2,k_2+1}, a_{3,k_3}, a_{4,k_4}; \boldsymbol{\delta}) - C(a_{1,k_1+1}, a_{2,k_2}, a_{3,k_3}, a_{4,k_4}; \boldsymbol{\delta})$$
$$+C(a_{1,k_1}, a_{2,k_2}, a_{3,k_3}, a_{4,k_4}; \boldsymbol{\delta}).$$

Outcomes	Year			
	1979	1980	1981	1982
0=Low	14 (0.122)	18 (0.157)	14 (0.122)	18 (0.157)
1=Medium	69 (0.600)	73 (0.635)	72 (0.626)	70 (0.609)
2=High	32 (0.278)	24 (0.209)	29 (0.252)	27 (0.235)

Table 1.6 *Ordinal stress data. Univariate marginal (and relative) frequencies for group living <5mi from Three Mile Island.*

Margin	$\hat{\rho}_N$ (SE)	$\hat{\rho}_S$	biv. obs.	biv. exp.
1,2	0.79 (0.06)	0.63	7 7 0 11 54 4 0 12 20	9.1 4.8 0.0 8.7 54.3 6.0 0.1 13.9 18.0
1,3	0.70 (0.07)	0.57	3 11 0 11 51 7 0 10 22	6.8 7.0 0.1 7.0 51.7 10.4 0.2 13.3 18.5
2,3	0.81 (0.06)	0.64	8 10 0 6 57 10 0 5 19	9.5 8.5 0.0 4.5 57.1 11.4 0.0 6.4 17.5
1,4	0.65 (0.08)	0.53	6 8 0 11 50 8 1 12 19	7.5 6.4 0.2 10.1 48.9 10.0 0.4 14.7 16.9
2,4	0.64 (0.08)	0.50	10 8 0 7 52 14 1 10 13	8.6 9.0 0.3 9.1 50.9 13.0 0.2 10.1 13.7
3,4	0.84 (0.05)	0.68	10 4 0 8 57 7 0 9 20	10.1 3.9 0.0 7.8 57.0 7.1 0.0 9.1 19.9

Table 1.7 *Ordinal stress data. Polychoric correlations $\hat{\rho}_N$, Spearman rank correlations with adjustment for ties, bivariate empirical marginal tables and bivariate expected frequencies based on fitted discretized Gaussian; the row and column labels (0,1,2) of the two-way tables are omitted. Note that latent polychoric correlations are larger in absolute values than actual correlations.*

Model	MLE	-Log-lik	AIC	BIC
unstr. Gauss.	0.79, 0.71, 0.81, 0.67, 0.65, 0.84	323.6	659.3	675.7
exch. Gauss.	$\hat{\rho} = 0.74$	330.4	662.8	665.5
Galambos	$\hat{\theta} = 1.46$	328.2	658.4	661.1

Table 1.8 *Ordinal stress data. Comparisons of log-likelihood and AIC/BIC values, with univariate parameters fixed at empirical values.*

4-tuple	Observed	Expected		
		unstr. Gauss.	exch. Gauss.	Galambos
0000	2	4.8	4.7	3.4
0010	2	0.9	1.7	1.6
0011	3	2.3	1.2	1.4
0101	1	0.2	0.7	0.9
0110	2	1.0	1.2	1.4
0111	4	2.6	2.4	3.0
1000	5	2.2	1.7	1.6
1001	1	1.3	1.2	1.4
1010	1	0.7	2.0	2.2
1011	4	4.1	4.2	4.8
1100	3	2.2	1.2	1.4
1101	2	1.1	2.4	3.0
1110	2	4.8	4.2	4.9
1111	38	35.7	34.9	36.7
1112	4	3.8	3.7	2.6
1121	2	3.0	4.5	3.3
1122	3	3.8	1.8	1.5
1211	2	2.1	2.7	1.6
1221	1	1.6	1.4	1.1
1222	1	1.9	1.6	1.6
2111	4	7.0	5.9	4.9
2112	3	2.0	2.4	2.0
2121	1	1.2	2.9	2.5
2122	4	3.2	3.3	3.2
2210	1	0.1	0.0	0.0
2211	2	3.0	1.8	1.4
2221	5	3.4	2.5	2.4
2222	12	10.8	10.7	12.8
others	0	4.1	6.0	6.3

Table 1.9 *Ordinal stress data for 4 years 1979-1982 following an accident at Three Mile Island, < 5 mi. group. Tabulated are four-tuples with non-zero frequencies, and expected frequencies for several models (unstructured Gaussian, exchangeable Gaussian and Galambos copulas).*

Table 1.8 has the parameter estimates of dependence parameters, negative log-likelihood and AIC/BIC values (equations in (1.11)) for three models: discretized Gaussian with unstructured correlation matrix, discretized Gaussian with exchangeable correlation matrix, and the Galambos copula. This allows the check of the effect of unstructured versus structured dependence and tail symmetric Gaussian versus tail asymmetric Galambos. The MLE of the latent correlation parameters in the unstructured correlation model are almost the same within two decimal places as the polychoric correlations which are based on separate bivariate margins. This demonstrates estimation consistency of bivariate marginal likelihoods and the 4-variate likelihood.

Based on AIC and BIC values for three models, the Galambos copula is the best fit of the three models considered here. Actually, other exchangeable copula models were fitted and the Galambos copula was best among these for AIC/BIC. Table 1.9 also has model-based expected frequencies for the different 4-vectors of ordinal responses. These also show that the Galambos copula fits a little better in the joint lower and upper corners. The sample size $n = 115$ is not large for 4-variate discrete data, so the exchangeable dependence model seems acceptable for matching observed frequencies. With larger sample sizes, one would generally try to fit a dependence structure with more parameters and the models developed in Chapter 3 can be used.

1.6   Copula models versus alternative multivariate models

An interpretation of the copula representation in Section 1.3 is that for statistical modeling, the copula approach has an advantage of having univariate margins of different types and the dependence structure can be modeled separately from univariate margins. The approach is shown in the examples in Section 1.5.

For example, one can have $Y_j \sim t_{\nu_j}$ with different shape parameters ν_j, or $Y_j \sim$ Pareto(α_j, σ_j) with different shape parameters α_j. The multivariate t and Pareto distributions presented in Section 1.2 have the requirements of constant ν and α respectively.

But the use of copulas is only one way to develop multivariate models. Depending on the context, other approaches such as use of latent variables, random effects, mixtures, stochastic operators, or conditional specifications might be more appropriate.

In statistical modeling in general, mathematical convenience plays a large role, especially when there is no stochastic or physical mechanism that leads to natural models. There is a quote attributed to George E. P. Box that says "All models are wrong, though some are useful." This applies to the use of the multivariate normal distribution as well as copula models. Statistical models are often a means to make inference about such quantities as quantiles, tail probabilities or conditional expectations, so models with flexible dependence and tail behavior can be desirable. The adequacy of statistical models does depend on their intended uses.

1.7   Terminology for multivariate distributions with U(0,1) margins

Early research in multivariate non-normal distributions developed mostly without copulas, but the concept of the copula was implicit in the literatures on multivariate extremes and multivariate distributions with specified univariate margins. Some history and alternative terms for the class of multivariate distributions with univariate $U(0, 1)$ margins are:

- copula: coined in Sklar (1959), has the characterization result but no mention of applications;
- *uniform representation*: in Kimeldorf and Sampson (1975, 1978) for bivariate, introduced for ease of investigation of dependence concepts and properties of bivariate distributions that are independent of the univariate margins;
- *dependence function*: in Deheuvels (1978), Galambos (1987) and used in multivariate extreme theory;
- *standard form*: in Cook and Johnson (1981) with applications.

So what is now commonly referred to as a copula was independently discovered in the 1970s and 1980s. In the fifteen year period of 1997–2012, there were more and more applications in the scientific literature, especially in insurance and finance where modeling of the joint tail is important. The increase in computer speed is a factor in the applications of copulas; some of the high-dimensional copula models in use now would have been difficult to consider in 1997. Books with copula applications include McNeil et al. (2005) for quantitative risk management, actuarial science and finance, Denuit et al. (2005) for actuarial science and dependent risks, Salvadori et al. (2007) for extremes in nature and the environment; Cherubini et al. (2004) and Cherubini et al. (2012) for finance.

1.8   Copula constructions and properties

To understand properties of copulas, there are many concepts of dependence, tail properties and tail asymmetries that are needed. These are presented in Chapter 2 before the copula construction methods in Chapter 3. Chapter 4 has summaries of many useful parametric copula families that can be used in construction of high-dimensional copulas. Chapter 5 covers inferential and diagnostic techniques for dependence models with copulas. Chapter 6 is concerned with numerical methods and algorithms for efficient copula computations. Chapter 7 has a variety of applications of copula models with data sets in different areas. Chapter 8 has more theory and background for results in Chapters 2 to 6.

Chapter 2

Basics: dependence, tail behavior and asymmetries

Classical multivariate statistics is based on the multivariate Gaussian distribution. This however, is mostly mathematical convenience, because the multivariate Gaussian distribution generally cannot be justified as a stochastic or physical mechanism. If variables can be considered as coming from sums or averages of many unmeasured quantities, then maybe the Central Limit Theorem provides a justification via an approximation.

The dependence for the multivariate Gaussian distribution is summarized in the correlation matrix. Correlation is not the best measure of dependence in general; it is a measure of linear dependence, and might not reach ± 1 for non-Gaussian random variables with perfect dependence (monotone relationship). For two continuous variables, extreme types of dependence are perfect positive dependence (or comonotonicity) and perfect negative dependence (or countermonotonicity).

Embrechts et al. (2002) mentioned several fallacies concerning dependence, two of which are repeated below.

- Fallacy 1. Marginal distributions and correlation determine the joint distribution.
- Fallacy 2. Given marginal distributions F_1 and F_2 for X_1 and X_2 respectively, all linear correlations between -1 and 1 can be attained through suitable specification of the joint distribution.

The above two statements do not hold, and there are many concepts of dependence that are needed to study multivariate distributions in general.

The chapter has basic material on concepts needed for construction of multivariate distributions and analyzing their dependence and tail properties. Simple examples are used to illustrate many of the concepts. Section 2.1 has basic results on multivariate cumulative distribution functions and their conditional distributions, and mixture models based on conditional independence. Section 2.2 has results on Laplace transforms of positive random variables because Laplace transforms appear in the multivariate cumulative distribution function (cdf) of several mixture model constructions. Section 2.3 has the basics of univariate extreme value theory as a primer for multivariate extremes. Related are ideas of univariate tail heaviness in Section 2.4. Section 2.5 has the basic result on the probability integral transform which is used for converting copulas to multivariate distributions and vice versa.

Basic properties of multivariate Gaussian and t, or more generally elliptical distributions, are covered in Sections 2.6 and 2.7 as there will be construction methods for multivariate models that include these as special cases.

Measures and concepts of dependence that are used for multivariate non-Gaussian distributions and copulas include: Fréchet classes and bounds, the concordance ordering, measures of monotone association (such as Kendall's tau, Spearman's rho, correlation of normal scores, Blomqvist's beta), positive quadrant/orthant dependence, dependence from stochastically increasing, total positivity of order 2, max infinite-divisibility, and tail dependence. Section 2.8 has the most useful dependence concepts for deriving dependence properties of copulas. Sections 2.9 and 2.10 have results for Fréchet classes and bounds given univariate margins or subsets of multivariate margins. Sections 2.11 and 2.12 have basic results on orderings of dependence and measures of monotone assocation. Section 2.13 is concerned with tail dependence.

The concordance ordering is a way to compare multivariate distributions with the same set of univariate margins so that we can decide if one multivariate distribution represents more dependence than another. For parametric families of multivariate distributions, a parameter is more interpretable if the joint distribution has more dependence as the parameter increases. Parametric families of multivariate distributions are usually constructed via stochastic representations and the simplest way to display the distribution is usually through the cdf — and then the density is obtained by differentiation (for continuous variables) or finite differences (for discrete variables). Often dependence and tail properties are obtained based on stochastic representations.

Multivariate distributions when represented through copulas need not have reflection symmetry (symmetry of joint upper and lower tails) such as for Gaussian and elliptical distributions. Concepts and measures of tail asymmetry and permutation asymmetry are covered in Sections 2.14 and 2.15. The concept of tail order, in Section 2.16, can also be used to quantify tail asymmetry.

Tail dependence is linked to the existence of a nontrivial limiting multivariate extreme value distribution. To understand how different copula families can differ in strength of dependence in the tails, the tail order in Section 2.16 is also used to quantify joint tails of copulas when there may not be the strongest tail dependence for multivariate extremes — this concept also has its origins in the extreme value literature. Other concepts that help with analysis of strength of dependence in the tails of copulas include semi-correlations of normal scores in Section 2.17, boundary conditional cdfs in Section 2.19, and related conditional tail expectations in Section 2.20.

Section 2.18 has basic results on tail dependence functions that are useful for analyzing tail dependence of mixture models. They also are used in some conditional tail expectations in Section 2.20 when there is tail dependence. Section 2.21 is concerned with tail comonotonicity, the strongest form of tail dependence.

Section 2.22 contains a chapter summary.

2.1 Multivariate cdfs and their conditional distributions

This section summarizes

- necessary conditions for multivariate cdfs;
- the decomposition of a multivariate distribution into absolutely continuous, singular and discrete parts;
- conditional distributions obtained from a given multivariate distribution;
- some basic operations on distribution functions such as the power of a cdf and mixtures.

2.1.1 Conditions for multivariate cdfs

Multivariate cdfs have additional necessary conditions in comparison to univariate cdfs, so that it is harder to write down arbitrary multivariate functions that correspond to cdfs.

The only basic requirements of univariate cdfs are monotonicity, right-continuity and endpoint constraints. If $G : \mathbb{R} \to [0,1]$ is a univariate cdf, then it is increasing (non-decreasing) from $G(-\infty) = 0$ to $G(\infty) = 1$; increasing includes flat segments and jumps like for cdf of discrete random variable. It then follows that the univariate survival function $\overline{G} = 1 - G$ is decreasing (non-increasing) from $\overline{G}(-\infty) = 1$ to $\overline{G}(\infty) = 0$.

Next we state the necessary and sufficient conditions for a right-continuous function F

on $\mathbb{R}^2$ to be a bivariate cdf; these are the basic properties of a bivariate cdf F. The conditions are:

1. $\lim_{x_j \to -\infty} F(x_1, x_2) = 0$, $j = 1, 2$;
2. $\lim_{x_j \to \infty \forall j} F(x_1, x_2) = 1$;
3. (rectangle inequality or 2-increasing) for all (a_1, a_2), (b_1, b_2) with $a_1 < b_1$, $a_2 < b_2$,

$$\Delta_{a_1}^{b_1} \Delta_{a_2}^{b_2} F := F(b_1, b_2) - F(a_1, b_2) - F(b_1, a_2) + F(a_1, a_2) \geq 0. \tag{2.1}$$

(i) If F has second-order derivatives, then condition 3 is equivalent to $\partial^2 F/\partial x_1 \partial x_2 \geq 0$.

(ii) Let $a_2 \to -\infty$ in (2.1); then $F(b_1, b_2) - F(a_1, b_2) \geq 0$ and F is increasing in the first variable. Similarly, from letting $a_1 \to -\infty$, F is increasing in the second variable. That is, 2-increasing implies increasing in each variable.

(iii) Conditions 1 and 2, combined with (ii), imply that $0 \leq F \leq 1$.

(iv) The univariate margins F_1, F_2 of $F(x_1, x_2)$ are obtained by letting $x_2 \to \infty$ and $x_1 \to \infty$, respectively.

The conditions or basic properties for a bivariate cdf extend to multivariate cdfs. Necessary and sufficient conditions for a right-continuous function F on $\mathbb{R}^d$ to be a multivariate cdf are:

1. $\lim_{x_j \to -\infty} F(x_1, \ldots, x_d) = 0$, $j = 1, \ldots, d$;
2. $\lim_{x_j \to \infty \forall j} F(x_1, \ldots, x_d) = 1$;
3. (rectangle inequality or d-increasing) for all $(a_1, \ldots, a_d)$, $(b_1, \ldots, b_d)$ with $a_i < b_i$, $i = 1, \ldots, d$,

$$\Delta_{a_1}^{b_1} \cdots \Delta_{a_d}^{b_d} F := \sum_{i_1=1}^{2} \cdots \sum_{i_d=1}^{2} (-1)^{i_1 + \cdots + i_d} F(x_{1 i_1}, \ldots, x_{d i_d}) \geq 0, \tag{2.2}$$

where $x_{j1} = a_j$, $x_{j2} = b_j$.

(i) If F has dth-order derivatives, then condition 3 is equivalent to $\partial^d F/\partial x_1 \cdots \partial x_d \geq 0$.

(ii) Let $a_2, \ldots, a_d \to -\infty$ in (2.2); then $F(b_1, b_2, \ldots, b_d) - F(a_1, b_2, \ldots, b_d) \geq 0$ and F is increasing in the first variable. Similarly, by symmetry, F is increasing in the remaining variables.

(iii) Let S be a non-empty subset of $\{1, \ldots, d\}$. The margin F_S of $F(\boldsymbol{x})$ is obtained by letting $x_i \to \infty$ for $i \notin S$.

(iv) If $(X_1, \ldots, X_d) \sim F$, then the rectangle condition is equivalent to

$$\mathbb{P}(a_j < X_j \leq b_j, \ j = 1, \ldots, d) \geq 0, \quad a_i < b_i \ \forall i.$$

(v) If F is a multivariate discrete cdf, then all probability masses are computed via (2.2).

For multivariate distributions, note that the multivariate survival function is not one minus the cdf. If $F \in \mathcal{F}(F_1, \ldots, F_d)$, the Fréchet class of multivariate distributions with univariate margins $F_1, \ldots, F_d$, and $\boldsymbol{X} \sim F$, then the *multivariate survival function* is defined as $\overline{F}(\boldsymbol{x}) = \mathbb{P}(X_1 > x_1, \ldots, X_d > x_d)$. By the inclusion-exclusion probability law,

$$\overline{F}(\boldsymbol{x}) = 1 - \sum_{j=1}^{d} F_j(x_j) + \sum_{S \subset \{1, \ldots, d\}, |S| \geq 2} (-1)^{|S|} F_S(x_i, i \in S), \quad \boldsymbol{x} \in \mathbb{R}^d,$$

where $|S|$ is the cardinality of S. If $F = C$ is the copula of $(U_1, \ldots, U_d)$, then the survival function of the copula is

$$\overline{C}(\boldsymbol{u}) = 1 - \sum_{j=1}^{d} u_j + \sum_{S \subset \{1, \ldots, d\}, |S| \geq 2} (-1)^{|S|} C_S(u_i, i \in S), \quad \boldsymbol{u} \in [0, 1]^d.$$

The survival or reflected copula $\widehat{C}$ is the distribution of the vector $(1 - U_1, \ldots, 1 - U_d)$ of reflected uniform random variables, and

$$\widehat{C}(\boldsymbol{u}) = \overline{C}(\mathbf{1}_d - \boldsymbol{u}) = 1 - \sum_{j=1}^{d}(1 - u_j) + \sum_{S \subset \{1,\ldots,d\}, |S| \geq 2} (-1)^{|S|} C_S(1 - u_i, i \in S), \quad \boldsymbol{u} \in [0,1]^d.$$

For the bivariate case,

$$\widehat{C}(u,v) = u + v - 1 + C(1 - u, 1 - v), \quad 0 \leq u \leq 1, \ 0 \leq v \leq 1.$$

2.1.2 Absolutely continuous and singular components

Random variables in statistical modeling could be discrete or continuous or mixed discrete-continuous. An example of the third type is a non-negative random variable with a mass at 0 but is otherwise continuous (a concrete example is the daily amount of rainfall).

The cdf of a discrete random variable is a step function with positive jumps; the cdf of a continuous random variable is continuous and the cdf of a mixed discrete-continuous has a mixture of continuous increasing segments and positive jump discontinuities.

Because a univariate cdf is increasing and right-continuous, it has right derivatives defined everywhere. Similarly, because of satisfying the rectangle condition, a multivariate cdf $F(x_1, \ldots, x_d)$ has a mixed derivative $f = \frac{\partial^d F}{\partial x_1 \cdots \partial x_d}$ everywhere if this is considered as a right derivative. A cdf F is *absolutely continuous* if it is the antiderivative of its mixed right derivative:

$$F(x_1, \ldots, x_d) = \int_{-\infty}^{x_1} \cdots \int_{-\infty}^{x_d} f(y_1, \ldots, y_d) \, dy_1 \cdots dy_d, \quad \boldsymbol{x} \in \mathbb{R}^d. \tag{2.3}$$

If $f \equiv 0$, then F is a step function and the corresponding random vector is purely discrete; in this case (2.3) definitely does not hold. It is possible for a continuous cdf F not to satisfy (2.3). Such univariate examples are not relevant for statistical modeling, but practical multivariate examples arise when there are functional relationships in the components of the random vector. A concrete example is given in Example 2.1 below.

The general result is that a cdf (univariate or multivariate) is a mixture of three components:

$$F = p_{\text{disc}} F_{\text{disc}} + p_{\text{sing}} F_{\text{sing}} + p_{\text{ac}} F_{\text{ac}},$$

where $F_{\text{disc}}, F_{\text{sing}}, F_{\text{ac}}$ are cdfs from respectively the *discrete, singular and absolutely continuous components*, with corresponding probabilities $p_{\text{disc}}, p_{\text{sing}}, p_{\text{ac}}$ such that $p_{\text{disc}} + p_{\text{sing}} + p_{\text{ac}} = 1$. The first discrete component $p_{\text{disc}} F_{\text{disc}}$ comes from the point masses of F and the third absolutely continuous component comes from integrating the mixed dth-order right derivative f of F, that is, $p_{\text{ac}} F_{\text{ac}}(\boldsymbol{x}) = \int_{-\infty}^{x_1} \cdots \int_{-\infty}^{x_d} f(\boldsymbol{y}) \, d\boldsymbol{y}$. The singular component comes from differencing: $p_{\text{sing}} F_{\text{sing}} = F - p_{\text{disc}} F_{\text{disc}} - p_{\text{ac}} F_{\text{ac}}$, or it can sometimes be identified from the functional relationship. The decomposition could also be written in terms of survival functions.

Example 2.1 (Bivariate Marshall-Olkin exponential with singular component). Let $X_1 = \min\{Z_1, Z_{12}\}$, $X_2 = \min\{Z_2, Z_{12}\}$ where Z_1, Z_2, Z_{12} are independent exponential random variables with rate parameters $\eta_1, \eta_2, \eta_{12}$ respectively. Then

$$\overline{F}(x_1, x_2) = \mathbb{P}(X_1 > x_1, X_2 > x_2) = \exp\{-\eta_1 x_1 - \eta_2 x_2 - \eta_{12}(x_1 \vee x_2)\}, \ x_1 \geq 0, \ x_2 \geq 0,$$

so that

$$\begin{aligned}
f(x_1, x_2) &= \overline{F}(x_1, x_2)\big[\eta_1 + \eta_{12} I(x_1 > x_2)\big]\big[\eta_2 + \eta_{12} I(x_2 > x_1)\big] \\
&= \begin{cases} \eta_2(\eta_1 + \eta_{12})e^{-\eta_2 x_2} e^{-(\eta_1 + \eta_{12})x_1}, & x_1 > x_2, \\ \eta_1(\eta_2 + \eta_{12})e^{-\eta_1 x_1} e^{-(\eta_2 + \eta_{12})x_2}, & x_1 < x_2. \end{cases}
\end{aligned}$$

Note that $f(x_1, x_2)$ is not well-defined (or unique) for $x_1 = x_2 > 0$, because the limits from $x_1 \downarrow x_2$ and $x_1 \uparrow x_2$ could be different. The absolutely continuous component $p_a \overline{F}_a$ can be obtained from $\int_{x_1}^{\infty} \int_{x_2}^{\infty} f(y_1, y_2) \mathrm{d}y_2 \mathrm{d}y_1$. Let $\eta_\bullet = \eta_1 + \eta_2 + \eta_{12}$. For $x_1 > x_2$, by splitting the integral into two parts, one gets:

$$\int_{y_1 = x_1}^{\infty} \int_{y_2 = x_2}^{y_1} f(y_1, y_2) \mathrm{d}y_2 \mathrm{d}y_1 + \int_{y_1 = x_1}^{\infty} \int_{y_2 = y_1}^{\infty} f(y_1, y_2) \mathrm{d}y_2 \mathrm{d}y_1$$

$$= \int_{x_1}^{\infty} (\eta_1 + \eta_{12}) [e^{-\eta_2 x_2} - e^{-\eta_2 y_1}] e^{-(\eta_1 + \eta_{12}) y_1} \mathrm{d}y_1 + \int_{x_1}^{\infty} \eta_1 e^{-(\eta_1 + \eta_2 + \eta_{12}) y_1} \mathrm{d}y_1$$

$$= e^{-\eta_2 x_2} e^{-(\eta_1 + \eta_{12}) x_1} - (\eta_1 + \eta_{12}) \eta_\bullet^{-1} e^{-\eta_\bullet x_1} + \eta_1 \eta_\bullet^{-1} e^{-\eta_\bullet x_1}.$$

Similarly, there is a symmetric relation for $x_1 < x_2$. Hence

$$p_{ac} \overline{F}_{ac}(x_1, x_2) = \begin{cases} e^{-\eta_2 x_2} e^{-(\eta_1 + \eta_{12}) x_1} - \eta_{12} \eta_\bullet^{-1} e^{-\eta_\bullet x_1}, & x_1 > x_2, \\ e^{-\eta_1 x_1} e^{-(\eta_2 + \eta_{12}) x_2} - \eta_{12} \eta_\bullet^{-1} e^{-\eta_\bullet x_2}, & x_1 < x_2, \end{cases}$$

and $p_{ac} = 1 - \eta_{12}/\eta_\bullet$ by letting $x_1, x_2 \to 0^+$. The singular component comes from the event $\{X_1 = X_2\}$ and $p_{sing} = \mathbb{P}(X_1 = X_2) = \mathbb{P}(Z_{12} < Z_1, Z_{12} < Z_2) = \eta_{12}/\eta_\bullet$. The singular component could be obtained directly as:

$$\overline{F}_{sing}(x_1, x_2) = \exp\{-\eta_\bullet (x_1 \vee x_2)\},$$

because $[Z_{12} | Z_{12} < Z_1, Z_{12} < Z_2]$ is exponential with rate $\eta_\bullet$. It can be checked that $p_{ac} \overline{F}_{ac} + p_{sing} \overline{F}_{sing} = \overline{F}$.

2.1.3 Conditional cdfs

Conditional cdfs of multivariate distributions and copulas are needed for simulation and for construction methods such as vines; both involves sequences or sets of conditional cdfs.

We introduce some notation for partial derivatives to be applied for conditional cdfs. Let h be a differentiable function of order 1. Then $\mathrm{D}_j h$ is a differential operator notation for the first order partial derivative with respect to the jth argument of h, that is, $\mathrm{D}_j h(\boldsymbol{x}) = \frac{\partial h}{\partial x_j}$. If h is also a differentiable function of order 2, then $\mathrm{D}_{jk}^2 h$ is a differential operator notation for a second order partial derivative, that is, $\mathrm{D}_{jk}^2 h(\boldsymbol{x}) = \frac{\partial^2 h}{\partial x_j \partial x_k}$, where j, k can be distinct or the same. This notation extends to higher order derivatives.

We will define the conditional distributions of a multivariate cdf through the differential operator notation. Let S_1, S_2 be two non-overlapping non-empty subsets of $\{1, \dots, d\}$.

Let $(X_1, \dots, X_d) \sim F$, where $F \in \mathcal{F}(F_1, \dots, F_d)$. If $X_1, \dots, X_d$ are all discrete, then conditional distributions of the form $\mathbb{P}(X_j \le x_j, j \in S_1 | X_k = x_k, k \in S_2)$ are defined from conditional probability applied to events. If $X_1, \dots, X_d$ are all continuous random variables, and $F_1, \dots, F_d$ are absolutely continuous with respective densities $f_1, \dots, f_d$, then the conditional cdfs are defined via limits.

If $S_2 - \{k\}$, then

$$F_{S_1 | k}(\boldsymbol{x}_{S_1} | x_k) := \lim_{\epsilon \to 0^+} \frac{\mathbb{P}(X_j \le x_j, j \in S_1, \, x_k \le X_k < x_k + \epsilon)}{\mathbb{P}(x_k \le X_k < x_k + \epsilon)} = \frac{\partial F_{S_1 \cup \{k\}} / \partial x_k}{f_k(x_k)}.$$

Similarly, the conditional survival function is:

$$\overline{F}_{S_1 | k}(\boldsymbol{x}_{S_1} | x_k) := \lim_{\epsilon \to 0^+} \frac{\mathbb{P}(X_j > x_j, j \in S_1, \, x_k \le X_k < x_k + \epsilon)}{\mathbb{P}(x_k \le X_k < x_k + \epsilon)} = \frac{-\partial \overline{F}_{S_1 \cup \{k\}} / \partial x_k}{f_k(x_k)}.$$

For the special case where $\boldsymbol{X} = \boldsymbol{U}$ is a vector of $U(0,1)$ random variables and $F = C$ is a copula, this conditional distribution is

$$C_{S_1|k}(\boldsymbol{u}_{S_1}|u_k) = \partial C_{S_1 \cup \{k\}}(\boldsymbol{u}_{S_1}, u_k)/\partial u_k.$$

For $d = 2$ with $S_1 = \{1\}$, $S_2 = \{2\}$, the conditional cdfs are denoted as $F_{1|2} = (D_2 F)/f_2$ and $C_{1|2} = D_2 C = \partial C/\partial u_2$.

If the cardinality of S_2 is greater than or equal to 2, then the definition of the conditional cdf is:

$$F_{S_1|S_2}(\boldsymbol{x}_{S_1}|\boldsymbol{x}_{S_2}) = \lim_{\epsilon \to 0} \frac{\mathbb{P}(X_j \leq x_j, j \in S_1; x_k \leq X_k < x_k + \epsilon, k \in S_2)}{\mathbb{P}(x_k \leq X_k < x_k + \epsilon, k \in S_2)} = \frac{\frac{\partial^{|S_2|} F_{S_1 \cup S_2}}{\prod_{k \in S_2} \partial x_k}}{f_{S_2}(\boldsymbol{x}_{S_2})},$$

provided F_{S_2} is absolutely continuous. Similarly, the conditional survival function is:

$$\overline{F}_{S_1|S_2}(\boldsymbol{x}_{S_1}|\boldsymbol{x}_{S_2}) = (-1)^{|S_2|} \frac{\partial^{|S_2|} \overline{F}_{S_1 \cup S_2}/\prod_{k \in S_2} \partial x_k}{f_{S_2}(\boldsymbol{x}_{S_2})}, \tag{2.4}$$

if $\overline{F}_{S_2}$ is absolutely continuous. If F_{S_2} is not absolutely continuous and has a singular component, then the above equation might be valid for points $\boldsymbol{x}_{S_2}$ where $f_{S_2}(\boldsymbol{x}_{S_2})$ is well defined; otherwise $F_{S_1|S_2}$ or $\overline{F}_{S_1|S_2}$ can be obtained from an understanding of the singular component. See Examples 2.3 and 2.4 below.

For the special case where $\boldsymbol{X} = \boldsymbol{U}$ is a vector of $U(0,1)$ random variables and $F = C$ is a copula, this conditional distribution is

$$C_{S_1|S_2}(\boldsymbol{u}_{S_1}|\boldsymbol{u}_{S_2}) = \frac{\partial^{|S_2|} C_{S_1 \cup S_2}(\boldsymbol{u}_{S_1}, \boldsymbol{u}_{S_2})/\prod_{k \in S_2} \partial u_k}{c_{S_2}(\boldsymbol{u}_{S_2})}, \tag{2.5}$$

if the marginal copula C_{S_2} is absolutely continuous with density c_{S_2}.

Example 2.2 (Bivariate Marshall-Olkin exponential continued). In Example 2.1, X_1 and X_2 are exponential random variables with rate parameters $\eta_1 + \eta_{12}$ and $\eta_2 + \eta_{12}$ respectively. By differentiation of the survival function with respect to x_2,

$$\begin{aligned}
\overline{F}_{1|2}(x_1|x_2) &= \frac{\overline{F}(x_1, x_2)[\eta_2 + \eta_{12} I(x_1 < x_2)]}{(\eta_2 + \eta_{12}) e^{(\eta_2 + \eta_{12}) x_2}} \\
&= \begin{cases} e^{-\eta_1 x_1}, & 0 \leq x_1 < x_2, \\ \frac{\eta_2}{\eta_2 + \eta_{12}} e^{-\eta_1 x_1 - \eta_{12}(x_1 - x_2)}, & x_1 \geq x_2. \end{cases}
\end{aligned}$$

This has a jump discontinuity of size $\eta_{12}/(\eta_2 + \eta_{12})$ when $x_1 = x_2$.

Example 2.3 (Trivariate Marshall-Olkin exponential). Let $X_1 = \min\{Z_1, Z_{12}, Z_{13}, Z_{123}\}$, $X_2 = \min\{Z_2, Z_{12}, Z_{23}, Z_{123}\}$, $X_3 = \min\{Z_3, Z_{13}, Z_{23}, Z_{123}\}$, where Z_S are independent exponential random variables with rate parameters η_S when S is a non-empty subset of $\{1, 2, 3\}$. Then

$$\overline{F}(x_1, x_2, x_3) = \mathbb{P}(X_1 > x_1, X_2 > x_2, X_3 > x_3) = \exp\left\{-\sum_S \eta_S(\vee_{j \in S} x_j)\right\}, \quad x_j > 0 \, \forall j,$$

and the $(2, 3)$ bivariate margin is

$$\overline{F}_{23}(x_2, x_3) = \exp\{-(\eta_2 + \eta_{12}) x_2 - (\eta_3 + \eta_{13}) x_3 - (\eta_{23} + \eta_{123})(x_2 \vee x_3)\}.$$

$\overline{F}_{23}$ is partly singular and its density is

$$f_{23}(x_2, x_3) = \begin{cases} (\eta_3 + \eta_{13})(\eta_2 + \eta_{12} + \eta_{23} + \eta_{123})e^{-(\eta_3+\eta_{13})x_3}e^{-(\eta_2+\eta_{12}+\eta_{23}+\eta_{123})x_2}, & x_2 > x_3, \\ (\eta_2 + \eta_{12})(\eta_3 + \eta_{13} + \eta_{23} + \eta_{123})e^{-(\eta_2+\eta_{12})x_2}e^{-(\eta_3+\eta_{13}+\eta_{23}+\eta_{123})x_3}, & x_2 < x_3. \end{cases}$$

Note that this density is not well defined for $x_2 = x_3$. $\overline{F}_{1|23}$ is valid based on (2.4) for $x_2 \neq x_3$. For $x_2 = x_3 = x$, the conditional cdf is based on $\{X_2 = X_3 = Z_{23}$ or $Z_{123}\}$:

$$\mathbb{P}(X_2 = X_3 = Z_{23} < Z_{123}|X_2 = X_3 = Z_{23} \wedge Z_{123}) = \eta_{23}/(\eta_{23} + \eta_{123}),$$
$$\mathbb{P}(X_1 > x_1|Z_{23} = x < \min\{Z_2, Z_3, Z_{12}, Z_{13}, Z_{123}\}) = e^{-\eta_1 x_1}e^{-(\eta_{12}+\eta_{13}+\eta_{123})(x_1 \vee x)},$$
$$\mathbb{P}(X_1 > x_1|Z_{123} = x < \min\{Z_2, Z_3, Z_{12}, Z_{13}, Z_{23}\}) = e^{-\eta_1 x_1}I(x_1 < x),$$
$$\overline{F}_{1|23}(x_1|x, x) = \frac{\eta_{23}}{\eta_{23} + \eta_{123}}e^{-\eta_1 x_1}e^{-(\eta_{12}+\eta_{13}+\eta_{123})(x_1 \vee x)} + \frac{\eta_{123}}{\eta_{23} + \eta_{123}}e^{-\eta_1 x_1}I(x_1 < x);$$

this has a jump discontinuity of size $\frac{\eta_{123}}{\eta_{23}+\eta_{123}}e^{-\eta_1 x}$ when $x_1 = x$.

Example 2.4 (Trivariate mixture of independence and perfect dependence). Let (U_1, U_2, U_3) be a random vector such that (i) U_1, U_2, U_3 are independent $U(0, 1)$ random variables with probability $0 < \pi < 1$ and (ii) $U_1 = U_2 = U_3 = U$ with probability $1 - \pi$ where $U \sim U(0, 1)$. Then each U_j has a $U(0, 1)$ distribution. The joint distribution or copula of U is

$$C(\boldsymbol{u}) = \pi u_1 u_2 u_3 + (1 - \pi)\min\{u_1, u_2, u_3\}, \quad \boldsymbol{u} \in [0, 1]^3.$$

The $(2, 3)$ marginal copula is $C_{23}(u_2, u_3) = \pi u_2 u_3 + (1 - \pi)\min\{u_2, u_3\}$, and it is partly singular; its density is $c_{23}(u_2, u_3) = \pi$. Equation (2.5) is valid only if $u_2 \neq u_3$, in which case it becomes

$$C_{1|23}(u_1|u_2, u_3) = \frac{\pi u_1}{\pi} = u_1;$$

this is essentially the conditional cdf based on the "independent" component. The conditional cdf $C_{1|23}$ when $u_2 = u_3 = u$, should be based on the event $\{U_2 = U_3 = U \in [u, u+\epsilon)\}$ with $\epsilon \to 0^+$ and this leads to $C_{1|23}(u_1|u, u) = I(u_1 > u)$.

2.1.4 Mixture models and conditional independence models

One cannot just write down a family of functions and expect it to satisfy the necessary conditions for a multivariate cdf in Section 2.1.1. As seen earlier in Section 1.2 and with more details in Chapter 3, multivariate distributions will be constructed through stochastic representations and limits.

A general approach to get a new multivariate distribution from parametric families is through mixture families, which is a special stochastic representation. For example, the multivariate Pareto and t_ν families in Section 1.2 are derived in this way. Let $\boldsymbol{X} = (X_1, \ldots, X_d)$ and let $\boldsymbol{Q} = \boldsymbol{q}$ be a random variable or vector. If $F_{\boldsymbol{X}|\boldsymbol{Q}}(\boldsymbol{x}|\boldsymbol{q})$ is defined for each $\boldsymbol{q}$, then the joint cdf of $\boldsymbol{X}$ is

$$F_{\boldsymbol{X}}(\boldsymbol{x}) = \int F_{\boldsymbol{X}|\boldsymbol{Q}}(\boldsymbol{x}|\boldsymbol{q})\,dF_{\boldsymbol{Q}}(\boldsymbol{q}). \tag{2.6}$$

Model (2.6) is called a *conditional independence* model if $X_1, \ldots, X_d$ are conditionally independent given $\boldsymbol{Q} = \boldsymbol{q}$ for every $\boldsymbol{q}$; then

$$F_{\boldsymbol{X}}(\boldsymbol{x}) = \int \prod_{j=1}^{d} F_{X_j|\boldsymbol{Q}}(x_j|\boldsymbol{q})\,dF_{\boldsymbol{Q}}(\boldsymbol{q}). \tag{2.7}$$

Some copula models in subsequent chapters are obtained via (2.6) or (2.7).

2.1.5   Power of a cdf or survival function

If F is univariate cdf, then so is F^q for any $q > 0$. If $\overline{F}$ is a univariate survival function, then so is $\overline{F}^q$ for any $q > 0$.

Any power of a univariate cdf is a cdf and F could be continuous or discrete. This follows because the only conditions for F to be a cdf are that it is increasing and right-continuous with $F(-\infty) = 0$ and $F(\infty) = 1$, and the only conditions for $\overline{F}$ to be a survival function are that it is decreasing and right-continuous with $\overline{F}(-\infty) = 1$ and $\overline{F}(\infty) = 0$. These properties are preserved for a positive power of a cdf and survival function.

In following the terminology of Marshall and Olkin (2007), for the family $G(\cdot; q) = F^q$, q is called a *resilience* parameter, and for the family $\overline{G}(\cdot; q) = \overline{F}^q$, q is called a *frailty* parameter. Note that F^q is increasing stochastically as q increases, because $1 - F^q(x)$ is increasing in q for any x, and $\overline{F}^q$ is decreasing stochastically as q increases because $\overline{F}^q(x)$ is decreasing in q for any x. This explains the terminology of resilience and frailty when F is a life distribution with support on $[0, \infty)$; for F^q, the lifetime is stochastically larger (more resilient) as q increases, and for $\overline{F}^q$, the lifetime is stochastically smaller (more frail) as q increases. As $q \to 0^+$, F^q converges in distribution to a degenerate distribution at the left endpoint of support of F (this could be $-\infty$), and as $q \to \infty$, F^q converges in distribution to a degenerate distribution at the right endpoint of support of F (this could be ∞). Similarly, as $q \to 0^+$, $\overline{F}^q$ converges in distribution to a degenerate distribution at the right endpoint of support of F and as $q \to \infty$, $\overline{F}^q$ converges in distribution to a degenerate distribution at the left endpoint of support of F.

The power of multivariate cdf is not necessary a cdf for dimensions $d \geq 2$. This can be seen for the bivariate case for an absolutely continuous cdf F with density f. If $F^q(x_1, x_2)$ is a cdf for all $q > 0$, then its mixed derivative

$$\frac{\partial^2 F^q}{\partial x_1 \partial x_2} = q F^{q-2} \Big[F \frac{\partial^2 F}{\partial x_1 \partial x_2} + (q-1) \frac{\partial F}{\partial x_1} \frac{\partial F}{\partial x_2} \Big]$$

should be non-negative for all x_1, x_2. With $q \to 0^+$, this would require

$$F \cdot f - \frac{\partial F}{\partial x_1} \frac{\partial F}{\partial x_2} \geq 0.$$

The theorems in Section 8.3 imply that if all positive powers of a multivariate cdf are cdfs, then this corresponds to a strong positive dependence condition.

For now, we provide some summaries of results for products of cdfs or survival functions as stochastic representations for the max/min operators and contrast these with results from convolutions (sum operator).

1. The product of two multivariate cdfs is another cdf. If F, G are d-variate cdfs such that $\boldsymbol{X} = (X_1, \ldots, X_d) \sim F$ and $\boldsymbol{Y} = (Y_1, \ldots, Y_d) \sim G$ with $\boldsymbol{X} \perp\!\!\!\perp \boldsymbol{Y}$, then the vector of coordinate-wise maxima $(X_1 \vee Y_1, \ldots, X_d \vee Y_d) \sim FG$.

2. From the above, any positive integer power of a multivariate cdf is a cdf.

3. (Sum-)infinitely divisible: $\boldsymbol{X}$ is *infinitely* divisible if for every positive integer n, there exists i.i.d. $\boldsymbol{X}_1^{(n)}, \ldots, \boldsymbol{X}_n^{(n)}$ such that $\boldsymbol{X} \stackrel{d}{=} \boldsymbol{X}_1^{(n)} + \cdots + \boldsymbol{X}_n^{(n)}$. If φ is characteristic function or moment generating function or LT of $\boldsymbol{X}$, then $\varphi^{1/n}$ is the corresponding function for $\boldsymbol{X}_1^{(n)}$.

4. Max-infinitely divisible: $\boldsymbol{X}$ is *max-infinitely divisible (max-id)* if for every positive integer n, there exists i.i.d. $\boldsymbol{X}_1^{(n)}, \ldots, \boldsymbol{X}_n^{(n)}$ such that $\boldsymbol{X} \stackrel{d}{=} \boldsymbol{X}_1^{(n)} \vee \cdots \vee \boldsymbol{X}_n^{(n)}$ (coordinate-wise maxima). If G is the cdf of $\boldsymbol{X}$, then $G^{1/n}$ is the cdf for $\boldsymbol{X}_1^{(n)}$. A multivariate cdf G is max-id if G^q is a cdf for all $q > 0$.

5. Min-infinitely divisible: $\boldsymbol{X}$ is *min-infinitely divisible (min-id)* if for every positive integer

n, there exists i.i.d. $\boldsymbol{X}_1^{(n)}, \ldots, \boldsymbol{X}_n^{(n)}$ such that $\boldsymbol{X} \overset{d}{=} \boldsymbol{X}_1^{(n)} \wedge \cdots \wedge \boldsymbol{X}_n^{(n)}$ (coordinate-wise minima). If $\overline{G}$ is the survival function of $\boldsymbol{X}$, then $\overline{G}^{1/n}$ is the survival function for $\boldsymbol{X}_1^{(n)}$. A multivariate cdf G is min-id if $\overline{G}^q$ is a survival function for all $q > 0$

6. From max-id and min-id to copulas: Let G be a continuous cdf G with copula $C(\boldsymbol{u}) = G(G_1^{-1}(u_1), \ldots, G_d^{-1}(u_d))$. If a continuous cdf G is max-id, then its copula is max-id. If a continuous cdf G is min-id, then $\overline{C}^q$ is a survival function for all $q > 0$, and its reflected copula $\widehat{C}(\boldsymbol{u}) = \overline{C}(\mathbf{1}_d - \boldsymbol{u})$ is max-id.

2.2   Laplace transforms

In this section, Laplace transforms are defined for non-negative random variables. If X is a non-negative random variable, then its Laplace transform (LT) or Laplace-Stiltjes transform is defined as:

$$\psi(s) = \psi_X(s) := \mathbb{E}[e^{-sX}] = \int_{[0,\infty)} e^{-sx} \mathrm{d}F_X(x), \quad s > 0. \qquad (2.8)$$

Since $|e^{-sx}| \leq 1$ for $x \geq 0$ and $s \geq 0$, the LT always exist for non-negative random variables. Hence, for non-negative random variables, it is more convenient to use than the moment generating function.

Some useful properties of LTs are given in Feller (1971) and Marshall and Olkin (2007). Properties that are used for construction and analysis of copula families are summarized below.

- If ψ is the LT of a non-negative random variable X, then $\psi(0) = 1$, ψ is bounded by 1 and strictly decreasing.
- A probability distribution with support on $[0, \infty)$ is uniquely determined by its LT (Feller (1971), p. 430).
- ψ is *completely monotone*, meaning that it has continuous derivatives of all orders and the derivatives alternate in sign, i.e., $(-1)^i \psi^{(i)}(s) \geq 0$ for $i = 1, 2, \ldots$ and $s > 0$, where $\psi^{(i)}$ is the ith derivative. The proof of complete monotonicity is via differentiation under the integral (see Feller (1971), p. 233), in which case, $(-1)^i \psi^{(i)}(s) = \mathbb{E}[X^i e^{-sX}]$ for $i = 1, 2, \ldots$.
- Bernstein's theorem (Bernstein (1928)): A function $\psi : (0, \infty) \to [0, 1]$ is the LT of a non-negative random variable iff it is completely monotone and $\psi(0) = 1$.
- LTs are survival functions on $[0, \infty)$ corresponding to the class of scale mixtures of exponential survival functions; this follows from (2.8) with F_X being the mixing distribution for the exponential rate parameter.
- If X has positive mass p_0 at 0, then (2.8) becomes

$$\psi(s) = p_0 + \int_{(0,\infty)} e^{-sx} \mathrm{d}F_X(x)$$

and $\psi(\infty) = \lim_{s \to \infty} \psi(s) = p_0$. Hence $\psi(\infty) = 0$ only if $p_0 = 0$.

- If $\psi(\infty) = p_0$, the functional inverse $\psi^{-1} : [p_0, 1] \to [0, \infty)$ is strictly decreasing and satisfies $\psi^{-1}(p_0) = \infty$, $\psi^{-1}(1) = 0$.

For applications of LTs for construction of copulas, it is necessary that $\psi(\infty) = p_0 = 0$ so that ψ^{-1} has (closed) domain $[0, 1]$, $\psi^{-1}(0) = \infty$ and $\exp\{-\psi^{-1}(F(x))\}$ is a proper cdf when F is a univariate cdf. Hence it is convenient to define:

$$\mathcal{L}_\infty = \{\psi : [0, \infty) \to [0, 1] \mid \psi(\infty) = 0, \ (-1)^i \psi^{(i)}(s) \geq 0, s > 0, \ i = 1, 2 \ldots\}. \qquad (2.9)$$

A random variable X is infinitely divisible if for every positive integer n, there exists i.i.d. $X_1^{(n)}, \ldots, X_n^{(n)}$ such that $X \stackrel{d}{=} X_1^{(n)} + \cdots + X_n^{(n)}$. If X is a non-negative random variable with LT ψ and it is infinitely divisible, then

$$\psi(s) = \mathbb{E}[e^{-sX}] = \mathbb{E}[e^{-s(X_1^{(n)} + \cdots + X_n^{(n)})}] = \left\{\mathbb{E}[e^{-s(X_1^{(n)})}]\right\}^n$$

so that the LT of $X_1^{(n)}$ is $\psi^{1/n}$. If $Z_{n,m}$ is the sum of m i.i.d. replicates of $X_1^{(n)}$, then its LT is $\psi^{m/n}$. Since positive rational numbers are dense in the positive reals and the continuity theorem applies to LTs, ψ^r is a LT for all $r > 0$ if ψ is the LT of an infinitely divisible non-negative random variable.

A LT is said to be *infinitely divisible* if it is the LT of a (non-negative) infinitely divisible distribution or random variable. In the case, there is a parameter $\theta > 0$ such that $\psi(s) = e^{-\theta \Psi(x)}$ and Ψ is called the Bernstein function. The function $\Psi : [0, \infty) \to [0, \infty)$ satisfies $\Psi(0) = 0$, $\Psi(\infty) = \infty$, Ψ is increasing, and $-\Psi$ is completely monotone. We define the following class of completely monotone functions:

$$\mathcal{L}_\infty^* = \{\omega : [0, \infty) \to [0, \infty) \mid \omega(0) = 0, \ \omega(\infty) = \infty, \ (-1)^{i-1}\omega^{(i)} \geq 0, \ i = 1, 2, \ldots\}. \quad (2.10)$$

The class $\mathcal{L}_\infty^*$ is important in the construction of some multivariate copulas based on one or more LTs.

The property of $-\log \psi \in \mathcal{L}_\infty^*$ for a LT ψ means that ψ is the LT of an infinitely divisible random variable. This is stated on p. 450 of Feller (1971) and summarized as Theorem 8.81.

If a non-negative random variable X has finite moments $\mathbb{E}(X^i)$ for $i = 1, 2, \ldots$, then its LT has a (convergent) infinite Taylor series in a neighborhood of 0:

$$\psi(s) = \mathbb{E}[e^{-sX}] = 1 + \sum_{i=1}^\infty s^i \psi^{(i)}(0)/i! = 1 + \sum_{i=1}^\infty s^i (-1)^i \mathbb{E}[X^i]/i!,$$

and the distribution F_X is uniquely determined from its moments (Feller (1971), pp. 233–234).

If a non-negative random variable X does not have finite expectations for all positive powers, then it is useful to define the maximal non-negative moment degree as $M_X = \sup\{m \geq 0 : \mathbb{E}(X^m) < \infty\}$; see Section 2.4 for its link to tail heaviness. If $k < M_X < k+1$ where k is a non-negative integer and $|\psi^{(k)}(0) - \psi^{(k)}(s)|$ is regularly varying at 0^+, then Theorem 8.88 says that in a neighborhood of 0, $\psi(s)$ has a Taylor expansion up to the kth term and then the next order term is s^{M_X}.

2.3   Extreme value theory

This section has a few results to (a) contrast extreme value limit theory for maxima/minima and central limit theory for means/averages, and (b) show a method to derive extreme value copulas.

Standard extreme value theory assumes data are i.i.d. samples of maxima or minima (each maxima is the maximum of i.i.d. or weakly dependent random variables). From the limiting extreme value distributions, tail distributions are derived which can be applied to the subset of a random sample that exceeds a high threshold. That is, extreme value theory can be used to estimate extreme quantiles (such as for the distribution of log-returns). The assumption for this theory is a density with exponential or inverse polynomial tail. This theory might give more reliable estimates of tail probabilities and extreme quantiles, because likelihood inference based on all data is dominated by the data values in the middle and inferences about the tail involves "extrapolation."

We next state the probabilistic framework to compare different limit laws from extremes and averages. Let $X_1, X_2, \ldots$ be i.i.d. random variables with distribution F. Let

$$S_n = X_1 + \cdots + X_n,$$
$$M_n = \max\{X_1, \ldots, X_n\} = X_1 \vee \cdots \vee X_n,$$
$$L_n = \min\{X_1, \ldots, X_n\} = X_1 \wedge \cdots \wedge X_n.$$

Central limit theory consists of the study of conditions for the existence of sequences $\{a_n\}$ and $\{b_n\}$ so that $(S_n - a_n)/b_n$ converges in distribution to a non-degenerate distribution. Extreme value theory consists of the study of conditions for the existence of sequences $\{a_n\}$ and $\{b_n\}$ so that $(M_n - a_n)/b_n$ or $(L_n - a_n)/b_n$ converges in distribution to a non-degenerate distribution.

For central limit theory, if F (or X_i) has finite variance, the limiting distribution is Gaussian, and if F has infinite variance, the limiting distribution (if it exists) is a stable distribution.

For extreme value theory, $L_n = -\max\{-X_1, \ldots, -X_n\}$ so that any result for maxima can be converted to a result on minima. Hence, usually results are presented for maxima.

Example 2.5 Extreme value limits are shown for a few univariate distributions with different types of tails.

1. Exponential distribution with $F(x) = 1 - e^{-x}$ for $x \geq 0$, and $F^{-1}(p) = -\log(1-p)$ for $0 < p < 1$:

$$\mathbb{P}\left(\frac{M_n - a_n}{b_n} \leq x\right) = \mathbb{P}(M_n \leq a_n + b_n x) = \left[1 - e^{-a_n - b_n x}\right]^n, \quad a_n + b_n x > 0.$$

Choose $a_n = \log n = F^{-1}(1 - n^{-1})$ and $b_n = 1$. Then

$$\mathbb{P}(M_n \leq a_n + b_n x) = \left[1 - n^{-1} e^{-x}\right]^n \to \exp\{-e^{-x}\}, \quad -\infty < x < \infty.$$

This is the Gumbel or extreme value distribution for maxima.

2. Fréchet distribution (of the reciprocal of a Weibull random variable). Let $M_n = \max\{X_1, \ldots, X_n\}$, where $X_i = Y_i^{-1}$ with Y_i i.i.d. Weibull$(\gamma, 1)$, i.e., $\mathbb{P}(X_i \leq x) = e^{-1/x^\gamma}$ for $x > 0$ with $\gamma > 0$. Take $a_n = 0$ and $b_n = n^{1/\gamma}$. Then

$$\mathbb{P}(M_n/n^{1/\gamma} \leq x) = \left[\exp\{-(n^{1/\gamma} x)^{-\gamma}\}\right]^n = \exp\{-x^{-\gamma}\}, \quad x > 0.$$

So the Fréchet distribution is an extreme value limit for maxima; this was also the extreme value limit for the Pareto distribution in Example 1.6.

3. Beta$(1, \beta)$ distribution with cdf $1 - (1-x)^\beta$ for $0 \leq x \leq 1$, where $\beta > 0$:

$$\mathbb{P}\left(\frac{M_n - a_n}{b_n} \leq x\right) = \left[1 - (1 - a_n - b_n x)^\beta\right]^n, \quad \text{if } 0 \leq a_n + b_n x \leq 1.$$

Take $a_n = 1$ and $b_n = n^{-1/\beta}$ to get $[1 - n^{-1}(-x)^\beta]^n \to \exp\{-(-x)^\beta\}$, for $x \leq 0$. This is an example of a Weibull distribution on $(-\infty, 0]$.

4. Location-scale property. Suppose $(M_n - a_n)/b_n \to_d H(\cdot)$ or $\mathbb{P}((M_n - a_n)/b_n \leq x) \to H(x) \; \forall x$, then

$$\frac{M_n - a_n}{b_n/b^*} + a^* = \frac{M_n - a_n + a^* b_n/b^*}{b_n/b^*} \to_d H\left(\frac{\cdot - a^*}{b^*}\right),$$

$$\mathbb{P}\left(\frac{M_n - a_n + a^* b_n/b^*}{b_n/b^*} \leq y\right) = \mathbb{P}\left(\frac{M_n - a_n}{b_n} \leq \frac{y - a^*}{b^*}\right) \to H\left(\frac{y - a^*}{b^*}\right).$$

Hence if a distribution is an extreme value limit, so are all of its location-scale transforms.

The three-types (Fisher-Tippet) theorem for extreme value limits says that possible limits for maxima, up to location-scale changes, are the following.

1. $H_W(x; \xi) = \exp\{-(-x)^{-1/\xi}\}$, $x \le 0$, $\xi = -1/\beta < 0$ (Weibull).
2. $H_G(x) = \exp\{-e^{-x}\}$, $-\infty < x < \infty$, $\xi = 0$ (Gumbel).
3. $H_F(x; \xi) = \exp\{-x^{-1/\xi}\}$, $x > 0$, $\xi = 1/\gamma > 0$ (Fréchet).

By negation of random variables, the only possible limits for minima, up to location-scale changes, are the following.

1. $H_W^*(x; \xi) = 1 - H_W(-x) = 1 - \exp\{-x^{-1/\xi}\}$, $x \ge 0$, $\xi < 0$.
2. $H_G^*(x) = 1 - H_G(-x) = 1 - \exp\{-e^x\}$, $-\infty < x < \infty$, $\xi = 0$.
3. $H_F^*(x; \xi) = 1 - H_F(-x) = 1 - \exp\{-(-x)^{-1/\xi}\}$, $x < 0$, $\xi > 0$.

By making a location-scale change, all three types can be combined into a single family with "tail heaviness" or tail index parameter ξ. The family is called the Generalized extreme value (GEV) distribution; it is also known as the von Mises or Jenkinson distribution. With $[y]_+ := \max\{0, y\}$, the GEV distribution is:

$$H_{GEV}(x; \xi) = \exp\{-[1 + \xi x]_+^{-1/\xi}\}, \quad -\infty < x < \infty, \; -\infty < \xi < \infty$$
$$= \begin{cases} \exp\{-[1 + \xi x]^{-1/\xi}\}, & x \le -1/\xi, \; \xi < 0, \\ \exp\{-e^{-x}\}, & -\infty < x < \infty, \; \xi = 0, \\ \exp\{-[1 + \xi x]^{-1/\xi}\}, & x > -1/\xi, \; \xi > 0. \end{cases}$$

The three-parameter location-scale family is $H_{GEV}((\cdot - \mu)/\sigma; \xi)$. A larger ξ means a heavier tail; see Section 2.4.

Definition 2.1 (Maximum domain of attraction or MDA). Suppose F is a distribution such that there exist sequences $\{a_n\}$ and $\{b_n\}$ so that $F^n(a_n + b_n x)$ converges in distribution to a non-degnerate H. F is said to be in the *maximum domain of attraction* of the Gumbel distribution, or MDA(Gumbel), if H is the Gumbel extreme value distribution. Similarly, F is in MDA(Fréchet;α) if H is the Fréchet distribution with parameter $\xi = \alpha^{-1}$ and in MDA(Weibull;β) if H is the Weibull distribution with parameter $\xi = -\beta^{-1}$.

From Galambos (1987), Resnick (1987) and Embrechts et al. (1997), sufficient conditions for MDA are the following.

1. For MDA(Gumbel), for all $-\infty < y < \infty$, $\lim_{t \to \infty} \overline{F}(t + y\,a(t))/\overline{F}(t) = e^{-y}$, where $a(\cdot)$ is a positive auxiliary function. The auxiliary function can be chosen as the mean residual life function $a(t) = \int_t^\infty \{\overline{F}(x)/\overline{F}(t)\}\,dx$.
2. For MDA(Fréchet;γ), F has upper support point ∞ and for all $x > 0$, $\overline{F}(tx)/\overline{F}(t) \to x^{-\gamma}$ as $t \to \infty$ (this is same as $\overline{F}$ is regularly varying at ∞ with index $-\gamma$).
3. For MDA(Weibull;β), F has finite upper support point v and for all $x > 0$, $F^*(x) = F(v - x^{-1})$ satisfies $\overline{F}^*(tx)/\overline{F}^*(t) \to x^{-\beta}$ as $t \to \infty$.

For MDA(Gumbel), the condition is also necessary.

2.4   Tail heaviness

There are several concepts that relate to tail heaviness of univariate densities with unbounded support; all involve the speed that the density goes to zero at $\pm\infty$. The concepts are maximal non-negative moment degree and regular variation, as well as maximum/minimum domain of attraction.

Definition 2.2 (Maximal non-negative moment). For a positive random variable $X \sim F$ the *maximal non-negative moment degree* is

$$M_X = M_F = \sup\{m \geq 0 : \mathbb{E}(X^m) < \infty\}. \qquad (2.11)$$

M_X is 0 if no moments exist and M_X is ∞ if all moments exist. A smaller value of M_X means that F has a heavier tail at ∞.

Definition 2.3 (Regular variation). A measurable function $g : \mathbb{R}_+ \to \mathbb{R}_+$ is *regularly varying* at ∞ with index α (written $g \in \mathrm{RV}_\alpha$) if for any $t > 0$,

$$\lim_{x \to \infty} \frac{g(xt)}{g(x)} = t^\alpha. \qquad (2.12)$$

For the lower limit at 0^+, if for any $t > 0$, $\lim_{x \to 0^+} g(xt)/g(x) = t^\alpha$, then g is regularly varying at 0^+ and denoted by $g \in \mathrm{RV}_\alpha(0^+)$. If (2.12) holds with $\alpha = 0$ for any $t > 0$, then g is said to be *slowly varying* at ∞, written as $g \in \mathrm{RV}_0$. Similarly, $\mathrm{RV}_0(0^+)$ is defined. A regularly varying function g can be written as $g(x) = x^\alpha \ell(x)$ where $\ell(x)$ is a slowly varying function.

Example 2.6 (Slowly varying).

1. Consider the Burr survival function with scale parameter equal to 1, that is, $\overline{F}(x) = (1 + x^\varsigma)^{-\alpha} = x^{-\varsigma\alpha}(1 + x^{-\varsigma})^{-\alpha}$ for $x > 0$. Then $\overline{F} \in \mathrm{RV}_{-\varsigma\alpha}$ since $(tx)^{-\varsigma\alpha}/x^{-\varsigma\alpha} = t^{-\varsigma\alpha}$ and $\ell(x) = (1 + x^{-\varsigma})^{-\alpha}$ is slowly varying because $\lim_{x \to \infty}[1 + (xt)^{-\varsigma}]^{-\alpha}/(1 + x^{-\varsigma})^{-\alpha} = 1$.

2. $(-\log u)^q \in \mathrm{RV}_0(0^+)$ for any $q \neq 0$, because as $u \to 0^+$,

$$\left(\frac{-\log ut}{-\log u}\right)^q = \left(1 + \frac{-\log t}{-\log u}\right)^q \sim 1 + q \cdot \frac{-\log t}{-\log u} \to 1,$$

   for any $t > 0$.

3. $\exp\{-\zeta(-\log u)^p\} \in \mathrm{RV}_0(0^+)$ for any $\zeta \neq 0$ and $0 < p < 1$, because as $u \to 0^+$,

$$\frac{\exp\{-\zeta(-\log ut)^p\}}{\exp\{-\zeta(-\log u)^p\}} = \exp\left\{-\zeta(-\log u)^p\left[\left(1 + \frac{-\log t}{-\log u}\right)^p - 1\right]\right\}$$

$$\sim \exp\{\zeta p(-\log t)(-\log u)^{p-1}\} \to 1,$$

   for any $t > 0$. $\qquad \square$

If a positive random variable has survival function $\overline{F} = 1 - F$ which is regularly varying with tail index $-\alpha$, then $\overline{F} \in \mathrm{RV}_{-\alpha}$, F is MDA(Fréchet;α) and the maximal non-negative moment degree is α. If F has density f which is regularly varying with index $-\alpha - 1$, then $\overline{F} \in \mathrm{RV}_{-\alpha}$, etc., as in the preceding sentence. The maximal non-negative moment degree could be $M_F = \alpha$ without F having a regularly varying tail.

If a positive random variable has survival function $\overline{F} = 1 - F$ that is regularly varying with index $-\infty$, then F is MDA(Gumbel) and the maximal non-negative moment degree is $M_F = \infty$. Also, exponential tails imply MDA(Gumbel) and $M_F = \infty$; and many common distributions with subexponential tails are in MDA(Gumbel); see Example 3.3.35 of Embrechts et al. (1997) and Goldie and Resnick (1988) for distributions that are both subexponential and in MDA(Gumbel).

For more on theory of regular variation and extreme value theory, references are Bingham et al. (1987), Resnick (1987, 2007) and de Haan and Ferreira (2006).

2.5   Probability integral transform

In this section, we summarize results on the probability integral transform that are useful for simulating random variables and also for the transformations to get $U(0,1)$ or other margins. The results are also implicit in the construction and use of copulas.

- If $U \sim U(0,1)$ and F is a univariate cdf with F^{-1} being the (generalized) inverse or quantile function, then $X = F^{-1}(U) \sim F$. The random variable X can be continuous or discrete.
- If $X \sim F$ is a continuous random variable, then $F(X) \sim U(0,1)$.
- If F_j is a continuous cdf for $j = 1, 2$ and $X_1 \sim F_1$, then $F_2^{-1}(F_1(X_1)) = F_2^{-1} \circ F_1(X_1) \sim F_2$ (this follows from the above two items).
- If F_j is a continuous cdf for $j = 1, 2$ and $X_1 \sim F_1$, then $F_2^{-1}(1 - F_1(X_1)) \sim F_2$.

2.6   Multivariate Gaussian/normal

In this section, we present basic results for the multivariate Gaussian distribution, including conditional distributions and parametrization via a set of partial correlations.

The multivariate Gaussian distribution can be constructed based on the property that any linear combination of independent Gaussian random variables is Gaussian. Hence a few different linear combination of same set of independent Gaussian random variables will be dependent, and have a joint multivariate Gaussian distribution. A characterization of a multivariate Gaussian random vector $X = (X_1, \ldots, X_d)$ is that any linear combination $\sum_{i=1}^{d} a_i X_i$, for real constants $\{a_i\}$, is univariate Gaussian.

The basics of the construction are given in the following itemized list.

- For the construction of d-variate Gaussian with $d \geq 2$, let $Z = (Z_1, \ldots, Z_d)^\top$ be a vector of independent $N(0,1)$ random variables, let $A = (a_{ij})$ be a $d \times d$ non-singular matrix with $AA^\top = \Sigma$, and let $\mu = (\mu_1, \ldots, \mu_d)^\top$ be a vector in $\mathbb{R}^d$. Then $X = \mu + AZ$ is multivariate Gaussian with mean vector μ, covariance matrix Σ and

$$f_{X-\mu}(y) = f_Z(A^{-1}y) \cdot |A^{-1}| = (2\pi)^{-d/2}|A|^{-1}\exp\{-\tfrac{1}{2}y^\top(AA^\top)^{-1}y\}, \quad y \in \mathbb{R}^d,$$

$$\phi_d(x; \mu, \Sigma) = f_X(x) = (2\pi)^{-d/2}|\Sigma|^{-\frac{1}{2}}\exp\{-\tfrac{1}{2}(x-\mu)^\top\Sigma^{-1}(x-\mu)\}, \quad x \in \mathbb{R}^d. \quad (2.13)$$

The latter is the joint density of X. The derivation is via the linear transformation with Jacobian $|A|^{-1} = |\Sigma|^{-1/2}$. The density (2.13) is denoted by $N_d(\mu, \Sigma)$.

- In non-matrix form, $X = \mu + AZ$ becomes $X_i = \sum_{j=1}^{d} a_{ij}Z_j + \mu_i$, $i = 1, \ldots, d$. Then $\text{Cov}(X_i, X_k) = \text{Cov}(\sum_{j=1}^{d} a_{ij}Z_j, \sum_{j=1}^{d} a_{kj}Z_j) = \sum_{j=1}^{d} a_{ij}a_{kj}$ which is the inner product if the ith and kth rows of A or the (i, k) element of $AA^\top$. Hence $AA^\top = \text{Cov}(X) = \Sigma_X = \Sigma$ is the covariance matrix of X.

- For a given positive definite matrix Σ, there are many non-singular matrices A such that $AA^\top = \Sigma$. If A is lower triangular (or upper triangular), then $\Sigma = AA^\top$ is the Cholesky decomposition of Σ; A is unique subject to positive diagonal elements.

- If the $d \times d$ matrix A used for the linear combination is singular, then $\Sigma = AA^\top$ is also singular. There is a $N_d(\mu, \Sigma)$ distribution in this case, but it is not absolutely continuous and does not have a density (with respect to Lebesgue measure). If Σ is non-negative definite and singular, then $N_d(\mu, \Sigma)$ is a singular multivariate Gaussian distribution.

Basic properties for marginal and conditional distributions are as follows.

- Let $X \sim N_d(\mu, \Sigma)$. Then $b^\top X$ is univariate Gaussian for all $b \in \mathbb{R}^d$ ($b \neq 0$).

- Let $X \sim N_d(\mu, \Sigma)$, let B be a $r \times d$ matrix ($1 \leq r \leq d$) with rank$(B)=r$. Then $BX \sim N_r(B\mu, B\Sigma B^\top)$. In particular, all of the marginal distributions of X (dimensions $2, \ldots, d-1$) are multivariate Gaussian.

- Let $X \sim N_d(\mu, \Sigma)$. Suppose $\text{Cov}(X_i, X_j) = 0$ for all $i \neq j$, $i \in I$, $j \notin I$, where I is a non-empty proper subset of $\{1, \ldots, d\}$. Then $(X_i, i \in I)$ and $(X_j, j \notin I)$ are mutually independent subvectors.

- Write $X = \begin{pmatrix} X_1 \\ X_2 \end{pmatrix}$, where X_1 is $d_1 \times 1$ and X_2 is $d_2 \times 1$, with $d = d_1 + d_2$. If $X \sim$ $N_d(\mu, \Sigma)$, the mean vector and covariance matrix are partitioned as $\mu = \begin{pmatrix} \mu_1 \\ \mu_2 \end{pmatrix}$ and $\Sigma = \begin{pmatrix} \Sigma_{11} & \Sigma_{12} \\ \Sigma_{21} & \Sigma_{22} \end{pmatrix}$.

(a) If Σ is positive definite, then

$$\det(\Sigma) = \det(\Sigma_{11}) \det(\Sigma_{22} - \Sigma_{21}\Sigma_{11}^{-1}\Sigma_{12}) = \det(\Sigma_{11}) \det(\Sigma_{22\cdot1}) \qquad (2.14)$$

and

$$\Sigma^{-1} = \begin{pmatrix} \Sigma_{11}^{-1} + \Sigma_{11}^{-1}\Sigma_{12}\Sigma_{22\cdot1}^{-1}\Sigma_{21}\Sigma_{11}^{-1} & -\Sigma_{11}^{-1}\Sigma_{12}\Sigma_{22\cdot1}^{-1} \\ -\Sigma_{22\cdot1}^{-1}\Sigma_{21}\Sigma_{11}^{-1} & \Sigma_{22\cdot1}^{-1} \end{pmatrix}, \qquad (2.15)$$

where

$$\Sigma_{22\cdot1} = \Sigma_{22} - \Sigma_{21}\Sigma_{11}^{-1}\Sigma_{12}. \qquad (2.16)$$

(b) The covariance matrix of $[X_2 \mid X_1 = x_1]$ for any x_1 with the dimension of X_1 is $\Sigma_{22\cdot1}$. Also,

$$[X_2 \mid X_1 = x_1] \sim N_{d_2}\left(\mu_2 + \Sigma_{21}\Sigma_{11}^{-1}(x_1 - \mu_1), \Sigma_{22\cdot1}\right). \qquad (2.17)$$

The conditional mean vector is $\mu_{2|1}(x_1) = \mu_2 + \Sigma_{21}\Sigma_{11}^{-1}(x_1 - \mu_1)$ and it is linear in x_1.

- For $d = 2$ with $d_1 = d_2 = 1$, the preceding result becomes:
(a) $\Sigma_{22\cdot1} = \sigma_{22} - \sigma_{12}^2/\sigma_{11} = \sigma_2^2 - \sigma_{12}^2/\sigma_1^2 = \sigma_2^2(1 - \rho_{12}^2)$.
(b) $\mu_{2|1}(x_1) = \mu_2 + \sigma_{12}\sigma_1^{-2}(x_1 - \mu_1) = \mu_2 + \sigma_2\rho_{12} \cdot (x_1 - \mu_1)/\sigma_1$.

- To demonstrate of the technique of decomposition of a multivariate density, an outline of the proof is given, first for bivariate $d = 2$ and then $d > 2$.

   Bivariate: With $\rho = \rho_{12}$, the determinant and inverse of $R = \begin{pmatrix} 1 & \rho \\ \rho & 1 \end{pmatrix}$ are $\det(R) = (1 - \rho^2)$ and $R^{-1} = (1 - \rho^2)^{-1} \begin{pmatrix} 1 & -\rho \\ -\rho & 1 \end{pmatrix}$. The bivariate Gaussian density $\phi_2(x; \mu, \Sigma) = \phi_2(x_1, x_2; \mu_1, \mu_2, \sigma_1, \sigma_2, \rho)$ can be written as

$$\frac{1}{(2\pi)\sigma_1\sigma_2(1 - \rho^2)^{1/2}} \exp\{-\tfrac{1}{2}(1 - \rho^2)^{-1}[z_1^2 + z_2^2 - 2\rho z_1 z_2]\},$$

where $z_1 = (x_1 - \mu_1)/\sigma_1$ and $z_2 = (x_2 - \mu_2)/\sigma_2$. The term in the exponent simplifies to:

$$(1-\rho^2)^{-1}\{z_1^2+z_2^2-2\rho z_1 z_2\} = z_1^2+(1-\rho^2)^{-1}\{\rho^2 z_1^2+z_2^2-2\rho z_1 z_2\} = z_1^2+(1-\rho^2)^{-1}(z_2-\rho z_1)^2.$$

Hence

$$\phi_2(x; \mu, \sigma_1, \sigma_2, \rho) = \sigma_1^{-1}\phi(z_1) \cdot \sigma_2^{-1}(1 - \rho^2)^{-1/2}(2\pi)^{-1/2} \exp\{-\tfrac{1}{2}(1 - \rho^2)^{-1}(z_2 - \rho z_1)^2\}$$
$$= \sigma_1^{-1}\phi(z_1) \cdot \sigma_2^{-1}(1 - \rho^2)^{-1/2}\phi\big((z_2 - \rho z_1)/(1 - \rho^2)^{1/2}\big)$$

and

$$\frac{(z_2 - \rho z_1)}{(1 - \rho^2)^{1/2}} = \sigma_2^{-1}(1 - \rho^2)^{-1/2}\left(x_2 - \mu_2 - \frac{\rho\sigma_2(x_1 - \mu_1)}{\sigma_1}\right).$$

Since in general,

$$f_{X_1,X_2}(x_1, x_2) = f_{X_1}(x_1) f_{X_2|X_1}(x_2|x_1),$$

we can identify $f_{X_2|X_1}(x_2|x_1)$ as a $N(\mu_2 + \rho\sigma_2(x_1 - \mu_1)/\sigma_1, \sigma_2^2(1 - \rho^2))$ density.

Multivariate: Partition $\boldsymbol{\Sigma}^{-1}$ as $\begin{pmatrix} \boldsymbol{\Sigma}^{11} & \boldsymbol{\Sigma}^{12} \\ \boldsymbol{\Sigma}^{21} & \boldsymbol{\Sigma}^{22} \end{pmatrix}$, and let $\boldsymbol{z}_1 = \boldsymbol{x}_1 - \boldsymbol{\mu}_1$, $\boldsymbol{z}_2 = \boldsymbol{x}_2 - \boldsymbol{\mu}_2$. With the partitioning, then using (2.15) the exponent of the multivariate Gaussian density is

$$\begin{aligned}
(\boldsymbol{x} - \boldsymbol{\mu})^\top \boldsymbol{\Sigma}^{-1}(\boldsymbol{x} - \boldsymbol{\mu}) &= \boldsymbol{z}_1^\top \boldsymbol{\Sigma}^{11} \boldsymbol{z}_1 + \boldsymbol{z}_1^\top \boldsymbol{\Sigma}^{12} \boldsymbol{z}_2 + \boldsymbol{z}_2^\top \boldsymbol{\Sigma}^{21} \boldsymbol{z}_1 + \boldsymbol{z}_2^\top \boldsymbol{\Sigma}^{22} \boldsymbol{z}_2 \\
&= \boldsymbol{z}_1^\top \boldsymbol{\Sigma}_{11}^{-1} \boldsymbol{z}_1 + \boldsymbol{z}_1^\top \boldsymbol{\Sigma}_{11}^{-1} \boldsymbol{\Sigma}_{12} \boldsymbol{\Sigma}_{22\cdot1}^{-1} \boldsymbol{\Sigma}_{21} \boldsymbol{\Sigma}_{11}^{-1} \boldsymbol{z}_1 - \boldsymbol{z}_1^\top \boldsymbol{\Sigma}_{11}^{-1} \boldsymbol{\Sigma}_{12} \boldsymbol{\Sigma}_{22\cdot1}^{-1} \boldsymbol{z}_2 - \boldsymbol{z}_2^\top \boldsymbol{\Sigma}_{22\cdot1}^{-1} \boldsymbol{\Sigma}_{21} \boldsymbol{\Sigma}_{11}^{-1} \boldsymbol{z}_1 \\
&\quad + \boldsymbol{z}_2^\top \boldsymbol{\Sigma}_{22\cdot1}^{-1} \boldsymbol{z}_2 \\
&= \boldsymbol{z}_1^\top \boldsymbol{\Sigma}_{11}^{-1} \boldsymbol{z}_1 + (\boldsymbol{z}_2 - \boldsymbol{\Sigma}_{21} \boldsymbol{\Sigma}_{11}^{-1} \boldsymbol{z}_1)^\top \boldsymbol{\Sigma}_{22\cdot1}^{-1} (\boldsymbol{z}_2 - \boldsymbol{\Sigma}_{21} \boldsymbol{\Sigma}_{11}^{-1} \boldsymbol{z}_1). \tag{2.18}
\end{aligned}$$

Also, using (2.14), the multiplicative constant term of the density is

$$(2\pi)^{d/2}|\boldsymbol{\Sigma}|^{1/2} = (2\pi)^{d_1/2}|\boldsymbol{\Sigma}_{11}|^{1/2} \cdot (2\pi)^{d_2/2}|\boldsymbol{\Sigma}_{22\cdot1}|^{1/2}.$$

Hence

$$\phi_d(\boldsymbol{x}; \boldsymbol{\mu}, \boldsymbol{\Sigma}) = \phi_{d_1}(\boldsymbol{x}_1; \boldsymbol{\mu}_1, \boldsymbol{\Sigma}_{22}) \cdot \phi_{d_2}(\boldsymbol{x}_2; \boldsymbol{\mu}_{2|1}(\boldsymbol{x}_1), \boldsymbol{\Sigma}_{22\cdot1}).$$

Since in general,

$$f_{\boldsymbol{X}_1, \boldsymbol{X}_2}(\boldsymbol{x}_1, \boldsymbol{x}_2) = f_{\boldsymbol{X}_1}(\boldsymbol{x}_1) f_{\boldsymbol{X}_2|\boldsymbol{X}_1}(\boldsymbol{x}_2|\boldsymbol{x}_1),$$

we can identify $f_{\boldsymbol{X}_2|\boldsymbol{X}_1}(\boldsymbol{x}_2|\boldsymbol{x}_1)$ as a $N_{d_2}(\boldsymbol{\mu}_{2|1}(\boldsymbol{x}_1), \boldsymbol{\Sigma}_{22\cdot1})$ density.

For the multivariate Gaussian copula, we can assume $\boldsymbol{\mu} = \boldsymbol{0}$ and $\boldsymbol{\Sigma}$ is a positive definite correlation matrix $\boldsymbol{R} = (\rho_{jk})$. Let $\Phi_d(\cdot; \boldsymbol{R})$ be the d-variate cdf for a $N_d(\boldsymbol{0}, \boldsymbol{R})$ random vector. With Φ denoting the univariate standard Gaussian cdf, the copula is:

$$\Phi_d\big(\Phi^{-1}(u_1), \ldots, \Phi^{-1}(u_d); \boldsymbol{R}\big), \quad 0 < u_1, \ldots, u_d < 1.$$

Next, partial correlations are defined. These are very important for vine copulas, as they are generalization of multivariate Gaussian copulas when the parametrization of $\boldsymbol{R}$ is a $d(d - 1)/2$-vector consisting $d - 1$ correlations and $(d - 1)(d - 2)/2$ partial correlations that are algebraically independent.

As indicated above, conditional distributions of multivariate Gaussian have means that depend on values of the conditioning variables and covariance matrices that are constant over values of the conditioning variables. When the conditional covariance matrix is scaled to have a diagonal of 1s, then it is a conditional correlation matrix, and the off-diagonal elements are partial correlations.

Suppose $d \geq 3$. For (2.16) and (2.17), take $d_1 = d - 2$ so that the first subvector has dimension $d - 2$ and the second subvector has dimension 2. From the 2×2 covariance matrix $\boldsymbol{\Sigma}_{22;1}$, we can obtain the partial correlation $\rho_{d-1,d;1\cdots d-2}$. In this case, with $\boldsymbol{a}_{d-1} = (\sigma_{1,d-1}, \ldots, \sigma_{d-2,d-1})^\top$ and $\boldsymbol{a}_d = (\sigma_{1,d}, \ldots, \sigma_{d-2,d})^\top$, then

$$\boldsymbol{\Sigma}_{22\cdot1} = \begin{pmatrix} \sigma_{d-1,d-1} - \boldsymbol{a}_{d-1}^\top \boldsymbol{\Sigma}_{11}^{-1} \boldsymbol{a}_{d-1} & \sigma_{d-1,d} - \boldsymbol{a}_{d-1}^\top \boldsymbol{\Sigma}_{11}^{-1} \boldsymbol{a}_d \\ \sigma_{d,d-1} - \boldsymbol{a}_d^\top \boldsymbol{\Sigma}_{11}^{-1} \boldsymbol{a}_{d-1} & \sigma_{dd} - \boldsymbol{a}_d^\top \boldsymbol{\Sigma}_{11}^{-1} \boldsymbol{a}_d \end{pmatrix}.$$

Hence the conditional correlation is

$$\rho_{d-1,d;1\cdots d-2} = \frac{\sigma_{d-1,d} - \boldsymbol{a}_{d-1}^\top \boldsymbol{\Sigma}_{11}^{-1} \boldsymbol{a}_d}{\left[\sigma_{d-1,d-1} - \boldsymbol{a}_{d-1}^\top \boldsymbol{\Sigma}_{11}^{-1} \boldsymbol{a}_{d-1}\right]^{1/2} \left[\sigma_{dd} - \boldsymbol{a}_d^\top \boldsymbol{\Sigma}_{11}^{-1} \boldsymbol{a}_d\right]^{1/2}}. \tag{2.19}$$

If $\boldsymbol{\Sigma} = \boldsymbol{R}$ is a correlation matrix, then in the above, $\sigma_{d-1,d} = \rho_{d-1,d}$ and $\sigma_{d-1,d-1} = $

$\sigma_{dd} = 1$. Other partial correlations can be obtained by permuting indices and working with submatrices of $\boldsymbol{\Sigma}$.

For $d = 3$, (2.19) has a simple form assuming $\boldsymbol{\Sigma}$ is a correlation matrix because then $\boldsymbol{\Sigma}_{11}$ is the scalar 1, $\boldsymbol{a}_{d-1} = \rho_{12}$ and $\boldsymbol{a}_d = \rho_{13}$. Therefore,

$$\rho_{23;1} = [\rho_{23} - \rho_{12}\rho_{13}] \Big/ \sqrt{(1 - \rho_{12}^2)(1 - \rho_{13}^2)},$$

$$\rho_{13;2} = [\rho_{13} - \rho_{12}\rho_{23}] \Big/ \sqrt{(1 - \rho_{12}^2)(1 - \rho_{23}^2)},$$

$$\rho_{12;3} = [\rho_{12} - \rho_{13}\rho_{23}] \Big/ \sqrt{(1 - \rho_{13}^2)(1 - \rho_{23}^2)}.$$

The latter two are obtained by permuting indices. Futhermore, as an alternative to (2.19), there is a recursion formula (Anderson (1958)) for partial correlations from lower order partial correlations; see (6.6). For example,

$$\rho_{34;12} = \frac{\rho_{34;1} - \rho_{23;1}\rho_{24;1}}{\sqrt{(1 - \rho_{23;1}^2)(1 - \rho_{24;1}^2)}}.$$

The recursion equation has been implemented in Algorithm 26 (in Chapter 6) if it is desired to efficiently compute all possible partial correlations in a set of variables.

An important idea is that there are sequences of correlations and partial correlations that are algebraically independent, leading to alternative parametrization of the correlation matrix that avoid the constraint of positive definiteness. For $d = 3$, $(\rho_{12}, \rho_{13}, \rho_{23;1}) \in (-1,1)^3$, $(\rho_{12}, \rho_{23}, \rho_{13;2}) \in (-1,1)^3$, and $(\rho_{13}, \rho_{23}, \rho_{12;3}) \in (-1,1)^3$ are parametrizations of the correlation matrix with algebraically independent parameters. This extends to higher dimensions; see Example 3.10 for $d = 4$. For dimension d, the general pattern for algebraic independence of the $d(d-1)/2$-vector is that there are $d - 1$ correlations, $d - 2$ partial correlations conditioning on one variable, ..., $d - 1 - k$ partial correlations conditioning on k variables for $k = 2, \ldots, d-2$. A special case with simpler notation has correlations $\rho_{j-1,j}$ for $j = 2, \ldots, d$ and partial correlations $\rho_{j-\ell,j;(j-\ell+1):(j-1)}$ for $j = \ell + 1, \ldots, d$ and $\ell = 2, \ldots, d - 1$.

In Kurowicka and Cooke (2006), this parametrization is called a partial correlation vine, of which we give more details in Sections 3.9.6 and 3.9.7.

2.7 Elliptical and multivariate t distributions

In this section, we state some general basic results on stochastic representations of spherical and elliptical distributions, and then go to the special case of multivariate t distributions. Many more properties of these distributions are given in Fang et al. (1990).

A spherical d-dimensional vector $\boldsymbol{X}$ arises from a mixture of uniform distributions on surfaces of d-dimensional hyperspheres with different radii. If $\boldsymbol{X}$ has a density, then the contours of the density are surfaces of hyperspheres, that is,

$$f_{\boldsymbol{X}}(\boldsymbol{x}) = \varphi_d(\boldsymbol{x}^\top \boldsymbol{x}), \quad \boldsymbol{x} \in \mathbb{R}^d, \tag{2.20}$$

for a function φ_d, which is called the *generator*. If $\boldsymbol{\Sigma}$ is a non-negative definite symmetric $d \times d$ matrix and $\boldsymbol{A}$ is a $d \times d$ matrix such that $\boldsymbol{A}\boldsymbol{A}^\top = \boldsymbol{\Sigma}$, then $\boldsymbol{Y} = \boldsymbol{A}\boldsymbol{X}$ has an elliptical distribution. If $\boldsymbol{X}$ has the density in (2.20) and $\boldsymbol{\Sigma}$ is positive definite, then $\boldsymbol{Y}$ has the density

$$f_{\boldsymbol{Y}}(\boldsymbol{y}) = |\boldsymbol{\Sigma}|^{-1/2} \varphi_d(\boldsymbol{y}^\top \boldsymbol{\Sigma}^{-1} \boldsymbol{y}), \quad \boldsymbol{y} \in \mathbb{R}^d.$$

Let $\boldsymbol{X} = (X_1, \ldots, X_d)^\top$ have a spherical distribution. Useful results about spherical distributions include the following.

- If F_{X_j} is a possible marginal distribution of a spherical distribution for any dimension $d \geq 2$, then it is a scale mixture of independent $N(0,1)$ random variables (Kelker (1970)). The stochastic representation is:

$$\boldsymbol{X} = \sqrt{Q}\,(Z_1,\ldots,Z_d)^T, \quad Z_j \sim N(0,1),\ Q > 0,\ Z_1,\ldots,Z_d, Q \text{ independent}. \quad (2.21)$$

- Note that if $Z_1,\ldots,Z_d$ are i.i.d. $N(0,1)$, then

$$(Z_1,\ldots,Z_d) = \sqrt{W_d}\,(V_{d1},\ldots,V_{dd}), \quad (V_{d1},\ldots,V_{dd}) \perp\!\!\!\perp W_d,$$

where $\boldsymbol{V}_d = (V_{d1},\ldots,V_{dd})^\top$ is uniform on the surface of the d-dimensional hypersphere and $W_d \sim \chi_d^2$ or $\text{Gamma}(d/2, \text{scale} = 2)$.

- If $\boldsymbol{X}$ is a scale mixture of Gaussian as in (2.21), then

$$\boldsymbol{X} = \sqrt{QW_d}\,\boldsymbol{V}_d = R_d \boldsymbol{V}_d, \quad R_d = \sqrt{QW_d}, \quad W_d \perp\!\!\!\perp Q \perp\!\!\!\perp \boldsymbol{V}_d.$$

This shows that $X_j \overset{d}{=} V_k \sqrt{QW_k}$ for any $k = 1, 2, \ldots$, where V_k has distribution that is the margin of a uniform distribution on the surface of the unit hypersphere in dimension k. By Theorem 4.21 of Joe (1997),

$$f_{V_k}(v) = [B(\tfrac{1}{2}, \tfrac{1}{2}(k-1))]^{-1}(1-v^2)^{(k-3)/2}, \quad -1 < v < 1. \quad (2.22)$$

- More generally, for a d-dimensional spherical distribution that need not be a scale mixture of Gaussian, the stochastic representation is

$$\boldsymbol{X} = R\boldsymbol{V}_d, \quad R \geq 0,\ R \perp\!\!\!\perp \boldsymbol{V}_d,$$

and if X_j has a density, it is

$$f_{X_j}(x) = \int_{|x|}^{\infty} r^{-1} f_{V_d}(x/r)\,dF_R(r), \quad -\infty < x < \infty,$$

using (2.22) with $k = d$. If R has a density, the joint density of $\boldsymbol{X}$ is:

$$f_{\boldsymbol{X}}(\boldsymbol{x}) = f_R\big([\boldsymbol{x}^\top \boldsymbol{x}]^{1/2}\big)/([\boldsymbol{x}^\top \boldsymbol{x}]^{(d-1)/2} s_d) =: \varphi_d(\boldsymbol{x}^\top \boldsymbol{x}),$$

where $s_d = 2\pi^{d/2}/\Gamma(d/2)$ is the surface area of the d-dimensional unit hypersphere. With integration from spherical coordinates, this implies

$$F_R(r) = s_d \int_0^r \varphi_d(t^2)\,t^{d-1}dt, \quad r \geq 0,$$

so that a necessary condition on φ_d for dimension d is that

$$\int_0^{\infty} \varphi_d(t^2)\,t^{d-1}dt = \tfrac{1}{2} \int_0^{\infty} \varphi_d(w)\,w^{d/2-1}dw < \infty.$$

Next consider a d-dimensional elliptical random vector $\boldsymbol{Y}$ centered at $\boldsymbol{0}$ with non-negative definite matrix $\boldsymbol{\Sigma} = \boldsymbol{A}\boldsymbol{A}^T$. The stochastic representation is:

$$\boldsymbol{Y} \overset{d}{=} R\boldsymbol{A}\boldsymbol{V}_d,$$

where the radial random variable $R \geq 0$ is independent of $\boldsymbol{V}_d$.

Due to Lemmas 2.6 and 5.3 of Schmidt (2002), the radial random variable R depends on the dimension d of the elliptical distribution, and there exists a constant $k^* > 0$ such that

$$R_m \overset{d}{=} k^* R_d \sqrt{B}, \quad B \sim \text{Beta}(\tfrac{1}{2}m, \tfrac{1}{2}(d-m)), \quad 1 \leq m < d,$$

and B is independent of R_d. As an example, $R_d^2 \sim \chi_d^2$ for multivariate normal and the product of independent χ_d^2 and $\text{Beta}(\tfrac{1}{2}m, \tfrac{1}{2}(d-m))$ random variables leads to a χ_m^2 random variable (a result which can be verified directly from properties of Gamma and Beta random variables).

We next derive the t_ν density for shape parameter $\nu > 0$; the proof for the multivariate t_ν density is similar.

For the scale mixture of Gaussian, the stochastic representation is

$$Y_{\mu,\sigma} = \mu + \sigma\sqrt{Q}\,Z, \quad Z \sim N(0,1), \quad \sigma > 0, -\infty < \mu < \infty,$$

where Q is a positive random variable independent of Z, μ is the location parameter, and σ is the scale parameter. To get the t_ν location-scale family with $\nu > 0$, let $W = 1/Q \sim \text{Gamma}(\nu/2, \text{scale} = 2/\nu)$ (with mean 1) or $W = S^2/\nu$ where $S^2 \sim \text{Gamma}(\nu/2, \text{scale} = 2)$. Let Y be the standard version with $\mu = 0$ and $\sigma = 1$. Then for $-\infty < y < \infty$,

$$\mathbb{P}(Y \leq y) = \int_0^\infty \mathbb{P}(Z \leq y\sqrt{W} \mid W = w)\, f_W(w)\, dw$$

$$= \int_0^\infty \mathbb{P}(Z \leq y\sqrt{w})\, f_W(w)\, dw = \int_0^\infty \Phi(y\sqrt{w})\, f_W(w)\, dw,$$

and the density is

$$f_Y(y) = \int_0^\infty w^{1/2}\phi(y\sqrt{w})\, f_W(w)\, dw.$$

Next, substitute the Gamma density:

$$f_Y(y) = \int_0^\infty w^{1/2}\frac{1}{\sqrt{2\pi}}\exp\{-\tfrac{1}{2}y^2 w\}\frac{w^{\nu/2-1}}{\Gamma(\nu/2)}(\tfrac{1}{2}\nu)^{\nu/2}e^{-\nu w/2}\, dw$$

$$= (\tfrac{1}{2}\nu)^{\nu/2}\frac{1}{\sqrt{2\pi}\,\Gamma(\nu/2)}\int_0^\infty w^{(\nu+1)/2-1}e^{-w(\nu+y^2)/2}\, dw$$

$$= (\tfrac{1}{2}\nu)^{\nu/2}\frac{1}{\sqrt{2\pi}\,\Gamma(\nu/2)}\cdot\Gamma([\nu+1]/2)\frac{2^{(\nu+1)/2}}{(\nu+y^2)^{(\nu+1)/2}}$$

$$= \frac{\Gamma([\nu+1]/2)}{\sqrt{\pi\nu}\,\Gamma(\nu/2)}\cdot\frac{1}{(1+y^2/\nu)^{(\nu+1)/2}} =: t_{1,\nu}(y). \tag{2.23}$$

Note that (2.23) has tails of the form $O(|y|^{-(\nu+1)})$ as $y \to \pm\infty$.

The d-variate t density with shape parameter $\nu > 0$, and positive definite covariance matrix Σ is derived in a similar way as a scale mixture of a multivariate Gaussian random vector. A stochastic representation is:

$$\boldsymbol{Y} = (Y_1, \ldots, Y_d)^\top = \sqrt{W^{-1}}\,\boldsymbol{Z},$$

where $\boldsymbol{Z} \sim N_d(\boldsymbol{0}, \boldsymbol{\Sigma})$ and $W > 0$ is independent of $\boldsymbol{Z}$. The cdf and density of $\boldsymbol{Y}$ are:

$$\mathbb{P}(\boldsymbol{Y} \leq \boldsymbol{y}) = \int_0^\infty \mathbb{P}(\boldsymbol{Z} \leq \boldsymbol{y}\sqrt{W} \mid W = w)\, f_W(w)\, dw$$

$$= \int_0^\infty \mathbb{P}(\boldsymbol{Z} \leq \boldsymbol{y}\sqrt{w})\, f_W(w)\, dw = \int_0^\infty \Phi_d(\boldsymbol{y}\sqrt{w}; \boldsymbol{\Sigma})\, f_W(w)\, dw,$$

$$f_{\boldsymbol{Y}}(\boldsymbol{y}) = \int_0^\infty w^{d/2}\phi_d(\boldsymbol{y}\sqrt{w}; \boldsymbol{\Sigma})\, f_W(w)\, dw, \quad \boldsymbol{y} \in \mathbb{R}^d. \tag{2.24}$$

Substituting $W \sim \text{Gamma}(\nu/2, \text{scale} = 2/\nu)$, then (2.24) becomes

$$\begin{aligned}
f_{\boldsymbol{Y}}(\boldsymbol{y}) &= \int_0^\infty w^{d/2} \frac{1}{(2\pi)^{d/2}|\boldsymbol{\Sigma}|^{1/2}} \exp\{-\tfrac{1}{2}\boldsymbol{y}^\top \boldsymbol{\Sigma}^{-1}\boldsymbol{y}w\} \frac{w^{\nu/2-1}}{\Gamma(\nu/2)} (\tfrac{1}{2}\nu)^{\nu/2} e^{-\nu w/2}\, \mathrm{d}w \\
&= (\tfrac{1}{2}\nu)^{\nu/2} \frac{1}{(2\pi)^{d/2}\Gamma(\nu/2)|\boldsymbol{\Sigma}|^{1/2}} \int_0^\infty w^{(\nu+d)/2-1} \exp\{-\tfrac{1}{2}w(\nu + \boldsymbol{y}^\top \boldsymbol{\Sigma}^{-1}\boldsymbol{y})\}\, \mathrm{d}w \\
&= (\tfrac{1}{2}\nu)^{\nu/2} \frac{1}{(2\pi)^{d/2}\Gamma(\nu/2)|\boldsymbol{\Sigma}|^{1/2}} \cdot \Gamma([\nu+d]/2) \frac{2^{(\nu+d)/2}}{(\nu + \boldsymbol{y}^\top \boldsymbol{\Sigma}^{-1}\boldsymbol{y})^{(\nu+d)/2}} \\
&= |\boldsymbol{\Sigma}|^{-1/2} \frac{\nu^{\nu/2}\Gamma((\nu+d)/2)}{\Gamma(\nu/2)\pi^{d/2}} (\nu + \boldsymbol{y}^\top \boldsymbol{\Sigma}^{-1}\boldsymbol{y})^{-(\nu+d)/2} \\
&= |\boldsymbol{\Sigma}|^{-1/2} \frac{\Gamma((\nu+d)/2)}{\Gamma(\nu/2)\,[\pi\nu]^{d/2}} (1 + \boldsymbol{y}^\top \boldsymbol{\Sigma}^{-1}\boldsymbol{y}/\nu)^{-(\nu+d)/2} =: t_{d,\nu}(\boldsymbol{y}; \boldsymbol{\Sigma}). \quad (2.25)
\end{aligned}$$

The derivation of the conditional densities is similar to that for multivariate Gaussian. Suppose the random vector $\boldsymbol{Y}$ is partitioned as $(\boldsymbol{Y}_1^\top, \boldsymbol{Y}_2^\top)^\top$ with the covariance matrix partitioned as $\boldsymbol{\Sigma} = \begin{pmatrix} \boldsymbol{\Sigma}_{11} & \boldsymbol{\Sigma}_{12} \\ \boldsymbol{\Sigma}_{21} & \boldsymbol{\Sigma}_{22} \end{pmatrix}$. Let the dimension of $\boldsymbol{Y}_1$ be d_1 and the dimension of $\boldsymbol{Y}_2$ be $d_2 = d - d_1$. The conditional density of $[\boldsymbol{Y}_2|\boldsymbol{Y}_1 = \boldsymbol{y}_1]$ is:

$$t_{d,\nu}(\boldsymbol{y}_1, \boldsymbol{y}_2; \boldsymbol{\Sigma})/t_{d_1,\nu}(\boldsymbol{y}_1; \boldsymbol{\Sigma}_{11}). \quad (2.26)$$

Let $\boldsymbol{\Sigma}_{22\cdot1} = \boldsymbol{\Sigma}_{22} - \boldsymbol{\Sigma}_{21}\boldsymbol{\Sigma}_{11}^{-1}\boldsymbol{\Sigma}_{12}$. From (2.18), an identity is:

$$\boldsymbol{y}^\top \boldsymbol{\Sigma}^{-1}\boldsymbol{y} = \boldsymbol{y}_1^\top \boldsymbol{\Sigma}_{11}^{-1}\boldsymbol{y}_1 + (\boldsymbol{y}_2 - \boldsymbol{\Sigma}_{21}\boldsymbol{\Sigma}_{11}^{-1}\boldsymbol{y}_1)^\top \boldsymbol{\Sigma}_{22\cdot1}^{-1}(\boldsymbol{y}_2 - \boldsymbol{\Sigma}_{21}\boldsymbol{\Sigma}_{11}^{-1}\boldsymbol{y}_1).$$

Let $\boldsymbol{y}_{2\cdot1} = \boldsymbol{y}_2 - \boldsymbol{\Sigma}_{21}\boldsymbol{\Sigma}_{11}^{-1}\boldsymbol{y}_1$. Then (2.26) becomes:

$$\begin{aligned}
&|\boldsymbol{\Sigma}_{22\cdot1}|^{-1/2} \frac{\Gamma((\nu+d)/2)}{\Gamma((\nu+d_1)/2)\,[\pi\nu]^{(d-d_1)/2}} \frac{(1 + [\boldsymbol{y}_1^\top\boldsymbol{\Sigma}_{11}^{-1}\boldsymbol{y}_1 + \boldsymbol{y}_{2\cdot1}^\top\boldsymbol{\Sigma}_{22\cdot1}^{-1}\boldsymbol{y}_{2\cdot1}]/\nu)^{-(\nu+d)/2}}{(1 + \boldsymbol{y}_1^\top\boldsymbol{\Sigma}_{11}^{-1}\boldsymbol{y}_1/\nu)^{-(\nu+d_1)/2}} \\
&= |\boldsymbol{\Sigma}_{22\cdot1}|^{-1/2} \frac{\Gamma((\nu+d)/2)}{\Gamma((\nu+d_1)/2)\,\pi^{(d-d_1)/2}} \frac{(\nu + \boldsymbol{y}_1^\top\boldsymbol{\Sigma}_{11}^{-1}\boldsymbol{y}_1 + \boldsymbol{y}_{2\cdot1}^\top\boldsymbol{\Sigma}_{22\cdot1}^{-1}\boldsymbol{y}_{2\cdot1})^{-(\nu+d)/2}}{(\nu + \boldsymbol{y}_1^\top\boldsymbol{\Sigma}_{11}^{-1}\boldsymbol{y}_1)^{-(\nu+d_1)/2}} \\
&= |\boldsymbol{\Sigma}_{22\cdot1}|^{-1/2} \frac{\Gamma((\nu+d)/2)}{\Gamma((\nu+d_1)/2)\,\pi^{(d-d_1)/2}} \frac{(1 + [\boldsymbol{y}_{2\cdot1}^\top\boldsymbol{\Sigma}_{22\cdot1}^{-1}\boldsymbol{y}_{2\cdot1}]/[\nu + \boldsymbol{y}_1^\top\boldsymbol{\Sigma}_{11}^{-1}\boldsymbol{y}_1])^{-(\nu+d)/2}}{(\nu + \boldsymbol{y}_1^\top\boldsymbol{\Sigma}_{11}^{-1}\boldsymbol{y}_1)^{(d-d_1)/2}} \\
&= |\boldsymbol{\Sigma}_{22\cdot1}|^{-1/2} \frac{\Gamma((\nu+d)/2)}{\Gamma((\nu+d_1)/2)\,\pi^{d_2/2}(\nu+d_1)^{d_2/2}} \frac{(\nu+d_1)^{d_2/2}}{(\nu + \boldsymbol{y}_1^\top\boldsymbol{\Sigma}_{11}^{-1}\boldsymbol{y}_1)^{d_2/2}} \\
&\quad \cdot \left(1 + \frac{(\nu+d_1)[\boldsymbol{y}_{2\cdot1}^\top\boldsymbol{\Sigma}_{22\cdot1}^{-1}\boldsymbol{y}_{2\cdot1}]/[\nu + \boldsymbol{y}_1^\top\boldsymbol{\Sigma}_{11}^{-1}\boldsymbol{y}_1]}{\nu+d_1}\right)^{-(\nu+d)/2}.
\end{aligned}$$

So the conditional density is d_2-variate t in $\boldsymbol{y}_2$ with shape parameter $\nu+d_1$, positive definite matrix parameter $\boldsymbol{\Sigma}_{22;1}[\nu + \boldsymbol{y}_1^\top\boldsymbol{\Sigma}_{11}^{-1}\boldsymbol{y}_1]/(\nu+d_1)$ and location parameter $\boldsymbol{\Sigma}_{21}\boldsymbol{\Sigma}_{11}^{-1}\boldsymbol{y}_1$.

For the special case when component 2 has dimension 1 ($d_1 = d - 1$), let $y_{2\cdot1} = y_d - \boldsymbol{\Sigma}_{21}\boldsymbol{\Sigma}_{11}^{-1}\boldsymbol{y}_1$. The conditional density of $[Y_d|(Y_1,\ldots,Y_{d-1}) = \boldsymbol{y}_1]$ is:

$$\boldsymbol{\Sigma}_{22\cdot1}^{-1/2} \frac{\Gamma((\nu+d)/2)}{\Gamma((\nu+d-1)/2)\,\pi^{1/2}} \frac{(\nu + \boldsymbol{y}_1^\top\boldsymbol{\Sigma}_{11}^{-1}\boldsymbol{y}_1 + \boldsymbol{\Sigma}_{22\cdot1}^{-1}y_{2\cdot1}^2)^{-(\nu+d)/2}}{(\nu + \boldsymbol{y}_1^\top\boldsymbol{\Sigma}_{11}^{-1}\boldsymbol{y}_1)^{-(\nu+d-1)/2}}. \quad (2.27)$$

This is univariate t with shape parameter $\nu + d - 1$, location parameter $\boldsymbol{\Sigma}_{21}\boldsymbol{\Sigma}_{11}^{-1}\boldsymbol{y}_1$ and scale parameter $\{\boldsymbol{\Sigma}_{22;1}[\nu + \boldsymbol{y}_1^\top\boldsymbol{\Sigma}_{11}^{-1}\boldsymbol{y}_1]/(\nu+d-1)\}^{1/2}$.

For the bivariate case, take (2.25) with $d = 2$, and let $\boldsymbol{R} = \begin{pmatrix} 1 & \rho \\ \rho & 1 \end{pmatrix}$. With parametrization via $-1 < \rho < 1$,

$$t_{2,\nu}(\boldsymbol{y};\rho) = (1-\rho^2)^{-1/2} \frac{\Gamma((\nu+2)/2)}{\Gamma(\nu/2)[\pi\nu]^{1/2}} \left(1 + \frac{y_1^2 + y_2^2 - 2\rho y_1 y_2}{\nu(1-\rho^2)}\right)^{-(\nu+2)/2}.$$

From (2.27) with $d = 2$, the conditional density of Y_2 given $Y_1 = y_1$ is a location-scale transform of the $t_{\nu+1}$ density with location parameter ρy_1, scale parameter $\sqrt{(1-\rho^2)(\nu+y_1^2)/(\nu+1)}$, leading to:

$$(1-\rho^2)^{-1/2} \frac{\Gamma((\nu+2)/2)}{\Gamma((\nu+1)/2)\,\pi^{1/2}} \frac{[\nu + y_1^2 + (y_2 - \rho y_1)^2/(1-\rho^2)]^{-(\nu+2)/2}}{(\nu+y_1^2)^{-(\nu+1)/2}}.$$

2.8 Multivariate dependence concepts

For study of multivariate distributions beyond the classical multivariate Gaussian, there are many important dependence concepts. The basic ones, which are used in subsequent chapters are presented here.

Bivariate versions of these dependence concepts are repeated from Chapter 2 of Joe (1997); not all multivariate versions are included here. Relations among the dependence concepts are stated as theorems in Section 8.3.

2.8.1 Positive quadrant and orthant dependence

Let $\boldsymbol{X} = (X_1, X_2)$ be a bivariate random vector with cdf F. $\boldsymbol{X}$ or F is *positive quadrant dependent* (PQD) if

$$\mathbb{P}(X_1 > a_1, X_2 > a_2) \geq \mathbb{P}(X_1 > a_1)\,\mathbb{P}(X_2 > a_2) \quad \forall a_1, a_2 \in \mathbb{R}. \tag{2.28}$$

Condition (2.28) is equivalent to

$$\mathbb{P}(X_1 \leq a_1, X_2 \leq a_2) \geq \mathbb{P}(X_1 \leq a_1)\,\mathbb{P}(X_2 \leq a_2) \quad \forall a_1, a_2 \in \mathbb{R}. \tag{2.29}$$

The reason (2.28) or (2.29) is a positive dependence concept is that X_1 and X_2 are more likely to be large together or to be small together compared with X_1' and X_2', where $X_1 \stackrel{d}{=} X_1'$, $X_2 \stackrel{d}{=} X_2'$, and $X_1' \perp\!\!\!\perp X_2'$. Reasoning similarly, $\boldsymbol{X}$ or F is *negative quadrant dependent* (NQD) if the inequalities in (2.28) and (2.29) are reversed.

For the multivariate extension, let $\boldsymbol{X}$ be a random d-vector ($d \geq 2$) with cdf F. $\boldsymbol{X}$ or F is *positive upper orthant dependent* (PUOD) if

$$\mathbb{P}(X_i > a_i,\ i = 1, \ldots, d) \geq \prod_{i=1}^{d} \mathbb{P}(X_i > a_i) \quad \forall \mathbf{a} \in \mathbb{R}^d, \tag{2.30}$$

and $\boldsymbol{X}$ or F is *positive lower orthant dependent* (PLOD) if

$$\mathbb{P}(X_i \leq a_i,\ i = 1, \ldots, d) \geq \prod_{i=1}^{d} \mathbb{P}(X_i \leq a_i) \quad \forall \mathbf{a} \in \mathbb{R}^d. \tag{2.31}$$

If both (2.30) and (2.31) hold, then $\boldsymbol{X}$ or F is *positive orthant dependent* (POD). Note that for the multivariate extension, (2.30) and (2.31) are not equivalent. For specific cdfs, only one of these inequalities might be easier to establish directly, but a proof might be feasible via a stochastic representation.

Intuitively, (2.30) means that $X_1, \ldots, X_d$ are more likely simultaneously to have large values, compared with a vector of independent random variables with the same corresponding univariate margins. If the inequalities in (2.30) and (2.31) are reversed, then the concepts of *negative lower orthant dependence* (NLOD), *negative upper orthant dependence* (NUOD) and *negative orthant dependence* (NOD) result.

2.8.2 Stochastically increasing positive dependence

Stochastically increasing is a concept of positive dependence that implies positive quadrant or orthant dependence. There are results on copulas that rely on this stronger dependence condition.

Let $\boldsymbol{X} = (X_1, X_2)$ be a bivariate random vector with cdf $F \in \mathcal{F}(F_1, F_2)$. X_2 is *stochastically increasing* (SI) in X_1 or $\{F_{2|1}(\cdot|x_1)\}$ is *stochastically increasing* as x_1 increases if

$$\mathbb{P}(X_2 > x_2 \mid X_1 = x_1) = 1 - F_{2|1}(x_2|x_1) \uparrow x_1 \ \forall x_2. \tag{2.32}$$

By reversing the roles of the indices of 1 and 2, one has X_1 SI in X_2 or $F_{1|2}$ SI. Similarly, X_2 is *stochastically decreasing* in X_1 if $\{F_{2|1}(\cdot|x_1)\}$ is stochastically decreasing as x_1 increases.

The reason (2.32) is a positive dependence condition is that X_2 is more likely to take on larger values as X_1 increases.

Multivariate versions of the above are the following.

(a) The random vector $(X_1, \ldots, X_d)$ is *positively* dependent through conditional stochastic ordering if $\{X_i : i \neq j\}$ conditional on $X_j = x$ is stochastically increasing as x increases, for all $j = 1, \ldots, d$.

(b) The random vector $(X_1, \ldots, X_d) \sim F$ is *conditionally increasing in sequence* (CIS) if for $i = 2, \ldots, d$, X_i is stochastically increasing in $X_1, \ldots, X_{i-1}$ or, for all x_i, $\mathbb{P}(X_i > x_i \mid X_j = x_j, j = 1, \ldots, i-1)$ is increasing in $x_1, \ldots, x_{i-1}$. The multivariate distribution F is also said to be CIS.

(c) The random vector $(X_1, \ldots, X_d) \sim F$ is *conditionally increasing* if X_j is stochastically increasing in $\boldsymbol{X}_S$ for all S such that $S \subset \{1, \ldots, d\} \backslash \{j\}$. The multivariate distribution F is also said to be conditionally increasing.

2.8.3 Right-tail increasing and left-tail decreasing

Let $\boldsymbol{X} = (X_1, X_2)$ be a bivariate random vector with cdf $F \in \mathcal{F}(F_1, F_2)$. X_2 is *right-tail increasing* (RTI) in X_1 if

$$\mathbb{P}(X_2 > x_2 \mid X_1 > x_1) = \overline{F}(x_1, x_2)/\overline{F}_1(x_1) \uparrow x_1 \ \forall x_2. \tag{2.33}$$

Similarly, X_2 is *left-tail decreasing* (LTD) in X_1 if

$$\mathbb{P}(X_2 \leq x_2 \mid X_1 \leq x_1) = F(x_1, x_2)/F_1(x_1) \downarrow x_1 \ \forall x_2. \tag{2.34}$$

The reason that (2.33) and (2.34) are positive dependence conditions is that, for (2.33), X_2 is more likely to take on larger values as X_1 increases, and, for (2.34) X_2 is more likely to take on smaller values as X_1 decreases. Reversing the directions of the monotonicities leads to negative dependence conditions.

2.8.4 Associated random variables

Let $\boldsymbol{X}$ be a random d-vector. $\boldsymbol{X}$ is (positively) *associated* if the inequality

$$\mathbb{E}[g_1(\boldsymbol{X}) \, g_2(\boldsymbol{X})] \geq \mathbb{E}[g_1(\boldsymbol{X})] \, \mathbb{E}[g_2(\boldsymbol{X})] \tag{2.35}$$

holds for all real-valued functions g_1, g_2 which are increasing (in each component) and are such that the expectations in (2.35) exist. Intuitively, this is a positive dependence condition for $\boldsymbol{X}$ because it means that two increasing functions of $\boldsymbol{X}$ have positive covariance whenever the covariance exists.

It may appear impossible to check this condition of association directly given a cdf F for $\boldsymbol{X}$. Where association of a random vector can be established, it is usually done by making use of a stochastic representation for $\boldsymbol{X}$. One important consequence of the association condition is that it implies the positive orthant dependence condition; see Theorem 8.6.

2.8.5 *Total positivity of order 2*

A non-negative function g on A^2, where $A \subset \mathbb{R}$, is *totally positive of order 2* (TP$_2$) if for all $x_1 < y_1$, $x_2 < y_2$, with $x_1, x_2, y_1, y_2 \in A$,

$$g(x_1, x_2)\, g(y_1, y_2) \geq g(x_1, y_2)\, g(y_1, x_2). \tag{2.36}$$

The "order 2" part of the definition comes from writing the difference $g(x_1, x_2)\, g(y_1, y_2) - g(x_1, y_2)\, g(y_1, x_2)$ as the determinant of a square matrix of order 2. Total positivity of higher orders involves the non-negativity of determinants of larger square matrices. If the inequality in (2.36) is reversed then g is *reverse rule of order 2* (RR$_2$).

For a bivariate cdf F with density f, three notions of positive dependence are: (i) f is TP$_2$; (ii) F is TP$_2$; (iii) $\overline{F}$ is TP$_2$. The reasoning behind (i) as a positive dependence condition is that for $x_1 < y_1$, $x_2 < y_2$, $f(x_1, x_2)\, f(y_1, y_2) \geq f(x_1, y_2)\, f(y_1, x_2)$ means that it is more likely to have two pairs with components matching low–low and high–high than two pairs with components matching low–high and high–low. Similarly, f RR$_2$ is a negative dependence condition.

Theorem 8.5 says that f TP$_2$ implies both F and $\overline{F}$ TP$_2$, and either F TP$_2$ or $\overline{F}$ TP$_2$ implies that F is PQD. Hence both (ii) and (iii) are positive dependence conditions. A direct explanation for (ii) as a positive dependence condition is as follows. The condition of F TP$_2$ is given by:

$$F(x_1, x_2)\, F(y_1, y_2) - F(x_1, y_2)\, F(y_1, x_2) \geq 0, \ \forall x_1 < y_1, \ x_2 < y_2.$$

This is equivalent to:

$$
\begin{aligned}
&F(x_1, x_2)\, [F(y_1, y_2) - F(y_1, x_2) - F(x_1, y_2) + F(x_1, x_2)] \\
&- [F(x_1, y_2) - F(x_1, x_2)]\, [F(y_1, x_2) - F(x_1, x_2)] \geq 0, \\
&\qquad \forall\, x_1 < y_1, \ x_2 < y_2.
\end{aligned}
\tag{2.37}
$$

If $(X_1, X_2) \sim F$, then the inequality in (2.37) is the same as

$$
\begin{aligned}
&\mathbb{P}(X_1 \leq x_1, X_2 \leq x_2)\, \mathbb{P}(x_1 < X_1 \leq y_1, x_2 < X_2 \leq y_2) \\
&- \mathbb{P}(X_1 \leq x_1, x_2 < X_2 \leq y_2)\, \mathbb{P}(x_1 < X_1 \leq y_1, X_2 \leq x_2) \geq 0,
\end{aligned}
$$

for all $x_1 < y_1$ and $x_2 < y_2$. This has an interpretation as before for low–low and high–high pairs versus low–high and high–low pairs. Similarly, the inequality resulting from $\overline{F}$ TP$_2$ can be written in the form of (2.37) with the survival function $\overline{F}$ replacing F.

2.9 Fréchet classes and Fréchet bounds, given univariate margins

Let $\mathcal{F}(F_1, \ldots, F_d)$ be the Fréchet class of d-variate distributions with given univariate margins $F_1, \ldots, F_d$, which could be continuous or discrete. This class is non-empty because it includes $F = \prod_{j=1}^{d} F_j$ which is the multivariate distribution of d independent random variables, the jth one with cdf F_j. The Fréchet bounds are the pointwise maxima and minima: $\min_{F \in \mathcal{F}} F(x_1, \ldots, x_d)$ and $\max_{F \in \mathcal{F}} F(x_1, \ldots, x_d)$, $\boldsymbol{x} = (x_1, \ldots, x_d) \in \mathbb{R}^d$.

For $F \in \mathcal{F}(F_1, F_2)$, Fréchet bounds are:

$$F^-(x_1, x_2) = \max\{0, F_1(x_1) + F_2(x_2) - 1\} \leq F(x_1, x_2) \leq \min\{F_1(x_1), F_2(x_2)\} = F^+(x_1, x_2). \tag{2.38}$$

The proof of inequality comes from simple set inequalities:

$$\mathbb{P}(A_1) + \mathbb{P}(A_2) - 1 \leq \mathbb{P}(A_1) + \mathbb{P}(A_2) - \mathbb{P}(A_1 \cup A_2) = \mathbb{P}(A_1 \cap A_2) \leq \min\{\mathbb{P}(A_1), \mathbb{P}(A_2)\},$$

with

$$A_1 = \{X_1 \le x_1\}, \quad A_2 = \{X_2 \le x_2\}, \quad X_1 \sim F_1, \ X_2 \sim F_2.$$

The upper bound F^+ and lower bound F^- are distributions in $\mathcal{F}(F_1, F_2)$ for any F_1, F_2 whether they are discrete, continuous or mixed.

For $F \in \mathcal{F}(F_1, \ldots, F_d)$, Fréchet bounds are:

$$\max\{0, F_1(x_1) + \cdots + F_d(x_d) - (d-1)\} \le F(\boldsymbol{x}) \le \min\{F_1(x_1), \ldots, F_d(x_d)\} = F^+(\boldsymbol{x}).$$

The upper bound F^+ is a distribution in $\mathcal{F}(F_1, \ldots, F_d)$ in general, but not the lower bound. It follows from Fréchet (1935) that the lower bound is pointwise sharp over $F \in \mathcal{F}(F_1, \ldots, F_d)$.

The Fréchet upper bound can represent perfect positive dependence if $F_1, \ldots, F_d$ are all continuous. The Fréchet lower bound can represent perfect negative dependence if $d = 2$ and F_1, F_2 are continuous. The bounds can also represent perfect dependence sometimes in the discrete case, but not with any generality; see Example 2.9.

If $F_1, \ldots, F_d$ are continuous and $X_j \sim F_j$, $j = 1, \ldots, d$, then the Fréchet upper bound corresponds to *comonotonic* random variables with $X_j = F_j^{-1}(F_1(X_1))$, $j = 2, \ldots, d$, i.e., *perfect positive dependence with $F_j^{-1} \circ F_1$ all monotone increasing.*

Comonotonic random variables (with finite second moments) do not have a correlation of 1, unless they belong to same location-scale family: $F_j^{-1} \circ F_1$ linear with positive slope.

Example 2.7 Suppose X_1, X_2 are comonotonic. If $X_1, X_2 \sim$ Exponential(1), then $X_2 = F_2^{-1}(F_1(X_1)) = X_1$ since $F_1 = F_2$. If $X_1 \sim$ Pareto$(\alpha, 1)$ and $X_2 \sim$ Exponential(1), then $X_2 = F_2^{-1}(F_1(X_1)) = -\log[(1+X_1)^{-\alpha}] = \alpha \log(1+X_1)$ is a monotone increasing functional relation but the correlation of X_1, X_2 is **not 1**.

If F_1, F_2 are continuous and $X_j \sim F_j$, $j = 1, 2$, then the Fréchet lower bound corresponds to *countermonotonic* random variables with $X_2 = F_2^{-1}(1 - F_1(X_1))$, i.e., *perfect negative dependence with $F_2^{-1}(1 - F_1(\cdot))$ monotone decreasing.*

Example 2.8 Suppose X_1, X_2 are countermonotonic. If $X_1, X_2 \sim$ Exponential(1), then $X_2 = F_2^{-1}(1 - F_1(X_1)) = F_2^{-1}(e^{-X_1}) = -\log[1 - e^{-X_1}]$ is a monotone decreasing functional relation but the correlation of X_1, X_2 is **not** -1.

The next example illustrates the Fréchet bounds in (2.38) for bivariate binary distributions and shows that in general these bounds do not correspond to perfect dependence or comonotonic/countermonotonic distributions.

Example 2.9 Let F_j correspond to Bernoulli(π_j) for $j = 1, 2$. Consider (Y_1, Y_2) as a bivariate binary pair so that it has a bivariate Bernoulli or binary distribution. Let $\mathbb{P}(Y_1 = 0, Y_2 = 0) = p_{00}$, $\mathbb{P}(Y_1 = 0, Y_2 = 1) = p_{01}$, $\mathbb{P}(Y_1 = 1, Y_2 = 0) = p_{10}$, $\mathbb{P}(Y_1 = 1, Y_2 = 1) = p_{11}$, $\mathbb{P}(Y_1 = 1) = p_{1+} = \pi_1$, and $\mathbb{P}(Y_2 = 1) = p_{+1} = \pi_2$. The bivariate probabilities are summarized in Table 2.1.

$Y_1 \backslash Y_2$	0	1	
0	p_{00}	p_{01}	$p_{0+} = 1 - \pi_1$
1	p_{10}	p_{11}	$p_{1+} = \pi_1$
	$p_{+0} = 1 - \pi_2$	$p_{+1} = \pi_2$	1

Table 2.1: *Bivariate Bernoulli distribution*

The bivariate Bernoulli distribution can be parametrized with two univariate parameters π_1, π_2 and one bivariate parameter p_{11}, leading to the correlation

$$\rho = (p_{11} - \pi_1\pi_2) \big/ \sqrt{\pi_1(1-\pi_1)\pi_2(1-\pi_2)}.$$

The extreme distributions in $\mathcal{F}(F_1, F_2)$ are in Table 2.2. The Fréchet upper bound is comonotonic (perfect positive dependence) only if $\pi_1 = \pi_2$ and the Fréchet lower bound is countermonotonic (perfect negative dependence) only if $\pi_1 + \pi_2 = 1$. From the Fréchet bound inequalities, $\max\{0, \pi_1 + \pi_2 - 1\} \le p_{11} \le \min\{\pi_1, \pi_2\}$, so that

$$\max\left\{-\sqrt{\frac{\pi_1\pi_2}{\overline{\pi}_1\overline{\pi}_2}}, -\sqrt{\frac{\overline{\pi}_1\overline{\pi}_2}{\pi_1\pi_2}}\right\} \le \rho \le \sqrt{\frac{\pi_{\min}(1-\pi_{\max})}{\pi_{\max}(1-\pi_{\min})}},$$

where $\overline{\pi}_j = 1 - \pi_j$, $j = 1, 2$, $\pi_{\min} = \min\{\pi_1, \pi_2\}$, and $\pi_{\max} = \max\{\pi_1, \pi_2\}$. Hence the range of the correlation depends on the univariate margin parameters π_1 and π_2, whereas the odds ratio is ∞ (respectively 0) for the Fréchet upper (lower) bound and the tetrachoric correlation is ± 1 at the bounds.

$Y_1 \backslash Y_2$	0	1	
0	$1 - \pi_2$	$\pi_2 - \pi_1$	$1 - \pi_1$
1	0	π_1	π_1
	$1 - \pi_2$	π_2	1

$Y_1 \backslash Y_2$	0	1	
0	$1 - \pi_1$	0	$1 - \pi_1$
1	$\pi_1 - \pi_2$	π_2	π_1
	$1 - \pi_2$	π_2	1

$Y_1 \backslash Y_2$	0	1	
0	$1 - \pi_1 - \pi_2$	π_2	$1 - \pi_1$
1	π_1	0	π_1
	$1 - \pi_2$	π_2	1

$Y_1 \backslash Y_2$	0	1	
0	0	$1 - \pi_1$	$1 - \pi_1$
1	$1 - \pi_2$	$\pi_1 + \pi_2 - 1$	π_1
	$1 - \pi_2$	π_2	1

Table 2.2 *Bivariate Bernoulli distributions that are Fréchet upper bounds (top row) and Fréchet lower bounds (bottom row). Upper left: $\pi_1 \le \pi_2$, correlation $\sqrt{\pi_1(1-\pi_2)/[\pi_2(1-\pi_1)]}$, odds ratio ∞, tetrachoric correlation 1; Upper right: $\pi_1 \ge \pi_2$, correlation $\sqrt{\pi_2(1-\pi_1)/[\pi_1(1-\pi_2)]}$, odds ratio ∞, tetrachoric correlation 1; Lower left: $\pi_1 + \pi_2 \le 1$, correlation $-\sqrt{\pi_1\pi_2/[(1-\pi_1)(1-\pi_2)]}$, odds ratio 0, tetrachoric correlation -1; Lower right: $\pi_1 + \pi_2 \ge 1$, correlation $-\sqrt{(1-\pi_1)(1-\pi_2)/[\pi_1\pi_2]}$, odds ratio 0, tetrachoric correlation -1.*

The *tetrachoric correlation* for a bivariate binary distribution F is the parameter ρ of the bivariate Gaussian copula C so that $C(F_1, F_2; \rho) = F$. This implies that $p_{00} = C(1 - \pi_1, 1 - \pi_2; \rho)$ and $p_{11} = \overline{C}(1 - \pi_1, 1 - \pi_2; \rho) = C(\pi_1, \pi_2; \rho)$. For the upper left of Table 2.2,

$$\pi_1 = \Phi(\min\{\Phi^{-1}(\pi_1), \Phi^{-1}(\pi_2)\}) = \Phi_2(\Phi^{-1}(\pi_1), \Phi^{-1}(\pi_2); 1),$$

and for the lower right,

$$\begin{aligned}
\pi_1 + \pi_2 - 1 = \pi_2 - (1 - \pi_1) &= \pi_2 - \min\{1 - \pi_1, \pi_2\} = \pi_2 - \Phi_2(\Phi^{-1}(1 - \pi_1), \Phi^{-1}(\pi_2); 1) \\
&= \pi_2 - \Phi_2(-\Phi^{-1}(\pi_1), \Phi^{-1}(\pi_2); 1) = \pi_2 - [\pi_2 - \Phi_2(\Phi^{-1}(\pi_1), \Phi^{-1}(\pi_2); -1)] \\
&= \Phi_2(\Phi^{-1}(\pi_1), \Phi^{-1}(\pi_2); -1).
\end{aligned}$$

The above derivations make use of the following: if $Z \sim \mathrm{N}(0, 1)$, then $\mathbb{P}(Z \le -a_1, Z \le a_2) = \mathbb{P}(-Z \ge a_1, Z \le a_2) = \mathbb{P}(Z \le a_2) - \mathbb{P}(-Z \le a_1, Z \le a_2) = \mathbb{P}(Z \le a_2) - \Phi_2(a_1, a_2; -1)$.

The above shows that the correlation is not a good dependence measure to use for binary variables, and the odds ratio and tetrachloric correlation are better measures; see Chaganty and Joe (2006) for ranges of correlation matrices for multivariate binary distributions. Hence any statistical method for multivariate binary observations based on the correlation matrix is potentially flawed. $\qquad\square$

Next we discuss the Fréchet class of copulas when F_j $(j = 1, \ldots, d)$ are U$(0, 1)$ cdfs. One member of this class corresponds to independence and the copula is denoted as $C^\perp(\boldsymbol{u}) = C^\perp_{1:d}(\boldsymbol{u}) = \prod_{j=1}^{d} u_j$ for $\boldsymbol{u} \in [0, 1]^d$. $C^\perp$ will be referred to as the *independence copula*; it is called the *product copula* in Nelsen (1999). If $(U_1, \ldots, U_d) \sim C^\perp$ is a vector of U$(0, 1)$ random variables, then $U_1 \perp\!\!\!\perp U_2 \perp\!\!\!\perp \cdots \perp\!\!\!\perp U_d$.

For the bivariate case, bounds for a copula C are:

$$C^-(u_1, u_2) := \max\{u_1 + u_2 - 1, 0\} \leq C(u_1, u_2) \leq C^+(u_1, u_2) := \min\{u_1, u_2\},$$

for $0 \leq u_1, u_2 \leq 1$. For d-variate with $d \geq 2$, the upper bound is $C^+(\boldsymbol{u}) = C^+_{1:d}(\boldsymbol{u}) = \min\{u_1, \ldots, u_d\}$ with $\boldsymbol{u} \in [0, 1]^d$. The upper bound C^+ is called the *comonotonicity or Fréchet upper bound* or Fréchet-Hoeffding upper bound copula. For $d = 2$, the lower bound C^- is called the *countermonotonicity or Fréchet lower bound* or Fréchet-Hoeffding lower bound copula.[1] Assuming $(U_1, \ldots, U_d)$ is a vector of U$(0, 1)$ random variables, for comonotonicity, $U_1 = \cdots = U_d$, and for bivariate countermonotonicity $U_1 = 1 - U_2$.

Note that Fréchet (1951) covered discrete and continuous margins in the bivariate case, and Hoeffding (1940) covered the bivariate case with continuous margins transformed to U$(-\frac{1}{2}, \frac{1}{2})$ margins. This explains why the distributions in (2.38) are called Fréchet bounds in general when the univariate margins can be discrete or continuous.

A more difficult result about the Fréchet lower bound was obtained in Dall'Aglio (1972); conditions were obtained in order that the Fréchet lower bound is a distribution for $d \geq 3$. In this case, the univariate margins must have a discrete component. Dall'Aglio's result is stated as Theorem 3.7 in Joe (1997); see also Section 8.4.

There are no countermonotonicity copulas for dimensions $d \geq 3$, but there are other copulas with extremal dependence besides the comonotonicity copulas. Let $1 \leq m < d$. If $U_1 \sim$ U$(0, 1)$ and $U_1 = \cdots = U_m = 1 - U_{m+1} = \cdots = 1 - U_d$, then the copula is

$$\begin{aligned}
\mathbb{P}(U_j \leq u_j, j = 1 \ldots, d) &= \mathbb{P}(U_1 \leq \min\{u_1, \ldots, u_m\}, 1 - U_1 \leq \min\{u_{m+1}, \ldots, u_d\}) \\
&= C^-\big(\min\{u_1, \ldots, u_m\}, \min\{u_{m+1}, \ldots, u_d\}\big) \\
&= C^-\big(C^+_{1:m}(u_1, \ldots, u_m), C^+_{1:(d-m)}(u_{m+1}, \ldots, u_d)\big); \qquad (2.39)
\end{aligned}$$

for the above equation $C^+_{1:1}(u) = u$.

2.10   Fréchet classes given higher order margins

In this section, we mention briefly some results for Fréchet classes given some higher-order margins. Pointwise upper and lower bounds can be obtained but in general even the pointwise upper bound as a function is not a multivariate distribution; see Chapter 3 of Joe (1997) for some results.

For sequential data analysis of multivariate data, one approach might be to fit univariate model F_j to the jth variable separately for $j = 1, \ldots, d$, and then find a multivariate model to describe the dependence. Copulas match with the Fréchet class $\mathcal{F}(F_1, \ldots, F_d)$. Because one can easily see dependence patterns in bivariate data through scatterplots, one might want to split the multivariate model into two steps: (a) fit bivariate (copula) models to each pair of variables, and (b) find a multivariate distribution compatible with the $\binom{d}{2}$ fitted bivariate models. Hence the Fréchet class of $\mathcal{F}(F_{jk}, 1 \leq j < k \leq d; F_j, 1 \leq j \leq d)$ is relevant and of interest. But this Fréchet class is intractable for analysis of compatibility and parametric models; see Section 8.4 for some results on checking non-compatibility.

---

[1]In Schweizer and Sklar (1983) and Nelsen (1999), the comonotonicity and countermonotonicity copulas are denoted as M and W respectively, but this notation is not convenient for this book as it takes up too many letters for copulas.

However, there are Fréchet classes of $(d-1)$ of the $\binom{d}{2}$ bivariate margins that are always non-empty and this is a key to the development of a class of multivariate distributions called vines. See Sections 3.7, 3.8 and 3.9.

If $\{(j_i, k_i) : i = 1, \ldots, d-1\}$ are pairs that satisfy a tree condition, then the Fréchet class $\mathcal{F}(F_{j_i k_i}, 1 \leq i \leq d; F_j, 1 \leq j \leq d)$ is always non-empty by exhibiting a joint distribution that satisfies a Markov or conditional independence property. Consider a graph with nodes labeled as $1, \ldots, d$ and draw an edge connecting j_i to k_i for $i = 1, \ldots, d$; the tree condition is that the graph is a tree (i.e., the graph has no cycles).

If the tree condition holds, then without loss of generality, one can permute the labels so that the edges have the form

$$\{(k_2, 2), (k_3, 3), \ldots, (k_d, d)\}, \quad k_j \in \{1, \ldots, j-1\}, \; j = 2, \ldots, d.$$

Note that $k_2 = 1$. With fixed univariate margins $F_1, \ldots, F_d$, and bivariate margins $F_{k_j, j} \in \mathcal{F}(F_{k_j}, F_j)$ for $j = 2, \ldots, d$, one can construct a multivariate distribution $F_{1:d}$ based on a Markov tree that is compatible with these margins.

We outline the details in the case that F_i is absolutely continuous with density f_i for $i = 1, \ldots, d$ and $F_{k_j, j}$ is absolutely continuous with conditional density $f_{j|k_j}$ for $j = 2, \ldots, d$. The compatible Markov tree density is

$$f_{1:d} = f_1 \prod_{j=2}^{d} f_{j|1:(j-1)} = f_1 \prod_{j=2}^{d} f_{j|k_j}. \tag{2.40}$$

If $d = 3$, then $k_3 = 1$ or 2. In the former case, $f_{1:3} = f_1 f_{2|1} f_{3|1}$ has specified margins f_{12} and f_{13} by respectively integrating out the third and second variables, and in the latter case, $f_{1:3} = f_1 f_{2|1} f_{3|2} = f_2 f_{1|2} f_{3|2}$ has specified bivariate margins f_{12} and f_{23} by respectively integrating out the third and first variables.

Suppose $d \geq 4$ and inductively that (2.40) is valid for up to $d - 1$. That is, $f_{1:d} = f_{1:(d-1)} f_{d|k_d}$ and $f_{1:(d-1)}$ has the specified bivariate marginal densities $f_{k_j, j}$ for $j = 2, \ldots, d-1$. By integrating out the dth variable, $f_{1:d}$ has marginal density $f_{1:(d-1)}$ so there is consistency for $f_{k_j, j}$ with $2 \leq j \leq d-1$. Let $S_d = \{1, \ldots, d-1\} \backslash k_d$. Then

$$\int f_{1:(d-1)}(\boldsymbol{x}_{1:(d-1)}) \, f_{d|k_d}(x_d|x_{k_d}) \, \mathrm{d}\boldsymbol{x}_{S_d} = f_{k_d}(x_{k_d}) \, f_{d|k_d}(x_d|x_{k_d}) = f_{k_d, d}(x_{k_d}, x_d).$$

So the last bivariate margin is also as specified.

2.11   Concordance and other dependence orderings

For $\mathcal{F}(F_1, \ldots, F_d)$, the set of distributions with given univariate margins $F_1, \ldots, F_d$, there are several ways to decide of one member of this class has more dependence than another member. This section summarizes such dependence orderings.

Definition 2.4 (Yanagimoto and Okamoto (1969), Tchen (1980) for bivariate; Joe (1990b) for multivariate). Let F, G be in $\mathcal{F}(F_1, \ldots, F_d)$. Then F *is less concordant than* G, *written* $F \prec_c G$ if $F \leq G$ pointwise and $\overline{F} \leq \overline{G}$ pointwise. That is, a random vector with distribution G has a larger probability of being small together or being large together, than a random vector with distribution F. Also write $\boldsymbol{X} \prec_c \boldsymbol{Y}$ if $F \prec_c G$ and $\boldsymbol{X} \sim F$, $\boldsymbol{Y} \sim G$.

For bivariate $F, G \in \mathcal{F}(F_1, F_2)$ and $F \leq G$ implies $\overline{F} \leq \overline{G}$, so only one of these inequalities needs to be shown. Recall that $\overline{F} = 1 - F_1 - F_2 + F$ for bivariate distributions. Example 2.10 has some details on establishing the concordance ordering for a bivariate parametric copula family.

If F, G are derived via stochastic representations, then this might be used to prove the $\prec_c$ ordering if it applies. Otherwise, because F and $\overline{F}$ are obtained from each other through the inclusion-exclusion principle, only one of these may have a simpler form, making one of $F \leq G$ or $\overline{F} \leq \overline{G}$ easier to establish.

Hence we also provide definitions for orderings based on upper orthants, lower orthants or bivariate margins.

Definition 2.5 Let $F, G \in \mathcal{F}(F_1, \ldots, F_d)$.

(a) G is *more PLOD (positive lower orthant dependent)* than F, written $F \prec_{cL} G$, if

$$F(\boldsymbol{x}) \leq G(\boldsymbol{x}) \quad \forall \boldsymbol{x} \in \mathbb{R}^d.$$

(b) G is *more PUOD (positive upper orthant dependent)* than F, written $F \prec_{cU} G$, if

$$\overline{F}(\boldsymbol{x}) \leq \overline{G}(\boldsymbol{x}) \quad \forall \boldsymbol{x} \in \mathbb{R}^d.$$

(c) G is *more pairwise concordant* than F, written $F \prec_c^{pw} G$, if, for all $1 \leq j < k \leq d$,

$$F_{jk}(x_j, x_k) \leq G_{jk}(x_j, x_k) \quad \forall (x_j, x_k) \in \mathbb{R}^2,$$

where F_{jk}, G_{jk} are the (j, k) bivariate margins of F, G respectively.

Example 2.10 (Bivariate concordance).

We show that the bivariate MTCJ and Galambos copula families, introduced in Section 1.2 are increasing in concordance, and that the boundaries are $C^\perp$ and C^+. The concordance property of the Galambos family follows because it is the extreme value limit of the MTCJ family, so we just directly show the concordance property of the MTCJ family.

The increasing in concordance property holds if

$$C_{\mathrm{MTCJ}}(u, v; \theta) = (u^{-\theta} + v^{-\theta} - 1)^{-1/\theta} \uparrow \theta \ \forall \ 0 < u, v < 1.$$

Take the partial derivative of $\log C$ with respect to θ to get:

$$\theta^{-2} \log(u^{-\theta} + v^{-\theta} - 1) + \theta^{-1} \frac{u^{-\theta} \log u + v^{-\theta} \log v}{(u^{-\theta} + v^{-\theta} - 1)} \geq 0$$

$$\Longleftrightarrow (u^{-\theta} + v^{-\theta} - 1) \log(u^{-\theta} + v^{-\theta} - 1) \geq u^{-\theta} \log u^{-\theta} + v^{-\theta} \log v^{-\theta}$$

$$\Longleftrightarrow 1 \log 1 + (s + t - 1) \log(s + t - 1) \geq s \log s + t \log t \ \forall s, t \geq 1.$$

The latter follows from a convexity/majorization inequality (Marshall et al. (2011)). If $a_1 + a_2 = b_1 + b_2$ and (a_1, a_2) is more spread out than (b_1, b_2) in the majorization ordering and g is convex, then

$$g(a_1) + g(a_2) \geq g(b_1) + g(b_2);$$

this is applied with $a_1 = 1$, $a_2 = s+t-1$, $b_1 = s$, $b_2 = t$ and $g(x) = x \log x$. An indirect proof of concordance follows from Theorem 8.22 by using the property that the MTCJ family is an Archimedean copula family based on the gamma LT.

For the limit of $\log C_{\mathrm{MTCJ}}$ as $\theta \to 0^+$: applying l'Hopital's rule,

$$\lim_{\theta \to 0^+} \frac{-\log(u^{-\theta} + v^{-\theta} - 1)}{\theta} = \lim_{\theta \to 0^+} \frac{[u^{-\theta} \log u + v^{-\theta} \log v]/(u^{-\theta} + v^{-\theta} - 1)}{1} = \log(uv).$$

After exponentiation, this is the independence copula $C^\perp(u, v) = uv$.

For the limit of C_{MTCJ} as $\theta \to \infty$:

$$(u^{-\theta} + v^{-\theta} - 1)^{-1/\theta} = u[1 + (u/v)^\theta - u^\theta]^{-1/\theta} = v[(v/u)^\theta + 1 - v^\theta]^{-1/\theta}.$$

For $0 < u < v < 1$, the middle term converges to u as $\theta \to \infty$, and for $0 < v < u < 1$, the right term converges to v as $\theta \to \infty$. This is the comonotonicity copula $C^+(u, v) = \min\{u, v\}$.

For the Galambos copula:

$$C(u, v; \theta) = C_{\text{Galambos}}(u, v; \theta) = uv \exp\{[(-\log u)^{-\theta} + (-\log v)^{-\theta}]^{-1/\theta}\}, \quad \theta > 0,$$

we obtain the limits as θ goes to the lower and upper boundaries. For the limit as $\theta \to 0^+$, $C(u, v; \theta) = uv \exp\{2^{-\infty}\} = uv$. Since

$$C(u, v; \theta) = uv \exp\left\{(-\log u)\left[1 + \left(\frac{-\log u}{-\log v}\right)^\theta\right]^{1/\theta}\right\} = uv \exp\left\{(-\log v)\left[1 + \left(\frac{-\log v}{-\log u}\right)^\theta\right]^{1/\theta}\right\},$$

the limit as $\theta \to \infty$ is $C^+(u, v) = \min\{u, v\}$. For example, $v < u$ implies $\left(\frac{-\log u}{-\log v}\right)^\theta \to 0$ as $\theta \to \infty$. $\qquad\square$

Sometimes a stronger dependence ordering can be established in the bivariate case. That is, an ordering of F, G that implies the concordance ordering of F and G. One such ordering is the stochastic increasing SI ordering; this is useful to establish some dependence results in Joe (1997) and in Chapter 8.

Definition 2.6 (more SI ordering; Yanagimoto and Okamoto (1969), Schriever (1987), Fang and Joe (1992); bivariate more (monotone) regression dependent or more stochastic increasing ordering). Let $F, G \in \mathcal{F}(F_1, F_2)$. Suppose that $F_{2|1}(x_2|x_1)$ and $G_{2|1}(x_2|x_1)$ are continuous in x_2 for all x_1. Then $G_{2|1}$ is *more SI* than $F_{2|1}$ (written $F \prec_{\text{SI}} G$ or $F_{2|1} \prec_{\text{SI}} G_{2|1}$) if $h(x_1, x_2) = G_{2|1}^{-1}[F_{2|1}(x_2|x_1)|x_1]$ is increasing in x_1. (Note that h is increasing in x_2 since, for each fixed x_1, it is a composition of increasing functions.)

In Joe (1997), this was renamed as more SI because $F_1 F_2 \prec_{\text{SI}} F$ if $F \in \mathcal{F}(F_1, F_2)$ is such that $F_{2|1}$ is stochastic increasing. Fang and Joe (1992) have examples showing that the $\prec_{\text{SI}}$ ordering holds for some 1-parameter copula families. There is no known multivariate version of this ordering.

Another dependence ordering is the supermodular ordering (called lattice-superadditive ordering in Joe (1990b)). It has been much studied in the 1990s; see Denuit et al. (2005) and Shaked and Shanthikumar (2007).

Definition 2.7 (Supermodular ordering). Let $F, G \in \mathcal{F}(F_1, F_2)$. F is smaller in the supermodular ordering, written $F \prec_{SM} G$, if $\int \varphi \, dF \leq \int \varphi \, dG$ when the integrals exist for all supermodular functions φ that satisfy

$$\varphi(\boldsymbol{x} \vee \boldsymbol{y}) + \varphi(\boldsymbol{x} \wedge \boldsymbol{y}) \geq \varphi(\boldsymbol{x}) + \varphi(\boldsymbol{y}), \quad \forall \, \boldsymbol{x}, \boldsymbol{y} \in \mathbb{R}^d.$$

The supermodular ordering has applications in actuarial risk analysis (Chapter 6 of Denuit et al. (2005)), but there are no results for copulas in this book that are based on this ordering,

There are other orderings in the literature that have names of the dependence concepts introduced in Section 2.8. However some of them are not related to the concordance ordering; see Colangelo (2008) and Chapter 9 of Shaked and Shanthikumar (2007).

2.12 Measures of bivariate monotone association

The term *monotone association* refers to dependence where if one variable increases then the other tends to increase (or decrease); that is, there is a roughly a monotone, not necessarily linear, relationship in terms of conditional expectation or conditional median of one variable given the other. A monotone relationship is a common form for two variables and many of the bivariate copula models are only suited for this type of dependence. Measures of bivariate monotone association, also called *measures of bivariate concordance*, are larger in absolute value if there is more probability near a monotone function relating two variables. They are better than correlation as dependence measures because they are invariant to strictly increasing transformations on the variables.

Scarsini (1984) developed criteria or axioms that measures of concordance should satisfy. See also Kimeldorf and Sampson (1989) for discussion of more general measures of positive dependence and Taylor (2007) for measures of multivariate concordance.

Definition 2.8 (Scarsini (1984)). A measure δ should satisfy the following for continuous random variables Y_1, Y_2.

(a) Domain: $\delta(Y_1, Y_2)$ can be defined for all random variables (including those without second moments).
(b) Symmetry: $\delta(Y_1, Y_2) = \delta(Y_2, Y_1)$.
(c) Coherence: $\delta(Y_1, Y_2) \le \delta(Y_1', Y_2')$ if $(F_{Y_1}(Y_1), F_{Y_2}(Y_2)) \prec_c (F_{Y_1'}(Y_1'), F_{Y_2'}(Y_2'))$ (that is, the copula of (Y_1, Y_2) is smaller than the copula of (Y_1', Y_2') in the concordance ordering).
(d) Range: $-1 \le \delta(Y_1, Y_2) \le 1$ (with $\delta(Y_1, Y_2) = 1$ for comonotonicity and $\delta(Y_1, Y_2) = -1$ for countermonotonicity).
(e) Independence: $\delta(Y_1, Y_2) = 0$ if $Y_1 \perp\!\!\!\perp Y_2$.
(f) Sign reversal: $\delta(-Y_1, Y_2) = -\delta(Y_1, Y_2)$;
(g) Continuity: if $(Y_{m1}, Y_{m2}) \sim F_m$ for $m = 1, 2, \ldots$, $(Y_1, Y_2) \sim F$, and $F_m \to_d F$, then $\lim_{m \to \infty} \delta(Y_{m1}, Y_{m2}) = \delta(Y_1, Y_2)$.
(h) Invariance: $\delta(Y_1, Y_2) = \delta(h_1(Y_1), h_2(Y_2))$ for strictly increasing functions h_1, h_2.

Scarsini (1984) shows that item (h) follows from items (a) to (g). Then from (f) and (h), $\delta(Y_1, Y_2) = -\delta(h(Y_1), Y_2)$ for h strictly decreasing. The proof of (h) is as follows: If $Y_1' = h_1(Y_1)$ and $Y_2' = h_2(Y_2)$ for strictly increasing functions h_1, h_2, then $C_{Y_1 Y_2} = C_{Y_1' Y_2'}$ where $C_{Y_1 Y_2}$ is the the copula of (Y_1, Y_2) and $C_{Y_1' Y_2'}$ is the copula of (Y_1', Y_2'). Hence $C_{Y_1 Y_2} \prec_c C_{Y_1' Y_2'} \prec_c C_{Y_1 Y_2}$ and (c) implies that $\delta(Y_1, Y_2) = \delta(Y_1', Y_2')$. However, for some measures, it might be easier to directly establish (h) and then (c) just needs to be verified for the case of $Y_1 \stackrel{d}{=} Y_1'$ and $Y_2 \stackrel{d}{=} Y_2'$.

Examples of measures of bivariate concordance are Kendall's tau (Kendall (1938)), Spearman's rank correlation (Spearman (1904)), Blomqvist's beta (Blomqvist (1950)) and the correlation of normal scores. Population and sample versions for continuous variables are given in different subsections below. There are several different ways to modify for discrete variables (or adjust for ties).

Note that Pearson's correlation $\rho(Y_1, Y_2)$ does not satisfy all of the axioms in Definition 2.8. Comments for Pearson's correlation and the axioms are the following:

(a) Domain: $\rho(Y_1, Y_2)$ exists only if Y_1 and Y_2 have finite second moments.
(b) Symmetry: yes.
(c) Coherence follows from Hoeffding's identity if $Y_1 \stackrel{d}{=} Y_1'$ and $Y_2 \stackrel{d}{=} Y_2'$. Let $(Y_1, Y_2) \sim F$, where $F \in \mathcal{F}(F_1, F_2)$, and suppose that the covariance of Y_1, Y_2 exists. Hoeffding's identity (Hoeffding (1940), Shea (1983)) is

$$\text{Cov}(Y_1, Y_2) = \int_{-\infty}^{\infty} \int_{-\infty}^{\infty} [\overline{F}(y_1, y_2) - \overline{F}_1(y_1)\overline{F}_2(y_2)] \, dy_2 dy_1.$$

For the proof of this identity, use the positive and negative parts of a random variable.
(d) Range: yes; 1 for comonotonicity: no; -1 for countermonotonicity: no.
(e) Independence: if Y_1 and Y_2 have finite second moments.

(f) Sign reversal: yes.

(g) Continuity: if finite second moments exist.

(h) Invariance to increasing transforms: no.

2.12.1 Kendall's tau

Let (Y_1, Y_2) and (Y_1', Y_2') be independent random pairs with continuous distribution $F = F_{12}$, copula C. The two pairs are *concordant* if $(Y_1 - Y_1')(Y_2 - Y_2') > 0$ and *discordant* if $(Y_1 - Y_1')(Y_2 - Y_2') < 0$. The population version of Kendall's tau is

$$
\begin{aligned}
\tau &= \mathbb{P}[(Y_1 - Y_1')(Y_2 - Y_2') > 0] - \mathbb{P}[(Y_1 - Y_1')(Y_2 - Y_2') < 0] \\
&= 2\mathbb{P}[(Y_1 - Y_1')(Y_2 - Y_2') > 0] - 1 = 4\mathbb{P}(Y_1 > Y_1', Y_2 > Y_2') - 1 \\
&= 4 \int_{[0,1]^2} C \, \mathrm{d}C - 1 = 4 \int F \, \mathrm{d}F - 1.
\end{aligned}
\tag{2.41}
$$

Let $C_{1|2}(u|v) = \frac{\partial C}{\partial v}(u, v)$ and $C_{2|1}(v|u) = \frac{\partial C}{\partial u}(u, v)$. Then, via integration by parts (see Fredricks and Nelsen (2007)),

$$
\int_{[0,1]^2} C(u, v) \, \mathrm{d}C(u, v) = \tfrac{1}{2} - \int_{[0,1]^2} C_{2|1}(v|u) \, C_{1|2}(u|v) \, \mathrm{d}u \mathrm{d}v,
$$

and

$$
\tau = 1 - 4 \int_{[0,1]^2} C_{2|1}(v|u) \, C_{1|2}(u|v) \, \mathrm{d}u \mathrm{d}v.
\tag{2.42}
$$

Note that (2.42) has an advantage as a numerical integration formula when the density c of C is unbounded.

The sample version for data (y_{i1}, y_{i2}), $i = 1, \ldots, n$, is

$$
\hat{\tau} = \frac{2}{n(n-1)} \sum_{1 \le i < j \le n} \left[I((y_{i1} - y_{j1})(y_{i2} - y_{j2}) > 0) - I((y_{i1} - y_{j1})(y_{i2} - y_{j2}) < 0) \right].
\tag{2.43}
$$

With no ties, this is the same as

$$
\hat{\tau} = \frac{4}{n(n-1)} \sum_{1 \le i < j \le n} I((y_{i1} - y_{j1})(y_{i2} - y_{j2}) > 0) - 1 = \frac{4}{n(n-1)} \sum_{1 \le i < j \le n} I_{ij} - 1,
$$

where $I_{ij} = I((y_{i1} - y_{j1})(y_{i2} - y_{j2}) > 0)$. As given above, the computation of $\hat{\tau}$, requires $O(n^2)$ operations. But there is a more efficient $O(n \log n)$ algorithm, due to Knight (1966), that is summarized in Algorithm 1 in Chapter 6; it includes an adjustment for ties. For a standard error (needed for an asymptotic confidence interval), a resampling method such as the delete-subset jackknife can be used.

In the case of a continuous bivariate distribution with copula C, as $n \to \infty$, $n^{-1}(\hat{\tau} - \tau)$ is asymptotically normal and, for the asymptotic variance,

$$
n \operatorname{Var}(\hat{\tau}) \to 16 \int [C + \overline{C}]^2 \mathrm{d}C - 4(\tau + 1)^2.
\tag{2.44}
$$

This can be proved from computing the covariance terms of the above I_{ij} as random variables, or as in Hoeffding (1948) using what is now known as Hoeffding's U-statistic formulation. (2.44) is a conversion of Hoeffding's result to a copula representation. When $C = C^\perp$, (2.44) becomes $4/9$, and when $C = C^+$ or C^-, (2.44) becomes 0. The numerically stable method to evaluate (2.44) is given in Section 6.7.

The axioms in Definition 2.8 are all satisfied for Kendall's tau $\tau = \tau(Y_1, Y_2) = \tau(C)$. Comments on the nontrivial properties are given below.

(c) Coherence: we use the identities in (2.52)–(2.53) given in Section 2.12.5 and the direct verification of invariance in (h). If $C_1 \prec_c C_2$, $\tau(C_2) - \tau(C_1)$ has the same sign as

$$\int C_2 \mathrm{d}C_2 - \int C_1 \mathrm{d}C_1 = \int (C_2 - C_1)\mathrm{d}C_2 + \int C_1 \mathrm{d}C_2 - \int C_1 \mathrm{d}C_1$$

$$= \int (C_2 - C_1)\mathrm{d}C_2 + \int (\overline{C}_2 - \overline{C}_1)\mathrm{d}C_1 \geq 0.$$

(d) Range $-1 \leq \tau(Y_1, Y_2) \leq 1$, with $\tau(Y_1, Y_2) = 1$ for comonotonicity and $\tau(Y_1, Y_2) = -1$ for countermonotonicity; this is based on the probablistic definition.

2.12.2   Spearman's rank correlation

With data (y_{i1}, y_{i2}), $i = 1, \ldots, n$, the sample version of Spearman's rank correlation or Spearman's rho is the correlation of the ranks r_{i1}, r_{i2}, where $r_{i1} = k$ if y_{i1} is kth smallest among $y_{11}, \ldots, y_{n1}$, and $r_{i2} = \ell$ if y_{i2} is ℓth smallest among $y_{12}, \ldots, y_{n2}$. If there are no ties,

$$\hat{\rho}_S = \frac{\sum_{i=1}^n r_{i1}r_{i2} - n[(n+1)/2]^2}{\sum_{i=1}^n \left[i - (n+1)/2\right]^2} = \frac{\sum_{i=1}^n r_{i1}r_{i2} - n[(n+1)/2]^2}{n(n^2-1)/12}.$$

Because $\hat{\rho}_S$ is defined based on ranks, and the data can be sorted into ranks with $O(n \log n)$ algorithms, $\hat{\rho}_S$ can be computed efficiently.

The population version, from the limit as $n \to \infty$, is:

$$\rho_S = 12 \int F_1(y_1)F_2(y_2)\,\mathrm{d}F(y_1, y_2) - 3 = 12 \int_{[0,1]^2} uv\,\mathrm{d}C(u,v) - 3 = \mathrm{Cor}[F_1(Y_1), F_2(Y_2)]$$

$$(2.45)$$

and this is the same as the correlation of the two uniform random variables $F_1(Y_1), F_2(Y_2)$ with $(Y_1, Y_2) \sim F \in \mathcal{F}(F_1, F_2)$. Since $\int_{[0,1]^2} uv\,\mathrm{d}C(u,v) = \mathbb{P}(U' \leq U, V' \leq V)$ where $U' \perp\!\!\!\perp V' \perp\!\!\!\perp (U,V)$ and $(U,V) \sim C$, $U' \sim \mathrm{U}(0,1)$, $V' \sim \mathrm{U}(0,1)$, then

$$\int_{[0,1]^2} uv\,\mathrm{d}C(u,v) = \int_{[0,1]^2} \overline{C}(u,v)\,\mathrm{d}u\mathrm{d}v = \int_{[0,1]^2} C(u,v)\,\mathrm{d}u\mathrm{d}v. \qquad (2.46)$$

The latter equality follows from substituting $\overline{C}(u,v) = 1 - u - v + C(u,v)$ and simplifying. Also

$$\int_{[0,1]^2} uv\,\mathrm{d}C(u,v) = \mathbb{E}[UV] = \mathbb{E}\{U\mathbb{E}[V|U]\} = \int_0^1 u \int_0^1 \overline{C}_{2|1}(v|u)\,\mathrm{d}v\,\mathrm{d}u$$

$$= \tfrac{1}{2} - \int_0^1 u \int_0^1 C_{2|1}(v|u)\,\mathrm{d}v\,\mathrm{d}u.$$

For numerical integration formulas,

$$\rho_S = 12 \int_{[0,1]^2} C(u,v)\,\mathrm{d}u\mathrm{d}v - 3 \qquad (2.47)$$

is better, or possibly

$$\rho_S = 3 - 12 \int_{[0,1]^2} u\,C_{2|1}(v|u)\,\mathrm{d}u\mathrm{d}v \qquad (2.48)$$

depending on the copula family.

Next, we discuss the asymptotics of $\hat{\rho}_S$. Asymptotically, $\hat{\rho}_S = \tilde{\rho}_S + O(n^{-1})$, where

$$\tilde{\rho}_S = 12 \cdot \frac{1}{n(n-1)(n-2)} \sum_{i \neq j, i \neq k, j \neq k} I_{ijk} - 3, \quad I_{ijk} = I(y_{i1} < y_{k1}, y_{j2} < y_{k2}). \quad (2.49)$$

For a standard error, a resampling method such as the delete-subset jackknife can be used.

In the case of a continuous bivariate distribution with copula C, as $n \to \infty$, $n^{-1}(\hat{\rho}_S - \rho_S)$ is asymptotically normal and, for the asymptotic variance,

$$n \operatorname{Var}(\hat{\rho}_S) \to 144 \int_0^1 \int_0^1 H(u, v) \, dC(u, v) - 9(\rho_S + 3)^2, \quad (2.50)$$

$$H(u, v) = u^2 v^2 + 2uv g_1(u) + 2uv g_2(v) + g_1^2(u) + g_2^2(v) + 2g_1(u)g_2(v),$$

$$g_1(u) = \int_0^1 \overline{C}(u, w) \, dw = \tfrac{1}{2} - u + \int_0^1 C(u, w) \, dw,$$

$$g_2(v) = \int_0^1 \overline{C}(x, v) \, dx = \tfrac{1}{2} - v + \int_0^1 C(x, v) \, dx.$$

This can be proved from computing the covariance terms of the I_{ijk} in (2.49) as random variables, or with the U-statistic formulation in Hoeffding (1948). (2.50) is a conversion of Hoeffding's result to a copula representation. When $C = C^{\perp}$, (2.50) becomes 1 and when $C = C^+$ or C^-, (2.50) becomes 0. The numerically stable method to evaluate (2.50) is given in Section 6.7.

The axioms in Definition 2.8 are all satisfied for Spearman's rank correlation $\rho_S = \rho_S(Y_1, Y_2) = \rho_S(C)$. Comments on the nontrivial properties are given below.

(c) Coherence: the invariance property (h) to increasing transforms can be verified directly, and if $C_1 \prec_c C_2$, then $\rho_S(C_2) - \rho_S(C_1) = 12 \int_{[0,1]^2} [\overline{C}_2(u, v) - \overline{C}_1(u, v)] du dv \geq 0$ using (2.46).

(d) Range $-1 \leq \rho_S(Y_1, Y_2) \leq 1$, with $\rho_S(Y_1, Y_2) = 1$ for comonotonicity and $\rho_S(Y_1, Y_2) = -1$ for countermonotonicity: this is based on the stochastic representations.

2.12.3 Blomqvist's beta

Blomqvist (1950) developed a quadrant measure of dependence that is now commonly called Blomqvist's β. It is also called the *medial correlation coefficient*. With $(U_1, U_2) \sim C$, it is defined as:

$$\beta = 4C(\tfrac{1}{2}, \tfrac{1}{2}) - 1 = 4\overline{C}(\tfrac{1}{2}, \tfrac{1}{2}) - 1 = 2\mathbb{P}((U_1 - \tfrac{1}{2})(U_2 - \tfrac{1}{2}) > 0) - 1.$$

As in the preceding section, let r_{i1}, r_{i2} be ranks of the data (y_{i1}, y_{i2}) for $i = 1, \ldots, n$. The most efficient sample version is

$$\hat{\beta} = 2n^{-1} \sum_{i=1}^n I([r_{i1} - \tfrac{1}{2} - n/2][r_{i2} - \tfrac{1}{2} - n/2] \geq 0) - 1.$$

This is exactly 1 for a perfectly positively dependent sample and -1 for a perfectly negatively dependent sample.

Converting the result in Blomqvist (1950), under the assumption that C is absolutely continuous, as $n \to \infty$,

$$n^{1/2}(\hat{\beta} - \beta) \to N(0, 1 - \beta^2)), \quad (2.51)$$

where the asymptotic variance is $1 - \beta^2 = 4\beta'(1 - \beta')$ with $\beta' = (1 + \beta)/2$. Hence the asymptotic variance can be estimated with the plug-in method.

The axioms in Definition 2.8 are all satisfied for Blomqvist's beta $\beta = \beta(Y_1, Y_2)$. Comments on the nontrivial properties are given below.

(c) Coherence: this follows after direct verification of property (h) on invariance to increasing transforms.

(e) Range $-1 \leq \beta(Y_1, Y_2) \leq 1$, with $\beta(Y_1, Y_2) = 1$ for comonotonicity and $\beta(Y_1, Y_2) = -1$ for countermonotonicity: because $C^+(\frac{1}{2}, \frac{1}{2}) = \frac{1}{2}$ and $C^-(\frac{1}{2}, \frac{1}{2}) = 0$.

2.12.4   Correlation of normal scores

If $(Y_1, Y_2) \sim F \in \mathcal{F}(F_1, F_2)$ with F_1, F_2 continuous, the population version of the correlation of normal scores is defined as:

$$\rho_N = \text{Cor}\left[\Phi^{-1}(F_1(Y_1)), \; \Phi^{-1}(F_2(Y_2))\right] = \text{Cor}\left[\Phi^{-1}(U_1), \; \Phi^{-1}(U_2)\right]$$

with $U_1 = F_1(Y_1)$ and $U_2 = F_2(Y_2)$, so that ρ_N depends only on the copula C of (Y_1, Y_2).

As defined in Section 2.12.2, let r_{i1}, r_{i2} be ranks of the data (y_{i1}, y_{i2}) for $i = 1, \ldots, n$. The sample version is the sample correlation after the normal score transform of ranks; that is, $\hat{\rho}_N$ is the sample correlation of the pairs $\Phi^{-1}([r_{i1} - \frac{1}{2}]/n), \Phi^{-1}([r_{i2} - \frac{1}{2}]/n)$. The asymptotic variance of $\hat{\rho}_N$ is given in Bhuchongkul (1964), but its form is not simple.

The axioms in Definition 2.8 can be verified in a similar way to Spearman's rank correlation.

2.12.5   Auxiliary results for dependence measures

One result for the proof of concordance ordering for Kendall's tau and Spearman's ρ_S is the following. Let $(X_1, X_2) \sim G$ and $(Y_1, Y_2) \sim F$ be two independent pairs of continuous random variables. Then (with dF as a Stieltjes integral, or replacing dF with density f)

$$\mathbb{P}(X_1 \leq Y_1, X_2 \leq Y_2) = \int_{-\infty}^{\infty} \int_{-\infty}^{\infty} \mathbb{P}(X_1 \leq y_1, X_2 \leq y_2) \, dF(y_1, y_2)$$

$$= \int_{-\infty}^{\infty} \int_{-\infty}^{\infty} G(y_1, y_2) \, dF(y_1, y_2) = \int \int G \, dF, \qquad (2.52)$$

$$\mathbb{P}(X_1 \leq Y_1, X_2 \leq Y_2) = \int_{-\infty}^{\infty} \int_{-\infty}^{\infty} \mathbb{P}(x_1 \leq Y_1, x_2 \leq Y_2) \, dG(x_1, x_2)$$

$$= \int_{-\infty}^{\infty} \int_{-\infty}^{\infty} \overline{F}(x_1, x_2) \, dG(x_1, x_2) = \int \int \overline{F} \, dG. \qquad (2.53)$$

Kendall's τ and Spearman's ρ_S for the bivariate Gaussian distribution $\Phi_2(\cdot; \rho)$ can be obtained in closed form from Theorem 8.53 with $\beta_0(\rho) = \Phi_2(0, 0; \rho) = \frac{1}{4} + (2\pi)^{-1} \arcsin(\rho)$. In this case, $\tau = 4\mathbb{P}(Z_1' < Z_1, Z_2' < Z_2) - 1$ where $(Z_1, Z_2), (Z_1', Z_2')$ are independent pairs from $\Phi_2(\cdot; \rho)$. Note that $\mathbb{P}(Z_1' < Z_1, Z_2' < Z_2) = \mathbb{P}(Y_1 < 0, Y_2 < 0)$ where $(Y_1, Y_2) = (Z_1' - Z_1, Z_2' - Z_2)$ is bivariate Gaussian with mean $(0, 0)$, and covariance matrix $\begin{pmatrix} 2 & 2\rho \\ 2\rho & 2 \end{pmatrix}$, or correlation matrix $\begin{pmatrix} 1 & \rho \\ \rho & 1 \end{pmatrix}$. Hence

$$\tau = 4\beta_0(\rho) - 1 = (2/\pi) \arcsin(\rho) \quad \text{and} \quad \beta = \tau.$$

Also

$$\rho_S = 12 \int \Phi(z_1)\Phi(z_2)\,\phi_2(z_1, z_2; \rho)\, dz_1 dz_2 - 3 = 12\mathbb{P}(Z_1'' < Z_1, Z_2'' < Z_2) - 3,$$

where $(Z_1, Z_2), (Z_1'', Z_2'')$ are independent pairs from $\Phi_2(\cdot; \rho)$ and $\Phi_2(\cdot; 0)$ respectively (that is, $Z_1'' \perp\!\!\!\perp Z_2'' \perp\!\!\!\perp (Z_1, Z_2)$). Similarly to above, $\mathbb{P}(Z_1'' < Z_1, Z_2'' < Z_2) = \mathbb{P}(X_1 < 0, X_2 < 0)$ where $(X_1, X_2) = (Z_1'' - Z_1, Z_2'' - Z_2)$ is bivariate Gaussian with mean $(0, 0)$, and covariance matrix $\begin{pmatrix} 2 & \rho \\ \rho & 2 \end{pmatrix}$, or correlation matrix $\begin{pmatrix} 1 & \rho/2 \\ \rho/2 & 1 \end{pmatrix}$. Hence

$$\rho_S = 12\beta_0(\tfrac{1}{2}\rho) - 3 = (6/\pi)\arcsin(\rho/2).$$

For bivariate t_ν (and bivariate absolutely continuous elliptical distributions), Kendall's tau is always the same as for bivariate Gaussian (Lindskog et al. (2003)), but Spearman's ρ_S will change with ν for fixed ρ.

Let $(Z_1, Z_2) \perp\!\!\!\perp (Z_1', Z_2') \perp\!\!\!\perp S \perp\!\!\!\perp S'$ where $(Z_1, Z_2) \overset{d}{=} (Z_1', Z_2') \sim \Phi_2(\cdot; \rho)$, $S \overset{d}{=} S' \sim F_S$ with $F_S(0) = 0$. For bivariate t_ν, $S = \sqrt{W}$ with $W \sim \text{Gamma}(\nu/2, \text{scale} = 2/\nu)$. Let $(Y_1, Y_2) = (Z_1, Z_2)/S$ and $(Y_1', Y_2') = (Z_1', Z_2')/S'$. Then with $R = S'/S$,

$$\mathbb{P}(Y_1' < Y_1, Y_2' < Y_2) = \mathbb{P}(Z_1' < Z_1 S'/S, Z_2' < Z_2 S'/S) = \mathbb{P}(Z_1' < Z_1 R, Z_2' < Z_2 R)$$
$$= \int_0^\infty \mathbb{P}(Z_1' < Z_1 r, Z_2' < Z_2 r)\, dF_R(r) = \int_0^\infty \beta_0(\rho)\, dF_R(r) = \beta_0(\rho),$$

because $(Z_1' - Z_1 r, Z_2' - Z_2 r)^\top \sim N_2\left((0,0)^\top, \begin{pmatrix} 1 + r^2 & \rho(1 + r^2) \\ \rho(1 + r^2) & 1 + r^2 \end{pmatrix}\right)$ with correlation parameter ρ. From the preceding, $\tau = 4\beta_0(\rho) - 1 = (2/\pi)\arcsin(\rho)$ for all bivariate $t_{2,\nu}(\rho)$ and more generally bivariate elliptical with correlation parameter ρ. Similarly, $\beta = 4\mathbb{P}(X_1 < 0, X_2 < 0) - 1 = (2/\pi)\arcsin(\rho)$.

Another result is on the comparison of ρ_S and τ. Capéraà and Genest (1993) prove that Spearman's ρ_S is larger than Kendall's τ for positively dependent random variables Y_1, Y_2 under the positive dependence condition: Y_2 right-tail increasing in Y_1 and Y_2 left-tail decreasing in Y_1 (or reversed roles of Y_1, Y_2). A simpler proof is given in Fredricks and Nelsen (2007).

2.12.6 Magnitude of asymptotic variance of measures of associations

Asymptotic variance formulas for $\hat{\tau}, \hat{\rho}_S, \hat{\beta}$ are given in (2.44), (2.50), (2.51). It is useful to know something about their magnitude. In the case of comonotonicity or countermonotonicity, the sample versions of the dependence measure is 1 (if they are defined in the best way). Hence the asymptotic variance should be smaller with strong positive or negative dependence. From numerical evaluations for some bivariate parametric copula families in Chapter 4, sometimes the asymptotic variance seems to be maximized at $C^\perp$ and otherwise the asymptotic variance is maximized at some $C(\cdot; \delta)$ where the parameter δ represents weak dependence (Kendall's tau between 0 and 0.1).

Some typical orders of magnitudes are shown in Table 2.3. Among the families in this table. for the Gumbel and MTCJ families, the asymptotic variance increases for a while as the parameter moves away from $C^\perp$ and then decreases. For the Frank copula, the asymptotic variance decreases as the parameter moves away from $C^\perp$ in a positive or negative direction. The largest asymptotic variance among these three families for $\hat{\tau}$ is 0.457 for Gumbel, and the largest for $\hat{\rho}_S$ is 1.005 for MTCJ.

τ	$\lim_{n\to\infty} n \, \mathrm{Var}(\hat{\tau})$			$\lim_{n\to\infty} n \, \mathrm{Var}(\hat{\rho}_S)$			$\lim_{n\to\infty} n \, \mathrm{Var}(\hat{\beta})$		
	Frank	Gumbel	MTCJ	Frank	Gumbel	MTCJ	Frank	Gumbel	MTCJ
0.0	0.444	0.444	0.444	1.000	1.000	1.000	1.000	1.000	1.000
0.1	0.433	0.457	0.450	0.960	0.978	0.986	0.987	0.991	0.991
0.2	0.401	0.442	0.436	0.851	0.842	0.857	0.950	0.961	0.961
0.4	0.284	0.345	0.351	0.490	0.535	0.526	0.800	0.841	0.837
0.4	0.284	0.345	0.351	0.490	0.535	0.527	0.800	0.841	0.837
0.5	0.211	0.274	0.286	0.299	0.366	0.387	0.692	0.749	0.738
0.6	0.139	0.198	0.213	0.142	0.196	0.220	0.565	0.637	0.613
0.7	0.077	0.123	0.138	0.044	0.079	0.094	0.425	0.504	0.465
0.8	0.033	0.059	0.070	0.007	0.018	0.024	0.282	0.353	0.304
1.0	0.000	0.000	0.000	0.000	0.000	0.000	0.000	0.000	0.000

Table 2.3 *Asymptotic variance of the sample version of Kendall's tau, Spearman's rank correlation and Blomqvist's beta for three parametric copula families; dependence parameters are chosen to attain some given τ values. Here, an asymptotic variance of $0.444 = 4/9$ for $C^\perp$ means that the SE of $\hat{\tau}$ is around $\sqrt{0.444/300} = 0.038$ and $\sqrt{0.444/1000} = 0.021$ for sample sizes of $n = 300$ and 1000 respectively.*

2.12.7 Measures of association for discrete/ordinal variables

For discrete response variables, there are versions of Kendall's τ and Spearman's ρ_S with adjustments for ties; see for example Agresti (1984). Because there is not necessarily a median value for a discrete random variable with equal probabilities below and above the median, there is not a version of Blomqvist's β for two discrete variables. The use of a bivariate Gaussian copula or latent bivariate Gaussian distribution leads to the polychoric correlation (Olsson (1979)) as the discrete counterpart of ρ_N. Binary response is a special case of ordinal response, and then an additional dependence measure is the (log) odds ratio.

Let two discrete response variables be denoted as y_1, y_2. Suppose the observation vectors are (y_{i1}, y_{i2}) for $i = 1, \dots, n$. The two variables can be summarized in a two-way table of proportions (p_{st}) or counts (n_{st}). Suppose the distinct discrete values are the integers $m_1, \dots, m_2$; for simpler notation, we assume the range of integers is the same for the two variables, but the definitions given below do not depend on this.

A measure based on concordant and discordant pairs is the *Goodman-Kruskal gamma*, which is defined as $\gamma = (p_{\mathrm{conc}} - p_{\mathrm{disc}})/(p_{\mathrm{conc}} + p_{\mathrm{disc}})$ where p_{conc} is the proportion of concordance and p_{disc} is the proportion of discordance in the two-way table (p_{st}); that is,

$$p_{\mathrm{conc}} = 2 \sum_{s=m_1}^{m_2-1} \sum_{t=m_1}^{m_2-1} p_{st} \Big\{ \sum_{s'>s, t'>t} p_{s't'} \Big\}, \quad p_{\mathrm{disc}} = 2 \sum_{s=m_1}^{m_2-1} \sum_{t=m_1+1}^{m_2} p_{st} \Big\{ \sum_{s'>s, t'<t} p_{s't'} \Big\}.$$

The sample version of γ is $\hat{\gamma} = [n_{\mathrm{conc}} - n_{\mathrm{disc}}]/[n_{\mathrm{conc}} + n_{\mathrm{disc}}]$, where n_{conc}=number of strictly concordant pairs and n_{disc}=number of strictly discordant pairs. Equivalently, use sample proportions $\hat{p}_{\mathrm{conc}}, \hat{p}_{\mathrm{disc}}$ in place of $p_{\mathrm{conc}}, p_{\mathrm{disc}}$ in the definition of γ.

A better version of τ with discrete data is *Kendall's τ_b*. The sample version is

$$\hat{\tau}_b = \frac{n_{\mathrm{conc}} - n_{\mathrm{disc}}}{\sqrt{[\binom{n}{2} - T_1][\binom{n}{2} - T_2]}},$$

where T_1 is the number of pairs tied on variable 1 only, and T_2 is the number of pairs tied on variable 2 only. The population version for the table (p_{st}) is

$$\tau_b = \frac{p_{\mathrm{conc}} - p_{\mathrm{disc}}}{\sqrt{(1 - \sum_s p_{s+}^2)(1 - \sum_t p_{+t}^2)}},$$

where (p_{s+}) and (p_{+t}) are the row and column margins of (p_{st}). Because $p_{\text{conc}} + p_{\text{disc}}$ cannot be larger than $1 - \sum_s p_{s+}^2$ and $1 - \sum_t p_{+t}^2$, then $|\gamma| \geq |\tau_b|$.

Spearman's rho for a two-way table is the correlation based on ranks with ties. For the two-way table of counts, let $n_{++} = n$ be the total, $(n_{m_1+}, \ldots, n_{m_2+})$ be the row margin vector, $(n_{+m_1}, \ldots, n_{+m_2})$ be the column margin vector, then *Spearman's rho with tied ranks* is

$$\hat{\rho}_S = \frac{\sum_{s=m_1}^{m_2} \sum_{t=m_1}^{m_2} n_{st}(r_{1s} - \bar{r})(r_{2t} - \bar{r})}{\left[\sum_{s=m_1}^{m_2} n_{s+}(r_{1s} - \bar{r})^2 \cdot \sum_{t=m_1}^{m_2} n_{+t}(r_{2t} - \bar{r})^2\right]^{1/2}},$$

$\bar{r} = (n_{++} + 1)/2$ is the average rank, $r_{1s} = \frac{1}{2}[(n_{m_1+} + \cdots + n_{s-1,+} + 1) + (n_{m_1+} + \cdots + n_{s+})]$, $r_{2t} = \frac{1}{2}[(n_{+m_1} + \cdots + n_{+,t-1} + 1) + (n_{+m_1} + \cdots + n_{+t})]$. For the population version ρ_S, let (p_{s+}) and (p_{+t}) be the marginal pmfs with corresponding cdfs F_1 and F_2. For variable 1, let the score for category s be $u_{1s} = [F_1(s-1) + F_1(s)]/2$, and for variable 2, let the score for category t be $u_{2t} = [F_2(t-1) + F_2(t)]/2$. Then

$$\rho_S = \frac{\sum_{s=m_1}^{m_2} \sum_{t=m_1}^{m_2} p_{st}(u_{1s} - \frac{1}{2})(u_{2t} - \frac{1}{2})}{\left[\sum_{s=m_1}^{m_2} p_{s+}(u_{1s} - \frac{1}{2})^2 \cdot \sum_{t=m_1}^{m_2} p_{+t}(u_{2t} - \frac{1}{2})^2\right]^{1/2}}.$$

Next, we give definitions of versions of ρ_N for ordinal variables.

Let $\hat{\zeta}_{1s} = \Phi^{-1}([n_{m_1+} + \cdots + n_{s+}]/n)$ (with $s = m_1, \ldots, m_2$) be the cutpoints for the latent $N(0,1)$ scale for variable 1 and let $\hat{\zeta}_{2t} = \Phi^{-1}([n_{+m_1} + \cdots + n_{+t}]/n)$ (with $t = m_1, \ldots, m_2$) be the cutpoints for the latent $N(0,1)$ scale for variable 2. Let $\hat{\zeta}_{1,m_1-1} = \hat{\zeta}_{2,m_1-1} = -\infty$. The *polychoric correlation*, based on the two-step estimation procedure, is defined as

$$\hat{\rho}_N = \underset{\rho}{\text{argmax}} \sum_{s=m_1}^{m_2} \sum_{t=m_1}^{m_2} n_{st} \log\{\Phi_2(\hat{\zeta}_{1s}, \hat{\zeta}_{2t}; \rho) - \Phi_2(\hat{\zeta}_{1,s-1}, \hat{\zeta}_{2t}; \rho)$$
$$- \Phi_2(\hat{\zeta}_{1s}, \hat{\zeta}_{2,t-1}; \rho) + \Phi_2(\hat{\zeta}_{1,s-1}, \hat{\zeta}_{2,t-1}; \rho)\}.$$

With $\zeta_{1s} = \Phi^{-1}(F_1(s))$ and $\zeta_{2t} = \Phi^{-1}(F_2(t))$, the population version is

$$\rho_N = \underset{\rho}{\text{argmax}} \sum_{s=m_1}^{m_2} \sum_{t=m_1}^{m_2} p_{st} \log\{\Phi_2(\zeta_{1s}, \zeta_{2t}; \rho) - \Phi_2(\zeta_{1,s-1}, \zeta_{2t}; \rho)$$
$$- \Phi_2(\zeta_{1s}, \zeta_{2,t-1}; \rho) + \Phi_2(\zeta_{1,s-1}, \zeta_{2,t-1}; \rho)\}.$$

It is called the *tetrachoric correlation* when both response variables are binary (e.g., $m_1 = 0$, $m_2 = 1$).

The polyserial correlation is similarly defined where one variable is continuous and the other is ordinal. Let the continuous variable be y_1 and let the ordinal/discrete variable be y_2. Suppose the observation vectors are (y_{i1}, y_{i2}) for $i = 1, \ldots, n$. Let $\{z_{i1}\}$ be the normal score transform for y_1 and suppose values of y_2 are the integer values $m_1, \ldots, m_2$. Let $\hat{\zeta}_{2t}$ be defined as above. Then the *polyserial correlation* is

$$\hat{\rho}_N = \underset{\rho}{\text{argmax}} \sum_{i=1}^{n} \log\left\{\phi(z_{i1}) \left[\Phi\left(\frac{\hat{\zeta}_{2,y_{i2}} - \rho z_{i1}}{(1-\rho^2)^{1/2}}\right) - \Phi\left(\frac{\hat{\zeta}_{2,y_{i2}-1} - \rho z_{i1}}{(1-\rho^2)^{1/2}}\right)\right]\right\},$$

and it is called a *biserial correlation* when y_2 is binary.

2.13   Tail dependence

Tail dependence is a measure of strength of dependence in the joint lower or joint upper tail of a multivariate distribution. It is presented in this section mainly for bivariate distributions, but the idea extends to multivariate.

The concept of bivariate tail dependence relates to the amount of dependence in the upper-quadrant tail or lower-quadrant tail of a bivariate distribution. It is a concept that is relevant to dependence in extreme values (which depends mainly on the tails) and in the derivation of multivariate extreme value distributions. It was presented in Joe (1993) as a way to differentiate different parametric bivariate copula families that interpolated $C^{\perp}$ and C^{+}.

The tail dependence coefficient is derived as a conditional probability so that it is between 0 and 1. Hence the condition (d) in Section 2.12 for a measure of monotone dependence is not satisfied. The lower bound of 0 is achieved for independence and countermonotonicity and other bivariate distributions.

The tail dependence coefficient is invariant to increasing transformations, so we present it first for bivariate copulas. If a bivariate copula C is such that

$$\lim_{u \to 1^-} \overline{C}(u,u)/(1-u) = \lambda_U$$

exists, then C has *upper tail dependence* if $\lambda_U \in (0,1]$ and no upper tail dependence if $\lambda_U = 0$. Similarly, if

$$\lim_{u \to 0^+} C(u,u)/u = \lambda_L$$

exists, C has *lower tail dependence* if $\lambda_L \in (0,1]$ and no lower tail dependence if $\lambda_L = 0$. We refer to λ_U and λ_L as the *tail dependence coefficients* whether they are positive or zero.

The derivation as a conditional probability is as follows. Suppose $(U_1, U_2) \sim C$. Then

$$\lambda_U = \lim_{u \to 1^-} \mathbb{P}(U_1 > u \mid U_2 > u) = \lim_{u \to 1^-} \mathbb{P}(U_2 > u \mid U_1 > u).$$

A similar expression holds for λ_L. If $\lambda_U > 0$ ($\lambda_L > 0$), there is a positive probability that one of U_1, U_2 takes values greater (less) than u given that the other is greater (less) than u, for u arbitrarily close to 1 (0).

If $(Y_1, Y_2) \sim F \in \mathcal{F}(F_1, F_2)$, where F is continuous, the non-copula forms of the tail dependence coefficients are:

$$\lambda_U = \lim_{u \to 1^-} \mathbb{P}\big(Y_2 > F_2^{-1}(u) \mid Y_1 > F_1^{-1}(u)\big) = \lim_{u \to 1^-} \mathbb{P}\big(Y_1 > F_1^{-1}(u) \mid Y_2 > F_2^{-1}(u)\big)$$

$$= \lim_{u \to 1^-} \mathbb{P}\big(Y_1 > F_1^{-1}(u), Y_2 > F_2^{-1}(u)\big)/(1-u),$$

$$\lambda_L = \lim_{u \to 0^+} \mathbb{P}\big(Y_2 \le F_2^{-1}(u) \mid Y_1 \le F_1^{-1}(u)\big) = \lim_{u \to 0^+} \mathbb{P}\big(Y_1 \le F_1^{-1}(u) \mid Y_2 \le F_2^{-1}(u)\big)$$

$$= \lim_{u \to 0^+} \mathbb{P}\big(Y_1 \le F_1^{-1}(u), Y_2 \le F_2^{-1}(u)\big)/u.$$

For a multivariate distribution with tail dependence, one can consider the set of bivariate tail dependence coefficients $(\lambda_{U,jk} : 1 \le j < k \le d)$ or $(\lambda_{L,jk} : 1 \le j < k \le d)$, or define multivariate tail dependence measures (Alink et al. (2007), Li (2008b)) as:

$$\lambda_U = \lim_{u \to 0^+} \overline{C}((1-u)\mathbf{1}_d)/u, \quad \lambda_L = \lim_{u \to 0^+} C(u\mathbf{1}_d)/u.$$

The interpretation as a conditional probability is as follows. If $(U_1, \ldots, U_d) \sim C$, then

$$\lambda_U = \lim_{u \to 0^+} \mathbb{P}(U_2 > 1-u, \ldots, U_d > 1-u \mid U_1 > 1-u),$$

$$\lambda_L = \lim_{u \to 0^+} \mathbb{P}(U_2 \le u, \ldots, U_d \le u \mid U_1 \le u).$$

Regarding Definition 2.8 for properties of measures of monotone assocation, not all are satisfied for the upper (or lower) tail dependence parameter $\lambda = \lambda(Y_1, Y_2)$, Axioms (a), (b), (c), (e), (g), (h) do hold (respectively for domain, symmetry, coherence, independence, continuity and invariance to strictly increasing transforms). The remaining axioms do not hold as explained below.

(d) Range: lower bound is 0 (because it is a conditional probability) even for negative dependence, and in particular, $\lambda(Y_1, Y_2) = 0$ for countermonotonicity. $\lambda(Y_1, Y_2) = 1$ for comonotonicity, but it can also be 1 for non-comonotonicity, see Section 2.21.

(f) Sign reversal: doesn't hold because $\lambda \geq 0$.

The empirical measure of tail dependence for data doesn't really exist because of the limit; the best that can be done are estimation procedures such as those in Dobrić and Schmid (2005) and Frahm et al. (2005). Also, for each pair of variables, a scatterplot (after transform to normal scores or uniform scores) can be inspected for tail dependence, and the empirical conditional proportions of $[u, 1]^2$ for some u near 1 and of $[0, u]^2$ for some u near 0 can be computed. An alternative to assess strength of dependence in the tails for data is the use of semi-correlations (Section 2.17) and other tail-weighted dependence measures.

For the bivariate Gaussian distribution or copula, $\lambda_U = \lambda_L = 0$ for any $-1 < \rho < 1$ for the correlation parameter. See Example 5.32 of McNeil et al. (2005) or Sibuya (1960) for a proof of the lack of tail dependence for the bivariate Gaussian copula. The bivariate t distribution or copula does have tail dependence, and one can prove $\lambda_U = \lambda_L = 0$ for bivariate Gaussian from the limit of the tail dependence parameter for the $t_\nu(\rho)$ distribution as $\nu \to \infty$.

Hence for models where the property of tail dependence is desired, non-Gaussian copulas with tail dependence are needed. For example, if there is tail dependence for joint risks in insurance or finance, then use of the Gaussian copula will underestimate the joint tail probability. The relevance of this is shown in the next example.

Example 2.11 Joint tail probabilities for multivariate Gaussian versus t_4. Consider $d = 5$ financial assets, and suppose they have the equicorrelation parameter $\rho = 0.5$. We compare a tail probability for multivariate Gaussian versus t_4. The probability that all 5 returns are below their 5th percentile (corresponding to an extreme simultaneous loss) is: $7.485 \times 10^{-5} = 1/13360$ for multivariate Gaussian, and $5.353 \times 10^{-4} = 1/1868$ for multivariate t_4.

Assuming 260 trading days per year, this means the extreme simultaneous loss would occur once every 13360/260=51 years assuming multivariate Gaussian and once every 1868/260=7.2 years assuming multivariate t_4. This is a big difference and shows that tail dependence can have a large impact on tail inferences. □

We next show a technique to compute the tail dependence parameter that is not a direct application of the definition. The definition of tail dependence is easy to apply only if the copula cdf has a simple form. If the copula cdf involves integrals and the conditional distributions are tractable, then the following calculation method (Demarta and McNeil (2005)) works. For example, it can be used for bivariate t distributions.

If Y_1, Y_2 are exchangeably dependent with $F_1 = F_2$, then

$$\lambda_L = 2 \lim_{y \to -\infty} \mathbb{P}(Y_2 \leq y \mid Y_1 = y) = 2 \lim_{y \to -\infty} F_{2|1}(y|y),$$

$$\lambda_U = 2 \lim_{y \to \infty} \mathbb{P}(Y_2 \geq y \mid Y_1 = y) = 2 \lim_{y \to \infty} \overline{F}_{2|1}(y|y).$$

This is a special case of Theorem 8.57.

By matching up with a $t_{\nu+1}$ density, $[Y_2 \mid Y_1 = y_1]$ has distribution that is a location-scale transform of the $t_{\nu+1}$ density (see Section 2.7):

$$[Y_2 \mid Y_1 = y_1] \sim t_{\nu+1}(\mu(y_1) = \rho y_1, \sigma(y_1)), \quad \sigma^2(y) = (1 - \rho^2)(\nu + y^2)/(\nu + 1).$$

Because of permutation symmetry in y_1, y_2, and reflection symmetry (invariance to $(y_1, y_2) \to (-y_1, -y_2)$ for the density) the upper/lower tail dependence parameter of $t_{2,\nu}(\rho)$ distribution is

$$\lambda = \lambda_L = \lambda_U = 2 \lim_{y \to -\infty} \mathbb{P}(Y_2 \le y | Y_1 = y)$$

and

$$\mathbb{P}(Y_2 \le y | Y_1 = y) = \mathbb{P}\big(\mu(y) + \sigma(y)Y \le y\big) = T_{1,\nu+1}\big([y - \rho y]/\sigma(y)\big),$$

where $Y \sim t_{\nu+1}$. Taking the limit as $y \to -\infty$, and factoring out $y/|y|$, leads to

$$\mathbb{P}(Y_2 \le y | Y_1 = y) = T_{1,\nu+1}\left(-\frac{(1-\rho)\sqrt{\nu+1}}{\sqrt{1-\rho^2}}\right) = T_{1,\nu+1}\left(-\sqrt{\frac{(\nu+1)(1-\rho)}{(1+\rho)}}\right)$$

and hence

$$\lambda = 2T_{1,\nu+1}\left(-\sqrt{\frac{(\nu+1)(1-\rho)}{(1+\rho)}}\right).$$

The bivariate t distribution extends to multivariate t, and the multivariate t copula has tail dependence. If $R = (\rho_{jk})_{1 \le j < k \le d}$ is the correlation matrix parameter, then the tail dependence parameter of the (j, k) margin is

$$\lambda_{jk} = 2T_{1,\nu+1}\left(-\sqrt{\frac{(\nu+1)(1-\rho_{jk})}{(1+\rho_{jk})}}\right).$$

For any $j < k$, $\lambda_{jk} = \lambda_{U,jk} = \lambda_{L,jk}$. The more general result for elliptical distributions is that tail dependence occurs if the random variable $\sqrt{Q}$ in (2.21) has a density that has a heavy enough tail, for example, a regularly varying tail. See the theorems in 8.8 for tail properties of elliptical distributions.

2.14 Tail asymmetry

For univariate distributions, one can compare the upper and lower tails to assess the tail asymmetry. There is a similar comparison for the joint upper and lower tails of multivariate distributions.

Definition 2.9 (Reflection or central symmetry/asymmetry). Let C be a copula such that $(U_1, \ldots, U_d) \sim C$, and define $\widehat{C}$ as the *reflected copula* such that $(1 - U_1, \ldots, 1 - U_d) \sim \widehat{C}$. *Reflection or central symmetry* holds if $C \equiv \widehat{C}$.

If $C(u\mathbf{1}_d) \ge \widehat{C}(u\mathbf{1}_d)$ for all $0 < u < u_0$ with some $0 < u_0 \le \frac{1}{2}$, then the copula has more probability in the lower tail and we say that there is *reflection asymmetry with skewness to lower tail*. If the inequality is reversed leading to $C(u\mathbf{1}_d) \le \widehat{C}(u\mathbf{1}_d)$ or all $0 < u < u_0$ with some $0 < u_0 \le \frac{1}{2}$, then the copula has more probability in the upper tail and we say that there is *reflection asymmetry with skewness to upper tail*.

In the literature on multivariate distributions, such as in Dharmadhikari and Joag-dev (1988), the definition of *central symmetry* is that a random d-vector $\boldsymbol{X}$ as centrally symmetric (about the origin) if $\mathbb{P}(\boldsymbol{X} \in A) = \mathbb{P}(-\boldsymbol{X} \in A) = \mathbb{P}(\boldsymbol{X} \in -A)$ for all Borel sets $A \in \mathbb{R}^d$; equivalently $\mathbb{P}(X_1 \le x_1, \ldots, X_d \le x_d) = \mathbb{P}(X_1 \ge -x_1, \ldots, X_d \ge -x_d)$ for all $\boldsymbol{x} \in \mathbb{R}^d$. Converting this to a property of a vector $\boldsymbol{U} = (U_1, \ldots, U_d)$ of U(0, 1) random variables which has a central point $\frac{1}{2}\mathbf{1}_d$, $\boldsymbol{U} \sim C$ is centrally symmetric (about $\frac{1}{2}\mathbf{1}_d$) if $\boldsymbol{U} \stackrel{d}{=} \mathbf{1}_d - \boldsymbol{U}$ or $C(\boldsymbol{u}) = \overline{C}(\mathbf{1}_d - \boldsymbol{u}) = \widehat{C}(\boldsymbol{u})$ for all $\boldsymbol{u} \in (0, 1)^d$. The term *reflection symmetric* is slightly more convenient in order to refer in general to $\widehat{C}$ as the reflected copula (of the reflection $\mathbf{1}_d - \boldsymbol{U}$).

If the copula density exists, then reflection symmetry implies that for $c(u_1, \ldots, u_d) = \partial^d C(u_1, \ldots, u_d)/\partial u_1 \cdots \partial u_d$, then

$$c(u_1, \ldots, u_d) = c(1 - u_1, \ldots, 1 - u_d), \quad \boldsymbol{u} \in [0, 1]^d.$$

In Nelsen (1999), the condition $C \equiv \widehat{C}$ in Definition 2.9 is called *radial symmetry*.[2]

As examples, multivariate Gaussian and t copulas are reflection symmetric, and the Frank copula family is the only bivariate reflection symmetric Archimedean family (but the multivariate Frank copula based on the logarithmic series Laplace transform is not reflection symmetric for dimensions $d \geq 3$). All non-boundary multivariate extreme value copulas (for maxima) are reflection asymmetric with skewness toward the joint upper tail.

The definition of skewness to the lower or upper tail is generally hard to check analytically, so we use the form of the tail of $C(u\mathbf{1}_d)$ and $\widehat{C}(u\mathbf{1}_d)$ as $u \to 0^+$ in Section 2.16 to assess tail asymmetry of copulas that satisfy some of the dependence concepts in Section 2.8.

2.15 Measures of bivariate asymmetry

In this section, we present and discuss some simple measures of bivariate asymmetry for bivariate copulas.

We define measures as a function of a bivariate copula C or a random pair (U, V) with $(U, V) \sim C$. There are two types of asymmetry that are most relevant for bivariate data analysis:

(i) lack of reflection symmetry: the difference of the distributions of (U, V) and the reflection $(1 - U, 1 - V)$, or the difference of $C(u, v)$ and $\widehat{C}(u, v) = u + v - 1 + C(1 - u, 1 - v)$;

(ii) lack of permutation symmetry or exchangeability: the difference of the distributions of (U, V) and the permutation (V, U), or the difference of $C_{12}(u, v) = C(u, v)$ and $C_{21}(u, v) = C(v, u)$.

The simplest one-parameter bivariate copula families are permutation symmetric but not necessarily reflection symmetric. Most of the commonly used two-parameter bivariate copula families are also permutation symmetric.

Measures of permutation asymmetry are studied in Klement and Mesiar (2006) and Nelsen (2007); measures of reflection asymmetry are studied in Dehgani et al. (2013) and Rosco and Joe (2013). The various measures for bivariate reflection asymmetry in Rosco and Joe (2013) can be adapted for permutation asymmetry, as given here. Measures of reflection asymmetry include the following.

- An L_∞ distance $\sup_{0 \leq u, v \leq 1} |C(u, v) - [u + v - 1 + C(1 - u, 1 - v)]|$ or its L_p counterpart $\{\int_{[0,1]^2} |C(u, v) - \widehat{C}(u, v)|^p du dv\}^{1/p}$ with $p \geq 1$.

- $\mathbb{E}[(U + V - 1)^3]$ measuring the skewness of the random variable $U + V - 1$ (which has mean 0), or its extension to other powers $\mathbb{E}[|U + V - 1|^k \operatorname{sign}(U + V - 1)]$ with positive-valued $k > 1$.

---

[2] Pages 11 and 15 of du Sautoy (2008) have the sentences: *"The honeybee likes the pentagonal symmetry of honeysuckle, the hexagonal shape of the clematis, and the highly radial symmetry of the daisy and sunflower. The bumblebee prefers mirror symmetry, such as the symmetry of the orchid, pea or foxglove."* and *"Now I hop along the beach kangaroo-fashion, and my two feet create a pattern with simple reflection. When I spin in the air and land facing the other way, I get a pattern with two lines of reflectional symmetry."* For dimension $d = 2$, radial symmetry makes me think of something like the preceding, and reflection symmetry as mentioned in the preceding matches mirror symmetry along the line $u_1 + u_2 = 1$ only if the copula is permutation symmetric. The transformations $U_j \to 1 - U_j$, which preserves marginal U(0, 1), are called reflections in the study of multivariate concordance, such as in Taylor (2007).

- Quantile-based skewness measure for $U + V - 1$:

$$\zeta_{U+V-1}(p) = \frac{Q(1-p) - 2Q(\frac{1}{2}) + Q(p)}{Q(1-p) - Q(p)}, \quad 0 < p < \frac{1}{2},$$

where $Q = Q_{U+V-1}$ be the quantile function of $U+V-1$. The value $p = 0.05$ is suggested in Rosco and Joe (2013) to achieve some sensitivity to the tails.

There are analogous measures of permutation asymmetry.

- An L_∞ distance $\sup_{0 \le u,v \le 1} |C(u,v) - C(v,u)|$ or its L_p counterpart.
- $\mathbb{E}[(U - V)^3]$ measuring the skewness of the random variable $U - V$ (which has mean 0), or its extension to other powers $\mathbb{E}[|U - V|^k \text{sign}(U - V)]$ with positive-valued $k > 1$. A formula that is valid even if C has a singular component is the following:

$$\mathbb{E}[(U - V)^3] = 3\mathbb{E}[UV^2 - U^2V] = 6\int_{[0,1]^2} [v\overline{C}(u,v) - u\overline{C}(u,v)]dudv$$

$$= 6\int_{[0,1]^2} [vC(u,v) - uC(u,v)]dudv. \tag{2.54}$$

- Quantile-based skewness measure for $U - V$:

$$\zeta_{U-V}(p) = \frac{Q(1-p) - 2Q(\frac{1}{2}) + Q(p)}{Q(1-p) - Q(p)}, \quad 0 < p < \frac{1}{2},$$

where $Q = Q_{U-V}$ is the quantile function of $U - V$. The choice of $p = 0.05$ achieves some sensitivity to the tails in the upper left or lower right corners.

The third moment skewness measures for reflection and permutation asymmetry take values between $-27/256$ and $27/256 = 0.1055$; and the L_∞ distance measures take values between 0 and $1/3$. The quantile-based skewness measures for reflection and permutation asymmetry take values between -1 and 1.

The L_∞ or L_p distances are non-negative. The sample version of them using the empirical distribution as an estimate of C are always non-negative, so that the distribution of the sample version of the measure has support on 0 to some upper bound (that depends on p), even if the copula is reflection or permutation symmetric. Getting an interval estimate is not straightforward, but Genest et al. (2012) have a test for permutation asymmetry based on the L_∞ distance of the empirical copula.

The skewness measures have simpler forms for data, because point estimators are straightforward to obtain and standard errors can be obtained through resampling methods such as bootstrapping. The skewness measures can be positive or negative so that they contain information on the direction of skewness. The quantile-based skewness measures for permutation asymmetry are more sensitive to bivariate copulas with positive dependence and no support in either the upper left corner or lower right corner of the unit square — these types of copulas can arise from bivariate distributions with a stochastic representation of $(X_1, X_2) = (Z_0 + Z_1, Z_0 + Z_2)$, or $(X_1, X_2) = (Z_0 \wedge Z_1, Z_0 \wedge Z_2)$, where Z_0, Z_1, Z_2 are independent non-negative random variables and one of Z_1, Z_2 is identically zero. See Sections 4.28 and 4.14 for some constructions of this form that are in use as statistical models.

For reflection asymmetry, there are also measures of tail asymmetry based on the difference of upper and lower semi-correlations (see Section 2.17). If $\varrho_U(C)$ is an upper tail-weighted measure of dependence, and $\varrho_L(C)$ is an analogous lower tail-weighted measure of dependence, then the difference $\Delta(C) = \varrho_U(C) - \varrho_L(C)$ is a measure of tail asymmetry, and $\Delta(C) = 0$ for copulas C that are reflection symmetric. Tail-weighted measures of asymmetry are useful for distinguishing different families of bivariate copulas that have different degrees of upper and lower tail order, as defined in Section 2.16.

2.16   Tail order

In this section, we discuss the *tail order*[3] as a concept for the strength of dependence in the joint tails of a multivariate distribution. By using it to compare opposite tails, it is also useful to assess tail or reflection asymmetry.

The concept originates in Ledford and Tawn (1996) (for multivariate extremes) and in Heffernan (2000) for bivariate copulas, but it is studied in much more depth in Hua and Joe (2011). The tail order has value ≥ 1, with larger values indicating less dependence in the joint tail. It is the reciprocal of the tail coefficient η in Ledford and Tawn (1996); η is called the residual dependence index in Hashorva (2010) and extreme residual coefficient in de Haan and Zhou (2011).

Definition 2.10 Let C be a d-dimensional copula. If there exists $\kappa_L(C) > 0$ and some $\ell(u)$ that is slowly varying at 0^+ (i.e., $\ell(tu)/\ell(u) \sim 1$ as $u \to 0^+$ for all $t > 0$) such that

$$C(u\mathbf{1}_d) \sim u^{\kappa_L(C)} \ell(u), \quad u \to 0^+,$$

then $\kappa_L(C)$ is called the *lower tail order* of C and $\Upsilon_L(C) = \lim_{u \to 0^+} \ell(u)$ is the lower tail order parameter. By reflection, the *upper tail order* is defined as $\kappa_U(C)$ such that

$$\overline{C}((1-u)\mathbf{1}_d) \sim u^{\kappa_U(C)} \ell^*(u), \quad u \to 0^+,$$

for some slowly varying function $\ell^*(u)$. The upper tail order parameter is then $\Upsilon_U(C) = \lim_{u \to 0^+} \ell^*(u)$.

With $\kappa = \kappa_L$ or κ_U, we further classify into the following.

- *Intermediate tail dependence*: for $1 < \kappa < d$ in dimension d, or $\kappa = 1$ and $\Upsilon = 0$.
- *Strong tail dependence*: $\kappa = 1$ with $\Upsilon > 0$.
- *Tail orthant independence* (or *tail quadrant independence*): $\kappa = d$ in dimension $d > 2$ (or $d = 2$), and the slowly varying function is (asymptotically) a constant

Some properties are the following.

- If $\kappa_L(C) = 1$ (respectively $\kappa_U(C) = 1$) and $\ell(u) \not\to 0$, then the usual definition of lower (respectively upper) tail dependence in Section 2.13 obtains with $\lambda_L = \Upsilon_L$ or $\lambda_U = \Upsilon_U$.
- $\kappa_L(C) = \kappa_U(C) = d$ for the d-dimensional independence copula $C^\perp$.
- It is not possible for $\kappa_U < 1$ or $\kappa_L < 1$, but it is possible for $\kappa_L(C)$ and $\kappa_U(C)$ to be $> d$ for copulas with some negative dependence.
- For the bivariate countermonotonicity copula, $\kappa_L(C)$ and $\kappa_U(C)$ can be considered as $+\infty$ because $C(u, u)$ and $\overline{C}(1 - u, 1 - u)$ are zero for $0 < u < \frac{1}{2}$.

It is possible for $\kappa_U = 1$ with $\lambda_U = 0$ with a slowly varying function $\ell^*(u)$ with $\lim_{u \to 0^+} \ell^*(u) = 0$. An example of this is the copula of the bivariate gamma convolution distribution; see Section 4.28.

To determine an approximation to κ_L and κ_U numerically, one can compute

$$\frac{\log C(u\mathbf{1}_d)}{\log u} \quad \text{and} \quad \frac{\log \overline{C}((1-u)\mathbf{1}_d)}{\log u},$$

for some small positive u values. Analytic examples are given in Example 2.12 below.

---

[3]Although not defined in Joe (1997), the tail order was implicitly used in two places: (i) in Theorem 4.16 for a result on lower tail dependence for a copula constructed from a mixture of max-id; (ii) in Section 7.1.7 to show that the trivariate (and multivariate) Frank copula is not tail or reflection symmetric.

2.16.1   Tail order function and copula density

In this subsection, study of the behavior of copula densities in the corners is accomplished via tail order functions. Assume C has lower tail order κ_L and upper tail order κ_U. Extending Definition 2.10 so that the limit is not necessarily along the main diagonal, define

$$b(\boldsymbol{w}) = \lim_{u \to 0^+} \frac{C(uw_1, \ldots, uw_d)}{u^{\kappa_L} \ell(u)}, \quad \boldsymbol{w} \in \mathbb{R}_+^d,$$

$$b^*(\boldsymbol{w}) = \lim_{u \to 0^+} \frac{\overline{C}(1 - uw_1, \ldots, 1 - uw_d)}{u^{\kappa_U} \ell^*(u)}, \quad \boldsymbol{w} \in \mathbb{R}_+^d,$$

as the lower and upper tail order functions respectively, when the limits exist.

Sometimes partial derivatives and the density have a simpler form than the copula cdf. We would like to know what tail properties will be inherited if we take partial derivatives of the copula. For example, for the lower tail, if

$$C(uw_1, \ldots, uw_d) \sim u^{\kappa_L} \ell(u)\, b(\boldsymbol{w}), \quad u \to 0^+, \; \boldsymbol{w} \in \mathbb{R}_+^d,$$

then we want to differentiate both sides of the above with respect to the w_j's to get:

$$u \frac{\partial C(u\boldsymbol{w})}{\partial w_j} \sim u^{\kappa_L} \ell(u) \frac{\partial b(\boldsymbol{w})}{\partial w_j}, \quad u \to 0^+, \; j = 1, \ldots, d,$$

and higher order derivatives up to:

$$u^d \frac{\partial^d C(u\boldsymbol{w})}{\partial w_1 \cdots \partial w_d} \sim u^{\kappa_L} \ell(u) \frac{\partial^d b(\boldsymbol{w})}{\partial w_1 \cdots \partial w_d}, \quad u \to 0^+. \tag{2.55}$$

A sufficient condition is finiteness near the corner and ultimate monotonicity of partial derivatives of the copula (that is, $\partial C/\partial u_j$ is ultimately monotone in u_j at 0^+, and similar for higher order derivatives). A proof is similar to that of Theorem 1.7.2 (Monotone density theorem) in Bingham et al. (1987). Parallel results apply for the joint upper tail.

Example 2.12 As an example of using the density to get the tail order, consider a multivariate Gaussian copula with positive definite correlation matrix $\boldsymbol{\Sigma}$ which satisfies $C_\Phi(u\mathbf{1}_d) \sim u^\kappa \ell(u) = h^* u^\kappa (-\log u)^\zeta$, $u \to 0^+$, where h^* is a constant. Then (as can be shown directly with the monotone density theorem), this would be equivalent to $c_\Phi(u\mathbf{1}_d) \sim h u^{\kappa-d}(-\log u)^\zeta$, $u \to 0^+$, where h is another constant. Thus, with ϕ_d for the multivariate Gaussian density,

$$1 = \lim_{u \to 0^+} \frac{c_\Phi(u\mathbf{1}_d)}{h\, u^{\kappa-d}(-\log u)^\zeta} = \lim_{u \to 0^+} \frac{\phi_d\left(\Phi^{-1}(u)\mathbf{1}_d; \boldsymbol{\Sigma}\right)}{h\, \phi^d(\Phi^{-1}(u))\, u^{\kappa-d}(-\log u)^\zeta}$$

$$= \lim_{z \to -\infty} \frac{\phi_d\left(z\mathbf{1}_d; \boldsymbol{\Sigma}\right)}{h\, \phi^d(z)\, [\Phi(z)]^{\kappa-d}\, [-\log(\Phi(z))]^\zeta} = \lim_{z \to -\infty} \frac{\phi_d\left(z\mathbf{1}_d; \boldsymbol{\Sigma}\right)}{h\, \phi^\kappa(z)\, |z|^{d-\kappa}\, [-\log(\phi(z)/|z|)]^\zeta}. \tag{2.56}$$

The above makes use of $\Phi(z) \sim \phi(z)/|z|$ as $z \to -\infty$. Since the exponent terms dominate the numerator and denominator of (2.56), to cancel the exponent terms, a necessary condition is that $\kappa = \mathbf{1}_d \boldsymbol{\Sigma}^{-1} \mathbf{1}_d^\top$, which turns out to be the tail order of the copula C_Φ. Also, to cancel the term of $|z|$ in (2.56), we need that $d - \kappa + 2\zeta = 0$, so $\zeta = (\kappa - d)/2$.

Example 2.13 Some bivariate examples of tail order and tail order functions are given here. They include tail dependence, intermediate tail dependence, and tail quadrant independence in the lower tail.

- For the bivariate Gaussian as a special case of preceding Example 2.12, $\kappa_L = \kappa_U = 2/(1+\rho)$ so there is intermediate tail dependence for $0 < \rho < 1$.
- For the bivariate MTCJ copula derived in Section 1.2,

$$C(u, v; \delta) = (u^{-\delta} + v^{-\delta} - 1)^{-1/\delta},$$

with density $c(u, v; \delta) = (1 + \delta)[uv]^{-\delta-1}(u^{-\delta} + v^{-\delta} - 1)^{-2-1/\delta}$. The reflected/survival bivariate copula is

$$\widehat{C}(u, v; \delta) = u + v - 1 + C(1 - u, 1 - v) = u + v - 1 + [(1 - u)^{-\delta} + (1 - v)^{-\delta} - 1]^{-1/\delta}.$$

The lower tail order of the reflected copula is the same as the upper tail order of the original copula. As $u \to 0^+$,

$$C(uw_1, uw_2; \delta) \sim u(w_1^{-\delta} + w_2^{-\delta})^{-1/\delta} =: ub(w_1, w_2) \Rightarrow \kappa_L = 1, \ \lambda_L = b(1, 1) = 2^{-1/\delta}.$$

A direct lower tail expansion of the density leads to:

$$c(uw_1, uw_2; \delta) \sim u^{-1}(1 + \delta)(w_1 w_2)^{-\delta-1}(w_1^{-\delta} + w_2^{-\delta})^{-2-1/\delta} = u^{-1}\frac{\partial^2 b(w_1, w_2)}{\partial w_1 \partial w_2}.$$

Also,

$$\begin{aligned}
\widehat{C}(u, u; \delta) &\sim 2u - 1 + [2(1 - u)^{-\delta} - 1]^{-1/\delta} \\
&\sim 2u - 1 + [2(1 + \delta u + \tfrac{1}{2}\delta(1 + \delta)u^2) - 1]^{-1/\delta} \\
&= 2u - 1 + [1 + 2\delta u + \delta(1 + \delta)u^2]^{-1/\delta} \\
&\sim 2u - 1 + [1 - 2u - (1 + \delta)u^2 + \tfrac{1}{2}(-\delta)^{-1}(-\delta^{-1} - 1)(2\delta u)^2] \\
&= (1 + \delta)u^2,
\end{aligned}$$

so that $\kappa_U = 2$.

- For the bivariate Galambos copula derived in Section 1.2,

$$C(u, v; \delta) = uv \exp\{(-\log u)^{-\delta} + (-\log v)^{-\delta})^{-1/\delta}\},$$

the reflected/survival bivariate copula is

$$\widehat{C}(u, v; \delta) = u + v - 1 + (1 - u)(1 - v) \exp\{(-\log[1 - u])^{-\delta} + (-\log[1 - v])^{-\delta})^{-1/\delta}\}.$$

As $u \to 0^+$,

$$\begin{aligned}
C(uw_1, uw_2; \delta) &= u^2 w_1 w_2 \exp\{(-\log u)2^{-1/\delta}[1 - \tfrac{1}{2}\delta(-\log w_1 w_2)/(-\log u)]^{-1/\delta}\} \\
&\sim u^{2-2^{-1/\delta}}(w_1 w_2)^{1-2^{-1/\delta-1}} =: u^{2-2^{-1/\delta}}b(w_1, w_2),
\end{aligned}$$

so that $\kappa_L = 2 - 2^{-1/\delta} \in (1, 2)$ for $\delta > 0$, and is increasing in δ. From the copula density in Section 4.9, it can be directly shown that as $u \to 0^+$,

$$c(uw_1, uw_2; \delta) \sim u^{-2^{-1/\delta}}(w_1 w_2)^{-2^{-1/\delta}-1} \cdot (1 - 2^{-1/\delta} + 2^{-2-2/\delta}) = u^{-2^{-1/\delta}}\frac{\partial^2 b(w_1, w_2)}{\partial w_1 \partial w_2}.$$

Next, as $u \to 0^+$,

$$\begin{aligned}
\widehat{C}(u, u; \delta) &= 2u - 1 + (1 - u)^2 \exp\{[2(-\log(1 - u))^{-\delta}]^{-1/\delta}\} \\
&\sim 2u - 1 + (1 - u)^2 \exp\{[2u^{-\delta}]^{-1/\delta}\} \\
&= 2u - 1 + (1 - u)^2 \exp\{2^{-1/\delta}u\} \\
&\sim 2u - 1 + (1 - 2u)(1 + 2^{-1/\delta}u) \sim 2^{-1/\delta}u,
\end{aligned}$$

so that $\kappa_U = 1$ and $\lambda_U = 2^{-1/\delta} \in (0, 1)$ for $\delta > 0$.

- For the bivariate Frank copula in Section 4.5, $C(u, v; \delta) = \widehat{C}(u, v; \delta)$ is reflection symmetric, with

$$C(u, v; \delta) = -\delta^{-1} \log\big([e^{-\delta u} + e^{-\delta v} - e^{-\delta}) - e^{-\delta u} e^{-\delta v}]/(1 - e^{-\delta})\big),$$

As $u \to 0^+$,

$$\begin{aligned}
C(uw_1, uw_2; \delta) &= -\delta^{-1} \log\big([-e^{-\delta} + e^{-\delta uw_1} + e^{-\delta uw_2} - e^{-\delta u(w_1 + w_2)}]/(1 - e^{-\delta})\big) \\
&= -\delta^{-1} \log\big([1 - e^{-\delta} + \tfrac{1}{2} u^2 \delta^2 (-2w_1 w_2)]/(1 - e^{-\delta})\big) \\
&= -\delta^{-1} \log\big(1 - \delta^2 u^2 w_1 w_2/(1 - e^{-\delta})\big) \sim \delta u^2 w_1 w_2/(1 - e^{-\delta}) =: u^2 b(w_1, w_2),
\end{aligned}$$

so that $\kappa_L = \kappa_U = 2$ and $\Upsilon = \delta/(1 - e^{-\delta})$. From the copula density in Section 4.5, it can be directly shown that as $u \to 0^+$,

$$c(uw_1, uw_2; \delta) \sim \delta/(1 - e^{-\delta}) = u^0 \frac{\partial^2 b(w_1, w_2)}{\partial w_1 \partial w_2}.$$

for any $w_1 > 0$, $w_2 > 0$.   □

There are some general results on tail orders for Archimedean and extreme value copulas in Hua and Joe (2011); see some of the theorems in Chapter 8.

We next mention the relevance of (2.55) with $d = 2$ for plots of contour densities. For simple bivariate copula families, with positive dependence, the copula density satisfies (with tail orders κ_L, κ_U, and slowly varying functions ℓ, ℓ^*):

$$\begin{aligned}
c(u, u) &\sim u^{\kappa_L - 2} \ell(u), \quad u \to 0^+, \\
c(u, u) &\sim (1 - u)^{\kappa_U - 2} \ell^*(1 - u), \quad u \to 1^-.
\end{aligned}$$

So for strong lower tail dependence ($\kappa_L = 1$) or intermediate lower tail dependence ($1 < \kappa_L < 2$), the copula asymptotes to ∞ at the lower corner $(0, 0)$. Analogously, for strong upper tail dependence ($\kappa_U = 1$) or intermediate upper tail dependence ($1 < \kappa_U < 2$), the copula asymptotes to ∞ at the lower corner $(1, 1)$. The density at the $(0, 0)$ and $(1, 1)$ corners are bounded only for tail quadrant independence or tail negative dependence ($\kappa_L \geq 2$, $\kappa_U \geq 2$).

One can similarly define tail orders for the $(0, 1)$ and $(1, 0)$ corners. For the positively dependent bivariate copulas with densities (with no singular component) in Chapter 4, the tail order at the $(0, 1)$ and $(1, 0)$ corners is 2 except for the bivariate t_ν copula where it is 1. This means that bivariate t_ν copula density asymptotes to ∞ at all four corners of $[0, 1]^2$.

Because the copula density asymptotes to ∞ for at least one corner for many commonly used bivariate copula families, we recommend the contour plots of copula densities be obtained with N$(0, 1)$ margins.

2.17   Semi-correlations of normal scores for a bivariate copula

Because it is useful to plot data after transforms to normal scores to assess tail dependence and asymmetry, we can further consider some bivariate summary values associated with these plots. In this section, we introduce semi-correlations which can be applied to all pairs of variables in a multivariate data set.

Let C be a bivariate copula, with density c and conditional cdfs $C_{2|1}$ and $C_{1|2}$. Suppose $(Z_1, Z_2) \sim C(\Phi, \Phi)$ and $(U_1, U_2) \sim C$. From Section 2.12.4, the correlation of the normal scores is:

$$\begin{aligned}
\rho_N = \mathrm{Cor}[\Phi^{-1}(U_1), \Phi^{-1}(U_2)] &= \mathrm{Cor}(Z_1, Z_2) \\
&= \int_{-\infty}^{\infty} \int_{-\infty}^{\infty} z_1 z_2 \, \phi(z_1) \, \phi(z_2) \, c(\Phi(z_1), \Phi(z_2)) \, \mathrm{d}z_1 \mathrm{d}z_2.
\end{aligned} \tag{2.57}$$

For the bivariate Gaussian copula with correlation ρ, clearly $\rho_N = \rho$. The upper and lower semi-correlations (of normal scores) are defined as:

$$\rho_N^+ = \mathrm{Cor}[Z_1, Z_2 | Z_1 > 0, Z_2 > 0],$$
$$\rho_N^- = \mathrm{Cor}[Z_1, Z_2 | Z_1 < 0, Z_2 < 0].$$

(2.58)

The sample versions $\hat{\rho}_N^-, \hat{\rho}^+$ are sample correlations in the joint lower and upper quadrants of the two variables after the normal score transform. Semi-correlations, and more generally correlations of truncated bivariate distributions, have been used in Ang and Chen (2002) and Gabbi (2005).

Similar to Spearman's ρ_S, the dependence measure ρ_N satisfies all of the desired properties for a measure of monotone association (Section 2.12). The upper and lower semi-correlations do not satisfy all of the properties; for example, the concordance ordering property doesn't hold.

Note that $\rho_N^+ = \rho_N^-$ for any reflection symmetric copula. For the bivariate Gaussian copula, there is a closed form formula. The first two conditional moments of bivariate Gaussian are derived in Tallis (1961) based on the moment generating function of a truncated multivariate Gaussian distribution, and recursions for higher order moments are derived in Shah and Parikh (1964). Here we just summarize the moments for the case of truncation at $(0, 0)$. For non-negative integers r, s, let

$$v_{r,s}(\rho) = v_{s,r}(\rho) = \mathbb{E}[Z_1^r Z_2^s | Z_1 > 0, Z_2 > 0; \rho]; \quad (Z_1, Z_2)^\top \sim \mathrm{N}_2 \left(\mathbf{0}, \begin{pmatrix} 1 & \rho \\ \rho & 1 \end{pmatrix} \right).$$

For bivariate Gaussian, by Theorem 8.53,

$$\beta_0 = \beta_0(\rho) = \mathbb{P}(Z_1 > 0, Z_2 > 0; \rho) = \tfrac{1}{4} + (2\pi)^{-1} \arcsin(\rho).$$

Furthermore,

$$v_{1,0} = (1 + \rho)/[2\beta_0 \sqrt{2\pi}\,],$$
$$v_{2,0} = 1 + \rho\sqrt{1 - \rho^2}\,/[2\pi\beta_0],$$
$$v_{1,1} = [v_{2,0} - 1 + \rho^2]/\rho = \rho + \sqrt{1 - \rho^2}\,/[2\pi\beta_0].$$

The (lower or upper) semi-correlation, assuming bivariate Gaussian, is

$$\mathrm{bvnsemic}(\rho) = \mathrm{Cor}(Z_1, Z_2 | Z_1 > 0, Z_2 > 0; \rho) = \frac{v_{1,1}(\rho) - v_{1,0}^2(\rho)}{v_{2,0}(\rho) - v_{1,0}^2(\rho)}.$$

(2.59)

In general, (2.58) can be expressed as integrals. Let $\beta_0 = C(\tfrac{1}{2}, \tfrac{1}{2}) = \overline{C}(\tfrac{1}{2}, \tfrac{1}{2})$, so that $\beta = 4\beta_0 - 1$ is Blomqvist's beta. Assuming C is permutation symmetric, the expressions in (2.58) can each be expressed in terms of three integrals:

$$\beta_0 \mathbb{E}[Z_1 | Z_1 > 0, Z_2 > 0] = \int_0^\infty z\phi(z)\, \overline{C}_{2|1}(\tfrac{1}{2} | \Phi(z))\, \mathrm{d}z,$$

$$\beta_0 \mathbb{E}[Z_1^2 | Z_1 > 0, Z_2 > 0] = \int_0^\infty z^2\phi(z)\, \overline{C}_{2|1}(\tfrac{1}{2} | \Phi(z))\, \mathrm{d}z,$$

$$\beta_0 \mathbb{E}[Z_1 Z_2 | Z_1 > 0, Z_2 > 0] = \int_0^\infty \int_0^\infty z_1 z_2 \phi(z_1)\phi(z_2)\, c(\Phi(z_1), \Phi(z_2))\, \mathrm{d}z_1 \mathrm{d}z_2;$$

$$\beta_0 \mathbb{E}[Z_1 | Z_1 < 0, Z_2 < 0] = \int_{-\infty}^0 z\phi(z)\, C_{2|1}(\tfrac{1}{2} | \Phi(z))\, \mathrm{d}z = \int_0^\infty z\phi(z)\, C_{2|1}(\tfrac{1}{2} | \Phi(-z))\, \mathrm{d}z,$$

$$\beta_0 \mathbb{E}[Z_1^2 | Z_1 < 0, Z_2 < 0] = \int_0^\infty z^2\phi(z)\, C_{2|1}(\tfrac{1}{2} | \Phi(-z))\, \mathrm{d}z,$$

$$\beta_0 \mathbb{E}[Z_1 Z_2 | Z_1 < 0, Z_2 < 0] = \int_0^\infty \int_0^\infty z_1 z_2 \phi(z_1)\phi(z_2)\, c(\Phi(-z_1), \Phi(-z_2))\, \mathrm{d}z_1 \mathrm{d}z_2.$$

If C is not permutation symmetric, there are additional expressions involving $C_{1|2}$.

Table 2.4 has semi-correlations for several bivariate copula families with different tail behavior; the families are bivariate Gaussian, Plackett, Frank, MTCJ, Gumbel, t_3 and BB1. Note that the lower and upper semi-correlations easily distinguish these copulas. The Plackett and Frank copulas have the same tail order of 2 but different coefficients for $C(u, u; \delta) \sim \zeta(\delta)u^2$ as $u \to 0^+$, where δ is the copula parameter. BB1 has two parameters (θ, δ) that can control lower and upper tail dependence, and different combinations of (θ, δ) lead to the same ρ_N and selected values of (θ, δ) are based on λ_U, ρ_N.

Standard deviations of the distributions of $\hat{\rho}_N^-, \hat{\rho}_N^+, \hat{\rho}_N^- - \hat{\rho}_N^+$ for $n = 600$ were computed via Monte Carlo simulation for the copula families with $\rho_N = 0.4$ and 0.6. The standard deviations are smaller for more dependence (compare the asymptotic variance for the empirical version of Kendall's τ, etc., in Table 2.3). These standard deviations can be adjusted for other sample sizes with the square root rule, to get an idea of their sampling variability.

copula	param.	ρ_N^-	ρ_N^+	λ_L	λ_U	SE($\hat{\rho}_N^-$)	SE($\hat{\rho}_N^+$)	SE($\hat{\rho}_N^- - \hat{\rho}_N^+$)
$\rho_N = 0.4$								
Gaussian	0.400	0.196	0.196			0.08	0.08	0.11
Plackett	3.755	0.189	0.189			0.07	0.07	0.10
Frank	2.768	0.139	0.139			0.07	0.07	0.10
MTCJ	0.697	0.463	0.039	0.370		0.07	0.07	0.10
Gumbel	1.350	0.132	0.415		0.329	0.08	0.07	0.10
t(3)	0.412	0.428	0.428	0.266	0.266	0.07	0.07	0.10
BB1	0.484,1.080	0.393	0.164	0.266	0.100	0.07	0.08	0.11
BB1	0.273,1.179	0.300	0.279	0.116	0.200	0.08	0.08	0.11
BB1	0.061,1.306	0.175	0.386	0.0002	0.300	0.08	0.07	0.11
$\rho_N = 0.6$								
Gaussian	0.600	0.356	0.356			0.07	0.07	0.09
Plackett	8.543	0.371	0.371			0.06	0.06	0.09
Frank	4.852	0.269	0.269			0.06	0.06	0.09
MTCJ	1.446	0.679	0.098	0.619		0.04	0.07	0.08
Gumbel	1.701	0.260	0.579		0.497	0.07	0.05	0.09
t(3)	0.614	0.540	0.540	0.383	0.383	0.06	0.06	0.08
BB1	1.146,1.080	0.638	0.197	0.571	0.100	0.05	0.07	0.09
BB1	0.702,1.239	0.549	0.345	0.451	0.250	0.06	0.07	0.09
BB1	0.271,1.475	0.405	0.489	0.177	0.400	0.07	0.06	0.09

Table 2.4 *Lower and upper semi-correlations given $\rho_N = 0.4$ and $\rho_N = 0.6$ for some 1-parameter and 2-parameter copula families. Note that λ_L or λ_U is zero if it is not included above. The bivariate Gaussian copula has intermediate tail dependence, the Gumbel copula has intermediate lower tail dependence, the Plackett and Frank have tail quadrant independence and the MTCJ copula has upper tail quadrant independence. On the right-hand side, the SDs of the distribution of $\hat{\rho}_N^-, \hat{\rho}_N^+, \hat{\rho}_N^- - \hat{\rho}_N^+$ are based on simulation with sample size $n = 600$; these can adjusted for other n's with the square root rule. For comparison, a theoretical calculation for $C^\perp$ using the delta method leads to $SD(\hat{\rho}_N^-) \approx 2\eta(1 - \eta^2)^{-1}[(2\eta^2 - 1)/n]^{1/2} = 2.296n^{-1/2}$ with $\eta = (2/\pi)^{1/2}$, or about $0.133, 0.093, 0.073$ for $n = 300, 600, 1000$ respectively; and so $SD(\hat{\rho}_N^- - \hat{\rho}_N^+) \approx 0.187, 0.133, 0.103$ for $n = 300, 600, 1000$ respectively. The SDs from simulations with $C^\perp$ were a little smaller. In comparison with variances for sample versions of τ, ρ_S, the SDs with dependence should also be smaller, as the SDs become 0 for C^+.*

Table 2.5 shows that $\rho_N^+ = \rho_N^-$ gets larger as ν decreases for the bivariate t_ν copula.

	$\rho = 0.4$				$\rho = 0.7$			
ν	τ	ρ_S	ρ_N	ρ_N^+	τ	ρ_S	ρ_N	ρ_N^+
3	0.262	0.369	0.389	0.422	0.494	0.662	0.686	0.601
8	0.262	0.379	0.398	0.288	0.494	0.676	0.698	0.518
13	0.262	0.381	0.399	0.253	0.494	0.679	0.699	0.497
18	0.262	0.382	0.400	0.238	0.494	0.680	0.700	0.488
23	0.262	0.383	0.400	0.222	0.494	0.681	0.700	0.482
100	0.262	0.384	0.400	0.199	0.494	0.683	0.700	0.467
∞	0.262	0.385	0.400	0.196	0.494	0.683	0.700	0.463

Table 2.5 *Bivariate t_ν: Comparison of measures of association and the semi-correlation as a function of ν for $\rho = 0.4$ and $\rho = 0.7$.*

More general are correlations of normal scores with other truncated tail regions, for example,

$$\rho_N^+(a) = \mathrm{Cor}[Z_1, Z_2 | Z_1 > a, Z_2 > a],$$
$$\rho_N^-(a) = \mathrm{Cor}[Z_1, Z_2 | Z_1 < -a, Z_2 < -a],$$

where $a \geq 0$. With $a \approx 1$, the resulting $\rho_N^-(a), \rho_N^+(a)$ match the tail dependence and tail order parameters of the bivariate copula families in Table 2.4. However the sample version with data would not be estimated as well with a larger threshold, because the truncation regions become smaller; that is, the standard deviations of the sampling distributions would be larger than those in Table 2.4.

2.18 Tail dependence functions

The tail dependence parameters measure the strength of dependence in the joint lower and joint upper corner along the main diagonal. Tail dependence functions summarize the dependence along other rays that go to $\mathbf{0}_d$ or $\mathbf{1}_d$ for a d-variate copula. Tail dependence functions can be helpful for the following.

- Deriving the limiting extreme value distribution when there is tail dependence; for example, Nikoloulopoulos et al. (2009) for the multivariate t_ν distribution, and Padoan (2011) for a multivariate skew-t distribution.
- Obtaining bounds on the tail dependence parameters for multivariate copulas such as vine and factor models which are obtained via a mixture of conditional distributions; see Joe et al. (2010), Krupskii and Joe (2013).

The tail dependence function has been used in multivariate extremes and tail expansions such as Jaworski (2006), Alink et al. (2007) and Klüppelberg et al. (2008). It is also related to multivariate regular variation; see Li and Sun (2009) and Li (2013). In multivariate extreme value theory. different functions can represent dependence in the tails; Section 6.4 of Falk et al. (2010) has a different tail dependence function.

In this section, we summarize some properties of tail dependence functions and their links to the exponent of the multivariate extreme value copula; the exposition is based on Joe et al. (2010).

Consider the d-dimensional copula C of $(U_1, \ldots, U_d)$, with reflected copula

$$\widehat{C}(v_1, \ldots, v_d) := \mathbb{P}\{1 - U_1 \leq v_1, \ldots, 1 - U_d \leq v_d\} = \overline{C}(1 - v_1, \ldots, 1 - v_d). \tag{2.60}$$

Let $I_d = \{1, \ldots, d\}$. The lower and upper tail dependence functions, denoted by $b(\cdot; C)$ and $b^*(\cdot; C)$ are defined as:

$$b(\boldsymbol{w}; C) := \lim_{u \to 0^+} u^{-1} \mathbb{P}\{U_i \leq uw_i, i \in I_d\} = \lim_{u \to 0^+} u^{-1} C(uw_i, i \in I_d), \forall \boldsymbol{w} \in \mathbb{R}_+^d;$$

$$b^*(\boldsymbol{w}; C) := \lim_{u \to 0^+} u^{-1} \mathbb{P}\{U_i > 1 - uw_i, i \in I_d\} \lim_{u \to 0^+} u^{-1} \overline{C}(1 - uw_i, i \in I_d), \forall \boldsymbol{w} \in \mathbb{R}_+^d.$$

Note that the multivariate regular variation property is usually assumed to ensure the existence of such limits.

The link to tail dependence is the following.

Definition 2.11 (Alternative definition of tail dependence).

(i) The copula C is said to be *lower tail dependent* if $b = b(\cdot; C)$ is non-zero. In this case, the lower tail dependence parameter is $b(\boldsymbol{1}_d)$.

(ii) The copula C is said to be *upper tail dependent* if $b^* = b^*(\cdot; C)$ is non-zero. In this case, the upper tail dependence parameter is $b^*(\boldsymbol{1}_d)$.

For any $\emptyset \neq S \subset I_d$ with $|S| \geq 2$, let C_S denote the copula of the $|S|$-dimensional margin $\{U_i, i \in S\}$, then $b_S(w_i, i \in S; C_S)$ and $b_S^*(w_i, i \in S; C_S)$ denote, respectively, the lower and upper tail dependence functions of C_S. For $|S| = 1$, say $S = \{j\}$, define $b_j(w_j) = b_j^*(w_j) = w_j$. From reflection, $b(\boldsymbol{w}; \widehat{C}) = b^*(\boldsymbol{w}; C)$ for any copula C, so it suffices to state results for the lower tail: $b(\cdot; C)$. For shorthand, write

$$C_S(u\boldsymbol{w}_S) \sim u \, b_S(\boldsymbol{w}_S), \quad \boldsymbol{w}_S = (w_i, i \in S),$$

where the copula argument of the b function is omitted unless needed for clarity.

Functions related to the tail dependence function are:

$$a(\boldsymbol{w}; C) = \lim_{u \to 0^+} u^{-1} \mathbb{P}\{U_i \leq uw_i, \text{ for some } i \in I_d\}$$

$$= \lim_{u \to 0^+} u^{-1} \big(1 - \mathbb{P}\{U_i > uw_i, \forall i \in I_d\}\big), \tag{2.61}$$

$$a^*(\boldsymbol{w}; C) = \lim_{u \to 0^+} u^{-1} \big(1 - \mathbb{P}\{U_i \leq 1 - uw_i, \forall i \in I_d\}\big),$$

$$a_S(\boldsymbol{w}_S; C_S) = \lim_{u \to 0^+} u^{-1} \big(1 - \mathbb{P}\{U_i > uw_i, \forall i \in S\}\big), \tag{2.62}$$

$$a_S^*(\boldsymbol{w}_S; C_S) = \lim_{u \to 0^+} u^{-1} \big(1 - \mathbb{P}\{U_i \leq 1 - uw_i, \forall i \in S\}\big),$$

where $\emptyset \neq S \subset I_d$.

The tail dependence functions share some similar properties to those of distribution functions.

(i) $b(\boldsymbol{w})$ is grounded, i.e., $b(w_i, i \in I_d) = 0$ if at least one w_i, $i \in I_d$, is zero.

(ii) $b(\boldsymbol{w})$ is d-monotone, i.e., satisfies the rectangle inequality.

(iii) $b(\boldsymbol{w})$ is homogeneous of order 1, i.e., $b(s\boldsymbol{w}) = sb(\boldsymbol{w})$ for any $s \geq 0$.

(iv) $b(\boldsymbol{w}) = 0$ for all $\boldsymbol{w} \geq \boldsymbol{0}$ iff $b(\boldsymbol{z}) = 0$ for some positive vector $\boldsymbol{z} = (z_1, \ldots, z_d)$.

(v) $a(\cdot)$ is homogeneous of order 1. For $1 \leq j \leq d$, $a_{I_d \setminus \{j\}}(w_i, i \neq j) = \lim_{w_j \to 0^+} a(\boldsymbol{w})$. Consequently, by iteration, $a_j(w_j) = \lim_{w_i \to 0^+, i \neq j} a(\boldsymbol{w}) = w_j$, $1 \leq j \leq d$.

Unless some conditions hold, one cannot necessarily get $b_S(\boldsymbol{w}_S)$ from $b(w_i, i \in I_d)$ by letting $w_j \to \infty$ for $j \notin S$. Sometimes this holds and sometimes it doesn't; for example, it fails to hold for the tail dependence functions of the multivariate t_ν distribution. Further examples and also other properties of the derivatives related to conditional tail dependence functions are given in Section 8.9.

Other useful properties of tail dependence functions and (2.61)–(2.62) are given below; the argument of C or C_S is omitted in the $b(\cdot)$ and $a(\cdot)$ functions.

1. Since b is differentiable almost surely and homogeneous of order 1, Euler's formula on homogeneous functions implies that

$$b(\boldsymbol{w}) = \sum_{j=1}^{d} w_j \frac{\partial b(\boldsymbol{w})}{\partial w_j}, \ \forall \boldsymbol{w} \in \mathbb{R}_+^d,$$

where the partial derivatives $\partial b / \partial w_j$ are homogeneous of order zero and bounded.

2. The $a(\cdot)$ function can be obtained from $\{b_S : S \subset I_d\}$:

$$a(\boldsymbol{w}) = \lim_{u \to 0^+} \sum_{S \subset I_d, S \neq \emptyset} (-1)^{|S|-1} u^{-1} \mathbb{P}\{U_i \leq u w_i, i \in S\} = \sum_{S \subset I_d, S \neq \emptyset} (-1)^{|S|-1} b_S(w_i, i \in S).$$

$$(2.63)$$

The above derivation follows from the inclusion-exclusion principle for expressing the probability of a union of events in terms of probabilities of intersections of subsets of the events. It follows from (2.63) that the set of tail dependence functions $\{b_S(\cdot; C_S)\}$ uniquely determine $a(\cdot)$.

3. There is a converse to (2.63):

$$
\begin{aligned}
b(\boldsymbol{w}) &= \lim_{u \to 0} u^{-1}\big(1 - \mathbb{P}\{U_i > u w_i, \text{ for some } i \in I_d\}\big) \\
&= \lim_{u \to 0} u^{-1}\Big(1 - \sum_{S \subset I_d, S \neq \emptyset} (-1)^{|S|-1} \mathbb{P}\{U_i > u w_i, i \in S\}\Big) \\
&= \lim_{u \to 0} u^{-1} \sum_{S \subset I_d, S \neq \emptyset} (-1)^{|S|-1}\big(1 - \mathbb{P}\{U_i > u w_i, i \in S\}\big) \\
&= \lim_{u \to 0} u^{-1} \sum_{S \subset I_d, S \neq \emptyset} (-1)^{|S|-1} \mathbb{P}\{U_i \leq u w_i, \text{ for some } i \in S\} \\
&= \sum_{S \subset I_d, S \neq \emptyset} (-1)^{|S|-1} a_S(w_i, i \in S).
\end{aligned}
$$

$$(2.64)$$

Similarly, for a margin indexed by S,

$$b_S(\boldsymbol{w}_S) = \sum_{S' \subset S, S' \neq \emptyset} (-1)^{|S'|-1} a_{S'}(w_i, i \in S').$$

That is, the function $a(\cdot)$ uniquely determines the tail dependence functions $\{b_S(\cdot; C_S) : S \subset I_d\}$.

4. To summarize, for the lower tail, the copula of $(1 - U_1, \ldots, 1 - U_d)$ is the reflected copula $\widehat{C}$ of C as given in (2.60), and

$$
\begin{aligned}
C(u\boldsymbol{w}) &= \mathbb{P}\{U_i \leq u w_i, \forall i \in I_d\} \sim u b(\boldsymbol{w}), \quad u \to 0^+, \\
\overline{C}(u\boldsymbol{w}) &= \widehat{C}(\boldsymbol{1}_d - u\boldsymbol{w}) = \mathbb{P}\{1 - U_i \leq 1 - u w_i, \forall i \in I_d\} \sim 1 - u\, a(\boldsymbol{w}), \quad u \to 0^+. \quad (2.65)
\end{aligned}
$$

Analogously, for the upper tail,

$$
\begin{aligned}
\widehat{C}(u\boldsymbol{w}) &= \mathbb{P}\{1 - U_i \leq u w_i, \forall i \in I_d\} \sim u b^*(\boldsymbol{w}), \quad u \to 0^+, \\
C(\boldsymbol{1}_d - u\boldsymbol{w}) &= \mathbb{P}\{U_i \leq 1 - u w_i, \forall i \in I_d\} \sim 1 - u\, a^*(\boldsymbol{w}), \ u \to 0^+. \quad (2.66)
\end{aligned}
$$

We next show that the (limiting) extreme value copulas of C can be conveniently expressed in terms of the $a(\cdot), a^*(\cdot)$ functions.

Let $(U_{i1}, \ldots, U_{id})$, $i = 1, 2, \ldots$, be an infinite random sample from C. The *lower extreme value limit* of C, which is the same as the upper extreme value limit of $\widehat{C}$, is given by

$$\lim_{n \to \infty} \widehat{C}^n(u_1^{1/n}, \ldots, u_d^{1/n})$$

$$= \lim_{n \to \infty} \mathbb{P}\big\{\big[\max_{i=1,\ldots,n}(1 - U_{i1})\big] \leq u_1^{1/n}, \ldots, \big[\max_{i=1,\ldots,n}(1 - U_{id})\big] \leq u_d^{1/n}\big\}.$$

For sufficiently large n, $u_j^{1/n} = \exp\{n^{-1} \log u_j\} \sim 1 + n^{-1} \log u_j$, so that with $\tilde{u}_j = -\log u_j$,

$$\widehat{C}^n(u_1^{1/n}, \ldots, u_d^{1/n}) \sim \widehat{C}^n(1 - n^{-1}\tilde{u}_1, \ldots, 1 - n^{-1}\tilde{u}_d)$$

$$\sim \big[1 - n^{-1} a(\tilde{u}_1, \ldots, \tilde{u}_d)\big]^n \to \exp\{-a(\tilde{u}_1, \ldots, \tilde{u}_d)\}, \qquad (2.67)$$

where the second approximation in (2.67) comes from (2.65) by substituting $u = n^{-1}$, $w_j = \tilde{u}_j$ $(j = 1, \ldots, d)$. If C has lower tail dependence, the lower extreme value copula C_{LEV} of C is

$$C_{\mathrm{LEV}}(u_1, \ldots, u_d) = \lim_{n \to \infty} \widehat{C}^n(u_1^{1/n}, \ldots, u_d^{1/n}) = \exp\{-a(-\log u_1, \ldots, -\log u_d)\},$$

and $a(\boldsymbol{w}) \leq \sum_{j=1}^d w_j$ with some pointwise inequalities. Since $b(\boldsymbol{w}; \widehat{C}) = b^*(\boldsymbol{w}; C)$ and $a(\boldsymbol{w}; \widehat{C}) = a^*(\boldsymbol{w}; C)$, if C has upper tail dependence, the upper extreme value copula C_{UEV} of C is

$$C_{\mathrm{UEV}}(u_1, \ldots, u_d) = \lim_{n \to \infty} C^n(u_1^{1/n}, \ldots, u_d^{1/n}) = \exp\{-a^*(-\log u_1, \ldots, -\log u_d)\}.$$

Hence the functions a, a^* are also the exponent functions of the limiting extreme value copulas.

These calculations are the generalization of the derivation of the Galambos copula in Example 1.6. We next show that the approach here is equivalent to the direct definition of the multivariate extreme value limit. The direct approach is based on $(X_{i1}, \ldots, X_{id})$, $i = 1, 2 \ldots$, be an infinite random sample from a d-variate distribution $F_{1:d}$. Let $M_{jn} = \max\{X_{1j}, \ldots, X_{nj}\}$ for $j = 1, \ldots, d$. Suppose there are location constants m_{jn} and positive scale constants s_{jn} such that $(M_{jn} - m_{jn})/s_{jn}$ converges in distribution to a member of the generalized extreme value family, for $j = 1, \ldots, d$; see Section 2.3. Then, assuming the joint upper tail of $F_{1:d}$ is well behaved,

$$\mathbb{P}\{(M_{jn} - m_{jn})/s_{jn} \leq z_j, j = 1, \ldots, d\} \to H_{1:d}(z_1, \ldots, z_d),$$

where $H_{1:d}$ is a multivariate extreme value distribution. By converting $H_{1:d}$ to a copula, a multivariate extreme value copula obtains.

Given a copula C with upper tail dependence function a^*, we show that the resulting extreme value copula is $C_{\mathrm{UEV}}(\boldsymbol{u}) = \exp\{-a^*(-\log u_1, \ldots, -\log u_d)\}$ for $F_{1:d} = C(F_1, \ldots, F_d)$ when F_j are all Pareto(α) with $\alpha > 0$ or when F_j are all U$(0, 1)$. For Pareto(α) with cdf $F_j(x) = 1 - (1 + x)^{-\alpha}$ for $x > 0$, one can take $m_{jn} = -1$ and $s_{jn} = n^{1/\alpha}$, and for U$(0, 1)$, one can take $m_{jn} = 1$ and $s_{jn} = n^{-1}$ (see Example 2.5 for the latter).

For the multivariate Pareto(α) distribution $C(F_1, \ldots, F_d)$, with $z_j > 0$ for $j = 1, \ldots, d$, and using (2.66), then for large n,

$$\mathbb{P}\{(M_{jn} - m_{jn})/s_{jn} \leq z_j, j = 1, \ldots, d\} = \mathbb{P}\{M_{jn} \leq n^{1/\alpha} z_j - 1, j = 1, \ldots, d\}$$

$$= C^n(1 - z_1^{-\alpha}/n, \ldots, 1 - z_d^{-\alpha}/n) \sim \big[1 - n^{-1} a^*(z_1^{-\alpha}, \ldots, z_d^{-\alpha})\big]^n$$

$$\to \exp\{-a^*(z_1^{-\alpha}, \ldots, z_d^{-\alpha})\} = H_{1:d}(z_1, \ldots, z_d).$$

The univariate margins of this multivariate extreme value distribution are all Fréchet with parameter α. To get a copula from $H_{1:d}$, set $u_j = \exp\{-z_j^{-\alpha}\}$ or $-\log u_j = z_j^{-\alpha}$.

For the multivariate distribution $C(F_1,\ldots,F_d)$ with F_j being $U(0,1)$, and with $z_j \le 0$ for $j = 1,\ldots,d$, then for large n,

$$\mathbb{P}\{(M_{jn} - m_{jn})/s_{jn} \le z_j, j = 1,\ldots,d\} = \mathbb{P}\{M_{jn} \le 1 + n^{-1}z_j, j = 1,\ldots,d\}$$
$$= C^n(1 + z_1/n,\ldots,1 + z_d/n) \sim [1 - n^{-1}a^*(-z_1,\ldots,-z_d)]^n$$
$$\to \exp\{-a^*(-z_1,\ldots,-z_d)\} = H_{1:d}(z_1,\ldots,z_d).$$

The univariate margins of this multivariate extreme value distribution are all Weibull with parameter 1. To get a copula from $H_{1:d}$, set $u_j = \exp\{z_j\}$ or $-\log u_j = -z_j$.

In both cases, the same limiting extreme value copula C_{UEV} arises. That is, C_{UEV} depends on a multivariate distribution $F_{1:d} = C(F_1,\ldots,F_d)$ only through C, and a^* is associated with C. Note that $C^n(u_1^{1/n},\ldots,u_d^{1/n})$ is the copula of

$$F_{1:d}^n(m_{1n} + s_{1n}z_1,\ldots,m_{dn} + s_{dn}z_d) = \mathbb{P}\{(M_{jn} - m_{jn})/s_{jn} \le z_j, j = 1,\ldots,d\}.$$

Next is an example to illustrate the use of the tail dependence functions to obtain the limiting extreme value copula.

Example 2.14 (Archimedean copula based on Sibuya LT and its Gumbel extreme value limit). Let $C(\boldsymbol{u}) = \psi(\psi^{-1}(u_1) + \cdots + \psi^{-1}(u_d))$, with LT $\psi(s) = 1 - (1 - e^{-s})^{1/\delta}$, $\delta > 1$. From Theorem 8.33, this has upper tail dependence and

$$\widehat{C}(u\boldsymbol{w}) \sim u\,b^*(\boldsymbol{w};C) = u\,b(\boldsymbol{w},\widehat{C}), \quad C(\boldsymbol{1}_d - u\boldsymbol{w}) \sim 1 - u\,a^*(\boldsymbol{w};C), \ u \to 0^+.$$

For $d = 2,3$:

$$\widehat{C}_{12}(v_1,v_2) = 1 - (1 - v_1) - (1 - v_2) + \psi(\psi^{-1}(1 - v_1) + \psi^{-1}(1 - v_2)),$$
$$\widehat{C}_{123}(v_1,v_2,v_3) = 1 - (1 - v_1) - (1 - v_2) - (1 - v_3) + \psi(\psi^{-1}(1 - v_1) + \psi^{-1}(1 - v_2))$$
$$+ \psi(\psi^{-1}(1 - v_1) + \psi^{-1}(1 - v_3)) + \psi(\psi^{-1}(1 - v_2) + \psi^{-1}(1 - v_3))$$
$$- \psi(\psi^{-1}(1 - v_1) + \psi^{-1}(1 - v_2) + \psi^{-1}(1 - v_3)).$$

Since $\psi^{-1}(t) = -\log\{1 - (1 - t)^\delta\}$, then $\psi^{-1}(1 - uw) = -\log\{1 - (uw)^\delta\} \sim (uw)^\delta$ as $u \to 0^+$; and $\psi(s) \sim 1 - s^{1/\delta}$ as $s \to 0^+$ or $\psi(u^\delta W) \sim 1 - (u^\delta W)^{1/\delta} = 1 - uW^{1/\delta}$ as $u \to 0^+$ with $W > 0$.

For the upper tail dependence function of C or lower tail dependence function of $\widehat{C}$, take the tail approximation $\widehat{C}(u\boldsymbol{w}) \sim u\,b(\boldsymbol{w};\widehat{C})$. For $d = 2,3$:

$$\widehat{C}_{12}(uw_1, uw_2) \sim u(w_1 + w_2) - 1 + \psi(u^\delta(w_1^\delta + w_2^\delta)) \sim u(w_1 + w_2) - u(w_1^\delta + w_2^\delta)^{1/\delta}$$
$$= u[w_1 + w_2 - (w_1^\delta + w_2^\delta)^{1/\delta}] =: ub(w_1,w_2;\widehat{C}_{12}),$$
$$\widehat{C}_{123}(uw_1, uw_2, uw_3) \sim u(w_1 + w_2 + w_3) - 2 + [1 - u(w_1^\delta + w_2^\delta)^{1/\delta}] + [1 - u(w_1^\delta + w_3^\delta)^{1/\delta}]$$
$$+ [1 - u(w_2^\delta + w_3^\delta)^{1/\delta}] - [1 - u(w_1^\delta + w_2^\delta + w_3^\delta)^{1/\delta}]$$
$$= u[w_1 + w_2 + w_3 - (w_1^\delta + w_2^\delta)^{1/\delta} - (w_1^\delta + w_3^\delta)^{1/\delta} - (w_2^\delta + w_3^\delta)^{1/\delta}$$
$$+ (w_1^\delta + w_2^\delta + w_3^\delta)^{1/\delta}]$$
$$=: ub(w_1, w_2, w_3; \widehat{C}_{123}).$$

From these, using the inclusion-exclusion principle, one gets:

$$a(w_1, w_2; \widehat{C}_{12}) = (w_1^\delta + w_2^\delta)^{1/\delta},$$
$$a(w_1, w_2, w_3; \widehat{C}_{123}) = (w_1^\delta + w_2^\delta + w_3^\delta)^{1/\delta},$$
$$a(\boldsymbol{w}; \widehat{C}_{1:d}) = (w_1^\delta + \cdots + w_d^\delta)^{1/\delta} = a^*(\boldsymbol{w}, C_{1:d}).$$

More simply, as $u \to 0^+$,

$$C_{1:d}(\mathbf{1}_d - u\mathbf{w}) \sim 1 - u\, a^*(\mathbf{w}; C_{1:d}),$$

$$\psi\Big(\sum_{j=1}^{d} \psi^{-1}(1 - uw_j)\Big) \sim \psi\Big(\sum_{j=1}^{d} u^\delta w_j^\delta\Big) \sim 1 - u(w_1^\delta + \cdots + w_d^\delta)^{1/\delta} = 1 - u\, a^*(\mathbf{w}; C_{1:d}).$$

This corresponds to the Gumbel copula in dimension d:

$$C_{\mathrm{UEV}}(u_1, \ldots, u_d) = \exp\{-A(-\log u_1, \ldots, -\log u_d; \delta)\},$$

where $A(\mathbf{w}; \delta) = [\sum_{j=1}^{d} w_j^\delta]^{1/\delta}$, $\delta > 1$.

Sometimes the $a(\cdot), a^*(\cdot)$ functions are easier to obtain and sometimes the $b(\cdot), b^*(\cdot)$ functions are easier. But then one can convert from one to the other via (2.63) and (2.64). □

Next we discuss derivatives of $b(\cdot)$ leading to conditional tail dependence functions.

Assuming some regular variation conditions on the lower tail of C, then b has the same order of derivatives as C. That is, we assume C has continuous derivatives to the dth order, so that the limit operation and differentiation are commutative. Then, as $u \to 0^+$, with $u_j = uw_j$,

$$\frac{\partial C}{\partial u_1}(u\mathbf{w}) = \frac{\partial}{\partial w_1} C(u\mathbf{w}) \cdot \frac{\partial w_1}{\partial u_1} \sim u \frac{\partial b(\mathbf{w})}{\partial w_1} \cdot u^{-1} = \frac{\partial b(\mathbf{w})}{\partial w_1},$$

$$\frac{\partial^2 C}{\partial u_1 \partial u_2}(u\mathbf{w}) = \frac{\partial^2}{\partial w_1 \partial w_2} C(u\mathbf{w}) \cdot \frac{\partial w_1}{\partial u_1} \frac{\partial w_2}{\partial u_2} \sim u \frac{\partial^2 b(\mathbf{w})}{\partial w_1 \partial w_2} \cdot u^{-2} = u^{-1} \frac{\partial^2 b(\mathbf{w})}{\partial w_1 \partial w_2}, \quad (2.68)$$

$$\vdots$$

$$\frac{\partial^d C}{\partial u_1 \cdots \partial u_d}(u\mathbf{w}) \sim u^{-(d-1)} \frac{\partial^d b(\mathbf{w})}{\partial w_1 \cdots \partial w_d}.$$

Now, we define conditional tail dependence functions. They appear in Theorem 8.68 on tail dependence of regular vines.

Let $S = \{k_1, k_2, \ldots, k_m\}$ be a subset of $\{1, \ldots, d\}$ with cardinality m of at least 2, and let C_S be the corresponding margin of C. Let $(w_{k_1}, \ldots, w_{k_m}) \in \mathbb{R}_+^m$. If C_S has multivariate lower tail dependence, then

$$b_{k_1|k_2\cdots k_m}(w_{k_1}|w_{k_2}, \ldots, w_{k_m}) = \lim_{u \to 0^+} C_{k_1|k_2\cdots k_m}(uw_{k_1}|uw_{k_2}, \ldots, uw_{k_m})$$

is not the zero function. If C_S has multivariate upper tail dependence, then

$$b^*_{k_1|k_2\cdots k_m}(w_{k_1}|w_{k_2}, \ldots, w_{k_m}) = \lim_{u \to 0^+} C_{k_1|k_2\cdots k_m}(1 - uw_{k_1}|1 - uw_{k_2}, \ldots, 1 - uw_{k_m})$$

is not the zero function.

It is shown in Joe et al. (2010) that the conditional tail dependence functions of form $b_{j|T}$ and $b^*_{j|T}$ with $\emptyset \neq T \subset I_d \backslash \{j\}$ can be obtained from the derivatives of margins of the multivariate tail dependence functions b and b^*. If $b_{k_1, k_2, \ldots, k_m}$ and $b_{k_2, \ldots, k_m}$ are lower tail dependence functions, then (see Theorem 8.58)

$$b_{k_1|k_2\cdots k_m}(w_{k_1}|w_{k_2}, \ldots, w_{k_m}) = \frac{\partial^{m-1} b_{k_1, k_2, \ldots, k_m}}{\partial w_{k_2} \cdots \partial w_{k_m}} \Big/ \frac{\partial^{m-1} b_{k_2, \ldots, k_m}}{\partial w_{k_2} \cdots \partial w_{k_m}}.$$

For $m = 2$, since $b_{k_2}(w) = w$, this simplifies to $b_{k_1|k_2}(w_{k_1}|w_{k_2}) = \frac{\partial b_{k_1, k_2}}{\partial w_{k_2}}$.

Remark 2.1 Note the following properties of tail dependence functions and conditional tail dependence functions as some of the arguments go to ∞.

- Bivariate case: By definition, $b(w, 1) = \lim_{u \to 0+} C(uw, u)/u \leq \lim_{u \to 0+} u/u = 1$ for $w > 1$, and $b(w, 1)$ is increasing in w. Hence $0 < b(\infty, 1) \leq 1$ if tail dependence exists.
- Bivariate conditional: $b_{1|2}(w|1) = \lim_{u \to 0+} C_{1|2}(uw|u)$ is increasing in w and $0 < b_{1|2}(\infty|1) \leq 1$ if tail dependence exists.
- Multivariate: $b(w_1, \ldots, w_{d-1}, 1) = \lim_{u \to 0+} C(uw_1, \ldots, uw_{d-1}, u)/u \leq \lim_{u \to 0+} u/u = 1$ for $w_j > 1$ ($j = 1, \ldots, d-1$), and $b(w_1, \ldots, w_{d-1}, 1)$ is increasing. Hence $0 < b(\infty, \ldots, \infty, 1) \leq 1$ if tail dependence exists. A similar conclusion holds if 1 is in another position instead of the dth position and the rest are ∞.
- Multivariate conditional: $b_{1:(d-1)|d}(\boldsymbol{w}_{1:(d-1)}|1) = \lim_{u \to 0+} C_{1:(d-1)|d}(uw_1, \ldots, uw_{d-1}|u)$ is increasing in w_j ($j = 1, \ldots, d-1$) and $0 < b_{1:(d-1)|d}(\infty, \ldots, \infty|1) \leq 1$ if tail dependence exists.

Example 2.15 We illustrate the various tail dependence functions using the trivariate MTCJ copula $C(u_1, u_2, u_3; \delta) = (u_1^{-\delta} + u_2^{-\delta} + u_3^{-\delta} - 2)^{-1/\delta}$ for $\delta > 0$. As $u \to 0^+$,

$$C(uw_1, uw_2, uw_3; \delta) \sim ub(w_1, w_2, w_3), \quad b(w_1, w_2, w_3) = (w_1^{-\delta} + w_2^{-\delta} + w_3^{-\delta})^{-1/\delta}.$$

For the C_{12} margins, a similar calculation leads to $b_{12}(w_1, w_2) = (w_1^{-\delta} + w_2^{-\delta})^{-1/\delta}$. Some partial derivatives are:

$$\frac{\partial C}{\partial u_1} = \left(u_1^{-\delta} + u_2^{-\delta} + u_3^{-\delta} - 2\right)^{-1/\delta-1} u_1^{-\delta-1},$$

$$\frac{\partial^2 C}{\partial u_1 \partial u_2} = (1 + \delta)\left(u_1^{-\delta} + u_2^{-\delta} + u_3^{-\delta} - 2\right)^{-1/\delta-2} u_1^{-\delta-1} u_2^{-\delta-1},$$

$$\frac{\partial^3 C}{\partial u_1 \partial u_2 \partial u_3} = (1 + \delta)(1 + 2\delta)\left(u_1^{-\delta} + u_2^{-\delta} + u_3^{-\delta} - 2\right)^{-1/\delta-3} u_1^{-\delta-1} u_2^{-\delta-1} u_3^{-\delta-1}.$$

With $\boldsymbol{w} = (w_1, w_2, w_3)$, it is straightforward to check that as $u \to 0^+$,

$$\frac{\partial C(u\boldsymbol{w}; \delta)}{\partial u_1} \sim \left(w_1^{-\delta} + w_2^{-\delta} + w_3^{-\delta}\right)^{-1/\delta-1} w_1^{-\delta-1} = \frac{\partial b}{\partial w_1},$$

$$\frac{\partial^2 C(u\boldsymbol{w}; \delta)}{\partial u_1 \partial u_2} \sim u^{-1}(1 + \delta)\left(w_1^{-\delta} + w_2^{-\delta} + w_3^{-\delta}\right)^{-1/\delta-2} (w_1 w_2)^{-\delta-1} \sim u^{-1} \frac{\partial^2 b}{\partial w_1 \partial w_2},$$

$$\frac{\partial^3 C(u\boldsymbol{w}; \delta)}{\partial u_1 \partial u_2 \partial u_3} \sim u^{-2}(1 + \delta)(1 + 2\delta)\left(w_1^{-\delta} + w_2^{-\delta} + w_3^{-\delta}\right)^{-1/\delta-3} (w_1 w_2 w_3)^{-\delta-1}$$

$$\sim u^{-2} \frac{\partial^3 b}{\partial w_1 \partial w_2 \partial w_3}.$$

Next we illustrate conditional tail dependence functions. A direct calculation yields

$$C_{3|12}(u_3|u_1, u_2; \delta) = \frac{\left(u_1^{-\delta} + u_2^{-\delta} + u_3^{-\delta} - 2\right)^{-1/\delta-2}}{\left(u_1^{-\delta} + u_2^{-\delta} - 1\right)^{-1/\delta-2}}.$$

Therefore,

$$b_{3|12}(w_3|w_1, w_2) = \lim_{u \to 0} C_{3|12}(uw_3|uw_1, uw_2; \delta)$$

$$= \frac{\left(w_1^{-\delta} + w_2^{-\delta} + w_3^{-\delta}\right)^{-1/\delta-2}}{\left(w_1^{-\delta} + w_2^{-\delta}\right)^{-1/\delta-2}} = \left(1 + \frac{w_3^{-\delta}}{w_1^{-\delta} + w_2^{-\delta}}\right)^{-1/\delta-2}.$$

Note that $b_{3|12}$ is homogeneous of order 0. It can be checked directly that

$$b_{3|12}(w_3|w_1, w_2) = \frac{\partial^2 b(w_1, w_2, w_3)}{\partial w_1 \partial w_2} \Big/ \frac{\partial^2 b_{12}(w_1, w_2)}{\partial w_1 \partial w_2}.$$

2.19 Strength of dependence in tails and boundary conditional cdfs

In Cooke et al. (2011b) with a study of dependence of two aggregate sums of indicators of extreme events, with an increasing number of summands, the strength of limiting dependence depended on boundary conditional cdfs of copulas. This motivated the further study of boundary conditional cdfs as an alternative indicator of strength of dependence in tails; rigorous details for bivariate copulas are given in Hua and Joe (2014). The boundary conditional cdfs for a bivariate copula are: $C_{1|2}(\cdot|1)$, $C_{2|1}(\cdot|1)$, $C_{1|2}(\cdot|0)$, $C_{2|1}(\cdot|0)$.

The tail orders for a bivariate copula involve only the upper corner near $(1,1)$ or the lower corner near $(0,0)$. In contrast, the properties of $C_{1|2}(\cdot|1)$ depend on the copula near the top edge of the unit square, and similarly the properties of $C_{2|1}(\cdot|1)$, $C_{1|2}(\cdot|0)$, $C_{2|1}(\cdot|0)$ depend respectively on the copula near the right edge, bottom edge, and left edge of the unit square. Because in general a copula can be pieced together from bivariate distributions with uniform margins on different non-overlapping subsquares of the unit square, the form of $C_{1|2}(\cdot|1)$ can be quite varied as a mix of continuous and discrete components.

With a positive dependence condition such as stochastically increasing, the boundary conditional cdf can have one of the following three forms:

(i) the boundary conditional cdf degenerate at conditioning value (e.g., $C_{1|2}(\cdot|1)$ degenerate at 1 with $C_{1|2}(1^-|1) = 0$; or $C_{1|2}(\cdot|0)$ degenerate at 0 with $C_{1|2}(0|0) = 1$);

(ii) the boundary conditional cdf has positive but not unit mass at the conditioning value (e.g., $0 < C_{1|2}(1^-|1) < 1$, or $0 < C_{1|2}(0|0) < 1$);

(iii) the boundary conditional cdf has no mass at conditioning value (e.g., $1 - C_{1|2}(1^-|1) = 0$, or $0 = C_{1|2}(0|0)$).

With this categorization, case (i) represents the strongest strength of dependence in the joint tail, case (ii) represents intermediate strength of dependence and case (iii) represents weak dependence in the joint tail. Theorem 8.12 links tail dependence to either (i) or (ii) when the stochastically increasing positive dependence condition holds. The form of the boundary conditional cdfs also affect the asymptotic form of some conditional tail expectations, as shown in Section 2.20. For some common classes of bivariate copulas, Section 4.1 summarizes some links of tail order and boundary conditional cdfs.

2.20 Conditional tail expectation for bivariate distributions

In this section, we show how the tail of the copula affects that behavior of $\mathbb{E}(X_1|X_2 > t)$ and $\mathbb{E}(X_1|X_2 = t)$ as $t \to \infty$ when X_1, X_2 have a common univariate cdf $F_1 = F_2 = F$ with support on $[0, \infty)$. We refer to these two conditional expectations as conditional tail expectations (CTEs) as $t \to \infty$. Under some conditions, if the copula of (X_1, X_2) has upper tail dependence, the CTE is asymptotically $O(t)$; if the copula of (X_1, X_2) has upper quadrant tail independence, the CTE is asymptotically $O(1)$; and if the copula of (X_1, X_2) has upper intermediate tail independence, the CTE is asymptotically $O(t^\gamma \ell(t))$ for some $0 < \gamma < 1$ and slowly varying function $\ell(t)$.

We will demonstrate these results with examples to show the type of transformation of integrands used to get the asymptotic rates. General results are derived rigorously in Hua and Joe (2014). The techniques should extend to the multivariate setting where the tail conditional expectation of a subset of variables is obtained given a tail event on the remaining variables.

For additional notation, let F_{12} be the joint cdf of (X_1, X_2), and C be the copula of (X_1, X_2) so that $F_{12} = C(F_1, F_2)$. Then

$$\overline{F}_{12} = \overline{C}(F_1, F_2) = \overline{F}_1 + \overline{F}_2 - 1 + C(F_1, F_2) = \widehat{C}(\overline{F}_1, \overline{F}_2),$$

where $\widehat{C}(u_1, u_2) = u_1 + u_2 - 1 + C(1 - u_1, 1 - u_2)$ is the reflected copula.

We assume that the tail of $\overline{F}$ is not too heavy, so that all conditional expectations exist as well as $\mathbb{E}(X_j) = \int_0^\infty \overline{F}(x)\,\mathrm{d}x$ for $j = 1, 2$. The conditional expectations are:

$$\mathbb{E}(X_1|X_2 > t) = \int_0^\infty \frac{\overline{F}_{12}(x,t)}{\overline{F}(t)}\,\mathrm{d}x = \int_0^\infty \frac{\widehat{C}(\overline{F}(x),\overline{F}(t))}{\overline{F}(t)}\,\mathrm{d}x, \qquad (2.69)$$

$$\mathbb{E}(X_1|X_2 = t) = \int_0^\infty \overline{F}_{1|2}(x|t)\,\mathrm{d}x = \int_0^\infty \widehat{C}_{1|2}(\overline{F}(x)|\overline{F}(t))\,\mathrm{d}x. \qquad (2.70)$$

Note that upper tail conditions on C correspond to lower tail conditions on the reflected copula $\widehat{C}$.

Let $u = \overline{F}(x)$ and $v = \overline{F}(t)$ where $v \to 0^+$ as $t \to \infty$. Assume $\widehat{C}$ has continuous second order derivatives. The simplest case to handle is when the expansions:

$$\widehat{C}(u,v) \sim \widehat{C}(u,0) + v\,\widehat{C}_{1|2}(u|0) + O(v^2) = v\,\widehat{C}_{1|2}(u|0) + O(v^2), \quad v \to 0^+,$$
$$\widehat{C}_{1|2}(u|v) \sim \widehat{C}_{1|2}(u|0) + O(v), \quad v \to 0^+,$$

are valid. Then (2.69) and (2.70) both become

$$\mathbb{E}(X_1|X_2 > t) \sim \mathbb{E}(X_1|X_2 = t) \sim \int_0^\infty \widehat{C}_{1|2}(\overline{F}(x)|0)\,\mathrm{d}x, \quad t \to \infty; \qquad (2.71)$$

this can be valid only if $\widehat{C}_{1|2}(\cdot|0)$ has no mass at 0, and $\widehat{C}_{1|2}(\overline{F}(x)|0)$ is the survival function of a random variable with finite mean. For copulas C with upper tail dependence (so that $\widehat{C}$ has lower tail dependence) and satisfying the positive dependence condition of SI, $\widehat{C}_{1|2}(0|0) \geq \lambda_U$ and then (2.71) is invalid. However, (2.71) holds for the bivariate Frank copula in (4.7) with dependence parameter δ; this has tail quadrant independence and $\widehat{C}_{1|2}(u|0) = (1 - e^{-\delta u})/(1 - e^{-\delta})$ for $0 < u < 1$ with $\widehat{C}_{1|2}(u|0) \sim O(u)$ as $u \to 0^+$.

For the case that C has upper tail dependence, we show an alternative expansion based on $\hat{b}$, the lower tail dependence function of $\widehat{C}$, and $\hat{b}_{1|2}(w_1|w_2) = \partial\hat{b}(w_1, w_2)/\partial w_2$. The expansion is a little different depending on the univariate survival function $\overline{F}$ but the technique is roughly the same. An outline with some details for Pareto(α) ($\alpha > 1$) and exponential margins is given in the next example.

Example 2.16 (CTEs that are $O(t)$ as $t \to \infty$). If $\overline{F}(x) = (1+x)^{-\alpha}$ for $x > 0$ with $\alpha > 1$, then asymptotically (2.69) and (2.70) become:

$$\mathbb{E}(X_1|X_2 > t) = (1+t)^\alpha \int_0^\infty \widehat{C}\big((1+x)^{-\alpha}, (1+t)^{-\alpha}\big)\,\mathrm{d}x$$

$$= (1+t)^\alpha t \int_0^\infty \widehat{C}\big((1+tw)^{-\alpha}, (1+t)^{-\alpha}\big)\,\mathrm{d}w$$

$$\sim t^{\alpha+1} \int_0^\infty \widehat{C}\big(t^{-\alpha}w^{-\alpha}, t^{-\alpha}\big)\,\mathrm{d}w \sim t \int_0^\infty \hat{b}(w^{-\alpha}, 1)\,\mathrm{d}w,$$

$$\mathbb{E}(X_1|X_2 = t) = \int_0^\infty \widehat{C}_{1|2}\big((1+x)^{-\alpha}|(1+t)^{-\alpha}\big)\,\mathrm{d}x = t \int_0^\infty \widehat{C}_{1|2}\big((1+tw)^{-\alpha}|(1+t)^{-\alpha}\big)\,\mathrm{d}w$$

$$\sim t \int_0^\infty \widehat{C}_{1|2}\big(t^{-\alpha}w^{-\alpha}|t^{-\alpha}\big)\,\mathrm{d}w \sim t \int_0^\infty \hat{b}_{1|2}(w^{-\alpha}|1)\,\mathrm{d}w.$$

The tail dependence functions in Section 2.18 are used in the above.

If the tail of $\overline{F}$ is in the domain of attraction of the Gumbel extreme value distribution instead of regularly varying, the asymptotic result is different but involves $\hat{b}$. We demonstrate this with $\overline{F}(x) = e^{-x}$ for $x > 0$. Asymptotically (2.69) and (2.70) become:

$$\mathbb{E}(X_1|X_2 > t) = e^t \int_0^\infty \widehat{C}(e^{-x}, e^{-t})\, dx = e^t \int_{-t}^\infty \widehat{C}(e^{-(y+t)}, e^{-t})\, dy$$

$$\sim e^t t \int_{-1}^\infty \widehat{C}(e^{-(tw+t)}, e^{-t})\, dw \sim t \int_{-1}^\infty \hat{b}(e^{-tw}, 1)\, dw$$

$$\sim t \int_{-1}^0 \hat{b}(\infty, 1)\, dw = t\,\hat{b}(\infty, 1),$$

$$\mathbb{E}(X_1|X_2 = t) = \int_0^\infty \widehat{C}_{1|2}(e^{-x}|e^{-t})\, dx = \int_{-t}^\infty \widehat{C}_{1|2}(e^{-(y+t)}|e^{-t})\, dy$$

$$\sim t \int_{-1}^\infty \widehat{C}_{1|2}(e^{-(tw+t)}|e^{-t})\, dw \sim t \int_{-1}^\infty \hat{b}_{1|2}(e^{-tw}|1)\, dw$$

$$\sim t \int_{-1}^0 \hat{b}_{1|2}(\infty|1)\, dw = t\,\hat{b}_{1|2}(\infty|1),$$

since $\hat{b}(0, 1) = 0$, $\hat{b}_{1|2}(0|1) = 0$, $\hat{b}(\infty, 1) \le 1$, and $\hat{b}_{1|2}(\infty|1) \le 1$ (see Remark 2.1).   □

Next is an example where the CTE is $O(t^\gamma)$ with $0 < \gamma < 1$. The transformations inside the integral and the method of approximation is different from the preceding Example 2.16.

Example 2.17 (CTEs that are $O(t^\gamma)$ as $t \to \infty$ with $0 < \gamma < 1$). We take C to be the reflected Gumbel copula or $\widehat{C}$ to be bivariate Gumbel (Section 4.8). We show some details for Pareto(α) and Weibull margins.

With $\widehat{C}(u, v) = \exp\{-[(-\log u)^\delta + (-\log v)^\delta]^{1/\delta}\}$ for $\delta > 1$,

$$\widehat{C}_{1|2}(u|v) = v^{-1}\widehat{C}(u, v) \cdot \left[1 + \left(\frac{-\log u}{-\log v}\right)^{1/\delta - 1}\right],$$

and it can be shown that $\widehat{C}_{1|2}(u|0) = 1$ for $0 < u \le 1$, that is, $\widehat{C}_{1|2}(\cdot|0)$ is degenerate at 0. Let $A(z_1, z_2) = (z_1^\delta + z_2^\delta)^{1/\delta}$ with derivative $A_2(z_1, z_2) := \partial A(z_1, z_2)/\partial z_2 = (1 + [z_1/z_2]^\delta)^{1/\delta - 1}$. The transforms inside the integrals of the conditional expectations will make use of

$$\widehat{C}(v^s, v) = v^{A(s,1)}, \quad \widehat{C}_{1|2}(v^s|v) = v^{A(s,1)-1} A_2(s, 1). \qquad (2.72)$$

For Pareto(α) with $\alpha > 1$, let $y = \alpha \log(1 + x)$, $x = e^{y/\alpha} - 1$, $dx = \alpha^{-1} e^{y/\alpha} dy$, $T = \alpha \log(1 + t)$. Then, asymptotically (2.69) and (2.70) become:

$$\mathbb{E}(X_1|X_2 > t) = \alpha^{-1} \int_0^\infty e^T \widehat{C}(e^{-y}, e^{-T})\, e^{y/\alpha} dy = \alpha^{-1} T \int_0^\infty e^T \widehat{C}(e^{-sT}, e^{-T})\, e^{sT/\alpha} ds$$

$$= \alpha^{-1} T \int_0^\infty e^T e^{-TA(s,1)} e^{sT/\alpha} ds = \alpha^{-1} T \int_0^\infty e^{-Tg(s;T)} ds, \qquad (2.73)$$

$$g(s; T) := g(s) = -1 - s/\alpha + A(s, 1) = -1 - s/\alpha + (s^\delta + 1)^{1/\delta},$$

$$\mathbb{E}(X_1|X_2 = t) = \frac{T}{\alpha} \int_0^\infty \widehat{C}_{1|2}(e^{-sT}|e^{-T})\, e^{sT/\alpha} ds = \frac{T}{\alpha} \int_0^\infty e^{-T[A(s,1)-1]} A_2(s, 1)\, e^{sT/\alpha} ds$$

$$= \alpha^{-1} T \int_0^\infty A_2(s, 1)\, e^{-Tg(s;T)} ds. \qquad (2.74)$$

In (2.73), g has a unique minimum in $(0, \infty)$ and is convex since $g'(s) = -\alpha^{-1} + (1 + s^{-\delta})^{1/\delta - 1}$

and $g''(s) = (\delta - 1)(1 + s^{-\delta})^{1/\delta - 2} s^{-\delta - 1} > 0$. The root of g' is $s_0 = (\alpha^{\delta/(\delta - 1)} - 1)^{-1/\delta}$. Also $g(0) = 0$, $g(\infty) = \infty$, $g'(0) = -\alpha^{-1}$ and $\alpha g(s_0) = -\alpha + (\alpha^{\delta/(\delta - 1)} - 1)^{1 - 1/\delta} \in (-1, 0)$. With these properties for g, the Laplace approximation can be applied to (2.73) to get (as $T \to \infty$):

$$\frac{T}{\alpha} \int_0^\infty \exp\{-Tg(s)\} ds \sim \frac{T}{\alpha} \int_0^\infty \exp\{-T[g(s_0) + (s - s_0)g'(s_0) + \tfrac{1}{2}(s - s_0)^2 g''(s_0)]\} ds$$

$$= \alpha^{-1} e^{-Tg(s_0)} T \int_0^\infty \exp\{-\tfrac{1}{2}Tg''(s_0)(s - s_0)^2\} ds$$

$$\sim \alpha^{-1/2}(1 + t)^{-\alpha g(s_0)} \sqrt{2\pi[g''(s_0)]^{-1} \log(1 + t)},$$

and similarly, in (2.74),

$$\alpha^{-1} T \int_0^\infty A_2(s, 1) \exp\{-Tg(s)\} ds \sim \alpha^{-1/2} A_2(s_0, 1)(1 + t)^{-\alpha g(s_0)} \sqrt{2\pi[g''(s_0)]^{-1} \log(1 + t)}.$$

Hence both conditional expectations are $O(t^{-\alpha g(s_0)}[\log t]^{1/2})$, as $t \to \infty$.

Next for the Weibull survival function $\overline{F}(x) = \exp\{-x^\beta\}$ with $x > 0$ and $\beta > 0$, let $T = -\log \overline{F}(t) = t^\beta$, $y = -\log \overline{F}(x) = x^\beta$ and $x = y^{1/\beta}$. Substituting into (2.69) leads to:

$$\mathbb{E}(X_1|X_2 > t) = \beta^{-1} \int_0^\infty e^T \widehat{C}(e^{-y}, e^{-T}) y^{1/\beta - 1} dy = \beta^{-1} T^{1/\beta} \int_0^\infty e^T \widehat{C}(e^{-sT}, e^{-T}) s^{1/\beta - 1} ds$$

$$= \beta^{-1} T^{1/\beta} \int_0^\infty \exp\{T[1 - (1 + s^\delta)^{1/\delta}]\} s^{1/\beta - 1} ds.$$

Let $z = (1 + s^\delta)^{1/\delta} - 1$, $s = m(z) = [(z + 1)^\delta - 1]^{1/\delta}$, with $m'(z) = [(z + 1)^\delta - 1]^{1/\delta - 1}(z + 1)^{\delta - 1}$. Then

$$\mathbb{E}(X_1|X_2 > t) = \beta^{-1} T^{1/\beta} \int_0^\infty e^{-Tz}[m(z)]^{1/\beta - 1} m'(z) \, dz$$

$$= \beta^{-1} T^{1/\beta} \int_0^\infty e^{-Tz}[(z + 1)^\delta - 1]^{1/(\beta\delta) - 1}(z + 1)^{\delta - 1} dz.$$

Let $\Upsilon(z) = [(z + 1)^\delta - 1]^{1/(\beta\delta) - 1}(z + 1)^{\delta - 1}$. This is bounded by an exponential function over all $z > 0$ and it is analytical for $0 < z < 1$, behaving like $(\delta z)^{1/(\beta\delta) - 1}$ for z near 0. By a slightly modified version of Watson's lemma (pp 20–21 of Wong (1989)), the integral behaves like:

$$\Gamma(\beta^{-1}\delta^{-1})\beta^{-1} T^{1/\beta} \delta^{1/(\beta\delta) - 1} T^{-1/(\beta\delta)} = O(t^{1 - 1/\delta}), \quad t \to \infty. \tag{2.75}$$

Similarly, in (2.70).

$$\mathbb{E}(X_1|X_2 = t) = \beta^{-1} T^{1/\beta} \int_0^\infty e^{-T[(1 + s^\delta)^{1/\delta} - 1]}(1 + s^\delta)^{1/\delta - 1} s^{1/\beta - 1} ds$$

$$= \beta^{-1} T^{1/\beta} \int_0^\infty e^{-Tz}[(z + 1)^\delta - 1]^{1/(\beta\delta) - 1} dz.$$

By Watson's lemma, the integral behaves like (2.75).

As the upper order for a bivariate reflected Gumbel copula is $\kappa_U = 2^{1/\delta}$, a larger δ implies a stronger degree of upper intermediate tail dependence. This observation is consistent to the pattern of $O(t^{1 - 1/\delta})$ for $\mathbb{E}[X_1|X_2 > t]$ and $\mathbb{E}[X_1|X_2 = t]$. As $\delta \to 1$ for the independence copula, the rate is $O(1)$, and as $\delta \to \infty$ for comonotonicity, the rate is $O(t)$. $\square$

The above approach can work more generally if something like (2.72) is a good enough approximation to use within Laplace's method or Watson's lemma. Note that conditional tail variance calculations could be done in a similar way, using

$$\text{Var}(X_1|X_2 > t) = \mathbb{E}(X_1^2|X_2 > t) - \mathbb{E}^2(X_1|X_2 > t),$$
$$\text{Var}(X_1|X_2 = t) = \mathbb{E}(X_1^2|X_2 = t) - \mathbb{E}^2(X_1|X_2 = t).$$

The complication is that the leading terms, for example, of $\mathbb{E}(X_1^2|X_2 > t)$ and $\mathbb{E}^2(X_1|X_2 > t)$, might cancel, so one must keep track of the validity of further terms in the tail expansions.

2.21 Tail comonotonicity

This section is concerned with tail comonotonicity which corresponds to an extreme form of tail dependence with the tail dependence parameter equal to 1. The copula need not be the comonotonicity copula C^+, that is, it can be absolutely continuous and satisfy tail comonotonicity. Tail comonotonic copulas can be used to get conservative bounds on joint tail probabilities.

Definition 2.12 A random vector $\boldsymbol{X}$ is said to be *upper tail comonotonic* if $\boldsymbol{X}$ has a copula C and its reflected copula $\widehat{C}$ satisfies

$$\lim_{u \to 0^+} \frac{\widehat{C}(uw_1, \ldots, uw_d)}{u} = \min\{w_1, \ldots, w_d\}, \quad w_i, \ldots, w_d \in [0, \infty);$$

the copula C is said to be an upper tail comonotonic copula. $\boldsymbol{X}$ is said to be *lower tail comonotonic* if $\boldsymbol{X}$ has a copula C that satisfies $\lim_{u \to 0^+} C(u\boldsymbol{w})/u = \min\{w_1, \ldots, w_d\}$, $w_i \in [0, \infty)$; the copula C is said to be a lower tail comonotonic copula.

In the extreme value literature (e.g., Resnick (2007)), the concept is called *asymptotic dependence*.

Some basic properties are the following. For a copula C, $\lambda_U(C) = 1$ if and only if the upper tail dependence function exists and $b^*(\boldsymbol{w}) = \min\{w_1, \ldots, w_d\}$. In parallel, $\lambda_L(C) = 1$ if and only if the lower tail dependence function exists and $b(\boldsymbol{w}) = \min\{w_1, \ldots, w_d\}$. If C is a d-variate copula, then any bivariate marginal copula C_{jk}, $j \neq k$, is upper (respectively lower) tail comonotonic if and only if C is upper (respectively lower) tail comonotonic.

Hua and Joe (2012b) have several ways to construct absolutely continuous tail comonotonic copulas. Cheung (2009) has a related definition of upper (lower) comonotonicity which requires the upper (lower) tail to satisfy $C(\boldsymbol{u}) = \min\{u_1, \ldots, u_d\}$ in the rectangular region beyond a threshold. Hence upper comonotonicity implies upper tail comonotonicity in the sense of Definition 2.12 but not the converse.

2.22 Summary for analysis of properties of copulas

In this chapter, we have covered some basic tools needed for the construction of multivariate distributions with different dependence structures in Chapter 3. Chapter 8 has further results (theorems and examples) concerning the dependence and tail behavior concepts in this chapter, when applied to general copula constructions or specific parametric copula families in Chapter 4. Many of the copula families are used in the data analysis examples in Chapter 7.

Chapter 3

Copula construction methods

This chapter presents copula construction methods, covering families with exchangeable dependence, conditional independence given latent variables or factor structural dependence, up to general unstructured dependence. The number of parameters is of the order of the dimension up to the square of the dimension. There is an emphasis on stochastic representations and interpretation of the dependence structures. Theorems for verifying dependence and tail properties of copula families are in Chapter 8.

Section 3.1 has an overview of dependence structures and flexibility of different copula construction methods. This includes desirable properties of copula families for applications and a discussion of which properties can be satisfied from different construction methods.

Section 3.2 has the construction of mixtures of powers of a univariate cdf or survival function, leading to Archimedean copulas based on Laplace transforms. Section 3.3 has Archimedean copulas based on the Williamson transform. Section 3.4 has copulas with hierarchical dependence including hierarchical Archimedean copulas. Section 3.5 has copulas based on the construction of mixtures of power of max-id distributions, and this includes hierarchical Archimedean as a special case. Section 3.6 adds the power parameter for multivariate distributions that are max-id and studies some limiting distributions.

Because simple constructions such as Archimedean and mixture of max-id cannot lead to flexible dependence, there are approaches to construct multivariate copulas based on a set of bivariate copulas. As background, Section 3.7 discusses the difficulties of working with the Fréchet class given bivariate marginal copulas, so that this approach is not practically feasible. Section 3.8 discusses Fréchet classes given two marginal multivariate distributions which have variables in common and introduces mixtures of conditional distributions. The vine pair-copula construction in Section 3.9 makes use of sequentially mixing conditional distributions and is a flexible approach involving a sequence of bivariate copulas, with most of them applied to pairs of univariate conditional distributions. Section 3.10 has factor copulas which are constructed with vine copulas rooted at latent variables, with response variables that can be continuous, discrete or mixed. Subsection 3.10.4 has results on the multivariate t copula with the correlation matrix having a factor dependence structure.

To get to high-dimensional copulas, it might be desirable to combine copulas for smaller subsets of variables that might be more homogeneous. Section 3.11 discusses constructions that build on copulas for groups of variables.

The aforementioned constructions can be used in different ways. Section 3.12 has copula models that can be considered as extensions of some Gaussian structural equation models. Section 3.13 has a comparison of different classes of graphical models, including truncated vines and factor copula models. Section 3.14 has stationary Markov, q-dependent and mixed Markov/q-dependent time series models that can be considered as extensions of Gaussian autoregressive moving-average time series models. Section 3.15 discusses multivariate extreme value limiting distributions, which can be obtained from multivariate copulas with tail dependence. In particular, some examples of multivariate extreme value copulas derived from factor copulas are obtained in Section 3.16.

Section 3.17 summarizes some other miscellaneous mixture approaches to get copula models. Section 3.18 has a summary of operations that can be used to obtain additional copula families. Some of these operations for bivariate copulas might lead to suitable candidates to use with the pair-copula construction.

Section 3.19 contains a chapter summary.

3.1 Overview of dependence structures and desirable properties

There are many ways to develop parametric multivariate models from a set of random variables. For likelihood inference, it would be desirable to have either the multivariate distribution or density (or both) have a closed form or a readily computable form. Furthermore, to go from a multivariate distribution to a copula, the univariate inverse cdfs should also be tractable/computable. Also, it would be desirable that the stochastic representation is a plausible data generation method and that the parameters of the model have simple interpretations. The suitable dependence structures depend on the context of applications.

In this section, we outline methods to derive multivariate distributions with different dependence structures ranging from simple to complex. Copulas are obtained from the multivariate distributions via Sklar's theorem. Let $(X_1, \ldots, X_d)$ be dependent random variables with joint cdf F. By deriving F and then obtaining the univariate margins F_j and inverse cdf F_j^{-1}, the copula C can be obtained.

The simplest dependence structure is exchangeable with positive dependence or conditional i.i.d. The general form is

$$F(x_1, \ldots, x_d) = \int \prod_{j=1}^{d} G(x_j; q) \, \mathrm{d}F_Q(q), \quad \boldsymbol{x} \in \mathbb{R}^d.$$

The stochastic representation is that $X_1, \ldots, X_d$ given $Q = q$ are conditional i.i.d. with the univariate distribution $G(\cdot; q)$. The survival function is

$$\overline{F}(x_1, \ldots, x_d) = \int \prod_{j=1}^{d} \overline{G}(x_j; q) \, \mathrm{d}F_Q(q), \quad \boldsymbol{x} \in \mathbb{R}^d.$$

For tractability, the most convenient choices are $G(x; q) = G^q(x; 1)$ in which case $q > 0$ is a resilience parameter, or $\overline{G}(x; q) = \overline{G}^q(x; 1)$ in which case $q > 0$ is a frailty parameter; this leads to the Archimedean copulas. For the survival function, $\overline{G}(\cdot; 1)$ being Exponential or Weibull are special cases. If $G(\cdot; q)$ is normal with mean 0 and standard deviation $q^{-1} > 0$, then spherical distributions that are inverse scale mixtures of $N(0, 1)$ are obtained.

DeFinetti's theorem (see Mai and Scherer (2012a)) implies that conditional i.i.d. and extendible to any dimension is the same as exchangeable with positive dependence. For exchangeable with possible negative dependence (and giving up the property of conditional i.i.d.), models include $F(x_1, \ldots, x_d) = \int G(x_1, \ldots, x_d; r) \, \mathrm{d}F_R(r)$ and $\overline{F}(x_1, \ldots, x_d) = \int \overline{G}(x_1, \ldots, x_d; r) \, \mathrm{d}F_R(r)$, where $G(\cdot; r)$ corresponds to a uniform distribution on the surface of a d-dimensional hypersphere (with radius r) or simplex (with sum r). The latter leads Archimedean copulas with possible negative dependence. Note that the maximum amount of exchangeable negative dependence decreases as the dimension d increases.

Exchangeable dependence is a strong assumption, and it would not be expected to hold in most practical situations. The condition can be relaxed in several ways: one is partial exchangeability where there are mutually exclusive subsets of variables such that the variables within any subset are considered exchangeable; another is conditional independence (independence without being identically distributed after conditioning).

Hierarchical dependence will refer to partial exchangeability where variables can be grouped into subsets as in hierarchical clustering. Clusters of variables that are farther apart have less dependence, and for two clusters, there is a constant association for the intercluster pairs of variables. Hierarchical Archimedean copulas can be obtained with this dependence structure. For the case of two clusters, say indexed in disjoint sets J_1, J_2 with union $\{1, \ldots, d\}$, a stochastic representation of the multivariate distribution is

$$F(x_1, \ldots, x_d) = \int \left\{ \int \prod_{i \in J_1} G_1^{q_1}(x_i) \mathrm{d}F_{Q_1}(q_1; q) \right\} \cdot \int \left\{ \int \prod_{j \in J_2} G_2^{q_2}(x_j) \mathrm{d}F_{Q_2}(q_2; q) \right\} \mathrm{d}F_Q(q).$$

Conditional independence models based on $p \geq 1$ latent variables have the form

$$F(x_1, \ldots, x_d) = \int \prod_{j=1}^{d} G_j(x_j | v_1, \ldots, v_p) \, \mathrm{d}F_{V_1, \ldots, V_p}(v_1, \ldots, v_p), \quad \boldsymbol{x} \in \mathbb{R}^d.$$

This can be a reasonable model when there are latent variables that can explain the dependence in the observed variables. When $p = 1$, $G_j(x_j | v_1)$ can be specified via a copula for (X_j, V_1). For $p \geq 2$, one possible specification for $G_j(x_j | v_1, \ldots, v_p)$ is via a vine distribution (described below).

For the bivariate case, one mainly wants distributions and copulas that can cover a range of dependence from perfect positive dependence to independence, and maybe extending to perfect negative dependence. This can be done with one of the above constructions. Then it would be nice if one could construct multivariate distributions given specified bivariate margins. However, compatibility conditions in general for bivariate margins are much more difficult than something like the constraint of positive definiteness for a correlation matrix.

We will present two ways to get multivariate distributions where there is a bivariate copula associated with (j, k) for every $1 \leq j < k \leq d$. One approach involves a mixture of products of powers of bivariate distributions, but it requires that the bivariate distributions used for the mixing are max-infinitely divisible (max-id); an advantage of this approach is that closed form multivariate distributions can be obtained. Another approach involves a set or sequence of mixtures of conditional distributions where one starts by mixing univariate marginal distributions to get F_{jk} for $d-1$ pairs (j, k) (that satisfy a tree condition) and then continues inductively by mixing univariate conditional distributions of form $F_{j|S}, F_{k|S}$ to get $F_{jk|S}$ for the $(d-1)(d-2)$ remaining (j, k) pairs. Multivariate Gaussian distributions can be obtained in this way, and then the parametrization of $d(d-1)/2$ parameters involves $d-1$ correlations and $(d-1)(d-2)/2$ partial correlations, in such a way that the parameter space is all of $(-1, 1)^{d(d-1)/2}$. That is, this parametrization has parameters that are algebraically independent and gets around the constraint of positive definiteness for $\binom{d}{2}$ correlations, and it is also the formulation that can extend to the vine copula or pair-copula construction. The classical approach of correlations and linear combinations does not extend beyond elliptical distributions, which include multivariate Gaussian, because elliptical distributions and multivariate stable distributions are the main multivariate families that have closure under linear combinations, and stable laws might not have second order moments.

The vine copula approach, derived from a sequence of mixing conditional distributions, is characterized by a sequence of bivariate copulas that are applied to univariate marginal or conditional distributions. It can lead to multivariate distributions that (a) have a wide range of dependence and tail behavior, (b) can be used to approximate multivariate distributions and (c) apply to continuous or discrete variables.

For smaller value of the dimension d, parametric vine copulas have a parameter associated to each (j, k) with $1 \leq j < k \leq d$. For larger values of d and high-dimensional copula modeling, truncated vine models with $O(d)$ parameters are more practical and are the copula analogue of (parsimonious) multivariate Gaussian models with structured correlation matrices. For truncated vine models, there are some conditional independence relations among the observed variables. For the factor copula models mentioned above, there is conditional independence among observed variables, given latent variables. Combined truncated vine-factor models have latent variables and some conditional independence relations conditional on latent variables in combination with observed variations; they can be considered as nonlinear structural equation models.

Multivariate extreme value distributions or copulas are suitable as models for multivariate maxima or minima. With exponential (respectively Fréchet) margins, the class of distributions have the stochastic representation of dependent exponential (Fréchet) random variables for which any weighted minimum (maximum) of the variables has an exponential (Fréchet) distribution. However, this leads to Pickands' representation which is based on a finite measure in a simplex; it is difficult to assess the form of dependence from the measure. To create parametric families of multivariate extreme value copulas with different dependence structures, we will instead take extreme value limits from any of the preceding constructions; useful parametric families obtain if we start with multivariate distributions with tail dependence.

The emphasis in this chapter is on construction methods of multivariate models from which copulas can be obtained. Constructions based on mixtures, latent variables and limits are important because one cannot write out functional forms and expect them to be families of multivariate cdfs with a wide range of dependence. In the literature on multivariate copulas, there are families that could be considered as "perturbations of independence." They have the form $C(u_1, \ldots, u_d) = \left(\prod_{j=1}^{d} u_j \right) \cdot \{1 + \theta g(u_1, \ldots, u_d)\}$ for $\boldsymbol{u} \in [0,1]^d$, where g satisfies some properties in order that univariate margins are $U(0,1)$ and θ lies in a small interval about 0, so that the mixed dth order derivative is non-negative. These types of copulas will not be discussed in this chapter, because their limited range of dependence means that they are not good candidates for models for data.

Details for constructions mentioned above are given in the remaining sections of this chapter, starting with simpler constructions that have exchangeable and partially exchangeable dependence structures, and then moving to construction methods that can accommodate more general dependence structures.

To conclude this section, we discuss some desirable properties of copula models, and summarize which of them can be satisfied for the various constructions.

class	A1	A2	A3	A4	B	C1	C2	D
Gaussian	yes	yes	no	no	yes	no	yes	yes
multivariate t_ν	no	yes	yes	no	yes	no	yes	yes
Archimedean	yes	no	no	bivariate	yes	yes	yes	yes
mixture of max-id	yes	yes	bivariate	bivariate	yes	yes	yes	partly
vine PCC	yes	yes	yes	yes	no	no	yes	yes
factor	yes	yes	yes	yes	yes	no	partly	yes

Table 3.1 *Satisfaction or not of properties for various constructions: (A1) inclusion of independence and comonotonicity copulas; (A2) flexible and wide range of dependence; (A3) flexible tail dependence; (A4) flexible tail asymmetries; (B) closure property under marginalization; (C1) closed-form cdf; (C2) closed-form density; (D) ease of simulation. "Bivariate" in a cell means that the property holds for bivariate but not general multivariate.*

Table 3.1 has desirable properties for parametric multivariate copula families, updated from Section 4.1 of Joe (1997). The closure property under marginalization could be weakened to the bivariate margins belonging to the same parametric family. A closed-form cdf is useful if the response variables are discrete and a closed-form density is useful for continuous response variables.

There is no known parametric multivariate family that satisfies all of the desirable properties, in which case one must decide on the relative importance of the properties in particular applications. The use of multivariate models can be considered as an approach to get inferences on quantities like rectangle and joint tail probabilities. For applications requiring upper and lower tail dependence, multivariate t copulas and vine/factor copulas are generally the best, and the latter can have reflection asymmetry. Further explanations are given below.

1. Multivariate Gaussian copulas do not have tail dependence or tail asymmetry, and the lack of a closed form cdf can be a problem for applications with multivariate discrete data. However, a regular vine construction for discrete variables might provide a good approximation, and for something like multivariate probit models, composite likelihood (Section 5.6) could be used instead of full likelihood.

2. Multivariate t copulas are suitable if tail dependence and approximate reflection symmetry hold. Property A1 is not satisfied because independence copula does not result when all of the ρ's are 0 (zero ρ's correspond to uncorrelatedness but not independence). Also, there is the lack of a closed form cdf.

3. There are parametric families of bivariate Archimedean copulas that can have different types of tail asymmetries so that these would suitable for use with vines. Exchangeable multivariate Archimedean copulas have limited applications but can be used for sensitivity analyses of probability calculations.

4. Mixtures of max-id copulas have flexible positive dependence but not flexible tail dependence. A closed form cdf means that this family can be convenient for positively dependent discrete/ordinal variables. Also extreme value limits from this class lead to some closed form multivariate extreme value copulas with flexible dependence (i.e., a parameter associated to each bivariate margin).

5. Vine pair-copula constructions (PCC) are not closed under margins and do not have closed form cdf. However, they have lots of flexibility and members of the class should provide good approximations when trivariate and higher-order margins have conditional distributions with copulas that do not vary much over different values of the conditioning variables. For multivariate discrete data, the lack of a closed form cdf is not a problem as sequences of conditional distributions can still be used; the joint pmf has a simple algorithmic form.

6. Factor copulas (structured or unstructured) are closed under margins but do not have closed form cdf. However, for discrete or continuous variables, the probability mass/density function can be computed based on the p-dimensional numerical integrals if there are p latent variables; so $p = 1, 2$ are easily feasible and $p = 3$ is possible. Some structural factor copula models with p latent variables can sometimes be handled efficiently without p-dimensional numerical integrals. Similar to multivariate Gaussian factor models, there can be issues of near-identifiability but these do not have much effect on joint tail probabilities.

3.2   Archimedean copulas based on frailty/resilience

This section describes a simple method to get what are commonly called Archimedean copulas. These copulas have an exchangeable dependence structure. The bivariate members of this class can be useful for multivariate construction methods in later sections that are based on a set of bivariate copulas. The family of copulas can be considered as a generalization of the copula family that results from a Gamma scale mixture of Weibull.

Let $G_1, \ldots, G_d$ be univariate cdfs. Let Q be a positive random variable with Laplace transform (LT) $\psi_Q : [0, \infty) \to [0, 1]$. Let $X_1, \ldots, X_d$ be dependent random variables that are conditionally independent given $Q = q$. If

$$[X_j \mid Q = q] \sim \overline{G}_j^q, \quad q > 0, \tag{3.1}$$

then Q is called a frailty random variable, and if

$$[X_j \mid Q = q] \sim G_j^q, \quad q > 0, \tag{3.2}$$

then Q is called a resilience random variable. We are following the construction and terminology of Marshall and Olkin (1988, 2007). Here, we assume that $Q = 0$ with zero probability, otherwise for frailty, $\overline{G}_j^0(x) = I_{(-\infty,\infty)}(x)$ corresponds to a defective random variable with mass at ∞ (if $\overline{G}_j$ has support on all of $\mathbb{R}$), and $G_j^0(x) = I_{(-\infty,\infty)}(x)$ corresponds to a defective random variable with mass at $-\infty$ (if G_j has support on all of $\mathbb{R}$). For (3.1), the joint survival is

$$\overline{F}(x_1, \ldots, x_d) = \int_0^\infty \overline{G}_1^q(x_1) \cdots \overline{G}_d^q(x_d) \, dF_Q(q) = \psi_Q\big(-\log \overline{G}_1(x_1) - \cdots - \log \overline{G}_d(x_d)\big),$$

with univariate survival functions

$$\overline{F}_j(x_j) = \psi_Q\big(-\log \overline{G}_j(x_j)\big), \quad j = 1, \ldots, d,$$

and inverse functions

$$\overline{F}_j^{-1}(u_j) = \overline{G}_j^{-1}\big(\exp\{-\psi_Q^{-1}(u_j)\}\big), \quad 0 < u_j < 1, \ j = 1, \ldots, d;$$

the survival copula, as in (1.10), is

$$C(\boldsymbol{u}) = \overline{F}\big(\overline{F}_1^{-1}(u_1), \ldots, \overline{F}_d^{-1}(u_d)\big) = \psi_Q\Big(\sum_{j=1}^d \psi_Q^{-1}(u_j)\Big), \quad \boldsymbol{u} \in [0,1]^d.$$

Similarly, for (3.2), the joint cdf is

$$F(x_1, \ldots, x_d) = \int_0^\infty G_1^q(x_1) \cdots G_d^q(x_d) \, dF_Q(q) = \psi_Q\big(-\log G_1(x_1) - \cdots - \log G_d(x_d)\big), \tag{3.3}$$

with univariate cdfs

$$F_j(x_j) = \psi_Q\big(-\log G_j(x_j)\big), \quad j = 1, \ldots, d,$$

and inverse functions

$$F_j^{-1}(u_j) = G_j^{-1}\big(\exp\{-\psi_Q^{-1}(u_j)\}\big), \quad 0 < u_j < 1, \ j = 1, \ldots, d;$$

the copula, via (1.9), is

$$C(\boldsymbol{u}) = F\big(F_1^{-1}(u_1), \ldots, F_d^{-1}(u_d)\big) = \psi_Q\Big(\sum_{j=1}^d \psi_Q^{-1}(u_j)\Big), \quad \boldsymbol{u} \in [0,1]^d,$$

the same as from the frailty representation.

For the resilience representation, note that $G_j = \exp\{-\psi_Q^{-1}(F_j)\}$, and if $F_j(x) = x$ for $0 < x < 1$, then $G_j(x) = \exp\{-\psi_Q^{-1}(x)\}$, $0 < x < 1$, for $j = 1, \ldots, d$. The condition that Q has no mass at 0 or $\lim_{s \to \infty} \psi_Q(s) = 0$ is important. If to the contrary, $\mathbb{P}(Q = 0) = \pi_0 = \lim_{s \to \infty} \psi_Q(s)$, then F_j cannot take values in $(0, \pi_0)$, that is, F_j is not continuous and the above copula equation is not valid.

The above construction leads to the functional form that is called an Archimedean copula (the name comes via Archimedean axiom, see Nelsen (2006)). This multivariate family was introduced in Kimberling (1974) without the stochastic representation. Consider

$$C_\psi(\boldsymbol{u}) = \psi\Big(\sum_{j=1}^{d} \psi^{-1}(u_j)\Big), \quad \boldsymbol{u} \in [0,1]^d, \tag{3.4}$$

This is a valid copula for any d whenever $\psi \in \mathcal{L}_\infty$ where $\mathcal{L}_\infty$ is the class of LTs of non-negative random variables with no mass at 0 (so $\psi(\infty) = 0$). This mixture representation implies positive dependence. In the next section, conditions are given for (3.4) to be a valid d-variate copula when ψ is not a LT.

Note that (3.4) is permutation symmetric $C_\psi(u_{i_1}, \ldots, u_{i_d}) = C_\psi(\boldsymbol{u})$ for any permutation $(i_1, \ldots, i_d)$ of $(1, \ldots, d)$ so that it is a copula of exchangeable U(0,1) random variables. If $G_1 = \cdots = G_d$ in (3.3), then the multivariate distribution is also permutation symmetric and the random variables $X_1, \ldots, X_d$ which satisfy (3.2) are conditional i.i.d. and exchangeable with positive dependence.

In Nelsen (1999), Genest and MacKay (1986) and other references, the bivariate form of the Archimedean copula is:

$$C_\varphi(u_1, u_2) = \xi^{-1}(\xi(u_1) + \xi(u_2)), \quad 0 \le u_1 \le 1, \ 0 \le u_2 \le 1, \tag{3.5}$$

and ξ is called a generator. The necessary and sufficient conditions for this to be a bivariate copula are: $\xi : [0,1] \to [0,\infty)$ decreasing, $\xi(0) = \infty$, $\xi(1) = 0$ and ξ convex. Some copulas of form (3.5) that don't come from LTs can have negative dependence.

However, the frailty and resilience interpretation of (3.4), following Marshall and Olkin (1988, 2007), is more useful to understand dependence and tail properties; see Section 8.5. It is also the form used when considering extensions such as hierarchical Archimedean copulas and those based on mixture of max-id distributions (Sections 3.4 and 3.5).

Common 1-parameter LTs that lead to bivariate Archimedean copulas that interpolate the independence copula $C^\perp$ and the comonotonicity copula C^+ are the gamma, positive stable, Sibuya and logarithmic series families of LTs; see the Appendix and Sections 4.5–4.8. The LT families are parametrized so that the copula is increasing in dependence as the parameter increases. There are many other parametric LT families with one or more parameters that also interpolate $C^\perp$ and C^+; see Sections 4.17–4.26. Note that (3.4) is more tractable if the inverse LT also has simple form. There are also many parametric LT families, but more useful Archimedean families obtain only if $C^\perp$ and C^+ are included and there is a concordance ordering with respect to at least one parameter.

We next summarize some results for conditional distributions of (3.4).

Suppose $d \ge 3$ and let $C_{1\cdots d-1|d}(u_1, \ldots, u_{d-1}|u_d) = \partial C_\psi(u_1, \ldots, u_d)/\partial u_d$ be the conditional distribution given $U_d = u_d$.

$$C_{1\cdots d-1|d}(u_1, \ldots, u_{d-1}|u_d) = \frac{\psi'\Big(\sum_{j=1}^{d-1} \psi^{-1}(u_j) + \varsigma\Big)}{\psi'(\varsigma)}, \tag{3.6}$$

where $\varsigma = \psi^{-1}(u_d)$. The jth $(1 \le j \le d-1)$ margin is

$$F_{j|d}(u_j|u_d) = \psi'(\psi^{-1}(u_j) + \varsigma)/\psi'(\varsigma) = h(\psi^{-1}(u_j) + \varsigma)/h(\varsigma) =: v_j,$$

with $h = -\psi'$ which is monotone increasing. Hence $\psi^{-1}(u_j) = h^{-1}(v_j h(\varsigma)) - \varsigma$ for $1 \le j \le d-1$, and the copula of (3.6) is:

$$C_{1\cdots d-1;d}(v_1, \ldots, v_{d-1}; \varsigma) = h\Big(\sum_{j=1}^{d-1} [h^{-1}(v_j h(\varsigma)) - \varsigma] + \varsigma\Big) \Big/ h(\varsigma);$$

this is an Archimedean copula $\varphi(\varphi^{-1}(v_1; \varsigma) + \cdots + \varphi^{-1}(v_{d-1}; \varsigma); \varsigma)$ with $s = \varphi^{-1}(v; \varsigma) = h^{-1}(vh(\varsigma)) - \varsigma$ and $v = \varphi(s; \varsigma) = h(\varsigma + s)/h(\varsigma)$.

Similarly, when conditioning on two or more variables, the copulas of conditional distributions of Archimedean copulas are Archimedean. We summarized the result for conditioning on two variables; from this the general pattern can be seen. Let $d \geq 4$ and let $(U_1, \ldots, U_d)$ be a random vector with copula C_ψ. Let $a_j = \psi^{-1}(u_j)$ for $j = d - 1$ and $j = d$, and let $\varsigma = a_{d-1} + a_d$. The conditional distribution given $U_{d-1} = u_{d-1}, U_d = u_d$ is:

$$C_{1 \cdots d-2 | d-1, d}(u_1, \ldots, u_{d-2} | u_{d-1}, u_d) = \frac{\partial^2 C(u_1, \ldots, u_d)/\partial u_{d-1} \partial u_d}{\psi''(a_{d-1} + a_d)/[\psi'(a_{d-1})\,\psi'(a_d)]}$$

$$= \frac{\psi''\left(\sum_{j=1}^{d-2} \psi^{-1}(u_j) + \varsigma\right)}{\psi''(\varsigma)} \tag{3.7}$$

By differentiation, with $h = \psi''$, and $0 < u_d \leq 1$, (3.7) has jth $(1 \leq j \leq d - 2)$ margin $F_{j|d-1,d}(u_j | u_{d-1}, u_d) = h(\psi^{-1}(u_j) + \varsigma)/h(\varsigma) =: v_j$, and h is monotone increasing. Hence $\psi^{-1}(u_j) = h^{-1}(v_j h(\varsigma)) - \varsigma$ for $1 \leq j \leq d - 2$ and the copula of the conditional distribution is:

$$C_{1 \cdots d-2; d-1, d}(v_1, \ldots, v_{d-2}; \varsigma) = h\left(\sum_{j=1}^{d-2} [h^{-1}(v_j h(\varsigma)) - \varsigma] + \varsigma\right) \Big/ h(\varsigma);$$

this is an Archimedean copula $\varphi(\varphi^{-1}(v_1; \varsigma) + \cdots + \varphi^{-1}(v_{d-2}; \varsigma); \varsigma)$ with $s = \varphi^{-1}(v; \varsigma) = h^{-1}(vh(\varsigma)) - \varsigma$ and $v = \varphi(s; \varsigma) = h(\varsigma + s)/h(\varsigma)$. The pattern extends to conditional distributions of Archimedean copulas with three or more conditioning variables. For $d > m \geq 2$, the conditional distribution of $[U_1, \ldots, U_m | U_{m+1} = u_{m+1}, \ldots, U_d = u_d]$ is

$$C_{1 \cdots m | m+1 \cdots d}(u_1, \ldots, u_m | u_{m+1}, \ldots, u_d) = \frac{\psi^{(d-m)}\left(\psi^{-1}(u_1) + \cdots + \psi^{-1}(u_m) + \varsigma\right)}{\psi^{(d-m)}(\varsigma),} \tag{3.8}$$

where $\varsigma = \sum_{k=m+1}^{d} \psi^{-1}(u_k)$; it has univariate margins

$$F_{j|(m+1):d}(u_j) = \psi^{(d-m)}\left(\psi^{-1}(u_j) + \varsigma\right) \Big/ \psi^{(d-m)}(\varsigma), \quad j = 1, \ldots, m,$$

and the copula of (3.8) is Archimedean with LT $\varphi_{d-m}(\cdot; \varsigma)$ coming from the $(d - m)$th derivative of ψ:

$$\varphi_{d-m}(s; \varsigma) = \frac{(-1)^{d-m}\psi^{(d-m)}(s + \varsigma)}{(-1)^{d-m}\psi^{(d-m)}}(\varsigma), \quad s \geq 0.$$

For $d > m \geq 2$, and starting from $(U_1, \ldots, U_d) \sim C_\psi$, the conditional distribution of $[U_1, \ldots, U_m | U_{m+1} \leq u_{m_1}, \ldots, U_d \leq u_d]$ is also Archimedean. With calculations similar to the above, it is straightforward to show that this conditional distribution is

$$\psi\left(\psi^{-1}(u_1) + \cdots + \psi^{-1}(u_m) + \varsigma\right)/\psi(\varsigma)$$

and its copula is

$$C^*_{1 \cdots m}(v_1, \ldots, v_m; u_{m+1}, \ldots, u_d) = \frac{\psi\left(\psi^{-1}[v_1 \psi(\varsigma)] + \cdots + \psi^{-1}[v_m \psi(\varsigma)] - (m - 1)\varsigma\right)}{\psi(\varsigma)}, \tag{3.9}$$

where $\varsigma = \sum_{k=m+1}^{d} \psi^{-1}(u_k)$. The copula (3.9) is Archimedean with LT

$$\varphi^*(s; \varsigma) = \psi(s + \varsigma)/\psi(\varsigma); \quad s \geq 0.$$

Other properties on Archimedean copulas are given in Section 8.5.

3.3 Archimedean copulas based on Williamson transform

The function form (3.4) for Archimedean copulas leads to a valid copula in dimension d for some functions ψ that need not be Laplace transforms of positive random variables.

A sufficient condition for (3.4) to be a copula in dimension d is that derivatives of ψ exist and alternate in sign up to order d, that is, $(-1)^k \psi^{(k)} \geq 0$, $k = 1, \ldots, d$. In this case, all mixed derivatives of (3.4) up to order d are non-negative and the copula density of (3.4) is

$$c_\psi(\boldsymbol{u}) = \frac{\partial^d C_\psi(\boldsymbol{u})}{\partial u_1 \cdots \partial u_d} = \psi^{(d)}\Big(\sum_{j=1}^d \psi^{-1}(u_j)\Big) \cdot \prod_{j=1}^d (\psi^{-1})'(u_j), \quad \boldsymbol{u} \in (0,1)^d.$$

Note that $(\psi^{-1})' \leq 0$. If $(-1)^{d-2}$ times the $(d-2)$th derivative is continuous, decreasing and convex, and the left and right one-sided $(d-1)$th derivatives are not the same everywhere, then $C_\psi(\boldsymbol{u}) = \psi(\sum_{j=1}^d \psi^{-1}(u_j))$ is a valid distribution function that has a singular component (and is not absolutely continuous).

For dimension d, the necessary and sufficient condition for ψ is that ψ is d-times monotone; see Malov (2001) and McNeil and Nešlehová (2009). The class of functions that are d-times monotone is studied in Williamson (1956).

Definition 3.1 The class of d-monotone generators for d-variate Archimedean copulas is:[1]

$$\mathcal{L}_d = \{\psi : [0, \infty) \to [0, 1] \mid \psi(0) = 1,\ \psi(\infty) = 0,\ (-1)^j \psi^{(j)} \geq 0,\ j = 1, \ldots, d-2,$$
$$(-1)^{d-2} \psi^{(d-2)} \text{ decreasing and convex}\}. \tag{3.10}$$

With $d \to \infty$, the class $\mathcal{L}_\infty$ in (2.9) of LTs of positive random variables is obtained; hence for $\psi \in \mathcal{L}_\infty$, $\psi(\infty) = 0$ and ψ is completely monotone.

If $\psi \in \mathcal{L}_\infty$, then $C_\psi(\boldsymbol{u}) = \psi(\sum_{j=1}^d \psi^{-1}(u_j))$ in (3.4), and it has support on all of the interior of $(0,1)^d$ because the cdf comes from a conditional independence model and

$$\mathbb{P}(a_j < U_j < b_j, j = 1, \ldots, d) = \int_0^\infty \prod_{j=1}^d [G^q(b_j) - G^q(a_j)] \mathrm{d}F_Q(q),$$

where $G(u) = \exp\{-\psi^{-1}(u)\}$ is strictly increasing and Q is a positive random variable with LT ψ. However. if $\psi \in \mathcal{L}_d \backslash \mathcal{L}_\infty$, it is possible that C_ψ has support on only a subset of $(0,1)^d$.

For $\psi \in \mathcal{L}_d$, Williamson (1956) proved that there is an increasing function H that is bounded below such that $\psi(s) = \int_0^\infty (1 - xs)_+^{d-1} \mathrm{d}H(x)$, where $(y)_+ = \max\{0, y\}$. With the boundary conditions $\psi(0) = 1$ and $\psi(\infty) = 0$, H can be taken as the cdf of a positive random variable. If R is a non-negative random variable with cdf F_R, McNeil and Nešlehová (2010) define the Williamson d-transform of F_R as

$$(\mathcal{W}_d F_R)(s) := \int_0^\infty \Big(1 - \frac{s}{r}\Big)_+^{d-1} \mathrm{d}F_R(r), \quad s \geq 0, \tag{3.11}$$

and proved that the inverse Williamson d-transform is

$$(\mathcal{W}_d^{-1}\psi)(t) = 1 - \sum_{k=0}^{d-2} (-1)^k t^k \psi^{(k)}(t)/k! - (-1)^{d-1} t^{d-1} \psi_+^{(d-1)}(t)/(d-1)!, \quad 0 \leq t \leq 1.$$

where $\psi_+^{(d-1)}$ is a right derivative.

---

[1] Although this class was used in Joe (1997), the form of $\mathcal{L}_d$ was not precise in the last two derivatives. The form used in this definition with a reference to Williamson (1956) was given in Joe and Ma (2000) for d-variate Archimedean copulas.

If $Q_d = (d-1)/R$, then (3.11) can be written as

$$(\mathcal{W}_d F_R)(s) = \int_0^\infty \left(1 - \frac{qs}{d-1}\right)_+^{d-1} \mathrm{d}F_{Q_d}(q), \quad s \geq 0;$$

this is closer in form to a LT.

McNeil and Nešlehová (2010) showed that an Archimedean copula with $\psi \in \mathcal{L}_d$ can be represented as the survival copula for a random vector:

$$\boldsymbol{X} = (X_1, \ldots, X_d) \stackrel{d}{=} R \times (S_1, \ldots, S_d), \tag{3.12}$$

where $R \perp\!\!\!\perp (S_1, \ldots, S_d)$, R is a positive random variable (satisfying $F_R(0) = 0$) with $\psi = \mathcal{W}_d F_R$ and $(S_1, \ldots, S_d)$ is uniformly distributed on the simplex $\{\boldsymbol{s} \in \mathbb{R}_+^d : \sum_i s_i = 1\}$.

With the stochastic representation (3.12), and $\boldsymbol{x} = (x_1, \ldots, x_d)$ with $x_j > 0$ for all j,

$$\begin{aligned}
\overline{F}_{\boldsymbol{X}}(\boldsymbol{x}) &= \int_0^\infty \mathbb{P}(S_1 > x_1/r, \ldots, S_d > x_d/r) \, \mathrm{d}F_R(r) \\
&= \int_0^\infty \left[1 - \frac{x_1 + \cdots + x_d}{r}\right]_+^{d-1} \mathrm{d}F_R(r) \\
&= \psi(x_1 + \cdots + x_d).
\end{aligned} \tag{3.13}$$

The equality (3.13) makes use of probability calculations for Dirichlet distributions (Theorem 5.4 of Fang et al. (1990)); alternatively a geometric approach is outlined below. The univariate survival functions are $\overline{F}_j(x_j) = \psi(x_j)$, $x_j > 0$, for $j = 1, \ldots, d$ and the copula of $\boldsymbol{X}$ is $C_\psi(\boldsymbol{u}) = \psi(\sum_{j=1}^d \psi^{-1}(u_j))$.

If $\psi \in \mathcal{L}_\infty$ is the LT of a positive random variable Q, then Proposition 1 of McNeil and Nešlehová (2010) implies that

$$R \stackrel{d}{=} E_d/Q,$$

where $E_d \perp\!\!\!\perp Q$ and E_d has the Erlang(d) or Gamma($d, 1$) distribution.

Proposition 4.1 of McNeil and Nešlehová (2009) says that $C_\psi(\boldsymbol{u}_{1:d})$ is absolutely continuous iff F_R is absolutely continuous where $\psi = \mathcal{W}_d F_R$. Also $\psi \in \mathcal{L}_{d+1}$ implies $C_\psi(\boldsymbol{u}_{1:d})$ is absolutely continuous.

Technical details for (3.13)

The result that is needed is that if $(S_1, \ldots, S_d)$ is uniform on the unit d-simplex $\{\boldsymbol{a} \in \mathbb{R}_+^d : \sum_i a_i = 1\}$, and $\boldsymbol{s} = (s_1, \ldots, s_d) \in \mathbb{R}_+^d$ with $0 \leq \sum_{i=1}^d s_i < 1$, then

$$\mathbb{P}(S_j > s_j, j = 1, \ldots, d) = (1 - s_\bullet)^{d-1}, \quad s_\bullet = \sum_{i=1}^d s_i.$$

For $d = 2$, this can be easily obtained by drawing a picture of the triangle with vertices $(0,0)$, $(1,0)$, $(0,1)$. The survival probability depends on $\boldsymbol{s}$ only through $s_\bullet$ because the region $\{S_1 > s_1, S_2 > s_2\}$ corresponds to a line segment from $(s_1, 1 - s_1)$ to $(1 - s_2, s_2)$ and this has length that is $\sqrt{2}(1 - s_\bullet)$ in comparison to $\sqrt{2}$ for the line segment from $(1,0)$ to $(0,1)$.

For $d \geq 3$, geometrically $\mathbb{P}\{S_1 > s_1, \cdots, S_d > s_d\}$ depends only on $s_\bullet$ and is the same as the relative volume of the sub-simplex with d vertices $(s_1, s_2, \ldots, s_{d-1}, 1 - s_\bullet + s_d)$, $(s_1, s_2, \ldots, 1 - s_\bullet + s_{d-1}, s_d), \ldots, (1 - s_\bullet + s_1, s_2, \ldots, s_{d-1}, s_d)$. The squared distance between consecutive vertices is $2(1 - s_\bullet)^2$, so the simplex is regular with identical faces. The simplex lies in a $(d-1)$-dimensional space. Hence the relative volume of this sub-simplex to the d vertices of unit d-simplex is

$$\left[\frac{\sqrt{2}(1 - s_\bullet)}{\sqrt{2}}\right]^{d-1} = (1 - s_\bullet)^{d-1}.$$

3.4 Hierarchical Archimedean and dependence

We define a hierarchical dependence structure to refer to a structure where (a) groups of variables are exchangeable, (b) pairs of variables in two separate groups have a common "dependence" with the strength of dependence being larger if the groups are closer. Mai and Scherer (2012a) study a dependence structure for copulas that is hierarchical and extendible, and their definition of hierarchical dependence is in terms of sigma algebras. This type of dependence might be plausible for portfolio credit risk, such as multiple obligors grouped by credit rating and industrial sector (Whelan (2004), Hering et al. (2010)).

Joe (1993) has a construction of this form with all bivariate margins being Archimedean, and then by extreme value limits there are also multivariate extreme value copulas with this dependence structure. Other copula constructions can also have hierarchical dependence, and the correlation matrix of a multivariate Gaussian or elliptical copula can have this structure.

In this hierarchical dependence structure for Archimedean copulas, a variable has the same "dependence" with all variables that are clustered more closely. So this dependence structure is quite restrictive. An example is in Figure 3.4 showing a dendogram as in hierarchical clustering (Everitt (1974)).

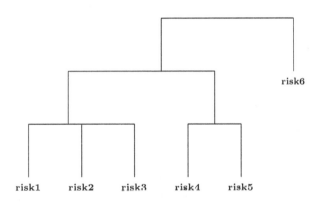

Figure 3.1 *Hierarchical Archimedean structure with 4 dependence parameters and 6 variables; one parameter for within group1=(risk1,risk2,risk3), one for within group2=(risk4,risk5), one for between group1 and group2, and one for risk6 with the union of group1 and group2.*

The general multivariate result is notationally complex, so we first indicate the pattern and conditions from the trivariate and 1-level hierarchical extensions of Archimedean copulas. These are now called *hierarchical Archimedean* or *nested Archimedean* copulas. The trivariate generalization is:

$$
\begin{aligned}
C(\boldsymbol{u}) &= \psi(\psi^{-1} \circ \varphi[\varphi^{-1}(u_1) + \varphi^{-1}(u_2)] + \psi^{-1}(u_3)) \tag{3.14} \\
&= \psi(\psi^{-1}[C_{12}(u_1, u_2)] + \psi^{-1}(u_3)) \\
&= \int_0^\infty \exp\{-q\psi^{-1}(C_{12}(u_1, u_2))\} \exp\{-q\psi^{-1}(u_3)\} \, \mathrm{d}F_Q(q), \quad \boldsymbol{u} \in [0, 1]^3,
\end{aligned}
$$

where $\psi, \varphi \in \mathcal{L}_\infty$, and C_{12} is a bivariate Archimedean with LT φ, and Q has LT ψ. For (3.14) to be valid as a stochastic representation, $H(u_1, u_2) = \exp\{-\psi^{-1}(C_{12}(u_1, u_2))\}$ must be max-id.

For the extension to d-variate with variables indexed in $S_1 = \{1, \ldots, m\}$ in group 1, and variables indexed in $S_2 = \{m+1, \ldots, d\}$ in group 2 $(2 \leq m \leq d-1)$, we can write:

$$C(\boldsymbol{u}) = \psi\big(\psi^{-1}[C_{1\cdots m}(u_1, \ldots, u_m)] + \psi^{-1}[C_{m+1\cdots d}(u_{m+1}, \ldots, u_d)]\big)$$

$$= \int_0^\infty e^{-q\psi^{-1}[C_{S_1}(u_1,\ldots,u_m)]} e^{-q\psi^{-1}[C_{S_2}(u_{m+1},\ldots,u_d)]} \, dF_Q(q), \quad \boldsymbol{u} \in [0,1]^d, \quad (3.15)$$

where $C_{S_1}(u_1, \ldots, u_m) = \varphi_1(\sum_{j=1}^m \varphi_1^{-1}(u_j))$, $C_{S_2}(u_{m+1}, \ldots, u_d) = \varphi_2(\sum_{j=m+1}^d \varphi_2^{-1}(u_j))$, and $\varphi_1, \varphi_2 \in \mathcal{L}_\infty^*$. For (3.15) to be valid as a stochastic representation, $H_i = \exp\{-\psi^{-1}(C_{S_i})\}$ must be max-id for $i = 1, 2$.

From the representation, the univariate margins of H_i are all $\exp\{-\psi^{-1}\}$ so that the copula K_i of H_i has the form of an Archimedean copula with generator $\zeta_i = \exp\{-\psi^{-1} \circ \varphi_i\}$ for $i = 1, 2$. From Theorem 8.31, this is well-defined for any d if $\omega_i = -\log \zeta_i = \psi^{-1} \circ \varphi_i \in \mathcal{L}_\infty^*$, where $\mathcal{L}_\infty^*$ is the class of completely monotone functions defined in (2.10).

See Example 3.2 later in this chapter for examples where this condition on ω_i is satisfied when $\psi, \varphi_1, \varphi_2$ all belong to the same parametric LT family.

If (3.14) and (3.15) can be shown to be max-id, then the form of (3.15) can be iterated with C_{S_1} and C_{S_2} having the same form in lower dimensions. See Theorem 8.32 on the conditions for (3.15) to be max-id if C_{S_1} and C_{S_2} are Archimedean copulas, so that (3.15) can be used in the next level of hierarchical nesting.

We describe the general hierarchical Archimedean copula in a form that is like hierarchical clustering and somewhat like h-extendible copulas in Mai and Scherer (2012a); in particular see Figure 2 of the latter.

For the iterative extension of (3.15), let $S_{1,1}, S_{1,2}$ be the first split into two non-empty subsets. If a subset has cardinality of 1 or 2, it need not be split any more. A subset of cardinality 3 or more can be split further if the corresponding set of variables is not exchangeable. We use the following notation to indicate further splits.

- Let $S_{2,1}, \ldots, S_{2,N_2}$ be the subsets after the second split, where $2 < N_2 \leq 2^2$. If $S_{1,a}$ is split, then the two non-empty subsets are say S_{2,a_1}, S_{2,a_2} with $S_{2,a_1} \cup S_{2,a_2} = S_{1,a}$. If $S_{1,a}$ is not split, then $S_{1,a} = S_{2,b}$ for some b.

- Let $S_{l,1}, \ldots, S_{l,N_l}$ be the subsets after the lth split, where $l < N_l \leq 2^l$, for $l = 2, \ldots, L$. If $S_{l-1,a}$ is split, then the two non-empty subsets are say S_{l,a_1}, S_{l,a_2} with $S_{l,a_1} \cup S_{l,a_2} = S_{l-1,a}$. If $S_{l-1,a}$ is not split, then $S_{l-1,a} = S_{l,b}$ for some b.

- The maximum L is $d-2$ in which case $N_l = l+1$.

Let $\varphi_{0,1} = \psi$. Let $\varphi_{1,1}, \varphi_{1,2}$ be the LTs associated to $S_{1,1}, S_{1,2}$ and let $\varphi_{l,1}, \ldots, \varphi_{l,N_l}$ be the LTs associated to $S_{l,1}, \ldots, S_{l,N_l}$. If a subset $S_{l,a}$ has size 1, then $\varphi_{l,a}$ is not needed. With this notation, the extension of (3.15) satisfies:

$$C(\boldsymbol{u}) = \psi\big(\psi^{-1}(C_{S_{1,1}}) + \psi^{-1}(C_{S_{1,2}})\big), \qquad (3.16)$$

$$C_{S_{l,a}} = \varphi_{l,a}\big(\varphi_{l,a}^{-1}(C_{S_{l+1,a_1}}) + \varphi_{l,a}^{-1}(C_{S_{l+1,a_2}})\big), \quad S_{l+1,a_1} \cup S_{l+1,a_2} = S_{l,a}, \; l \geq 1,$$

$$\varphi_{l,a}^{-1} \circ \varphi_{l+1,a_k} \in \mathcal{L}_\infty^*, \text{ for } \ell \geq 0, \; |S_{l+1,a_k}| > 1, \text{ and } k = 1, 2,$$

$$C_{S_{l,a}}(u_j; j \in S_{l,a}) = \varphi_{l,a}\Big(\sum_{j \in S_{l,a}} \varphi_{l,a}^{-1}(u_j)\Big), \quad \text{if } S_{l,a} \text{ is not split.}$$

We describe some properties of the hierarchical Archimedean copulas in (3.15) and (3.16). Each bivariate margin is Archimedean. For the (j,k) margin, the LT is $\varphi = \varphi_{l,a}$ associated with the smallest subset $S_{l,a}$ (largest l) that contains both j, k. If no subset contains both j, k, then the LT is ψ. The condition of $\varphi_{l,a}^{-1} \circ \varphi_{l+1,a_k} \in \mathcal{L}_\infty^*$ means that if (j, k_1) is nested in a smaller subset than (j, k_2), then the (j, k_1) bivariate margin is more concordant than the (j, k_2) margin; see Corollary 8.24. For (3.15), this means that the bivariate (j, k_1) margin has more dependence than (j, k_2) margin if $1 \leq j, k_1 \leq m$ and $k_2 > m$, or if $j, k_1 > m$ and $k_2 \leq m$.

In dimension d, if the smallest subsets all have cardinality 1 or 2, the number of distinct bivariate Archimedean copula margins is $d - 1$ (out of $d(d - 1)/2$ bivariate margins) and this is the most possible. An example for (3.15) with $d = 4$ and $m = 2$ in which case the $d - 1 = 3$ LTs are $\psi, \varphi_1, \varphi_2$. The $(1, 2)$ bivariate margin has LT φ_1, the $(3, 4)$ bivariate margin has LT φ_2, and the remaining four bivariate margins have LT ψ.

There have been several recent papers on hierarchical Archimedean copulas, including McNeil (2008), Hofert (2008), Hering et al. (2010), Savu and Trede (2010), Hofert (2011), and Okhrin et al. (2013). Joe (1993) obtained conditions on $\varphi_{l,a}^{-1} \circ \varphi_{l+1,a_k}$ from the stochastic mixture representation, and listed some examples where the bivariate margins are all in the same parametric Archimedean family, and Hofert (2008) obtained more parametric examples with this property. Hering et al. (2010) use the stochastic representation and have parametric families for the infinitely divisible LTs of the form $\exp\{-\varphi_{l,a}^{-1} \circ \varphi_{l+1,a_k}\}$ without requiring that all bivariate margins be in the same parametric Archimedean family.

The Gumbel copula is an Archimedean and extreme value copula, and it extends to the hierarchical Archimedean form. Joe (1993) also mentions the hierarchical form of the Galambos copula (Section 4.9); the hierarchical Galambos extreme value copula can be obtained as a lower extreme value limit (Section 3.15) of hierarchical Archimedean copulas that have lower tail dependence.

Other hierarchical Archimedean copulas based on 2-parameter LT families (in the Appendix) are indicated in the next example.

Example 3.1 (Hierarchical Archimedean and 2-parameter LT families).
Suppose φ_1, φ_2 are in the same 2-parameter LT family with respective parameters (θ, δ_1) and (θ, δ_2) with $\delta_1 \leq \delta_2$. Let $\omega = \varphi_1^{-1} \circ \varphi_2$. We will show for the cases below that $\omega \in \mathcal{L}_\infty^*$ and then from the above theory, we can construct hierarchical Archimedean copulas. The extra parameter θ allows the construction of parametric hierarchical Archimedean copulas with a wider variety of tail behavior than is available with 1-parameter LT families. If θ is different for φ_1, φ_2, then the condition on ω is not tractable to check.

- Mittag-Leffler LT family, $\varphi_i(s) = (1 + s^{1/\delta_i})^{-1/\theta}$ with $\varphi_i^{-1}(t) = (t^{-\theta} - 1)^{\delta_i}$; $\omega(s) = s^\gamma$, with $\gamma = \delta_1/\delta_2 < 1$, is in $\mathcal{L}_\infty^*$. As a special trivariate case, let $\psi(s) = (1 + s)^{-1/\theta}$ with $\delta_1 = 1$ and let $\varphi(s) = (1 + s^{1/\delta})^{-1/\theta}$ with $\delta = \delta_2 \geq 1$ and $\gamma = 1/\delta$. Then (3.14) becomes

$$C_{123}(\boldsymbol{u}; \theta, \delta) = \left\{ \left[(u_1^{-\theta} - 1)^\delta + (u_2^{-\theta} - 1)^\delta \right]^{1/\delta} + u_3^{-\theta} \right\}^{-1/\theta}, \quad \theta > 0, \delta \geq 1. \quad (3.17)$$

The $(1, 3)$ and $(2, 3)$ margins are (4.10) and the more concordant $(1, 2)$ margin is (4.53). If we substitute the univariate Pareto survival margins $\overline{F}_j(x) = (1 + x_j)^{-1/\theta} = u_j$ for $j = 1, 2, 3$, then the resulting trivariate survival function is:

$$\overline{F}_{123}(x_1, x_2, x_3) = \left[(x_1^\delta + x_2^\delta)^{1/\delta} + x_3 + 1 \right]^{-1/\theta}, \quad x_1, x_2, x_3 \geq 0; \quad (3.18)$$

this example was initially obtained by writing a trivariate Pareto extension of Example 1.3 with hierarchical structure.

- LT family F (gamma stopped gamma), $\varphi_i(s) = [1 + \delta_i^{-1} \log(1 + s)]^{-1/\theta}$ with $\theta > 0$, $\delta_i > 0$: $\omega(s) = (1 + s)^\gamma - 1$, with $\gamma = \delta_1/\delta_2 < 1$, is in $\mathcal{L}_\infty^*$.
- LT family G (positive stable stopped gamma), $\varphi_i(s) = \exp\{-[\delta_i^{-1} \log(1 + s)]^{1/\theta}\}$ with $\theta \geq 1$, $\delta_i > 0$: $\omega(s) = (1 + s)^\gamma - 1$, with $\gamma = \delta_1/\delta_2 < 1$.
- LT family H (Sibuya stopped positive stable), $\varphi_i(s) = 1 - [1 - \exp\{-s^{1/\delta_i}\}]^{1/\theta}$ with $\theta \geq 1$, $\delta_i > 1$: $\omega(s) = s^\gamma$, with $\gamma = \delta_1/\delta_2 < 1$.
- LT family I (Sibuya stopped gamma), $\varphi_i(s) = 1 - [1 - (1 + s)^{-1/\delta_i}]^{1/\theta}$ with $\theta \geq 1$ and $\delta_i > 0$: $\omega(s) = (1 + s)^\gamma - 1$, with $\gamma = \delta_1/\delta_2 < 1$.

3.5   Mixtures of max-id

A multivariate cdf G is *max-infinitely* divisible (max-id) if G^q is a cdf for all $q > 0$. Analogously, a multivariate survival function $\overline{G}$ is *min-infinitely divisible (min-id)* if $\overline{G}^q$ is a survival function for all $q > 0$. Max-infinite and min-infinite divisibility are strong positive dependence conditions; max-id is satisfied by all extreme value copulas and both max-id/min-id are satisfied by bivariate Archimedean copulas C_ψ with $(\log \psi)' \leq 0$ and $(\log \psi)'' \geq 0$; in particular for those based on a LT ψ.

The idea of mixing max-id or min-id distributions to form copulas is studied in detail in Joe and Hu (1996), but earlier ideas of such mixtures appear in Marshall and Olkin (1990).

Let K be a max-id d-variate copula, and let $\psi = \psi_Q \in \mathcal{L}_\infty$ be the LT of a resilience random variable. Then

$$F_{\psi, K} = \int_0^\infty K^q \mathrm{d}F_Q(q) = \psi(-\log K)$$

is a multivariate distribution, and it has copula

$$C_{\psi, K}(\boldsymbol{u}) = \psi\Big(-\log K\big(\exp\{-\psi^{-1}(u_1)\}, \ldots, \exp\{-\psi^{-1}(u_d)\}\big)\Big), \quad \boldsymbol{u} \in [0,1]^d. \qquad (3.19)$$

Note that min-id does not lead to a new construction. If K is min-id, so that $\overline{K}^q$ is a survival function for all $q > 0$, then

$$\overline{F}_{\psi, K} = \int_0^\infty \overline{K}^q \mathrm{d}F_Q(q) = \psi(-\log \overline{K})$$

is a multivariate survival function, and then

$$\overline{C}_{\psi, \overline{K}}(\boldsymbol{u}) = \psi\Big(-\log \overline{K}\big(\exp\{-\psi^{-1}(u_1)\}, \ldots, \exp\{-\psi^{-1}(u_d)\}\big)\Big), \quad \boldsymbol{u} \in [0,1]^d,$$

is the survival function of a copula, and the copula is

$$\widehat{C}_{\psi, \overline{K}}(\boldsymbol{u}) = \psi\Big(-\log \overline{K}\big(\exp\{-\psi^{-1}(1-u_1)\}, \ldots, \exp\{-\psi^{-1}(1-u_d)\}\big)\Big)$$
$$= \psi\Big(-\log \widehat{K}\big(\exp\{-\psi^{-1}(u_1)\}, \ldots, \exp\{-\psi^{-1}(u_d)\}\big)\Big) = C_{\psi, \widehat{K}}(\boldsymbol{u}),$$

where $\widehat{K}$ is the reflected copula of K and is max-id.

This construction generalizes Archimedean copulas to copula families with more flexible positive dependence than exchangeable, and they are advantageous to use if copula cdfs with closed analytic forms are desirable. For bivariate with $d = 2$, there are lots of choices for ψ and K. Some 2-parameter bivariate copula families are constructed in this way, one parameter δ for K and one parameter θ for ψ. This includes the BB1–BB7 families in Joe (1997), originally from Joe and Hu (1996).

For $d \geq 2$, there are results in Chapter 4 of Joe (1997) with sufficient conditions are a multivariate copula to be max-id. Max-id copulas are positively dependent (Theorems 8.7 and 8.5) and hence the mixture (3.19) is also positively dependent. However, different notions of positive dependence in Chapter 8 require different assumptions for the proofs. As K increases in the PLOD ordering with ψ fixed, then so does $C_{\psi, K}$.

Special cases are given next to show that flexibility of the construction in (3.19).

- Suppose $d = 2$ and K is bivariate Archimedean with LT ζ. Then (3.19) becomes:

$$C_{\psi, K}(u_1, u_2) = \psi\big(-\log \zeta\big[\zeta^{-1}(e^{-\psi^{-1}(u_1)}) + \zeta^{-1}(e^{-\psi^{-1}(u_2)})\big]\big) = \varphi\big(\varphi^{-1}(u_1) + \varphi^{-1}(u_2)\big).$$

This is a bivariate Archimedean copula with generator

$$\varphi(s) = \psi(-\log \zeta(s)), \quad \varphi^{-1}(t) = \zeta^{-1}(e^{-\psi^{-1}(t)}). \tag{3.20}$$

By construction, this is positively dependent and hence $-\log \varphi$ is log concave, and $\varphi \in \mathcal{L}_2$ (see Theorem 8.84). By Theorem 8.81, φ is a LT if ζ is the LT of an infinitely divisible distribution (or $-\log \zeta \in \mathcal{L}_\infty^*$).

- Suppose $d = 2$ and $K(v_1, v_2) = e^{-A(-\log v_1, -\log v_2)}$ is a bivariate extreme value copula with exponent function A (see Section 3.15). Then (3.19) becomes:

$$C_{\psi,K}(u_1, u_2) = \psi(A(\psi^{-1}(u_1), \psi^{-1}(u_2))).$$

This matches the form of the bivariate Archimax copula in Capéraà et al. (2000); this form is valid with $\psi \in \mathcal{L}_2$.

- Max-id copulas are positively dependent, so the boundary copulas in this class are $K = C^\perp$ and $K = C^+$, and they are both max-id. Their use in (3.19) leads to $C_{\psi,\text{independence}} = C_\psi$, an Archimedean copula, and $C_{\psi,\text{comonotonicity}} = C^+$.

- If H_{ij} is a max-id bivariate copula for $1 \le i < j \le d$, then the product $\prod_{1 \le i < j \le d} H_{ij}$ is a multivariate cdf, leading to max-id copula

$$K(\boldsymbol{v}) = \prod_{1 \le i < j \le d} H_{ij}(v_i^{1/(d-1)}, v_j^{1/(d-1)}).$$

Then (3.19) becomes

$$C_{\psi,K}(\boldsymbol{u}) = \psi\Big(-\sum_{i<j} \log H_{ij}(\exp\{-\psi^{-1}(u_i)/[d-1]\}, \exp\{-\psi^{-1}(u_j)/[d-1]\})\Big).$$

Each bivariate marginal copula is more concordant than the bivariate Archimedean copula C_ψ. An interpretation is that the LT ψ leads to a minimal level of (pairwise) dependence, and the copulas H_{ij} add some individual pairwise dependence beyond the global dependence.

- In the product, H_{ij} could be included only for a subset of pairs. For example with $d = 3$, if H_{12} and H_{32} are max-id bivariate copulas, then $K(v_1, v_2, v_3) = H_{12}(v_1, v_2^{1/2}) H_{32}(v_3, v_2^{1/2})$ is a trivariate max-id copula, and (3.19) leads to a trivariate copula with bivariate margins

$$C_{j2}(u_j, u_2) = \psi\Big(-\log H_{j2}(e^{-\psi^{-1}(u_j)}, e^{-0.5\psi^{-1}(u_2)}) + \tfrac{1}{2}\psi^{-1}(u_2)\Big), \quad j = 1, 3, \tag{3.21}$$

and $C_{13}(u_1, u_3) = \psi(\psi^{-1}(u_1) + \psi^{-1}(u_3))$ is a bivariate Archimedean copula.
If $H_{12} = H_{32} = H$, then $C_{12} = C_{32}$, and $C_{13} \prec_c C_{12}$, because H max-id implies that it is PQD and then

$$-\log H(e^{-\psi^{-1}(u)}, e^{-0.5\psi^{-1}(v)}) + \tfrac{1}{2}\psi^{-1}(v) \le \psi^{-1}(u) + \psi^{-1}(v).$$

Note that (3.21) is example of a bivariate copula that is not permutation symmetric in general. Even if H is permutation symmetric, C_{32} is not permutation symmetric.

- For comparison, if H_{12}, H_{32} are max-id bivariate copulas, and $H_{13}(u_1, u_3) = u_1 u_3$, then $K(v_1, v_2, v_3) = H_{12}(v_1^{1/2}, v_2^{1/2}) H_{32}(v_3^{1/2}, v_2^{1/2}) (v_1 v_3)^{1/2}$ is a trivariate max-id copula, and (3.19) leads to a trivariate copula with bivariate margins:

$$C_{j2} = \psi\Big(-\log H_{j2}(e^{-0.5\psi^{-1}(u_j)}, e^{-0.5\psi^{-1}(u_2)}) + \tfrac{1}{2}\psi^{-1}(u_j) + \tfrac{1}{2}\psi^{-1}(u_2)\Big), \quad j = 1, 3,$$

and $C_{13}(u_1, u_3) = \psi(\psi^{-1}(u_1) + \psi^{-1}(u_3))$ is a bivariate Archimedean copula. If $H_{12} = H_{23} = H$ is permutation symmetric, then $C_{12} = C_{32} = C_{23}$, and $C_{13} \prec_c C_{12}$, then

$$C_{\psi,K}(\boldsymbol{u}) = \psi\Big(\sum_{j \in \{1,3\}} \Big[-\log H\big(e^{-0.5\psi^{-1}(u_j)}, e^{-0.5\psi^{-1}(u_2)}\big) + \tfrac{1}{2}\psi^{-1}(u_j) \Big] \Big). \qquad (3.22)$$

This $C_{\psi,K}$ is a suitable copula for the transition of a stationary time series of Markov order 2, when there is more dependence for measurements at nearer time points. An example of such a use is in Section 11.5 of Joe (1997).

- For a 4-variable copula suitable for the transition of a stationary time series of Markov order 3, let H_1, H_2 be two permutation symmetric max-id bivariate copulas, and take

$$K(\boldsymbol{v}) = H_1(v_1^{1/3}, v_2^{1/3})\, H_1(v_2^{1/3}, v_3^{1/3})\, H_1(v_3^{1/3}, v_4^{1/3})\, H_2(v_1^{1/3}, v_3^{1/3})\, H_2(v_2^{1/3}, v_4^{1/3})\, (v_1 v_4)^{1/3}.$$

Its bivariate margins are $K_{12} = K_{23} = K_{34} = H_1(u^{1/3}, v^{1/3})(uv)^{2/3}$, $K_{13} = K_{24} = H_2(u^{1/3}, v^{1/3})(uv)^{2/3}$, and $K_{14}(u, v) = uv$. If $H_2 \prec_c H_1$, then $K_{13} \prec_c K_{12}$.

- For another trivariate example, let K_{12} be a max-id bivariate copula, so that it is the bivariate margin of $K(v_1, v_2, v_3) = K_{12}(v_1, v_2)\, v_3$, which is a max-id trivariate copula. Then (3.19) leads to a trivariate copula

$$C_{\psi,K}(\boldsymbol{u}) = \psi\Big(-\log K_{12}(e^{-\psi^{-1}(u_1)}, e^{-\psi^{-1}(u_2)}) + \psi^{-1}(u_3) \Big) \qquad (3.23)$$

with bivariate margins $C_{13}(u, v) = C_{23}(u, v) = \psi(\psi^{-1}(u) + \psi^{-1}(v))$. This is less concordant than $C_{12}(u, v) = \psi\big(-\log K_{12}(e^{-\psi^{-1}(u_1)}, e^{-\psi^{-1}(u_2)}) \big)$ because K_{12} is max-id and PQD. Next take a special case where $K_{12}(u_1, u_2) = \zeta(\zeta^{-1}(u_1) + \zeta^{-1}(u_2))$, a bivariate Archimedean copula based on LT $\zeta \in \mathcal{L}_\infty$. This K_{12} is max-id due to Theorems 8.28 and 8.5. Then (3.23) becomes

$$C_{\psi,K}(\boldsymbol{u}) = \psi\Big(-\log \zeta\big[\zeta^{-1}(e^{-\psi^{-1}(u_1)}) + \zeta^{-1}(e^{-\psi^{-1}(u_2)}) \big] + \psi^{-1}(u_3) \Big)$$

$$= \psi\Big(\psi^{-1} \circ \varphi[\varphi^{-1}(u_1) + \varphi^{-1}(u_2)] + \psi^{-1}(u_3) \Big), \qquad (3.24)$$

where $\varphi^{-1}(t), \varphi(s)$ are given in (3.20). For (3.24), $C_{13} = C_{23}$ are Archimedean copulas based on ψ and and C_{12} is an Archimedean copula based on φ. This has the form of a hierarchical Archimedean in Section 3.4.

- The preceding example extends to $K(\boldsymbol{v}) = K_{1 \cdots m}(v_1, \ldots, v_m)\, K_{m+1 \cdots d}(v_{m+1}, \ldots, v_d)$, where $2 \leq m \leq d - 2$, $d \geq 2$, if $K_{1 \cdots m}$ and $K_{m+1 \cdots d}$ are both max-id. If both of these are Archimedean copulas, then (3.19) has a hierarchical Archimedean form. This type of copula might be a reasonable model if the variables can be separated into two clusters that are conditionally independent given a resilience (or frailty) random variable.

- If K is an extreme value copula, then it is max-id. If $\psi(s) = \exp\{-s^{1/\theta}\}$ is the positive stable LT with parameter $\theta > 1$ and K is a d-variate extreme value copula with $-\log K(\boldsymbol{v}) = A_K(\boldsymbol{x})$ with $x_j = -\log v_j$, then (3.19) is an extreme value copula

$$\psi\big(A_K[\psi^{-1}(u_1), \ldots, \psi^{-1}(u_d)] \big) = \exp\{ -A_K^{1/\theta}[(-\log u_1)^\theta, \ldots, (-\log u_d)^\theta] \}$$

with exponent function

$$A_C(\boldsymbol{x}) = A_K^{1/\theta}(x_1^\theta, \ldots, x_d^\theta), \qquad \theta \geq 1.$$

When K is derived from a product of bivariate extreme value copulas, this construction leads to some closed form multivariate extreme value copulas — families MM1, MM3 in Joe and Hu (1996) and Joe (1997) — with a different dependence parameter for each bivariate margin; see Section 4.8.

- In order to have a form that is closed under margins, K can be the copula from a product of bivariate max-id copulas H_{ij} and univariate beta distributions $v_i^{\gamma_i}$, leading to

$$K(\boldsymbol{v}) = \prod_{1 \le i < j \le d} H_{ij}(v_i^{1/(d-1+\gamma_i)}, v_j^{1/(d-1+\gamma_j)}) \prod_{i=1}^{d} v_i^{\gamma_i/(d-1+\gamma_i)},$$

where $\gamma_i \ge 0$, $i = 1, \ldots, d$.

- If ψ is a LT of a positive integer-valued random variable, then (3.19) is a copula for any K, that is, the requirement of max-id on K is not needed. It is mentioned in Section 4.4 of Joe (1997) when the functional form (3.19) can be a copula without ψ being a LT or K being max-id.

Example 3.2 (Links among LT families).
Let's take a closer look at (3.23) and (3.24) when ψ is in the positive stable, gamma, Sibuya and logarithmic series LT families. This will show more links among the LT families, and how (3.24) is a generalization of the hierarchical Archimedean construction.

Suppose ψ, φ are in the same LT family with respective parameters $\theta_1 < \theta_2$ and let $\gamma = \theta_1/\theta_2 < 1$. If (3.24) is satisfied, so that $\varphi = \psi(-\log \zeta)$ for a LT ζ and $\zeta(s) = e^{-\omega(s)}$, $\omega = \psi^{-1} \circ \varphi$.

The four special cases are the following.

- Positive stable LT, $\exp\{-s^{1/\theta}\}$: $\omega(s) = s^\gamma$ is easily shown to be in $\mathcal{L}_\infty^*$. Note that $e^{-\omega(s)}$ is again a positive stable LT.
- Sibuya LT, $1 - (1 - e^{-s})^{1/\theta}$: $\omega(s) = -\log\{1 - (1 - e^{-s})^\gamma\}$. Since $e^{-\omega(s)}$ is again a Sibuya LT, which is infinitely divisible, then $\omega \in \mathcal{L}_\infty^*$ by Theorem 8.81.
- Gamma LT, $(1 + s)^{-1/\theta}$: $\omega(s) = (1 + s)^\gamma - 1$ is easily shown to be in $\mathcal{L}_\infty^*$. Hence $e^{-\omega(s)}$ is the LT of an infinitely divisible random variable, and it corresponds to LT family L (exponentially tilted positive stable) in the Appendix.
- Logarithmic series LT, $-\theta^{-1}\log[1 - (1 - e^{-\theta})e^{-s}]$:

$$\omega(s) = -\log\{(1 - [1 - (1 - e^{-\theta_2})e^{-s}]^\gamma)/(1 - e^{-\theta_1})\}.$$

It is shown in the Appendix of Joe (1997) that $e^{-\omega(s)}$ is the LT of an infinitely divisible random variable, so that $\omega \in \mathcal{L}_\infty^*$. With a reparametrization, $e^{-\omega(s)}$ corresponds to LT family J (generalized Sibuya) in the Appendix.

Example 3.3 (Functional form with K not max-id).
If ψ is a logarithmic series LT with parameter $\theta > 0$, then K in (3.23) could be a Frank copula based on $\zeta \in \mathcal{L}_2$ being an extension of the logarithmic series LT with positive or negative parameter γ. The $(1, 2)$ bivariate margin of (3.24) is Archimedean with generator $\varphi = \psi(-\log \zeta) \in \mathcal{L}_2$:

$$\varphi(s) = -\theta^{-1}\log\{1 + \gamma^{-1}(1 - e^{-\theta})\log[1 - (1 - e^{-\gamma})e^{-s}]\},$$

$$\varphi^{-1}(t) = -\log\left[\frac{1 - e^{-\gamma(1 - e^{-\theta t})/(1 - e^{-\theta})}}{1 - e^{-\gamma}}\right], \quad \vartheta = 1 - e^{-\theta},$$

$$C_{12}(u, v) = -\frac{1}{\theta}\log\left\{1 + \frac{\vartheta}{\gamma}\log\left[1 - \frac{(1 - e^{-\gamma(1 - e^{-\theta u})/(1 - e^{-\theta})})(1 - e^{-\gamma(1 - e^{-\theta v})/(1 - e^{-\theta})})}{1 - e^{-\gamma}}\right]\right\}.$$

If K is a Frank copula, then the $(1, 2)$ margin of (3.24), as shown above, is not a Frank copula, but the other two bivariate margins are. If $\gamma < 0$, the above $(1, 2)$ bivariate margin C_{12} can be less concordant than the common $(1, 3)$ and $(2, 3)$ bivariate margins (Frank with parameter $\theta > 0$). Contrast this with the trivariate hierarchical Archimedean copula construction where the two bivariate margins that are common are less concordant than the third.

The above special cases and examples show that the construction (3.19) has some flexibility of dependence and closed form cdfs. Hence mixture of max-id copulas can be convenient as models for low-dimensional multivariate discrete data. The construction leads to some 2-parameter positively dependent bivariate copula families with two parameters that can allow for more flexible upper and lower tail behavior; the most useful parametric families of this type are included in Chapter 4. Also, it leads to copula families with more flexible dependence structures than hierarchical Archimedean copulas, but not as much as vine copulas (Section 3.9).

From members of (3.19) with upper or lower tail dependence, some limiting extreme value copulas have closed form cdf and some flexible dependence beyond exchangeable or partially exchangeable; see (4.17), (4.18) and (4.65) which were originally in Joe and Hu (1996) and Joe (1997).

Other properties for the family of mixtures of max-id copulas are given in Section 8.7.

3.6 Another limit for max-id distributions

If K is a max-id copula, then

$$C(\boldsymbol{u}; \eta) = K^{1/\eta}(u_1^\eta, \ldots, u_d^\eta), \quad \boldsymbol{u} \in [0.1]^d,$$

is a copula for all $\eta > 0$. If K is an extreme value copula, $C(\cdot; \eta) = K$ for all $\eta > 0$. If K is not an extreme value copula, then $\eta \to 0^+$ corresponds to the extreme value limit. For a max-id copula that is not an extreme value copula, the limit as $\eta \to \infty$ could also be interesting. Results of this theory are some 3-parameter copula families that are summarized in Section 4.31.

In the bivariate case, if K is max-id then K has TP$_2$ cdf so that it is left-tail decreasing (see Theorems 8.7 and 8.5); the stronger dependence property of SI cannot be proved in general from the max-id assumption.

If K has upper tail dependence, then as $\eta \to 0^+$, $C(\cdot; \eta)$ converges to an extreme value copula that is not $C^\perp$, and $\lim_{\eta \to \infty} C(\cdot; \eta)$ might have weaker dependence (such as $C^\perp$). If K does not have upper tail dependence, then $\lim_{\eta \to 0^+} C(\cdot; \eta) = C^\perp$, and there is a possibility that $C(\cdot; \eta)$ is increasing in concordance as η increases.

For bivariate Archimedean copulas, $\lim_{\eta \to \infty} C(\cdot; \eta)$ is determined by $\lim_{s \to \infty} \psi'(s)/\psi(s)$ or the behavior of $\psi(s)$ as $s \to \infty$. If the limiting ratio is positive, then $\lim_{\eta \to \infty} C(\cdot; \eta) = C^\perp$ (e.g., Frank with logarithmic series LT, Joe/B5 with Sibuya LT). If the limiting ratio is zero, then $\lim_{\eta \to \infty} C(\cdot; \eta) = C^+$ (e.g., MTCJ with gamma LT, or Gumbel with positive stable LT). For Archimedean copulas, the conditions for a limit of $C^\perp$ or C^+ are similar to those for classification of the lower tail order (see Theorem 8.37).

For a copula K that is a reflection of an Archimedean copula, then $\lim_{\eta \to \infty} C(\cdot; \eta)$ depends on the behavior $1 - \psi(s)$ as $s \to 0^+$; the conditions are similar to those for upper tail behavior of Archimedean copulas (see Theorems 8.33 and 8.35).

A summary for bivariate Archimedean and reflected Archimedean copulas is given Table 3.2 followed by some derivations. For notation, we use:

$$C_{\psi,P}(u,v;\eta) = K^{1/\eta}(u^\eta, v^\eta) \text{ if } K(u,v) = \psi(\psi^{-1}(u) + \psi^{-1}(v)), \tag{3.25}$$

$$C_{\psi,R}(u,v;\eta) = K^{1/\eta}(u^\eta, v^\eta) \text{ if } K(u,v) = u + v - 1 + \psi(\psi^{-1}(1-u) + \psi^{-1}(1-v)). \tag{3.26}$$

ψ at ∞	ψ at 0	$\psi^{1/\eta}[\psi^{-1}(u^\eta) + \psi^{-1}(v^\eta)]$		$\{u^\eta + v^\eta - 1 + \psi[\psi^{-1}(1-u^\eta)$ $+\psi^{-1}(1-v^\eta)]\}^{1/\eta}$	
		$\eta \to 0^+$	$\eta \to \infty$	$\eta \to 0^+$	$\eta \to \infty$
(a)	(i)	$C^\perp$	$C^\perp$	$C^\perp$	$C^\perp$
(a)	(ii)	$C^\perp$	$C^\perp$	$C^\perp$	C_{mix}
(a)	(iii)	$\mathrm{Gumbel}(\xi^{-1})$	$C^\perp$	$C^\perp$	C^+
(b)	(i)	$C^\perp$	$\mathrm{Gumbel}(\gamma^{-1})$	$C^\perp$	$C^\perp$
(b)	(ii)	$C^\perp$	$\mathrm{Gumbel}(\gamma^{-1})$	$C^\perp$	C_{mix}
(b)	(iii)	$\mathrm{Gumbel}(\xi^{-1})$	$\mathrm{Gumbel}(\gamma^{-1})$	$C^\perp$	C^+
(c)	(i)	$C^\perp$	C^+	$\mathrm{Galambos}(\zeta^{-1})$	$C^\perp$
(c)	(ii)	$C^\perp$	C^+	$\mathrm{Galambos}(\zeta^{-1})$	C_{mix}
(c)	(iii)	$\mathrm{Gumbel}(\xi^{-1})$	C^+	$\mathrm{Galambos}(\zeta^{-1})$	C^+

Table 3.2 *The lower tail behavior of $\psi(\psi^{-1}(u) + \psi^{-1}(v))$ depends on the the upper tail behavior of the LT ψ; this is separated into 3 cases: (a) $\psi(s) \sim a_1 e^{-a_2 s}$ as $s \to \infty$; (b) $\psi(s) \sim a_1 s^b e^{-a_2 s^\gamma}$ as $s \to \infty$ where $0 < \gamma < 1$ and $a_1 > 0$, $a_2 > 0$; (c) $\psi(s) \sim a_1 s^{-\zeta}$ as $s \to \infty$ where $\zeta > 0$ and $a_1 > 0$. The upper tail behavior of $\psi(\psi^{-1}(u) + \psi^{-1}(v))$ depends on the the lower tail behavior of the LT ψ; this is separated into 3 cases: (i) $1 - \psi(s) \sim a_1's - a_2's^2$ as $s \to 0^+$ where $a_1' > 0$, $a_2' > 0$; (ii) $1 - \psi(s) \sim a_1's - a_\kappa s^\kappa$ as $s \to 0^+$ where $1 < \kappa < 2$ and $a_1' > 0$, $a_\kappa > 0$; (iii) $1 - \psi(s) \sim a_0 s^\xi$ as $s \to 0^+$ where $0 < \xi < 1$. Within this table, Gumbel refers the bivariate Gumbel copula, Galambos refers the bivariate Galambos copula, and $C_{\mathrm{mix}}(u,v;\kappa) = (uv)^{\kappa-1}(u \wedge v)^{2-\kappa} = (C^\perp)^{\kappa-1}(C^+)^{2-\kappa}$ for $1 < \kappa < 2$ is a reparametrization of (4.37). The limits in columns 3 and 5 follow from Theorems 8.33 and 8.34.*

If the 1-parameter copula family $C(\cdot; \delta)$ has no upper tail dependence and conditions are satisfied so $C^{1/\eta}(u^\eta, v^\eta; \delta) \to C^+$ as $\eta \to \infty$, then possibly a 2-parameter family that is increasing in concordance in two parameters might result. Table 3.2 shows that there are choices of $C(\cdot; \delta)$ from Archimedean or reflected Archimedean copulas.

We next give an outline of limiting results in columns 4 and 6 of Table 3.2.

Column 4 of Table 3.2. (Derivations for the limit of $C_{\psi,P}$).
For the limit as $\eta \to \infty$, apply l'Hopital's rule to get:

$$\lim_{\eta \to \infty} \frac{\log \psi[\psi^{-1}(u^\eta) + \psi^{-1}(v^\eta)]}{\eta}$$
$$= \lim_{\eta \to \infty} \frac{\psi'[\psi^{-1}(u^\eta) + \psi^{-1}(v^\eta)]}{\psi[\psi^{-1}(u^\eta) + \psi^{-1}(v^\eta)]} \cdot \left\{ (\psi^{-1})'(u^\eta) u^\eta \log u + (\psi^{-1})'(v^\eta) v^\eta \log v \right\}$$
$$= \lim_{\eta \to \infty} \frac{\psi'(s_1 + s_2)}{\psi(s_1 + s_2)} \cdot \left\{ \frac{\psi(s_1)}{\psi'(s_1)} \log u + \frac{\psi(s_2)}{\psi'(s_2)} \log v \right\}, \tag{3.27}$$

where $s_1 := \psi^{-1}(u^\eta) \to \infty$, $s_2 := \psi^{-1}(v^\eta) \to \infty$ as $\eta \to \infty$.
We separate into three cases (a), (b), (c) as in Table 3.2.

(a) If $-\lim_{s \to \infty} \psi'(s)/\psi(s) = r_\infty > 0$, then the limit in (3.27) is $\log(uv)$ and the limit of $C(\cdot; \eta)$ is $C^\perp$ as $\eta \to \infty$. This holds if LT has the form $\psi(s) \sim a_1 e^{-a_2 s}$ as $s \to \infty$ and $\psi'(s) \sim [a_1 e^{-a_2 s}]'$ for some $a_1, a_2 > 0$.

(b) If $\psi(s) \sim T(s) = a_1 s^b \exp\{-a_2 s^\gamma\}$ and $\psi'(s) \sim T'(s)$ as $s \to \infty$, where $0 < \gamma < 1$, then $-\psi'(s)/\psi(s) \sim a s^{\gamma-1}$ where $a = a_2 \gamma$. The limit of ψ'/ψ is 0 as $s \to \infty$ but the rate is slower than $O(s^{-1})$. The limit of (3.27) is a positively dependent copula that is the Gumbel copula with parameter $\gamma^{-1} > 1$.
Proof: For large η, (3.27) is approximately:

$$(s_1 + s_2)^{\gamma-1}\{s_1^{1-\gamma} \log u + s_2^{1-\gamma} \log v\}. \tag{3.28}$$

The condition on ψ implies that $\psi^{-1}(t) \sim [a_2^{-1}(-\log t)]^{1/\gamma} \ell^*(-\log t)$ as $t \to 0^+$, where ℓ^* is slowly varying. Hence, as $\eta \to \infty$, (3.28) converges to:

$$-[(-\log u)^{1/\gamma} + (-\log v)^{1/\gamma}]^{\gamma-1} \{(-\log u)^{1/\gamma} + (-\log v)^{1/\gamma}\} = -[(-\log u)^{1/\gamma} + (-\log v)^{1/\gamma}]^\gamma.$$

The exponential of this is the Gumbel copula with parameter γ^{-1}.

(c) If $-\lim_{s\to\infty} \psi'(s)/\psi(s) = 0$, then the limit depends on the rate of ψ'/ψ for 3 different arguments.

If $\psi(s) \in RV_{-\zeta}$ as $s \to \infty$ with $\zeta > 0$ and $-\psi'(s)/\psi(s) = O(s^{-1})$, then the limit of (3.27) is C^+. Note that $\psi(s) \in RV_{-\zeta}$ with monotone tail implies $\psi'(s) \in RV_{-\zeta-1}$ (Theorem 8.90) and $-\psi'(s)/\psi(s) = O(s^{-1})$.

Proof: : Let $-\psi'(s)/\psi(s) = as^{-1}$ as $s \to \infty$ for some $a > 0$. Then for large η, (3.27) is approximately:

$$(s_1 + s_2)^{-1}\{s_1 \log u + s_2 \log v\} = (1 + s_2/s_1)^{-1}\{\log u + s_2 s_1^{-1} \log v\}$$
$$= (1 + s_1/s_2)^{-1}\{\log v + s_1 s_2^{-1} \log u\}. \tag{3.29}$$

The condition on ψ implies $\psi^{-1}(t) \in RV_{-1/\zeta}$ as $t \to 0^+$. For $u < v$, then $s_2 s_1^{-1} = \psi^{-1}(v^\eta)/\psi^{-1}(u^\eta) \approx (u/v)^{\eta/\zeta}$ and (3.29) has a limit of $\log u$. Similarly, the limit of (3.29) is $\log v$ if $u > v$.

The different possibilities occur with the following four 1-parameter LT families.

A. Positive stable: $\psi(s) = \exp\{-s^{1/\theta}\}$, $\theta > 1$: $-\psi'(s)/\psi(s) \sim \theta^{-1} s^{1/\theta-1}$ as $s \to \infty$ (exponent is between -1 and 0).
B. Gamma $\psi(s) = (1+s)^{-1/\theta}$, $\theta > 0$: $-\psi'(s)/\psi(s) \sim \theta^{-1}(1+s)^{-1} \sim \theta^{-1} s^{-1}$ as $s \to \infty$.
C. Sibuya: $\psi(s) = 1 - (1-e^{-s})^{1/\theta}$: $-\psi'(s)/\psi(s) \sim \theta^{-1}(1-e^{-s})^{1/\theta-1} e^{-s}/[\theta^{-1} e^{-s}] \to 1$ as $s \to \infty$.
D. Logarithmic series: $\psi(s) = -\theta^{-1} \log[1-(1-e^{-\theta})e^{-s}]$: $-\psi'(s)/\psi(s) \sim [1-(1-e^{-\theta})e^{-s}]^{-1} \to 1$ as $s \to \infty$.

Column 6 of Table 3.2. (Derivations for the limit of $C_{\psi,R}$).

We want the limit as $\eta \to \infty$ of

$$\{u^\eta + v^\eta - 1 + \psi[\psi^{-1}(1-u^\eta) + \psi^{-1}(1-v^\eta)]\}^{1/\eta}. \tag{3.30}$$

We separate into three cases (i), (ii), (iii) as in Table 3.2.

(i) If $1 - \psi(s) \approx a_1 s - a_2 s^2 + o(s^2)$ as $s \to 0^+$ with $a_1 > 0$, then the limit of (3.30) is $C^\perp$.
Proof: $\psi^{-1}(1-t) \sim t/a_1 + b_2 t^2 + o(t^2)$ as $t \to 0^+$, where $b_2 = a_2/a_1^3$. As $\eta \to \infty$, (3.30) becomes approximately

$$\{u^\eta + v^\eta - 1 + \psi[(u^\eta/a_1) + (v^\eta/a_1) + b_2 u^{2\eta} + b_2 v^{2\eta}]\}^{1/\eta}$$
$$\sim [u^\eta + v^\eta - u^\eta - v^\eta - a_1 b_2 u^{2\eta} - a_1 b_2 v^{2\eta} + a_2(u^\eta/a_1 + v^\eta/a_1)^2]^{1/\eta}$$
$$\sim [(a_2/a_1^2 - a_1 b_2)(u^{2\eta} + v^{2\eta}) + 2a_2 a_1^{-1} u^\eta v^\eta]^{1/\eta} = (2a_2 a_1^{-1})^{1/\eta} uv \to uv.$$

(ii) If $1 - \psi(s) \sim a_1 s - a_\kappa s^\kappa + o(s^\kappa)$ as $s \to 0^+$ with $a_1 > 0$, $a_\kappa > 0$, and $1 < \kappa < 2$, then the limit of (3.30) is $C_{\text{mix}}(u,v;\kappa) = (uv)^{\kappa-1}(u \wedge v)^{2-\kappa} = (C^\perp)^{\kappa-1}(C^+)^{2-\kappa}$.
Proof: : $\psi^{-1}(1-t) \sim t/a_1 + b_\kappa t^\kappa + o(t^\kappa)$ as $t \to 0^+$, where $b_\kappa = a_\kappa/a_1^{1+\kappa}$. As $\eta \to \infty$,

(3.30) becomes approximately

$$\left\{u^\eta + v^\eta - 1 + \psi[(u^\eta/a_1) + (v^\eta/a_1) + b_\kappa u^{\kappa\eta} + b_\kappa v^{\kappa\eta}]\right\}^{1/\eta}$$
$$\sim \left[-a_1 b_\kappa(u^{\kappa\eta} + v^{\kappa\eta}) + a_\kappa(u^\eta/a_1 + v^\eta/a_1)^\kappa\right]^{1/\eta}$$
$$= (a_\kappa a_1^{-\kappa})^{1/\eta}\left[(u^\eta + v^\eta)^\kappa - u^{\kappa\eta} - v^{\kappa\eta}\right]^{1/\eta}$$
$$= (a_\kappa a_1^{-\kappa})^{1/\eta}u^\kappa\left[(1 + (v/u)^\eta)^\kappa - 1 - (v/u)^{\kappa\eta}\right]^{1/\eta}$$
$$= (a_\kappa a_1^{-\kappa})^{1/\eta}v^\kappa\left[(1 + (u/v)^\eta)^\kappa - 1 - (u/v)^{\kappa\eta}\right]^{1/\eta}.$$

If $u > v$, the limit is

$$(a_\kappa a_1^{-\kappa})^{1/\eta}u^\kappa\left[\kappa(v/u)^\eta - (v/u)^{\kappa\eta}\right]^{1/\eta} \to u^{\kappa-1}v,$$

and if $u < v$, the limit is $v^{\kappa-1}u$.

(iii) If $1 - \psi(s) \sim a_0 s^\xi$ as $s \to 0^+$ with $0 < \xi < 1$, then the limit of (3.30) is C^+.
   Proof: $\psi^{-1}(1 - t) \sim (t/a_0)^{1/\xi}$ as $t \to 0^+$, and as $\eta \to \infty$, (3.30) is approximately

$$\left\{u^\eta + v^\eta - a_0[(u^\eta/a_0)^{1/\xi} + (v^\eta/a_0)^{1/\xi}]^\xi\right\}^{1/\eta} = \left\{u^\eta + v^\eta - [u^{\eta/\xi} + v^{\eta/\xi}]^\xi\right\}^{1/\eta}$$
$$= u\left\{1 + (v/u)^\eta - [1 + (v/u)^{\eta/\xi}]^\xi\right\}^{1/\eta} = v\left\{1 + (u/v)^\eta - [1 + (u/v)^{\eta/\xi}]^\xi\right\}^{1/\eta}.$$

If $u > v$, then

$$u[(v/u)^\eta - \xi(v/u)^{\eta/\xi}]^{1/\eta} \to v, \quad \eta \to \infty,$$

and similarly the limit is u if $u < v$. □

For bivariate copulas other than Archimedean or reflected Archimedean, or for multivariate copulas K that are max-id, it can be shown the limit of $C(\boldsymbol{u}; \eta) = K^{1/\eta}(u_1^\eta, \ldots, u_d^\eta)$ as $\eta \to \infty$ is an extreme value copula. In Table 3.2, the Gumbel and C_{mix} are extreme value copulas.

The most interesting 3-parameter families to come from (3.25) and (3.26) are summarized in Section 4.31.4 with the BB7 and reflected BB1 copulas for K respectively. For another example, the Frank copula with positive dependence satisfies the category (a)(i), in Table 3.2, so that $C^\perp$ obtains as $\eta \to 0^+$ and as $\eta \to \infty$. Numerical evaluations show that the most dependence occurs for a η value larger than 1. A more interesting example that is not Archimedean is given below.

Example 3.4 The bivariate Gaussian distribution/copula with positive dependence is max-id, so one can consider the family: $C(u, v; \eta) = \Phi_2^{1/\eta}(\Phi(u^\eta), \Phi(v^\eta); \rho)$ for $\eta > 0$. Because the bivariate Gaussian distribution does not have tail dependence, $\lim_{\eta \to 0+} C(u, v; \eta) = uv$ for $0 < u, v < 1$. Numerically, where $C(u, v; \eta)$ can be computed before η gets too large, $C(u, v; \eta)$ appears to be increasing in concordance as η increases. For the other limit, $\lim_{\eta \to \infty} \Phi_2^{1/\eta}(\Phi(u^\eta), \Phi(v^\eta); \rho)$ can be obtained using two applications of l'Hopital's rule and the second order approximation $\Phi^{-1}(p) \sim -a_p + a_p^{-1} \log[a_p(2\pi)^{1/2}]$ as $p \to 0^+$, where $a_p = (-2\log p)^{1/2}$. The limiting extreme value copula is $C(u, v; \rho) = \exp\{-A(-\log u, -\log v; \rho)\}$, where

$$A(x, y; \rho) = \begin{cases} y, & 0 \leq x/(x+y) \leq \rho^2/(1+\rho^2), \\ (x + y - 2\rho[xy]^{1/2})/(1 - \rho^2), & \rho^2/(1+\rho^2) \leq x/(x+y) \leq 1/(1+\rho^2), \\ x, & 1/(1+\rho^2) \leq x/(x+y) \leq 1. \end{cases}$$

Note that $A(1, 1; \rho) = 2/(1 + \rho)$ and the tail dependence parameter is $2 - A(1, 1; \rho) = 2\rho/(1 + \rho)$ which increases from 0 to 1 as ρ increases from 0 to 1.

3.7 Fréchet class given bivariate margins

The idea of dependence modeling with copulas is to specify univariate marginal models as a first step and then proceed to the dependence structure. For the dependence structure, bivariate analysis is typical to get an idea of the strength of general and tail dependence in pairs. It might be desirable to fit parametric bivariate copula families and then find a multivariate copula that has these fitted bivariate margins.

Then the Fréchet class $\mathcal{F}(F_{jk}, 1 \leq j < k \leq d; F_j, 1 \leq j \leq d)$ of given (continuous) univariate and bivariate margins is relevant. But this class is hard to study; there are no general results on when the set of $\binom{d}{2}$ bivariate margins or copulas are compatible with a d-variate distribution. Any such condition would be the generalization of the positive definite constraint on a correlation matrix. There are conditions in Joe (1997) for non-compatibility and other more general conditions are given in Section 8.4; also included there are examples to show compatibility constraints for some of the 1-parameter bivariate copula families in Chapter 4.

Molenberghs and Lesaffre (1994) have a construction based on an extension of the bivariate Plackett copula; the rectangle condition has to be checked numerically and some analysis is in Section 4.8 of Joe (1997). For the Molensbergh-Lesaffre construction, recursive equations must be solved and this is computationally demanding as the dimension d increases. However, it has been applied for discrete response variables.

3.8 Mixtures of conditional distributions

In this section, we show how Sklar's theorem applies to a set of univariate conditional distributions, all conditioned on variables in an index set S. Sequential mixtures of conditional distributions lead to the vine pair-copula construction in Section 3.9.

Shorthand notation used here includes: $\boldsymbol{x}_J = (x_j : j \in J)$ where J is a subset of $\{1, \ldots, d\}$; $(-\infty, \boldsymbol{x}_J) = \prod_{j \in J}(-\infty, x_j)$; $(-\infty, \boldsymbol{x}_J] = \prod_{j \in J}(-\infty, x_j]$.

Consider d random variables $X_1, \ldots, X_d$ with multivariate distribution F. Let S be a non-empty subset of $\{1, \ldots, d\}$, which will be the *conditioning set* of variables. Let T be a subset of S^c with cardinality of at least two, which will be the *conditioned set* of variables. With $M = S \cup T$, we can write

$$F_M(\boldsymbol{x}_M) = \int_{(-\infty, \boldsymbol{x}_S]} F_{T|S}(\boldsymbol{x}_T | \boldsymbol{y}_S) \, dF_S(\boldsymbol{y}_S); \qquad (3.31)$$

the conditional distribution $F_{T|S}(\cdot | \boldsymbol{x}_S)$ exists almost everywhere on a set $\mathcal{X} \subset \mathbb{R}^{|S|}$ with $\mathbb{P}(\boldsymbol{X}_S \in \mathcal{X}) = 1$.

Wherever $F_{T|S}(\cdot | \boldsymbol{x}_S)$ exists, it has univariate margins $F_{j|S}(x_j | \boldsymbol{x}_S) : j \in T$, so that Sklar's theorem implies that there is a copula $C_{T;S}(\cdot; \boldsymbol{x}_S)$ such that

$$F_{T|S}(\boldsymbol{x}_T | \boldsymbol{x}_S) = C_{T;S}(F_{j|S}(x_j | \boldsymbol{x}_S) : j \in T; \boldsymbol{x}_S);$$

note that this copula depends on $\boldsymbol{x}_S$. Here, $C_{T;S}$ is a $|T|$-dimensional copula applied to univariate margins $F_{j|S}(\cdot | \boldsymbol{x}_S)$; it is not a conditional distribution obtained from $C_{S \cup T}$ so the notation $C_{T|S}$ is inappropriate. Hence (3.31) can be represented as:

$$F_M(\boldsymbol{x}_M) = \int_{(-\infty, \boldsymbol{x}_S]} C_{T;S}(F_{j|S}(x_j | \boldsymbol{y}_S) : j \in T; \boldsymbol{y}_S) \, dF_S(\boldsymbol{y}_S). \qquad (3.32)$$

We next proceed with the absolutely continuous case and then the discrete case.

For the absolutely continuous case, we could convert univariate margins to $U(0,1)$. If

$F = C$ is a copula and $F_S = C_S$ is absolutely continuous with density c_S, then (3.32) can be written as:

$$C_M(\boldsymbol{u}_M) = \int_{(-\infty, \boldsymbol{u}_S]} C_{T;S}\big(C_{j|S}(u_j|\boldsymbol{v}_S) : j \in T; \, \boldsymbol{v}_S\big) \, c_S(\boldsymbol{v}_S) \, d\boldsymbol{v}_S. \tag{3.33}$$

In the above, $C_{j|S}$ is not a copula but is a conditional distribution from the copula $C_{S \cup \{j\}}$. If $T = \{i_1, i_2\}$ has cardinality 2, then the notation simplifies to:

$$C_{S \cup \{i_1, i_2\}}(\boldsymbol{u}_{S \cup \{i_1, i_2\}}) = \int_{(-\infty, \boldsymbol{u}_S]} C_{i_1, i_2;S}(C_{i_1|S}(u_{i_1}|\boldsymbol{v}_S), C_{i_2|S}(u_{i_2}|\boldsymbol{u}_S); \, \boldsymbol{v}_S) \, c_S(\boldsymbol{v}_S) \, d\boldsymbol{v}_S.$$
$$\tag{3.34}$$

If F is a multivariate Gaussian distribution of $\boldsymbol{Z}$, the copulas for the conditional distributions do not depend on the values $\boldsymbol{v}_S$ of the conditioning variables, that is

$$C_{T;S}(\cdot) = C_{T;S}(\cdot; \boldsymbol{v}_S) \tag{3.35}$$

depends just on the variables in the set $M = S \cup T$ (or the correlation matrix for $\boldsymbol{Z}_M$).

There are two ways to view the above. Firstly, (3.32) and (3.33) are representations of the marginal distributions F_M and C_M. Secondly, if copulas $C_{S \cup \{j\}}$, $j \in T$, are given, and $C_{T;S}$ is specified satisfying (3.35), then (3.33) is a method to construct a copula C_M from lower-dimensional copulas. In this case, (3.35) is called the *simplifying assumption*

If all of the univariate margins are discrete, then (3.32) could be converted to a similar equation with the joint pmf:

$$\mathbb{P}(X_j \le x_j, j \in T; X_i = x_i, i \in S) = C_{T;S}\big(F_{j|S}(x_j|\boldsymbol{x}_S) : j \in T; \boldsymbol{x}_S\big) \cdot f_S(\boldsymbol{x}_S),$$

where $\boldsymbol{x}_S$ is a possible value for $\boldsymbol{X}_S$ and the copula $C_{T;S}(\cdot; \boldsymbol{x}_S)$ is not unique. With the simplifying assumption, if $T = \{i_1, i_2\}$ has cardinality 2, then using the rectangle condition, the joint pmf is

$$\begin{aligned}
f_M(\boldsymbol{x}_M) &= f_M(x_{i_1}, x_{i_2}, \boldsymbol{x}_S) = f_S(\boldsymbol{x}_S) \cdot f_{i_1, i_2|S}(x_{i_1}, x_{i_2}|\boldsymbol{x}_S) \tag{3.36} \\
&= f_S(\boldsymbol{x}_S)\big\{ C_{T;S}\big(F_{i_1|S}(x_{i_1}|\boldsymbol{x}_S), F_{i_2|S}(x_{i_2}|\boldsymbol{x}_S)\big) - C_{T;S}\big(F_{i_1|S}(x_{i_1}^-|\boldsymbol{x}_S), F_{i_2|S}(x_{i_2}|\boldsymbol{x}_S)\big) \\
&\quad - C_{T;S}\big(F_{i_1|S}(x_{i_1}|\boldsymbol{x}_S), F_{i_2|S}(x_{i_2}^-|\boldsymbol{x}_S)\big) + C_{T;S}\big(F_{i_1|S}(x_{i_1}^-|\boldsymbol{x}_S), F_{i_2|S}(x_{i_2}^-|\boldsymbol{x}_S)\big) \big\},
\end{aligned}$$

where $x_{i_1}, x_{i_2}, \boldsymbol{x}_S$ are possible values of the discrete random variables, and $F_{i_j|S}(x^-|\boldsymbol{x}_S)$ refers to the left-hand limit of the cdf at a possible value x (that is, $f_{i_j|S}(x|\boldsymbol{x}_S) = F_{i_j|S}(x|\boldsymbol{x}_S) - F_{i_j|S}(x^-|\boldsymbol{x}_S)$).

Similarly, if some of the variables are discrete and others are continuous, (3.32) can still be applied. For example, if $T = \{i_1, i_2\}$ and only variable i_1 is discrete, then

$$\begin{aligned}
\mathbb{P}(X_{i_1} = x_{i_1}, X_{i_2} \le x_{i_2}, \boldsymbol{X}_S \le \boldsymbol{x}_S) &= \int_{(-\infty, \boldsymbol{x}_S]} \big[C_{i_1, i_2;S}\big(F_{i_1|S}(x_{i_1}|\boldsymbol{y}_S), F_{i_2|S}(x_{i_2}|\boldsymbol{y}_S)\big) \\
&\quad - C_{i_1, i_2;S}\big(F_{i_1|S}(x_{i_1}^-|\boldsymbol{y}_S), F_{i_2|S}(x_{i_2}|\boldsymbol{y}_S)\big) \big] \, dF_S(\boldsymbol{y}_S)
\end{aligned}$$

with the simplifying assumption.

In subsequent sections, the multivariate Gaussian example is imitated and copula models for sequences of conditional distributions are constructed with the *simplifying assumption* of (3.35).

3.9 Vine copulas or pair-copula constructions

This section shows some concrete applications of Section 3.8 to derive the vine copula or pair-copula construction. We start with a sequence of conditional distributions that is relevant for time series analysis, and then generalize it. The notation $i : j$ with integers $i < j$ is shorthand for the set $\{i, i+1, \ldots, j-1, j\}$.

Example 3.5 Equation (3.34) is to be applied iteratively to get a joint d-dimensional distribution or copula.

- For the first step, start with copulas or bivariate distributions on the sets $\{j, j+1\}$, $j = 1, \ldots, d-1$.
- Note that sets $\{j, j+1\}$ and $\{j+1, j+2\}$ have the integer $j+1$ in common; take $S = \{j+1\}$, $T = \{j, j+2\}$ and a bivariate copula $C_{j,j+2;j+1}$ applied to margins $C_{j|j+1}, C_{j+2|j+1}$ to get a trivariate copula $C_{j:(j+2)}$ $(j = 1, \ldots, d-2)$.
- The sets $j : (j+2)$ and $(j+1) : (j+3)$ have $\{j+1, j+2\}$ in common; take $S = \{j+1, j+2\}$, $T = \{j, j+3\}$ and a bivariate copula $C_{j,j+3;j+1,j+2}$ applied to margins $C_{j|j+1,j+2}, C_{j+3|j+1,j+2}$ to get a 4-variate copula $C_{j:(j+3)}$ $(j = 1, \ldots, d-3)$.
- Iterate this for $S = \{(j+1) : (j+\ell)\}$, $T = \{j, j+\ell+1\}$ and a bivariate copula $C_{j,j+\ell+1;(j+1):(j+\ell)}$ applied to margins $C_{j|(j+1):(j+\ell)}, C_{j+\ell+1|(j+1):(j+\ell)}$ to get a multivariate copula $C_{j:(j+\ell+1)}$ $(j = 1, \ldots, d-\ell-1, \ell = 2, \ldots, d-2)$.
- The last step has $S = 2 : (d-1)$ and $T = \{1, d\}$.

If the above procedure is used to decompose a multivariate Gaussian distribution with correlation matrix $\boldsymbol{R} = (\rho_{ij})$, then from Joe (1996a), the bivariate copula $C_{j,j+\ell+1;(j+1):(j+\ell)}$ is a bivariate Gaussian copula with parameter $\rho_{j,j+\ell+1;(j+1):(j+\ell)}$, which is a partial correlation based on $\boldsymbol{R}$.

3.9.1 Vine sequence of conditional distributions

Example 3.5 can be generalized to other sequences as follows. We assume there is a permutation of the variables that the variable j will be conditioned on variables $1, \ldots, j-1$ for $j = 2, \ldots, d$.

For variable or index j $(2 \le j \le d)$, let $P_{j-1} = (a_{1j}, \ldots, a_{j-1,j})$ be a permutation of $\{1, \ldots, j-1\}$ such that a_{1j} is first paired to j and then a_{2j} and finally $a_{j-1,j}$. In the ℓth step $(2 \le \ell \le j-1)$, $a_{\ell j}$ is paired to j conditional on (variables with) indices $a_{1j}, \ldots, a_{\ell-1,j}$. For $j = 2$, $a_{12} = 1$; for $j = 3$, there are two choices $P_2 = (1, 2)$ or $(2, 1)$. For $j \ge 4$, there will be constraints on the allowable permutations P_{j-1} that depend on the permutations chosen at stages $3, \ldots, j-1$.

The terms $[j, a_{1j}]$ and $[j, a_{\ell j} | a_{1j}, \ldots, a_{\ell-1,j}]$ $(1 < \ell < j)$ are used to summarize the sequence for index j, and the objects $[j, a_{\ell j} | a_{1j}, \ldots, a_{\ell-1,j}]$ will be considered as edges of a tree.

For $1 \le \ell < j \le d$, define set $S_{j,a_{\ell j}} = \{a_{1j}, \ldots, a_{\ell-1,j}\}$ as the conditioning set when $T = \{j, a_{\ell j}\}$ is the conditioned set, leading to the set $N_{j,\ell} = S_{j,a_{\ell j}} \cup \{j, a_{\ell j}\}$ (cardinality $\ell + 1$) that appears after the ℓth substep of the jth step. Note that $S_{j,a_{1j}} = \emptyset$ and the sequence starts with $S_{2,1} = \{1, 2\}$.

For set $N_{j,\ell}$, denote $e_1 = [j | a_{1j}, \ldots, a_{\ell-1,j}]$ and $e_2 = [a_{\ell j} | a_{1j}, \ldots, a_{\ell-1,j}]$ as the nodes or vertices of the edge $\boldsymbol{e} = [j, a_{\ell j} | a_{1j}, \ldots, a_{\ell-1,j}]$; for this edge $\boldsymbol{e}$, a bivariate copula $C_{j,a_{\ell j};a_{1j},\ldots,a_{\ell-1,j}}$ is assigned to couple with conditional univariate margins $F_{j|a_{1j},\ldots,a_{\ell-1,j}}$ and $F_{a_{\ell j}|a_{1j},\ldots,a_{\ell-1,j}}$.

If (3.34) is to be applied iteratively to a sequence $N_{j,\ell}$ for margins of a copula, then prior to $N_{j,\ell}$, $e_2 = [a_{\ell j} | a_{1j}, \ldots, a_{\ell-1,j}]$ must be constructable, which means that $\{a_{1j}, \ldots, a_{\ell-1,j}, a_{\ell j}\} = N_{j',\ell-1}$ where $j' = \max\{a_{1j}, \ldots, a_{\ell j}\}$. This is because (3.34) would require $C_{N_{j',\ell-1}}$ to exist before the conditional distribution $C_{a_{\ell j}|a_{1j},\ldots,a_{\ell-1,j}}$ can be obtained. If $\ell = 1$, $C_{a_{\ell j}|a_{1j},\ldots,a_{\ell-1}}(u_{a_{\ell j}}|\emptyset)$ is just considered as $u_{a_{1j}}$. So the requirement on the sequence $\{N_{j,\ell} : \ell = 1, \ldots, j-1; j = 3, \ldots, d\}$ imposes constraints on the permutations of $P_{j-1} = (a_{1j}, \ldots, a_{j-1,j}\}$, but a_{1j} can be arbitrarily equal to any of the preceeding positive integers at step j.

Example 3.6 We show a few specific examples of sequences.

1. (Increasing sequence) $(a_{1j}, \ldots, a_{j-1,j}) = (1, \ldots, j-1)$. This is always valid because $[j|1, \ldots, \ell-1]$ and $[\ell|1, \ldots, \ell-1]$ can form an edge since $N_{\ell,\ell-1} = \{1, \ldots, \ell\}$ appears before $N_{j,\ell} = \{1, \ldots, \ell, j\}$ for any $j > \ell$.

2. (Decreasing sequence) $(a_{1j}, \ldots, a_{j-1,j}) = (j-1, j-2, \ldots, 1)$. This corresponds to Example 3.5. $N_{j,\ell} = \{j, j-1, \ldots, j-\ell\}$, and $[j|j-1, \ldots, j-\ell+1]$ and $[j-\ell|j-1, \ldots, j-\ell+1]$ can form an edge since $N_{j-1,\ell-1} = \{j-1, \ldots, j-\ell\}$ appears before $N_{j,\ell}$.

3. $d = 5$ with $P_1 = (1)$, $P_2 = (2,1)$, $P_3 = (2,1,3)$, $P_4 = (3,2,1,4)$. This leads to the sequence of edges:

$$[12], [23], [13|2], [24], [14|2], [34|12], [35], [25|3], [15|23], [45|123],$$

and corresponding sequence of N's:

$$\{1,2\}, \{2,3\}, \{1,3,2\}, \{2,4\}, \{1,4,2\}, \{3,4,1,2\}, \{3,5\}, \{2,5,3\}, \{1,5,2,3\}, \{4,5,1,2,3\}.$$

Note with stage $j = 5$ that after $a_{15} = 3$, the only possible choice for a_{25} is 2 because $a_{25} = 1$ would require that $N = \{1,3\}$ exists earlier in the sequence and $a_{25} = 4$ would require that $N = \{3,4\}$ exists earlier in the sequence. □

There are many possible permutations at step j to meet all of the constraints. When the constraints are satisfied, the result is a vine $\mathcal{V}$ (Cooke (1997), Bedford and Cooke (2001, 2002)) that consists of a set of edges $\{e = [j, a_{\ell j}|S(j, a_{\ell j})] : \ell = 1, \ldots, j-1; j = 2, \ldots, d\}$. The two boundary cases of vines are in Example 3.6: case (1) is called a *C-vine or canonical vine* and case (2) is called a *D-vine or drawable vine*. Other vines are called *regular vines or R-vines*. The explanation of the term *vine* comes in the next subsection with a graphical representation of the trees and edges.

Vines have the following properties based on the above construction.

- For $\ell = 1, \ldots, d-1$, let $\mathcal{T}_\ell$ consist of edges e where the corresponding sets are $N_{j,\ell}$ with cardinality $\ell + 1$ for $j = \ell + 1, \ldots, d$. The number of edges in $\mathcal{T}_\ell$ is $d - \ell$ and integers $\ell + 1, \ldots, d$ each appearing as a conditioned variable on at least one edge of this set.
- The size of the conditioning set for any edge in $\mathcal{T}_\ell$ is $\ell - 1$ (the conditioning set is $\emptyset$ for $\ell = 1$).
- Every pair $\{i_1, i_2\}$ appears exactly once as a conditioned set.
- Each edge e in $\mathcal{T}_\ell$ (after a little relabeling) becomes a node in $\mathcal{T}_{\ell+1}$; the d nodes of $\mathcal{T}_1$ can be considered as $1, \ldots, d$.
- If edges are used to connect the nodes in $\mathcal{T}_\ell$, then the result is a *tree*, that is, a connected graph with no cycles.

Example 3.7 Consider case 3 in Example 3.6 in more detail. For the vine, $\mathcal{T}_1$ has edges $\{[12], [23], [24], [35]\}$; $\mathcal{T}_2$ has edges $\{[13|2], [14|2], [25|3]\}$ (the edge $[23]$ is converted as node $[3|2]$ for $[13|2]$ and $[2|3]$ for $[25|3]$ etc.; $\mathcal{T}_3$ has edges $\{[34|12], [15|23]\}$ and $\mathcal{T}_4$ has edge $\{[45|123]\}$. Suppose we assign 10 bivariate absolutely continuous copulas to these edges: $C_{12}, C_{23}, C_{24}, C_{35}, C_{13;2}, C_{14;2}, C_{25;3}, C_{34;12}, C_{15;23}, C_{45;123}$. With this vine sequence, we can construct a multivariate absolutely continuous distribution with univariate marginal cdfs $F_1, \ldots, F_5$ and densities $f_1, \ldots, f_5$ using the approach in Section 3.8.

(a) $F_{12} = C_{12}(F_1, F_2)$;
(b) $F_{23} = C_{23}(F_2, F_3)$;
(c) $F_{24} = C_{24}(F_2, F_4)$;
(d) $F_{35} = C_{35}(F_3, F_5)$;
(e) $F_{13|2} = C_{13;2}(F_{1|2}, F_{3|2})$, integrate with respect to f_2 to get F_{123};
(f) $F_{14|2} = C_{14;2}(F_{1|2}, F_{4|2})$, integrate with respect to f_2 to get F_{124};
(g) $F_{25|3} = C_{25;3}(F_{2|3}, F_{5|3})$, integrate with respect to f_3 to get F_{235};

(h) $F_{34|12} = C_{34;12}(F_{3|12}, F_{4|12})$, integrate with respect to f_{12} to get F_{1234};

(i) $F_{15|23} = C_{14;23}(F_{1|23}, F_{5|23})$, integrate with respect to f_{23} to get F_{1235};

(j) $F_{45|123} = C_{45;123}(F_{4|123}, F_{5|123})$, integrate with respect to f_{123} to get F_{12345}.

One can easily check that the conditional distribution arguments of each copula for trees T_2 to T_4 can be obtained from an earlier distribution in the sequence.

3.9.2   *Vines as graphical models*

In this section, we give the definition of a vine from Bedford and Cooke (2001) and Kurowicka and Cooke (2006), and show how vine distributions can be drawn as graphical models.

Definition 3.2 (Regular vine). V is a *regular vine on d elements*, with $\mathcal{E}(V) = \mathcal{E}_1 \cup \cdots \cup \mathcal{E}_{d-1}$ denoting the set of edges of V, if

1. $V = \{T_1, \ldots, T_{d-1}\}$ [consists of $d-1$ trees];

2. T_1 is a connected tree with nodes $\mathcal{N}_1 = \{1, \ldots, d\}$, and edges $\mathcal{E}_1$; for $\ell = 2, \ldots, d-1$, T_ℓ is a tree with nodes $\mathcal{N}_\ell = \mathcal{E}_{\ell-1}$ [edges in a tree becomes nodes in the next tree];

3. *(proximity)* for $\ell = 2, \ldots, d-1$, for $\{n_1, n_2\} \in \mathcal{E}_\ell$, $\#(n_1 \triangle n_2) = 2$, where $\triangle$ denotes symmetric difference and $\#$ denotes cardinality [nodes joined in an edge differ by two elements].

The *degree of a node* is defined as the number of edges attached to that node. A regular vine is called a *canonical* or *C-vine* if tree T_ℓ has a unique node of degree $d - \ell$ (the maximum degree) for $\ell = 1, \ldots, d-2$. A regular vine is called a *D-vine* if all nodes in T_1 have degree not higher than two.

The following result is due to Bedford and Cooke (2001). Let $V = \{T_1, \ldots, T_{d-1}\}$ be a regular vine, then

1. the number of edges is $d(d-1)/2$ [equals number of pairs of variables];

2. each conditioned set is a doubleton, each pair of variables occurs exactly once as a conditioned set [matching a bivariate distribution with univariate conditional margins];

3. if two edges have the same conditioning set, then they are the same edge.

Vines are typically displayed in two forms: (i) Figures 3.2 and 3.3, showing edges for one tree becoming nodes for the next tree, and the explanation of the term "vine" with the D-vine picture that looks somewhat like a grapevine, (ii) Figures 3.4 and 3.5, showing the trees of the vine separately.

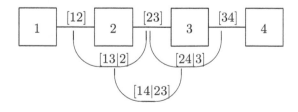

Figure 3.2: *A drawable D-vine with 4 variables and 3 trees*

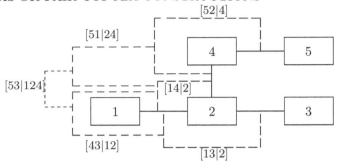

Figure 3.3: *A regular vine with 5 variables and 4 trees (and no intersecting edges)*

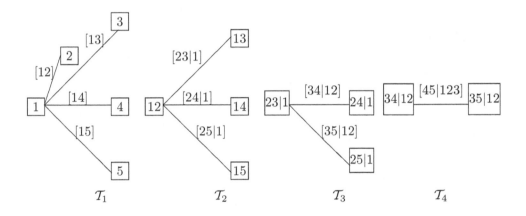

Figure 3.4: *A C-vine with 5 variables and 4 trees (separate T_j, $j = 1, \ldots, 4$).*

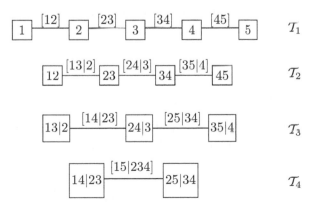

Figure 3.5: *A D-vine with 5 variables and 4 trees (separate T_j, $j = 1, \ldots, 4$).*

Example 3.8 In this example, we show the possible decisions of edges in the construction of a 5-dimension vine that is not the C-vine or D-vine. Shorthand notation without extra punctuation is used for edges and nodes. Consider the ten pairs:

$$
\begin{array}{ccccc}
 & & & & 15 \\
 & & 14 & & 25 \\
 & 13 & 24 & & 35 \\
12 & 23 & 34 & & 45 \\
\end{array}
$$

For first tree of vine, pick a pair from each column (always top pair means C-vine, always bottom pair means D-vine).

- Suppose Tree T_1 has nodes 1,2,3,4,5 and edges 12,23,24,35.
- 12,23,24,35 are now the nodes of Tree T_2. Consider pairing of these nodes. Possibilities are the following. (i) nodes 12 and 23: edge 13|2; (ii) nodes 12 and 24: edge 14|2; (iii) nodes 12 and 35: $\emptyset$; (iv) nodes 23 and 24: edge 34|2; (v) nodes 23 and 35: edge 25|3; (vi) nodes 24 and 35: $\emptyset$. So to get Tree T_2, choose 3 edges and avoid a cycle, say 13|2, 14|2, 25|3.
- 13|2, 14|2, 25|3 are now the nodes of Tree T_3. Consider pairing of these nodes. Possibilities are the following. (i) nodes 13|2 and 14|2: edge 34|12; (ii) nodes 13|2 and 25|3: edge 15|23; (iii) nodes 14|2 and 25|3: need 2 indices in the intersection. So to get Tree T_3, the only possibility is to have edges 34|12 and 15|23.
- Finally, for Tree T_4, the nodes 34|12 and 15|23 have edge 45|123.

For a vine copula applied to this vine, one needs to specify 10 bivariate copulas: $C_{12}, C_{23}, C_{24}, C_{35}, C_{13;2}, C_{14;2}, C_{25;3}, C_{34;12}, C_{15;23}, C_{45;123}$; see Example 3.7.

3.9.3 Vine distribution: conditional distributions and joint density

A *vine copula* or *pair-copula construction* is the result copula when a set of $\binom{d}{2}$ bivariate copulas are used sequentially in (3.34) leading to a d-dimensional copula, constructed to satisfy the simplifying assumption (3.35). If $[i_1, i_2|S(i_1, i_2)]$ is an edge, then the bivariate copula associated to this edge is denoted as $C_{i_1 i_2; S(i_1, i_2)}$ — or more simply $C_{i_1 i_2}$ if $S(i_1, i_2) = \emptyset$. If the univariate margins are not necessarily $U(0, 1)$, then the result is a *vine distribution*.

To avoid technicalities, we assume in this subsection that all of the $C_{i_1 i_2; S(i_1,i_2)}$ are absolutely continuous.

For the ℓth substep of the jth step of the vine construction, the copula $C_{j;a_{\ell j}:S(j,a_{\ell j})}$ is applied to (univariate) conditional distributions $C_{j|S(j,a_{\ell j})}$ and $C_{a_{\ell j}|S(j,a_{\ell j})}$. With $N_{j,\ell} = S(j, a_{\ell j}) \cup \{j, a_{\ell j}\}$, this results in:

$$
\begin{aligned}
C_{N_{j,\ell}}(\boldsymbol{u}_{N_{j,\ell}}) = \int_{[\boldsymbol{0}, \boldsymbol{u}_S]} & C_{j,a_{\ell j};S(j,a_{\ell j})} \big(C_{j|S(j,a_{\ell j})}(u_j|\boldsymbol{v}_S), \, C_{a_{\ell j}|S(j,a_{\ell j})}(u_{a_{\ell j}}|\boldsymbol{v}_S) \big) \\
& \cdot c_{S(j,a_{\ell j})}(\boldsymbol{v}_S) \, \mathrm{d}\boldsymbol{v}_S, \quad S = S(j, a_{\ell j}).
\end{aligned}
\tag{3.37}
$$

To get the conditional distribution of $C_{j|S(j,a_{\ell j}) \cup a_{\ell j}}$, differentiate (3.37) with respect to $(u_i : i \in S(j, a_{\ell j}) \cup a_{\ell j})$ and apply (2.5) to obtain:

$$
\begin{aligned}
C_{j|S(j,a_{\ell j}) \cup a_{\ell j}}&(u_j|\boldsymbol{u}_{S(j,a_{\ell j}) \cup a_{\ell j}}) \\
&= C_{j|a_{\ell j};S(j,a_{\ell j})} \big(C_{j|S(j,a_{\ell j})}(u_j|\boldsymbol{u}_{S(j,a_{\ell j})}) \mid C_{a_{\ell j}|S(j,a_{\ell j})}(u_{a_{\ell j}}|\boldsymbol{u}_{S(j,a_{\ell j})}) \big).
\end{aligned}
\tag{3.38}
$$

Note that $c_{S(j,a_{\ell j}) \cup a_{\ell j}}(\boldsymbol{u}_{S(j,a_{\ell j}) \cup a_{\ell j}})$ cancels from the numerator and denominator of the conditional cdf. The construction of the vine implies that the margin $C_{S(j,a_{\ell j}) \cup a_{\ell j}}$ was created earlier in the sequence. The equation (3.38) is a recursion equation for $l = 1, \ldots, j-1$. If

$P_{j-1} = (a_{1j}, \ldots, a_{j-1,j})$, then the recursion starts with $C_{j,a_{1j}}(u_j, u_{a_{1j}})$ (because $S(j, a_{1j}) = \emptyset$) and $C_{j,a_{2j};a_{1j}}(C_{j|a_{1j}}, C_{a_{2j}|a_{1j}})$ and ends with

$$C_{j|1:(j-1)}(u_j|\boldsymbol{u}_{1:(j-1)}) = C_{j|S(j,a_{j-1,j})\cup a_{j-1,j}}(u_j|\boldsymbol{u}_{S(j,a_{j-1,j})\cup a_{j-1,j}}) \tag{3.39}$$
$$= C_{j|a_{j-1,j};S(j,a_{j-1,j})}(C_{j|S(j,a_{j-1,j})}(u_j|\boldsymbol{u}_{S(j,a_{j-1,j})}) \mid C_{a_{j-1,j}|S(j,a_{j-1,j})}(u_{a_{j-1,j}}|\boldsymbol{u}_{S(j,a_{j-1,j})})).$$

The vine pair-copula construction does not lead to closed-form cdf for the d-variate copula, but the copula density has a simple form (Bedford and Cooke, 2001) with no integrals, assuming all of the bivariate copulas have closed form densities.

We go from (3.38) to show how the joint density $c_{1:d}$ is derived. The joint density can be decomposed as

$$c_{1:d} = c_{12} \cdot \prod_{j=3}^{d} c_{j|1:(j-1)}.$$

Differentiating (3.39) with respect to u_j leads to $c_{j|1:(j-1)}$. Let $S_{j\ell} = S(j, a_{\ell j})$. Applying the chain rule and using the recursion (3.38) lead to

$$c_{j|1:(j-1)}(u_j|u_{1:(j-1)}) = \prod_{\ell=j-1}^{1} c_{j,a_{\ell j};S_{j\ell}}(C_{j|S_{j\ell}}(u_j|\boldsymbol{u}_{S_{j\ell}}) \mid C_{a_{\ell j}|S_{j\ell}}(u_{a_{\ell j}}|\boldsymbol{u}_{S_{j\ell}})). \tag{3.40}$$

When substituted into the above expression for $c_{1:d}$, a more compact form of the joint density is

$$c_{1:d} = \prod_{[i_1,i_2|S(i_1,i_2)]\in\mathcal{E}(\mathcal{V})} c_{i_1,i_2;S(i_1,i_2)}(C_{i_1|S(i_1,i_2)}, C_{i_2|S(i_1,i_2)}). \tag{3.41}$$

This is summarized in Theorem 8.64 in a form applied to a density with cdf $F \in \mathcal{F}(F_1, \ldots, F_d)$:

$$f_{1:d} = \prod_{j=1}^{d} f_j \cdot \prod_{[i_1,i_2|S(i_1,i_2)]\in\mathcal{E}(\mathcal{V})} c_{i_1,i_2;S(i_1,i_2)}(F_{i_1|S(i_1,i_2)}, F_{i_2|S(i_1,i_2)}),$$

since $F_{k|S}(x_k|\boldsymbol{x}_S) = C_{k|S}(F_k(x_k)|F_i(x_i) : i \in S)$ and

$$f_{1:d}(\boldsymbol{x}) = c_{1:d}(F_1(x_1), \ldots, F_d(x_d)) \prod_{j=1}^{d} f_j(x_j).$$

Note the following special cases of (3.40).

- (C-vine) (3.40) becomes

$$c_{j|1:(j-1)} = \prod_{\ell=1}^{j-1} c_{j,\ell;1:(\ell-1)}(C_{j|1:(\ell-1)}, C_{\ell|1:(\ell-1)}).$$

- (D-vine) (3.40) becomes

$$c_{j|1:(j-1)} = \prod_{\ell=1}^{j-1} c_{j,j-\ell;(j-\ell+1):(j-1)}(C_{j|(j-\ell+1):(j-1)}, C_{j-\ell|(j-\ell+1):(j-1)}).$$

3.9.4　Vine array

The vine array, denoted as A, is a simple way to store the edges of a vine in a two-dimensional array. We will define the array with tree $\mathcal{T}_\ell$ ($\ell = 2, \ldots, d-1$) in row ℓ of a $d \times d$ matrix, where only the upper triangle is used. The array A closely follows the vine construction in Section 3.9.1. Column j has the permutation $P_{j-1} = (a_{1j}, \ldots, a_{j-1,j})$ of the previously added variables, used in the jth step of the construction. The diagonal of the vine array shows the order in which variables are added. We will also show the vine array in the form where the order need not be $1, 2 \ldots, d$.

As a concrete example, suppose $P_2 = (1,2)$, $P_3 = (1,2,3)$, $P_4 = (2,1,3,4)$, $P_5 = (2,1,4,3,5)$. This means that variable 3 is joined to variable 1 on $\mathcal{T}_1$ and the variable 2 in $\mathcal{T}_2$ leading to edges $[13], [23|1]$. Then variable 4 is joined to variables 1,2,3 in $\mathcal{T}_1, \mathcal{T}_2, \mathcal{T}_3$ respectively leading to edges $[14], [24|1], [34|12]$, etc. A vine array stores all of the conditional distributions, for example, with

$$
\begin{bmatrix}
- & 12 & 13 & 14 & 25 & 26 \\
 & - & 23|1 & 24|1 & 15|2 & 16|2 \\
 & & - & 34|12 & 35|12 & 46|12 \\
 & & & - & 45|123 & 36|124 \\
 & & & & - & 56|1234 \\
 & & & & & -
\end{bmatrix}, \quad
A = \begin{bmatrix}
1 & 1 & 1 & 1 & 2 & 2 \\
 & 2 & 2 & 2 & 1 & 1 \\
 & & 3 & 3 & 3 & 4 \\
 & & & 4 & 4 & 3 \\
 & & & & 5 & 5 \\
 & & & & & 6
\end{bmatrix}.
$$

The right-hand array above is a more compact way to store the information of the vine, from which the edges of the vine can be retrieved as given in the left-hand array.[2]

In applications to data, it is arbitrary which variable is labeled as variable j for $j = 1, \ldots, d$. The above vine array is still well defined if a permutation $(j_1, \ldots, j_6)$ of $(1, \ldots, 6)$ is applied to the elements of the array. For example. if the permutation is $(3, 4, 2, 6, 1, 5)$, mean the vine starts with the pair of variables 3,4, then variable 2 is joined to variable 3 in $\mathcal{T}_1$ and variable 4 in $\mathcal{T}_2$, etc., then the above arrays become:

$$
\begin{bmatrix}
- & 34 & 32 & 36 & 41 & 45 \\
 & - & 42|3 & 46|3 & 31|4 & 35|4 \\
 & & - & 26|34 & 21|43 & 65|43 \\
 & & & - & 61|432 & 25|436 \\
 & & & & - & 15|4362 \\
 & & & & & -
\end{bmatrix}, \quad
A = \begin{bmatrix}
3 & 3 & 3 & 3 & 4 & 4 \\
 & 4 & 4 & 4 & 3 & 3 \\
 & & 2 & 2 & 2 & 6 \\
 & & & 6 & 6 & 2 \\
 & & & & 1 & 1 \\
 & & & & & 5
\end{bmatrix}.
$$

In this more general version of the vine array $A = (a_{ij})$, the jth column above the diagonal has a permutation of the diagonal numbers $(a_{11}, \ldots, a_{j-1,j-1})$. Note that there are additional constraints on the jth column given previous columns to get a valid vine array; see Section 8.11.

As a check on understanding the notation of a vine array, one can verify that the left array below matches a vine but not the right array:

$$
\begin{bmatrix}
1 & 1 & 1 & 2 & 1 & 2 \\
 & 2 & 2 & 1 & 3 & 4 \\
 & & 3 & 3 & 2 & 1 \\
 & & & 4 & 4 & 3 \\
 & & & & 5 & 5 \\
 & & & & & 6
\end{bmatrix}, \quad
\begin{bmatrix}
1 & 1 & 1 & 1 & 4 & 5 \\
 & 2 & 2 & 3 & 3 & 3 \\
 & & 3 & 2 & 2 & 4 \\
 & & & 4 & 1 & 2 \\
 & & & & 5 & 1 \\
 & & & & & 6
\end{bmatrix}.
$$

---

[2]In the vine literature, sometimes the vine array is specified in the reverse order with the (ℓ, j) element for the ℓth tree and jth variable stored in position $(d+1-\ell, d+1-j)$.

The next property is that the conversion of vines to vine arrays is a one-to-many mapping. We show this with two representations of the D-vine on $1, 2, 3, 4, 5$:

$$\begin{bmatrix} 1 & 1 & 2 & 3 & 4 \\ & 2 & 1 & 2 & 3 \\ & & 3 & 1 & 2 \\ & & & 4 & 1 \\ & & & & 5 \end{bmatrix}, \quad \begin{bmatrix} 3 & 3 & 3 & 4 & 2 \\ & 4 & 4 & 3 & 3 \\ & & 2 & 2 & 4 \\ & & & 5 & 5 \\ & & & & 1 \end{bmatrix}.$$

3.9.5   Vines with some or all discrete marginal distributions

In this section, we show where the density of the vine is different or similar when some of all of the variables are discrete.

In the discrete case, for a vine $\mathcal{V}$ with edges $\{[i_1, i_2 | S(i_1, i_2)]\}$, (3.36) can be used recursively to get a joint pmf of $X_1, \ldots, X_d$. References are Panagiotelis et al. (2012) for the D-vine with all discrete variables, and Stöber et al. (2013a) for the R-vine with mixed discrete and continuous variables. Here, we show the key ideas and steps when the vine has been summarized via a vine array A.

For generic notation, let j, k be indices with $1 \leq k < j \leq d$ and let $S \subset \{1, \ldots, d\}$ be a subset of indices not including $\{j, k\}$. For tree $\mathcal{T}_\ell$, the notation below can be applied with $k = a_{\ell j}$ and $S = \{a_{1j}, \ldots, a_{\ell-1,j}\}$ for $j = \ell, \ldots, d$. The corresponding random variables/vectors are $X_k, X_j, \boldsymbol{X}_S$. Each variable can be either discrete or (absolutely) continuous and densities below are defined relative to an appropriate product measure involving 1-dimensional Lebesgue and counting measures. The relation

$$f_{j|S \cup \{k\}} = f_{kj|S} / f_{k|S} \tag{3.42}$$

holds for all combinations of discrete and continuous variables, include $S = \emptyset$, but the form of $f_{kj|S}$ depends on which of X_j, X_k are discrete.

If both are (absolutely) continuous, then using Sklar's theorem, there are copula densities $c_{kj;S}(\cdot; \boldsymbol{x}_S)$ such that

$$f_{kj|S}(x_k, x_j | \boldsymbol{x}_S) = c_{kj;S}(F_{k|S}(x_k | \boldsymbol{x}_S), F_{j|S}(x_j | \boldsymbol{x}_S); \boldsymbol{x}_S) \cdot f_{k|S}(x_k | \boldsymbol{x}_S) f_{j|S}(x_j | \boldsymbol{x}_S).$$

If $\widetilde{c}_{kj;S}(x_k, x_j; \boldsymbol{x}_S) := c_{kj;S}(F_{k|S}(x_k | \boldsymbol{x}_S), F_{j|S}(x_j | \boldsymbol{x}_S); \boldsymbol{x}_S)$, then from (3.42)

$$f_{j|S \cup \{k\}} = \widetilde{c}_{jk;S} f_{j|S}. \tag{3.43}$$

With the possibility of discrete variables for X_k, X_j, for shorthand notation, let $F_{i|S}^+ = F_{i|S}(x_i | \boldsymbol{x}_S)$ and $F_{i|S}^- = F_{i|S}(x_i^- | \boldsymbol{x}_S)$ for $i = k, j$.

If both X_k, X_j are discrete, then using Sklar's theorem, there are copulas $C_{kj;S}(\cdot; \boldsymbol{x}_S)$ (not necessarily unique) such that

$$\begin{aligned} f_{kj|S}(x_k, x_j | \boldsymbol{x}_S) &= \mathbb{P}(X_k = x_k, X_j = x_j | \boldsymbol{X}_S = \boldsymbol{x}_S) \\ &= C_{kj;S}(F_{k|S}^+, F_{j|S}^+; \boldsymbol{x}_S) - C_{kj;S}(F_{k|S}^-, F_{j|S}^+; \boldsymbol{x}_S) \\ &\quad - C_{kj;S}(F_{k|S}^+, F_{j|S}^-; \boldsymbol{x}_S) + C_{kj;S}(F_{k|S}^-, F_{j|S}^-; \boldsymbol{x}_S). \end{aligned}$$

Equation (3.43) holds in this case if we define

$$\begin{aligned} \widetilde{c}_{kj;S}(x_k, x_j; \boldsymbol{x}_S) := \big[&C_{kj;S}(F_{k|S}^+, F_{j|S}^+; \boldsymbol{x}_S) - C_{kj;S}(F_{k|S}^-, F_{j|S}^+; \boldsymbol{x}_S) - C_{kj;S}(F_{k|S}^+, F_{j|S}^-; \boldsymbol{x}_S) \\ &+ C_{kj;S}(F_{k|S}^-, F_{j|S}^-; \boldsymbol{x}_S) \big] \big/ \big[f_{k|S}(x_k | \boldsymbol{x}_S) f_{j|S}(x_j | \boldsymbol{x}_S) \big]. \end{aligned} \tag{3.44}$$

When only one of the variables X_k, X_j is discrete, the conditional distributions $C_{j|k;S} = D_1 C_{kj;S}$ or $C_{k|j;S} = D_2 C_{kj;S}$ appear in $\widetilde{c}_{kj;S}$.

If X_j is discrete and X_k is continuous, then using Sklar's theorem, there are copulas $C_{kj;S}(\cdot; \boldsymbol{x}_S)$ such that

$$f_{kj|S}(x_k, x_j|\boldsymbol{x}_S) = [C_{j|k;S}(F_{j|S}^+|F_{k|S}; \boldsymbol{x}_S) - C_{j|k;S}(F_{j|S}^-|F_{k|S}; \boldsymbol{x}_S)]f_{k|S}(x_k|\boldsymbol{x}_S). \qquad (3.45)$$

Equation (3.43) holds in this case if we define

$$\widetilde{c}_{kj;S}(x_k, x_j; \boldsymbol{x}_S) := \frac{C_{j|k;S}(F_{j|S}^+|F_{k|S}; \boldsymbol{x}_S) - C_{j|k;S}(F_{j|S}^-|F_{k|S}; \boldsymbol{x}_S)]}{f_{j|S}(x_j|\boldsymbol{x}_S)}. \qquad (3.46)$$

If X_k is discrete and X_j is continuous, then using Sklar's theorem, there are copulas $C_{kj;S}(\cdot; \boldsymbol{x}_S)$ such that

$$f_{kj|S}(x_k, x_j|\boldsymbol{x}_S) = C_{k|j;S}(F_{k|S}^+|F_{j|S}; \boldsymbol{x}_S) - C_{k|j;S}(F_{k|S}^-|F_{j|S}; \boldsymbol{x}_S)]f_{j|S}(x_j|\boldsymbol{x}_S).$$

Equation (3.43) holds in this case if we define

$$\widetilde{c}_{kj;S}(x_k, x_j; \boldsymbol{x}_S) := C_{k|j;S}(F_{k|S}^+|F_{j|S}; \boldsymbol{x}_S) - C_{k|j;S}(F_{k|S}^-|F_{j|S}; \boldsymbol{x}_S)$$
$$\overline{\qquad\qquad\qquad\qquad f_{k|S}(x_k|\boldsymbol{x}_S). \qquad\qquad\qquad}$$

The joint density (all variations of discrete and continuous variables) can be written as:

$$f_1 f_{2|1} f_{3|12} \cdots f_{d|1:(d-1)}. \qquad (3.47)$$

For a vine model with the simplifying assumption, copulas for the form $C_{kj;S}$ do not depend on $\boldsymbol{x}_S$. In all cases, $\widetilde{c}_{kj;S}$ involves $F_{k|S}, F_{j|S}$ computed in an earlier step.

For a C-vine, S is chosen to be a set of the form $1, \ldots, \ell - 1$ and for a D-vine, S is chosen to be a set of the form $\{m - \ell + 1, \ldots, m - 1\}$. More specific examples are the following.

- For a C-vine with $d = 5$, (3.47) would be computed as

$$f_{2|1} = \widetilde{c}_{12} f_2; \quad f_{3|12} = \widetilde{c}_{23;1} f_{3|1}, \quad f_{3|1} = \widetilde{c}_{13} f_3;$$
$$f_{4|123} = \widetilde{c}_{34;12} f_{4|12}, \quad f_{4|12} = \widetilde{c}_{24;1} \widetilde{c}_{14} f_4;$$
$$f_{5|1234} = \widetilde{c}_{45;123} f_{5|123}, \quad f_{5|123} = \widetilde{c}_{35;12} \widetilde{c}_{25;1} \widetilde{c}_{15} f_5.$$

- For a D-vine with $d = 5$, (3.47) would be computed as

$$f_{2|1} = \widetilde{c}_{12} f_2; \quad f_{3|12} = \widetilde{c}_{13;2} f_{3|2}, \quad f_{3|2} = \widetilde{c}_{23} f_3;$$
$$f_{4|123} = \widetilde{c}_{14;23} f_{4|23}, \quad f_{4|23} = \widetilde{c}_{24;3} \widetilde{c}_{34} f_4;$$
$$f_{5|1234} = \widetilde{c}_{15;234} f_{5|234}, \quad f_{5|234} = \widetilde{c}_{25;34} \widetilde{c}_{35;4} \widetilde{c}_{45} f_5.$$

- For the B1-vine (see Section 3.9.8) with $d = 5$, the vine array is $\begin{bmatrix} 1 & 1 & 1 & 1 & 2 \\ & 2 & 2 & 2 & 1 \\ & & 3 & 3 & 3 \\ & & & 4 & 4 \\ & & & & 5 \end{bmatrix}$, and (3.47) would be computed as

$$f_{2|1} = \widetilde{c}_{12} f_2; \quad f_{3|12} = \widetilde{c}_{23;1} f_{3|1}, \quad f_{3|1} = \widetilde{c}_{13} f_3;$$
$$f_{4|123} = \widetilde{c}_{34;12} f_{4|12}, \quad f_{4|12} = \widetilde{c}_{24;1} \widetilde{c}_{14} f_4;$$
$$f_{5|1234} = \widetilde{c}_{45;123} f_{5|123}, \quad f_{5|123} = \widetilde{c}_{35;12} \widetilde{c}_{15;2} \widetilde{c}_{25} f_5.$$

In extending the above examples, more generally for the vine array $A = (a_{\ell j})_{1 \le \ell \le j \le d}$,

$$f_{j|1:(j-1)} = f_{j|a_{1j}\cdots a_{j-1,j}} = \left\{ \prod_{\ell=j-1}^{2} \widetilde{c}_{a_{\ell j}j;a_{1j}\cdots a_{\ell-1,j}} \right\} \times \widetilde{c}_{a_{1j}j} \times f_j. \tag{3.48}$$

This result is given in Stöber et al. (2013a).

Even if there are some discrete variables, the combination of (3.47) and (3.48) shows that with some discrete variables, the Bedford and Cooke (2001) decomposition (Theorem 8.64) applies with the use of $\widetilde{c}_{jk;S}$ instead of $c_{jk;S}$.

In the iterative computation of the joint density $f_{1:d}$, the conditional cdfs $F_{k|S\cup\{j\}}, F_{j|S\cup\{k\}}$, with $k = a_{\ell j}$ and $S = \{a_{1j}, \ldots, a_{\ell-1,j}\}$, are computed after $C_{kj;S}$. In the expressions for $\widetilde{c}_{kj;S}$, note that $F_{j|S}, F_{k|S}$ are computed from the preceding tree. The conditional distributions for $\mathcal{T}_\ell$ are:

$$F_{j|S\cup\{k\}}(x_j|\boldsymbol{x}_{S\cup\{k\}}) = \mathbb{P}(X_j \le x_j|X_k = x_k, \boldsymbol{X}_S = \boldsymbol{x}_S)$$

$$= \begin{cases} C_{j|k;S}(F_{j|S}|F_{k|S}), & X_k \text{ continuous,} \\ [C_{kj;S}(F_{k|S}^+, F_{j|S}) - C_{kj;S}(F_{k|S}^-, F_{j|S})]/[F_{k|S}^+ - F_{k|S}^-], & X_k \text{ discrete.} \end{cases}$$

Similarly, $F_{k|S\cup\{j\}}(x_k|\boldsymbol{x}_{S\cup\{j\}})$ is obtained.

For the D-vine and other regular vines for multivariate discrete data, some algorithms implementing the recursions are given in Section 6.10.

3.9.6 Truncated vines

A vine is called a t-truncated vine if the copulas for trees $\mathcal{T}_{t+1}, \ldots, \mathcal{T}_{d-1}$ are all $C^\perp$; it is also called a vine truncated after tree $\mathcal{T}_t$. A 1-truncated vine is a Markov tree, and depends only on $C_{12}, C_{a_{13}3}, \ldots, C_{a_{1d}d}$ in terms of the vine array notation, and $C_{a_{j2}j;a_{j1}} = C^\perp$ for $j = 3, \ldots, d$ etc. A 2-truncated vine has second order dependence; it depends on the copulas $C_{a_{j1}j}$ for $j = 2, \ldots, d$ and $C_{a_{j2}j;a_{j1}}$ for $j = 3, \ldots, d$, with $C_{a_{j3}j;a_{j1}a_{j2}} = C^\perp$ for $j = 4, \ldots, d$ etc. For a t-truncated vine, the vine copula density for d continuous variables in (3.41) simplifies because $c_{i_1,i_2;S(i_1,i_2)} = 1$ (density for conditional independence) if the cardinality $|S(i_1,i_2)| \ge t$. When some variables are discrete, for a t-truncated vine with $|S(i_1,i_2)| \ge t$, $\widetilde{c}_{i_1,i_2;S(i_1,i_2)}$ based on (3.44), (3.45) and (3.46) also become 1. Hence from (3.48), there is a simplification to:

$$f_{j|a_{1j}\cdots a_{j-1}j} = f_{j|a_{1j}\cdots a_{tj}} = \left\{ \prod_{\ell=t\wedge(j-1)}^{2} \widetilde{c}_{ja_{\ell j};a_{1j}\cdots a_{\ell-1,j}} \right\} \times \widetilde{c}_{ja_{1j}} \times f_j.$$

for $j - 1 \ge t$ or $j \ge t + 1$. Then (3.47) becomes

$$f_1 f_{2|1} \prod_{j=3}^{d} f_{j|a_{1j}\cdots a_{t\wedge(j-1),j}}.$$

A truncated vine model implies that there are lots of conditional independence relations in the variables. This is indicated below for some Gaussian models with partial correlation truncated vines.

Example 3.9 Let $\boldsymbol{Z} \sim N_d(\boldsymbol{0}, \boldsymbol{R})$ where $\boldsymbol{R}$ is a non-singular correlation matrix. The (j, k) entry of the precision matrix (inverse of the covariance matrix) is opposite in sign to the partial correlation of Z_j, Z_k given $\{Z_i : i \in T(j,k)\}$, where $T(j,k) = \{1, \ldots, d\}\backslash\{j, k\}$. More specifically, if $\boldsymbol{R}^{-1} = (\rho^{jk})$, then $\rho^{jk} = -\sigma_{jk;T(j,k)}|\boldsymbol{R}^{(jk)}||\boldsymbol{R}|^{-1}$ for $j \neq k$, where $\sigma_{jk;T(j,k)}$ is the conditional covariance of Z_j, Z_k given $\{Z_i : i \in T(j,k)\}$ and $\boldsymbol{R}^{(jk)}$ is the submatrix of $\boldsymbol{R}$ with the jth and kth rows/columns removed (Whittaker (1990) or Exercise 2.19 of Joe (1997)). Gaussian models that are Markov trees in d variables have $d-1$ pairs that are connected with edges and $(d-1)(d-2)/2$ zeros in the upper triangle of the inverse correlation matrix. Similarly, 2-truncated Gaussian vines have $d-1$ pairs in tree T_1 and $d-2$ conditionally dependent pairs in tree T_2, and $(d-2)(d-3)/2$ zeros in the upper triangle of the inverse correlation matrix. More generally, ℓ-truncated Gaussian vines have $d-1$ pairs in tree T_1 and conditionally dependent pairs in tree $T_2. \ldots, T_\ell$, and $(d-\ell)(d-\ell-1)/2$ zeros in the upper triangle of the inverse correlation matrix.

A specific numerical example is given below to illustrate this. Consider a partial correlation vine with trees and edges as follows; T_1: $12, 23, 24, 35$; T_2: $13|2, 34|2, 25|3$; T_3: $14|23$,

$$45|23; \quad T_4: 15|234. \text{ The vine array is } A = \begin{bmatrix} 1 & 1 & 2 & 2 & 3 \\ & 2 & 1 & 3 & 2 \\ & & 3 & 1 & 4 \\ & & & 4 & 1 \\ & & & & 5 \end{bmatrix}.$$

Consider a 1-truncated R-vine with correlations $\rho_{12} = 0.4, \rho_{23} = 0.5, \rho_{24} = 0.6, \rho_{35} = 0.7$ for T_1. This leads to:

$$\boldsymbol{R} = \begin{pmatrix} 1.00 & 0.40 & 0.20 & 0.24 & 0.14 \\ 0.40 & 1.00 & 0.50 & 0.60 & 0.35 \\ 0.20 & 0.50 & 1.00 & 0.30 & 0.70 \\ 0.24 & 0.60 & 0.30 & 1.00 & 0.21 \\ 0.14 & 0.35 & 0.70 & 0.21 & 1.00 \end{pmatrix}, \ \boldsymbol{R}^{-1} = \begin{pmatrix} 1.1905 & -0.4762 & 0 & 0 & 0 \\ -0.4762 & 2.0863 & -0.6667 & -0.9375 & 0 \\ 0 & -0.6667 & 2.2941 & 0 & -1.3725 \\ 0 & -0.9375 & 0 & 1.5625 & 0 \\ 0 & 0 & -1.3725 & 0 & 1.9608 \end{pmatrix}.$$

The vine construction implies $\rho_{13;2} = 0$, $\rho_{34;2} = 0$, $\rho_{25;3} = 0$, $\rho_{14;23} = 0$, $\rho_{45;23} = 0$, $\rho_{15;234} = 0$. Then using the recursion formula for partial correlations

$$\rho_{14;235} \propto \rho_{14;23} - \rho_{15;23}\rho_{45;23} = 0,$$
$$\rho_{45;123} \propto \rho_{45;23} - \rho_{14;23}\rho_{15;23} = 0,$$
$$\rho_{45;23} \propto \rho_{45;2} - \rho_{34;2}\rho_{35;2} = 0 \Rightarrow \rho_{45;2} = 0,$$
$$\rho_{15;234} \propto \rho_{15;23} - \rho_{14;23}\rho_{45;23} = 0 \Rightarrow \rho_{15;23} = 0,$$
$$\rho_{13;24} \propto \rho_{13;2} - \rho_{14;2}\rho_{34;2} = 0,$$
$$\rho_{15;234} \propto \rho_{15;24} - \rho_{13;24}\rho_{35;24} = 0 \Rightarrow \rho_{15;24} = 0,$$
$$\rho_{13;245} \propto \rho_{13;24} - \rho_{15;24}\rho_{35;24} = 0.$$

Similarly,

$$\rho_{34;12} \propto \rho_{34;2} - \rho_{13;2}\rho_{14;2} = 0,$$
$$\rho_{45;123} \propto \rho_{45;12} - \rho_{34;12}\rho_{35;12} = 0 \Rightarrow \rho_{45;12} = 0,$$
$$\rho_{34;125} \propto \rho_{34;12} - \rho_{35;12}\rho_{45;12} = 0,$$
$$\rho_{15;23} \propto \rho_{15;3} - \rho_{12;3}\rho_{25;3} = 0 \Rightarrow \rho_{15;3} = 0,$$
$$\rho_{25;13} \propto \rho_{25;3} - \rho_{12;3}\rho_{15;3} = 0,$$
$$\rho_{45;123} \propto \rho_{45;13} - \rho_{24;13}\rho_{25;13} = 0 \Rightarrow \rho_{45;13} = 0,$$
$$\rho_{25;134} \propto \rho_{25;13} - \rho_{24;13}\rho_{45;13} = 0.$$

3.9.7   Multivariate distributions for which the simplifying assumption holds

In this section, we mention multivariate distributions for which (a) all conditional distributions have copulas that don't depend on the value of the conditioning variables, or (b) there is a vine sequence of conditional distributions for which conditional distributions have copulas that don't depend on the value of the conditioning variables. We refer to these below as Property (a) and Property (b) respectively. For Property (a), the multivariate copula of the distribution has a vine decomposition for any regular vine or equivalently the copula satisfies the simplifying assumption (3.35) for any regular vine. For Property (b), there is a vine decomposition for a specific regular vine or equivalently the copula satisfies the simplifying assumption for a specific vine.

It is well known that multivariate Gaussian distributions have conditional distributions that are univariate or lower-dimensional multivariate Gaussian (see Section 2.6). Because copulas are not affected by location and scale changes, the copula of conditional distributions of multivariate Gaussian given a particular subset of variables does not depend on the value of the conditioning variables.

The vine pair-copula construction can be considered as a generalization of multivariate Gaussian distributions which are parametrized in terms $d-1$ correlations and $(d-1)(d-2)/2$ partial correlations. In the case of a d-variate Gaussian distribution, there is a correlation associated with each of the $d-1$ edges in tree T_1 of the vine, and partial correlation associated with the $(d-1)(d-2)/2$ edges in trees T_2 to T_{d-1}. For a non-singular d-variate Gaussian distribution with correlation matrix $R = (\rho_{ij})$, $\{\rho_e : e \in \mathcal{E}(\mathcal{V})\}$ is a parametrization of the correlation matrix that is in the space $(-1,1)^{d(d-1)/2}$; the algebraic independence of the ρ_e follows from the algebraic independence of the bivariate copulas on the edges of a vine (and the bivariate copulas are taken as bivariate Gaussian to get the multivariate Gaussian copula). That is, this is a parametrization that consists of algebraically independent correlations and partial correlations, and avoids the positive definite constraints in the ρ_{ij}. The parametrization is called a partial correlation vine in Kurowicka and Cooke (2006). See Example 3.10 below for more details.

This parametrization is useful when one wants to consider parametric truncated vine models with $O(d)$ parameters and not $O(d^2)$ parameters.

Example 3.10 Some details of the vine decomposition and the determinant of a correlation matrix based on the vine are shown.

For the trivariate Gaussian distribution $F_{123} = \Phi_3(z_1, z_2, z_3; \Sigma)$ where Σ is a correlation matrix, the bivariate copula of $F_{23|1}(z_2, z_3|z_1)$ is

$$C_{23;1}(u_2, u_3) = \Phi_2(\Phi^{-1}(u_2), \Phi^{-1}(u_3); \rho_{23;1}), \quad \text{for any } z_1,$$

where

$$\rho_{23;1} = \frac{\rho_{23} - \rho_{12}\rho_{13}}{\sqrt{(1 - \rho_{12}^2)(1 - \rho_{13}^2)}}.$$

The range of $(\rho_{12}, \rho_{13}, \rho_{23;1})$ is $(-1,1)^3$ or equivalently $\rho_{12}, \rho_{13}, \rho_{23;1}$ are algebraically independent parameters. By permuting indices,

$$C_{13;2}(u_1, u_3) = \Phi_2(\Phi^{-1}(u_1), \Phi^{-1}(u_3); \rho_{13;2}), \quad \text{for any } z_2,$$
$$C_{12;3}(u_1, u_2) = \Phi_2(\Phi^{-1}(u_1), \Phi^{-1}(u_2); \rho_{12;3}), \quad \text{for any } z_3,$$

where

$$\rho_{13;2} = \frac{\rho_{13} - \rho_{12}\rho_{23}}{\sqrt{(1 - \rho_{12}^2)(1 - \rho_{23}^2)}}, \quad \rho_{12;3} = \frac{\rho_{12} - \rho_{13}\rho_{23}}{\sqrt{(1 - \rho_{13}^2)(1 - \rho_{23}^2)}}.$$

Hence the range of $(\rho_{12}, \rho_{23}, \rho_{13;2})$ or $(\rho_{13}, \rho_{23}, \rho_{12;3})$ is also $(-1,1)^3$.

Let

$$R = \begin{pmatrix} 1 & \rho_{12} & \rho_{13} \\ \rho_{12} & 1 & \rho_{23} \\ \rho_{13} & \rho_{23} & 1 \end{pmatrix}$$

be positive definite so that $1 - \rho_{jk}^2 > 0$ for all $j < k$ and $\det(\boldsymbol{R}) > 0$. With some algebra, it can be shown that

$$\det(\boldsymbol{R}) = (1 - \rho_{12}^2)(1 - \rho_{13}^2)(1 - \rho_{23;1}^2);$$

also this formula holds with permutation of indices.

For 4-variate Gaussian with cdf $F_{1234} = \Phi_4(\cdot; \boldsymbol{\Sigma})$, the parameter of the copula $C_{34;12}$ of $F_{34;12}$ is the partial correlation

$$\rho_{34;12} = \frac{\rho_{34;1} - \rho_{23;1}\rho_{24;1}}{\sqrt{(1 - \rho_{23;1}^2)(1 - \rho_{24;1}^2)}}.$$

Examples of partial correlation vines for $d = 4$ are

$$(\rho_{12}, \rho_{13}, \rho_{14}, \rho_{23;1}, \rho_{24;1}, \rho_{34;12}) \in (-1, 1)^6$$

and

$$(\rho_{12}, \rho_{23}, \rho_{34}, \rho_{13;2}, \rho_{24;3}, \rho_{14;23}) \in (-1, 1)^6.$$

From Kurowicka and Cooke (2006), with $\boldsymbol{R} = (\rho_{jk})_{1 \le j < k \le 4}$,

$$\begin{aligned}
\det(\boldsymbol{R}) &= (1 - \rho_{12}^2)(1 - \rho_{13}^2)(1 - \rho_{23;1}^2)(1 - \rho_{14}^2)(1 - \rho_{24;1}^2)(1 - \rho_{34;12}^2) \\
&= (1 - \rho_{12}^2)(1 - \rho_{23}^2)(1 - \rho_{13;2}^2)(1 - \rho_{34}^2)(1 - \rho_{24;3}^2)(1 - \rho_{14;23}^2);
\end{aligned}$$

there are similar expressions for the determinant from other vine decompositions.

The general partial correlation vine is $\{\rho_e : e \in \mathcal{E}(\boldsymbol{V})\}$ where $\mathcal{V} = (\mathcal{T}_1, \ldots, \mathcal{T}_{d-1})$ is a vine. The general determinant result for a partial correlation vine is

$$\log[\det(\boldsymbol{R})] = \sum_{e \in \mathcal{E}(\mathcal{V})} \log(1 - \rho_e^2).$$

For a truncated vine with zero partial correlations for trees $\mathcal{T}_{m+1}, \ldots, \mathcal{T}_{d-1}$, the determinant becomes

$$\log[\det(\boldsymbol{R})] = \sum_{e \in \mathcal{T}_1, \ldots, \mathcal{T}_m} \log(1 - \rho_e^2).$$

The determinant criterion is important when looking for a best truncated vine; see Section 6.17. $\square$

Multivariate t distributions have conditional distributions that are univariate or lower-dimensional multivariate t with a location and scale changes, and shape (or degree of freedom) parameter that is increased by the number of conditioning variables (see Section 2.7). Hence multivariate t copulas satisfy the simplifying assumption for any regular vine.

There are other multivariate distributions that satisfy Property (a) or (b). A list is given below; most of these are in the literature on multivariate distributions and others were obtained based on constructions in the preceding sections.

- The multivariate Pareto and Burr distributions in Examples 1.3 and 1.4 are closed under conditional distributions (the scale parameter depends on the values of the conditioning variables), as shown by Takahasi (1965). Hence the resulting multivariate copula, which is Archimedean based on a gamma LT, satisfies the simplifying assumption for any vine.
- The trivariate Pareto survival function in (3.18) and hierarchical Archimedean copula (3.17) is an example of a distribution that satisfies the simplifying assumption with the vine based on $\{13, 23, 12|3\}$ but not $\{12, 13, 23|1\}$ or $\{12, 23, 13|2\}$ if $\delta > 1$. It is easier to determine the copula $C_{12;3}$ from (3.18) than (3.17). The trivariate survival function is:

$$\overline{F}_{123}(x_1, x_2, x_3) = [(x_1^\delta + x_2^\delta)^{1/\delta} + x_3 + 1]^{-1/\theta}, \quad x_1, x_2, x_3 > 0, \ \theta > 0, \ \delta > 1.$$

The common univariate survival margin and density are $(1+x)^{-1/\theta}$ and $\theta^{-1}(1+x)^{-1/\theta-1}$ respectively. Hence

$$\overline{F}_{12|3}(x_1, x_2|x_3) = \frac{-\partial \overline{F}_{123}}{\partial x_3} \Big/ f_3(x_3) = \left\{ \left[\left(\frac{x_1}{1+x_3}\right)^\delta + \left(\frac{x_2}{1+x_3}\right)^\delta \right]^{1/\delta} + 1 \right\}^{-1/\theta-1},$$

and its copula does not depend on x_3 because $1 + x_3$ appears as a scale parameter. The copula $C_{12;3}$ is in the BB1 family (4.53) with parameter $(\theta/[1+\theta], \delta)$. Because of symmetry in x_1, x_2 the other two bivariate conditional distributions are the same, and

$$\overline{F}_{23|1}(x_2, x_3|x_1) = \left\{ \frac{(x_1^\delta + x_2^\delta)^{1/\delta} + x_3 + 1}{x_1 + 1} \right\}^{-1/\theta-1} (x_1^\delta + x_2^\delta)^{1/\delta-1} x_1^{\delta-1}.$$

The copula $C_{23;1}(\cdot; x_1)$ depends x_1 and cannot be written conveniently because $\overline{F}_{2|1}(\cdot|x_1)$ and $\overline{F}_{3|1}(\cdot|x_1)$ do not have closed form inverses.

- The preceding example can be extended to a trivariate copula (3.23) based on a mixture of max-id, if the max-id distribution is a bivariate extreme value copula. With ψ being a gamma LT with parameter $\theta > 0$ and $K_{12}(u, v) = \exp\{-A(-\log u, \log v)\}$ where A is homogeneous of order 1 and $A(x, 0) = A(0, x) = x$, then (3.23) becomes

$$C_{\psi,K}(\boldsymbol{u}) = \left[-\log K_{12}\left(e^{-(u_1^{-\theta}-1)}, e^{-(u_2^{-\theta}-1)}\right) + u_3^{-\theta} \right]^{-1/\theta} = \left[A\left(u_1^{-\theta}-1, u_2^{-\theta}-1\right) + u_3^{-\theta} \right]^{-1/\theta}.$$

Hence
$$C_{12|3}(u_1, u_2|u_3) = \left[u_3^\theta A\left(u_1^{-\theta} - 1, u_2^{-\theta} - 1\right) + 1 \right]^{-1/\theta-1},$$

and with $v_j = [u_3^\theta(u_j^{-\theta} - 1) + 1]^{-1/\theta-1}$ for $j = 1, 2$, the corresponding copula is

$$\begin{aligned} C_{12;3}(v_1, v_2; u_3) &= \left[u_3^\theta A\left(\{v_1^{-\theta/(1+\theta)} - 1\}u_3^{-\theta}, \{v_2^{-\theta/(1+\theta)} - 1\}u_3^{-\theta}\right) + 1 \right]^{-1/\theta-1} \\ &= \left[A\left(v_1^{-\theta/(1+\theta)} - 1, v_2^{-\theta/(1+\theta)} - 1\right) + 1 \right]^{-1/\theta-1}. \end{aligned}$$

Therefore $C_{12;3}$ does not depend on u_3. It can be checked for nontrivial A that $C_{23;1}$ depends on u_1 so that the simplifying assumption is not satisfied for all vines. The hierarchical Archimedean copula in (3.17) for the preceding example obtains as a special case with K_{12} being the Gumbel copula (4.14). Also note that the $(1, 2)$ margin of $C_{\psi,K}$ is more concordant than the $(1, 3)$ and $(2, 3)$ margins.

- The multivariate GB2 distribution (see Section 4.27) has conditional independence based on a gamma mixture. All of its conditional distributions depend only on the sum of the values of the conditioning variables in the scale parameter. Hence the multivariate GB2 copula satisfies the simplifying assumption for any regular vine.

- The Dirichlet distribution holds for $\boldsymbol{X} = (X_1, \ldots, X_d)$ if it has the stochastic representation $\boldsymbol{X} = (Z_1, \ldots, Z_d)/Z_\bullet$, where $Z_\bullet = Z_1 + \cdots + Z_d$ and the Z_j's are independent Gamma random variables with different shape parameters and a common scale parameter. Let S be a non-empty proper subset of $\{1, \ldots, d\}$ and let $T = \{1, \ldots, d\}\backslash S$. Then $[(X_j : j \in T)|(X_k = x_k : k \in S)]$ has a Dirichlet distribution after rescaling to have a sum of 1 (Chapter 40 of Johnson and Kotz (1972)). Hence the copula of the Dirichlet distribution satisfies the simplifying assumption for any vine.

- Proposition 3 of Hobæk Haff et al. (2010) has a construction based on a specific vine where conditional distributions are in a family with location and scale parameters depending on the values of conditioning variables. A condition is needed to satisfy Property (b).

It is proved in Stöber et al. (2013b) that among Archimedean copulas, only the ones based on the gamma LT and their extensions to negative dependence satisfy the simplifying assumption. Also Stöber et al. (2013b) prove that among elliptical distributions only the multivariate t_ν (with multivariate Gaussian as $\nu \to \infty$) and their extensions to elliptical distributions with finite support satisfy the simplifying assumption. The t distribution (see Section 2.7) is based on a scale mixture of Gaussian where the reciprocal of the random scale variable has a Gamma distribution. Hence there was similarity in the proofs of characterizations within the Archimedean and elliptical copulas.

Since Sklar's theorem states that the copula function is not unique in the discrete case, the simplifying assumption is potentially less restrictive in the discrete case than in the continuous case. Hence there are more multivariate discrete distributions (among simple parametric families) that can be decomposed as vines satisfying the simplifying assumption. We now investigate in more detail the restrictions that this assumption imposes.

Let j, k be distinct indices not in the subset of indices S. For a multivariate discrete distribution, $F_{jk|S}, F_{j|S}, F_{k|S}$ are discrete. Let the discrete random variables be $X_1, \ldots, X_d$ and let their support sets be $\mathcal{X}_1, \ldots, \mathcal{X}_d$ respectively.

For fixed $\boldsymbol{x}_S$, there is a copula $C_{jk;S}(\cdot; \boldsymbol{x}_S)$ such that $F_{jk|S}(x_j, x_k | \boldsymbol{x}_S) = C_{jk;S}(F_{j|S}(x_j|\boldsymbol{x}_S), F_{k|S}(x_k|\boldsymbol{x}_S); \boldsymbol{x}_S)$ and $C_{jk;S}(\cdot; \boldsymbol{x}_S)$ is unique on the domain $D(\boldsymbol{x}_S) = \{F_{j|S}(x|\boldsymbol{x}_S) : x \in \mathcal{X}_j\} \times \{F_{k|S}(x|\boldsymbol{x}_S) : x \in \mathcal{X}_k\}$. Under some conditions there is a copula C^* such that $C^* = C_{jk;S}(\cdot; \boldsymbol{x}_S)$ on the set $D(\boldsymbol{x}_S)$ for all $\boldsymbol{x}_S$; that is, the copulas $C_{jk;S}(\cdot; \boldsymbol{x}_S)$ can be "merged" so that

$$F_{jk|S}(x_j, x_k | \boldsymbol{x}_S) = C^*(F_{j|S}(x_j|\boldsymbol{x}_S), F_{k|S}(x_k|\boldsymbol{x}_S)), \quad \forall x_j, x_j, \boldsymbol{x}_S. \tag{3.49}$$

To determine if there is such a C^*, a linear programming problem can be set up.

If there is a copula C^*, then there are constraints $C^*(a_i, b_i) = p_i$, $i = 1, \ldots, N$, where the N equations in a_i, b_i, p_i arise from (3.49). Let $a_{(1)} < \cdots < a_{(N_1)}$ be the distinct a_i values and let $b_{(1)} < \cdots < b_{(N_2)}$ be the distinct b_i values. Let $a_{(0)} = b_{(0)} = 0$ and $a_{(N_1+1)} = b_{(N_2+1)} = 1$. Let

$$R_{jk} = C^*(a_{(j)}, b_{(k)}) - C^*(a_{(j-1)}, b_{(k)}) - C^*(a_{(j)}, b_{(k-1)}) + C^*(a_{(j-1)}, b_{(k-1)})$$

for $1 \leq j \leq N_1 + 1$, $1 \leq k \leq N_2 + 1$. The N constraints are $C^*(a_i, b_i) = p_i$, and also $C^*(a_{(j)}, b_{(N_2+1)}) = a_{(j)}$, $C^*(a_{(N_1+1)}, b_{(k)}) = b_{(k)}$, $C^*(a_{(0)}, b_{(k)}) = 0 = C^*(a_{(j)}, b_{(0)})$. The rectangle probabilities R_{jk} lead to $(N_1+1)(N_2+1)$ non-negativity constraints on $N_1 N_2 - N$ unknowns $\{C^*(a_{(j)}, b_{(k)}) : j = 1, \ldots, N_1, k = 1, \ldots, N_2\} \backslash \{C^*(a_i, b_i) : i = 1, \ldots, N\}$. In addition, there are other monotonicity constraints on $C^*(a_{(j)}, b_{(k)})$ relative to neighboring $C^*(a_i, b_i)$. If there is a non-negative rectangle solution, then there exists a valid distribution with support on $\{0, a_{(1)}, \ldots, a_{(N_1)}, 1\} \times \{0, b_{(1)}, \ldots, b_{(N_2)}, 1\}$. This can be extended nonuniquely to a compatible bivariate copula C^* (with support on all of $[0,1]^2$); a simple extension is one which is piecewise uniform over the rectangles $[a_{(j-1)}, a_{(j)}] \times [b_{(k-1)}, b_{(k)}]$ (page 14 of Joe (1997)) and other methods are given in Genest and Nešlehová (2007)). Given the constraints, linear programming or linear system reduction can be used to check for a non-negative solution where an arbitrary objective function is minimized or maximized.

Note that if $a_i < a_{i'}$ and $b_i < b_{i'}$ and $C^*(a_i, b_i) = p_i$, $C^*(a_{i'}, b_{i'}) = p_{i'}$, then it is straightforward to use set inequalities to get the necessary condition:

$$\max\{a_{i'} + b_{i'} - 1, p_i\} \leq p_{i'} \leq \min\{a_{i'}, b_{i'}, p_i + a_{i'} - a_i + b_{i'} - b_i\}. \tag{3.50}$$

This condition can be interpreted as the local dependence not being substantially different; this inequality might fail if p_i matches a Fréchet upper bound, p_i' matches a Fréchet lower bound and $a_{i'} - a_i$, $b_{i'} - b_i$ are quite small.

The idea of the next example is to show that the simplifying assumption is much easier to be satisfied for multivariate discrete distributions, and to show a technique for verifying the simplifying assumption. Essentially, it should hold if the strength of dependence in the discrete conditional distributions $F_{jk|S}(\cdot|\boldsymbol{x}_S)$ is not too different for different $\boldsymbol{x}_S$. Hence this property can lead to the construction of a multivariate discrete distribution for which the simplifying assumption is satisfied for a particular vine but not necessarily all vines.

Example 3.11 (Trivariate beta-binomial). For an example where the number of inequalities does not get too large, consider the following simple multivariate exchangeable model with support on $\{0, 1, \ldots, K\}^d$, $d = 1, 2, \ldots$. The mixing random variable $P \sim \text{Beta}(\alpha, \beta)$ and conditioned on $P = p$, the random variables $X_1, \ldots, X_d$ are independent Binomial(K, p). The Beta distribution mixing distribution has parameters $\alpha = \pi/\gamma$, $\beta = (1 - \pi)/\gamma$ where $0 < \pi < 1$, $\gamma > 0$, so that $\alpha + \beta = \gamma^{-1}$. The multivariate pmf is

$$
\begin{aligned}
f_{1:d}(x_1, \ldots, x_d) &= \int_0^1 \prod_{j=1}^d \binom{K}{x_j} p^{x_j}(1-p)^{K-x_j} \cdot \frac{p^{\pi/\gamma-1}(1-p)^{(1-\pi)/\gamma-1}}{B(\pi/\gamma, (1-\pi)/\gamma)} dp \\
&= \prod_{j=1}^d \binom{K}{x_j} \frac{B(\pi/\gamma + x_\bullet, (1-\pi)/\gamma) + Kd - x_\bullet)}{B(\pi/\gamma, (1-\pi)/\gamma)} \\
&= \prod_{j=1}^d \binom{K}{x_j} \frac{\prod_{i=0}^{x_\bullet-1}(\pi/\gamma + i) \cdot \prod_{i=0}^{Kd-x_\bullet-1}((1-\pi)/\gamma + i)}{\prod_{i=0}^{Kd-1}(\gamma^{-1} + i)} \\
&= \left\{ \prod_{j=1}^d \binom{K}{x_j} \right\} \frac{\prod_{i=0}^{x_\bullet-1}(\pi + i\gamma) \cdot \prod_{i=0}^{Kd-x_\bullet-1}((1-\pi) + i\gamma)}{\prod_{i=0}^{Kd-1}(1 + i\gamma)},
\end{aligned}
$$

where $x_\bullet = x_1 + \cdots + x_d$ and $x_j \in \{0, 1, \ldots, K\}$ for $j = 1, \ldots, d$. The density is valid even for $d = 1$, in which case:

$$
f_1(x) = \binom{K}{x} \frac{\prod_{i=0}^{x-1}(\pi + i\gamma) \cdot \prod_{i=0}^{K-x-1}((1-\pi) + i\gamma)}{\prod_{i=0}^{K-1}(1 + i\gamma)}, \quad x = 0, \ldots, K.
$$

For beta-binomial with $d = 3$ and $K = 2$, some specific examples are given with different amounts of overlaps in the marginal cdf values of the bivariate conditional distributions.

(a) $\pi = 0.4$, $\gamma = 1$: the conditional distributions $F_{12|3}(\cdot|x_3)$ are shown in the table below, with conditional univariate marginal distributions in the columns labeled as 2 and in the row labeled as 2.

$i_1 \backslash i_2$	$F_{12|3}(i_1, i_2|0)$			$F_{12|3}(i_1, i_2|1)$			$F_{12|3}(i_1, i_2|2)$		
	0	1	2	0	1	2	0	1	2
0	0.670	0.765	0.780	0.191	0.308	0.347	0.025	0.058	0.080
1	0.765	0.919	0.953	0.308	0.580	0.720	0.058	0.179	0.320
2	0.780	0.953	1.000	0.347	0.720	1.000	0.080	0.320	1.000

The correlation of each of the three conditional distributions is 0.4. The univariate marginal values of a's (also b's because of exchangeability) are:

$$(a_1, \ldots, a_6) = (0.080, 0.320, 0.347, 0.720, 0.780, 0.953).$$

There is no overlap of these cdf values from the three bivariate conditional distributions, so it suffices to check the condition (3.50) with $(a_i, a_{i'}) = (b_i, b_{i'})$ equal to $(0.320, 0.347)$ and $(0.720, 0.780)$ with a values occuring from $F_{12|3}(\cdot|x_3)$ with different x_3. The required conditions are:

$$
\begin{aligned}
\max\{0.347 + 0.347 - 1, \, 0.179\} &= C^*(0.320, 0.320)\} \le C^*(0.347, 0.347) = 0.191 \\
&\le \min\{0.347, \, 0.179 + 2(0.347 - 0.320)\}, \\
\max\{0.780 + 0.780 - 1, \, 0.580\} &= C^*(0.720, 0.720)\} \le C^*(0.780, 0.780) = 0.670 \\
&\le \min\{0.780, \, 0.580 + 2(0.780 - 0.720)\}.
\end{aligned}
$$

This is fine, because $0.179 \le 0.191 \le 0.233$ and $0.580 \le 0.670 \le 0.700$.

(b) $\pi = 0.3$, $\gamma = 0.7$: the conditional distributions are in the table below.

$i_1 \backslash i_2$	$F_{12\|3}(i_1, i_2\|0)$			$F_{12\|3}(i_1, i_2\|1)$			$F_{12\|3}(i_1, i_2\|2)$		
	0	1	2	0	1	2	0	1	2
0	0.679	0.776	0.790	0.226	0.356	0.395	0.045	0.100	0.132
1	0.776	0.929	0.960	0.356	0.642	0.772	0.100	0.281	0.452
2	0.790	0.960	1.000	0.395	0.772	1.000	0.132	0.452	1.000

The correlation of each of the three conditional distributions is 0.36842. The univariate marginal values are: $(a_1, \ldots, a_6) = (b_1, \ldots, b_6) = (0.132, 0.395, 0.452, 0.772, 0.790, 0.960)$. There is overlap in the range of marginal cdf values from the bivariate conditional distributions given $y = 1$ and given $y = 2$. Conditions (3.50) for the first and second conditional distributions, with more significant digits, lead to:

$$\max\{0.79032 + 0.79032 - 1, 0.64243\} = C(0.77151, 0.77151)\} \leq C(0.79032, 0.79032)$$
$$= 0.67940 \leq \min\{0.79032, 0.64243 + 2(0.79032 - 0.77151)\}.$$

This is satisfied because $0.64243 \leq 0.67940 \leq 0.68005$. For the overlapping second and third conditional distributions, the monotonicity and non-negative rectangle constraints can be reduced to the following because of symmetry:

$$0.2265 + 0.0453 - 2z_1 \geq 0, \qquad\qquad 0.2810 + 0.2265 - 2z_3 \geq 0,$$
$$z_3 + z_1 - 0.2265 - 0.1003 \geq 0, \qquad\qquad z_4 + z_3 - 0.2810 - 0.3559 \geq 0,$$
$$0.3559 + 0.1003 - z_3 - z_2 \geq 0, \qquad\qquad 0.6424 + 0.2810 - 2z_4 \geq 0,$$
$$z_1 = C^*(0.395, 0.132) \geq 0.0453, \qquad\qquad z_3 = C^*(0.452, 0.395) \geq 0.2265,$$
$$z_2 = C^*(0.772, 0.132) \in [0.1003, 0.1317], \qquad z_4 = C^*(0.772, 0.452) \geq 0.3559.$$

These constraints can be seen in Figure 3.6. The linear programming solution from $\min z_1$ given the constraints leads to: $(z_1, z_2, z_3, z_4) = (0.073, 0.100, 0.254, 0.383)$. From this, one can get consistent values for $C^*(a_5, a_j)$ and $C^*(a_6, a_j)$ for $j = 1, \ldots, 6$.

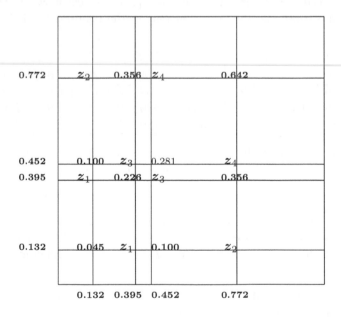

Figure 3.6 *Grid of the bivariate copula to show the unknowns z_1, z_2, z_3, z_4, the constraints, and the specified values for C^*.*

(c) $\pi = 0.4$, $\gamma = 0.5$: the conditional distributions are in the table below.

$i_1 \backslash i_2$	$F_{12\mid 3}(i_1, i_2\mid 0)$			$F_{12\mid 3}(i_1, i_2\mid 1)$			$F_{12\mid 3}(i_1, i_2\mid 2)$		
	0	1	2	0	1	2	0	1	2
0	0.516	0.649	0.672	0.183	0.310	0.352	0.042	0.099	0.132
1	0.649	0.874	0.928	0.310	0.605	0.748	0.099	0.289	0.468
2	0.672	0.928	1.000	0.352	0.748	1.000	0.132	0.468	1.000

The correlation of each of the three conditional distributions is $1/3$. The univariate marginal values are: $(a_1, \ldots, a_6) = (b_1, \ldots, b_6) = (0.132, 0.352, 0.468, 0.672, 0.748, 0.928)$. There is overlap from the bivariate conditional distributions given $x_3 = 1$ with those given $x_3 = 0$ and given $x_3 = 2$. With the symmetry, one can set up a linear programming problem with variables:

$$z_1 = C^*(0.352, 0.132),\ z_2 = C^*(0.672, 0.132),\ z_3 = C^*(0.748, 0.132),\ z_4 = C^*(0.928, 0.132)$$
$$z_5 = C^*(0.468, 0.352),\ z_6 = C^*(0.672, 0.352),\ z_7 = C^*(0.928, 0.352)$$
$$z_8 = C^*(0.672, 0.468),\ z_9 = C^*(0.748, 0.468),\ z_{10} = C^*(0.928, 0.468),$$
$$z_{11} = C^*(0.748, 0.672),$$
$$z_{12} = C^*(0.928, 0.748).$$

Given the constraints from monotonicities and non-negative rectangle probabilities of the copula cdf, the inequality constraints can be reduced to the following because of symmetry (a diagram similar to Figure 3.6 can be drawn):

$$0.183 + 0.042 - 2z_1 \geq 0,$$
$$z_5 + z_1 - 0.183 - 0.099 \geq 0,$$

$$z_6 + 0.099 - z_5 - z_2 \geq 0, \qquad z_3 - z_2 \geq 0,$$
$$0.310 + z_2 - z_6 - z_3 \geq 0, \qquad z_4 - z_3 \geq 0,$$
$$z_7 + z_3 - 0.310 - z_4 \geq 0, \qquad z_6 - z_2 \geq 0,$$
$$0.289 + 0.183 - 2z_5 \geq 0, \qquad z_6 - z_5 \geq 0,$$
$$z_8 + z_5 - 0.289 - z_6 \geq 0, \qquad z_7 - z_4 \geq 0,$$
$$z_9 + z_6 - z_8 - 0.310 \geq 0, \qquad z_8 - z_6 \geq 0,$$
$$z_{10} + 0.310 - z_9 - z_7 \geq 0, \qquad z_9 - z_8 \geq 0,$$
$$0.516 + 0.289 - 2z_8 \geq 0, \qquad z_{10} - z_7 \geq 0,$$
$$z_{11} + z_8 - 0.516 - z_9 \geq 0, \qquad z_{10} - z_9 \geq 0,$$
$$0.649 + z_9 - z_{11} - z_{10} \geq 0, \qquad z_{11} - z_9 \geq 0,$$
$$0.605 + 0.516 - 2z_{11} \geq 0, \qquad z_2 - 0.99 \geq 0,$$
$$z_{12} + z_{11} - 0.605 - 0.649 \geq 0,$$
$$0.874 + 0.605 - 2z_{12} \geq 0.$$

The linear programming solution from min z_1 subject to the constraints leads to:

$$(z_1, \ldots, z_{12}) = (0.046, 0.099, 0.099.0.099, 0.236, 0.310, 0.310, 0.363, 0.363, 0.363, 0.516, 0.738).$$

□

In the above, we have listed some examples of multivariate distributions that satisfy the simplifying assumption. At the other extreme, we can consider multivariate distributions that are furthest from satisfying the simplifying assumption. With a fairly natural stochastic representation and dependence structure, one such class are factor models based on positive random variables in a convolution-closed family; see Section 4.28. For all conditional distributions, the conditional dependence can vary from conditional independence to the conditional Fréchet upper bound as the value of the conditioning variables increase. Hence the simplifying assumption is far from holding for any regular vine.

For the special case of a trivariate distribution for the gamma factor model, Stöber et al.

(2013b) assessed the closeness of vine approximations, and concluded that typically sample sizes in the hundreds would be needed to distinguish the trivariate distribution from the best vine approximation.

To summarize this subsection, multivariate distributions can be decomposed as a sequence of conditional distributions in many ways; in general the simplifying assumption is not satisfied but might be a good approximation. Vine copula models in applications, such as in Chapter 7, are considered as approximations for the purpose of making inferences about quantities of interest. Because it is known that some factor models can deviate the most from the simplifying assumption, factor copula models in Sections 3.10 and 3.11 should also be considered for comparisons, especially when dependence can be explained by some latent variables. More discussion and comparisons are given in Section 3.13.

3.9.8   Vine equivalence classes

The idea of vine equivalence classes is implicit in the preceding subsections. If the vine is a "graph," then relabeling the nodes for T_1 and applying the same relabeling to trees $T_2, \ldots, T_{d-1}$ does not change the dependence structure. Counting equivalence classes is a way to understand the number of distinct d-dimensional vine structures.

Vines in the same equivalence classes can be matched after a permutation of indices. Denote a permutation P of $\{1, \ldots, d\}$ as $P(1 \cdots d) = k_1 \cdots k_d$ where components of P are summarized as $P(i) = k_i$. For a subset S, we use the notation $P(S) = \{P(j) : j \in S\}$.

Definition 3.3 (Equivalence classes of vines). Consider a regular vine written in the form of a set of $\binom{d}{2}$ conditional distributions:

$$\{[\{i_1, i_2\} | S_\ell(i_1, i_2)] : 1 \le i_1 < i_2 \le d, \text{ with } d - \ell - 1 \text{ pairs at } T_{\ell+1}, \ell = 0, \ldots, d - 2\}.$$

The subscript on S is used to clarify the size ℓ of the conditioning set. Let P be a permutation. Then the regular vines $\{[\{i_1, i_2\} \mid S_\ell(i_1, i_2)]\}$ and $\{[\{P(i_1), P(i_2)\} \mid P(S_\ell(i_1, i_2))]\}$ are said to be in the same equivalence class.

Alternatively, two regular vines that are summarized as $\{[\{i_1, i_2\} \mid S_\ell(i_1, i_2)]\}$ and $\{[\{i'_1, i'_2\} \mid S_\ell(i'_1, i'_2)]\}$ are in the *same equivalence class* if there is a permutation P such that $\{[\{P(i_1), P(i_2)\} \mid P(S_\ell(i_1, i_2))]\}$ and $\{[\{i'_1, i'_2\} \mid S_\ell(i'_1, i'_2)]\}$ are the same. Note that in the set notation, the order within the subset of conditioned variables or within the subset of conditioning variables doesn't matter. Also, the order of the listing of the distributions doesn't matter.

We will now use shorthand notation, without punctuation, for the edges. For $d = 3$, we can assume that 12 is a pair at tree T_1, then the second pair is either 23 (D-vine) or 13 (C-vine). The D-vine has edges 12, 23, 13|2 in standard form, and the C-vine has edges 12, 13, 23|1 in standard form. For the C-vine, permute $\pi(123) = 213$ to get $\{21, 23, 13|2\}$, which is the same as the D-vine. Note that for this vine in the graph representation, T_1 has one node of degree 2 and two nodes of degree 1.

For $d = 4$, for tree T_1, two nodes can have degree 2 and the other two have degree 1 (similar to the D-vine), or one node can have degree 3 and the other three have degree 1 (similar to the C-vine). This leads to two equivalence classes: the C-vine and D-vine, as shown below.

- D-vine: 12, 23, 34, 13|2, 24|3, 14|23. Note that there is no alternative conditioning after T_1 since the margins 12 and 34 cannot be paired to have a common conditioning index.
- C-vine: 12, 13, 14, 23|1, 24|1, 34|12 in standard form,
- An alternative conditioning starting from $\{12, 13, 14\}$ at T_1 is 12, 13, 14, 23|1, 34|1, 24|13. Permute $P(1234) = 1324$ to get 13, 12, 14, 32|1, 24|1, 34|12, so this is now a C-vine in standard form.

- A second alternative conditioning starting from $\{12, 13, 14\}$ at T_1 is 12, 13, 14, 24|1, 34|1, 23|14. Permute $P(1234) = 1432$ to get 14, 13, 12, 42|1, 32|1, 43|12, and this is also a C-vine in standard form.

Using the previous enumeration method starting from

$$
12 \quad \begin{matrix} 13 \\ 23 \end{matrix} \quad \begin{matrix} 14 \\ 24 \\ 34 \end{matrix} \quad \begin{matrix} 15 \\ 25 \\ 35 \\ 45 \end{matrix}
$$

there are 6 equivalence classes for $d = 5$. The enumeration starts by noting that there are three possibilities for tree T_1, (a) three nodes with degree 2 and the other two with degree 1 (similar to the D-vine), or (b) one node with degree 3, one node with degree 2 and the other three with degree 1, or (c) one node with degree 4 and the other four with degree 1 (as in the first tree of the C-vine).

(a) D-vine 12, 23, 34, 45, 13|2, 24|3, 35|4, 14|23, 25|34, 15|234 (the T_1 degrees of the nodes are 1,2,2,2,1 and there are unique trees T_2, T_3, T_4).

(b) 12, 13, 14, 25 (the T_1 degrees of the nodes are 3,2,1,1,1). T_2 possibilities are 23|1, 24|1, 34|1, 15|2, of which 2 of first 3 can be chosen (as they form a cycle). Variables 3 and 4 are symmetric so there are two cases for T_2:

(i) 23|1, 24|1, 15|2, leading to T_3 possibilities 34|12, 35|12, 45|123 — 2 cases for next step: (i′) 34|12, 35|12, 45|123 or (i″) 35|12, 45|12, 34|125,

or, (ii) 23|1, 34|1, 15|2, leading to the remainder 24|13, 35|12, 45|123.

(c) 12, 13, 14, 15 (T_1 degrees of the nodes are 4,1,1,1,1). T_2 possibilities are 23|1, 24|1, 34|1, 25|1, 35|1, 45|1 — can choose 3 of 6 without a cycle.

(i) 23|1, 24|1, 25|1 (C-vine-like) leads to T_3 possibilities 34|12, 35|12, 45|12, and the remaining step is a C-vine for any choices, say 34|12, 35|12, 45|123, or

(ii) 23|1, 24|1, 35|1 (which is a D-vine 4-2-3-5 given 1), so following are 34|12, 25|13, 45|123.

The 6 equivalence classes for 5-dimensional vines can be summarized as in the following table, where the four equivalence classes between the C-vine and D-vine are denoted as B0, B1, B2 and B3.

D : 12, 23, 34, 45,	13|2, 24|3, 35|4,	14|23, 25|34,	15|234	
B1: 12, 13, 14, 25,	23|1, 24|1, 15|2,	34|12, 35|12,	45|123	
B2: 12, 13, 14, 25,	23|1, 24|1, 15|2,	35|12, 45|12,	34|125	
B3: 12, 13, 14, 25,	23|1, 34|1, 15|2,	24|13, 35|12,	45|123	
C : 12, 13, 14, 15,	23|1, 24|1, 25|1,	34|12, 35|12,	45|123	
B0: 12, 13, 14, 15,	23|1, 24|1, 35|1,	34|12, 25|13,	45|123	

The number of equivalence classes grows exponentially as d increases. The analysis and count of the number of equivalence classes of vines is given in Joe et al. (2011); see also Theorem 8.63.

3.9.9 Historical background of vines

This subsection mentions some history and the early references; more details are in Cooke et al. (2011a). The name "vine" comes from Cooke (1997), and the general development of regular vines as in Definition 3.2, is given in Bedford and Cooke (2001, 2002). The special case of D-vine, with no indication of the generalization, is in Joe (1994, 1996a, 1997), and was constructed as a sequence of mixtures of conditional distributions. Joe (1996a) mentions that the vine density involves no integrals but did not develop the vine copula density result of Bedford and Cooke (2001), because the initial interest was in the extreme value limit.

When there is tail dependence in the vine copula, the extreme value limit does involve integrals.

3.10   Factor copula models

Factor copula models are conditional independence models where observed variables are conditionally independent given one or more latent variables. These models extend the multivariate Gaussian model where the $d \times d$ correlation matrix $\boldsymbol{R}$ has the factor correlation structure if $\boldsymbol{R} = \boldsymbol{A}\boldsymbol{A}^\top + \boldsymbol{\Psi}^2$ where $\boldsymbol{A}$ is $d \times p$ with $1 \leq p < d$ and $\boldsymbol{\Psi}^2$ is a diagonal matrix with elements $\psi_1^2, \ldots, \psi_d^2$. For multivariate Gaussian with standardization to $N(0, 1)$ margins, the stochastic representation for observed variables $Z_1, \ldots, Z_d$, latent variables $W_1, \ldots, W_p$ and residual variables $\epsilon_1, \ldots, \epsilon_d$ is

$$ Z_j = \alpha_{j1} W_1 + \cdots + \alpha_{jp} W_d + \psi_j \epsilon_j, \quad j = 1, \ldots, d. $$

The matrix $\boldsymbol{A} = (\alpha_{jk})$ is called the matrix of loadings. It is not unique because $(\boldsymbol{A}\boldsymbol{P})(\boldsymbol{P}^\top \boldsymbol{A}^\top) = \boldsymbol{A}\boldsymbol{A}^\top$ if $\boldsymbol{P}$ is an orthogonal $p \times p$ matrix.

For continuous response variables, a general factor copula model is presented in Section 3.10.1; the 1-factor copula is given in Joe (2011b) and a general p-factor copula is developed in Krupskii and Joe (2013) together with numerical implementation of the likelihood for $p = 1$ and $p = 2$. For ordinal response variables, the factor copula model presented in Section 3.10.2 is studied in Nikoloulopoulos and Joe (2014) in the context of item response.

Also Klüppelberg and Kuhn (2009) consider a model with multivariate t_ν, where the correlation matrix has factor correlation structure. This model is useful but is not a special case of the factor copula model; more discussion about this is in Section 3.10.4.

3.10.1   Continuous response

We assume that all copulas are absolutely continuous and have densities, so that the log-likelihood for continuous data will involve the density of the factor copula, after fitting univariate margins and converting to $U(0, 1)$.

In the p-factor copula model, the dependent uniform random variables $U_1, \ldots, U_d$ are assumed to be conditionally independent given p latent variables $V_1, \ldots, V_p$. Without loss of generality, we can assume V_i are i.i.d. $U(0, 1)$. Let the conditional cdf of U_j given $V_1, \ldots, V_p$ be denoted by $F_{j|V_1,\ldots,V_p} = C_{j|V_1,\ldots,V_p}$. Then,

$$ C(\boldsymbol{u}) = \int_{[0,1]^p} \prod_{j=1}^d C_{j|V_1,\ldots,V_p}(u_j|v_1,\ldots,v_p) \, \mathrm{d}v_1 \cdots \mathrm{d}v_p, \quad \boldsymbol{u} \in [0,1]^d. \qquad (3.51) $$

Any conditional independence model, based on p independent latent variables, can be put in this form after transforms to $U(0, 1)$ random variables. Hence, the dependence structure of $\boldsymbol{U}$ is then defined through conditional distributions $C_{1|V_1,\ldots,V_p}, \ldots, C_{d|V_1,\ldots,V_p}$. We will call (3.51) a *factor copula model*, after $C_{j|V_1,\ldots,V_p}$ are expressed appropriately in terms of a sequence of bivariate copulas that link the observed variables U_j and the latent variables V_k. Some of the bivariate copulas are applied to conditional distributions.

In the statistical finance literature, there are alternative factor copula models with different specification of $F_{j|V_1,\ldots,V_p}$; see McNeil et al. (2005) (Section 9.7.2), Hull and White (2004) and Oh and Patton (2012). These all have a linear latent structure and are not as general as the models presented in this section.

We next show details for the case of $p = 1$. For $j = 1, \ldots, d$, denote the joint cdf and density of (V_1, U_j) by $C_{V_{1j}}$ and $c_{V_{1j}}$ respectively. The conditional distribution of U_j given

$V_1 = v$ is $C_{j|V_1}(u_j|v) = \partial C_{V_1 j}(v, u_j)/\partial v$. With $p = 1$, equation (3.51) becomes:

$$C(\boldsymbol{u}) = \int_0^1 \prod_{j=1}^d C_{j|V_1}(u_j|v_1) \, dv_1, \quad \boldsymbol{u} \in [0,1]^d. \tag{3.52}$$

If the density of C_{j,V_1} exists, then $\frac{\partial}{\partial u} C_{j|V_1}(u|v_1) = \frac{\partial^2}{\partial v_1 \partial u} C_{V_1 j}(v_1, u) = c_{V_1 j}(v_1, u)$. The existence of the densities $c_{V_1 j}$ implies by differentiation of (3.52) that the density of the 1-factor copula is

$$c(\boldsymbol{u}) = \frac{\partial^d C(\boldsymbol{u})}{\partial u_1 \ldots \partial u_d} = \int_0^1 \prod_{j=1}^d c_{V_1 j}(v_1, u_j) \, dv_1, \quad \boldsymbol{u} \in (0,1)^d, \tag{3.53}$$

assuming regularity conditions for interchanging the order of differentiation and integration are satisfied. In this model, dependence is defined by d bivariate linking copulas $C_{V_1 1}, \ldots, C_{V_1 d}$; there are no constraints amongst these bivariate copulas. Note that any conditional independence model for absolutely continuous random variables, conditioned on one latent variable, can be written in this form. For a parametric model (3.52), assume that there is a parameter θ_{j1} for $C_{V_1 j}$ for $j = 1, \ldots, d$.

Archimedean copulas based on LTs in $\mathcal{L}_\infty$ are special cases of 1-factor copula models with $C_{V_1 j}$ common for all j (the resulting joint copula has a simple form even if the $C_{V_1 j}$ do not). Also, as shown below, when the $C_{V_1 j}(\cdot; \theta_{j1})$ are all bivariate Gaussian copulas, then (3.53) becomes the copula of the multivariate Gaussian distribution with a 1-factor correlation matrix.

A main advantage of model (3.52) is that it allows for different types of tail dependence structure. As it was shown in Joe et al. (2010) and Joe (2011b), if all bivariate linking copulas are lower (upper) tail dependent then all bivariate margins of U are also lower (upper) tail dependent respectively. Thus, with appropriately chosen linking copulas asymmetric dependence structure as well as tail dependence can be easily modeled. Also if each $C_{V_1 j}$ is reflection symmetric, then the joint copula (3.52) is reflection symmetric.

For the special case of bivariate Gaussian linking copulas, let C_{j,V_1} be the bivariate Gaussian copula with correlation θ_{j1} for $j = 1, \ldots, d$. Then $C_{V_1 j}(v, u) = \Phi_2(\Phi^{-1}(v), \Phi^{-1}(u); \theta_{j1})$ and

$$C_{j|V_1}(u|v) = \Phi\left(\frac{\Phi^{-1}(u) - \theta_{j1}\Phi^{-1}(v)}{(1 - \theta_{j1}^2)^{1/2}}\right).$$

For (3.52), let $u_j = \Phi(z_j)$ to get a multivariate distribution with $N(0,1)$ margins. Then

$$F(z_1, \ldots, z_d) := C(\Phi(z_1), \ldots, \Phi(z_d)) = \int_0^1 \prod_{j=1}^d \Phi\left(\frac{z_j - \theta_{j1}\Phi^{-1}(v_1)}{(1 - \theta_{j1}^2)^{1/2}}\right) dv_1$$

$$= \int_{-\infty}^\infty \left\{\prod_{j=1}^d \Phi\left(\frac{z_j - \theta_{j1}w}{(1 - \theta_{j1}^2)^{1/2}}\right)\right\} \cdot \phi(w) \, dw.$$

Hence this model is the same as a multivariate Gaussian model with a 1-factor correlation structure because this multivariate cdf comes from the representation:

$$Z_j = \theta_{j1}W + (1 - \theta_{j1}^2)^{1/2}\epsilon_j, \quad j = 1, \ldots, d,$$

where $W, \epsilon_1, \ldots, \epsilon_d$ are i.i.d. $N(0,1)$ random variables. The correlation of $Z_j, Z_{j'}$ is $\theta_{j1}\theta_{j'1}$ for $j \neq j'$.

If $C_{V_1 j}$ is the t copula with correlation θ_{j1}, and shape parameter ν_i for $j = 1, \ldots, d$, then $c(u_1, \ldots, u_d)$ is not the multivariate t copula density.

We next show details for $p = 2$. Let $C_{V_1 j}$ be the copula of (V_1, U_j) as before. Also let $C_{V_2 j; V_1}$ be the copula for margins $C_{j|V_1} = C_{U_j|V_1}$ and $C_{V_2|V_1}$, and let $c_{V_2 j; V_1}$ be its density. We make the simplifying assumption that the copula for $C_{U_j|V_1}(\cdot|v_1)$ and $C_{V_2|V_1}(\cdot|v_1)$ does not depend on v_1; this is the same assumption used in the vine pair-copula construction; because of the latent variables, the use of the simplifying assumption is not strong. Note that $C_{V_2|V_1}$ is the $U(0,1)$ cdf since we assume that V_2 is independent of V_1. Then the independence of V_1, V_2 implies

$$C_{j|V_1,V_2}(u|v_1, v_2) = \mathbb{P}(U_j \leq u|V_1 = v_1, V_2 = v_2) = \frac{\partial}{\partial v_2}\mathbb{P}(U_j \leq u, V_2 \leq v_2|V_1 = v_1)$$

$$= \frac{\partial}{\partial v_2}C_{V_2 j; V_1}\big(v_2, C_{j|V_1}(u|v_1)\big) = C_{j|V_2; V_1}\big(C_{j|V_1}(u|v_1)|v_2\big),$$

where $C_{j|V_2; V_1}(x|v) = \partial C_{V_2 j; V_1}(v, x)/\partial v$. Then (3.51) becomes:

$$C(\boldsymbol{u}) = \int_0^1 \int_0^1 \prod_{j=1}^d C_{j|V_1,V_2}(u_j|v_1, v_2)\, dv_1 dv_2 = \int_0^1 \int_0^1 \prod_{j=1}^d C_{j|V_2; V_1}\big(C_{j|V_1}(u_j|v_1)|v_2\big)\, dv_1 dv_2.$$

$$(3.54)$$

By differentiation with respect to $u_1, \ldots, u_d$, (3.54) implies that the 2-factor copula density is

$$c(\boldsymbol{u}) = \int_0^1 \int_0^1 \prod_{j=1}^d \big\{c_{V_2 j; V_1}\big(v_2, C_{j|V_1}(u_j|v_1)\big) \cdot c_{V_1 j}(v_1, u_j)\big\}\, dv_1 dv_2. \qquad (3.55)$$

The dependence structure is defined through $2d$ bivariate linking copulas $C_{V_1 1}, \ldots, C_{V_1 d}$, $C_{V_2 1; V_1}, \ldots, C_{V_2 d; V_1}$; there are no constraints amongst these $2d$ bivariate copulas. For a parametric model (3.54), assume that there is a parameter θ_{j1} for $C_{V_1 j}$ and a parameter θ_{j2} for $C_{V_2 j; V_1}$ for $j = 1, \ldots, d$. Clearly, this is an extension of the 1-factor copula model and different types of dependence and tail behavior can be accommodated.

This model includes the 2-factor multivariate Gaussian model as a special case. Suppose $C_{V_1 j}$ and $C_{V_2 j; V_1}$ are Gaussian copulas with correlations $\theta_{j1} = \alpha_{j1}$ and $\theta_{j2} = \alpha_{j2}/(1-\alpha_{j1}^2)^{1/2}$ respectively, $j = 1, \ldots, d$. Here α_{j2} is a correlation of $Z_j = \Phi(U_j)$ and $W_2 = \Phi(V_2)$ so that the independence of V_1, V_2 implies that θ_{j2} is the partial correlation of Z_j and W_2 given $W_1 = \Phi(V_1)$ (in general $\rho_{ZW_2; W_1} = [\rho_{ZW_2} - \rho_{ZW_1}\rho_{W_2 W_1}]/[(1-\rho_{ZW_1}^2)(1-\rho_{W_2 W_1})]^{1/2}$). Then, using the above conditional distribution of the bivariate Gaussian copula,

$$C_{j|V_2; V_1}\big(C_{j|V_1}(u|v_1)|v_2\big) = \Phi\left(\left[\frac{\Phi^{-1}(u) - \alpha_{j1}\Phi^{-1}(v_1)}{(1-\alpha_{j1}^2)^{1/2}} - \theta_{j2}\Phi^{-1}(v_2)\right] \Big/ (1-\theta_{j2}^2)^{1/2}\right)$$

$$= \Phi\left(\frac{\Phi^{-1}(u) - \alpha_{j1}\Phi^{-1}(v_1) - \theta_{j2}(1-\alpha_{j1}^2)^{1/2}\Phi^{-1}(v_2)}{[(1-\alpha_{j1}^2)(1-\theta_{j2}^2)]^{1/2}}\right).$$

With $z_j = \Phi(u_j)$, $j = 1, \ldots, d$, the cdf for the 2-factor model becomes

$$F(z_1, \ldots, z_d) := C(\Phi(z_1), \ldots, \Phi(z_d))$$

$$= \int_{-\infty}^{\infty} \int_{-\infty}^{\infty} \prod_{j=1}^d \Phi\left(\frac{z_j - \alpha_{j1}w_1 - \theta_{j2}(1-\alpha_{j1}^2)^{1/2}w_2}{[(1-\alpha_{j1}^2)(1-\theta_{j2}^2)]^{1/2}}\right) \cdot \phi(w_1)\phi(w_2)\, dw_1 dw_2.$$

Hence this model is the same as a multivariate Gaussian model with a 2-factor correlation structure because this multivariate cdf comes from the representation:

$$Z_j = \alpha_{j1}W_1 + \alpha_{j2}W_2 + [(1-\alpha_{j1}^2)(1-\theta_{j2}^2)]^{1/2}\epsilon_j, \quad j = 1, \ldots, d,$$

where $W_1, W_2, \epsilon_1, \ldots, \epsilon_d$ are i.i.d. N(0,1) random variables. The correlation of $Z_j, Z_{j'}$ is $\alpha_{j1}\alpha_{j'1} + \alpha_{j2}\alpha_{j'2}$ for $j \neq j'$, and the $d \times 2$ loading matrix is $\boldsymbol{A} = (\alpha_{jk})_{1 \leq j \leq d, 1 \leq k \leq 2}$.

The factor copula model can be straightforwardly extended to $p > 2$ factors and it becomes an extension of the p-factor multivariate Gaussian distribution with a correlation matrix $\boldsymbol{\Sigma}$ that has the p-factor structure, that is, $\boldsymbol{\Sigma} = \boldsymbol{A}\boldsymbol{A}^\top + \boldsymbol{\Psi}^2$, where $\boldsymbol{A}$ is a $d \times p$ matrix of loadings and $\boldsymbol{\Psi}^2$ is a diagonal matrix of residual variances. The main difference is that for the factor copula model, the parameters for the second to pth factors are partial correlations $\rho_{Z_j W_k; W_1 \cdots W_{k-1}}$ for $j = 1, \ldots, d$ and $k = 2, \ldots, p$. The advantage of this parametrization is that all of these partial correlations and the $\rho_{Z_j W_1}$ are algebraically independent in the interval $(-1, 1)$, and the partial correlation parametrization is the one that can extend to the factor copula, by replacing each correlation with factor 1 by a bivariate copula, and each partial correlation for factors 2 to p with a bivariate copula applied to conditional distributions. Note that the p-factor copula density involves a p-dimensional integral in general when the bivariate linking copulas are not all Gaussian.

We next provide some details on how the parameters of bivariate Gaussian linking copulas are related to the matrix of loadings $\boldsymbol{A}$ in the classical factor model with p factors. Consider the model,

$$Z_j = \sum_{i=1}^p \alpha_{ji} W_i + \psi_j \epsilon_j, \quad j = 1, \ldots, d, \tag{3.56}$$

where $W_1, \ldots, W_p, \epsilon_1, \ldots, \epsilon_d$ are i.i.d. N(0,1) random variables. Let the matrix of loadings be $\boldsymbol{A} = (A_{ji})$; one possibility has $A_{ji} = \alpha_{ji}, i = 1, \ldots, p, j = 1, \ldots, d$. This matrix is unique up to orthogonal transformations. The unconditional distribution of (Z_j, W_1) is given by a bivariate Gaussian distribution with correlation $\mathrm{Cor}(Z_j, W_1) = \alpha_{j1}$, which follows from (3.56). Similarly, for $k = 2, \ldots, p$, the conditional distribution $F_{Z_j, W_k; W_1, \ldots, W_{k-1}}$ is a bivariate Gaussian distribution with correlation

$$
\begin{aligned}
\rho_{Z_j, W_k; W_1, \ldots, W_{k-1}} &= \frac{\mathrm{Cov}(Z_j, W_k | W_1, \ldots, W_{k-1})}{[\mathrm{Var}(Z_j | W_1, \ldots, W_{k-1}) \, \mathrm{Var}(W_k | W_1, \ldots, W_{k-1})]^{1/2}} \\
&= \frac{\alpha_{jk}}{(1 - \alpha_{j1}^2 - \ldots - \alpha_{j,k-1}^2)^{1/2}}.
\end{aligned}
\tag{3.57}
$$

As a result, C_{Z_j, W_1} is a bivariate copula with correlation α_{j1} and $C_{Z_j, W_k; W_1, \ldots, W_{k-1}}$ is a bivariate Gaussian copula with correlation $\rho_{Z_j, W_k; W_1, \ldots, W_{k-1}}$ as given in (3.57).

With a general $2 \leq p < d$ and assuming the simplifying assumption, we note that due to independence $V_1, \ldots, V_k$ (for $1 \leq k \leq p$), (3.51) becomes

$$
\begin{aligned}
C_{j|V_1, \ldots, V_k}(u|v_1, \ldots, v_k) &= \frac{\partial \mathbb{P}(U_j \leq u, V_k \leq v_k | V_1 = v_1, \ldots, V_{k-1} = v_{k-1})}{\partial v_k} \\
&= \frac{\partial C_{j, V_k; V_1, \ldots, V_{k-1}}(C_{j|V_1, \ldots, V_{k-1}}(u|v_1, \ldots, v_{k-1}), v_k)}{\partial v_k} \\
&= C_{j|V_k; V_1, \ldots, V_{k-1}}(C_{j|V_1, \ldots, V_{k-1}}(u|v_1, \ldots, v_{k-1})|v_k).
\end{aligned}
\tag{3.58}
$$

This recursion formula for $C_{j|V_1, \ldots, V_p}(u|v_1, \ldots, v_p)$ can be further expanded to express this conditional probability in terms of bivariate linking copulas $C_{j|V_k; V_1, \ldots, V_{k-1}}$ for $k \leq p$. The density can be then obtained by differentiating the cdf and applying the chain rule.

The factor copula models are very general latent variable models. They are also equivalent to truncated C-vines rooted at the latent variables; the view as C-vines means that one can obtain the joint density of the observed and unobserved variables through the copula density representation of Bedford and Cooke (2001) and then integrate the latent variables to get the density of the observed variables. In particular, the factor copula model that satisfies the simplifying assumption with p latent variables $V_1, \ldots, V_p$ can be represented as a C-vine copula model for $(V_1, \ldots, V_p, U_1, \ldots, U_d)$ rooted at the latent variables and truncated at the pth level (all copulas at higher levels are independence copulas). By integrating over latent variables, one gets that the p-factor copula density is a p-dimensional integral in general when the bivariate linking copulas are not all Gaussian.

We use the notation of Section 3.9 to show the edges of the trees of the vine based on observed variables, denoted as $1, \ldots, d$, and latent variables, denoted as $V_1, \ldots, V_p$. For the 1-factor model, or 1-truncated C-vine rooted at V_1, the tree T_1 has edges $[V_1, 1], \ldots, [V_1, d]$. For the 2-factor model, or 2-truncated C-vine rooted at V_1 in T_1 and V_2 in T_2, then T_1 has edges $[V_1, V_2], [V_1, 1], \ldots, [V_1, d]$ and T_2 has edges $[V_2, 1 | V_1], \ldots, [V_2, d | V_1]$. For the p-factor model, T_1 has edges $[V_1, V_2], \ldots, [V_1, V_p], [V_1, 1], \ldots, [V_1, d]$; for $2 \leq \ell < p$, T_ℓ has edges $[V_\ell, V_{\ell+1} | V_1, \ldots, V_{\ell-1}], \ldots, [V_\ell, V_p | V_1, \ldots, V_{\ell-1}], [V_\ell, 1 | V_1, \ldots, V_{\ell-1}], \ldots, [V_\ell, d | V_1, \ldots, V_{\ell-1}]$; and T_p has edges $[V_p, 1 | V_1, \ldots, V_{p-1}], \ldots, [V_p, d | V_1, \ldots, V_{p-1}]$.

For the p-factor model, let the bivariate linking copulas and densities for T_ℓ be $C_{V_\ell j; V_1 \cdots V_{\ell-1}}$ and $c_{V_\ell j; V_1 \cdots V_{\ell-1}}$ for $j = 1, \ldots, d$. Using Theorem 8.64, the integrand or the joint copula density of $V_1, \ldots, V_p, U_1, \ldots, U_d$ is

$$\prod_{j=1}^{d} c_{V_1 j}(v_1, u_j) \cdot \prod_{\ell=2}^{p} \prod_{j=1}^{d} c_{V_\ell j; V_1 \cdots V_{\ell-1}} \left(v_\ell, C_{j | V_1 \cdots V_{\ell-1}}(u_j | v_1, \ldots, v_{\ell-1}) \right),$$

where $C_{j | V_1 \cdots V_{\ell-1}}$ are obtained by recursion using (3.58).

This viewpoint of factor models as vines with latent variables leads to the extension to a subset of structural equation models (SEMs) or structural covariance models given in Section 3.12. It also helps in proving dependence properties of the factor copula model (3.51). Section 8.13 has theorems concerning tail and dependence properties for factor copulas constructed via vines. Some basic properties are summarized below.

- In general, if the bivariate linking copulas are not Gaussian, then (3.51) with (3.58) is identifiable with dp copulas, whereas with Gaussian copulas and $pd - p(p-1)/2 > 0$, the loadings matrix A can be rotated to $A^* = AP$ with $pd - p(p-1)/2$ parameters via $p(p-1)/2$ Givens rotations. For example, for $d \geq 5$ and the 2-factor model with $2d \leq d(d-1)/2$, there are $2d - 1$ parameters in the Gaussian model, whereas the 2-factor copula in general can accommodate $2d$ distinct bivariate linking copulas. In specific applications with a parametric version of (3.54), if the log-likelihood is quite flat, then one of the linking copulas for the second factor can be set to $C^\perp$.

- The factor copula models are closed under margins and extendible, that is, the same form of model results when deleting or adding a variable. Note that this property doesn't hold in general for vine copulas without latent variables.

- For the 1-factor model (3.52), if the $C_{j | V_1}$ are stochastically increasing for all j, then the bivariate margins $C_{jj'}$ are positive quadrant dependent for all $j \neq j'$.

- For the 2-factor model (3.54), if $C_{V_1 j}, C_{V_2 j; V_1}$ have conditional distributions $C_{j | V_1}, C_{j | V_2; V_1}$ that are stochastically increasing for all j, then the bivariate margins $C_{jj'}$ are positive quadrant dependent for all $j \neq j'$.

- For the 1-factor model (3.52), if the $C_{V_1 j}$ have lower (upper) tail dependence for all j, then the bivariate margins $C_{jj'}$ have lower (upper) tail dependence for all $j \neq j'$.

- For the 2-factor model (3.54), if the $C_{V_1 j}$ have lower (upper) tail dependence for all j and $C_{j | V_2; V_1}(t_{j0} | 0) > 0$ when $t_{j0} = \lim_{w \to 0+} \lim_{u \to 0+} C_{j | V_1}(u | wu)$ for all j, then the bivariate margins $C_{jj'}$ have lower (upper) tail dependence for all $j \neq j'$.

3.10.2 Discrete ordinal response

In this section, some details of the factor copula model are given for ordinal response variables; this assumes conditional independence of ordinal variables $Y_1, \ldots, Y_d$ given latent variables. The main application, with many ordinal variables that follow a factor structure, consists of item response; see Nikoloulopoulos and Joe (2014).

For simpler presentation, we assume the category labels for $Y_1, \ldots, Y_d$ are the same and in the set $\{0, 1 \ldots, K - 1\}$. Let the cutpoints in the uniform $U(0, 1)$ scale for the jth item/variable be $a_{j,k}$, $k = 1, \ldots, K - 1$, with $a_{j,0} = 0$ and $a_{j,K} = 1$. These correspond to $a_{j,k} = \Phi(\zeta_{j,k})$, where $\zeta_{j,k}$ are cutpoints in the normal $N(0, 1)$ scale.

For the 1-factor model, let V_1 be a latent variable, which we assumed to be standard uniform (without loss of generality). From Sklar's theorem, there is a bivariate copula $C_{V_1 j}$ such that $\mathbb{P}(V_1 \leq v, Y_j \leq y) = C_{V_1 j}(v, F_j(y))$ for $0 < v < 1$ where F_j is the cdf of Y_j; note that F_j is a step function with jumps at $0, \ldots, K - 1$ and $F_j(y) = a_{j,y+1}$. Then it follows that

$$F_{j|V_1}(y|v) := \mathbb{P}(Y_j \leq y|V_1 = v) = \frac{\partial C_{V_1 j}(v, F_j(y))}{\partial v}; \qquad (3.59)$$

let $C_{j|V_1}(a|v) = \partial C_{V_1 j}(v, a)/\partial v$ with previous notation.

The pmf for the 1-factor model is:

$$\pi_{1:d}(\boldsymbol{y}) = \int_0^1 \prod_{j=1}^d \mathbb{P}(Y_j = y_j|V_1 = v) \, dv = \int_0^1 \prod_{j=1}^d \left\{ C_{j|V_1}(F_j(y_j)|v) - C_{j|V_1}(F_j(y_j - 1)|v) \right\} dv$$

$$= \int_0^1 \prod_{j=1}^d \left\{ C_{j|V_1}(a_{j,y_j+1}|v) - C_{j|V_1}(a_{j,y_j}|v) \right\} dv = \int_0^1 \prod_{j=1}^d f_{V_1 j}(v, y_j) \, dv, \qquad (3.60)$$

where $f_{V_1 j}(v, y) = C_{j|V_1}(a_{j,y+1}|v) - C_{j|V_1}(a_{j,y}|v)$ is the joint density of V_1 and Y_j.

For the 2-factor model, consider two latent variables V_1, V_2 that are, without loss of generality, independent $U(0, 1)$ random variables. Let $C_{V_1 j}$ be defined as in the 1-factor model, and let $C_{V_2 j; V_1}$ be a bivariate copula such that

$$\mathbb{P}(V_2 \leq v_2, Y_j \leq y|V_1 = v_1) = C_{V_2 j; V_1}(v_2, F_{j|V_1}(y|v_1)), \quad 0 < v_1, v_2 < 1, \ y = 0, \ldots, K - 1,$$

where $F_{j|V_1}$ is given in (3.59). Here we are making the simplifying assumption that the copula for the univariate conditional distributions $F_{V_2|V_1} = F_{V_2}$ and $F_{j|V_1}$ does not depend on v_1; this is a model assumption, as by Sklar's theorem there exist such bivariate copulas that in general depend on $v_1 \in (0, 1)$. Then for $0 < v_1, v_2 < 1$,

$$\mathbb{P}(Y_j \leq y|V_1 = v_1, V_2 = v_2) = \frac{\partial}{\partial v_2} \mathbb{P}(V_2 \leq v_2, Y_j \leq y|V_1 = v_1)$$

$$= \frac{\partial}{\partial v_2} C_{V_2 j; V_1}(v_2, F_{j|V_1}(y|v_1)) = C_{j|V_2; V_1}(F_{j|V_1}(y|v_1)|v_2),$$

where $C_{j|V_2; V_1}(a|v) = \partial C_{V_2 j; V_1}(v, a)/\partial v$.

The pmf for the 2-factor model is:

$$\pi_{1:d}(\boldsymbol{y}) = \int_0^1 \int_0^1 \prod_{j=1}^d \mathbb{P}(Y_j = y_j|V_1 = v_1, V_2 = v_2) \, dv_1 dv_2$$

$$= \int_0^1 \int_0^1 \left\{ \prod_{j=1}^d C_{j|V_2; V_1}(F_{j|V_1}(y_j|v_1)|v_2) - C_{j|V_2; V_1}(F_{j|V_1}(y_j - 1|v_1)|v_2) \right\} dv_1 dv_2$$

$$= \int_0^1 \int_0^1 \prod_{j=1}^d f_{V_2 j|V_1}(v_2, y_j|v_1) \, dv_1 dv_2, \qquad (3.61)$$

where $f_{V_2 j|V_1}(v_2, y|v_1) = C_{j|V_2; V_1}(F_{j|V_1}(y|v_1)|v_2) - C_{j|V_2; V_1}(F_{j|V_1}(y - 1|v_1)|v_2)$. The idea in the derivation of this 2-factor model can be extended to $p \geq 3$ factors or latent variables; the pmf then involves p-dimensional integrals.

For parametric 1-factor and 2-factor models, we let $C_{V_1 j}$ and $C_{V_2 j; V_1}$ be parametric bivariate copulas, say with parameters θ_{j1} and θ_{j2} respectively. For the set of all parameters, let $\boldsymbol{\theta} = (a_{jk}, j = 1, \ldots, d, k = 1, \ldots, K - 1; \theta_{11}, \ldots, \theta_{d1})$ for the 1-factor model and $\boldsymbol{\theta} = (a_{jk}, j = 1, \ldots, d, k = 1, \ldots, K - 1; \theta_{11}, \ldots, \theta_{d1}, \theta_{12}, \ldots, \theta_{d2})$ for the 2-factor model. By assuming V_1, V_2 to be independent random variables, we model the joint distribution of (V_1, V_2, Y_j) in terms of two bivariate copulas rather than one trivariate copula. This model allows for selection of $C_{V_1 j}$ and $C_{V_2 j; V_1}$ independently among a variety of parametric copula families, i.e., there are no constraints in the choices of parametric copulas $\{C_{V_1 j}, C_{V_2 j; V_1} : j = 1, \ldots, d\}$.

We next show what happens when all the bivariate copulas are Gaussian. The resulting model in this case is the same as the discretized multivariate Gaussian model with factor structure. Let $U_1, \ldots, U_d, V_1, V_2$ be $U(0, 1)$ random variables. Let $\Phi_2(\cdot; \rho)$ be bivariate Gaussian cdf with correlation ρ and standard normal margins. Suppose $U_j = \Phi(Z_j)$ for $j = 1, \ldots, d$, $V_1 = \Phi(W_1)$, $V_2 = \Phi(W_2)$, where $W_1, W_2, Z_1, \ldots, Z_d$ are jointly multivariate Gaussian, and W_1, W_2 are independent. The parameter θ_{j1} for $C_{V_1 j}$ is the correlation parameter of W_1, Z_j, and $C_{V_1 j}$ is the distribution of $(\Phi(W_1), \Phi(Z_j))$. Hence,

$$C_{V_1 j}(v, a) = \Phi_2(\Phi^{-1}(v), \Phi^{-1}(a); \theta_{j1}) \quad \text{and} \quad C_{j|V_1}(a|v) = \Phi\Big(\frac{\Phi^{-1}(a) - \theta_{j1}\Phi^{-1}(v)}{(1 - \theta_{j1}^2)^{1/2}}\Big).$$

Let $F_j(y) = a_{j, y+1} = \Phi(\zeta_{j, y+1})$. Then, for $j = 1, \ldots, d$,

$$C_{j|V_1}(a_{j, y_j+1}|v) = \Phi\Big(\frac{\zeta_{j, y_j+1} - \theta_{j1}\Phi^{-1}(v)}{(1 - \theta_{j1}^2)^{1/2}}\Big), \tag{3.62}$$

and the pmf (3.60) for the 1-factor model becomes

$$\int_0^1 \prod_{j=1}^d \left\{ \Phi\Big(\frac{\zeta_{j, y_j+1} - \theta_{j1}\Phi^{-1}(v)}{(1 - \theta_{j1}^2)^{1/2}}\Big) - \Phi\Big(\frac{\zeta_{j, y_j} - \theta_{j1}\Phi^{-1}(v)}{(1 - \theta_{j1}^2)^{1/2}}\Big) \right\} dv$$

$$= \int_{-\infty}^\infty \prod_{j=1}^d \left\{ \Phi\Big(\frac{\zeta_{j, y_j+1} - \theta_{j1}w}{(1 - \theta_{j1}^2)^{1/2}}\Big) - \Phi\Big(\frac{\zeta_{j, y_j} - \theta_{j1}w}{(1 - \theta_{j1}^2)^{1/2}}\Big) \right\} \cdot \phi(w)\, dw.$$

Hence this model is a discretized multivariate Gaussian model with a 1-factor correlation matrix $\boldsymbol{R} = (\rho_{jj'})$ with $\rho_{jj'} = \theta_{j1}\theta_{j'1}$ for $j \neq j'$.

For the 2-factor model with Gaussian copulas, the parameter θ_{j2} for $C_{V_2 j; V_1}$ is the partial correlation of W_2, Z_j given W_1, so that $\theta_{j2} = \alpha_{j2}/(1 - \theta_{j1}^2)^{1/2}$, where α_{j2} is the correlation of W_2, Z_j. Note that $C_{V_2 j; V_1}$ is the copula of $[W_2, Z_j | W_1 = w]$ for any $-\infty < w < \infty$. Also $\alpha_{j1} = \theta_{j1}$ is the correlation of W_1, Z_j and parameter of $C_{V_1 j}$. Hence,

$$C_{V_2 j; V_1}(v, a) = \Phi_2(\Phi^{-1}(v), \Phi^{-1}(a); \theta_{j2}) \quad \text{and} \quad C_{j|V_2; V_1}(a|v) = \Phi\Big(\frac{\Phi^{-1}(a) - \theta_{j2}\Phi^{-1}(v)}{(1 - \theta_{j2}^2)^{1/2}}\Big).$$

Therefore $F_{j|V_1}(y_j|v_1)$ is given as in (3.62) with $v = v_1$, and

$$C_{j|V_2; V_1}(F_{j|V_1}(y_j|v_1)|v_2) = \Phi\left(\left[\frac{\zeta_{j, y_j+1} - \theta_{j1}\Phi^{-1}(v_1)}{(1 - \theta_{j1}^2)^{1/2}} - \theta_{j2}\Phi^{-1}(v_2) \right] \Big/ (1 - \theta_{j2}^2)^{1/2} \right)$$

$$= \Phi\left(\frac{\zeta_{j, y_j+1} - \theta_{j1}\Phi^{-1}(v_1) - \theta_{j2}(1 - \theta_{j1}^2)^{1/2}\Phi^{-1}(v_2)}{[(1 - \theta_{j1}^2)(1 - \theta_{j2}^2)]^{1/2}} \right).$$

The pmf (3.61) for the 2-factor model, with $\alpha_{j2} = \theta_{j2}(1 - \theta_{j1}^2)^{1/2}$, becomes

$$\int \int \prod_{j=1}^{d} \left\{ \Phi\left(\frac{\zeta_{j,y_j+1} - \theta_{j1}w_1 - \alpha_{j2}w_2}{[(1 - \theta_{j1}^2)(1 - \theta_{j2}^2)]^{1/2}}\right) - \Phi\left(\frac{\zeta_{j,y_j} - \theta_{j1}w_1 - \alpha_{j2}w_2}{[(1 - \theta_{j1}^2)(1 - \theta_{j2}^2)]^{1/2}}\right) \right\} \cdot \phi(w_1)\phi(w_2) \, dw_1 dw_2.$$

Hence this model is a discretized multivariate Gaussian model with a 2-factor correlation matrix $\boldsymbol{R} = (\rho_{jj'})$ with $\rho_{jj'} = \theta_{j1}\theta_{j'1} + \theta_{j2}\theta_{j'2}[(1 - \theta_{j1}^2)(1 - \theta_{j'1}^2)]^{1/2} = \alpha_{j1}\alpha_{j'1} + \alpha_{j2}\alpha_{j'2}$ for $j \neq j'$.

The dependence properties at the end of Section 3.10.1 extend to ordinal response. For the 1-factor model, if $C_{j|V_1}$ are stochastically increasing for all j, then $(Y_j, Y_{j'})$ is positive quadrant dependent for all $j \neq j'$. For the 2-factor model, if $C_{V_1 j}, C_{V_2 j; V_1}$ have conditional distributions $C_{j|V_1}, C_{j|V_2; V_1}$ that are stochastically increasing for all j, then $(Y_j, Y_{j'})$ is positive quadrant dependent for all $j \neq j'$.

3.10.3 Mixed continuous and ordinal response

In this subsection, we show the extension to a mix of continuous and ordinal responses. Assume that J_1 is a set of indices for continuous variables and J_2 is a set of indices for ordinal variables on integer values. Then the densities (or likelihoods) for the d variables for the 1-factor and 2-factor models are given below. The pattern extends to the p-factor model.

- 1-factor: If the latent variable is V_1, and the cutpoints are $\{a_{j,k}\}$ for $j \in J_2$ on the $U(0, 1)$ scale, the joint density is

$$f(u_j, j \in J_1; y_j, j \in J_2) = \int_0^1 \prod_{j \in J_1} c_{j,V_1}(u_j, v) \cdot \prod_{j \in J_2} \left\{ C_{j|V_1}(a_{j,y_j+1}|v) - C_{j|V_1}(a_{j,y_j}|v) \right\} dv.$$

- 2-factor: If the latent variables are V_1, V_2, and the cutpoints are $\{a_{j,k}\}$ for $j \in J_2$ on the $U(0, 1)$ scale, the joint density is

$$f(u_j, j \in J_1; y_j, j \in J_2) = \int_0^1 \int_0^1 \prod_{j \in J_1} \left\{ c_{V_2 j; V_1}(C_{j|V_1}(u_j|v_1), v_2) \cdot c_{j,V_1}(u_j, v_1) \right\}$$

$$\cdot \prod_{j \in J_2} \left\{ C_{j|V_2; V_1}(F_{j|V_1}(y_j|v_1)|v_2) - C_{j|V_2; V_1}(F_{j|V_1}(y_j - 1|v_1)|v_2) \right\} dv_1 dv_2,$$

where $F_{j|V_1}(y)$ is given in (3.59).

3.10.4 t-Copula with factor correlation matrix structure

A multivariate t_ν copula has a correlation matrix which can have a factor correlation structure; this is studied in Klüppelberg and Kuhn (2009). This multivariate t_ν factor model is not the same as the factor copula model in (3.51) with bivariate t_ν linking copulas, but it is useful to consider because the log-likelihood does not involve integrals.

Equations for the multivariate t_ν factor log-likelihood can be obtained based on formulas for the inverse and determinant of a correlation matrix with factor structure; these are summarized from Krupskii (2014).

Let $\boldsymbol{A} = (\alpha_{jk})$ be a $d \times p$ matrix of loadings where $1 \leq p < d$. Let $\boldsymbol{R} = \boldsymbol{A}\boldsymbol{A}^\top + \boldsymbol{\Psi}^2$, where $\boldsymbol{\Psi}^2$ is a diagonal matrix with elements $\psi_1^2, \ldots, \psi_d^2$. We will generalize to $\boldsymbol{\Sigma} = \boldsymbol{A}\boldsymbol{A}^\top + \boldsymbol{D}$ where $\boldsymbol{D} = \text{diag}(D_{11}, \ldots, D_{dd})$. Then with $\boldsymbol{I}_p$ being the $p \times p$ identity matrix,

$$\det(\boldsymbol{\Sigma}) = \det(\boldsymbol{A}\boldsymbol{A}^\top + \boldsymbol{D}) = \det(\boldsymbol{D}) \cdot \det(\boldsymbol{I}_p + \boldsymbol{A}^\top \boldsymbol{D}^{-1} \boldsymbol{A}) = \prod_{j=1}^{d} D_{jj} \cdot \det(\boldsymbol{I}_p + \boldsymbol{A}^\top \boldsymbol{D}^{-1} \boldsymbol{A}),$$

$$(3.63)$$

and
$$\Sigma^{-1} = D^{-1} - D^{-1}A(I_p + A^\top D^{-1}A)^{-1}A^\top D^{-1}.$$

Note that $I_p + A^\top D^{-1}A$ is a $p \times p$ matrix so that, if $p \ll d$, it is much faster to compute the determinant and inverse of this $p \times p$ matrix than the $d \times d$ matrix Σ. Equation (3.63) is a special case of Theorem 18.1.1 of Harville (1997).

The bi-factor structural correlation matrix is a special case of the loading matrix A with $p = m + 1$ based on m non-overlapping groups of variables; see Section 3.11.1. A can be written as

$$\begin{pmatrix} \phi_1 & \alpha_1 & 0 & \cdots & 0 \\ \phi_2 & 0 & \alpha_2 & \cdots & 0 \\ \vdots & \vdots & \vdots & \ddots & \vdots \\ \phi_m & 0 & 0 & \cdots & \alpha_m \end{pmatrix}, \tag{3.64}$$

where the gth group J_g (block of rows) has size d_g for $g = 1, \ldots, m$. The elements of the diagonal matrix Ψ^2 are $1 - \phi_{gj}^2 - \alpha_{gj}^2$ for j in the gth group. In (3.63), $\det(D) = \det(\Psi^2)$ simplifies to $\prod_{g=1}^m \prod_{j \in J_g} (1 - \phi_{gj}^2 - \alpha_{gj}^2)$.

3.11   Combining models for different groups of variables

If there are many variables that are divided into groups, a modeling approach might be to first fit lower-dimensional copulas separately by groups and then combine these into a higher-dimensional copula. Different combination approaches are presented in this section.

Combining distributions of groups of variables cannot be done in generality because there is no general construction method for multivariate distributions with given marginal distributions that are not univariate; see some (negative) results in Marco and Ruiz-Rivas (1992) and Genest et al. (1995b). However, if there is a latent variable such that groups of variables are conditionally independent given the latent variable, then a joint copula model can be constructed.

Suppose the variables form m groups with d_g variables in the gth group J_g, and the total number of variables is $d_1 + d_2 + \cdots + d_m = d$. Suppose there is a latent variable V_0 that explains the dependence among different groups. Then a general copula form, with conditional independence of groups of variables given $V_0 \sim U(0,1)$, is:

$$C(u) = \int_0^1 \left\{ \prod_{g=1}^m C_g(u_{J_g}; v_0) \right\} dv_0, \quad u \in [0,1]^d, \tag{3.65}$$

where $C_g(u_{J_g}; v_0)$ is the copula for the gth group when the latent variable has value v_0. The joint copula density is

$$c(u) = \int_0^1 \left\{ \prod_{g=1}^m c_g(u_{J_g}; v_0) \right\} dv_0, \quad u \in (0,1)^d.$$

One approach with group conditional independence comes from mixing powers of max-id distributions on groups. Suppose the latent variable is $Q = F_Q^{-1}(V_0)$ where $V_0 \sim U(0,1)$ and F_Q has support on the positive reals. Let $\psi = \psi_Q$ be the LT of Q. Let $F_{Y_{J_g}}^q$ be the joint distribution of the gth group given $Q = q$ and let C_g be the copula of $F_{Y_{J_g}}$. Then the joint distribution is $\int_0^\infty \prod_{g=1}^m \left\{ F_{Y_{J_g}}^q(y_{J_g}) \right\} dF_Q(q)$ and its copula is

$$C(u) = \psi \left(-\sum_{g=1}^m \log C_g \left(e^{-\psi^{-1}(u_j)}, j \in J_g \right) \right), \quad u \in [0,1]^d.$$

Sections 8.3 and 8.5 have conditions for Archimedean and other copulas that are max-id. Methods that don't depend on max-id are given in several subsections below.

3.11.1 Bi-factor copula model

The copula version of the confirmatory factor model with some structural zero loadings is a special case of (3.65). In this case, there is a latent variable V_g for group g ($g = 1, \ldots, m$), and variables in the gth group are conditionally independent given V_0 and V_g. Then the joint copula is:

$$\int_0^1 \prod_{g=1}^m \left\{ \int \prod_{j \in J_g} C_{j|V_g,V_0}(u_j|v_g, v_0) \, dv_m \right\} dv_0. \tag{3.66}$$

The within group copula could be a 1-factor copula (given V_0) with cdf

$$C_{j|V_g,V_0}(u_j|v_g, v_0) = C_{j|V_g;V_0}\big(C_{j|V_0}(u_j|v_0)|C_{V_g|V_0}(v_g|v_0)\big). \tag{3.67}$$

If $V_0, V_1, \ldots, V_m$ are mutually independent latent random variables, then the total number of bivariate linking copulas is $2d$. We call the combination of (3.66) and (3.67) a bi-factor copula model, and it is a special case of a structured factor copula model; see Krupskii and Joe (2014). This can also be considered as a 2-truncated vine rooted at the latent variables.

The model with (3.66) and (3.67) is the bi-factor model of Holzinger and Swineford (1937), and Gibbons and Hedeker (1992) when all of the linking copulas are bivariate Gaussian. Assume the the observed variables have been standardized to have mean 0 and variance 1. Let $W_0, W_1, \ldots, W_m$ be the independent latent $N(0,1)$ random variables, W_0 is a global latent variable and W_g ($g = 1, \ldots, m$) is a latent variable for the gth group of variables. The stochastic representation of the bi-factor model is:

$$Y_j = \phi_{j1} W_0 + \phi_{j2} W_g + \psi_j \epsilon_j, \quad j \in J_g, \tag{3.68}$$

where $\phi_{j1} = \rho_{j,W_0}$ is the correlation of Y_j with W_0, $\rho_{j,W_g;W_0} = \phi_{j2}/(1 - \phi_{j1}^2)^{1/2}$ is the partial correlation of Y_j with W_g given W_0 when $j \in J_g$, and $\psi_j^2 = 1 - \phi_{j1}^2 - \phi_{j2}^2$.

The number of parameters in the Gaussian bi-factor structure is $2d - N_1 - N_2$, where N_1 is the number of groups of size 1 and N_2 is the number of groups of size 2. For a group g of size 1 with variable j, W_g is absorbed with ϵ_j because ϕ_{j2} would not be identifiable. For a group g of size 2 with variables j_1, j_2, the parameters ϕ_{j_12} and ϕ_{j_22} appear only in the correlation for variables j_1, j_2 and this correlation is $\phi_{j_11}\phi_{j_21} + \phi_{j_12}\phi_{j_22}$. Since only the product $\phi_{j_12}\phi_{j_22}$ appears, one of ϕ_{j_12}, ϕ_{j_22} can be taken as 1 without loss of generality. For the bi-factor copula with non-Gaussian linking copulas, near non-identifiability can occur when there are groups of size 2; in this case, one of the linking copulas to the group latent variable can be fixed (say at C^+) for a group of size 2.

The bi-factor model can be extended to a tri-factor model (Krupskii (2014)), where each group is divided into two or more subgroups. The number of bivariate linking copulas is $3d$, with each variable linking to (i) the global latent factor, (ii) a latent factor that is specific to the group that the variable is in, and (iii) a latent factor that is specific to the subgroup that the variable is in. For the Gaussian version, the loading matrix has $1 + m + m'$ columns where $m' = \sum_{g=1}^m h_g$ and h_g is the number of subgroups for group g; the first column of the loading matrix has loadings for all variables and the column for a specific group (subgroup) has non-zero values only for the variables in this group (subgroup).

3.11.2 Nested dependent latent variables

An alternative way to have distinct latent variables for different groups is a 1-truncated vine with $m + 1$ latent variables $V_0, V_1, \ldots, V_m$, and bivariate linking copulas:

$$C_{V_0V_1}, \ldots, C_{V_0V_m}; \quad \text{and } C_{U_jV_g} \text{ for } j \in J_g, \ g = 1, \ldots, m.$$

We call this a *nested factor copula model*; see Krupskii and Joe (2014).

For the special case of multivariate Gaussian with zero means and unit variances, let $W_0, W_1', \ldots, W_m'$ be the dependent latent $N(0,1)$ random variables, W_0 is a global latent variable and $W_g' = \beta_g W_0 + (1 - \beta_g^2)^{1/2} W_g$ for the latent variables for group g. Here $W_1, \ldots, W_g$ are i.i.d. $N(0,1)$, independent of W_0. The stochastic representation is:

$$Y_j = \beta_{jg} W_g' + \psi_j \epsilon_j, \quad j \in J_g,$$
$$W_g' = \beta_g W_0 + (1 - \beta_g^2)^{1/2} W_g, \quad g = 1, \ldots, m,$$

where $\psi_j = 1 - (1 - \beta_{jg}^2)^{1/2}$, or

$$Y_j = \beta_{jg} \beta_g W_0 + \beta_{jg} (1 - \beta_g^2)^{1/2} W_g + \psi_j \epsilon_j, \quad j \in J_g.$$

Hence for multivariate Gaussian, this is a special case of (3.68) with $\phi_{j1} = \beta_{jg} \beta_g$ and $\phi_{j2} = \beta_{jg} (1 - \beta_g^2)^{1/2}$ for $j \in J_g$.

The number of parameters in the Gaussian nested factor structure is $d + m$ for $m \geq 3$ and $d + 1$ for two groups, because in the case of two groups, $\beta_1 \beta_2$ occur only as a product in the correlation of variables in different groups.

This nested factor model can be extended to a factor model with one extra level of nesting where each group is divided into two or more subgroups; see Krupskii (2014). The number of bivariate linking copulas is $m + m' + d$, where m' is defined in the preceding subsection.

3.11.3   Dependent clusters with conditional independence

If one has copula models for different groups or clusters, then a larger multivariate copula model can be built assuming conditional independence of different groups of variables given aggregation variables for the groups. The main idea of this construction is in Brechmann (2013) and Brechmann (2014).

Consider m non-overlapping groups of absolutely continuous random variables with $\boldsymbol{X}_g = (X_j : j \in J_g)$ as the vector of variables in group J_g. For $g = 1, \ldots, m$, let $V_g = h_g(\boldsymbol{X}_g)$ be a scalar aggregation random variable and let $j_g \in J_g$ be such that $(X_j : j \in J_g \backslash \{j_g\}, V_g)$ and $\boldsymbol{X}_g$ form a 1-1 mapping. Suppose the V_g are dependent and $f_{\boldsymbol{V}}$ is the density of $\boldsymbol{V} = (V_1, \ldots, V_m)$. Consider the conditional independence model with

$$\boldsymbol{X}_1 \perp\!\!\!\perp \cdots \perp\!\!\!\perp \boldsymbol{X}_m | V_1, \ldots, V_m, \tag{3.69}$$

and suppose

$$[\boldsymbol{X}_g | V_1, \ldots, V_m] \stackrel{d}{=} [\boldsymbol{X}_g | V_g] \tag{3.70}$$

(conditional distribution of $\boldsymbol{X}_g$ given the m aggregation variables is the same as the conditional distribution of $\boldsymbol{X}_g$ given V_g). If $f_{\boldsymbol{X}_g}$ is the density for group g, then joint density of $\boldsymbol{X}_1, \ldots, \boldsymbol{X}_m$ is

$$f_{\boldsymbol{X}_1, \ldots, \boldsymbol{X}_m}(\boldsymbol{x}_1, \ldots, \boldsymbol{x}_m) = \frac{f_{\boldsymbol{V}}(v_1, \ldots, v_m)}{\prod_{g=1}^m f_{V_g}(v_g)} \prod_{g=1}^m f_{\boldsymbol{X}_g}(\boldsymbol{x}_g), \quad v_g = h_g(\boldsymbol{x}_g) \text{ for all } g. \tag{3.71}$$

For simpler notation, we show the proof based on two groups $\boldsymbol{X}_1 = (X_1, \ldots, X_{d_1})$ and $\boldsymbol{X}_2 = (X_{d_1+1}, \ldots, X_{d_1+d_2})$, with a 1-1 mapping of $(X_1, \ldots, X_{d_1})$ to $(X_1, \ldots, X_{d_1-1}, V_1)$ and a 1-1 mapping of $(X_{d_1+1}, \ldots, X_{d_1+d_2})$ to $(X_{d_1+1}, \ldots, X_{d_1+d_2-1}, V_2)$.

In general, for a 1-1 mapping $\boldsymbol{z} = (z_1, \ldots, z_k) \mapsto \boldsymbol{y} = (y_1, \ldots, y_k)$ with $(y_1, \ldots, y_k) =$

$(s_1(\boldsymbol{z}), \ldots, s_k(\boldsymbol{z}))$ for functions $s_1, \ldots, s_k$, the relationship of the densities, based on the Jacobian, is

$$f_{\boldsymbol{Z}}(\boldsymbol{z}) = f_{\boldsymbol{Y}}(s_1(\boldsymbol{z}), \ldots, s_k(\boldsymbol{z})) \left| \frac{\partial(y_1, \ldots, y_k)}{\partial(z_1, \ldots, z_k)} \right|. \tag{3.72}$$

From (3.69) and (3.70),

$$f_{\boldsymbol{X}_1, \boldsymbol{X}_2 | V_1, V_2} = f_{\boldsymbol{X}_1 | V_1} \cdot f_{\boldsymbol{X}_2 | V_2}.$$

From (3.72), (i) with $\boldsymbol{Z} = \boldsymbol{X}_1$, $\boldsymbol{Y} = (X_1, \ldots, X_{d_1-1}, V_1)$, $k = d_1$, $s_1, \ldots, s_{d_1-1}$ being identity functions and $s_{d_1} = h_1$, and (ii) with $\boldsymbol{Z} = \boldsymbol{X}_2$, $\boldsymbol{Y} = (X_{d_1}, \ldots, X_{d_1+d-2-1}, V_2)$, $k = d_2$, $s_1, \ldots, s_{d_2-1}$ being identity functions and $s_{d_2} = h_2$, then

$$f_{\boldsymbol{X}_1}(\boldsymbol{x}_1) = f_{X_1, \ldots, X_{d_1-1}, V_1} \cdot \frac{\partial h_1(\boldsymbol{x}_1)}{\partial x_{d_1}},$$

$$f_{\boldsymbol{X}_2}(\boldsymbol{x}_2) = f_{X_{d_1+1}, \ldots, X_{d_1+d_2-1}, V_2} \cdot \frac{\partial h_2(\boldsymbol{x}_2)}{\partial x_{d_1+d_2}}.$$

Hence

$$f_{X_1, \ldots, X_{d_1-1} | V_1} = \frac{f_{\boldsymbol{X}_1}(\boldsymbol{x}_1)}{f_{V_1} \cdot \partial h_1(\boldsymbol{x}_1)/\partial x_{d_1}},$$

$$f_{X_{d_1+1}, \ldots, X_{d_1+d_2-1} | V_2} = \frac{f_{\boldsymbol{X}_2}(\boldsymbol{x}_2)}{f_{V_2} \cdot \partial h_2(\boldsymbol{x}_2)/\partial x_{d_1+d_2}},$$

$$f_{\boldsymbol{X}_1, \boldsymbol{X}_2} = f_{X_1, \ldots, X_{d_1-1}, V_1, X_{d_1+1}, \ldots, X_{d_1+d_2-1}, V_2} \cdot \frac{\partial h_1(\boldsymbol{x}_1)}{\partial x_{d_1}} \frac{\partial h_2(\boldsymbol{x}_2)}{\partial x_{d_1+d_2}}$$

$$= f_{V_1 V_2} f_{X_1, \ldots, X_{d_1-1} | V_1} f_{X_{d_1+1}, \ldots, X_{d_1+d_2-1} | V_2} \cdot \frac{\partial h_1(\boldsymbol{x}_1)}{\partial x_{d_1}} \frac{\partial h_2(\boldsymbol{x}_2)}{\partial x_{d_1+d_2}}$$

$$= \frac{f_{V_1 V_2}}{f_{V_1} f_{V_2}} f_{\boldsymbol{X}_1}(\boldsymbol{x}_1) f_{\boldsymbol{X}_2}(\boldsymbol{x}_2).$$

Note that if $V_1 \perp\!\!\!\perp \cdots \perp\!\!\!\perp V_m$, then the groups of random variables are mutually independent. If the V_g are dependent, then the random variables in different groups are dependent as given by the density (3.71).

Special cases of (3.71) include the following.

1. When $\boldsymbol{X}_g = \boldsymbol{U}_g$ is a multivariate random vector distributed with copula C_g for $g = 1, \ldots, m$ and $V_g = C_g(\boldsymbol{U}_g)$ is the multivariate probability integral transform via the copula cdf, then V_g has distribution $F_{V_g} = F_K(\cdot; C_g)$ which is a Kendall function (see Section 8.14). This aggregation through copula cdfs is studied in detail in Brechmann (2014) and is called the hierarchical Kendall copula. It is most convenient to use when each $\boldsymbol{U}_g$ consists of exchangeable random variables that have a cdf that is an Archimedean copula. The convenience is the simplicity of F_{V_g} in this case.

2. Related ideas on hierarchical risk aggregation are in Arbenz et al. (2012).

3. Clustered multivariate Gaussian distributions arise as a special case when $\boldsymbol{X}_g$ is multivariate Gaussian and each V_g is a sum (and hence having a Gaussian distribution); $(V_1, \ldots, V_m)$ can have a joint multivariate normal distribution. Further details of this case are given in Example 3.12 below, when $m = 2$ and each of $\boldsymbol{X}_1, \boldsymbol{X}_2$ consists of exchangeable Gaussian random variables.

4. Clustered multivariate t distributions arise as a special case when $\boldsymbol{X}_g$ is multivariate t_{ν_g} and each V_g is a sum (and hence has univariate t_{ν_g} distribution); $(V_1, \ldots, V_m)$ can have a joint multivariate t_ν copula.

Example 3.12 For the case of multivariate Gaussian with two exchangeable groups, we make some comparisons below, where the within group correlations are ρ_1, ρ_2 for the two groups. Let the group sizes be d_1 and d_2.

- Bi-factor with two exchangeable groups J_1, J_2. The stochastic representation is: $X_j = \phi_g V_0 + \gamma_g V_g + \psi_g \epsilon_j$ for $j \in J_g$, where $\psi_g^2 = 1 - \phi_g^2 - \gamma_g^2$. The within group correlation is $\rho_g = \phi_g^2 + \gamma_g^2 \geq 0$ for $g = 1, 2$, the between group correlation is $\varrho = \phi_1 \phi_2$ and $|\varrho| \leq (\rho_1 \rho_2)^{1/2}$.

- Oblique factor with two exchangeable groups J_1, J_2. The stochastic representation is: $X_j = \gamma_g V_g + \psi_g \epsilon_j$ for $j \in J_g$, where $\psi_g^2 = 1 - \gamma_g^2$. Let $\rho_{V_1 V_2}$ be the correlation of V_1, V_2. The within group correlation is $\rho_g = \gamma_g^2 \geq 0$ for $g = 1, 2$, and the between group correlation is $\varrho = \gamma_1 \gamma_2 \rho_{V_1 V_2}$ and $|\varrho| \leq (\rho_1 \rho_2)^{1/2}$.

- Conditional independence as in (3.71) with summation as the aggregation function. Suppose $J_1 = \{1, \ldots, d_1\}$ and $J_2 = \{d_1 + 1, \ldots, d_1 + d_2\}$. Let $V_1 = X_1 + \cdots + X_{d_1}$ and $V_2 = X_{d_1 + 1} + \cdots + X_{d_1 + d_2}$. Suppose the correlation of V_1, V_2 is $\rho_{V_1 V_2} \in (-1, 1)$. Let $d = d_1 + d_2$. The between group correlation is:

$$
\begin{aligned}
\varrho = \mathrm{Cor}(X_1, X_d) &= \mathrm{Cov}(X_1, X_d) = \mathrm{Cov}(\mathbb{E}[X_1 | V_1], \mathbb{E}[X_d | V_2]) = \mathrm{Cov}(d_1^{-1} V_1, d_2^{-1} V_2) \\
&= d_1^{-1} d_2^{-1} \rho_{V_1, V_2} [d_1 + d_1(d_1 - 1)\rho_1]^{1/2} [d_2 + d_2(d_2 - 1)\rho_2]^{1/2} \\
&= \rho_{V_1, V_2} \frac{[1 + (d_1 - 1)\rho_1]^{1/2} [1 + (d_2 - 1)\rho_2]^{1/2}}{d_1^{1/2} d_2^{1/2}}.
\end{aligned}
$$

Note that if $\rho_{V_1 V_2} > 0$, $\rho_1 > 0$ and $\rho_2 > 0$, then $\varrho \geq (\rho_1 \rho_2)^{1/2}$ occurs if

$$
\rho_{V_1 V_2} \geq \frac{(d_1 \rho_1)^{1/2} (d_2 \rho_2)^{1/2}}{[1 + (d_1 - 1)\rho_1]^{1/2} [1 + (d_2 - 1)\rho_2]^{1/2}},
$$

for example, if $d_1 = d_2 = 4$ and $\rho_1 = \rho_2 = 0.5$, then this occurs for $\rho_{V_1 V_2} \geq 0.8$. Hence (3.71) can lead to a higher between group correlation than the bi-factor and oblique factor models; also the within group correlations ρ_1, ρ_2 could be negative.

3.12   Nonlinear structural equation models

In the multivariate Gaussian case, truncated vines and factor models lead to structured correlation matrices. A theory that covers both is structural equation or covariance models. In the setting of dependence models where univariate margins are considered in the first stage, the dependence is summarized in structured correlation matrices.

Gaussian structural equation models (SEMs) can be presented as very general and they include common factor, bi-factor and other structured factor models as special cases; see for example, Bollen (1989), Browne and Arminger (1995), and Mulaik (2009). In this context of non-experimental data, SEMs can be used to get parsimonious structures without trying to infer causation. Latent variable models are reasonable in many applications as an explanation of the dependence in the observed variables, and SEMs can provide some flexibility in the dependence conditioned on the latent variables. With observational data, there need not be a specific correlation structure proposed in advance, but rather a class of plausible parsimonious structures can be fit and compared.

To show the main ideas, we will present some specific Gaussian structural equation models, write them with a parametrization of partial correlations that are algebraically independent, and then find the vine representation. For the vine copula extension, each correlation can be replaced by a bivariate copula, and each partial correlation can be replaced by a bivariate copula applied to univariate conditional distributions as margins. Nonlinear structural equations can be written for the vine copula model. Some data examples with Gaussian SEMs are given in Sections 7.11 and 7.12.

Consider the following linear representation for d variables that have a multivariate Gaussian distribution:

$$Z_j = \boldsymbol{\phi}_j^\top \boldsymbol{Z} + \boldsymbol{\delta}_j^\top \boldsymbol{W} + \psi_j \epsilon_j, \quad j = 1, \ldots, d, \qquad (3.73)$$

where $\epsilon_j \sim N(0,1)$ is independent of Z_i, $i \neq j$, and $\boldsymbol{W} = (W_1, \ldots, W_p)^\top$, the ϵ_j's are mutually independent, and ψ_j is chosen such that $\text{Var}(Z_j) = 1$. The variables in $\boldsymbol{W}$ are latent factors of $\boldsymbol{Z}$ and have loading vector $\boldsymbol{\delta}_j \in \mathbb{R}^p$ for variable Z_j. The vector $\boldsymbol{\phi}_j = (\phi_{j1}, \ldots, \phi_{jd})' \in \mathbb{R}^d$, specifies between-variable dependence of the components of $\boldsymbol{Z}$. Assume that $\phi_{jj} = 0$ and $\boldsymbol{\phi}_j$, $j = 1, \ldots, d$, have been chosen so that the model is well-defined.

In other words, we assume that Z_j is explained by its relationship to other components of $\boldsymbol{Z}$ as well as to a set of common latent factors $\boldsymbol{W}$. The idiosyncratic variance is given through the ϵ_j. Different specific choices of $\boldsymbol{\phi}_j$ and $\boldsymbol{\delta}_j$, with $O(d)$ non-zero values, lead to parsimonious and well-defined parameterizations of the correlation matrix $\boldsymbol{R} = (\rho_{ij})_{i,j=1,\ldots,d}$ of $\boldsymbol{Z}$. If $\boldsymbol{\phi}_j = \boldsymbol{0}$ for all j, then a factor model obtains. If $\boldsymbol{\delta}_j = \boldsymbol{0}$ for all j, then a truncated vine model obtains.

Other structures also permit a representation in terms of correlations and partial correlations that are algebraically independent.

In the simplest case, we have one factor W and one neighbor $Z_{k(j)}$ of each variable Z_j, $j = 1, \ldots, d$, in a Markov tree, corresponding to a 1-truncated vine model. For $d \geq 5$, model (3.73) can therefore be stated as

$$\begin{aligned} Z_1 &= \delta_1 W + \psi_1 \epsilon_1, \\ Z_2 &= \phi_2 Z_1 + \delta_2 W + \psi_2 \epsilon_2, \\ Z_j &= \phi_j Z_{k(j)} + \delta_j W + \psi_j \epsilon_j, \quad j = 3, \ldots, d, \end{aligned} \qquad (3.74)$$

where, without loss of generality, $[1, 2]$ is an edge of the Markov tree and $\phi_1 = 0$. The correlations between Z_j and W are denoted by $\rho_{j,W}$, and $\psi_j = \sqrt{1 - \phi_j^2 - \delta_j^2 - 2\phi_j \delta_j \rho_{k(j),W}}$. The model (3.74) is studied in Brechmann and Joe (2014a) and maximum likelihood estimation of the parameters is implemented with a modified expectation-maximization algorithm,

Using shorthand notation k_j for $k(j)$ and with $\rho_{jk;W}$ as the partial correlation of variables j, k given W, we get that

$$\begin{aligned} \rho_{j,W} &= \text{Cov}(Z_j, W) = \phi_j \rho_{k_j,W} + \delta_j, \quad j = 1, \ldots, d; \\ 1 &= \text{Var}(Z_j) = \phi_j^2 + \delta_j^2 + 2\phi_j \delta_j \rho_{k_j,W} + \psi_j^2, \quad j = 1, \ldots, d; \\ \rho_{jk_j;W} &[(1 - \rho_{j,W}^2)(1 - \rho_{k_j,W}^2)]^{1/2} = \text{Cov}(Z_j, Z_{k_j}|W) = \phi_j \text{Var}(Z_{k_j}|W) \\ &= \phi_j (1 - \rho_{k_j,W}^2), \quad j = 2, \ldots, d. \end{aligned}$$

Given the ϕ_j and δ_j, the above equations can be used to solve for $\rho_{j,W}$ and $\rho_{jk_j;W}$. Also, the loading parameters δ_j and the between-variable parameters ϕ_j can be expressed in terms of correlations and partial correlations as

$$\begin{aligned} \phi_j &= \rho_{jk_j;W}(1 - \rho_{j,W}^2)^{1/2}/(1 - \rho_{k_j,W}^2)^{1/2}, \quad j = 2, \ldots, d; \\ \delta_j &= \rho_{j,W} - \phi_j \rho_{k_j,W}, \quad j = 1, \ldots, d; \\ \psi_j^2 &= 1 - \phi_j^2 - \delta_j^2 - 2\phi_j \delta_j \rho_{k_j,W} = (1 - \rho_{j,W}^2)(1 - \rho_{jk_j;W}^2). \quad j = 1, \ldots, d. \end{aligned} \qquad (3.75)$$

There is a 1-1 mapping from $(\delta_1, \ldots, \delta_d, \phi_2, \ldots, \phi_d)$ to $(\rho_{1,W}, \ldots, \rho_{d,W}, \{\rho_{jk(j);W}, j = 2, \ldots, d\})$. Assuming that correlations $\rho_{k,W}$ are given, the full correlation matrix may then be obtained as well by interpreting model (3.74) as a 2-truncated vine. We thus have a correlation matrix parameterization $\boldsymbol{R} = \boldsymbol{R}(\delta_1, \ldots, \delta_d, \phi_2, \ldots, \phi_d) = \boldsymbol{R}(\rho_{1,W}, \ldots, \rho_{d,W}, \{\rho_{jk(j);W}, j = 2, \ldots, d\})$ in terms of $2d - 1$ parameters, for $d \geq 5$.

To show a nonlinear SEM extension, we first show some details for a specific residual Markov tree with $d = 5$ and $k(2) = 1$, $k(3) = 2$, $k(4) = 2$, $k(5) = 3$, before the general case. Consider

$$
\begin{aligned}
Z_1 &= \delta_1 W + \psi_1 \epsilon_1, \\
Z_2 &= \phi_2 W + \delta_2 Z_1 + \psi_2 \epsilon_2, \\
Z_3 &= \phi_3 W + \delta_3 Z_2 + \psi_3 \epsilon_3, \\
Z_4 &= \phi_4 W + \delta_4 Z_2 + \psi_4 \epsilon_4, \\
Z_5 &= \phi_5 W + \delta_5 Z_3 + \psi_5 \epsilon_5.
\end{aligned}
\tag{3.76}
$$

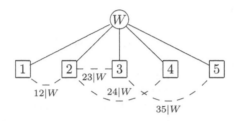

Figure 3.7 *Combined path diagram and truncated vine for model (3.76) with Z_j replaced by j for $j = 1, \ldots, 5$. The dashed edges indicate the partial correlations given W for the Markov tree conditional dependence of the variables after adjusting for W.*

For the general copula version of (3.76), the correlation $\rho_{j,W}$ in (3.75) is replaced by a bivariate copula $C_{j,W}$ for $j = 1, \ldots, d$ and the partial correlation $\rho_{jW;k(j)}$ is replaced by a bivariate copula $C_{j,k(j);W}$ for $j = 2, \ldots, d$. This d-dimensional copula model with $Z_1, \ldots, Z_d$ (and cdfs F_{Z_j}) is based on a 2-truncated vine with edges $[Z_j, W]$ ($j = 1, \ldots, d$) for tree $\mathcal{T}_1$ and edges $[Z_j, W | Z_{k(j)}]$ ($j = 2, \ldots, d$) for tree $\mathcal{T}_2$. See Figure 3.7 for a vine diagram the combines with the path diagram used for SEMs. From (3.38), for (3.76) with $d = 5$,

$$
\begin{aligned}
F_{Z_j|W}(z_j|w) &= C_{j|W}\big(F_{Z_j}(z_j)|F_W(w)\big), \quad j = 1, \ldots, 5, \\
F_{Z_2|Z_1,W}(z_2|z_1, w) &= C_{2|1;W}\big(F_{Z_2|W}(z_2|w) \mid F_{Z_1|W}(z_1|w)\big), \\
F_{Z_3|Z_2,W}(z_3|z_2, w) &= C_{3|2;W}\big(F_{Z_3|W}(z_3|w) \mid F_{Z_2|W}(z_2|w)\big), \\
F_{Z_4|Z_2,W}(z_4|z_2, w) &= C_{4|2;W}\big(F_{Z_4|W}(z_4|w) \mid F_{Z_2|W}(z_2|w)\big), \\
F_{Z_5|Z_3,W}(z_5|z_3, w) &= C_{5|3;W}\big(F_{Z_5|W}(z_5|w) \mid F_{Z_3|W}(z_3|w)\big).
\end{aligned}
$$

Let $P_1, \ldots, P_d$ be independent $U(0,1)$ random variables, and let $W, \epsilon_1, \ldots, \epsilon_d$ be $N(0,1)$ random variables with $\epsilon_j = \Phi^{-1}(P_j)$. Based on the conditional method for simulation, a nonlinear SEM, with $Z_j \sim N(0,1)$, (3.76) has representation:

$$
\begin{aligned}
Z_1 &= F_{Z_1|W}^{-1}\big(\Phi(\epsilon_1)|W\big), \\
Z_2 &= F_{Z_2|Z_1,W}^{-1}\big(\Phi(\epsilon_2)|Z_1, W\big), \\
Z_3 &= F_{Z_3|Z_2,W}^{-1}\big(\Phi(\epsilon_3)|Z_2, W\big), \\
Z_4 &= F_{Z_4|Z_2,W}^{-1}\big(\Phi(\epsilon_4)|Z_2, W\big), \\
Z_5 &= F_{Z_5|Z_3,W}^{-1}\big(\Phi(\epsilon_5)|Z_3, W\big).
\end{aligned}
$$

More generally, for $j = 2, \ldots, d$,

$$F_{Z_j|Z_{k(j)},W}(z_j|z_{k(j)}, w) = C_{j|k(j);W}\big(F_{Z_j|W}(z_j|w) \mid F_{Z_{k(j)}|W}(z_{k(j)}|w)\big).$$

and

$$Z_j = F^{-1}_{Z_j|Z_{k(j)},W}\big(\Phi(\epsilon_j)|Z_{k(j)}, W\big).$$

If $C_{j,W}$ is a Gaussian copula with parameter $\rho_{j,W}$ and $C_{j,k(j);W}$ is a Gaussian copula with parameter $\rho_{j,k(j);W}$, then this becomes model (3.74).

Let $U = (U_1, \ldots, U_d)$ be a multivariate $U(0,1)$ random vector with $U_j = \Phi(Z_j)$ for $j = 1, \ldots, d$, and let $V = \Phi(W)$. Then, from the truncated vine and Theorem 8.64, the copula density is

$$c_U(u) = \int_0^1 \prod_{j=1}^d c_{j,W}(u_j, v) \cdot \prod_{j=2}^d c_{j,k(j);W}\big(C_{j|W}(u_j|v), C_{k(j)|W}(u_{k(j)}|v)\big)\, \mathrm{d}v.$$

The above procedure can be applied to any Gaussian SEM for which the dependence structure can be converted to a truncated partial correlation vine. The resulting model can be considered as a nonlinear SEM.

3.13   Truncated vines, factor models and graphical models

Whittaker (1990) has a Gaussian graphical model that is based on sparseness of the inverse correlation matrix. In this section, we make some comparisons of this with vines and path diagrams of structural equation models.

Consider d jointly Gaussian variables summarized in a $d \times d$ correlation matrix R. If the correlation matrix has the form of a truncated partial correlation vine, then there are many zeros in the inverse correlation matrix, but the parametrization in terms of the partial correlation vine is more convenient for extensions to copula models because of the algebraic independence of parameters in the partial correlation vine. If the inverse correlation matrix R^{-1} has some zeros indicating zero partial correlation of two variables given the remainder (e.g., $\rho_{12;3\cdots d} = 0$ implies that variables 1 and 2 are conditionally independent given variables 3 to d if the joint distribution is Gaussian), then the remaining non-zero partial correlations given $d - 2$ conditioning variables are not algebraically independent.

The algorithms for finding an approximate dependence structure based on a truncated partial correlation vine are more convenient than trying to determining an approximation based on the inverse correlation matrix. Also interpretations of partial correlations in a vine are similar when variables are added or deleted, but the set of relevant partial correlations represented by the inverse correlation matrix changes when variables are added or deleted,

General correlation matrices based on factor models do not have zeros in the inverse correlation matrix. Factor models are often reasonable if the dependence in observed variables can be explained based on unobserved (latent) variables. The common dependence on latent variables implies that two observed variables are conditionally dependent given other observed variables. Structural equation models are based on extensions of common factor models, and latent variables play an important role. These models can be represented as graphical models through path diagrams (see Figure 3.7 for an example) to show where conditional independence or dependence (usually given latent variables) occurs.

Vines can approximate but do not include general copulas with exchangeable dependence. Factor models include positive exchangeable dependence as a special case. The combination of truncated vines and structured factor models lead to a wide class of parsimonious dependence structures for high-dimensional applications.

The various types of factor copula models in the preceding sections are closed under margins whereas vine copulas without latent variables are not closed under margins; the closure property helps in numerical estimation for large d.

For applications, one should consider the dependence structures that can come from truncated vines, factor models and their generalizations, making use of context with possible latent variables where relevant. Theorems with tail and dependence properties of vine copulas are given in Section 8.12. One result is that to get upper (lower) tail dependence in all bivariate margins, it suffices that all bivariate copulas in tree $\mathcal{T}_1$ are upper (lower) tail dependent. Similarly, for factor copulas, to get upper (lower) tail dependence in all bivariate margins, it suffices that all bivariate linking copulas to the first latent variable are upper (lower) tail dependent.

The context of an application can help in choosing a vine, if there are no obvious latent variables to explain the dependence. D-vines are more natural if there is a time or linear spatial order in variables. C-vines are more natural if there are leading variables that influence others. Otherwise, intermediate regular vines between the boundary cases of C-vines and D-vines might be better. With large dimension d, truncated vines will have fewer parameters, and also could be better than graphical models based on high order conditional independencies.

In general, copulas based on truncated R-vines would be considered as good approximations to multivariate distributions; for a given multivariate distribution, there could be more than one R-vine approximation that perform equally well. There are different algorithms (see Section 6.17) for finding good R-vine approximations; the space is too large for an exhaustive search of vines and parametric pair-copula families that could be assigned to each edge. For applications, one should check the approximating R-vine for an interpretation in the context of the variables; if there is no simple interpretation, then the R-vine should just be considered a model that fits well.

Because there are many vine approximations possible for a multivariate distribution, a conjecture is that the best vines are such that conditional dependence varies little as functions of values of the conditioning variables. This conjecture could be checked with some multivariate model constructions that are plausible for applications by generating simulated data and comparing different vine copula fits.

3.14 Copulas for stationary time series models

In this section, we present stochastic representations for time series, with continuous response variable, that extend several stationary Gaussian time series, in particular, Markov order p time series as extension of Gaussian AR(p) with p being a positive integer, q-dependent time series as extension of Gaussian MA(q) with q being a positive integer, and mixed p-Markov/q-dependent as extension of Gaussian ARMA(p, q). Chapter 8 of Joe (1997) only has the extensions of AR(p) and MA(1). With copula-based transition probabilities, the extension to non-stationary times series with covariates can incorporate the covariates into univariate margins that are regression models.

Based on the stochastic representations and assuming stationarity, we also derive the representations in terms of appropriate $(p+1)$-variate and $(q+1)$-variate copulas that have some constraints in terms of margins.

Copula-based Markov time series and other ways to use copulas in time series models have been studied and applied in Darsow et al. (1992), Chen and Fan (2006), Biller (2009), Ibragimov (2009), and Beare (2010).

For all of the models given in this section, $\{\epsilon_s\}$ is a sequence of continuous i.i.d. innovation random variables with cdf F_ϵ, $\{Y_t\}$ is a sequence of observed random variables such that ϵ_s is independent of Y_t for $t \leq s - 1$, and for the ARMA extension, $\{W_t\}$ is a sequence of unobserved random variables such that ϵ_s is independent of $\{W_t : t \leq s - 1\}$.

Markov

For Markov order 1, the general form is $Y_t = h(\epsilon_t; Y_{t-1})$ where $h(x; y)$ can be assumed to be increasing in x. For Markov order p with $p \geq 2$, the general form is $Y_t = h(\epsilon_t; (Y_{t-p}, \ldots, Y_{t-1}))$ where $h(x; \boldsymbol{y})$ is increasing in x. It is clear that Y_t depends on the past only through the p previous observations.

If we assume that the Y_t are continuous random variables with a common cdf F_Y, and the ϵ_s are continuous, then because of the possibility of applying probability integral transforms, we can assume without loss of generality that the ϵ_s are U(0, 1) random variables and $Y_t = F_Y^{-1}(U_t)$ where the U_t are U(0, 1). So we assume that $Y_t = U_t \sim$ U(0, 1) in order to discuss the dependence structure.

For $p = 1$, we can take $h(x; u) = C_{2|1}^{-1}(x|u)$ for $0 < x < 1$ and $0 < u < 1$ where C_{12} is an arbitrary bivariate copula, and C_{12} is the copula of (U_{t-1}, U_t). For $p \geq 2$, $h(x; \boldsymbol{u}) = C_{p+1|1:p}^{-1}(x|\boldsymbol{u})$ where $C_{1:(p+1)}$ is a $(p+1)$-variate copula with margins that satisfies $C_{12} = C_{1+i,2+i}$ for $i = 1, \ldots, p-1$ and $C_{1:m} = C_{(1+i):(m+i)}$ for $i = 1, \ldots, p+1-m$, $m = 2, \ldots, p$, and $C_{1:(p+1)}$ is the copula of $(U_{t-p}, \ldots, U_t)$. To get copulas with these marginal constraints, one possibility comes from the mixture of max-id copula (Section 3.5) with permutation symmetric bivariate max-id copulas H_{jk} that satisfy $H_{jk} = H_{j+i,k+i}$. Another possibility is a D-vine on $p + 1$ variables: the copulas for tree T_1 are all the same as C_{12}, that is, $C_{12} = C_{23} = \cdots = C_{p,p+1}$; the copulas for tree T_2 are all the same as $C_{13;2}$, that is, $C_{13;2} = \cdots = C_{p-1,p+1;p}$; and so on up to tree T_p with copula $C_{1,p+1;2:p}$.

If $C_{1:(p+1)}$ is a Gaussian copula with correlation matrix that is Toeplitz with lag m serial correlation ρ_m, then the Gaussian AR(p) time series model results after U_t are transformed to Gaussian.

q-dependent

For 1-dependent, the general model is $Y_t = g(\epsilon_t; \epsilon_{t-1})$, where $g(x; v)$ can be assumed to be increasing in x. For q-dependent with $q \geq 2$, the general model is $Y_t = g(\epsilon_t; (\epsilon_{t-q}, \ldots, \epsilon_{t-1}))$, where $g(x; \boldsymbol{v})$ can be assumed to be increasing in x. It is clear the $Y_{t'}$ and Y_t are independent if $|t - t'| > q$.

If we assume that the Y_t are continuous random variables with a common cdf F_Y, and the ϵ_s are continuous, then we can assume without loss of generality that F_ϵ and F_Y are U(0, 1). Hence we replace Y_t by U_t in order to present the dependence structure.

For $q - 1$, we can take $g(x; v) = K_{2|1}^{-1}(x|v)$ for $0 < x < 1$ and $0 < v < 1$, where K_{12} is an arbitrary bivariate copula and then K_{12} is the copula of (ϵ_{t-1}, U_t) because with the independence of $\epsilon_{t-1}, \epsilon_t$,

$$\mathbb{P}(\epsilon_{t-1} \leq v, U_t \leq u) = \mathbb{P}(\epsilon_{t-1} \leq v, g(\epsilon_t; \epsilon_{t-1}) \leq u)$$
$$= \int_0^v \mathbb{P}(g(\epsilon_t; z) \leq u) \, \mathrm{d}z = \int_0^v \mathbb{P}(\epsilon_t \leq g^{-1}(u; z)) \, \mathrm{d}z$$
$$= \int_0^v g^{-1}(u; z) \, \mathrm{d}z = \int_0^v K_{2|1}(u|z) \, \mathrm{d}z = K_{12}(v, u),$$

for $0 < v < 1$ and $0 < u < 1$. For $q \geq 2$, we can take $g(x; \boldsymbol{v}) = K_{q+1|1:q}^{-1}(x|\boldsymbol{v})$ where $K_{1:(q+1)}$ is a $(q+1)$-variate copula with margin $K_{1:q}$ being an independence copula and then $K_{1:(q+1)}$ is the copula of $(\epsilon_{t-q}, \ldots, \epsilon_{t-1}, U_t)$, with $U_t = g(\epsilon_t; (\epsilon_{t-q}, \ldots, \epsilon_{t-1}))$. The proof is as follows. With $0 < v_i < 1$ for $i = 1, \ldots, q$ and $0 < u < 1$,

$$\mathbb{P}(\epsilon_{t-q} \leq v_1, \ldots, \epsilon_{t-1} \leq v_q, U_t \leq u) = \int_0^{v_1} \cdots \int_0^{v_q} \mathbb{P}(g(\epsilon_t; \boldsymbol{z}) \leq u) \, \mathrm{d}\boldsymbol{z}$$
$$= \int_0^{v_1} \cdots \int_0^{v_q} \mathbb{P}(\epsilon_t \leq g^{-1}(u; \boldsymbol{z})) \, \mathrm{d}\boldsymbol{z} = \int_0^{v_1} \cdots \int_0^{v_q} g^{-1}(u; \boldsymbol{z}) \, \mathrm{d}\boldsymbol{z} = K_{1:(q+1)}(\boldsymbol{v}_{1:q}, u)$$

if $g^{-1}(u; \boldsymbol{z}) = K_{q+1|1:q}(u|\boldsymbol{z})$.

For $q \geq 2$, $K_{1:(q+1)}$ satisfying the marginal constraint can be obtained by D-vines. For $q = 2$, K_{12} is an independence copula and $K_{23}, K_{13;2}$ are general; for the absolutely continuous case, with densities denoted as k, then $k_{123}(v_1, v_2, u) = k_{23}(v_2, u) \, k_{13;2}(v_1, K_{3|2}(u|v_2))$. For $q \geq 3$, $K_{j-1,j}$ ($2 \leq j \leq q$), $K_{j-2,j;j-1}$ ($3 \leq j \leq q$), $\ldots, K_{1q;2:(q-1)}$ are independence copulas, and $K_{q,q+1}, K_{q-1,q+1;q}, \ldots, K_{1,q+1;2:q}$ are general; for the absolutely continuous case, the density is

$$k_{1:(q+1)}(\boldsymbol{v}_{1:q}, u) = k_{q,q+1}(v_q, u) \cdot \prod_{j=q-1}^{1} k_{j,q+1;(j+1):q}\big[v_j, K_{q+1|(j+1):q}(u|\boldsymbol{v}_{(j+1):q})\big].$$

To get a Gaussian MA(q) time series model after transform to Gaussian margins, $K_{1:(q+1)}$ is a Gaussian copula with correlation matrix $\boldsymbol{R} = \begin{pmatrix} \boldsymbol{I}_q & \boldsymbol{r} \\ \boldsymbol{r}^\top & 1 \end{pmatrix}$ and the column vector $\boldsymbol{r} = (\rho_{1,q+1}, \ldots, \rho_{q,q+1})^\top$ is chosen so that $\boldsymbol{R}$ is positive definite.

Mixed Markov and q-dependent

For a general nonlinear analogue of ARMA($1, q$), we extend the Poisson ARMA($1, q$) time series model of McKenzie (1988) to get:

$$W_s = h(\epsilon_s; W_{s-1}), \quad Y_t = g(\epsilon_t; W_{t-q}, \{\epsilon_{t-i}, 1 \leq i < q\}), \tag{3.77}$$

where $h(x; \cdot)$ and $g(x; \cdot)$ can be assumed to be increasing in x. The function h drives the Markov model on the unobserved $\{W_s\}$ and Y_t depends on W_{t-q} which is independent of $\epsilon_t, \epsilon_{t-1}, \ldots, \epsilon_{t-q+1}, \epsilon_{t-q}$.

By comparing with the pure Markov and pure q-dependent copula-based models, in the case of continuous response, we can assume that F_W, F_ϵ, F_Y are all U$(0, 1)$, and that $h(x; u) = C_{2|1}^{-1}(x|u)$ for a bivariate copula C_{12} and $g(x; \boldsymbol{z}) = K_{q+1|1:q}^{-1}(x|\boldsymbol{z})$ where $K_{1:(q+1)}$ satisfies the conditions given above. We again use $Y_t = U_t \sim$ U$(0, 1)$ to present the dependence structure,

For $p = q = 1$, $W_s = C_{2|1}^{-1}(\epsilon_s|W_{s-1})$ and $U_t = K_{2|1}^{-1}(\epsilon_t|W_{t-1})$. Note the special cases lead to the previous Markov order 1 and 1-dependent models. If $C_{12} = C^\perp$, then $W_s = \epsilon_s$ and $\{U_t\}$ is a 1-dependent sequence; and if $K_{12} = C^+$, then $U_t = W_{t-1} = C_{2|1}^{-1}(\epsilon_{t-1}|W_{t-2})$ is a Markov sequence.

Model (3.77) extends to mixed Markov order p with q-dependent:

$$W_s = h(\epsilon_s; (W_{s-p}, \ldots, W_{s-1})), \quad U_t = g(\epsilon_t; W_{t-q}, (\epsilon_{t-q+1}, \ldots, \epsilon_{t-1})),$$

where $h(v; \boldsymbol{u}) = C_{p+1|1:p}^{-1}(v|\boldsymbol{u})$ for a $(p + 1)$-variate copula $C_{1:(p+1)}$, satisfying the above conditions.

If $C_{1:(p+1)}$ and $K_{1:(q+1)}$ are Gaussian copulas satisfying the conditions in the preceding subsections, then the Gaussian ARMA(p, q) time series model is obtained.

Likelihood and joint density for continuous response

Suppose an observed time series is $\{y_t : t = 1, \ldots, n\}$. For a Markov model, the likelihood or joint density is a product of conditional densities based on $c_{1:(p+1)}(u_{t-p}, \ldots, u_t)$, where $u_i = F_Y(y_i; \eta)$ for a univariate distribution F_Y. For a q-dependent time series model, or for a mixed Markov/q-dependent model, the joint density is in general a multidimensional integral with dimension equal to the Markov order p. We show a few details below for the joint density of n consecutive observations that have been converted to U$(0, 1)$ marginally. This density for the dependence structure can be combined with a stationary marginal distribution F_Y for a continuous response.

The details of using a 1-1 transform and the Jacobian can be seen from the case of $p = 1$ and $q = 1$.

Let $g = K_{2|1}^{-1}$ and $h = C_{2|1}$, and let k_{12} be the density of K_{12}. The model is $W_s = h(\epsilon_s; W_{s-1})$, $U_t = g(\epsilon_t; W_{t-1})$. Given independent random variables $W_0, \epsilon_1, \epsilon_2, \ldots, \epsilon_n$, the stochastic sequence $U_1, W_1, U_2, W_1, \ldots, U_n, W_n$ can be obtained. Write

$$U_1 = g(\epsilon_1; W_0) =: a_1(\epsilon_1; W_0),$$
$$U_2 = g(\epsilon_2; W_1) = g(\epsilon_2; h(\epsilon_1; W_0)) =: a_2(\epsilon_1, \epsilon_2; W_0),$$
$$U_3 = g(\epsilon_3; W_2) = g(\epsilon_3; h[\epsilon_2; h(\epsilon_1; W_0)]) =: a_3(\epsilon_1, \epsilon_2, \epsilon_3; W_0),$$
$$\vdots = \vdots$$
$$U_n = g(\epsilon_n; W_{n-1}) = g(\epsilon_n; h[\epsilon_{n-1}; W_{n-2}]) =: a_n(\epsilon_1, \ldots, \epsilon_n; W_0).$$

Given W_0, there is a 1-1 transform from $(\epsilon_1, \ldots, \epsilon_n)$ to $(U_1, \ldots, U_n)$. The inverse transform is

$$\epsilon_1 = K_{2|1}(U_1|W_0) =: b_1(U_1; W_0),$$
$$\epsilon_2 = K_{2|1}(U_2|h(\epsilon_1; W_0)) =: b_2(U_1, U_2; W_0),$$
$$\epsilon_3 = K_{2|1}(U_3|h[\epsilon_2; h(\epsilon_1; W_0)]) =: b_3(U_1, U_2, U_3; W_0),$$
$$\vdots = \vdots$$
$$\epsilon_n = K_{2|1}(U_n|h[\epsilon_{n-1}; W_{n-2}]) =: b_n(U_1, \ldots, U_n; W_0).$$

For the non-random variable transformation, write $e_j = b_j(u_1, \ldots, u_j; w_0)$ and $w_j = h(e_j; w_{j-1}) = C_{2|1}(e_j|w_{j-1})$ for $j = 1, \ldots, n$. The Jacobian matrix is lower triangular, so its determinant is $|J(u_1, \ldots, u_n; w_0)| = \frac{\partial b_1}{\partial u_1} \frac{\partial b_2}{\partial u_2} \cdots \frac{\partial b_n}{\partial u_n}$. Note the b_2 might not be increasing in u_1, etc.

The joint pdf of $(U_1, \ldots, U_n)$ is

$$c_{U_1, \ldots, U_n}(\boldsymbol{u}) = \int_0^1 f_{\epsilon_1, \ldots, \epsilon_n, W_0}[b_1(u_1; w_0), \ldots, b_n(u_1, \ldots, u_n; w_0)] \, |J(u_1, \ldots, u_n; w_0)| \, dw_0$$

$$= \int_0^1 \prod_{j=1}^n \frac{\partial b_j}{\partial u_j} \, dw_0 = \int_0^1 \prod_{j=1}^n k_{12}(w_{j-1}, u_j) \, dw_0.$$

For $q \geq 1$ and $p \geq 0$, the copula density of $U_1, \ldots, U_n$ can be written as an integral with dimension equal to $\max\{p, q\}$. For $q = 0$, no integration is needed because the first $p + 1$ observations can be assume to have copula $C_{1:(p+1)}$.

With details similar to the case of $p = 1, q = 1$, some joint densities are the following.

- 1-dependent:

$$c_{U_1, \ldots, U_n}(\boldsymbol{u}) = \int_0^1 \prod_{j=1}^n k_{12}(e_{j-1}, u_j) \, de_0,$$

where $e_j := K_{2|1}(u_j|e_{j-1})$ for $j = 1, \ldots, n$.

- 2-dependent:

$$c_{U_1, \ldots, U_n}(\boldsymbol{u}) = \int_0^1 \int_0^1 \prod_{j=1}^n k_{123}(e_{j-2}, e_{j-1}, u_j) \, de_0 de_1,$$

with $e_j := K_{3|12}(u_j|e_{j-2}, e_{j-1})$ for $j = 1, \ldots, n$.

- $p = 1, q = 2$:

$$c_{U_1, \ldots, U_n}(\boldsymbol{u}) = \int_0^1 \int_0^1 \prod_{j=1}^n k_{123}(w_{j-2}, e_{j-1}, u_j) \, de_0 dw_{-1},$$

with $w_s = C_{2|1}^{-1}(e_s|w_{s-1})$ for $s = 1, \ldots, n$ and $e_j = K_{3|12}(u_j|w_{j-2}, e_{j-1})$ for $j = 1, \ldots, n$.

- $p = 2, q = 1$:

$$c_{U_1,\ldots,U_n}(\boldsymbol{u}) = \int_0^1 \int_0^1 \prod_{j=1}^n k_{12}(w_{j-1}, u_j)\, \mathrm{d}w_0 \mathrm{d}w_{-1},$$

with $w_s = C_{3|12}^{-1}(e_s|w_{s-2}, w_{s-1})$ for $s = 1, \ldots, n$, and $e_j = K_{2|1}(u_j|w_{j-1})$ for $j = 1, \ldots, n$.

For a stationary model with discrete margin F_Y, we need to transform from a U$(0,1)$ sequence U_t to Y_t via F_Y^{-1}. For $p = q = 1$, this leads to $W_s = C_{2|1}^{-1}(\epsilon_s|W_{s-1})$ for the unobserved Markov U$(0,1)$ sequence, $U_t = K_{2|1}^{-1}(\epsilon_t|W_{t-1})$ for a U$(0,1)$ sequence and $Y_t = F_Y^{-1}[K_{2|1}^{-1}(\epsilon_t|W_{t-1})] = F_Y^{-1}(U_t)$ as a sequence with distribution F_Y. The joint distribution of $(Y_1, \ldots, Y_n)$ involves an integral of dimension $n + 1$.

3.15　Multivariate extreme value distributions

Early uses of copulas (or dependence functions) and multivariate non-Gaussian distributions appear in the multivariate extreme value literature, for example, Deheuvels (1978) and Galambos (1987).

Let $(X_{i1}, \ldots, X_{id})$ be i.i.d. random vectors with distribution F, $i = 1, 2, \ldots$. Let $M_{jn} = \max_{1 \le i \le n} X_{ij}$, $j = 1, \ldots, d$, be the componentwise maxima.

Multivariate extreme value distributions come from non-degenerate limits (in law) of

$$\big((M_{1n} - a_{1n})/b_{1n}, \ldots, (M_{dn} - a_{dn})/b_{dn}\big),$$

where the b_{jn} are positive. If a limiting distribution exists, then each univariate margin must be in the GEV family (see Section 2.3). The multivariate limiting distribution can then be written in the form

$$C\big(H(z_1; \xi_1), \ldots, H(z_d; \xi_d)\big),$$

where $H(z_j; \xi_j)$ are GEV distributions and C is a multivariate copula.

The main result for multivariate extremes is due to de Haan and Resnick (1977) and Pickands (1981). It is usually presented in the form of max-stable multivariate distributions with univariate Fréchet margins as in de Haan and Resnick (1977) or min-stable survival functions with univariate exponential margins as in Pickands (1981). We will present the latter form.

Extreme value copulas satisfy further conditions:

$$C(u_1^r, \ldots, u_d^r) = C^r(\boldsymbol{u}), \ \forall r > 0, \ \boldsymbol{u} \in [0, 1]^d.$$

For the multivariate min-stable exponential survival form, the multivariate survival function $\overline{G}$ satisfies the condition

$$-\log \overline{G}(r\boldsymbol{x}) = -r \log \overline{G}(\boldsymbol{x}) \ \forall r > 0, \ \boldsymbol{x} \in [0, \infty)^d,$$

and $\overline{G}(0, \ldots, 0, x_j, 0, \ldots, 0) = e^{-x_j}$ $(j = 1, \ldots, d)$. With $A = -\log \overline{G}$, the copula is

$$C(\boldsymbol{u}) = \exp\{-A(-\log u_1, \ldots, -\log u_d)\},$$

where A maps $[0, \infty)^d \to [0, \infty)$ and has the following properties:

- $A(0, \ldots, 0, x_j, 0, \ldots, 0) = x_j$ for $j = 1, \ldots, d$;
- A is homogeneous of order 1;
- A is convex;
- $\max\{x_1, \ldots, x_d\} \le A(x_1, \ldots, x_d) \le x_1 + \cdots + x_d$ (boundaries are comonotonicity and independence).

Furthermore, the representation of de Haan and Resnick (1977) and Pickands (1981) is that

$$A(\boldsymbol{x}) = -\log \overline{G}(\boldsymbol{x}) = \int_{\mathcal{S}_d} [\max_{1 \le i \le d} (q_i x_i)] \, d\mu(\boldsymbol{q}), \quad x_j \ge 0, \; j = 1, \ldots, d, \tag{3.78}$$

where $\mathcal{S}_d = \{\boldsymbol{q} : q_i \ge 0, j = 1, \ldots, d, \sum_j q_j = 1\}$ is the d-dimensional unit simplex and μ is a finite measure on $\mathcal{S}_d$.

So unlike central limit theory which leads to multivariate Gaussian with a finite number of dependence parameters, the family of extreme value copulas is infinite-dimensional for each $d \ge 2$. For parametric inference, it might be desirable to find finite-dimensional parametric subfamilies that have a variety of dependence structures like those mentioned in Section 3.1.

If $(X_1, \ldots, X_d) \sim G$ such that G satisfies (3.78), then

$$\min\{X_1/w_1, \ldots, X_d/w_d\} \sim \text{Exponential}(\text{rate} = A(w_1, \ldots, w_d))$$

for all $(w_1, \ldots, w_d) \in (0, \infty)^d$, since

$$\mathbb{P}(X_j/w_j > z, j = 1, \ldots, d) = \exp\{-A(w_1 z, \ldots, w_d z)\} = \exp\{-z A(w_1, \ldots, w_d)\}.$$

That is, min-stable multivariate exponential distributions have the property of closure under weighted minima; compare with the property of closure under linear combinations for multivariate Gaussian.

Some multivariate extreme value families have been obtained from extreme value limits of multivariate distributions with tail dependence. Examples are the following.

1. Limit of multivariate Pareto (Galambos, 1975).

2. Limit of mixture of max-id copulas with tail dependence: some with closed form and flexible dependence structures (Joe and Hu, 1996).

3. Limit of multivariate t_ν copula (Nikoloulopoulos et al. (2009)): the correlation matrix $\boldsymbol{\Sigma}$ is a parameter, and boundary cases are Hüsler-Reiss copulas (as $\nu \to \infty$ and correlation parameters approach 1 at rate $(-\log n)^{-1}$ as $n \to \infty$) and Marshall-Olkin copulas (as $\nu \to 0^+$).

4. Limit of multivariate skew-t_ν distributions (Padoan, 2011).

5. Limit of D-vine copulas with tail dependence (Joe, 1994): the limiting extreme value copulas have integrals and so are computationally feasible in low dimensions .

There is theory from different copula constructions on whether there is upper or lower tail dependence; then the limiting upper or lower extreme value copula can be obtained; see Section 2.18.

For 1-truncated and 2-truncated R-vines with tail dependence, the limiting extreme value copulas are simplest when the vine is the C-vine. For a 1-truncated C-vine,

$$C_{1:d}(\boldsymbol{u}) = \int_0^{u_1} \left\{ \prod_{j=2}^d C_{j|1}(u_j|v) \right\} dv, \quad \boldsymbol{u} \in [0.1]^d,$$

so that

$$u^{-1} C_{1:d}(u\boldsymbol{w}) = \int_0^{w_1} \left\{ \prod_{j=2}^d C_{j|1}(uw_j|uw) \right\} dw.$$

Let $b_{j|1}(w_j|w) = \lim_{u \to 0^+} C_{j|1}(uw_j|uw)$ for $j = 1, \ldots, d$. Under the condition that $b_{j|1}(\cdot|w)$ is a proper distribution for all j, then the (lower) tail dependence function of $C_{1:d}$ is

$$b_{1:d}(\boldsymbol{w}) = \int_0^{w_1} \left\{ \prod_{j=2}^d b_{j|1}(w_j|w) \right\} dw, \quad \boldsymbol{w} \in \mathbb{R}_+^d.$$

With these conditions, marginal tail dependence functions are obtained by letting appropriate w_k increase to ∞. Then the exponent function A of the limiting extreme value copula can be obtained by inclusion-exclusion as in (2.63). For the 1-truncated D-vine,

$$C_{1:d}(\boldsymbol{u}) = \int_0^{u_1} \cdots \int_0^{u_{d-1}} \left\{ \prod_{j=2}^{d-1} c_{j-1,j}(v_{j-1}, v_j) \right\} C_{d|d-1}(u_d|v_{d-1})\, dv_1 dv_2 \cdots dv_{d-1}$$

$$= \int_0^{u_2} \cdots \int_0^{u_{d-1}} C_{1|2}(u_1|v_2) \left\{ \prod_{j=3}^{d-1} c_{j-1,j}(v_{j-1}, v_j) \right\} C_{d|d-1}(u_d|v_{d-1})\, dv_2 \cdots dv_{d-1},$$

$$C_{1:d}(u\boldsymbol{w}) = u \int_0^{w_2} \cdots \int_0^{w_{d-1}} C_{1|2}(uw_1|uv_2) \left\{ \prod_{j=3}^{d-1} u\, c_{j-1,j}(uv_{j-1}, uv_j) \right\}$$

$$\cdot C_{d|d-1}(uw_d|uv_{d-1})\, d\boldsymbol{v}_{2:(d-1)}.$$

Let $b_{j-1,j}$ be the tail dependence function of $C_{j-1,j}$, and let $b_{j|j-1}(w_j|w) = \lim_{u \to 0^+} C_{j|j-1}(uw_j|uw)$ and $b_{j-1|j}(w_j|w) = \lim_{u \to 0^+} C_{j-1|j}(uw_j|uw)$ be conditional tail dependence functions for $j = 2, \ldots, d$. Under the condition that $C_{j-1,j}$ has continuous density and $b_{j|j-1}(\cdot|w)$, $b_{j-1|j}(\cdot|w)$ are proper distributions for all j, as $u \to 0^+$, $C_{1|2}(uw_1|uv_2) \to b_{1|2}(w_1|v_2)$, $C_{d|d-1}(uw_d|uv_{d-1}) \to b_{d|d-1}(w_d|v_{d-1})$, $u\, c_{j-1,j}(uv_{j-1}, uv_j) \to (D_{12}^2 b_{j-1,j})(v_{j-1}, v_j)$ and the (lower) tail dependence function is

$$b_{1:d}(\boldsymbol{w}) = \int_0^{w_2} \cdots \int_0^{w_{d-1}} b_{1|2}(w_1|v_2) \left\{ \prod_{j=3}^{d-2} (D_{12}^2 b_{j-1,j})(v_{j-1}, v_j) \right\} b_{d|d-1}(w_d|v_{d-1})\, dv_2 \cdots dv_{d-1}.$$

For some truncated R-vines in between the C-vine and D-vine, the dimension of the integral can be reduced to be between 1 and $d-2$. With details similar to the above, Section 3.16 has the extreme value limit of copulas with factor structure and tail dependence.

In the multivariate extreme value literature, there are also other ways to derive extreme value models; one approach is through stationary max-stable random fields. Some references with various approaches are in Schlather (2002), Cooley et al. (2010), and Padoan (2013).

3.16 Multivariate extreme value distributions with factor structure

The multivariate Gumbel and Galambos copulas have exchangeable dependence. The t-EV copulas with more flexible dependence have been obtained based on the extreme value limit of the multivariate t_ν distribution; similar to the multivariate t_ν distribution, there is a parameter for each bivariate margin; also, the multivariate Hüsler-Reiss copula is obtained on the boundary.

For large dimension d, if the extremes are linked to dependence via a latent variable, one might want to have a multivariate extreme value distribution with $O(d)$ parameters and dependence in the form of a 1-factor model.

The theory of tail dependence functions in Section 2.18 can be applied to the 1-factor copula with bivariate linking copulas C_{Vj} for $j = 1, \ldots, d$; its cdf is

$$C(\boldsymbol{u}) = \int_0^1 \prod_{j=1}^d C_{j|V}(u_j|u_0)\, du_0, \quad \boldsymbol{u} \in [0,1]^d.$$

For $j = 1, \ldots, d$, assume lower tail dependence of C_{Vj} with tail dependence function $b_{jV}(w,v)$ that satisfies $b_{jV}(w, \infty) = w$; also for the conditional tail dependence function $b_{j|V}(w_j|v)$, suppose $b_{j|V}$ is a proper distribution on $[0, \infty)$. Then, using arguments similar to the proof of part (d) of Theorem 8.76, as $u \to 0^+$,

$$\frac{C(u\boldsymbol{w})}{u} = u^{-1} \int_0^1 \prod_{j=1}^d C_{j|V}(uw_j|u_0)\, \mathrm{d}u_0 = \int_0^{1/u} \prod_{j=1}^d C_{j|V}(uw_j|uw_0)\, \mathrm{d}w_0$$

$$\to \int_0^\infty \prod_{j=1}^d b_{j|V}(w_j|w_0)\, \mathrm{d}w_0.$$

The tail dependence function of the mixture is

$$b(\boldsymbol{w}) = \int_0^\infty \prod_{j=1}^d b_{j|V}(w_j|w_0)\, \mathrm{d}w_0, \quad \boldsymbol{w} \in \mathbb{R}_+^d,$$

and the tail dependence function of a margin S (with cardinality ≥ 2) is

$$b_S(\boldsymbol{w}_S) = \int_0^\infty \prod_{j \in S} b_{j|V}(w_j|w_0)\, \mathrm{d}w_0, \quad \boldsymbol{w}_S \in \mathbb{R}_+^{|S|}.$$

From (2.63), the exponent function of the lower extreme value limit is:

$$A(\boldsymbol{w}) = \sum_{j=1}^d w_j - \sum_{1 \leq i < j \leq d} b_{ij}(w_i, w_j) + \sum_{i_1 < i_2 < i_3} b_{i_1 i_2 i_3}(w_{i_1}, w_{i_2}, w_{i_3}) + \cdots + (-1)^d b(w_1, \ldots, w_d). \tag{3.79}$$

Note also that for a fixed j with $b_{jV}(w, \infty) = w$, then

$$\int_0^\infty b_{j|V}(w_j|w_0)\, \mathrm{d}w_0 = \int_0^\infty \frac{\partial b_{jV}(w_j, w_0)}{\partial v}\, \mathrm{d}w_0 = b_{jV}(w_j, v)\, |_{v=0}^\infty = w_j - 0 = w_j,$$

The above implies that $b_{j|V}(w_j|w_0) \to 0$ as $w_0 \to \infty$ at a fast enough rate so that $b_{j|V}(w_j|\cdot)$ is integrable. Assuming $b_{jV}(w, \infty) = w$ for all j, by combining the integrands of the $2^d - 1$ terms in A in (3.79), one gets[3]

$$A(\boldsymbol{w}) = \int_0^\infty \Big[1 \quad \prod_{j=1}^d \{1 \quad b_{j|V}(w_j|v)\}\Big]\, \mathrm{d}v, \quad \boldsymbol{w} \subset \mathbb{R}_+^d,$$

As $v \to \infty$, $1 - \prod_{j=1}^d \{1 - b_{j|V}(w_j|v)\} \sim \sum_{j=1}^d b_{j|V}(w_j|v)$ so that the integrand can be dominated by an integrable function.

When b_{jV} is the tail upper dependence function for the bivariate Gumbel (or respectively Galambos) copula with parameter $\delta_j > 1$ ($\delta_j > 0$) for $j = 1, \ldots, d$, then

$$b_{j|V,\mathrm{Gumbel}}(w|v) = 1 - (v^{\delta_j} + w^{\delta_j})^{1/\delta_j - 1} v^{\delta_j - 1} = 1 - (1 + [w/v]^{\delta_j})^{1/\delta_j - 1},$$

or

$$b_{j|V,\mathrm{Galambos}}(w|v) = (v^{-\delta_j} + w^{-\delta_j})^{-1/\delta_j - 1} v^{-\delta_j - 1} = (1 + [w/v]^{-\delta_j})^{-1/\delta_j - 1}.$$

For these functions, with w fixed, $b_{j|V,\mathrm{Gumbel}}(w|v) \sim (\delta_j^{-1} - 1)(w/v)^{\delta_j}$ as $v \to \infty$ and $b_{j|V,\mathrm{Galambos}}(w|v) \sim (w/v)^{\delta_j + 1}$ as $v \to \infty$.

3.17   Other multivariate models

In this section, we present some other multivariate models that have been converted to copulas. Before this, the first subsection below has some parallels between Archimedean copulas and elliptical distributions that provide a basis for extensions.

---

[3]Thanks to David Lee for the derivation.

3.17.1   Analogy of Archimedean and elliptical

Consider $R(Z_1, \ldots, Z_d)$, where $Z_1, \ldots, Z_d, R$ are mutually independent and R is a radial random variable. Elliptical distributions based on scale mixtures of Gaussian arise when the Z_j are $N(0,1)$ random variables and Archimedean copulas arise when the Z_j are exponential random variables.

To be more precise, spherical distributions with the same generator existing for all dimensions $d = 2, 3, \ldots$ imply that they are scale mixtures of Gaussian distributions with mean 0 with a radial random variable R that doesn't depend on d. Likewise, Archimedean copulas with the same generator existing for all dimensions $d = 2, 3, \ldots$ imply the copula comes from a rate mixture of exponential survival distributions. The survival function of the mixture is the same as the Laplace transform of the mixing distribution F_Q or rate random variable $Q = R^{-1}$, that does not depend on d.

A spherical distribution exists for a specific dimension d if there is a stochastic representation $R(S_1, \ldots, S_d)$ where R is a radial random variable, and $(S_1, \ldots, S_d)$ is uniform on the surface of the d-dimensional unit hypersphere. An Archimedean copula exists for a specific dimension d if there is a copula that comes from the stochastic representation $R(S_1, \ldots, S_d)$ where R is a radial random variable, and $(S_1, \ldots, S_d)$ is uniform on the surface of the d-dimensional unit simplex. The univariate marginal survival function of the mixture is the same as the Williamson d-monotone transform of the mixing distribution F_R.

McNeil and Nešlehová (2010) extend Archimedean copulas to Liouville copulas, by replacing a Dirichlet$(1, \ldots, 1)$ distribution (uniform on simplex) for $(S_1, \ldots, S_d)$ with a Dirichlet$(\alpha_1, \ldots, \alpha_d)$ distribution where the α_j are integers. The integer parameters are only needed to get some closed form expressions. See Section 3.17.2 for a few more details of this construction.

Going back to elliptical distributions, $(S_1, \ldots, S_d)$ is uniform on the surface of the d-dimensional unit hypersphere implies $S_1^2 + \cdots + S_d^2 = 1$ and $(T_1, \ldots, T_d) = (S_1^2, \ldots, S_d^2)$ is Dirichlet$(\frac{1}{2}, \ldots, \frac{1}{2})$. The proof of this statement is as follows. The area of surface of hypersphere is $2\pi^{d/2}/\Gamma(d/2)$. The Jacobian of $s \to s^{1/2} = t$ is $\frac{1}{2}t^{-1/2}$. There is a factor of 2^d from the symmetry and $s \to |s|$.

A summary of the parallels is given in Table 3.3.

Mixture 1	Property	Mixture 2
Spherical	conditional i.i.d.	Archimedean
(generator φ_d for dim. d):		(generator ψ for any d):
scale mixture of i.i.d. N(0,1)		scale mixture of i.i.d. exponential
$\downarrow$		$\downarrow$
Spherical (fixed d, φ_d)	not necessarily	Archimedean (fixed d, ψ_d):
scale mixture of uniform	conditional i.i.d.	scale mixture of uniform
on surface of sphere		on unit simplex
$\downarrow$		$\downarrow$
Elliptical via closure		No extension
under linear combinations		matching linearity

Table 3.3 *Some parallels of spherical distributions and Archimedean copulas. Spherical extends to elliptical for non-exchangeable dependence. There is nothing as general for Archimedean copulas; extensions with less flexible dependence are copulas that are hierarchial Archimedean or mixtures of max-infinite-divisible. The vine pair-copula construction is a more flexible dependence model that generalizes multivariate Gaussian after parametrization via partial correlation vines.*

3.17.2 Other constructions

Luo and Shevchenko (2010) define a multivariate t distribution where the univariate margins can have different shape (or degree of freedom) parameters. With location parameters of 0, the stochastic representation is:

$$(X_1, \ldots, X_d) = (Q_1^{1/2} Z_1, \ldots, Q_d^{1/2} Z_d), \quad Q_j^{-1} = 2\nu_j^{-1} F_\Gamma^{-1}(V; \nu_j/2) \text{ for } j = 1, \ldots, d,$$
$$V \sim U(0,1), \quad \boldsymbol{Z} = (Z_1, \ldots, Z_d) \sim N_d(\boldsymbol{0}, \boldsymbol{R}), \quad V \perp \boldsymbol{Z},$$

where $F_\Gamma(\cdot; \vartheta)$ is the Gamma cdf with shape parameter ϑ and scale parameter 1. Hence $Q_j^{-1} \sim \text{Gamma}(\nu_j/2, \text{scale} = 2/\nu_j)$ are dependent random variables and $Q_j^{1/2} Z_j \sim t_{\nu_j}$. The usual multivariate t_ν is a special case when $\nu_1 = \cdots = \nu_d = \nu$. The joint cdf is:

$$F_{\boldsymbol{X}}(\boldsymbol{x}; \boldsymbol{R}, \nu_1, \ldots, \nu_d) = \int_0^1 \Phi_d(x_1 q_1^{-1/2}, \ldots, x_d q_d^{-1/2}; \boldsymbol{R}) \, dv, \quad q_j^{-1} = 2\nu_j^{-1} F_\Gamma^{-1}(v; \nu_j/2)],$$

and its joint pdf is:

$$f_{\boldsymbol{X}}(\boldsymbol{x}; \boldsymbol{R}, \nu_1, \ldots, \nu_d) = \int_0^1 \phi_d(x_1 q_1^{-1/2}, \ldots, x_d q_d^{-1/2}; \boldsymbol{R}) \cdot \left(\prod_{j=1}^d q_j^{-1/2} \right) dv, \quad \boldsymbol{x} \in \mathbb{R}^d.$$

Hence, this can be converted to a copula whose density is a 1-dimensional integral:

$$c(\boldsymbol{u}; \boldsymbol{R}, \nu_1, \ldots, \nu_d) = \frac{f_{\boldsymbol{X}}\left(T_{\nu_1}^{-1}(u_1), \ldots, T_{\nu_d}^{-1}(u_d); \boldsymbol{R}, \nu_1, \ldots, \nu_d\right)}{\prod_{j=1}^d t_{\nu_j}\left(T_{\nu_j}^{-1}(u_j)\right)}, \quad \boldsymbol{u} \in (0,1)^d,$$

where T_ν and t_ν are the t_ν cdf and pdf respectively.

Prior to Luo and Shevchenko (2010), a grouped t-copula has been defined in Daul et al. (2003). The grouped t-copula obtains if the d variables are divided into m groups such that there is common degree of freedom parameter ν_g in group g for $g = 1, \ldots, m$.

Section 5.3.3 of McNeil et al. (2005) mentions the importance of having copulas with tail asymmetry for finance applications in the context of market risk. In order to mix multivariate Gaussian distributions and get multivariate distributions and copulas with tail asymmetry, Section 3.2.2 of McNeil et al. (2005) has a Gaussian mean-variance mixture construction with stochastic representation:

$$\boldsymbol{X} \stackrel{d}{=} \boldsymbol{m}(Q) + Q^{1/2} \boldsymbol{A} \boldsymbol{Z},$$

where (i) $\boldsymbol{Z} \sim N_k(\boldsymbol{0}, I_k)$; (ii) $Q \geq 0$ is a scalar random variable independent of $\boldsymbol{Z}$; (iii) $\boldsymbol{A}$ is a $d \times k$ matrix; and (iv) $\boldsymbol{m} : [0, \infty) \to \mathbb{R}^d$ is a measureable function. Elliptical distributions are a special case when $\boldsymbol{m} = \boldsymbol{0}$. A concrete specification is $\boldsymbol{m}(Q) = \boldsymbol{\mu} + Q\boldsymbol{\gamma}$ where $\boldsymbol{\mu}, \boldsymbol{\gamma}$ are vector parameters in $\mathbb{R}^d$. McNeil et al. (2005) refer to the resulting copulas as skewed normal mixture copulas, of which a special case is a skewed t copula or multivariate generalized hyperbolic copula, when Q has a generalized inverse Gaussian distribution.

Other skew-elliptical copulas with tail asymmetry arise from the skew-normal distribution of Azzalini (1985) and the skew-t distribution of Azzalini and Capitanio (2003). These multivariate distributions can be derived from a stochastic representation where the multivariate Gaussian or t random vector is conditioned on one variable being positive.

For the multivariate skew-t distribution of Azzalini and Capitanio (2003), $(Y_1, \ldots, Y_d)$ is distributed as $[(X_1, \ldots, X_d) \mid X_0 > 0]$, where $(X_0, X_1, \ldots, X_d)$ is jointly multivariate t with shape parameter ν and correlation matrix $\boldsymbol{R} = \begin{pmatrix} 1 & \boldsymbol{\delta}^\top \\ \boldsymbol{\delta} & \boldsymbol{\Omega} \end{pmatrix}$. Based on this stochastic representation, the multivariate skew-t density is

$$2 t_{d,\nu}(\boldsymbol{y}; \boldsymbol{\Omega}) \, T_{\nu+d}\left(\boldsymbol{\alpha}^\top \boldsymbol{y} \left[\frac{\nu + d}{\nu + \boldsymbol{y}^\top \boldsymbol{\Omega}^{-1} \boldsymbol{y}} \right]^{1/2} \right),$$

where $\boldsymbol{\alpha}^\top = (1 - \boldsymbol{\delta}^\top \boldsymbol{\Omega}^{-1} \boldsymbol{\delta})^{-1/2} \boldsymbol{\delta}^\top \boldsymbol{\Omega}^{-1}$. Its univariate margins are

$$2t_\nu(y_j) \, T_{\nu+1}\Big(\zeta_j y_j \Big[\frac{\nu+1}{\nu + y_j^2}\Big]^{1/2}\Big),$$

where $\zeta_j = (1 - \delta_j^2)^{-1/2} \delta_j$ can be interpreted as a skewness parameter and $\delta_j = \zeta_j / (1 + \zeta_j^2)^{1/2}$. The bivariate $(1, 2)$ marginal density, with $\rho = \rho_{12}$, has the form

$$2t_{2,\nu}(y_1, y_2; \rho) \, T_{2+\nu}\Big(\frac{(\beta_1 y_1 + \beta_2 y_2)}{\sigma} \Big[\frac{\nu+2}{\nu + (y_1^2 + y_2^2 - 2\rho y_1 y_2)/(1 - \rho^2)}\Big]^{1/2}\Big),$$

where

$$\begin{pmatrix} \beta_1 \\ \beta_2 \end{pmatrix} = (1 - \rho^2)^{-1} \begin{pmatrix} \delta_1 - \rho \delta_2 \\ \delta_2 - \rho \delta_1 \end{pmatrix},$$

and

$$\sigma^2 = \Big[1 - \frac{\delta_1^2 + \delta_2^2 - 2\rho \delta_1 \delta_2}{(1 - \rho^2)}\Big].$$

Note the constraint

$$\delta_1^2 + \delta_2^2 - 2\rho \delta_1 \delta_2 \le 1 - \rho^2$$

from the positive definiteness condition.

Copulas based on Liouville distributions are studied in McNeil and Nešlehová (2010). The idea is as follows. Archimedean copulas based on the Williamson transform (Section 3.3) are copulas for random vectors with the stochastic representation

$$\boldsymbol{X} = (X_1, \ldots, X_d) \overset{d}{=} R \times (S_1, \ldots, S_d),$$

where $R \perp\!\!\!\perp (S_1, \ldots, S_d)$, R is a positive random variable and $(S_1, \ldots, S_d)$ is uniformly distributed on the unit simplex. If the distribution of $(S_1, \ldots, S_d)$ is instead Dirichlet$(\alpha_1, \ldots, \alpha_d)$ on the unit simplex, then $\boldsymbol{X} \overset{d}{=} R(S_1, \ldots, S_d)$ has a Liouville distribution. The survival copula of $F_{\boldsymbol{X}}$ is called a Liouville copula.

The density for Dirichlet$(\alpha_1, \ldots, \alpha_d)$ is

$$f_D(s_1, \ldots, s_{d-1}; \alpha_1, \ldots, \alpha_d) = \frac{\Gamma(\alpha_\bullet)}{\prod_{j=1}^d \Gamma(\alpha_j)} \prod_{j=1}^{d-1} s_j^{\alpha_j - 1} \cdot [1 - s_1 - \ldots - s_{d-1}]^{\alpha_d - 1},$$

$$s_j \ge 0, j = 1, \ldots, d-1; \quad 1 - s_1 - \ldots - s_{d-1} \ge 0,$$

where $\alpha_\bullet = \sum_{j=1}^d \alpha_j$. If F_R has density f_R, the Liouville density (Section 6.1 of Fang et al. (1990)) is:

$$f_{\boldsymbol{X}}(\boldsymbol{x}) = \Big\{\prod_{j=1}^d \frac{x_j^{\alpha_j - 1}}{\Gamma(\alpha_j)}\Big\} \cdot \frac{\Gamma(\alpha_\bullet)}{x_\bullet^{\alpha_\bullet - 1}} f_R(x_\bullet), \quad x_\bullet = x_1 + \cdots + x_d, \; \boldsymbol{x} \in (0, \infty)^d.$$

This can be proved through the 1-1 transformation $\boldsymbol{X} \leftrightarrow (S_1, \ldots, S_{d-1}, R)$ with $R = X_1 + \cdots + X_d$ and $S_j = X_j / R$ for $j = 1, \ldots, d-1$. The univariate marginal cdfs are

$$F_{X_j}(x) = 1 - \int_x^\infty [1 - F_B(x/r; \alpha_j, \alpha_\bullet - \alpha_j)] \, f_R(r) \, dr, \quad x > 0,$$

where F_B is the cdf of the Beta distribution. The convenience of the Liouville copula based on f_R depends on the univariate cdfs and quantile functions having a simpler form. See

Section 4.27 for an example of a Liouville copula based on a multivariate GB2 distribution where the cdf involves the incomplete beta function.

For the case where α_j is a positive integer for $j = 1, \ldots, d$ (so that $\alpha_\bullet$ is an integer), and ψ is the Williamson $\alpha_\bullet$-transform of F_R, McNeil and Nešlehová (2010) show that

$$\overline{F}_{\boldsymbol{X}}(\boldsymbol{x}) = \sum_{i_1=0}^{\alpha_1-1} \cdots \sum_{i_d=0}^{\alpha_d-1} (-1)^{i_1+\cdots+i_d} \frac{\psi^{(i_1+\cdots+i_d)}(x_1 + \cdots + x_d)}{i_1! \cdots i_d!} \prod_{j=1}^{d} x_j^{i_j}, \quad \boldsymbol{x} \in \mathbb{R}_+^d.$$

The jth univariate margin is $F_j(x) = 1 - \sum_{i=0}^{\alpha_j-1} (-1)^i \psi^{(i)}(x)/i!$, for $x > 0$.

3.18 Operations to get additional copulas

As shown in Chapter 1 and the preceding section, any continuous multivariate distribution can be converted to a copula.

This chapter has covered some construction methods to get copula families with different types of dependence structures. Several high-dimensional copula constructions are based on a sequence of bivariate copulas. There are a variety of operations that can be used to get more parametric low-dimensional copula families, some with one or more additional parameters. Some of the operations were used within the construction methods.

One purpose of operations on copulas is to derive copula families with an extra parameter; for applications, as the sample size increases, log-likelihood analysis can better discriminate multi-parameter families. Especially useful to get extra flexibility for bivariate copulas in vines and factor models would be parametric copula families with asymmetric tails and (i) intermediate tail dependence in both tails, or (ii) tail dependence in one tail and intermediate tail dependence in the other.

Because there are many ways of deriving new parametric families of copulas, it is important to compare new families with previously defined families in terms of dependence and tail properties (concordance, tail dependence, tail order, tail asymmetry) and also with Kullback-Leibler sample sizes (Section 5.7). If a new parametric copula family has large Kullback-Leibler sample sizes (or small divergence) in comparison with an existing family, then it is better only if it is simpler to use.

A summary of some operations is given in the following list.

- Convex combination: If C_1 and C_2 are d-variate copulas, then so is $\pi C_1 + (1 - \pi)C_2$ for $0 \leq \pi \leq 1$.
- General mixture: If $C(\cdot; \eta)$ is a family of d-variate copulas indexed by η and $F_H(\eta)$ is a distribution, then $\int C(\boldsymbol{u}; \eta) \, \mathrm{d}F_H(\eta)$ is a copula.
- Maximum from two independent random vectors: let C_1 be a d-variate copula, and let C_2 be a m-variate copula if $m > 1$ or the U(0,1) cdf if $m = 1$. Suppose $1 \leq m \leq d$. if $\boldsymbol{U}_1 = (U_{11}, \ldots, U_{1d}) \sim C_1$ and $\boldsymbol{U}_2 = (U_{21}, \ldots, U_{2m}) \sim C_2$, with $\boldsymbol{U}_1 \perp\!\!\!\perp \boldsymbol{U}_2$. Then the copula of the componentwise maximum $(U_{11} \vee U_{21}, \ldots, U_{1m} \vee U_{2m}, U_{1,m+1}, \ldots, U_{1d})$ is $C_1(u_1^{1/2}, \ldots, u_m^{1/2}, u_{m+1}, \ldots, u_d) \, C_2(u_1^{1/2}, \ldots, u_m^{1/2})$. For $d = m = 2$, it becomes $C_1(u_1^{1/2}, u_2^{1/2}) \, C_2(u_1^{1/2}, u_2^{1/2})$. For $d = 2$ and $m = 1$, it becomes $C_1(u_1^{1/2}, u_2) \, u_1^{1/2}$.
- Maximum of independent multivariate beta random vectors: More generally, with C_1, C_2 as given in the preceding item, then for $0 < \gamma < 1$,

$$C_1(u_1^\gamma, \ldots, u_m^\gamma, u_{m+1}, \ldots, u_d) \, C_2(u_1^{1-\gamma}, \ldots, u_m^{1-\gamma})$$

is a $d-$variate copula.
- Asymmetrization via maximum of more general multivariate beta random vectors:

$$C_1(u_1^{\gamma_1}, \ldots, u_m^{\gamma_m}, u_{m+1}, \ldots, u_d) \, C_2(u_1^{1-\gamma_1}, \ldots, u_m^{1-\gamma_m}).$$

A special case in Genest et al. (1998) has $m = d = 2$, $C_2 = C^\perp$ leads to $C_1(u_1^{\gamma_1}, u_2^{\gamma_2}) u_1^{1-\gamma_2} u_2^{1-\gamma_2}$. This is mentioned in Section 5.4.3 of McNeil et al. (2005) as a way to perturb a bivariate copula to get some permutation asymmetry.

- Associated copulas from reflections $U_j \to 1 - U_j$ for $j \in S$, where S is a non-empty subset of $\{1, \ldots, d\}$.

- Symmetrizing operation: for a copula C with $(U_1, \ldots, U_d) \sim C$, let $\widehat{C}$ be the copula of $(1 - U_1, \ldots, 1 - U_d)$, and define $C_{\text{sym}} = \frac{1}{2}(C + \widehat{C})$. Then C_{sym} is a reflection symmetric copula. In the bivariate case, $C, \widehat{C}$ and C_{sym} have the same values of Blomqvist β and the same values of Spearman's rank correlation. But for Kendall's tau, $\tau(C_{\text{sym}}) = \tau(C)$ iff $\int (\widehat{C} - C) \, dC = 0$.

- In the bivariate case, the mixture of max-id can be used to get a family that combines two different copula families; one of these must be max-id and the other must be Archimedean. Choices for the max-id copula include common 1-parameter families such as Plackett and bivariate Gaussian in the positive dependence range. The copula $C_{\psi,K} = \psi(-\log K(e^{-\psi^{-1}(u;\theta)}, e^{-\psi^{-1}(v;\theta)}; \delta); \theta)$ includes $K(u, v; \delta)$ if $\psi(s; \theta)$ becomes e^{-s} at a boundary, and it includes $\psi(\psi^{-1}(u; \theta) + \psi^{-1}(v; \theta); \theta)$ if K becomes $C^\perp$ at a boundary. The BB1–BB7 copulas in Chapter 4 are constructed in this way from some of the 1-parameter copula families.

- For max-id copulas, the approach in Section 3.6 can be used to add an extra power parameter, that is, $C(\boldsymbol{u}; \eta) = K^{1/\eta}(u_1^\eta, \ldots, u_d^\eta)$ when K is max-id.

Of course, the usefulness of any of the operations applied to a parametric family of copulas depends on whether the resulting larger family is tractable, and preferably has interpretation in terms of dependence increasing as some parameters increase.

For special families such as extreme value copulas and Archimedean copulas, there are operations on the exponent function and LT respectively to preserve them. For extreme value copulas, the geometric mixture $C_1^\gamma C_2^{1-\gamma}$ corresponds to a convex combination of the exponent functions A_1, A_2 when written as $C_i(\boldsymbol{u}) = \exp\{-A_i(-\log u_1, \ldots, -\log u_d)\}$. For Archimedean copulas, there are several operations of LTs in $\mathcal{L}_\infty$ that lead to other LTs in $\mathcal{L}_\infty$. These operations on LTs have some analogies to some operators on life distributions on page 218 of Marshall and Olkin (2007).

- If $\psi \in \mathcal{L}_\infty$, then so is $\psi_\varsigma(s) = \psi(s + \varsigma)/\psi(\varsigma)$ for every $\varsigma > 0$. The complete monotonicity follows easily from this property for ψ and boundary conditions are easily shown to be satisfied. As an example, the positive stable LT $\psi(s) = \exp\{-s^{1/\theta}\}$ ($\theta \geq 1$), with $\alpha = \psi(\varsigma)$ and $\varsigma = (-\log \alpha)^\theta$ leads to

$$\psi_\varsigma(s) = \alpha^{-1} \exp\{-[s + (-\log \alpha)^\theta]^{1/\theta}\}, \quad s \geq 0.$$

This LT is used in Hougaard (1986) and elsewhere. After setting $\delta^{-1} = (-\log \alpha) > 0$ and $\alpha^{-1} = e^{1/\delta}$, this is the LT of the BB9 copula in Section 4.25.

Interpretations of ψ_ς are the following.

(a) If ψ is the LT of a random variable X with density f_X, then ψ_ς is the LT of (a random variable with density) $e^{-\varsigma x} f_X(x)/\psi(\varsigma)$ by writing

$$\psi_\varsigma(s) = \int_0^\infty e^{-sx} \frac{e^{-\varsigma x} f_X(x)}{\psi(\varsigma)} \, dx, \quad s \geq 0.$$

(b) In Section 3.2, this LT generator also comes from the new Archimedean copula obtained from a Archimedean copula conditioned on a lower rectangular region.

- Let ψ be a LT in $\mathcal{L}_\infty$ with finite mean. Then $\varphi(s) = \psi'(s)/\psi'(0)$ is a LT in $\mathcal{L}_\infty$ because $\varphi(0) = 1$, $\varphi(\infty) = 0$ and φ is completely monotone. Similarly if ψ is the LT of a random variable with finite mth moment where m is a positive integer, then $\varphi(s) = \psi^{(m)}(s)/\psi^{(m)}(0)$ is a LT in $\mathcal{L}_\infty$.

- Combining the two preceding cases, for an Archimedean copula based on LT ψ, conditioning given $U_{m+1} = u_{m+1}, \ldots, U_d = u_d$ with $2 \leq m < d$ leads to another Archimedean copula based on a normalization of the mth order derivative $\psi^{(m)}(s + \varsigma)/\psi^{(m)}(\varsigma)$ where $\varsigma > 0$ depends on $u_{m+1}, \ldots, u_d$; see Section 3.2 and Mesfioui and Quessy (2008).

- Let $\psi \in \mathcal{L}_\infty$. Consider

$$v(s) = \int_s^\infty \psi(t)\, dt \Big/ \int_0^\infty \psi(t)\, dt, \quad s \geq 0,$$

where the integral in the denominator is finite. Then $v(0) = 1$, $v(\infty) = 0$ and v is completely monotone, so that $v \in \mathcal{L}_\infty$. Examples of this construction are in Sections 4.11 and 4.31.1 with the positive stable and Mittag-Leffler LTs; see also Section A.1.4.

- $\varphi(s) = \psi(-\log \zeta(s))$ is a LT when ζ is an infinitely divisible LT and $\psi \in \mathcal{L}_\infty$ (Theorem 8.82); the functional inverse is $\varphi^{-1}(t) = \zeta^{-1}(\exp\{-\psi^{-1}(t)\}$. Several 2-parameter LT families in Section A.1.2 are obtained in this way. Choices of $\zeta(\cdot; \delta)$ include gamma, positive stable, exponentially tilted positive stable, leading respectively to $-\log \zeta$ being $\delta^{-1} \log(1 + s)$ for $\delta > 0$, $s^{1/\delta}$ for $\delta > 1$ and $(\delta^{-\vartheta} + s)^{1/\vartheta} - \delta^{-1}$ for $\vartheta > 1$ and $\delta > 0$. The latter is part of Theorem 8 in Hofert (2008).

- Vine copula models depend on flexible bivariate copula families, so we also mention results that lead to $\psi \in \mathcal{L}_2$ (and not necessarily $\mathcal{L}_\infty$). We convert published results stated with bivariate Archimedean copulas in the parametrization of (3.5) to that of (3.4). Operations (c)–(g) below are from Genest et al. (1998) and Lemmas 5–9 of Michiels and De Schepper (2012). An example of case (f) is $-\log \zeta(s) = (\delta^{-\vartheta} + s)^{1/\vartheta} - \delta^{-1}$ with additional parameter range $0 < \vartheta < 1$ and $\delta^{-1} \geq 1 - \vartheta$ (this additional range means that $\zeta \in \mathcal{L}_2$ but $\zeta \notin \mathcal{L}_\infty$; see pages 161–162 and family MB9E in Joe (1997)).

(a) Theorem 4.5.1 of Nelsen (1999): $\psi_\gamma(s) := \psi(s^{1/\gamma}) \in \mathcal{L}_2$ for $\gamma > 1$ if $\psi \in \mathcal{L}_2$; this is relaxed version $\psi(-\log \zeta)$ with ζ being the positive stable LT.

(b) Theorem 4.5.1 of Nelsen (1999): $\psi_\alpha(s) := [\psi(s)]^{1/\alpha} \in \mathcal{L}_2$ for $0 < \alpha < 1$ if $\psi \in \mathcal{L}_2$. If furthermore, ψ is twice differentiable and $\eta(s) = -\log \psi(s)$ is such that $\eta'' \leq 0$, then $\psi_\alpha \in \mathcal{L}_2$ for all $\alpha > 0$. For $\mathcal{L}_\infty$, suppose ψ is an infinitely divisible LT, then (i) $\psi^{1/\alpha}$ is a LT for all $\alpha > 0$, but otherwise $\psi^{1/\alpha}$ need not be a LT for $0 < \alpha < 1$; (ii) by Theorem 8.81, $\eta \in \mathcal{L}_\infty^*$ or $(-1)^{i-1}\eta^{(k)} \geq 0$ for $k = 1, 2, \ldots$. Hence for $\psi \in \mathcal{L}_2$, the relaxed condition for $\psi_\alpha \in \mathcal{L}_2$ for all $\alpha > 0$ is that $\eta \in \mathcal{L}_2^*$.

(c) (Right composition of ψ^{-1}). $\psi_g^{-1} = \psi^{-1} \circ g^{-1}$ with $\psi_g(s) := g[\psi(s)] \in \mathcal{L}_2$ if $\psi \in \mathcal{L}_2$ and g is convex and strictly increasing on $[0, 1]$ with $g(0) = 0$ and $g(1) = 1$. Note that $\psi_g(s) = h[-\log \psi(s)]$ if $h(t) = g(e^{-t})$ for $t \geq 0$ (and $h(0) = 1$, $h(\infty) = 0$).

(d) (Left composition of ψ^{-1}). $\psi_h^{-1} = h^{-1} \circ \psi^{-1}$ with $\psi_h(s) := \psi[h(s)] \in \mathcal{L}_2$ if $\psi \in \mathcal{L}_2$ and h is concave and strictly increasing on $[0, \infty)$ with $h(0) = 0$ and $h(\infty) = \infty$. This is a relaxation of $\psi(-\log \zeta)$ from $-\log \zeta \in \mathcal{L}_\infty^*$ for ζ being an infinitely divisible LT to a function $h \in \mathcal{L}_2^*$.

(e) (Scaling on ψ^{-1}). $\psi_\varsigma^{-1}(t) = \psi^{-1}[t\psi(\varsigma)] - \varsigma$ or $\psi_\varsigma(s) := \psi(s + \varsigma)/\psi(\varsigma) \in \mathcal{L}_2$ of $\psi \in \mathcal{L}_2$ for $\varsigma > 0$. This is a relaxed version of one of the above results for LTs in $\mathcal{L}_\infty$.

(f) (Composition via exponentiation on ζ^{-1}, ψ^{-1}). $\varphi^{-1} = \zeta^{-1}(\exp\{-\psi^{-1}\})$ or $\varphi(s) = \psi(-\log \zeta(s)) \in \mathcal{L}_2$ if $\psi, \zeta \in \mathcal{L}_2$ and $\psi'' \geq -\psi'$.

(g) (Linear combination on ψ_i^{-1}). If $\psi_1, \psi_2 \in \mathcal{L}_2$ and $\alpha_1, \alpha_2 > 0$, then ψ defined through $\psi^{-1}(t) = \alpha_1 \psi_1^{-1}(t) + \alpha_1 \psi_2^{-1}(t)$ is in $\mathcal{L}_2$. This does not lead to a general closed form expression for ψ.

3.19   Summary for construction methods

This chapter covers copula construction methods that have a variety of dependence structures. The dependence structures that can be adapted to high dimensions are mostly analogues of methods in classical multivariate (Gaussian) analysis.

Some of the constructions lead to copulas that have integrals in the density. Other constructions lead to copula models where there are recursive equations. Computationally, the low-dimensional integrals and recursions do not hinder likelihood inference. See Chapter 6 for numerical methods and algorithms.

Chapter 4

Parametric copula families and properties

This chapter lists parametric copula families that could be useful for dependence models; all except one include at least two of the independence copula $C^\perp$, the comonotonicity copula C^+ and the countermonotonicity copula C^-. In most sections, a bivariate family is presented followed by dependence and tail properties, multivariate extensions, and some historical background. The idea is to include many useful properties of parametric copula families that aid in model selection. This chapter is mainly aimed to be used as a reference in the style of Johnson and Kotz (1972).

Since many families include $C^\perp$ and C^+, the dependence and tail properties are useful and needed for distinguishing parametric families and understanding when they might be used. For parametric copula families that are members of Archimedean, elliptical, extreme value or mixture of max-id, properties can be obtained from theorems in Chapter 8.

Families that don't include C^+ have limited range of dependence and are not useful for statistical modeling. The only one of this type listed in this chapter is the Morgenstern or FGM copula; this is a simple copula that may be useful for illustration of dependence properties.

For the bivariate version of a copula family, the copula cdf, conditional cdf $C_{2|1}$ and copula density are given, as well as boundary conditional cdfs $C_{2|1}(\cdot|0)$, $C_{2|1}(\cdot|1)$ and other tail behavior. Dependence properties such as the concordance ordering, positive quadrant dependence, stochastically increasing and TP$_2$ are mentioned when they are satisfied. Also, if relevant, results are given for measures of dependence such as Blomqvist's β, Kendall's τ, Spearman's ρ_S, and lower and upper tail dependence parameters λ_L, λ_U or lower and upper tail order parameters κ_L, κ_U.

For a bivariate copula with well-behaved joint lower tail, tail behavior includes the following for the lower tail. If the lower tail order satisfies $\kappa_L \geq 1$ and

$$C(uw_1, uw_2) \sim u^{\kappa_L} \ell(u)\, b(w_1, w_2), \quad u \to 0^+,$$

where $\ell(u)$ is a slowly varying function and b is the lower tail order function. Under some regularity conditions (e.g., existing finite density in the interior of the unit square, ultimately monotone in the tail),

$$uC_{2|1}(uw_2|uw_1) \sim u^{\kappa_L} \ell(u) \frac{\partial b}{\partial w_1}(w_1, w_2), \quad u \to 0^+,$$

$$uC_{1|2}(uw_1|uw_2) \sim u^{\kappa_L} \ell(u) \frac{\partial b}{\partial w_2}(w_1, w_2), \quad u \to 0^+,$$

$$u^2 c(uw_1, uw_2) \sim u^{\kappa_L} \ell(u) \frac{\partial^2 b}{\partial w_1 \partial w_2}(w_1, w_2), \quad u \to 0^+.$$

Or equivalently,

$$C_{2|1}(uw_2|uw_1) \sim u^{\kappa_L-1}\ell(u)\frac{\partial b}{\partial w_1}(w_1,w_2), \quad u \to 0^+,$$

$$C_{1|2}(uw_1|uw_2) \sim u^{\kappa_L-1}\ell(u)\frac{\partial b}{\partial w_2}(w_1,w_2), \quad u \to 0^+,$$

$$c(uw_1,uw_2) \sim u^{\kappa_L-2}\ell(u)\frac{\partial^2 b}{\partial w_1 \partial w_2}(w_1,w_2), \quad u \to 0^+.$$

For $\kappa_L = 1$ with tail dependence parameter $0 < \lambda_L \le 1$, it can be assumed that $\ell \equiv 1$. For the case of tail quadrant independence if $C(uw_1, uw_2) = \zeta u^2 w_1 w_2$ as $u \to 0$ for a constant $\zeta > 0$, then $\kappa_L = 2$, $\ell(u) \equiv \zeta$, $b(w_1, w_2) = w_1 w_2$, and $\frac{\partial^2 b}{\partial w_1 \partial w_2} \equiv 1$.

For the upper tail, the tail expansions are similar with the lower tail expansion of the reflected or survival copula $\widehat{C}$. With upper tail order function b^* and upper tail order $\kappa_U \ge 1$, then under some regularity conditions, with a slowly varying function ℓ^*,

$$\widehat{C}(uw_1, uw_2) = \overline{C}(1-uw_1, 1-uw_2) \sim u^{\kappa_U}\ell^*(u)\,b^*(w_1,w_2), \quad u \to 0^+,$$

$$\widehat{C}_{2|1}(uw_2|uw_1) = \overline{C}_{2|1}(1-uw_2|1-uw_1) \sim u^{\kappa_U-1}\ell^*(u)\frac{\partial b^*}{\partial w_1}(w_1,w_2), \quad u \to 0^+,$$

$$\widehat{C}_{1|2}(uw_1|uw_2) = \overline{C}_{1|2}(1-uw_1|1-uw_2) \sim u^{\kappa_U-1}\ell^*(u)\frac{\partial b^*}{\partial w_2}(w_1,w_2), \quad u \to 0^+,$$

$$\hat{c}(uw_1, uw_2) = c(1-uw_1, 1-uw_2) \sim u^{\kappa_U-2}\ell^*(u)\frac{\partial^2 b^*}{\partial w_1 \partial w_2}(w_1,w_2), \quad u \to 0^+.$$

Note that if $1 \le \kappa_L < 2$ and the above hold, then $c(uw_1, uw_2)$ asymptotes to ∞ in the joint lower corner as $u \to 0^+$; similarly if $1 \le \kappa_U < 2$, then $c(1-uw_1, 1-uw_2)$ asymptotes to ∞ in the joint upper corner as $u \to 0^+$.

Similar tail expansions exist for multivariate copulas under some regularity conditions.

Section 4.1 has a summary of tail orders of the parametric families in Sections 4.3 to 4.31, and Section 4.2 has a summary of some general results on tail behavior of bivariate Archimedean and extreme value copulas. The bivariate versions of the families in Sections 4.3 to 4.31 are mostly in the elliptical, Archimedean, mixture of max-id or extreme value classes. Section 4.31 includes a few new parametric copula families, constructed based on some of the methods in Chapter 3. Section 4.32 has some further dependence comparisons of families that have comparable tail properties.

Section 4.33 contains a chapter summary.

4.1   Summary of parametric copula families

In this section, we provide a tabular summary of the bivariate version of the parametric copula families that are listed in this chapter, starting with 1-parameter families. The intention is that this table provides a summary from which the different families can be compared for tail asymmetry and strength of dependence in the tails.

The main features of Table 4.1 are: (a) the lower and upper tail orders (κ_L and κ_U), (b) indication of reflection symmetry or not, (c) indication of extendability or not to negative dependence. Properties (a) and (b) concern the amount of tail asymmetry, and property (c) needs to be considered for applications where not all variables are positively related (or conditionally positively related).

copula fam.	κ_L or λ_L	κ_U or λ_U	refl. sym.	neg. dep.
		1-parameter		
Gaussian	$\kappa_L = 2/(1+\rho)$	$\kappa_U = 2/(1+\rho)$	Y	Y
Plackett	$\kappa_L = 2$	$\kappa_U = 2$	Y	Y
Frank	$\kappa_L = 2$	$\kappa_U = 2$	Y	Y
reflected MTCJ	$\kappa_L = 2$	$\lambda_U = 2^{-1/\delta}$	N	Y*
Joe/B5	$\kappa_L = 2$	$\lambda_U = 2 - 2^{1/\delta}$	N	N
Gumbel (EV)	$\kappa_L = 2^{1/\delta}$	$\lambda_U = 2 - 2^{1/\delta}$	N	N
Galambos (EV)	$\kappa_L = 2 - 2^{-1/\delta}$	$\lambda_U = 2^{-1/\delta}$	N	N
Hüsler-Reiss (EV)	$\kappa_L = 2\Phi(\delta^{-1})$	$\lambda_U = 2 - 2\Phi(\delta^{-1})$	N	N
integ.pos.stableLT	$\kappa_L = 2^{1/\delta}$	$\kappa_U = 2 \wedge (1 + \delta^{-1})$	N	Y
inv.gammaLT	$\kappa_L = \sqrt{2}$	$\kappa_U = 1 \vee (\delta^{-1} \wedge 2)$	N	N
		2-parameter		
t_ν	$\lambda_L = \lambda_U$	$\lambda_U = 2T_{1,\nu+1}(-\sqrt{\frac{(\nu+1)(1-\rho)}{(1+\rho)}})$	Y	Y
t-EV	$\kappa_L = 2 - \lambda_U$	$\lambda_U = 2T_{1,\nu+1}(-\sqrt{\frac{(\nu+1)(1-\rho)}{(1+\rho)}})$	N	N
BB1	$\lambda_L = 2^{-1/(\delta\theta)}$	$\lambda_U = 2 - 2^{1/\delta}$	N	N
BB2	$\lambda_L = 1$	$\kappa_U = 2$	N	N
BB3	$\lambda_L = 1$	$\lambda_U = 2 - 2^{1/\theta}$	N	N
BB4	$\lambda_L = (2 - 2^{-1/\delta})^{-1/\theta}$	$\lambda_U = 2^{-1/\delta}$	N	N
BB5 (EV)	$\kappa_L = 2 - \lambda_U$	$\lambda_U = 2 - (2 - 2^{-1/\delta})^{1/\theta}$	N	N
BB6	$\kappa_L = 2^{1/\delta}$	$\lambda_U = 2 - 2^{1/(\delta\theta)}$	N	N
BB7	$\lambda_L = 2^{-1/\delta}$	$\lambda_U = 2 - 2^{1/\theta}$	N	N
BB8	$\kappa_L = 2$	$\kappa_U = 2$	N	N
BB9	$\kappa_L = 2^{1/\theta}$	$\kappa_U = 2$	N	N
BB10	$\kappa_L = 2$	$\kappa_U = 2$	N	N
		1- to 3-parameter, some new		
psSSiA	$\kappa_L = 2^{1/\theta}$	$\lambda_U = 2 - 2^{1/(\delta\theta)}$	N	N
gaSSiA	$\lambda_L = 2^{-1/\theta}$	$\lambda_U = 2 - 2^{1/\delta}$	N	N
integ.MitLefA	$\lambda_L = 2^{-1/(\vartheta\delta)}$	$\kappa_U = 1 + \delta^{-1}$	N	N
gamma factor	$1 < \kappa_L < 2$	$\lambda_U = 0, \kappa_U = 1$	N	N
BB1r.power	$\lambda_L = (2 - 2^{1/\delta})^{1/\eta}$	$\lambda_U = 2^{-1/(\delta\theta)}$	N	N
BB7.power	$\lambda_L = 2^{-1/(\delta\eta)}$	$\lambda_U = 2 - 2^{1/\theta}$	N	N
GB2 (Gaussian at boundary)	$\kappa_L = 2$	$\lambda_U > 0$	N	N

* extends to negative dependence with support being a proper subset of $(0,1)^2$.

Table 4.1 *Lower and upper tail dependence or order parameters for many of the bivariate parametric families in this chapter. EV in the left-hand column indicates an extreme value copula. The direction of tail asymmetry can be determined from this table; suitable copula families might be used in survival or reflected form to match tail asymmetry seen in particular data sets. New parametric copula families can be compared against this table for a comparison of tail properties. The last two columns contain the indicator of reflection symmetry, and the indicator of extendability to negative dependence and countermonotonicity*

.

Bivariate families with permutation asymmetry in Table 4.1 are GB2 and the gamma factor copulas. In addition, there are several asymmetric versions of the Gumbel and Galambos copula families. There are Liouville copulas (Section 3.17.2) that also have permutation asymmetry. Some rough analysis based on asymmetry measures seems to show that more extreme bivariate permutation asymmetry with positive dependence can be attained from copulas deriving from stochastic representations of the form $(X_1, X_2) = (Z_0 + Z_1, Z_0 + Z_2)$, or $(X_1, X_2) = (Z_0 \wedge Z_1, Z_0 \wedge Z_2)$, where Z_0, Z_1, Z_2 are independent non-negative random variables and one of Z_1, Z_2 is identically zero. These include asymmetric Gumbel/Galambos for the max operator $\wedge$ and the gamma factor model for the convolution operator.

The 1-parameter copula families cover a variety of different tail behavior when interpolating $C^{\perp}$ and C^{+}, and the 2-parameter copula families add some flexibility for joint lower and upper tail behavior. Nevertheless, as the sample size gets larger, the list in Table 4.1 could be inadequate for applications, especially the 1-parameter families. One can try to combine the copula families in Table 4.1 with operations in Section 3.18 to get parametric copula families with more parameters. For example, three-parameter copula families could have (i) parameters that indicate the strength of dependence in the middle, joint lower tail, joint upper tail; (ii) parameters to indicate strength of dependence in the middle, tail asymmetry and permutation asymmetry.

4.2   Properties of classes of bivariate copulas

Many of the parametric bivariate copula families in this chapter are Archimedean, elliptical, extreme value or mixture of max-id. In this section, a summary in Table 4.2 is given for the tail order (Section 2.16) and the form of boundary conditional cdf (Section 2.19) for bivariate Archimedean, extreme value and t_{ν} copulas. Tail inferences such the joint tail probabilities and conditional tail expectations can be quite different for copulas in different categories in Table 4.2. In addition, for bivariate mixtures of max-id copulas, a summary is given to indicate which dependence properties follow from general results and which must be determined individually.

tail order	cdf	copula family	condition
(a) $\kappa = 1$	(i)	Archimedean (lower)	$\psi \in \mathrm{RV}_{-\alpha}$
	(i)	Archimedean (upper)	$\psi'(0) = -\infty$
	(i)	Extreme value (upper)	$B'(0) = -1$ or $B'(1) = 1$
(a) $\kappa = 1$	(ii)	Extreme value (upper)	$B'(0) > -1$ or $B'(1) < 1$
	(ii)	t_{ν}	
(b) $1 < \kappa < 2$	(i)	Archimedean (lower)	$\psi(s) \sim a_1 s^q e^{-a_2 s^p}$, $s \to \infty$, with $0 < p < 1$
	(i)	Gaussian with $\rho > 0$	
	(i)	Extreme value (lower)	$B'(0) = -1$ or $B'(1) = 1$
(b) $1 < \kappa < 2$	(iii)	Extreme value (lower)	$B'(0) > -1$ or $B'(1) < 1$
	(iii)	Archimedean (upper)	$1 < M_{\psi} < 2$
(c) $\kappa = 2$	(iii)	Archimedean (lower)	$\psi(s) \sim a_1 s^q e^{-a_2 s^p}$, $s \to \infty$, with $p = 1$
(c) $\kappa = 2$	(iii)	Archimedean (upper)	$M_{\psi} > 2$

Table 4.2 *Table 2: Summary of tail order and boundary conditional cdfs for Archimedean, extreme value, and Gaussian/t copulas with positive dependence. Classification for the boundary conditional cdf are: (i) boundary conditional cdf degenerate at the conditioning value; (ii) boundary conditional cdf with positive but not unit mass at the conditioning value; (iii) boundary conditional cdf with no mass at the conditioning value, where the boundary conditional cdf refers to one of $C_{1|2}(\cdot|0), C_{1|2}(\cdot|1), C_{2|1}(\cdot|0), C_{2|1}(\cdot|1)$ depending on the context. For a bivariate extreme value copula of form $e^{-A(-\log u, -\log v)}$, $B(w) = A(w, 1-w)$. For a bivariate Archimedean copula, ψ is a LT in $\mathcal{L}_{\infty}$ and M_{ψ} is the maximal non-negative moment degree as given in (2.11)*

.

The bivariate families BB1–BB7 in Sections 4.17 to 4.23 were constructed from the mixture of max-id approach. The generic form is:

$$C_{\psi,K}(u,v) = \psi\big(-\log K(e^{-\psi^{-1}(u)}, e^{-\psi^{-1}(v)})\big), \tag{4.1}$$

where K is a bivariate max-id copula and ψ is a LT. If K has the form of an Archimedean copula with infinitely divisible LT ζ, then, in (4.1), C is an Archimedean copula with LT $\varphi(s;\theta,\delta) = \psi(-\log\zeta(s;\delta);\theta)$ and $\varphi^{-1}(t;\theta,\delta) = \zeta^{-1}(e^{-\psi^{-1}(t;\theta)};\delta)$. That is, if $K(x,y;\delta) = \zeta(\zeta^{-1}(x;\delta) + \zeta^{-1}(y;\delta);\delta)$, then

$$C(u,v;\theta,\delta) = \psi\big(-\log\zeta\big[\zeta^{-1}(e^{-\psi^{-1}(u)}) + \zeta^{-1}(e^{-\psi^{-1}(v)})\big]\big) = \varphi(\varphi^{-1}(u) + \varphi^{-1}(v)). \tag{4.2}$$

Two-parameter families obtain if $K(\cdot;\delta)$ is parametrized by δ and $\psi(\cdot;\theta)$ is parametrized by θ. If K is increasing in concordance as δ increases, then clearly C increases in concordance as δ increases with θ fixed. The concordance ordering for δ fixed and θ varying must be checked individually for each ψ, K combination — it has been established for BB1, BB2, BB4, BB5, BB6; for BB3 and BB7, and it only holds for a subset of the δ values.

4.3   Gaussian

The first copula family for which properties are listed is the multivariate Gaussian or normal copula. This is the copula used in classical multivariate statistics after individual variables have been transformed to Gaussian.

With $\boldsymbol{R}$ being a $d \times d$ correlation matrix, the multivariate copula cdf is:

$$C(\boldsymbol{u};\boldsymbol{R}) = \Phi_d\big(\Phi^{-1}(u_1),\ldots,\Phi^{-1}(u_d);\boldsymbol{R}\big), \quad \boldsymbol{u} \in [0,1]^d, \tag{4.3}$$

and the copula density is

$$c(\boldsymbol{u};\boldsymbol{R}) = \frac{\phi_d\big(\Phi^{-1}(u_1),\ldots,\Phi^{-1}(u_d);\boldsymbol{R}\big)}{\prod_{j=1}^d \phi\big(\Phi^{-1}(u_j)\big)}, \quad \boldsymbol{u} \in (0,1)^d,$$

when $\boldsymbol{R}$ is positive definite.

4.3.1   Bivariate Gaussian copula

When $d = 2$, the argument of the bivariate cdf is replaced by $\rho \in [-1,1]$, leading to:

$$C(u,v;\rho) = \Phi_2\big(\Phi^{-1}(u), \Phi^{-1}(v);\rho\big), \quad 0 < u,v < 1. \tag{4.4}$$

The conditional distribution is:

$$C_{2|1}(v|u;\rho) = \Phi\Big(\frac{\Phi^{-1}(v) - \rho\Phi^{-1}(u)}{\sqrt{1-\rho^2}}\Big).$$

The copula density for $-1 < \rho < 1$ is:

$$c(u,v;\rho) = \frac{\phi_2(\Phi^{-1}(u),\Phi^{-1}(v);\rho)}{\phi(\Phi^{-1}(u))\,\phi(\Phi^{-1}(v))} = (1-\rho^2)^{-1/2} \exp\Big\{-\tfrac{1}{2}(x^2 + y^2 - 2\rho xy)$$
$$(1-\rho^2)\Big\} \cdot \exp\{\tfrac{1}{2}(x^2 + y^2)\},$$

with $x = \Phi^{-1}(u)$, $y = \Phi^{-1}(v)$ and $0 < u,v < 1$.

Some properties of the bivariate Gaussian copula (4.4) are the following.

- Increasing in $\prec_c$, increasing in $\prec_{SI}$; also increasing in TP_2 ordering (Genest and Verret (2002)).

- C^- for $\rho = -1$, $C^\perp$ for $\rho = 0$, C^+ for $\rho = 1$.
- TP_2 density for $0 < \rho < 1$, RR_2 density for $-1 < \rho < 0$.
- Reflection symmetry.
- Same family with parameter $-\rho$ with a 90° rotation.
- Kendall's τ and Blomqvist's β: both $2\pi^{-1}\arcsin(\rho)$, or $\rho = \sin(\pi\tau/2) = \sin(\pi\beta/2)$.
- Spearman's $\rho_S = 6\pi^{-1}\arcsin(\rho/2)$, or $\rho = 2\sin(\pi\rho_S/6)$.
- Boundary conditional cdfs: $C_{2|1}(\cdot|0; \rho)$ is degenerate at 0 for $\rho > 0$ and degenerate at 1 for $\rho < 0$. $C_{2|1}(\cdot|1; \rho)$ is degenerate at 1 for $\rho > 0$ and degenerate at 0 for $\rho < 0$.
- Intermediate tail dependence for $\rho > 0$, with $\kappa_L = \kappa_U = 2/(1 + \rho)$ for $-1 < \rho < 1$.
- A non-standard upper extreme value limit leads to the Hüsler-Reiss family; see Section 4.10.

4.3.2 Multivariate Gaussian copula

Some properties of the multivariate Gaussian copula (4.3) are the following.

- Increasing in concordance as any ρ_{jk} increases (with constraints of positive definiteness).
- Multivariate tail order is $\kappa_L = \kappa_U = \mathbf{1}^\top \mathbf{R}^{-1} \mathbf{1}$.
- Multivariate TP_2 if the $\mathbf{R}^{-1}$ has non-diagonal elements that are non-positive, or equivalently partial correlations of two variables given the rest are all non-negative.
- The multivariate Hüsler-Reiss family is obtained from a non-standard upper extreme value limit.

4.4 Plackett

For $0 \le \delta < \infty$, the copula cdf is:

$$C(u, v; \delta) = \tfrac{1}{2}\eta^{-1}\{1 + \eta(u + v) - [(1 + \eta(u + v))^2 - 4\delta\eta uv]^{1/2}\}, \quad 0 \le u, v \le 1, \quad (4.5)$$

where $\eta = \delta - 1$. The conditional distribution is:

$$C_{2|1}(v|u; \delta) = \tfrac{1}{2} - \tfrac{1}{2}\frac{[\eta u + 1 - (\eta + 2)v]}{[(1 + \eta(u + v))^2 - 4\delta\eta uv]^{1/2}}.$$

The copula density is

$$c(u, v; \delta) = \frac{\delta[1 + \eta(u + v - 2uv)]}{[(1 + \eta(u + v))^2 - 4\delta\eta uv]^{3/2}}, \quad 0 < u, v < 1.$$

Some properties of (4.5) are the following.

- Increasing in $\prec_c$, increasing in $\prec_{SI}$.
- C^- as $\delta \to 0^+$, $C^\perp$ as $\delta \to 1$, C^+ as $\delta \to \infty$.
- Stochastically increasing positive dependence for $\delta > 1$; stochastically decreasing negative dependence for $\delta > 1$.
- Reflection symmetry.
- If (U, V) is in this family with parameter δ, then $(U, 1 - V)$ has a copula within the same family, since $u - C(u, 1 - v; \delta) = C(u, v; \delta^{-1})$.
- Constructed from the cross-product ratio:

$$\frac{C(u, v; \delta)[1 - u - v + C(u, v)]}{[u - C(u, v; \delta)][v - C(u, v; \delta)]} \equiv \delta, \quad (4.6)$$

Solving this quadratic equation leads to (4.5).
- Spearman's $\rho_S = (\delta + 1)/(\delta - 1) - [2\delta\log(\delta)]/(\delta - 1)^2$; see Mardia (1967).

- Blomqvist's $\beta = (\sqrt{\delta} - 1)/(\sqrt{\delta} + 1)$ and $\delta = [(1 + \beta)/(1 - \beta)]^2$.
- Boundary conditional cdfs: $C_{2|1}(v|0; \delta) = \delta v/[1 + (\delta - 1)v]$, $C_{2|1}(v|1; \delta) = v/[v + \delta(1 - v)]$.
- Tail order $\kappa_L = \kappa_U = 2$ and $C(u, u) \sim \delta u^2$ as $u \to 0^+$.
- Corners of copula density: $c(0, 0; \delta) = c(1, 1; \delta) = \delta$.
- For $\delta \geq 1$, $C(u, v; \delta)$ is max-id. This can be shown via Theorem 8.8 by showing that $C(u, v; \delta)\, c(u, v; \delta) - C_{2|1}(v|u; \delta)\, C_{1|2}(u|v; \delta) \geq 0$ with the help of symbolic manipulation software.

Multivariate extension

Molenberghs and Lesaffre (1994) have a construction based on extension of bivariate Plackett. The validity of the rectangle condition can only be checked numerically; some analysis is given in Section 4.8 of Joe (1997).

Recursive equations must be solved to get the multivariate cdf and this is computationally demanding as dimension d increases. Also one cannot easily simulate from the construction.

Origins and further notes

- The idea of Plackett (1965) is a continuous bivariate distribution $F \in \mathcal{F}(F_1, F_2)$ such that if $(X_1, X_2) \sim F$, then the odds ratio

$$\frac{\mathbb{P}(X_2 > x_2 | X_1 > x_1)/\mathbb{P}(X_2 \leq x_2 | X_1 > x_1)}{\mathbb{P}(X_2 > x_2 | X_1 \leq x_1)/\mathbb{P}(X_2 \leq x_2 | X_1 \leq x_1)}$$
$$= \frac{\mathbb{P}(X_1 > x_1, X_2 > x_2)\mathbb{P}(X_1 \leq x_1, X_2 \leq x_2)}{\mathbb{P}(X_1 > x_1, X_2 \leq x_2)\mathbb{P}(X_1 \leq x_1, X_2 > x_2)} \equiv \delta,$$

for all x_1, x_2. Converting to a copula with $X_1, X_2 \sim \mathrm{U}(0, 1)$, leads to the equation (4.6). This is a quadratic equation in C for any $0 < u, v < 1$, with solution given in (4.5), as shown in Mardia (1967).
- This copula is convenient for discretizing two continuous variables to binary so that a measure of dependence is constant (invariant to the cutpoints for the discretization). Note that there is no copula so that the discretization leads to a constant correlation (invariant to the cutpoints).

4.5 Copulas based on the logarithmic series LT

The bivariate copula family is presented first, followed by some properties and multivariate extensions.

4.5.1 Bivariate Frank

For $-\infty < \delta < \infty$, the copula cdf is:

$$C(u, v; \delta) = -\delta^{-1} \log\left(\frac{1 - e^{-\delta} - (1 - e^{-\delta u})(1 - e^{-\delta v})}{1 - e^{-\delta}}\right), \quad 0 \leq u, v \leq 1, \qquad (4.7)$$

The density is

$$c(u, v; \delta) = \frac{\delta(1 - e^{-\delta})\, e^{-\delta(u+v)}}{[1 - e^{-\delta} - (1 - e^{-\delta u})(1 - e^{-\delta v})]^2}, \quad 0 < u, v < 1.$$

The conditional cdf and its inverse are:

$$C_{2|1}(v|u;\delta) = e^{-\delta u}[(1-e^{-\delta})(1-e^{-\delta v})^{-1} - (1-e^{-\delta u})]^{-1},$$
$$C_{2|1}^{-1}(p|u;\delta) = -\delta^{-1}\log\{1 - (1-e^{-\delta})/[(p^{-1}-1)e^{-\delta u}+1]\}.$$

Some properties of (4.7) are the following.

- Increasing in $\prec_c$, increasing in $\prec_{SI}$.
- C^- as $\delta \to -\infty$, $C^\perp$ as $\delta \to 0^+$, C^+ as $\delta \to \infty$.
- Reflection symmetry $c(1-u, 1-v;\delta) = c(u,v;\delta)$ can be shown via algebra. Frank (1979) characterized this as the only reflection symmetric Archimedean copula.
- If (U,V) is in this family with parameter δ, then $(U, 1-V)$ has a copula within the same family, since $u - C(u, 1-v;\delta) = C(u,v;-\delta)$.
- For $\delta > 0$, TP_2 density, TP_2 cdf, TP_2 survival function, max-id, min-id.
- Archimedean with $\psi(s;\delta) = -\delta^{-1}\log[1 - (1-e^{-\delta})e^{-s}]$; this is the logarithmic series LT for $\delta > 0$. The inverse is $\psi^{-1}(t;\delta) = -\log[(1-e^{-\delta t})/(1-e^{-\delta})]$ for $0 \le t \le 1$.
- Kendall's $\tau = 1 + 4\delta^{-1}[D_1(\delta) - 1]$ and Spearman's $\rho_S = 1 + 12\delta^{-1}[D_2(\delta) - D_1(\delta)]$, where $D_k(x) = kx^{-k}\int_0^x t^k(e^t-1)^{-1}dt$ for $k = 1, 2$ (see Nelsen (1986) and Genest (1987)).
- Blomqvist's $\beta = 4\delta^{-1}\log[(2e^{-\delta/2} - 2e^{-\delta})/(1-e^{-\delta})] - 1$; an iterative method is needed to solve for δ given β.
- Tail order $\kappa_L = \kappa_U = 2$, and $C(u,u;\delta) \sim \delta(1-e^{-\delta})^{-1}u^2$ as $u \to 0^+$.
- Boundary conditional cdfs: $C_{2|1}(v|0;\delta) = (1-e^{-\delta v})/(1-e^{-\delta})$ for $0 < v < 1$, $C_{2|1}(v|1;\delta) = 1 - (e^{\delta} - e^{\delta v})/(e^{\delta} - 1)$ for $0 < v < 1$.
- Corners of copula density: $c(0,0,\delta) = c(1,1;\delta) = \delta/(1-e^{-\delta})$.

4.5.2 Multivariate Frank extensions

The multivariate Frank copula is:

$$C(\boldsymbol{u};\delta) = \psi\Big(\sum_{j=1}^d \psi^{-1}(u_j)\Big) = -\delta^{-1}\log\Big\{1 - \frac{\prod_j(1-e^{-\delta u_j})}{(1-e^{-\delta})^{d-1}}\Big\}, \quad \boldsymbol{u} \in [0,1]^d. \qquad (4.8)$$

Some of its properties are the following.

- Increasing in $\prec_c$ as δ increases. This follows from Corollary 8.24 since $\omega(s) = \psi^{-1}(\psi(s;\delta_2);\delta_1)$ is in $\mathcal{L}_\infty^*$ for $1 < \delta_1 < \delta_2$.
- Reflection asymmetric with skewness to the upper corner for dimension $d \ge 3$. The tail orders are $\kappa_L = \kappa_U = d$. In the joint lower tail $C(u\boldsymbol{1}_d) \sim u^d(\delta/[1-e^{-\delta}])^{d-1}$, and in the joint upper tail $\overline{C}((1-u)\boldsymbol{1}_d) \sim u^d(\delta/[1-e^{-\delta}])^{d-1}e^{-\delta(d-1)}\mathbb{E}[N^{d-1}]$ as $u \to 0^+$, where $N \sim \text{Geometric}(e^{-\delta})$ on the positive integers. The coefficient for the upper corner is larger for $\theta > 0$ because $\mathbb{E}[N^{d-1}] > e^{\delta(d-1)}$. Details of these calculations are instructive to show a method based on the representation from a conditional independence model; they are given in the subsection below on technical details.
- In Table 5.4 of Joe (1997), the range of $\delta < 0$ leading to Archimedean copulas based on non-LTs quickly decreases as d increases, and there is limited negative dependence for dimensions $d \ge 3$.
- Conditional distributions (given 1 or 2 variables) have copula in family BB10 (Archimedean with negative binomial LT).
- The trivariate hierarchical Archimedean extension is in Joe (1993). For the representation as a mixture of a bivariate max-id copula K, the copula K is Archimedean with the LT of family BB8. See Example 3.2.

As indicated above, there is a hierarchical Archimedean extension of the Frank copula, with the form given in Section 3.4. A bigger generalization comes from Section 3.5; a mixture of max-id copula $C_{\psi,K}$, with ψ logarithmic series with parameter $\theta > 0$ and K arbitrary d-variate, leading to

$$C_{\psi,K}(\boldsymbol{u}) = -\theta^{-1} \log\left[1 - (1 - e^{-\theta}) \log K\left(\frac{1 - e^{-\theta u_1}}{1 - e^{-\theta}}, \ldots, \frac{1 - e^{-\theta u_d}}{1 - e^{-\theta}}\right)\right].$$

Here, K is not required to be max-id, because the logarithmic series random variable has support on the positive integers; see Example 3.3.

Technical details

We show how to get the limits of $C(u\mathbf{1}_d)/u^d$ and $\overline{C}((1-u)\mathbf{1}_d)/u^d$ as $u \to 0^+$.

For the lower tail, $1 - e^{-\delta u} = \delta u + O(u^2)$, so that (4.8) becomes

$$C(u\mathbf{1}_d) \sim -\delta^{-1} \log\{1 - (1 - e^{-\delta})^{1-d}[\delta^d u^d + O(u^{d+1})]\} \sim u^d \delta^{d-1}/(1 - e^{-\delta})^{d-1}, \quad u \to 0^+.$$

For the upper tail,

$$\overline{C}((1-u)\mathbf{1}_d) = \int_0^\infty [1 - G^q(1-u)]^d \mathrm{d}F_Q(q), \tag{4.9}$$

where $G(v) = \exp\{-\psi^{-1}(v; \delta)\} = (1 - e^{-\delta v})/(1 - e^{-\delta})$ and F_Q is the logarithmic cdf with probability $p_i = (1 - e^{-\delta})^i/[i\delta]$ for $i = 1, 2, \ldots$. Let $N \sim \text{Geometric}(e^{-\delta})$ on the positive integers. Then (4.9) becomes

$$\overline{C}((1-u)\mathbf{1}_d) = \sum_{i=1}^\infty p_i\left[1 - (1 - e^{-\delta(1-u)})^i/(1 - e^{-\delta})^i\right]^d \sim u^d \sum_{i=1}^\infty p_i\left\{\frac{i\delta e^{-\delta}}{1 - c^{-\delta}}\right\}^d$$

$$= u^d(\delta/[1 - e^{-\delta}])^{d-1} e^{-\delta(d-1)} \sum_{i=1}^\infty i^{d-1} e^{-\delta}(1 - e^{-\delta})^{i-1}$$

$$= u^d(\delta/[1 - e^{-\delta}])^{d-1} e^{-\delta(d-1)} \mathbb{E}(N^{d-1}).$$

Since $h(x) = x^a$ is strictly convex for $a > 1$, $\mathbb{E}(N^{d-1})) > [\mathbb{E}(N)]^{d-1} = e^{\delta(d-1)}$ for $d > 2$ and $\delta > 0$. Hence the coefficient of u^d is larger for the upper tail.

Origins and further notes

- Frank (1979) obtained the bivariate copula as a function such that $C(u, v)$ and $C^\smallfrown(u, v) = u + v - C(u, v)$ are associative in the sense that $C(C(u, v), w) = C(u, C(v, w))$. Then $C(u, v) = 1 - u - v + C(1 - u, 1 - v)$, and this family has the only bivariate Archimedean copulas that are reflection symmetric.
- The bivariate family was further studied in Nelsen (1986) and Genest (1987).
- Marshall and Olkin (1988) include the positively dependent Frank subfamily as an example of a copula generated by the mixture (3.4). The LT for the mixture is included in their Example 4.3.
- The bivariate Frank copula is useful as a copula family with closed form cdf, and having both negative and positive dependence from countermonotonic to comonotonic. The multivariate Frank copula can be a convenient closed form cdf for positively dependent exchangeable random variables; see for example Meester and MacKay (1994).
- Because of tail quadrant independence, the bivariate Frank copula would seldom be a good choice for tree $\mathcal{T}_1$ of a vine or factor copula, especially when tail dependence of variables is desired. But it can be convenient for positive or negative conditional dependence in subsequent trees of a vine or factor copula.

4.6 Copulas based on the gamma LT

The bivariate copula family is presented first, followed by some properties and multivariate extensions.

4.6.1 Bivariate Mardia-Takahasi-Clayton-Cook-Johnson

For $0 \le \delta < \infty$, the copula cdf is:

$$C(u, v; \delta) = (u^{-\delta} + v^{-\delta} - 1)^{-1/\delta}, \quad 0 \le u, v \le 1. \tag{4.10}$$

The conditional cdf is

$$C_{2|1}(v|u; \delta) = [1 + u^{\delta}(v^{-\delta} - 1)]^{-1-1/\delta}$$

with inverse

$$C_{2|1}^{-1}(p|u; \delta) = [(p^{-\delta/(1+\delta)} - 1)u^{-\delta} + 1]^{-1/\delta}.$$

The density is

$$c(u, v; \delta) = (1 + \delta)[uv]^{-\delta-1}(u^{-\delta} + v^{-\delta} - 1)^{-2-1/\delta}, \quad 0 < u, v < 1.$$

Some properties of (4.10) are the following.

- Increasing in $\prec_c$, increasing in $\prec_{SI}$.
- $C^{\perp}$ as $\delta \to 0^+$, C^+ as $\delta \to \infty$.
- Archimedean with gamma LT $\psi(s; \delta) = (1+s)^{-1/\delta}$ and $\psi^{-1}(t; \delta) = t^{-\delta} - 1$ for $0 \le t \le 1$.
- For $\delta > 0$, TP_2 density, TP_2 cdf, TP_2 survival function, max-id, min-id.
- Extends to bivariate Archimedean with $\delta \ge -1$ if $\psi(s; \delta) = (1 + \delta s)_+^{-1/\delta}$. So this is $\psi(s; \delta) = [\max\{0, 1 + \delta s\}]^{-1/\delta}$ if $-1 \le \delta < 0$; this latter range has negative dependence and C^- obtains for $\delta = -1$. For the range with negative dependence, the density is 0 on the set $\{(u, v) : u^{-\delta} + v^{-\delta} < 1\}$, so the extension is not useful for statistical modeling.
- Kendall's $\tau = \delta/(\delta + 2)$ for $\delta \ge -1$, and $\delta = 2\tau/(1 - \tau)$.
- Blomqvist's $\beta = 4(2^{1+\delta} - 1)^{-1/\delta}$.
- Lower tail dependence, $\lambda_L = 2^{-1/\delta}$ and $\delta = (\log 2)/(-\log \lambda_L)$; the lower extreme value limit leads to the Galambos copula.
- Boundary conditional cdfs: $C_{2|1}(v|0; \delta) = 1$, and $C_{2|1}(v|1; \delta) = v^{1+\delta}$.
- Lower tail dependence function (same as upper tail of Galambos) is $b_L(w_1, w_2; \delta) = (w_1^{-\delta} + w_2^{-\delta})^{-1/\delta}$, and $\frac{\partial b_L^2}{\partial w_1 \partial w_2} = (\delta + 1)(w_1 w_2)^{-\delta-1}(w_1^{-\delta} + w_2^{-\delta})^{-1/\delta-2}$. So $c(u, u; \delta) \sim u^{-1} \frac{\partial b_L^2}{\partial w_1 \partial w_2}(1, 1; \delta) = u^{-1}(\delta + 1)2^{-1/\delta-2}$ as $u \to 0^+$.
- Upper tail: $\kappa_U = 2$, and as $u \to 0^+$, $\overline{C}(1 - uw_1, 1 - uw_2; \delta) \sim (1 + \delta)u^2 w_1 w_2$, so that $c(1 - u, 1 - u; \delta) \sim (1 + \delta)$.

4.6.2 Multivariate MTCJ extensions

The d-variate Archimedean copula family is:

$$C(\boldsymbol{u}; \delta) = [u_1^{-\delta} + \cdots + u_d^{-\delta} - (d - 1)]^{-1/\delta}, \quad \boldsymbol{u} \in [0, 1]^d. \tag{4.11}$$

Some properties are the following.

- Increasing in $\prec_c$ as δ increases. This follows from Corollary 8.24 since $\omega(s) = \psi^{-1}(\psi(s; \delta_2); \delta_1) = (1 + s)^{\delta_1/\delta_2} - 1$ is in $\mathcal{L}_{\infty}^*$ for $1 < \delta_1 < \delta_2$.
- The extension to range $-1/(d-2) \le \delta < 0$ is given in page 158 of Joe (1997) to illustrate how some Archimedean copula families can be extended to negative dependence; because support is not all of $(0, 1)^d$; this extension is not useful for statistical modeling.

- The trivariate hierarchical Archimedean extension is in Joe (1993). For the representation as a mixture of a bivariate max-id copula K, the copula K is Archimedean with the LT of family BB9. See Example 3.2.

A bigger generalization than hierarchical Archimedean comes from Section 3.5; a mixture of max-id copula $C_{\psi,K}$, with ψ gamma with parameter $\theta > 0$ and K max-id, leading to

$$C_{\psi,K}(\boldsymbol{u}) = \left[1 - \log K\left(e^{-(u_1^{-\theta}-1)}, \ldots, e^{-(u_d^{-\theta}-1)}\right)\right]^{-1/\theta}, \quad \boldsymbol{u} \in [0,1]^d.$$

This has lower tail dependence according to Theorem 8.51. If

$$K(\boldsymbol{v}) = \prod_{1 \le i < j \le d} H_{ij}(v_i^{1/(d-1)}, v_j^{1/(d-1)}; \delta_{ij}), \quad \boldsymbol{v} \in [0,1]^d,$$

is a product of Galambos copulas, then family MM2 in Joe (1997) is obtained; see (4.64).

Origins and further notes

There have been several independent discoveries of the form of this copula or multivariate distribution with different univariate margins. Hence this is the copula family with the most names used by different authors; they are indicated below.

- Mardia (1962) has a multivariate Pareto distribution with Pareto type I univariate margins, leading to copula (4.11).
- Takahasi (1965) derived the multivariate Burr distribution as a gamma mixture of Weibull survival functions, and Mardia's multivariate Pareto is a special case. Takahasi's derivation predates the frailty model, because the power of a Burr survival function is another Burr survival function. The Burr survival function includes Pareto of type II as a special case.
- Kimeldorf and Sampson (1975) obtained the bivariate uniform representation (their name for copula) by transforming the margins of Mardia's bivariate Pareto. Joe (1997) cites Kimeldorf and Sampson (1975) as the earliest reference for the bivariate copula (4.10).
- Galambos (1975) took the extreme value limit of the Mardia's multivariate Pareto distribution to get multivariate extreme value distributions with the copula in (4.20).
- Cook and Johnson (1981) obtained the multivariate copula by transforming Mardia's multivariate Pareto and Takahasi's multivariate Burr to a standard form (their name for copula). Their paper has one of the first plots to show tail asymmetry (or non-elliptical shaped density contours) and tail dependence with $N(0,1)$ margins,
- Clayton (1978) derived the bivariate distribution through a differential equation, and influenced the research on frailty models for multivariate survival data, and did not refer to the earlier papers on multivariate Pareto and Burr distributions. Oakes (1982, 1989) (and others) follow up on Clayton (1978) with further use of this copula for inference with bivariate survival data.
- The extension to negative dependence for the bivariate family is given in Genest and MacKay (1986); the subfamily with negative dependence (but not positive dependence) also appears in Ruiz-Rivas (1981).
- Galambos (1987), in Section 5.1, refers to (4.10) with parameter $\delta = 1$ and exponential survival margins as Mardia's distribution.
- Hutchinson and Lai (1990) refer to (4.10) as the Pareto copula and Johnson (1987) refer to it as Burr-Pareto-logistic, since it is also the copula of the bivariate logistic distribution of Gumbel (1961) when $\delta = 1$. Devroye (1986) refer to (4.11) as the Cook-Johnson copula.
- Genest and Rivest (1993) refer to (4.10) as the Clayton copula, and Genest and MacKay (1986) refer to (4.10) as the Cook-Johnson copula.

- Hougaard (2000) refers to this family as the gamma frailty model, or Clayton or Clayton-Oakes model.
- Charpentier and Juri (2006) characterize this copula family as being invariant to tail events; for example, $[(U, V)|U < a, V < b]$ has copula (4.10) for any $0 < a, b < 1$ if (U, V) has distribution (4.10). See also Juri and Wüthrich (2002).
- Joe et al. (2010) refer to the closure property of (4.11) under conditioning, as proved by Takahasi (1965). So reference to earlier authors in the name of the copula was to avoid an anachronism. By including Mardia-Takahasi in the name of this copula, this is consistent in following the practice used in other distributions such as Marshall-Olkin multivariate exponential, Galambos multivariate extreme value, etc., in which the original authors did not convert their multivariate distributions to copulas. (As indicated in Chapter 1, copulas only became popular much later.) The abbreviation to MTCJ follows the abbreviation style of FGM used for Farlie-Gumbel-Morgenstern; the initial C can also include Clayton for the bivariate copula.
- Durante and Sempi (2010) refer to the multivariate (4.11) as the Mardia-Takahasi-Clayton copula.

4.7   Copulas based on the Sibuya LT

The bivariate copula family is presented first, followed by some properties and multivariate extensions.

4.7.1   Bivariate Joe/B5

For $1 \le \delta < \infty$, the copula cdf is:

$$C(u, v; \delta) = 1 - \left([1 - u]^\delta + [1 - v]^\delta - [1 - u]^\delta[1 - v]^\delta\right)^{1/\delta}, \quad 0 \le u, v \le 1. \qquad (4.12)$$

The conditional cdf is $C_{2|1}(v|u; \delta) = [1 + (1 - v)^\delta(1 - u)^{-\delta} - (1 - v)^\delta]^{-1+1/\delta}[1 - (1 - v)^\delta]$. The density is

$$c(u, v; \delta) = \left(\overline{u}^\delta + \overline{v}^\delta - \overline{u}^\delta\overline{v}^\delta\right)^{-2+1/\delta}\overline{u}^{\delta-1}\overline{v}^{\delta-1}\left[\delta - 1 + \overline{u}^\delta + \overline{v}^\delta - \overline{u}^\delta\overline{v}^\delta\right],$$

where $\overline{u} = 1 - u$, $\overline{v} = 1 - v$ and $0 < u, v < 1$.

Because of the closeness of this copula family to the reflection of (4.10), $C_{2|1}^{-1}(p|u; \delta)$ can be computed via Newton-Raphson iterations with a starting value of

$$v_0 = 1 - \{([1 - p]^{-\delta^*/(1+\delta^*)} - 1)(1 - u)^{-\delta^*} + 1\}^{-1/\delta^*},$$

where $\delta^* = \delta - 1$.

Some properties of (4.12) are the following.
- Increasing in $\prec_c$, increasing in $\prec_{SI}$.
- $C^\perp$ for $\delta = 1$, C^+ as $\delta \to \infty$.
- Archimedean copula based on the Sibuya LT $\psi(s; \delta) = 1 - (1 - e^{-s})^{1/\delta}$, and $\psi^{-1}(t; \delta) = -\log[1 - (1 - t)^\delta]$ for $0 \le t \le 1$.
- For $\delta > 1$, TP$_2$ density, TP$_2$ cdf, TP$_2$ survival function, max-id, min-id, upper tail dependence.
- Kendall's tau

$$\tau = 1 + 2(2 - \delta)^{-1}[\text{digamma}(2) - \text{digamma}(2/\delta + 1)]$$

with limit $1 - \text{trigamma}(2)$ as $\delta \to 2$. [1] This is a simplification of a result of Schepsmeier

---

[1]digamma$(z) = \frac{d}{dz}\log\Gamma(z)$ and trigamma$(z) = \frac{d^2}{dz^2}\log\Gamma(z)$; the commonly used symbols are ψ, ψ' but we are using ψ here for the LT

(2010); it can also be obtained as a limit of the Kendall τ value of the BB7 copula as the second BB7 parameter goes to 0.

- Blomqvist's $\beta = 3 - 4[2(\frac{1}{2})^{\delta} - (\frac{1}{4})^{\delta}]^{1/\delta}$.
- Upper tail dependence with $\lambda_U = 2 - 2^{1/\delta}$ and $\delta = \log(2)/\log(2 - \lambda_U)$. The upper extreme value limit leads to the Gumbel copula.
- Boundary conditional cdfs: $C_{2|1}(v|0;\delta) = 1 - (1-v)^{\delta}$ and $C_{2|1}(\cdot|1;\delta)$ is degenerate at 1.
- Upper tail dependence function (same as Gumbel): $b_U(w_1, w_2; \delta) = w_1 + w_2 - (w_1^{\delta} + w_2^{\delta})^{1/\delta}$, $\frac{\partial b_U^2}{\partial w_1 \partial w_2} = (\delta - 1)(w_1 w_2)^{\delta - 1}(w_1^{\delta} + w_2^{\delta})^{1/\delta - 2}$. So $c(1 - u, 1 - u; \delta) \sim u^{-1} \frac{\partial b_U^2}{\partial w_1 \partial w_2}(1, 1; \delta) = u^{-1}(\delta - 1)2^{1/\delta - 2}$ as $u \to 0^+$.
- Lower tail: $\kappa_L = 2$ and $C(uw_1, uw_2; \delta) \sim \delta u^2 w_1 w_2$ as $u \to 0^+$, so that $c(0, 0; \delta) = \delta$.

4.7.2 Multivariate extensions with Sibuya LT

The d-variate Archimedean copula family is:

$$C(\boldsymbol{u}; \delta) = 1 - \left\{1 - \prod_{j=1}^{d}[1 - (1 - u_j)^{\delta}]\right\}^{1/\delta}, \quad \boldsymbol{u} \in [0,1]^d, \ \delta \geq 1. \tag{4.13}$$

Properties are the following.

- Increasing in $\prec_c$ as δ increases. This follows from Corollary 8.24 since $\omega(s) = \psi^{-1}(\psi(s; \delta_2); \delta_1)$ is in $\mathcal{L}_{\infty}^*$ for $1 < \delta_1 < \delta_2$.
- The hierarchical Archimedean extension is given in Joe (1993). For the representation as a mixture of power of an Archimedean copula K, the copula K is in this family, that is, (4.12) and (4.13). See Example 3.2.

A bigger generalization comes from Section 3.5; a mixture of max-id copula $C_{\psi,K}$, with ψ Sibuya with parameter $\theta > 1$ and K is arbitrary d-variate, leading to

$$C_{\psi,K}(\boldsymbol{u}) = 1 - \left[1 - \log K\left(1 - (1 - u_1)^{\theta}, \ldots, 1 - (1 - u_d)^{\theta}\right)\right]^{1/\theta}, \quad \boldsymbol{u} \in [0,1]^d.$$

Here, K is not required to be max-id, because the Sibuya random variable has support on the positive integers.

Origins and further notes

- The bivariate family was obtained as a special case of the two-parameter family BB8 (Section 4.24), and the BB8 family appears within the hierarchical Archimedean extension of the Frank copula.
- The Sibuya LT corresponds to the survival function $1 - (1 - e^{-x})^{\alpha}$, which is a power of the exponential cdf. When the power α is positive (and not just restricted to be in $(0,1)$), it is called the Verhulst distribution, on page 333 of Marshall and Olkin (2007).

4.8   Copulas based on the positive stable LT

The bivariate copula family is presented first, followed by some properties and multivariate extensions.

4.8.1 Bivariate Gumbel

The copula cdf is:

$$C(u, v; \delta) = \exp\{-([-\log u]^{\delta} + [-\log v]^{\delta})^{1/\delta}\}, \quad 0 \leq u, v \leq 1, \ 1 \leq \delta < \infty. \tag{4.14}$$

Let $x = -\log u$, $y = -\log v$, or $u = e^{-x}$, $v = e^{-y}$. Then

$$\overline{G}(x, y; \delta) = C(u, v; \delta) = \exp\{-[x^\delta + y^\delta]^{1/\delta}\}, \quad x \geq 0, y \geq 0, \quad 1 \leq \delta < \infty,$$

is a bivariate exponential survival function, with margins $\overline{G}_1 = \overline{G}_2$ and $\overline{G}_1(x) = e^{-x}$.

The conditional distribution is

$$C_{2|1}(v|u; \delta) = \overline{G}_{2|1}(y|x; \delta) = e^x \cdot \exp\{-[x^\delta + y^\delta]^{1/\delta}\}[x^\delta + y^\delta]^{1/\delta - 1}x^{\delta - 1},$$
$$= u^{-1}\exp\{-[x^\delta + y^\delta]^{1/\delta}\} \cdot [1 + (y/x)^\delta]^{1/\delta - 1},$$

and the inverse $\overline{G}_{2|1}^{-1}(p|x; \delta)$ means solving $\overline{G}_{2|1}(y|x; \delta) = p$ for y with fixed $x > 0$ and $0 < p < 1$. Letting $z = [x^\delta + y^\delta]^{1/\delta} \geq x$, then one can solve $e^{-z}z^{1-\delta}e^x x^{\delta-1} = p$ or

$$h(z) = z + (\delta - 1)\log z - [x + (\delta - 1)\log x - \log p] = 0 \qquad (4.15)$$

for z, given p, x, δ and finally get $y = (z_0^\delta - x^\delta)^{1/\delta}$, where z_0 is the root of (4.15). Since $h'(z) = 1 + (\delta - 1)z^{-1}$ has a simple form, the solution of (4.15) can be implemented with a modified Newton-Raphson numerical method.

The density is

$$c(u, v; \delta) = \exp\{-[x^\delta + y^\delta]^{1/\delta}\}[(x^\delta + y^\delta)^{1/\delta} + \delta - 1][x^\delta + y^\delta]^{1/\delta - 2}(xy)^{\delta - 1}(uv)^{-1}, \quad 0 < u, v < 1.$$

Some properties of (4.14) are the following.

- Increasing in $\prec_c$, increasing in $\prec_{\text{SI}}$.
- $C^\perp$ for $\delta = 1$, C^+ as $\delta \to \infty$.
- Archimedean copula with positive stable LT $\psi(s) = \psi(s; \delta) = \exp\{-s^{1/\delta}\}$, and $\psi^{-1}(t; \delta) = (-\log t)^\delta$ for $0 \leq t \leq 1$.
- TP$_2$ density (and hence TP$_2$ cdf and survival function).
- Min-id (or reflected copula $\widehat{C}$ is max-id).
- Extreme value copula with exponent: $A(x, y; \delta) = (x^\delta + y^\delta)^{1/\delta}$; $B(w; \delta) = A(w, 1-w; \delta) = [w^\delta + (1-w)^\delta]^{1/\delta}$. $B'(0; \delta) = -1$ and $B'(1; \delta) = 1$ for $\delta > 1$.
- Kendall's $\tau = (\delta - 1)/\delta$ and $\delta = 1/(1 - \tau)$.
- Upper tail dependence with parameter $\lambda_U = 2 - A(1, 1; \delta) = 2 - 2^{1/\delta}$ and $\delta = \log(2)/\log(2 - \lambda_U)$.
- Blomqvist's $\beta = 2^{2-A(1,1;\delta)} - 1 = 2^{\lambda_U} - 1 = 2^{2-2^{1/\delta}} - 1$, and $\delta = \log(2)/\log[2 - (\log(1 + \beta))/\log(2)]$
- Spearman's $\rho_S = 12\int_0^1 [1 + B(w; \delta)]^{-2}dw - 3$.
- Upper tail dependence function $b_U(w_1, w_2; \delta) = w_1 + w_2 - (w_1^\delta + w_2^\delta)^{1/\delta}$, and $\partial b_U^2/\partial w_1 \partial w_2 = (\delta - 1)(w_1 w_2)^{\delta - 1}(w_1^\delta + w_2^\delta)^{1/\delta - 2}$. So $c(1 - u, 1 - u; \delta) \sim u^{-1}\frac{\partial b_U^2}{\partial w_1 \partial w_2}(1, 1; \delta) = u^{-1}(\delta - 1)2^{1/\delta - 2}$ as $u \to 0^+$.
- Lower tail order and tail order function $\kappa_L = A(1, 1; \delta) = 2^{1/\delta}$; $b_L(w_1, w_2; \delta) = (w_1 w_2)^{\kappa_L/2}$, $\ell_L(u) = 1$; $\partial b_L^2/\partial w_1 \partial w_2 = (\kappa_L/2)^2(w_1 w_2)^{\kappa_L/2 - 1}$. So $c(u, u; \delta) \sim u^{\kappa_L - 2}\frac{\partial b_L^2}{\partial w_1 \partial w_2}(1, 1; \delta) = u^{\kappa_L - 2}(\kappa_L/2)^2$ as $u \to 0^+$.
- Boundary conditional cdfs: for $\delta > 1$, $C_{2|1}(v|0; \delta) = 1$ for $0 \leq v \leq 1$ (that is, degenerate at 0), and $C_{2|1}(v|1; \delta) = 0$ for $0 \leq v < 1$ with $C_{2|1}(1|1; \delta) = 1$ (that is, degenerate at 1).
- Additional properties are listed as part of the multivariate extension.

4.8.2 Multivariate Gumbel extensions

The permutation symmetric or exchangeable Archimedean copula, for $\delta \geq 1$, is:

$$C(\boldsymbol{u}; \delta) = \exp\left\{-\left[\sum_{j=1}^d [-\log u_j]^\delta\right]^{1/\delta}\right\} = \psi\left(\sum_{j=1}^d \psi^{-1}(u_j; \delta); \delta\right) = \exp\{-A(\boldsymbol{x}; \delta)\}, \qquad (4.16)$$

where $\psi(s;\delta)$ is the positive stable LT, $x_j = -\log u_j$ and $0 \le u_j \le 1$ $(j = 1, \ldots, d)$ and $A(\boldsymbol{x};\delta) = [\sum_{j=1}^{d} x_j^{\delta}]^{1/\delta}$.

There is a hierarchical Archimedean copula with partial symmetry or exchangeability, and bivariate Gumbel copulas for every bivariate margin. The number of dependence parameters for the d-variate version is $d - 1$. If a pair of variables is closer together in the hierarchy, then the pair has a larger value of the dependence parameter. With $x_j = -\log u_j$, for the trivariate case, the copula is:

$$C(u_1, u_2, u_3; \delta_1, \delta_2) = \exp\left\{-\left([(x_1^{\delta_1} + x_2^{\delta_1}]^{\delta_2/\delta_1} + x_3^{\delta_2})^{1/\delta_2}\right\},$$

$$= \psi_{\delta_2}\left(\psi_{\delta_2}^{-1} \circ \psi_{\delta_1}\left[\psi_{\delta_1}^{-1}(u_1) + \psi_{\delta_1}^{-1}(u_2)\right] + \psi_{\delta_2}^{-1}(u_3)\right), \quad 1 \le \delta_2 \le \delta_1,$$

where $\psi_{\delta} = \psi(\cdot;\delta)$ is the positive stable LT. The above symmetric Gumbel (4.16) obtains when $\delta_1 = \delta_2$.

A bigger generalization comes from Section 3.5; a mixture of max-id copula $C_{\psi,K}$, with ψ positive stable with parameter $\theta \ge 1$ and K max-id, can have some of the margins being bivariate Gumbel, and the copula is more PLOD than (4.16). If K is an extreme value copula with $-\log K(\boldsymbol{u}) = A_K(\boldsymbol{x})$, $x_j = -\log u_j$ $(j = 1, \ldots, d)$, then

$$C_{\psi,K}(\boldsymbol{u}) = \exp\{-A_K(\psi^{-1}(u_1), \ldots, \psi^{-1}(u_d))\} = \exp\{-A_K^{1/\delta}(x_1^{\delta}, \ldots, x_d^{\delta})\}, \quad \delta \ge 1,$$

is also an extreme value copula. If $K(\boldsymbol{u}) = \prod_{1 \le i < j \le d} H_{ij}(u_i^{1/(d-1)}, u_j^{1/(d-1)}; \delta_{ij})$ where each H_{ij} is a Gumbel or Galambos copula, then the families MM1 and MM3 in Joe (1997) are obtained. With $x_j = -\log u_j$, $j = 1, \ldots, d$, these are respectively

$$C_{\mathrm{MM1}}(\boldsymbol{u}) = \exp\left\{-\left[\sum_{i<j}\left\{\left(\frac{x_i^{\theta}}{d-1}\right)^{\delta_{ij}} + \left(\frac{x_j^{\theta}}{d-1}\right)^{\delta_{ij}}\right\}^{1/\delta_{ij}}\right]^{1/\theta}\right\}, \tag{4.17}$$

$$C_{\mathrm{MM3}}(\boldsymbol{u}) = \exp\left\{-\left[\sum_{j=1}^{d} x_j^{\theta} - \sum_{i<j}\left\{\left(\frac{x_i^{\theta}}{d-1}\right)^{-\delta_{ij}} + \left(\frac{x_j^{\theta}}{d-1}\right)^{-\delta_{ij}}\right\}^{-1/\delta_{ij}}\right]^{1/\theta}\right\}. \tag{4.18}$$

In both MM1 and MM3, $\theta \ge 1$; in MM1, $\delta_{ij} \ge 1$, and in MM3, $\delta_{ij} > 0$.

Properties of (4.16) and extensions are the following.

- Increasing in $\prec_c$ as δ increases. This follows from Corollary 8.24 since $\omega(s) = \psi^{-1}(\psi(s;\delta_2);\delta_1) = s^{\delta_1/\delta_2}$ is in $\mathcal{L}_{\infty}^*$ for $1 < \delta_1 < \delta_2$.
- The Gumbel copula family is the only one in the intersection of Archimedean and extreme value copula classes; see Genest and Rivest (1989).
- A representation of $A(\boldsymbol{x};\delta)$ as an integral over a simplex is given in Smith (1990b).
- The Gumbel copula with $\delta > 1$ is the upper extreme value limit of an Archimedean copula if the LT φ is such that $1 - \varphi(s) \in \mathrm{RV}_0(\delta^{-1})$ as $s \to 0^+$; this result appears in Genest and Rivest (1989) and Joe et al. (2010).
- The hierarchical Archimedean extension is given in Joe (1993). For the representation as a mixture of a max-id copula K, the copula K is in this family, that is, (4.14) and (4.16). See Example 3.2.

There are several asymmetric Gumbel copulas.

1. Take the multivariate Marshall-Olkin exponential distribution, and replace $\vee_{j \in S} x_j$ by $[\sum_{j \in S} x_j^{\delta}]^{1/\delta}$ for any subset S of $\{1, \ldots, d\}$. Then convert the multivariate survival function into a copula. See Sections 4.15.1 and 4.14.2.
2. See Section 4.15.2 for a bivariate asymmetric Gumbel copula based on the deHaan representation.

Origins and further notes

- The original source is Gumbel (1960b), where the multivariate distribution with extreme value margins is given.
- In the multivariate extreme value literature, this distribution is often called the logistic model; see Tiago de Oliveira (1980), Tawn (1990) and Coles (2001).
- In the multivariate survival literature, the following conditional independence model is called a positive stable frailty model (see Hougaard (2000)):

$$\overline{F}(\boldsymbol{x}; \delta) = \int \prod_{j=1}^{d} \overline{G}_j^q(x_j) \, dF_Q(q) = \psi\Big(-\sum_{j=1}^{d} \log \overline{G}_j(x_j); \delta\Big),$$

where F_Q is the cdf of a positive stable random variable with LT $\psi(s; \delta) = \exp\{-s^{1/\delta}\}$.
- The hierarchical Gumbel models have appeared as choice models in McFadden (1974).

4.9 Galambos extreme value

The bivariate copula family is presented first, followed by some properties and multivariate extensions.

4.9.1 Bivariate Galambos copula

The copula cdf is:

$$C(u, v; \delta) = uv \exp\{[(-\log u)^{-\delta} + (-\log v)^{-\delta}]^{-1/\delta}\}, \quad 0 \le u, v \le 1, \ 0 \le \delta < \infty. \quad (4.19)$$

Let $x = -\log u$, $y = -\log v$, or $u = e^{-x}$, $v = e^{-y}$. The density is

$$\begin{aligned}
c(u, v; \delta) = \ & [C(u, v; \delta)/uv] \cdot \big[1 - (x^{-\delta} + y^{-\delta})^{-1-1/\delta}(x^{-\delta-1} + y^{-\delta-1}) \\
& + (x^{-\delta} + y^{-\delta})^{-2-1/\delta}(xy)^{-\delta-1}\{1 + \delta + (x^{-\delta} + y^{-\delta})^{-1/\delta}\}\big],
\end{aligned}$$

where $x = -\log u$, $y = -\log v$ and $0 < u, v < 1$. The conditional cdf is:

$$C_{2|1}(v|u; \delta) = v \exp\{(x^{-\delta} + y^{-\delta})^{-1/\delta}\} \cdot \{1 - [1 + (x/y)^{\delta}]^{-1-1/\delta}\}.$$

For the conditional inverse cdf $C_{2|1}^{-1}(p|u; \delta)$, solve the equation $C_{2|1}(v|u; \delta) = p$ by finding the root y of

$$h(y) = \log p - \log C_{2|1}(v|u) = \log p + y - (x^{-\delta} + y^{-\delta})^{-1/\delta} - \log[1 - x^{-\delta-1}(x^{-\delta} + y^{-\delta})^{-1/\delta-1}]$$

and set $v = e^{-y}$. For Newton-Raphson iterations, use

$$h'(y) = 1 - y^{-\delta-1}(x^{-\delta} + y^{-\delta})^{-1/\delta-1} + \frac{(1+\delta)x^{-\delta-1}y^{-\delta-1}(x^{-\delta} + y^{-\delta})^{-1/\delta-2}}{[1 - x^{-\delta-1}(x^{-\delta} + y^{-\delta})^{-1/\delta-1}]}.$$

For u close to 1 with $(-\log u)^{\delta}$ and x being large, numerically the following can be done. Let $y = xr^{1/\delta}$ or $y^{\delta} = rx^{\delta}$ and solve for r in the equation:

$$h_1(r) = \log p + xr^{1/\delta} - x(1 + r^{-1})^{-1/\delta} - \log\{1 - (1 + r^{-1})^{-1/\delta-1}\}$$

with derivative

$$h_1'(r) = \delta^{-1}xr^{1/\delta-1} - \delta^{-1}x(1 + r^{-1})^{-1/\delta-1}r^{-2} + \frac{(1+\delta^{-1})(1 + r^{-1})^{-1/\delta-2}r^{-2}}{\{1 - (1 + r^{-1})^{-1/\delta-1}\}}.$$

Some properties of (4.19) are the following.

- Increasing in $\prec_c$, increasing in $\prec_{SI}$. with the latter proved with Theorem 2.14 in Joe (1997) involving much algebra and many inequalities.
- $C^\perp$ for $\delta \to 0^+$, C^+ as $\delta \to \infty$.
- Stochastic increasing positive dependence.
- Extreme value copula with exponent: $A(x,y;\delta) = x+y-(x^{-\delta}+y^{-\delta})^{-1/\delta}$ and $B(w;\delta) = A(w,1-w;\delta) = 1-[w^{-\delta}+(1-w)^{-\delta}]^{-1/\delta}$. $B'(0;\delta) = -1$ and $B'(1;\delta) = 1$ for $\delta > 0$.
- Upper tail dependence with $\lambda_U = 2 - A(1,1;\delta) = 2^{-1/\delta}$ and $\delta = (\log 2)/(-\log \lambda_U)$.
- Blomqvist's $\beta = 2^{2-A(1,1;\delta)} - 1 = 2^{2^{-1/\delta}} - 1$, and $\delta = \log(2)/\log[\log(2)/\log(1+\beta)]$.
- With $B'(w;\delta) = [1+\{\frac{1-w}{w}\}^\delta]^{-1/\delta-1} - [1+\{\frac{w}{1-w}\}^\delta]^{-1/\delta-1}$, Kendall's τ can be computed from (8.11).
- Spearman's $\rho_S = 12\int_0^1 [1+B(w;\delta)]^{-2} dw - 3$.
- Boundary conditional cdfs: $C_{2|1}(\cdot|0;\delta)$ is degenerate at 0, $C_{2|1}(\cdot|1;\delta)$ is degenerate at 1.
- Upper tail dependence function $b_U(w_1,w_2;\delta) = (w_1^{-\delta}+w_2^{-\delta})^{-1/\delta}$ and $\partial b_U^2/\partial w_1 \partial w_2 = (\delta+1)(w_1 w_2)^{-\delta-1}(w_1^{-\delta}+w_2^{-\delta})^{-1/\delta-2}$. Therefore $c(1-u,1-u;\delta) \sim u^{-1}\frac{\partial b_U^2}{\partial w_1 \partial w_2}(1,1;\delta) = u^{-1}(\delta+1)2^{-1/\delta-2}$ as $u \to 0^+$.
- Lower tail order and tail order function $\kappa_L = A(1,1;\delta) = 2 - 2^{-1/\delta}$; $b_L(w_1,w_2;\delta) = (w_1 w_2)^{\kappa_L/2}$, $\ell_L(u) = 1$; $\partial b_L^2/\partial w_1 \partial w_2 = (\kappa_L/2)^2(w_1 w_2)^{\kappa_L/2-1}$. Therefore $c(u,u;\delta) \sim u^{\kappa_L-2}\frac{\partial b_L^2}{\partial w_1 \partial w_2}(1,1;\delta) = u^{\kappa_L-2}(\kappa_L/2)^2$ as $u \to 0^+$.

4.9.2 Multivariate Galambos extensions

First we mention the extension with exchangeable dependence. The exponent function A in dimension d involves 2^{d-1} terms. The copula is:

$$C(\boldsymbol{u};\delta) = \exp\{-A(-\log u_1,\ldots,-\log u_d;\delta)\}, \quad \boldsymbol{u} \in [0,1]^d,$$

$$A(\boldsymbol{x};\delta) = \sum_{j=1}^d x_i - \sum_{1\le j<k\le d}(x_j^{-\delta}+x_k^{-\delta})^{-1/\delta} + \sum_{S:|S|>2}(-1)^{|S|-1}[\sum_{i\in S} x_i^{-\delta}]^{-1/\delta}, \quad (4.20)$$

where $S \subset \{1,\ldots,d\}$ in the third sum.

Properties of (4.20) are the following.

- The Galambos copula with $\delta > 0$ is the lower extreme value limit of an Archimedean copula (upper extreme value limit of the reflected copula) if the LT ψ is such that $\psi(s) \in \mathrm{RV}(\delta^{-1})$ as $s \to \infty$.
- There is an extension with hierarchical dependence from the lower extreme value limit of the hierarchical Archimedean copula based on gamma LTs.

Origins and further notes

- The multivariate extreme value distribution was obtained by Galambos (1975) as an extreme value limit of the multivariate Pareto distribution with (4.11) as a survival copula.
- In the extreme value literature, such as Coles (2001), this is called the negative logistic model.

4.10   Hüsler-Reiss extreme value

The bivariate copula family is presented first, followed by some properties and multivariate extensions.

4.10.1 Bivariate Hüsler-Reiss

Let Φ, ϕ be the standard normal cdf and pdf respectively, and let $x = -\log u$ and $y = -\log v$. For $\delta \geq 0$, the copula cdf is:

$$C(u,v;\delta) = \exp\{-x\Phi(\delta^{-1} + \tfrac{1}{2}\delta \log[x/y]) - y\Phi(\delta^{-1} + \tfrac{1}{2}\delta \log[y/x])\}, \ 0 \leq u, v \leq 1. \quad (4.21)$$

The density is

$$c(u,v;\delta) = \frac{C(u,v;\delta)}{uv} \cdot \left[\Phi(\delta^{-1} + \tfrac{1}{2}\delta\log[y/x])\,\Phi(\delta^{-1} + \tfrac{1}{2}\delta\log[x/y]) + \tfrac{1}{2}\delta y^{-1}\phi(\delta^{-1} + \tfrac{1}{2}\delta\log[x/y])\right],$$

for $0 < u, v < 1$. The conditional cdf is:

$$C_{2|1}(v|u;\delta) = C(u,v;\delta) \cdot u^{-1}\Phi(\delta^{-1} + \tfrac{1}{2}\delta\log[x/y]).$$

For the conditional inverse cdf $C_{2|1}^{-1}(p|u;\delta)$, one can solve

$$h(y) = x\{1 - \Phi(\delta^{-1} + \tfrac{1}{2}\delta\log[x/y])\} - y\Phi(\delta^{-1} + \tfrac{1}{2}\delta\log[y/x]) + \log\Phi(\delta^{-1} + \tfrac{1}{2}\delta\log[x/y]) - \log p = 0,$$

with

$$h'(y) = -\Phi(\delta^{-1} + \tfrac{1}{2}\delta\log[y/x]) - \tfrac{1}{2}y^{-1}\frac{\phi(\delta^{-1} + \tfrac{1}{2}\delta\log[x/y])}{\Phi(\delta^{-1} + \tfrac{1}{2}\delta\log[x/y])}.$$

Some properties of (4.21) are the following.

- Increasing in $\prec_c$, increasing in $\prec_{SI}$, with the latter proved with Theorem 2.14 in Joe (1997) and the inequality $\phi(z) + z\Phi(z) \geq 0$ for all $z < 0$.
- $C^{\perp}$ as $\delta \to 0^+$, C^+ as $\delta \to \infty$.
- Stochastic increasing positive dependence.
- Blomqvist's $\beta = 2^{2 - 2\Phi(\delta^{-1})} - 1$, and $\delta = \{\Phi^{-1}(1 - [\log(1+\beta)]/(2\log 2))\}^{-1}$.
- Extreme value copula with exponent function $A(x,y;\delta) = x\Phi(\delta^{-1} + \tfrac{1}{2}\delta\log[x/y]) + y\Phi(\delta^{-1} + \tfrac{1}{2}\delta\log[y/x])$. With $B(w;\delta) = A(w, 1-w;\delta)$,

$$B(w;\delta) = w\Phi\left(\delta^{-1} + \tfrac{1}{2}\delta\log\frac{w}{1-w}\right) + (1-w)\Phi\left(\delta^{-1} + \tfrac{1}{2}\delta\log\frac{1-w}{w}\right), \quad (4.22)$$

$$B'(w;\delta) = \Phi\left(\delta^{-1} + \tfrac{1}{2}\delta\log\frac{w}{1-w}\right) - \Phi\left(\delta^{-1} + \tfrac{1}{2}\delta\log\frac{1-w}{w}\right).$$

Hence $B'(0;\delta) = -1$ and similarly, $B'(1;\delta) = 1$ for $\delta > 0$.

- Upper tail dependence with $\lambda_U = 2 - A(1,1;\delta) = 2[1 - \Phi(\delta^{-1})]$ and $\delta = 1/\Phi^{-1}(1 - \lambda/2)$.
- With $B'(w;\delta)$ as above, Kendall's τ can be computed from (8.11).
- Spearman's $\rho_S = 12\int_0^1 [1 + B(w;\delta)]^{-2}dw - 3$.
- Boundary conditional cdfs: $C_{2|1}(\cdot|0;\delta)$ is degenerate at 0, $C_{2|1}(\cdot|1;\delta)$ is degenerate at 1.
- Lower tail order is $\kappa_L = 2\Phi(\delta^{-1})$.

4.10.2 Multivariate Hüsler-Reiss

The copula cdf of the multivariate extension is:

$$C(\boldsymbol{u}; \{\delta_{ij}\}_{1 \leq i,j \leq d}) = \exp\{-A(-\log u_1, \ldots, -\log u_d; \{\delta_{ij}\})\}, \quad \boldsymbol{u} \in [0,1]^d,$$

$$A(\boldsymbol{x};\delta) = \sum_{j=1}^{d} x_i + \sum_{S:|S| \geq 2} (-1)^{|S|-1} r_S\big(x_i, i \in S; \{\delta_{ij}\}_{i,j \in S}\big) \quad (4.23)$$

where the sum is over $S \subset \{1, \ldots, d\}$, and for $S = \{i_1, \ldots, i_s\} \subset I$, $|S| = s \geq 2$,

$$r_S(x_i, i \in S; \{\delta_{ij}\}_{i,j \in S}) = \int_0^{x_{i_s}} \overline{\Phi}_{s-1}\left(\delta_{i_j,i_s}^{-1} + \tfrac{1}{2}\delta_{i_j,i_s} \log \frac{y}{x_{i_j}}, 1 \leq j \leq s-1; \mathbf{\Gamma}_{S,i_s}\right) dy,$$

and $\overline{\Phi}_{s-1}(\cdot; \mathbf{\Gamma}_{S,j})$ is the survival function of the multivariate Gaussian distribution with correlation matrix $\mathbf{\Gamma}_{S,j} = (\rho_{i,k;j})$ whose (i,k) entry for $i, k \in S \backslash \{j\}$ equals to

$$\rho_{i,k;j} := \left(\delta_{ij}^{-2} + \delta_{kj}^{-2} - \delta_{ik}^{-2}\right) / [2\delta_{ij}^{-1}\delta_{kj}^{-1}],$$

and δ_{ii}^{-1} is defined as zero for all i. A more compact form of the exponent function A in (4.23) is given in Nikoloulopoulos et al. (2009) as

$$A(\boldsymbol{x}; \ \delta_{ij}, 1 \leq i, j \leq d) = \sum_{j=1}^d x_j \Phi_{d-1}\left(\delta_{ij}^{-1} + \tfrac{1}{2}\delta_{ij} \log \frac{x_j}{x_i}, i \in I_j; \mathbf{\Gamma}_{1:d,j}\right), \qquad (4.24)$$

where $I_j = \{1, \ldots, d\} \backslash \{j\}$ and $\mathbf{\Gamma}_{1:d,j} = \mathbf{\Gamma}_{S,j}$ when $S = \{1, \ldots, d\}$.

Origins and further notes

- The multivariate distribution was derived in Hüsler and Reiss (1989) as a non-standard extreme value limit of the multivariate Gaussian distribution. The limit is taken with the correlations approaching 1 at an appropriate rate (otherwise the limit has the independence copula, since the multivariate Gaussian distribution does not have tail dependence).
- From a max-stable process, Smith (1990b) derived the bivariate distribution in (4.21). Genton et al. (2011) derived multivariate distributions of the max-stable process and their result is the same as (4.24) after reparametrization.
- Hashorva (2006) has shown that the Hüsler-Reiss copula is also the limit for general elliptical distributions when a random radius has a distribution function in the Gumbel max-domain of attraction.
- In Nikoloulopoulos et al. (2009), the multivariate Hüsler-Reiss distribution is obtained as a limit of the multivariate t-EV distribution in Section 4.16 with the parameter $\nu \to \infty$ and correlation parameters approach 1 at an appropriate rate.
- This multivariate distribution may be convenient for some extreme value applications because like the multivariate Gaussian distribution it has a parameter for each bivariate margin. The d-dimensional density ($d \geq 3$) is not convenient to work with because it essentially involves a $(d-1)$-dimensional integral. However, for inference, the IFM approach (Section 5.5) can proceed with estimates of GEV parameters from separate univariate likelihoods followed by bivariate parameters from separate bivariate likelihoods. An alternative estimating equation approach is through composite likelihood (Section 5.6) with the constraint of the δ parameters included in the numerical optimization.

4.11   Archimedean with LT that is integral of positive stable

The Archimedean copula based on the LT formed as a normalized integral of the positive stable LT has interesting dependence and tail properties.

4.11.1   *Bivariate copula in Joe and Ma, 2000*

For $\delta > 0$, let $F_\Gamma(\cdot; \delta)$ and $F_\Gamma^{-1}(\cdot; \delta)$ be the cdf and quantile functions for the Gamma$(\delta, 1)$ distribution. The copula cdf is:

$$C(u, v; \delta) = 1 - F_\Gamma\left(\left\{[F_\Gamma^{-1}(1-u; \delta)]^\delta + [F_\Gamma^{-1}(1-v; \delta)]^\delta\right\}^{1/\delta}; \delta\right) \qquad (4.25)$$

$$= \psi(\psi^{-1}(u; \delta) + \psi^{-1}(v; \delta); \delta), \quad 0 \leq u, v \leq 1, \ \delta > 0,$$

where ψ is given below in (4.26). The conditional distribution is:

$$C_{2|1}(v|u;\delta) = \frac{\exp\left\{-\left([F_\Gamma^{-1}(1-u;\delta)]^\delta + [F_\Gamma^{-1}(1-v;\delta)]^\delta\right)^{1/\delta}\right\}}{\exp\{-F_\Gamma^{-1}(1-u;\delta)\}}.$$

Its inverse is

$$C_{2|1}^{-1}(p|u;\delta) = 1 - F_\Gamma(y;\delta), \quad \text{where } y = [(x - \log p)^\delta - x^\delta]^{1/\delta}, \ x = F_\Gamma^{-1}(1-u;\delta).$$

The density is:

$$c(u,v;\delta) = \frac{\Gamma^2(1+\delta)\,\psi''\left([F_\Gamma^{-1}(1-u;\delta)]^\delta + [F_\Gamma^{-1}(1-v;\delta)]^\delta;\delta\right)}{\exp\{-F_\Gamma^{-1}(1-u;\delta)\}\exp\{-F_\Gamma^{-1}(1-v;\delta)\}}, \quad 0 < u,v < 1,$$

where $\psi'(s;\delta) = -e^{-s^{1/\delta}}/\Gamma(1+\delta)$ and $\psi''(s;\delta) = \delta^{-1}s^{1/\delta-1}e^{-s^{1/\delta}}/\Gamma(1+\delta)$.

Some properties of (4.25) are the following.

- Increasing in $\prec_c$ (concordance); a proof is given in Joe and Ma (2000).
- C^+ as $\delta \to \infty$, $C^\perp$ for $\delta = 1$, C^- as $\delta \to 0^+$. Direct proofs of the bounds are non-trivial and given in the subsection below on technical details.
- TP$_2$ density if $\delta > 1$.
- Archimedean copula C_ψ, with generator

$$\psi(s;\delta) = 1 - F_\Gamma(s^{1/\delta};\delta) = \int_{s^{1/\delta}}^\infty z^{\delta-1}e^{-z}dz \ / \ \Gamma(\delta) = \int_s^\infty \exp\{-y^{1/\delta}\}dy \ / \ \Gamma(1+\delta).$$
(4.26)

  ψ is a LT for $\delta \geq 1$ and $\psi \in \mathcal{L}_2$ for $0 < \delta < 1$, and $\psi^{-1}(t;\delta) = [F_\Gamma^{-1}(1-t;\delta)]^\delta$. $-\log\psi$ is concave for $\delta > 1$ and is convex for $0 < \delta < 1$. The outline of the proof is in the technical subsection below. By Corollary 8.23, (4.25) is PQD for $\delta > 1$ and NQD for $0 < \delta < 1$.

- Kendall's $\tau = 1 - 4[\Gamma(1+\delta)]^{-2}\int_0^\infty se^{-2s^{1/\delta}}ds = 1 - 2^{2-2\delta}\delta\Gamma(2\delta)[\Gamma(1+\delta)]^{-2}$, using Theorem 8.26. This becomes $\tau = 1 - 2\pi^{-1/2}[\Gamma(1+\delta)]^{-1}\Gamma(\frac{1}{2}+\delta)$, using the Gamma function duplication formula, Therefore, as $\delta \to 0^+$, $\tau \to -1$. Using the Stirling approximation, $\tau \sim 1 - 2/\sqrt{\pi\delta}$ as $\delta \to \infty$. Combined with the $\prec_c$ ordering, this establishes the bounds of C^- and C^+.
- Blomqvist's $\beta = 3 - 4F_\Gamma(2^{1/\delta}\{F_\Gamma^{-1}(\frac{1}{2};\delta)\};\delta)$.
- The lower tail order is $\kappa_L = 2^{1/\delta}$, which is < 2 for $\delta > 1$ and > 2 for $0 < \delta < 1$. Note that $\kappa_L > 2$ indicates negative dependence, because positive dependence implies $\kappa_L \leq 2$. For $\delta \geq 1$, this follows from Theorem 8.37 because $\psi(s;\delta) = 1 - F_\Gamma(s^{1/\delta};\delta) \sim [\Gamma(\delta)]^{-1}s^{1-1/\delta}\exp\{-s^{-1/\delta}\}$ as $s \to \infty$ using $1 - F_\Gamma(y) \sim y^{\delta-1}e^{-y}/\Gamma(\delta)$ as $y \to \infty$. For $0 < \delta < 1$, a direct calculation leads to $C(u,u;\delta) \sim \ell(u)u^{2^{1/\delta}}$ as $u \to 0^+$, where $\ell(u)$ is a slowly varying function.
- The upper tail order is $\kappa_U = \min\{2, 1+\delta^{-1}\}$. For $\delta > 1$, $\kappa_U = 1+\delta^{-1}$ (follows from Theorems 8.88 and 8.35 because $\psi(s;\delta) \sim 1 - a_1 s + a_2 s^{1+1/\delta}$ as $s \to 0^+$ where $a_1 > 0$ and $a_2 > 0$). For $0 < \delta \leq 1$, $\kappa_U = 2$ (because $\psi''(0;\delta) = 0$ for $0 < \delta < 1$, $\psi''(0;\delta)$ is finite for $\delta = 1$, and $\psi(s;\delta) \sim 1 - a_1 s + O(s^2)$ as $s \to 0^+$ where $a_1 > 0$). This is intermediate upper tail dependence ($1 < \kappa_U < 2$) for $\delta > 1$.
- The tail order $\kappa_{NW} = \kappa_{SE}$ in the northwest or southeast corners of the unit square is 2 and $u - C(u;1-u;\delta) \sim \Gamma(\delta)u^2(-\log u)^{1-\delta}$ as $u \to 0^+$.
- The support is all of $(0,1)^2$ for any $\delta > 0$. For $\delta \geq 1$, since $\kappa_L = 2^{1/\delta} \leq 1+\delta^{-1} = \kappa_U \leq \kappa_{NW}$, the tail asymmetry is skewed to the lower tail relative to the upper tail. For $0 < \delta < 1$, since $\kappa_L = 2^{1/\delta} > 2 = \kappa_U \geq \kappa_{NW}$, and the tail asymmetry is skewed to the upper tail relative to the lower tail.

- Boundary conditional cdfs: $C_{2|1}(\cdot|0;\delta)$ is degenerate at 0 for $\delta > 1$, and $C_{2|1}(\cdot|0;\delta)$ is degenerate at 1 for $0 < \delta < 1$. Also, $C_{2|1}(v|1;\delta) = \exp\{-F_\Gamma^{-1}(1-v;\delta)\}$ from (8.9).
- Copula density at corners: For $\delta > 1$, $\kappa_L < 2$ and $\kappa_U < 2$ imply that $c(0,0;\delta)$ and $c(1,1;\delta)$ are infinite. For $0 < \delta \leq 1$, $c(0,0;\delta)$ and $c(1,1;\delta)$ are finite.
- Additional properties are listed as part of the multivariate extension.

Technical details

The second derivative $-\log\psi$ is obtained as well as the concordance ordering (bivariate case) and the bounds as $\delta \to 0^+$ and $\delta \to \infty$.

1. The sign of $-\log\psi$ is the same as the sign of $(\psi')^2 - \psi\psi''$ and for (4.26), this is equal in sign to

$$e^{-s^{1/\delta}} - \delta^{-1}s^{1/\delta-1}\int_s^\infty e^{-y^{1/\delta}}dy. \qquad (4.27)$$

If $0 < \delta < 1$, $s^{1/\delta-1} \leq y^{1/\delta-1}$ for $y > s$ and (4.27) is bounded below by

$$e^{-s^{1/\delta}} - \delta^{-1}\int_s^\infty y^{1/\delta-1}e^{-y^{1/\delta}}dy = 0.$$

For $\delta > 1$, 0 is an upper bound. Hence $(-\log\psi)''$ is ≥ 0 for $0 < \delta < 1$ and ≤ 0 for $\delta > 1$.

2. A simpler proof of concordance than that in the Appendix of Joe and Ma (2000) might be possible. Let $0 < \delta_1 < \delta_2$. By Theorem 8.22, it suffices to show for (4.26) that $\omega(s) = \psi^{-1}(\psi(s;\delta_1);\delta_2)$ is convex or the $\omega'(s)$ is increasing in $s > 0$. Let f_Γ be the density function for Gamma$(\delta,1)$. By the chain rule, with $t = \psi(s;\delta_1)$,

$$\omega'(s) = \frac{\psi'(s;\delta_1)}{\psi'\left(\psi^{-1}(\psi(s;\delta_1);\delta_2)\right)} = \frac{\psi'(\psi^{-1}(t;\delta_1);\delta_1)}{\psi'(\psi^{-1}(t;\delta_2);\delta_2)} \overset{\text{sign}}{=} \frac{\exp\{-F_\Gamma^{-1}(1-t;\delta_1)\}}{\exp\{-F_\Gamma^{-1}(1-t;\delta_2)\}}$$

$$= \exp\{F_\Gamma^{-1}(1-t;\delta_2) - F_\Gamma^{-1}(1-t;\delta_1)\}.$$

Hence ω' is increasing if $F_\Gamma^{-1}(u;\delta_2) - F_\Gamma^{-1}(u;\delta_1)$ is increasing in u or if $F_\Gamma^{-1}(F_\Gamma(x;\delta_1);\delta_2) - x$ is increasing in x. It remains to show this analytically.

3. Upper bound is C^+. Without loss of generality, in (4.25), assume $0 < u < v < 1$, so that $x = x_\delta = F_\Gamma^{-1}(1-u;\delta) \geq F_\Gamma^{-1}(1-v;\delta) = y = y_\delta$. For $\delta \geq 1$, $x \leq (x^\delta + y^\delta)^{1/\delta} = x[1 + (y/x)^\delta]^{1/\delta} \leq x \cdot 2^{1/\delta}$. Therefore

$$1 - F_\Gamma(2^{1/\delta}x_\delta;\delta) \leq C(u,v;\delta) \leq 1 - F_\Gamma(x_\delta;\delta) = u.$$

For the lower bound, $1 - F_\Gamma(2^{1/\delta}x_\delta;\delta) \sim 1 - F_\Gamma(x_\delta;\delta) - (2^{1/\delta} - 1)x_\delta f_\Gamma(x_\delta;\delta) \to u$ as $\delta \to \infty$.

4. Lower bound is C^-. Let $H(s;\delta) = F_\Gamma(s^{1/\delta};\delta) = \int_0^s \exp\{-y^{1/\delta}\}dy/\Gamma(1+\delta)$ for $s \geq 0$, with functional inverse $H^{-1}(p;\delta) = [F_\Gamma^{-1}(p;\delta)]^\delta$ $(0 \leq p \leq 1)$. As $\delta \to 0^+$, $H(\cdot;\delta) \to_d H(\cdot;0)$, the cdf of U$(0,1)$. The copula cdf (4.25) becomes $1 - H[H^{-1}(1-u;\delta) + H^{-1}(1-v;\delta);\delta]$ and as $\delta \to 0^+$, this converges to $1 - \min\{2 - u - v, 1\} = \max\{0, u + v - 1\} = C^-(u,v)$.

5. Tail order in the northwest or southeast corners of the unit square. Using Theorem 8.38,

$$-\psi'\left(\psi^{-1}(u;\delta);\delta\right) \cdot \psi^{-1}(1-u;\delta) \sim \Gamma(\delta)u^2(-\log u)^{1-\delta}.$$

The approximation comes from $F_\Gamma^{-1}(p;\delta) \sim -\log(1-p) + (\delta-1)\log[-\log(1-p)] - \log\Gamma(\delta)$ as $p \to 1^-$ and $F_\Gamma^{-1}(p;\delta) \sim [p\Gamma(\delta+1)]^{1/\delta}$ as $p \to 0^+$, so that

$$\psi^{-1}(1 - u;\delta) = [F_\Gamma^{-1}(u;\delta)]^\delta \sim u\Gamma(\delta+1),$$
$$-\psi'\left(\psi^{-1}(u;\delta);\delta\right) = \exp\{-F_\Gamma^{-1}(1-u;\delta)\}/\Gamma(\delta+1) \sim \delta^{-1}u(-\log u)^{1-\delta}.$$

4.11.2 Multivariate extension

For $\delta \geq 1$, the multivariate Archimedean extension is:

$$C(\boldsymbol{u}; \delta) = 1 - F_\Gamma \left(\left\{ \sum_{j=1}^{d} [F_\Gamma^{-1}(1 - u_j; \delta)]^\delta \right\}^{1/\delta}; \delta \right) \qquad (4.28)$$

- For $d \geq 3$, $\psi(s; \delta)$ in (4.26) can be used to get an Archimedean copula only for $\delta \geq 1$ (the region where ψ is a LT). This is because $\psi \notin \mathcal{L}_3$ for $0 < \delta < 1$; the third derivative is:

$$\psi'''(s; \delta) = \frac{s^{1/\delta - 2} e^{-s^{1/\delta}}}{\delta^2 \Gamma(1 + \delta)} \cdot \{1 - \delta - s^{1/\delta}\},$$

  and this cannot not be entirely non-positive if $0 < \delta < 1$.
- From Theorem 8.37, the lower tail order is $d^{1/\delta}$. From Theorem 8.88, the upper tail order is $1 + \delta^{-1}$. Since $2^{1/\delta} < 1 + \delta^{-1} < 3^{1/\delta}$ for $\delta > 1$, the direction of tail skewness is the lower tail for bivariate and the upper tail for $d \geq 3$.
- Since $\psi'(s; \delta)/\psi'(0; \delta)$ is the positive stable LT, using results in Section 3.2, the copula of $C_{12|3}(u_1, u_2|u_3; \delta)$ has the form of BB9 in Section 4.25.

Origins and further notes

- The multivariate family is derived in Joe and Ma (2000) as a min-stable multivariate $\overline{G}_0$ distribution when $\overline{G}_0$ is the gamma survival function.
- This family was used as an example of an Archimedean copula with intermediate tail in both the joint lower and upper tails in Hua and Joe (2011).
- The bivariate (4.25) is a one-parameter copula family that is permutation symmetric and reflection asymmetric, interpolates $C^-, C^\perp, C^+$, and has support on all of $(0, 1)^2$ for the whole parameter range. These properties make it a desirable bivariate parametric family for use in vines when the combination of reflection asymmetry and negative dependence is required. The common bivariate copula families with negative dependence are the Frank, Plackett, Gaussian and t_ν copulas, and all of these are reflection symmetric.

4.12 Archimedean based on LT of inverse gamma

Let $Y = X^{-1}$ have the inverse Gamma (IΓ) distribution, where $X \sim \text{Gamma}(\alpha, 1)$ for $\alpha > 0$. Then it is straightforward to derive that the maximal non-negative moment degree of Y is $M_Y = \alpha$. The LT of the inverse Gamma distribution is:

$$\psi(s; \delta) = \frac{2}{\Gamma(\alpha)} s^{\alpha/2} K_\alpha(2\sqrt{s}), \quad s \geq 0, \ \delta = \alpha^{-1} > 0, \qquad (4.29)$$

where K_α is the modified Bessel function of the second kind. Let $C_{1:d}(\boldsymbol{u}; \delta)$ be the Archimedean copula with the LT in (4.29). Because $\psi^{-1}(\cdot; \delta)$ does not have a closed form, this copula is not simple. For the bivariate copula density, the first two derivatives of ψ are needed: $\psi'(s; \delta) = -2s^{(\alpha-1)/2} K_{\alpha-1}(2s^{1/2})/\Gamma(\alpha)$, $\psi''(s; \delta) = 2s^{(\alpha-2)/2} K_{\alpha-2}(2s^{1/2})/\Gamma(\alpha)$. Hua and Joe (2011) showed that this copula has different tail properties, compared with other one-parameter Archimedean copula families, as summarized below.

- Numerically, it has been checked for the bivariate case that the concordance ordering is increasing as δ increases. $C^\perp$ obtains as $\delta \to 0^+$ and C^+ obtains as $\delta \to \infty$.
- For bivariate, Kendall's $\tau = \delta/(\delta + 2)$.
- With $\alpha = \delta^{-1}$, the upper tail order is $\max\{1, \min\{\alpha, d\}\}$. For the bivariate case, $\kappa_U = \max\{1, \min\{\alpha, 2\}\}$. That is, there is upper tail dependence for only part of the parameter range.

- The lower tail order is $\kappa_L = \sqrt{d}$; for bivariate, $\kappa_L = \sqrt{2}$.
- For bivariate, there is reflection asymmetry with skewness to the upper tail for $\delta > \sqrt{1/2}$ and skewness to the lower tail for $0 < \delta < \sqrt{1/2}$.

4.13 Multivariate t_ν

For notation, let $T_{1,\nu}$ or T_ν be univariate t cdf with parameter $\nu > 0$ and let $T_{d,\nu}(\cdot; \boldsymbol{\Sigma})$ for d-variate t with parameter $\nu > 0$ and covariance matrix $\boldsymbol{\Sigma}$. Similarly, let $t_{1,\nu}$ (or t_ν), and $t_{d,\nu}(\cdot; \boldsymbol{\Sigma})$ be the corresponding densities. If $\boldsymbol{\Sigma} = \boldsymbol{R}$ is a correlation matrix, then the univariate marginal cdfs of $T_{d,\nu}(\cdot; \boldsymbol{R})$ are all equal to $T_{1,\nu}$.

The univariate t_ν density is

$$t_{1,\nu}(y) = \frac{\Gamma((\nu+1)/2)}{\Gamma(\nu/2)\sqrt{\pi\nu}}(1 + y^2/\nu)^{-(\nu+1)/2}.$$

The univariate t_ν density with location and scale parameters is

$$t_{1,\nu}(x; \mu, \sigma) = \sigma^{-1}\frac{\Gamma((\nu+1)/2)}{\Gamma(\nu/2)\sqrt{\pi\nu}}\left(1 + \frac{[(x-\mu)/\sigma]^2}{\nu}\right)^{-(\nu+1)/2}. \tag{4.30}$$

The d-variate t_ν density with correlation matrix $\boldsymbol{R}$ is

$$t_{d,\nu}(\boldsymbol{y}; \boldsymbol{R}) = |\boldsymbol{R}|^{-1/2}\frac{\Gamma((\nu+d)/2)}{\Gamma(\nu/2)[\pi\nu]^{d/2}}(1 + \boldsymbol{y}^\top\boldsymbol{R}^{-1}\boldsymbol{y}/\nu)^{-(\nu+d)/2}, \quad \boldsymbol{y} \in \mathbb{R}^d;$$

see Section 2.7 for a derivation. Using the probability integral transform, the multivariate t_ν copula cdf is:

$$C(\boldsymbol{u}; \boldsymbol{R}, \nu) = T_{d,\nu}\big(T_{1,\nu}^{-1}(u_1), \cdots, T_{1,\nu}^{-1}(u_d); \boldsymbol{R}\big), \quad \boldsymbol{u} \in [0,1]^d, \tag{4.31}$$

and the multivariate t_ν copula density with positive definite $\boldsymbol{R}$ is:

$$c(\boldsymbol{u}; \boldsymbol{R}, \nu) = \frac{t_{d,\nu}\big(T_{1,\nu}^{-1}(u_1), \cdots, T_{1,\nu}^{-1}(u_d); \boldsymbol{R}\big)}{\prod_{j=1}^d t_{1,\nu}\big[T_{1,\nu}^{-1}(u_j)\big]}, \quad \boldsymbol{u} \in (0,1)^d.$$

The multivariate Gaussian copula obtains when $\nu \to \infty$.

For the bivariate case with $\boldsymbol{R} = \begin{pmatrix} 1 & \rho \\ \rho & 1 \end{pmatrix}$, we use the simpler notation $t_{2,\nu}(\cdot; \rho)$ for $t_{2,\nu}(\cdot; \boldsymbol{R})$. The bivariate density with $-1 < \rho < 1$ is

$$t_{2,\nu}(\boldsymbol{y}; \rho) = (1 - \rho^2)^{-1/2}\frac{\Gamma((\nu+2)/2)}{\Gamma(\nu/2)[\pi\nu]}\left(1 + \frac{y_1^2 + y_2^2 - 2\rho y_1 y_2}{\nu(1 - \rho^2)}\right)^{-(\nu+2)/2}. \tag{4.32}$$

From Section 2.7, the conditional distribution $[Y_2 \mid Y_1 = y_1]$ is a location-scale transform of the $t_{\nu+1}$ density, that is, (4.30) with $\mu(y_1) = \rho y_1$ and $\sigma^2(y) = (1 - \rho^2)(\nu + y^2)/(\nu + 1)$. The bivariate t_ν copula conditional distribution is:

$$C_{2|1}(v|u; \rho, \nu) = T_{\nu+1}\left(\frac{T_\nu^{-1}(v) - \rho T_\nu^{-1}(u)}{\sqrt{(1 - \rho^2)(\nu + [T_\nu^{-1}(u)]^2)/(\nu + 1)}}\right). \tag{4.33}$$

With $y_1 = T_\nu^{-1}(u)$, $y_2 = T_\nu^{-1}(v)$ and $0 < u, v < 1$, the bivariate t_ν copula density is:

$$c(u, v; \rho, \nu) = \frac{t_{2,\nu}\big(T_\nu^{-1}(u), T_\nu^{-1}(v); \rho\big)}{t_\nu\big(T_\nu^{-1}(u)\big)\, t_\nu\big(T_\nu^{-1}(u)\big)}$$

$$= \frac{1}{\sqrt{1 - \rho^2}}\frac{\Gamma((\nu+2)/2)\,\Gamma(\nu/2)}{\Gamma^2((\nu+1)/2)}\frac{\left(1 + \frac{y_1^2 + y_2^2 - 2\rho y_1 y_2}{\nu(1 - \rho^2)}\right)^{-(\nu+2)/2}}{\left(1 + \frac{y_1^2}{\nu}\right)^{-(\nu+1)/2}\left(1 + \frac{y_2^2}{\nu}\right)^{-(\nu+1)/2}}. \tag{4.34}$$

Some properties of the multivariate t copula (4.31) are the following.

- Increasing in $\prec_c$ as a special case of a general concordance result for elliptical distributions; see Theorem 8.52.
- Multivariate tail dependence indices of the form $\mathbb{P}(U_i > u, i = 1, \ldots, d \mid U_j > u)$ and $\mathbb{P}(U_i > u, i \notin J \mid U_j > u.j \in J)$, with $\emptyset \neq J \subset \{1, \ldots, d\}$, are given Chan and Li (2008), in terms of moments of the underlying multivariate Gaussian distribution.
- The (lower or upper) tail dependence function of (4.31) is given by

$$b(\boldsymbol{w}; \boldsymbol{R}, \nu) = \sum_{j=1}^{d} w_j T_{d-1,\nu+1}\left(\frac{\sqrt{\nu+1}}{\sqrt{1-\rho_{ij}^2}}\left[\rho_{ij} - \left(\frac{w_i}{w_j}\right)^{-1/\nu}\right], i \neq j; \boldsymbol{R}_j\right),$$

for all $\boldsymbol{w} = (w_1, \ldots, w_d) \in \mathbb{R}_+^d$ where $\boldsymbol{R}_j = (\rho_{k_1 k_2;j})_{k_1,k_2 \neq j}$ is a matrix of partial correlations given the jth variable. With $d = 2$, this simplifies to:

$$b(\boldsymbol{w}; \rho, \nu) = w_1 T_{\nu+1}\left(\frac{\sqrt{\nu+1}}{\sqrt{1-\rho^2}}\left[\rho - \left(\frac{w_2}{w_1}\right)^{-1/\nu}\right]\right) + w_2 T_{\nu+1}\left(\frac{\sqrt{\nu+1}}{\sqrt{1-\rho^2}}\left[\rho - \left(\frac{w_1}{w_2}\right)^{-1/\nu}\right]\right).$$

Further properties of the bivariate version $C(\cdot; \rho, \nu)$ of (4.31) with (4.32)–(4.34) are the following.

- C^- obtains as $\rho \to -1$, and C^+ obtains as $\rho \to 1$. When $\rho = 0$, the copula is not $C^\perp$.
- Positive quadrant dependent for $\rho > 0$ and negative quadrant dependent for $\rho < 0$.
- Kendall's τ and Blomqvist's β are both equal $2\pi^{-1} \arcsin(\rho)$, the same as for bivariate Gaussian.
- The tail dependence parameter is $\lambda = \lambda_L = \lambda_U = 2T_{\nu+1}\left(-\sqrt{(\nu+1)(1-\rho)/(1+\rho)}\right)$: for fixed $\rho \in (-1, 1)$, this is decreasing in ν with limit of $2\Phi(-\infty) = 0$ as $\nu \to \infty$ and $2T_1\left(-\sqrt{(1-\rho)/(1+\rho)}\right)$ as $\nu \to 0^+$.
- Consider the range of (β, λ). Since $T_1(x) = \frac{1}{2} + \pi^{-1} \arctan x$ is the Cauchy cdf with density $t_1(x) = \pi^{-1}(1+x^2)^{-1}$, the limit of λ as $\nu \to 0^+$ is

$$1 + 2\pi^{-1} \arctan\left(-\sqrt{(1-\rho)/(1+\rho)}\right) = 1 + 2\pi^{-1} \arcsin\left(-\sqrt{(1-\rho)/2}\right) = \frac{1}{2} + \pi^{-1} \arcsin \rho;$$

this identity is in Hua and Joe (2012a). Substituting $\rho = \sin(\pi\beta/2)$ leads to limit of $(1+\beta)/2$. That is, given $\beta \in (-1, 1)$, the maximum λ is $(1+\beta)/2$. Equivalently, given $\lambda \in (0, 1)$, then $\beta \geq 2\lambda - 1$.
- The range of λ as a function of ρ, as ν varies, is shown in Frahm et al. (2003) for bivariate t and elliptical copulas.
- The lower boundary conditional cdf is $C_{2|1}(v|0; \rho, \nu) = T_{\nu+1}(\rho\sqrt{\nu+1}/\sqrt{1-\rho^2}) =: p$ for $0 < v < 1$. That is, this has mass of p at 0 and mass of $1 - p$ at 1. By symmetry, the upper boundary cdf has mass p at 1 and $1 - p$ at 0.
- For $\rho > 0$, the conditional distributions $F_{2|1}(\cdot|x)$ of bivariate t_ν distribution are not stochastically increasing as x increases, because the conditional scale of bivariate t_ν is infinite ∞ as $x \to \pm\infty$.

4.14 Marshall-Olkin multivariate exponential

The bivariate copula family is presented first, followed by some properties and multivariate extensions.

4.14.1 Bivariate Marshall-Olkin

For the bivariate Marshall-Olkin exponential distribution, from Marshall and Olkin (1967a) and Example 2.1, a stochastic representation is:

$$X = \min\{Z_1/(1 - \pi_1), Z_{12}/\pi_1\}, \quad Y = \min\{Z_2/(1 - \pi_2), Z_{12}/\pi_2\},$$

where Z_1, Z_2, Z_{12} are independent Exponential(1) random variables, and $0 \leq \pi_1, \pi_2 \leq 1$. The survival function is:

$$\overline{G}(x, y; \pi_1, \pi_2) = e^{-(1-\pi_1)x} e^{-(1-\pi_2)y} e^{-\max\{\pi_1 x, \pi_2 y\}} = \exp\{-A(x, y; \pi_1, \pi_2)\}, \qquad (4.35)$$

for $x, y \geq 0$, where $A(x, y; \pi_1, \pi_2) = (1 - \pi_1)x + (1 - \pi_2)y + \max\{\pi_1 x, \pi_2 y\}$, and

$$\begin{aligned} B(w; \pi_1, \pi_2) &= A(w, 1 - w; \pi_1, \pi_2) = (1 - \pi_1)w + (1 - \pi_2)(1 - w) + \max\{\pi_1 w, \pi_2(1 - w)\} \\ &= \begin{cases} 1 - \pi_1 w, & 0 \leq w \leq \pi_2/(\pi_1 + \pi_2), \\ 1 - \pi_2(1 - w), & \pi_2/(\pi_1 + \pi_2) \leq w \leq 1. \end{cases} \end{aligned}$$

There is a singular component, and $B'(0; \pi_1, \pi_2) = -\pi_1$, $B'(1; \pi_1, \pi_2) = \pi_2$. At the trough, $B(\pi_2/(\pi_1 + \pi_2); \pi_1, \pi_2) = 1 - \pi_1\pi_2/(\pi_1 + \pi_2)$.

The parametrization is such that the univariate margins of (4.35) are exponential distributions with mean 1, and the (survival) copula cdf is:

$$C(u, v; \pi_1, \pi_2) = \min\{u^{\pi_1}, v^{\pi_2}\} u^{1-\pi_1} v^{1-\pi_2} = \begin{cases} uv^{1-\pi_2}, & u^{\pi_1} \leq v^{\pi_2}, \\ vu^{1-\pi_1}, & u^{\pi_1} > v^{\pi_2}, \end{cases} \qquad (4.36)$$

for $0 \leq u, v \leq 1$. The conditional distributions are:

$$C_{2|1}(v|u; \pi_1, \pi_2) = \begin{cases} v^{1-\pi_2}, & u^{\pi_1} \leq v^{\pi_2}, \\ (1 - \pi_1)vu^{-\pi_1}, & u^{\pi_1} > v^{\pi_2}. \end{cases}$$

$$C_{1|2}(u|v; \pi_1, \pi_2) = \begin{cases} (1 - \pi_2)uv^{-\pi_2}, & u^{\pi_1} \leq v^{\pi_2}, \\ u^{1-\pi_1}, & u^{\pi_1} > v^{\pi_2}. \end{cases}$$

For the permutation symmetric case of $\delta = \pi_1 = \pi_2$, the exponent function is $A(x, y; \delta, \delta) = (1 - \delta)(x + y) + \delta \max\{x, y\}$ and C is a geometric mixture of $C^{\perp}$ and C^+ with

$$C(u, v; \delta, \delta) = [\min\{u, v\}]^{\delta} [uv]^{1-\delta}, \quad 0 < u, v < 1; \qquad (4.37)$$

this case is also known as the Cuadras-Augé copula (Cuadras and Augé (1981)).

Another special case has $\pi_2 = 1$ and $\pi_1 = \pi \in [0, 1]$ leading to

$$C(u, v; \pi, 1) = \min\{u^{\pi}, v\} u^{1-\pi}, \quad 0 < u, v < 1, \ 0 \leq \pi \leq 1. \qquad (4.38)$$

This family can be used to compare the amount of bivariate non-exchangeability or permutation asymmetry.

Some properties of (4.36) are the following.

- (4.36) is increasing in the concordance ordering as π_1, π_2 increase. If $\pi_1 = 0$ or $\pi_2 = 0$, then $C^{\perp}$ obtains. If $\pi_1 = \pi_2 = 1$, then C^+ obtains.
- (4.36) is also the copula of

$$\left(\min\{\pi_1 Z_1/(1 - \pi_1), Z_{12}\}, \min\{\pi_2 Z_2/(1 - \pi_2), Z_{12}\}\right),$$

where Z_1, Z_2, Z_{12} are independent unit Exponential random variables. From Example 2.1, the mass of the singularity component is $\pi_1\pi_2/[\pi_1 + \pi_2 - \pi_1\pi_2]$.
- Kendall's $\tau = \pi_1\pi_2/[\pi_1 + \pi_2 - \pi_1\pi_2]$.
- Spearman's $\rho_S = 3\pi_1\pi_2/[2\pi_1 + 2\pi_2 - \pi_1\pi_2]$.
- Blomqvist's $\beta = 2^{\min\{\pi_1, \pi_2\}} - 1$.
- (4.36) appears as boundary of several extreme value copula families: asymmetric Gumbel/Galambos (Section 4.15) and t-EV (Section 4.16).

Some properties of (4.38) are the following.

- Increasing in $\prec_c$ with $C^\perp$ for $\pi = 0$ and C^+ for $\pi = 1$.
- Singularity on the set $\{(u, v) : v = u^\pi, 0 < u < 1\}$ with mass π for $0 < \pi \le 1$.
- The support is in the region $\{(u, v) : 0 < v < u^\pi < 1\}$ and the copula density is $(1-\pi)u^{-\pi}$ in this region.
- Kendall's $\tau = \pi$.
- Spearman's $\rho_S = 3\pi/(2 + \pi)$.
- Blomqvist's $\beta = 2^\pi - 1$.
- If $(U, V) \sim C(\cdot; \pi, 1)$, then from (2.54), the bivariate skewness coefficient $\mathbb{E}[(U - V)^3]$ can be evaluated as:

$$6 \int_0^1 \int_0^1 (v - u)C(u, v; \pi, 1)\, dv du = \frac{\pi(1 - \pi)}{2(1 + \pi)(3 + \pi)},$$

  and this is maximized with value 0.0253 at $\pi = [-3 + \sqrt{24}]/5 = 0.3798$. The maximum value of this bivariate skewness coefficient is $27/256 = 0.1055$,
- $\max_{u,v} |C(u, v; \pi, 1) - C(v, u, \pi, 1)| = \pi(1 - \pi)/[1 + \pi(1 - \pi)]^{1+1/[\pi(1-\pi)]}$ when $u = u^* = 1/[1 + \pi(1 - \pi)]^{1/[\pi(1-\pi)]}$ and $v = (u^*)^\pi$. The maximum of the L_∞ distance over different π occurs at $\pi = \frac{1}{2}$ leading to a value of $4^4/5^5 = 0.08192$; for comparison, the maximum L_∞ distance is $1/3$ over all bivariate copulas.
- For comparison, for the asymmetric Gumbel copula (Section 4.15.1) with a range of δ values, the skewness coefficient and L_∞ distance are tabulated below.

δ	skewness	L_∞
1.11	0.0015	0.0030
1.43	0.0059	0.0119
2.00	0.0117	0.0257
3.33	0.0184	0.0454
10.0	0.0231	0.0695
∞	0.0253	0.0819

Hence the most extreme bivariate permutation asymmetry is linked to a singular component.

4.14.2 Multivariate Marshall-Olkin exponential and extensions

The most general bivariate Marshall-Olkin exponential distribution involves three independent Exponential(1) random variables. The most general for d-variate involves $2^d - 1$ independent Exponential(1) random variables, with a random variable Z_S for each non-empty $S \subset \{1, \ldots, d\}$. Let $\mathcal{S} = 2^{\{1, \ldots, d\}} \setminus \emptyset$.

We next list the multivariate exponential distribution of Marshall and Olkin (1967b) and some of its extensions. Let $\xi_S > 0$ for each S and let $v_S = \sum_{T:S \subset T} \xi_T$. For $|S| = 1$, simplifying notation such as ξ_i and v_i will be used. For $S \in \mathcal{S}$, let Z_S be an exponential random variable with rate parameter ξ_S (mean ξ_S^{-1}). For $j = 1, \ldots, d$, let $X_j = \min\{Z_S : j \in S\}$.

The exponent A of the Marshall-Olkin multivariate exponential survival function $\overline{G} = e^{-A}$ is

$$A(\boldsymbol{x}; \{\xi_S, S \in \mathcal{S}\}) = \sum_{S \in \mathcal{S}} \xi_S \max_{i \in S} x_i = \sum_{S \in \mathcal{S}} (-1)^{|S|+1} v_S \min_{i \in S} x_i, \quad \boldsymbol{x} \in \mathbb{R}_+^d. \tag{4.39}$$

This can be rearranged to get

$$A(\boldsymbol{x}; \{\xi_S, S \in \mathcal{S}\}) = \sum_{S \in \mathcal{S}} \xi_S a_S(x_i, i \in S), \tag{4.40}$$

where

$$a_S(x_i, i \in S) = \sum_{T \subset S} (-1)^{|T|+1} [\min_{i \in T} x_i]. \tag{4.41}$$

Joe (1990a) has families of min-stable multivariate exponential survival functions in the form of (4.40), where a_S is replaced by something other than (4.41) or $a_S(\boldsymbol{x}_S) = \max_{i \in S} x_i$. These include

$$A(\boldsymbol{x}; \{\xi_S, S \in \mathcal{S}\}, \delta) = \sum_{S \in \mathcal{S}} \xi_S \Big(\sum_{i \in S} x_i^\delta \Big)^{1/\delta}, \quad \boldsymbol{x} \in \mathbb{R}_+^d. \tag{4.42}$$

where $1 \leq \delta \leq \infty$, with $\delta \to \infty$ in (4.42) leading to (4.39), and

$$A(\boldsymbol{x}; \{\xi_S, S \in \mathcal{S}\}, \delta) = \sum_{S \in \mathcal{S}} (-1)^{|S|+1} v_S \Big[\sum_{i \in S} x_i^{-\delta} \Big]^{-1/\delta} = \sum_{S \in \mathcal{S}} \xi_S a_S(x_i, i \in S; \delta), \quad \boldsymbol{x} \in \mathbb{R}_+^d. \tag{4.43}$$

where

$$a_S(x_i, i \in S; \delta) = \sum_{T \subset S} (-1)^{|T|+1} \Big[\sum_{i \in T} x_i^{-\delta} \Big]^{-1/\delta}$$

generalizes (4.41) and $0 \leq \delta \leq \infty$, with $\delta \to \infty$ in (4.43) leading to (4.39). Note that if $\delta = 1$ in (4.42) or if $\delta \to 0^+$ in (4.43), then $\overline{G}$ becomes a product of the univariate margins, i.e., $\exp\{-\sum_{i=1}^d v_i x_i\}$. Hence δ in both (4.42) and (4.43) is a global dependence parameter and ξ_S or v_S for $|S| \geq 2$ are parameters indicating the strength of dependence for the variables in S. The univariate survival functions for (4.39), (4.42) and (4.43) are $\overline{G}_j(x_j) = \exp\{-v_j x_j\}$, $j = 1, \ldots, d$, and each family has k-variate $(2 \leq k < d)$ marginal survival functions of the same form. By rescaling, one gets min-stable multivariate exponential survival functions with unit means for the univariate margins. Special cases of (4.42) and (4.43) include the Gumbel and Galambos families and their permutation-symmetric extensions (take $\xi_S = 0$ if $S \neq \{1, \ldots, d\}$ and $\xi_{\{1,\ldots,d\}} = 1$).

Other results for Marshall-Olkin copulas include the following.

- Li (2008a) and Li (2008b) has results for multivariate tail dependence of Marshall-Olkin copulas and related distributions.
- Mai and Scherer (2010) study the Pickands representation for Marshall-Olkin copulas.
- Chapter 3 of Mai and Scherer (2012b) has alternative parametrizations and properties of exchangeable and extendible Marshall-Olkin copulas.

4.15 Asymmetric Gumbel/Galambos copulas

In this section, asymmetric versions of the Gumbel and Galambos copula families (or min-stable multivariate exponential survival functions) are defined. The bivariate versions can be useful choices for edges of vines when there is the combination of (i) strong tail dependence in one joint tail, (ii) intermediate tail dependence in the other joint tail, and (iii) permutation asymmetry is desired.

4.15.1 Asymmetric Gumbel with Marshall-Olkin at boundary

For a bivariate asymmetric Gumbel survival function with the Marshall-Olkin exponential distribution as $\delta \to \infty$, let $\overline{G}(x, y; \delta, \pi_1, \pi_2) = \exp\{-A(x, y; \delta, \pi_1, \pi_2)\}$ be a bivariate exponential survival function with three-parameter copula family

$$C(u, v; \delta, \pi_1, \pi_2) = e^{-A(-\log u, -\log v; \delta, \pi_1, \pi_2)}, \quad 0 \leq u, v \leq 1, \ \delta \geq 1, \ 0 \leq \pi_1 < \pi_2 \leq 1, \tag{4.44}$$

where

$$A(x, y; \delta, \pi_1, \pi_2) = [\pi_1^\delta x^\delta + \pi_2^\delta y^\delta]^{1/\delta} + (1 - \pi_1)x + (1 - \pi_2)y,$$
$$B(w; \delta, \pi_1, \pi_2) = A(w, 1 - w; \delta, \pi_1, \pi_2) = [\pi_1^\delta w^\delta + \pi_2^\delta (1 - w)^\delta]^{1/\delta} + (\pi_2 - \pi_1)w + 1 - \pi_2.$$

This is one example of an asymmetric Gumbel bivariate distribution given in Tawn (1988). These can be derived as an extension of the bivariate Marshall-Olkin exponential survival function (4.35) by replacing $\max\{\pi_1 x, \pi_2 y\}$ by $[\pi_1^\delta x^\delta + \pi_2^\delta y^\delta]^{1/\delta}$ with $\delta \geq 1$. This general technique is also used in Joe (1990a) with the multivariate extension; see Section 4.14.2. If $\pi_1 = \pi_2 = \pi$, this is convex combination of the exponent functions for Gumbel and independence. Also, (4.44) is increasing in concordance as δ, π_1, π_2 increase.

4.15.2   Asymmetric Gumbel based on deHaan representation

An alternative representation, from de Haan (1984), for min-stable multivariate exponential survival functions has the exponent $A = -\log \overline{G}$ in the form:

$$A(\boldsymbol{x}) = \int_0^1 \left[\max_{1 \leq j \leq d} x_j g_j(v) \right] dv, \quad \boldsymbol{x} \in \mathbb{R}_+^d, \tag{4.45}$$

where the g_j are pdfs on $[0, 1]$ for $j = 1, \ldots, d$.

As shown in Joe (1993), the Gumbel and Galambos families in (4.14) and (4.19) can be unified with this form with

$$g_1(v) = (1 - \zeta)v^{-\zeta}, \quad g_2(v) = (1 - \eta)(1 - v)^{-\eta}, \quad 0 < v < 1,$$

with $0 < \zeta = \eta < 1$ for (4.14) and $\zeta = \eta < 0$ for (4.19). With ζ, η both less than 1, a two-parameter extension of (4.14) and (4.19) obtains. The magnitude of the difference of ζ and η measures asymmetry, and the average of ζ and η measures dependence. There is an obvious multivariate extension using more g_j of the same form in (4.45).

For $\zeta > 0$, $\eta > 0$, expansion of the integral in (4.45) leads to

$$A(x, y; \zeta, \eta) = (x + y)B(x/(x + y); \zeta, \eta), \tag{4.46}$$

where

$$B(w; \zeta, \eta) = wz^{1-\zeta} + (1 - w)(1 - z)^{1-\eta}, \quad 0 \leq w \leq 1, \tag{4.47}$$

and $z = z(w; \zeta, \eta)$ is the root of the equation

$$(1 - \zeta)w(1 - z)^\eta - (1 - \eta)(1 - w)z^\zeta = 0. \tag{4.48}$$

Some details are given in the subsection below. This model is called *bilogistic* in Smith (1990a) and Joe et al. (1992).

For $\zeta < 0$, $\eta < 0$, one obtains

$$A^*(x, y; \zeta, \eta) = (x + y)B^*(x/(x + y); \zeta, \eta), \tag{4.49}$$

where

$$B^*(w; \zeta, \eta) = 1 - wz^{1-\zeta} - (1 - w)(1 - z)^{1-\eta}, \quad 0 \leq w \leq 1, \tag{4.50}$$

and $z = z(w; \zeta, \eta)$ is the root of the equation

$$(1 - \zeta)wz^{-\zeta} - (1 - \eta)(1 - w)(1 - z)^{-\eta} = 0. \tag{4.51}$$

It is shown below that $B'(1; \zeta, \eta) = -B'(0; \zeta, \eta) = 1$ and $(B^*)'(1; \zeta, \eta) =$

$-(B^*)'(0; \zeta, \eta) = 1$ so that only the first asymmetric version provides alternative derivatives at the boundary.

The distributions from (4.46) are increasing in concordance as ζ or η decreases for $\zeta, \eta > 0$; C^+ obtains in the limit as $\zeta = \eta \to 0^+$ and $C^\perp$ obtains as $\zeta = \eta \to 1$. The distributions from (4.49) are also increasing in concordance as ζ or η increases; C^+ obtains in the limit as $\zeta = \eta \to 0^+$ and $C^\perp$ obtains as $\zeta = \eta \to -\infty$. $C^\perp$ obtains more generally as one of $\zeta, \eta > 0$ (or respectively, $\zeta, \eta < 0$) is fixed and the other approaches 1 (respectively, ∞). Different limits occur as one of the parameters approaches 0, for example, as $\eta \to 0^+$ in (4.46), or as $\eta \to 0^+$ in (4.49).

Technical details

1. When $\zeta = \eta \in (0, 1)$, then

$$A(x, y; \zeta, \eta) = \int_0^1 \max\{x(1-\zeta)v^{-\zeta}, \ y(1-\eta)(1-v)^{-\eta}\} \, dv = (x^{1/\zeta} + y^{1/\zeta})^\zeta,$$

corresponding to the Gumbel copula with parameter $\delta = 1/\zeta > 1$.
Proof: $xv^{-\zeta} \geq y(1-v)^{-\zeta} \iff v \leq [1 + (y/x)^{1/\zeta}]^{-1} = x^{1/\zeta}/[x^{1/\zeta} + y^{1/\zeta}] =: t$, and

$$A(x, y; \zeta, \eta) = (1-\zeta)\int_0^t xv^{-\zeta} dv + (1-\zeta)\int_t^1 y(1-v)^{-\zeta} dv = xt^{1-\zeta} + y(1-t)^{1-\zeta}.$$

2. $B(w) = B(w; \zeta, \eta) = wz^{1-\zeta} + (1-w)(1-z)^{1-\eta}$; $z = z(w)$ is the root of (4.48).
Proof: $w(1-\zeta)v^{-\zeta} \geq (1-w)(1-\eta)(1-v)^{-\eta} \iff v \leq z = z(w)$, and

$$B(w) = \int_0^1 \max\{w(1-\zeta)v^{-\zeta}, \ (1-w)(1-\eta)(1-v)^{-\eta}\} \, dv$$

$$= \int_0^z w(1-\zeta)v^{-\zeta} dv + \int_z^1 (1-w)(1-\eta)(1-v)^{-\eta} \, dv$$

$$= wz^{1-\zeta} + (1-w)(1-z)^{1-\eta}.$$

3. When $\zeta = \eta \in (-\infty, 0)$, then

$$A^*(x, y; \zeta, \eta) = \int_0^1 \max\{x(1-\zeta)v^{-\zeta}, \ y(1-\eta)(1-v)^{-\eta}\} \, dv = x + y - (x^{1/\zeta} + y^{1/\zeta})^\zeta,$$

corresponding to the Galambos copula with parameter $\delta = -1/\zeta > 0$.
Proof: $xv^{-\zeta} \geq y(1-v)^{-\zeta} \iff v \geq [1 + (y/x)^{1/\zeta}]^{-1} = x^{1/\zeta}/[x^{1/\zeta} + y^{1/\zeta}] =: t$, and

$$A^*(x, y; \zeta, \eta) = (1-\zeta)\int_0^t y(1-v)^{-\zeta} dv + (1-\zeta)\int_t^1 xv^{-\zeta} dv$$

$$= y[1 - (1-t)^{1-\zeta}] + x[1 - t^{1-\zeta}] = x + y - (x^{1/\zeta} + y^{1/\zeta})^\zeta.$$

4. $B^*(w) = B^*(w; \zeta, \eta) = 1 - wz^{1-\zeta} - (1-w)(1-z)^{1-\eta}$; $z = z(w)$ is the root of (4.51).
Proof: $w(1-\zeta)v^{-\zeta} \geq (1-w)(1-\eta)(1-v)^{-\eta} \iff v \geq z = z(w)$, and

$$B^*(w) = \int_0^1 \max\{w(1-\zeta)v^{-\zeta}, \ (1-w)(1-\eta)(1-v)^{-\eta}\} dv$$

$$= \int_0^z (1-w)(1-\eta)(1-v)^{-\eta} dv + \int_z^1 w(1-\zeta)v^{-\zeta} dv$$

$$= (1-w) - (1-w)(1-z)^{1-\eta} + w - wz^{1-\zeta}.$$

5. B' and $B^{*\prime}$ at boundaries of 0 and 1 for (4.47) and (4.50).
   Consider first (4.47). Note that $z(0) = 0$ and $z(1) = 1$.

$$B'(w) = z^{1-\zeta} - (1-z)^{1-\eta} + \left[w(1-\zeta)z^{-\zeta} - (1-w)(1-\eta)(1-z)^{-\eta} \right] z'$$

Then it can be shown that $z(w) \sim a_0 w^{1/\zeta}$ as $w \to 0^+$ where $a_0 = [(1-\zeta)/(1-\eta)]^{1/\zeta}$. Similarly, with $1 - z(w) \sim a_1(1-w)^{1/\eta}$ as $w \to 1^-$, where $a_1 = [(1-\eta)/(1-\zeta)]^{1/\eta}$. Because of the smoothness and boundedness of $z(w)$, $z'(w) \sim a_0 \zeta^{-1} w^{1/\zeta-1}$ as $w \to 0^+$, and $z'(w) \sim a_1 \eta^{-1}(1-w)^{1/\eta-1}$ as $w \to 1^-$. Hence as $w \to 0^+$,

$$B'(w) = O(w^{1/\zeta-1}) - 1 + O(w^{1/\zeta}) + [(1-\zeta)a_0^{-\zeta} - (1-\eta) + O(w^{1/\zeta}) + O(w)]a_0 \zeta^{-1} w^{1/\zeta-1},$$

or $B'(w) = -1 + O(w^{1/\zeta-1})$ and $B'(0) = -1$ since $0 < \zeta < 1$. Similarly, $B'(1) = 1$.
The analysis of the derivative of (4.50) is similar. Note that $z(0) = 1$ and $z(1) = 0$ for (4.51) and similar to above, $z(w) \sim 1 - a_0 w^{-1/\eta}$ and $z'(w) \sim a_0 \eta^{-1} w^{-1/\eta-1}$ as $w \to 0^+$ where $a_0 = [(1-\zeta)/(1-\eta)]^{-1/\eta}$, and $z(w) \sim a_1(1-w)^{-1/\zeta}$ and $z'(w) \sim -a_1 \zeta^{-1}(1-w)^{-1/\zeta-1}$ as $w \to 1^-$ where $a_1 = [(1-\eta)/(1-\zeta)]^{-1/\zeta}$. The derivative of (4.50) is

$$(B^*)'(w) = -z^{1-\zeta} + (1-z)^{1-\eta} + \left[-w(1-\zeta)z^{-\zeta} + (1-w)(1-\eta)(1-z)^{-\eta} \right] z'.$$

As $w \to 0^+$, $(B^*)'(w) = -1 + O(w^{-1/\eta})$ and $(B^*)'(0) = -1$ since $\eta < 0$. Similarly, $(B^*)'(1) = 1$.

6. The case of parameters with mixed signs, such as $0 < \zeta < 1$ and $\eta < 0$ is harder to analyze and the set for which each term is largest might not be an interval.
   For $0 < \zeta < 1$ and $\eta < 0$, some details of the B function are given below:

$$B(w) = \int_0^1 \max\{w(1-\zeta)v^{-\zeta}, \ (1-w)(1-\eta)(1-v)^{-\eta}\}dv,$$

$$w(1-\zeta) \ge (1-w)(1-\eta)v^{\zeta}(1-v)^{-\eta} \iff v \le z_1 \text{ or } v \ge z_2,$$

where $0 < z_1(w) < z_0 < z_2(w) < 1$ are the two roots of

$$-w(1-\zeta) + (1-w)(1-\eta)z^{\zeta}(1-z)^{-\eta} = 0,$$

provided $0 < w < w_0 < 1$, where $z_0 = \zeta/(\zeta - \eta) = \operatorname{argmax} z^{\zeta}(1-z)^{-\eta}$, and $w_0 = (1-\eta)z_0^{\zeta}(1-z_0)^{-\eta}/[1 - \zeta + (1-\eta)z_0^{\zeta}(1-z_0)^{-\eta}] \in [\frac{1}{2}, 1)$ satisfies

$$-w_0(1-\zeta) + (1-w_0)(1-\eta)z_0^{\zeta}(1-z_0)^{-\eta} = 0.$$

As $w \to 0^+$, $z_1(w) \to 0^+$ and $z_2(w) \to 1^-$.
For $0 < w < w_0$,

$$B(w) = \int_0^{z_1(w)} w(1-\zeta)v^{-\zeta}dv + \int_{z_1(w)}^{z_2(w)} (1-w)(1-\eta)v^{-\eta}dv + \int_{z_2(w)}^1 w(1-\zeta)v^{-\zeta}dv$$

$$= w\left[1 - z_2^{1-\zeta} + z_1^{1-\zeta} \right] + (1-w)\left[(1-z_1)^{1-\eta} - (1-z_2)^{1-\eta} \right],$$

and for $w_0 < w < 1$, $B(w) = \int_0^1 w(1-\zeta)v^{-\zeta}dv = w$. The trough value of B is smaller than w_0 and occurs at a value between $1 - w_0$ and w_0.
From analyzing the derivative B' near the boundaries, it can be shown that $B'(0) = -1$ and $B'(1) = 1$.

4.16   Extreme value limit of multivariate t$_\nu$

Let $T_\eta = T_{1,\eta}$, $t_\eta = t_{1,\eta}$ be the univariate t cdf and pdf with shape (or degree of freedom) parameter $\eta > 0$. Let $T_{m,\eta}(\cdot; \boldsymbol{R})$ be the m-variate t cdf with parameter η and correlation matrix $\boldsymbol{R}$. The name t-EV comes from Demarta and McNeil (2005), where the bivariate t-EV distribution is derived. The multivariate t-EV distribution is derived in Nikoloulopoulos et al. (2009).

The multivariate t$_\nu$-EV copula, obtained by the extreme value limit of the t$_\nu$ copula, is given by

$$C(u_1, \ldots, u_d; \boldsymbol{R}, \nu) = \exp\{-A(\boldsymbol{w}; \boldsymbol{R}, \nu)\}, \quad w_j = -\log u_j, \ j = 1, \ldots, d, \ \boldsymbol{u} \in [0,1]^d,$$

with exponent

$$A(\boldsymbol{w}; \boldsymbol{R}, \nu) = \sum_{j=1}^{d} w_j T_{d-1,\nu+1} \left(\frac{\sqrt{\nu+1}}{\sqrt{1 - \rho_{ij}^2}} \left[\left(\frac{w_i}{w_j}\right)^{-1/\nu} - \rho_{ij} \right], i \neq j; \boldsymbol{R}_j \right),$$

In the above, $\boldsymbol{R}_j$ is the correlation matrix with jth variable partialled out; that is, $\boldsymbol{R}_j = (\rho_{k_1 k_2; j})_{k_1, k_2 \neq j}$. For $d = 2$, $\boldsymbol{R}_1, \boldsymbol{R}_2$ are equal to the scalar 1.

4.16.1   Bivariate t-EV

For $d = 2$, with $\rho = \rho_{12} \in [-1, 1]$ and $\nu > 0$, the copula cdf is

$$C(u, v; \rho, \nu) = \exp\{-A(-\log u, -\log v; \rho, \nu)\} = \exp\{-(x+y)B(x/[x+y]; \rho, \nu)\}, \quad (4.52)$$

with $x = \log u, y = -\log v$, and $0 \leq u, v \leq 1$; also $B(w; \rho, \nu) = A(w, 1 - w; \rho, \nu)$ and

$$B(w; \rho, \nu) = w T_{\nu+1} \left(\frac{\sqrt{\nu+1}}{\sqrt{1-\rho^2}} \left[\left\{ \frac{w}{1-w} \right\}^{1/\nu} - \rho \right] \right) + (1-w) T_{\nu+1} \left(\frac{\sqrt{\nu+1}}{\sqrt{1-\rho^2}} \left[\left\{ \frac{1-w}{w} \right\}^{1/\nu} - \rho \right] \right).$$

For the conditional distributions and copula density, let $\zeta = (\nu+1)^{1/2}/(1-\rho^2)^{1/2}$, $z_{xy} = \zeta[\{x/y\}^{1/\nu} - \rho]$, and $z_{yx} = \zeta[\{x/y\}^{1/\nu} - \rho]$. Then

$$A(x, y; \rho, \nu) = x T_{\nu+1}(z_{xy}) + y T_{\nu+1}(z e_{yx}),$$

$$C_{2|1}(v|u; \rho, \nu) = u^{-1} \cdot C(u, v) \cdot \frac{\partial A(x, y; \rho, \nu)}{\partial x},$$

$$c(u, v; \rho, \nu) = (uv)^{-1} \cdot C(u, v) \cdot \left\{ \frac{\partial A(x, y; \rho, \nu)}{\partial x} \frac{\partial A(x, y; \rho, \nu; \rho, \nu)}{\partial y} - \frac{\partial^2 A(x, y; \rho, \nu)}{\partial x \partial y} \right\},$$

$$\frac{\partial A(x, y; \rho, \nu)}{\partial x} = T_{\nu+1}(z_{xy}) + t_{\nu+1}(z_{xy}) \cdot \zeta \nu^{-1}(x/y)^{1/\nu} - t_{\nu+1}(z_{yx}) \cdot \zeta \nu^{-1}(y/x)^{1/\nu+1},$$

$$\frac{\partial A(x, y; \rho, \nu)}{\partial y} = T_{\nu+1}(z_{yx}) + t_{\nu+1}(z_{yx}) \cdot \zeta \nu^{-1}(y/x)^{1/\nu} - t_{\nu+1}(z_{xy}) \cdot \zeta \nu^{-1}(x/y)^{1/\nu+1},$$

$$\frac{-\partial^2 A(x, y; \rho, \nu)}{\partial x \partial y} = t_{\nu+1}(z_{xy}) \cdot \zeta(\nu+1) \nu^{-2} y^{-1}(x/y)^{1/\nu} + t'_{\nu+1}(z_{xy}) \cdot \zeta^2 \nu^{-2} y^{-1}(x/y)^{2/\nu}$$

$$+ t_{\nu+1}(z_{yx}) \cdot \zeta(\nu+1) \nu^{-2} x^{-1}(y/x)^{1/\nu} + t'_{\nu+1}(z_{yx}) \cdot \zeta^2 \nu^{-2} x^{-1}(y/x)^{2/\nu},$$

where $t'_{\nu+1}(z) = -t_{\nu+1}(z) \cdot (\nu+2)(\nu+1)^{-1} z(1 + z^2/[\nu+1])^{-1}$ is the derivative of the $t_{\nu+1}$ density.

Some properties of the bivariate t-EV copula (4.52) are the following.

- The upper tail dependence parameter is $\lambda_U = 2T_{1,\nu+1}(-\sqrt{(\nu+1)(1-\rho)/(1+\rho)}\,)$, the same as the upper tail dependence parameter if the bivariate $t_\nu(\rho)$ copula. The lower tail order is $\kappa_L = 2 - \lambda_U$.
- The derivative of B is

$$B'(w;\rho,\nu) = T_{\nu+1}\Big(\frac{\sqrt{\nu+1}}{\sqrt{1-\rho^2}}\Big[\Big\{\frac{w}{1-w}\Big\}^{1/\nu} - \rho\Big]\Big)$$

$$+w\,t_{\nu+1}\Big(\frac{\sqrt{\nu+1}}{\sqrt{1-\rho^2}}\Big[\Big\{\frac{w}{1-w}\Big\}^{1/\nu} - \rho\Big]\Big)\cdot\nu^{-1}\Big(\frac{w}{1-w}\Big)^{1/\nu}[w(1-w)]^{-1}$$

$$-T_{\nu+1}\Big(\frac{\sqrt{\nu+1}}{\sqrt{1-\rho^2}}\Big[\Big\{\frac{1-w}{w}\Big\}^{1/\nu} - \rho\Big]\Big)$$

$$-(1-w)\,t_{\nu+1}\Big(\frac{\sqrt{\nu+1}}{\sqrt{1-\rho^2}}\Big[\Big\{\frac{1-w}{w}\Big\}^{1/\nu} - \rho\Big]\Big)\cdot\nu^{-1}\Big(\frac{1-w}{w}\Big)^{1/\nu}[w(1-w)]^{-1}.$$

As $w \to 0^+$, the first term becomes $T_{\nu+1}(-\rho\sqrt{(\nu+1)/(1-\rho^2)}\,)$, the third term becomes $-T_{\nu+1}(\infty) = -1$, the second term becomes 0, and the fourth term behaves like $w^{1/\nu}$. Hence $B'(0;\rho,\nu) > -1$ and similarly, $B'(1;\rho,\nu) < 1$ for $-1 < \rho < 1$.

4.17 Copulas based on the gamma stopped positive stable LT

The bivariate copula family presented in this section is a nice family to have for asymmetric lower and upper tail dependence, with λ_L, λ_U algebraically independent in the interval $(0,1)$. There are other 2-parameter families with this property but this is one for which the concordance ordering has been established for both parameters.

Because of its nice properties, more details on the properties are given for this family than other 2-parameter copula families constructed from 2-parameter LTs or mixture of max-id.

4.17.1   Bivariate BB1: Joe and Hu, 1996

In (4.1), let K be the Gumbel family and let ψ be the gamma LT family. Then the resulting two-parameter copula family of the form (4.2) is

$$C(u,v;\theta,\delta) = \Big\{1 + \big[(u^{-\theta}-1)^\delta + (v^{-\theta}-1)^\delta\big]^{1/\delta}\Big\}^{-1/\theta} \tag{4.53}$$

$$= \varphi(\varphi^{-1}(u) + \varphi^{-1}(v)), \quad 0 \le u,v \le 1,\ \theta > 0,\ \delta \ge 1,$$

where $\varphi(s) = \varphi(s;\theta,\delta) = (1 + s^{1/\delta})^{-1/\theta}$ (Mittag-Leffler LT or gamma stopped positive stable family in the Appendix) and $\varphi^{-1}(t;\theta,\delta) = (t^{-\theta}-1)^\delta$ for $0 \le t \le 1$.

Let $x = (u^{-\theta}-1)^\delta$ and $y = (v^{-\theta}-1)^\delta$; these are monotone decreasing transforms taking 0 to ∞ and 1 to 0. The inverse transforms are $u = (1 + x^{1/\delta})^{-1/\theta}$ and $v = (1 + y^{1/\delta})^{-1/\theta}$. Then $C(u,v;\theta,\delta) = \overline{G}(x,y;\theta,\delta) = [1 + (x+y)^{1/\delta}]^{-1/\theta}$, where $\overline{G}(x,y;\theta,\delta)$ is a bivariate survival function with univariate margins $\overline{G}_1 = \overline{G}_2$ and $\overline{G}_1(x) = [1 + x^{1/\delta}]^{-1/\theta}$, a Burr survival function on $[0,\infty)$, $g_1(x) = -\partial\overline{G}_1(x)/\partial x = (\delta\theta)^{-1}(1 + x^{1/\delta})^{-1/\theta-1}x^{1/\delta-1}$. The conditional cdf and copula density are:

$$C_{2|1}(v|u;\theta,\delta) = \frac{\partial\overline{G}}{\partial x}\cdot\frac{\partial x}{\partial u}, \qquad\qquad \frac{\partial x}{\partial u} = -\delta\theta(u^{-\theta}-1)^{\delta-1}u^{-\theta-1},$$

$$c(u,v;\theta,\delta) = \frac{\partial^2\overline{G}}{\partial x\partial y}\cdot\frac{\partial x}{\partial u}\cdot\frac{\partial y}{\partial v}.$$

For the inverse of the conditional cdf, solve for y or equivalently v in the equation:

$$C_{2|1}(v|u;\theta,\delta) = \overline{G}_{2|1}(y|x;\theta,\delta) = p, \quad \overline{G}_{2|1}(y|x;\theta,\delta) = \frac{-\partial \overline{G}(x,y;\theta,\delta)/\partial x}{g_1(x)}. \qquad (4.54)$$

Furthermore,

$$\overline{G}_{2|1}(y|x;\theta,\delta) = \frac{\frac{1}{\delta\theta}[1+(x+y)^{1/\delta}]^{-1/\theta-1}(x+y)^{1/\delta-1}}{\frac{1}{\delta\theta}(1+x^{1/\delta})^{-1/\theta-1}x^{1/\delta-1}},$$

$$C_{2|1}(v|u;\theta,\delta) = \{1+(x+y)^{1/\delta}\}^{-1/\theta-1}(x+y)^{1/\delta-1} \cdot x^{1-1/\delta}u^{-\theta-1},$$

$$c(u,v;\theta,\delta) = \{1+(x+y)^{1/\delta}\}^{-1/\theta-2}(x+y)^{1/\delta-2}[\theta(\delta-1)+(\theta\delta+1)(x+y)^{1/\delta}]$$
$$\cdot(xy)^{1-1/\delta}(uv)^{-\theta-1}.$$

For a starting value y_0 to solve (4.54) via Newton-Raphson iterations, a possible choice (based on large x, y) is from

$$[(x+y_0)^{1/\delta}]^{-1/\theta-1}(x+y_0)^{1/\delta-1} = p\delta\theta g_1(x)$$

or

$$y_0 = [p\delta\theta g_1(x)]^{-1/(1+1/[\delta\theta])} - x.$$

As $x \to 0^+$, say $x < 0.00001$, then $y \to 0^+$, and $y_0 \sim \min\{x[p^{-\delta/(\delta-1)}-1], 0.00001\}$ can be used as a starting point. As $x \to \infty$, say $x > 10000$, then the solution satisfies $y = rx \to \infty$, and $y_0 = rx$ with $r = p^{-\delta\theta/(1+\delta\theta)} - 1$ can be used as a starting point.

Some properties of the family of copulas (4.53) are the following.

- The MTCJ family is a subfamily when $\delta = 1$, and the Gumbel family obtains as $\theta \to 0^+$.
- $C^\perp$ obtains as $\theta \to 0^+$ and $\delta \to 1^+$, and C^+ obtains as $\theta \to \infty$ or $\delta \to \infty$.
- Concordance increases as δ increases because the Gumbel family is increasing in $\prec_c$.
- Concordance increases as θ increases by Theorem 8.22 because $\omega''(s) \le 0$, where $\omega(s) = \varphi^{-1}(\varphi(s;\theta_2,\delta);\theta_1,\delta) = [(1+s^{1/\delta})^\gamma - 1]^\delta$, with $\theta_1 < \theta_2$ and $\gamma = \theta_1/\theta_2 < 1$.
- The lower and upper tail dependence parameters are respectively $\lambda_L = 2^{-1/(\delta\theta)}$ and $\lambda_U = 2 - 2^{1/\delta}$. The range of (λ_L, λ_U) is $(0,1)^2$,
- Given tail parameters λ_U, λ_L:

$$\delta = [\log 2]/\log(2-\lambda_U), \quad \theta = [\log 2]/(-\delta\log\lambda_L) = \log(2-\lambda_U)/(-\log\lambda_L). \qquad (4.55)$$

- The extreme value limits from the lower and upper tails are the Galambos and Gumbel families, respectively.
- Kendall's $\tau = 1 - 2/[\delta(\theta+2)]$, by using $\varphi'(s;\theta,\delta) = -(\theta\delta)^{-1}(1+s^{1/\delta})^{-1/\theta-1}s^{1/\delta-1}$ in Theorem 8.26 with change of variables $w = s^{1/\delta}$ and $v = w/(1+w)$. This was originally derived in Schepsmeier (2010).
- Blomqvist's $\beta = 4\beta^* - 1$, where

$$\beta^* = \{1+2^{1/\delta}(2^\theta-1)\}^{-1/\theta} = \{1+(2-\lambda_U)(2^\theta-1)\}^{-1/\theta}. \qquad (4.56)$$

- Range of (β, λ_U): Given $\lambda_U \in (0,1)$ and the increasing in concordance in θ, the minimum and maximum values of β come from $4\beta^* - 1$ with $\theta \to 0^+$ and $\theta \to \infty$. This leads to the range $[2^{\lambda_U} - 1, 1]$ given λ_U.
  $\theta \to 0^+$: $\beta^* \sim \{1+(2-\lambda_U)\theta\log 2\}^{-1/\theta} \sim 2^{\lambda_U-2}$.
  $\theta \to \infty$: $\beta^* \sim \{1+(2-\lambda_U)2^\theta\}^{-1/\theta} \sim \frac{1}{2}$.

- Range of (β, λ_L): Given $\lambda_L \in (0, 1)$, the minimum and maximum values of β come from $4\beta^* - 1$ with $\delta \to 1^+$ and $\delta \to \infty$. With $\beta^* = \{1 + 2^{1/\delta}(2^{\log 2/(-\delta \log \lambda_L)} - 1)\}^{\delta \log \lambda_L/\log 2}$, this leads to the range $[\beta_{\min}(\lambda_L), 1]$ given λ_L, where

$$\beta_{\min}(\lambda) = 4\{2^{1-\log 2/\log \lambda} - 1\}^{\log \lambda/\log 2} - 1 \leq 2^{\lambda} - 1.$$

$\delta \to \infty$: $\beta^* \sim \{1 + \delta^{-1}\log 2^{-\log 2/\log \lambda_L}\}^{\delta \log \lambda_L/\log 2} \sim \frac{1}{2}$.

- Boundary conditional cdfs: $C_{2|1}(\cdot|0; \theta, \delta)$ is degenerate at 0, $C_{2|1}(\cdot|1; \theta, \delta)$ is degenerate at 1.

4.17.2 BB1: range of pairs of dependence measures

We consider range of two of out $(\beta, \lambda_L, \lambda_U)$ as given in (4.56) and (4.55). This can be compared with other bivariate copula families where λ_L, λ_U can independently take values in $(0, 1)$.

For below, let $\gamma = \beta^* = (\beta + 1)/4$ be in $(\frac{1}{4}, \frac{1}{2})$.

1. The range of (λ_L, λ_U) is $(0, 1)^2$.
2. The range of (β, λ_U) is $\lambda_U \leq 2 - (\log \gamma^{-1})/(\log 2)$.
   Proof: $\gamma = \{1 + (2 - \lambda_U)(2^{\theta} - 1)\}^{-1/\theta}$ means that if the (β, λ_U) combination is possible, then

$$\gamma^{-\theta} - 1 = (2 - \lambda_U)(2^{\theta} - 1)$$

or one can solve the following equation in $\theta \geq 0$:

$$h_U(\theta) = \log(2 - \lambda_U) + \log(2^{\theta} - 1) - \log(\gamma^{-\theta} - 1) = 0.$$

It can be shown with l'Hopital's rule that $h_U(0) = \log(2 - \lambda_U) + \log[(\log 2)/(\log \gamma^{-1})]$ and $h_U(\infty) = -\infty$, since

$$\lim_{\theta}(2^{\theta} - 1)/(\gamma^{-\theta} - 1) = \lim_{\theta}(2^{\theta}\log 2)/(\gamma^{-\theta}\log \gamma^{-1})$$

and this is $(\log 2)/(\log \gamma^{-1})$ for $\theta \to 0^+$ and 0 for $\theta \to \infty$. Also

$$h'_U(\theta) = \frac{2^{\theta}\log 2}{2^{\theta} - 1} - \frac{\gamma^{-\theta}\log \gamma^{-1}}{\gamma^{-\theta} - 1},$$

$$\theta h'_U(\theta) = (z_1 \log z_1)/(z_1 - 1) - (z_2 \log z_2)/(z_2 - 1), \quad z_1 = 2^{\theta} < z_2 = \gamma^{-\theta},$$

$$k(z) := (z \log z)/(z - 1) \uparrow z \geq 1, \quad k'(z) = (z - 1)^{-2}[z - 1 - \log z].$$

Hence $h'_U < 0$ and there is a root of h_U only if $h_U(0) \geq 0$, or if $(\log 2)(2 - \lambda_U)/(\log \gamma^{-1}) \geq 1$ or $\lambda_U \leq 2 - (\log \gamma^{-1})/(\log 2)$.

3. The range of (β, λ_L) is $\lambda_L \leq 2\gamma = \frac{1}{2}(1 + \beta)$.
   Proof: $\gamma = \{1 + 2^{1/\delta}(2^{\theta} - 1)\}^{-1/\theta}$ and $\lambda_L = 2^{-1/(\delta\theta)}$ means that if the (β, λ_L) combination is possible, then

$$\gamma^{-\theta} - 1 = \lambda_L^{-\theta}(2^{\theta} - 1)$$

or one can solve the following equation in $\theta \geq 0$:

$$h_L(\theta) = -\theta \log \lambda_L + \log(2^{\theta} - 1) - \log(\gamma^{-\theta} - 1) = 0.$$

From l'Hopital's rule, as above, $h_L(0) = \log[(\log 2)/(\log \gamma^{-1})] < 0$, and for large θ, $h_L(\theta) \approx \theta\{-\log \lambda_L + \log 2 + \log \gamma\}$. Hence $h_L(\infty) \geq 0$ if $\lambda_L \leq 2\gamma = \frac{1}{2}(1 + \beta)$, in which case h_L has a root.

4. Range of (β, λ_L) versus range of (β, λ_U). As a comparison,

$$2\gamma \geq 2 - (\log \gamma^{-1})/(\log 2), \quad \gamma \in [\frac{1}{4}, \frac{1}{2}].$$

That is, there is more range in (β, λ_L) than (β, λ_U). The upper bound of $2\gamma = \frac{1}{2}(\beta + 1)$ is the same as the bound from the $t_{\nu}(\rho)$ copula with a given Blomqvist's β that is positive.

4.17.3   Multivariate extensions of BB1

The Mittag-Leffler LT can be used to get a multivariate Archimedean copula. Hierarchical Archimedean copulas with Mittag-Leffler LTs can be defined if θ is fixed; see Example 3.1. Also this LT can be used for the mixture of max-id copula construction for more flexible dependence. These properties also hold for the extensions of BB2, BB3, BB5, BB6, BB7 below, as they are all Archimedean copulas based on 2-parameter LTs of the form $\varphi(s; \theta, \delta) = \psi(-\log \zeta(s; \delta); \theta)$.

Origins and further notes

- The BB1 copula as well as BB2–BB7 below come from the mixture of max-id construction of Joe and Hu (1996).
- In order to get asymmetric tail dependence for all pairs of variables, the BB1 copula is a good choice of a bivariate copula for the first tree of a vine pair-copula construction or for the bivariate linking copulas to the first latent variable in factor copulas.

4.18   Copulas based on the gamma stopped gamma LT

Details and properties are mainly given for the bivariate copula based on the 2-parameter LT family.

4.18.1   Bivariate BB2: Joe and Hu, 1996

In (4.1), let K be the MTCJ family and let ψ be the gamma LT family. Then the resulting two-parameter copula family of the form (4.2) is

$$C(u, v; \theta, \delta) = \left[1 + \delta^{-1} \log\left(e^{\delta(u^{-\theta}-1)} + e^{\delta(v^{-\theta}-1)} - 1\right)\right]^{-1/\theta} \tag{4.57}$$

$$= \varphi(\varphi^{-1}(u) + \varphi^{-1}(v)), \quad 0 \le u, v \le 1, \ \theta, \delta > 0,$$

where $\varphi(s) = \varphi(s; \theta, \delta) = [1 + \delta^{-1} \log(1 + s)]^{-1/\theta}$ (gamma stopped gamma LT in the Appendix) and $\varphi^{-1}(t; \theta, \delta) = \exp\{\delta(t^{-\theta} - 1)\} - 1$ for $0 \le t \le 1$.

Let $x = \exp\{\delta(u^{-\theta} - 1)\} - 1$ and $y = \exp\{\delta(v^{-\theta} - 1)\} - 1$; these are monotone decreasing transforms taking 0 to ∞ and 1 to 0. The inverse transforms are $u = [1 + \delta^{-1} \log(x + 1)]^{-1/\theta}$ and $v = [1 + \delta^{-1} \log(y+1)]^{-1/\theta}$. Then $C(u, v; \theta, \delta) = \overline{G}(x, y; \theta, \delta) = [1 + \delta^{-1} \log(x + y + 1)]^{-1/\theta}$, where $\overline{G}(x, y; \theta, \delta)$ is a bivariate survival function with univariate margins $\overline{G}_1 = \overline{G}_2$ and $\overline{G}_1(x) = [1 + \delta^{-1} \log(x + 1)]^{-1/\theta}$, a survival function on $[0, \infty)$, $g_1(x) = -\partial \overline{G}_1(x)/\partial x = (\delta\theta)^{-1}[1 + \delta^{-1} \log(x + 1)]^{-1/\theta-1}(x + 1)^{-1}$. The conditional cdf and copula density are:

$$C_{2|1}(v|u; \theta, \delta) = \frac{\partial \overline{G}}{\partial x} \cdot \frac{\partial x}{\partial u}, \qquad\qquad \frac{\partial x}{\partial u} = -\delta\theta(x + 1)u^{-\theta-1},$$

$$c(u, v; \theta, \delta) = \frac{\partial^2 \overline{G}}{\partial x \partial y} \cdot \frac{\partial x}{\partial u} \cdot \frac{\partial y}{\partial v}.$$

For the inverse of the conditional cdf, solve for y or equivalently v in the equation:

$$C_{2|1}(v|u; \theta, \delta) = \overline{G}_{2|1}(y|x; \theta, \delta) = p, \quad \overline{G}_{2|1}(y|x; \theta, \delta) = \frac{-\partial \overline{G}(x, y; \theta, \delta)/\partial x}{g_1(x)}. \tag{4.58}$$

Furthermore,

$$\overline{G}_{2|1}(y|x;\theta,\delta) = \frac{\frac{1}{\delta\theta}[1+\delta^{-1}\log(x+y+1)]^{-1/\theta-1}(x+y-1)^{-1}}{\frac{1}{\delta\theta}[1+\delta^{-1}\log(x+1)]^{-1/\theta-1}(x+1)^{-1}},$$

$$C_{2|1}(v|u;\theta,\delta) = \{1+\delta^{-1}\log(x+y+1)\}^{-1/\theta-1}(x+y+1)^{-1}\cdot(x+1)u^{-\theta-1},$$

$$c(u,v;\theta,\delta) = \{1+\delta^{-1}\log(x+y+1)\}^{-1/\theta-2}(x+y+1)^{-2}[1+\theta+\theta\delta+\theta\log(x+y+1)]$$
$$\cdot(x+1)(y+1)(uv)^{-\theta-1}.$$

For a starting value y_0 to solve (4.58) via Newton-Raphson iterations, a possible choice is to set $y = x$ in the nondominant term and get an equation for y_0:

$$[1+\delta^{-1}\log(2x+1)]^{-1/\theta-1}(x+y+1)^{-1} = p\delta\theta g_1(x) = p[1+\delta^{-1}\log(x+1)]^{-1/\theta-1}(x+1)^{-1}$$

or

$$y_0 = (x+1)p^{-1}\left\{\frac{1+\delta^{-1}\log(x+1)}{1+\delta^{-1}\log(2x+1)}\right\}^{1+1/\theta} - 1 + x,$$

provided the above is > 0. If x evaluates numerically as ∞, then suppose $y = xr$ and solve the equation $h(r) = 1 + \log(1+r)/[\delta(u^{-\theta}-1)] - [p(1+r)]^{-\theta/(1+\theta)} = 0$ for $r > 0$, and set $v = [u^{-\theta}+\delta^{-1}\log r]^{-1/\theta}$.

Similarly, if x or y evaluates as ∞ in C, $C_{2|1}$ and c, then let $\log r = \delta(v^{-\theta}-u^{-\theta})$. Then $\log(x+y+1) \sim \log(1+r) + \delta(u^{-\theta}-1)$, and the various functions become:

$$C(u,v;\theta,\delta) \sim \{u^{-\theta}+\delta^{-1}\log(1+r)\}^{-1/\theta},$$

$$C_{2|1}(v|u;\theta,\delta) \sim \{u^{-\theta}+\delta^{-1}\log(1+r)\}^{-1/\theta-1}(1+r)^{-1}u^{-\theta-1},$$

$$c(u,v;\theta,\delta) \sim \{u^{-\theta}+\delta^{-1}\log(1+r)\}^{-1/\theta-2}[1+\theta+\theta\log(1+r)+\theta\delta u^{-\theta}]\cdot\frac{r^{-1}(1+r)^{-2}}{(uv)^{\theta+1}}.$$

Some properties of the family of copulas (4.57) are the following.

- $C^{\perp}$ obtains as $\theta \to 0^+$, C^+ obtains as $\theta \to \infty$ or $\delta \to \infty$.

- The limit as $\delta \to 0^+$ is the MTCJ family.

- Concordance increases as δ increases because the MTCJ family is increasing in $\prec_c$.

- Concordance increases as θ increases by Theorem 8.22 because $\omega'' \le 0$, where $\omega(s) = \varphi^{-1}(\varphi(s;\theta_2,\delta);\theta_1,\delta) = \exp\{\delta([1+\delta^{-1}\log(1+s)]^{\gamma}-1)\}-1$, with $\theta_1 < \theta_2$ and $\gamma = \theta_1/\theta_2 < 1$.

- The lower tail dependence parameter is $\lambda_L = 1$, that is, this family is lower tail comonotonic. This property causes the numerical problems in evaluations referred to above.

- The upper tail order is $\kappa_U = 2$.

- In terms of tail order and dependence, this family is the most tail asymmetric possible with positive dependence.

- Boundary conditional cdfs: $C_{2|1}(\cdot|0;\theta,\delta)$ is degenerate at 0; with $y = \exp\{\delta(v^{-\theta}-1)\}-1$, $C_{2|1}(v|1;\theta,\delta) = \{1+\delta^{-1}\log(y+1)\}^{-1/\theta-1}(y+1)^{-1}$.

Further notes

Hua and Joe (2012b) use the (reflected) BB2 copula as a model to get conservative estimates of VaR and CTE for losses. To get conservative estimates in a multivariate setting, one can use the LT $\varphi(s;\theta,\delta)$ in a mixture of max-id copula. Then resulting multivariate copula has lower tail order function $b(\boldsymbol{w}) = \min\{w_1,\ldots,w_d\}$. The same comments also apply to the BB3 copula (in Section 4.19) and its LT.

An example of a multivariate application for conservative estimates is in Section 7.9.

4.19 Copulas based on the positive stable stopped gamma LT

Details and properties are mainly given for the bivariate copula based on the 2-parameter LT family.

4.19.1 Bivariate BB3: Joe and Hu, 1996

In (4.1), let K be the MTCJ family and let ψ be the positive stable LT family. Then the resulting two-parameter copula family of the form (4.2) is

$$C(u,v;\theta,\delta) = \exp\{-[\delta^{-1}\log(e^{\delta\tilde{u}^\theta} + e^{\delta\tilde{v}^\theta} - 1)]^{1/\theta}\} \tag{4.59}$$
$$= \varphi(\varphi^{-1}(u) + \varphi^{-1}(v)), \quad 0 \le u, v \le 1, \ \theta \ge 1, \delta > 0,$$

where $\tilde{u} = -\log u$, $\tilde{v} = -\log v$, $\varphi(s) = \varphi(s;\theta,\delta) = \exp\{-[\delta^{-1}\log(1+s)]^{1/\theta}\}$ (positive stable stopped gamma LT in the Appendix), and $\varphi^{-1}(t;\theta,\delta) = \exp\{\delta(-\log t)^\theta\} - 1\}$ for $0 \le t \le 1$.

Let $x = \exp\{\delta(-\log u)^\theta\} - 1$ and $y = \exp\{\delta(-\log v)^\theta\} - 1$; these are monotone decreasing taking 0 to ∞ and 1 to 0. The inverse transforms are $u = \exp\{[\delta^{-1}\log(x+1)]^{1/\theta}\}$ and $v = \exp\{[\delta^{-1}\log(y+1)]^{1/\theta}\}$. Then $C(u,v;\theta,\delta) = \overline{G}(x,y;\theta,\delta) = \exp\{-[\delta^{-1}\log(x+y+1)]^{1/\theta}\}$, where $\overline{G}(x,y;\theta,\delta)$ is a bivariate survival function with univariate margins $\overline{G}_1 = \overline{G}_2$ and $\overline{G}_1(x) = \exp\{-[\delta^{-1}\log(x+1)]^{1/\theta}\}$, a survival function on $[0,\infty)$, $g_1(x) = -\partial\overline{G}_1(x)/\partial x = \overline{G}_1(x) \cdot (\delta^{1/\theta}\theta)^{-1}[\log(x+1)]^{1/\theta-1}(x+1)^{-1}$. The conditional cdf and copula density are:

$$C_{2|1}(v|u;\theta,\delta) = \frac{\partial\overline{G}}{\partial x} \cdot \frac{\partial x}{\partial u}, \qquad \frac{\partial x}{\partial u} = -\delta\theta(x+1)(-\log u)^{\theta-1}/u,$$

$$c(u,v;\theta,\delta) = \frac{\partial^2\overline{G}}{\partial x\partial y} \cdot \frac{\partial x}{\partial u} \cdot \frac{\partial y}{\partial v}.$$

For the inverse of the conditional cdf, solve for y or equivalently v in the equation:

$$C_{2|1}(v|u;\theta,\delta) = \overline{G}_{2|1}(y|x;\theta,\delta) = p, \quad \overline{G}_{2|1}(y|x;\theta,\delta) = \frac{-\partial\overline{G}(x,y;\theta,\delta)/\partial x}{g_1(x)}. \tag{4.60}$$

Furthermore,

$$\overline{G}_{2|1}(y|x;\theta,\delta) = \frac{\overline{G}(x,y;\theta,\delta) \cdot \frac{1}{\delta^{1/\theta}\theta}[\log(x+y+1)]^{1/\theta-1}(x+y+1)^{-1}}{\overline{G}_1(x) \cdot \frac{1}{\delta^{1/\theta}\theta}[\log(x+1)]^{1/\theta-1}(x+1)^{-1}},$$

$$C_{2|1}(v|u;\theta,\delta) = \overline{G} \cdot [\log(x+y+1)]^{1/\theta-1}(x+y+1)^{-1} \cdot \delta^{1-1/\theta}(x+1)(-\log u)^{\theta-1}/u,$$

$$c(u,v;\theta,\delta) = \overline{G} \cdot [\log(x+y+1)]^{1/\theta-2}(x+y+1)^{-2}\{(\theta-1)\delta^{1/\theta} + \theta\delta^{1/\theta}\log(x+y+1)$$
$$+ [\log(x+y+1)]^{1/\theta}\} \cdot \delta^{2-2/\theta}(x+1)(y+1)(-\log u)^{\theta-1}(-\log v)^{\theta-1}/(uv).$$

For the solution to (4.60), Newton-Raphson iterations can proceed with

$$h(y) = \log(x+y+1) + (1-\theta^{-1})\log[\log(x+y+1)] + \delta^{-1/\theta}[\log(x+y+1)]^{1/\theta}$$
$$+ \log(p) + (\theta^{-1}-1)\log[\log(x+1)] - \delta^{-1/\theta}[\log(x+1)]^{1/\theta} - \log(x+1) = 0$$

and

$$h'(y) = \left\{1 + (1-\theta^{-1})[\log(x+y+1)]^{-1} + \delta^{-1/\theta}\theta^{-1}[\log(x+y+1)]^{1/\theta-1}\right\}(x+y+1)^{-1}.$$

If u is small enough (below e^{-1} with θ large), x evaluates numerically as ∞, but $x_l = \log(x) = \delta(-\log u)^\theta$ is finite. Then with $y = xr$ as $x \to \infty$, solve

$$h_\infty(r) = \log(r+1) + (1-\theta^{-1})\log[x_l + \log(r+1)] + \delta^{-1/\theta}[x_l + \log(r+1)]^{1/\theta}$$

$$+ \log(p) + (\theta^{-1} - 1) \log x_l - \delta^{-1/\theta} x_l^{1/\theta} = 0$$

for $r > 0$. The derivative is:

$$h'_\infty(r) = \left\{ 1 + (1 - \theta^{-1})[x_l + \log(r+1)]^{-1} + \delta^{-1/\theta}\theta^{-1}[x_l + \log(r+1)]^{1/\theta - 1} \right\}(r+1)^{-1}.$$

For the solution, set $v = \exp\{-\delta^{-1/\theta}[x_l + \log r]^{1/\theta}\}$.

If u is large enough (above e^{-1} with θ large), then x can evaluate numerically as 0 or very close to 0. In this case, $x \approx \delta(-\log u)^\theta$, and the solution $y \approx \delta(-\log v)^\theta$ will be small. With $y/x \to r$ as $x \to 0^+$, write $\log(1 + x + y) \approx x(1+r)$ and $\log(1+x) \approx x$. Change to solving the following

$$h_0(r) = xr + (1 - \theta^{-1})\log(r+1) + \delta^{-1/\theta}x^{1/\theta}(r+1)^{1/\theta} + \log(p) - \delta^{-1/\theta}x^{1/\theta} = 0$$

for $r > 0$. The derivative is:

$$h'_0(r) = x + (1 - \theta^{-1})(r+1)^{-1} + \delta^{-1/\theta}\theta^{-1}x^{1/\theta}(r+1)^{1/\theta - 1}.$$

For the solution, set $v = \exp\{-(rx/\delta)^{1/\theta}\}$.

Also if x or y numerically evaluates as ∞, then take

$$C(u, v; \theta, \delta) \sim \exp\{-[\delta^{-1}\log(1+r) + (-\log u)^\theta]^{1/\theta}\}, \quad r = \exp\{\delta[(-\log v)^\theta - (-\log u)^\theta]\},$$
$$C_{2|1}(v|u; \theta, \delta) \sim C \cdot (\log r)[\log(1+r)]^{-1}\delta(-\log u)^{\theta-1}u^{-1}.$$

If x or y numerically evaluates as 0 or very close to 0,

$$C(u, v; \theta, \delta) \sim \exp\{-[(-\log u)^\theta + (-\log v)^\theta]^{1/\theta}\}, \quad r = (-\log v)^\theta/(-\log u)^\theta,$$
$$C_{2|1}(v|u; \theta, \delta) \sim C \cdot (1+r)^{1/\theta - 1}[1 + \delta(-\log u)^\theta][1 + \delta(-\log u)^\theta + \delta(-\log v)^\theta]^{-1}u^{-1}.$$

Some properties of the family of copulas (4.59) are the following.

- The MTCJ family is a subfamily when $\theta = 1$, and the Gumbel family obtains as $\delta \to 0^+$.
- C^+ obtains as $\theta \to \infty$ or $\delta \to \infty$.
- The lower tail dependence parameter is $\lambda_L = 2^{-1/\delta}$ when $\theta = 1$ and $\lambda_L = 1$ when $\theta > 1$, so this family is tail comonotonic if $\theta > 1$. This tail comonotonicity property causes the numerical problems in evaluations referred to above.
- The upper tail dependence parameter is $\lambda_U = 2 - 2^{1/\theta}$, independent of δ. The upper extreme value limit is the Gumbel copula family.
- Concordance increases as δ increases because the MTCJ family is increasing in $\prec_c$.
- Concordance increases as θ increases for $0 < \delta \le e^2$ by Theorem 8.22 because $\omega'' \le 0$, where $\omega(s) = \varphi^{-1}(\varphi(s; \theta_2, \delta); \theta_1, \delta) = \exp\{\delta^{1-\gamma}[\log(1+s)]^\gamma\} - 1$ with $\theta_1 < \theta_2$ and $\gamma = \theta_1/\theta_2 < 1$. The concordance ordering in θ can fail for $\delta > e^2$. With a change of parametrization to (θ, α) with $\alpha = \delta^{1/\theta}$, the family of copulas has been shown to be increasing in concordance with both parameters θ and α.
- Boundary conditional cdfs: $C_{2|1}(\cdot|0; \theta, \delta)$ is degenerate at 0, $C_{2|1}(\cdot|1; \theta, \delta)$ is degenerate at 1.

4.20 Gamma power mixture of Galambos

This is a family that comes from a power mixture of the Galambos copulas. Details are given for the bivariate case, followed by multivariate extensions.

4.20.1   Bivariate BB4: Joe and Hu, 1996

In (4.1), let K be the Galambos family and let ψ be the gamma LT family. Then the resulting two-parameter copula family is

$$C(u, v; \theta, \delta) = \left(u^{-\theta} + v^{-\theta} - 1 - \left[(u^{-\theta} - 1)^{-\delta} + (v^{-\theta} - 1)^{-\delta}\right]^{-1/\delta}\right)^{-1/\theta}, \quad \theta \geq 0, \delta > 0, \quad (4.61)$$

for $0 \leq u, v \leq 1$.

Let $x = (u^{-\theta} - 1)^{-\delta}$ and $y = (v^{-\theta} - 1)^{-\delta}$; these are monotone increasing transforms from $[0, 1)$ to $[0, \infty)$. The inverse transforms are $u = (1 + x^{-1/\delta})^{-1/\theta}$ and $v = (1 + y^{-1/\delta})^{-1/\theta}$. Then $C(u, v; \theta, \delta) = G(x, y; \theta, \delta) = [1 + x^{-1/\delta} + y^{-1/\delta} - (x + y)^{1/\delta}]^{-1/\theta}$, where $G(x, y; \theta, \delta)$ is a bivariate cdf with univariate margins $G_1 = G_2$ and $G_1(x) = [1 + x^{-1/\delta}]^{-1/\theta}$, $g_1(x) = \partial G_1(x)/\partial x = (\delta\theta)^{-1}(1 + x^{-1/\delta})^{-1/\theta-1}x^{-1/\delta-1}$. The conditional cdf and copula density are:

$$C_{2|1}(v|u; \theta, \delta) = \frac{\partial G}{\partial x} \cdot \frac{\partial x}{\partial u}, \qquad \frac{\partial x}{\partial u} = \delta\theta(u^{-\theta} - 1)^{-\delta-1}u^{-\theta-1},$$

$$c(u, v; \theta, \delta) = \frac{\partial^2 G}{\partial x \partial y} \cdot \frac{\partial x}{\partial u} \cdot \frac{\partial y}{\partial v}.$$

For the inverse of the conditional cdf, solve for y or equivalently v in the equation:

$$C_{2|1}(v|u; \theta, \delta) = G_{2|1}(y|x; \theta, \delta) = p, \quad G_{2|1}(y|x; \theta, \delta) = \frac{\partial G(x, y; \theta, \delta)/\partial x}{g_1(x)}. \qquad (4.62)$$

Furthermore,

$$G_{2|1}(y|x; \theta, \delta) = \frac{\frac{1}{\delta\theta}[1 + x^{-1/\delta} + y^{-1/\delta} - (x + y)^{-1/\delta}]^{-1/\theta-1}\{x^{-1/\delta-1} - (x + y)^{-1/\delta-1}\}}{\frac{1}{\delta\theta}(1 + x^{-1/\delta})^{-1/\theta-1}x^{-1/\delta-1}},$$

$$C_{2|1}(v|u; \theta, \delta) = [1 + x^{-1/\delta} + y^{-1/\delta} - (x + y)^{-1/\delta}]^{-1/\theta-1}\{x^{-1/\delta-1} - (x + y)^{-1/\delta-1}\}\frac{x^{1+1/\delta}}{u^{1+\theta}},$$

$$c(u, v; \theta, \delta) = [1 + x^{-1/\delta} + y^{-1/\delta} - (x + y)^{-1/\delta}]^{-1/\theta-2} \cdot (xy)^{1+1/\delta}(uv)^{-\theta-1}$$
$$\cdot \Big[(\theta + 1)\{x^{-1/\delta-1} - (x + y)^{-1/\delta-1}\}\{y^{-1/\delta-1} - (x + y)^{-1/\delta-1}\}$$
$$+ \theta(1 + \delta)\{1 + x^{-1/\delta} + y^{-1/\delta} - (x + y)^{-1/\delta}\}(x + y)^{-1/\delta-2}\Big].$$

For $v = C_{2|1}^{-1}(p|u; \theta, \delta)$, solve (4.62) via

$$h(y) = -(\theta^{-1} + 1)\log[1 + x^{-1/\delta} + y^{-1/\delta} - (x + y)^{-1/\delta}] + \log\{x^{-1/\delta-1} - (x + y)^{-1/\delta-1}\}$$
$$+ (\delta^{-1} + 1)\log x - (1 + \theta)\log u - \log p = 0$$

to get y_0 and set $v = (1 + y_0^{-1/\delta})^{-1/\theta}$.

Because of the closeness of the Gumbel and Galambos copulas, this family is close to BB1.

Some properties of the family of copulas (4.61) are the following.

- The MTCJ family is obtained when $\delta \to 0^+$, and the Galambos family obtains as $\theta \to 0^+$.

- C^+ obtains as $\theta \to \infty$ or $\delta \to \infty$.

- The lower tail dependence parameter is $\lambda_L = (2 - 2^{-1/\delta})^{-1/\theta}$, and the upper tail dependence parameter is $\lambda_U = 2^{-1/\delta}$, independent of θ. The range of (λ_L, λ_U) is $(0, 1)^2$,

- Given tail parameters λ_U, λ_L: $\delta = [\log 2]/(-\log \lambda_U)$ and $\theta = [\log(2 - 2^{-1/\delta})]/(-\log \lambda_L) = \log(2 - \lambda_U)/(-\log \lambda_L)$.

- The lower extreme value limit leads to the min-stable bivariate exponential family $\exp\{-A(x, y; \theta, \delta)\}$, with

$$A(x, y; \theta, \delta) = x + y - [x^{-\theta} + y^{-\theta} - (x^{\theta\delta} + y^{\theta\delta})^{-1/\delta}]^{-1/\theta}; \tag{4.63}$$

the copula $C_{\text{LEV}}(u, v; \theta, \delta) = \exp\{-A(-\log u, -\log v; \theta, \delta)\}$, a two-parameter extension of the Galambos family.

- The upper extreme value limit is the Galambos family.

- Concordance increases as δ increases because the bivariate Galambos family used for K is increasing in $\prec_c$.

- Concordance increases as θ increase by showing that the $\partial \log C / \partial \theta \geq 0$: the derivative is split into the sum of two parts. For the first part, the non-negativity comes from $(u^{-\theta} + v^{-\theta} - 1 - z, z + 1)$ majorizing $(u^{-\theta}, v^{-\theta})$ where $z = [(u^{-\theta} - 1)^{-\delta} + (v^{-\theta} - 1)^{-\delta}]^{-1/\delta}$, because $z + 1 \leq \min\{u^{-\theta}, v^{-\theta}\}$. For the second part, the non-negativity $g(x)/x$ decreasing in $x \geq 0$, where $g(x) = x^{1+1/\delta}(1 + x^{-1/\delta}) \log(1 + x^{-1/\delta})$.

- Blomqvist's $\beta = 4\beta^* - 1$ and

$$\beta^* = \{2^{\theta+1} - 1 - 2^{-1/\delta}(2^\theta - 1)\}^{-1/\theta} = \{2^{\theta+1} - 1 - \lambda_U(2^\theta - 1)\}^{-1/\theta},$$

given λ_L, λ_U, this is the same β^* as in the BB1 family.

- Range of (β, λ_U): Given $\lambda_U \in (0, 1)$, the range of β is $[2^{\lambda_U} - 1, 1]$.

- Range of (β, λ_L): Given $\lambda_L \in (0, 1)$, the range of β is $[\beta_{\min}(\lambda_L), 1]$, where

$$\beta_{\min}(\lambda) = 4\{2^{1 - \log 2/\log \lambda} - 1\}^{\log \lambda/\log 2} - 1 \leq 2^\lambda - 1.$$

- Boundary conditional cdfs: $C_{2|1}(\cdot|0; \theta, \delta)$ is degenerate at 0, $C_{2|1}(\cdot|1; \theta, \delta)$ is degenerate at 1.

4.20.2 Multivariate extensions of BB4

By using the mixture of max-id construction (3.19), with ψ being a gamma LT and K being a product of bivariate Galambos copulas $K(\boldsymbol{u}) = \prod_{1 \leq i < j \leq d} H_{ij}(u_i^{1/(d-1)}, u_j^{1/(d-1)}; \delta_{ij})$, then the copula is:

$$C_{\text{MM2}}(\boldsymbol{u}; \theta, \{\delta_{ij}\}) = \left[\sum_{j=1}^d u_j^{-\theta} - (d-1) - \sum_{1 \leq i < j \leq d} (\hat{u}_i^{-\delta_{ij}} + \hat{u}_j^{-\delta_{ij}})^{-1/\delta_{ij}}\right]^{-1/\theta}, \tag{4.64}$$

$$\hat{u}_j = (u_j^{-\theta} - 1)/(d-1), \quad 0 < u_j < 1, \quad \theta > 0, \ \delta_{ij} > 0.$$

This is a multivariate copula where the lower tail dependence can vary with the bivariate margin. The lower extreme value limit copula is

$$C_{\text{MM8}}(\boldsymbol{u}; \theta, \{\delta_{ij}\}) = \exp\left\{-\left[x_1 + \cdots + x_d + \sum_{S \in \{1,\ldots,d\}, |S| \geq 2} (-1)^{|S|+1} r_S(x_i, i \in S)\right]\right\}, \tag{4.65}$$

$$r_S(x_i, i \in S) = \left[\sum_{i \in S} x_i^{-\theta} - \sum_{i < j, i \in S, j \in S} \left\{\left(\frac{x_i^{-\theta}}{d-1}\right)^{-\delta_{ij}} + \left(\frac{x_j^{-\theta}}{d-1}\right)^{-\delta_{ij}}\right\}^{-1/\delta_{ij}}\right]^{-1/\theta},$$

$$x_j = -\log u_j, \quad 0 \leq u_j \leq 1, \quad \theta > 0, \ \delta_{ij} > 0.$$

These two models are listed as MM2 and MM8 in Joe (1997). MM8 is the third example of a multivariate extreme value copula family with some flexible dependence (different tail dependence for different bivariate margins) and closed form copula cdf. The other two are (4.17) and (4.18).

4.21 Positive stable power mixture of Galambos

This is a family that comes from a power mixture of the Galambos copulas. Details are given for the bivariate case, followed by multivariate extensions.

4.21.1 Bivariate BB5: Joe and Hu, 1996

In (4.1), let K be the Galambos family and let ψ be the positive stable LT family. Then the two-parameter copula family is

$$C(u,v;\theta,\delta) = \exp\{-[x^\theta + y^\theta - (x^{-\theta\delta} + y^{-\theta\delta})^{-1/\delta}]^{1/\theta}\}, \quad 0 \le u, v \le 1, \ \theta \ge 1, \delta > 0, \ (4.66)$$

where $x = -\log u$, $y = -\log v$.

Write $C(u,v;\theta,\delta) = \overline{G}(x,y;\theta,\delta)$ in (4.66), where $\overline{G}(x,y;\theta,\delta)$ is a bivariate survival function with Exponential(1) margins: $\overline{G}_1 = \overline{G}_2$ and $\overline{G}_1(x) = e^{-x}$. Then

$$C_{2|1}(v|u;\theta,\delta) = \frac{-\partial\overline{G}}{\partial x} \cdot u^{-1},$$

$$c(u,v;\theta,\delta) = \frac{\partial^2\overline{G}}{\partial x\partial y} \cdot (uv)^{-1}.$$

For the inverse of the conditional cdf, solve for y or equivalently v in the equation:

$$C_{2|1}(v|u;\theta,\delta) = \overline{G}_{2|1}(y|x;\theta,\delta) = p, \quad \overline{G}_{2|1}(y|x;\theta,\delta) = \frac{-\partial G(x,y;\theta,\delta)/\partial x}{e^{-x}}. \quad (4.67)$$

Furthermore, with $t = x^\theta + y^\theta - (x^{-\theta\delta} + y^{-\theta\delta})^{-1/\delta}$,

$$\overline{G}_{2|1;\theta,\delta}(y|x) = \frac{\overline{G}(x,y;\theta,\delta) \cdot t^{1/\theta-1}\{x^{\theta-1} - (x^{-\theta\delta} + y^{-\theta\delta})^{-1/\delta-1}x^{-\theta\delta-1}\}}{e^{-x}},$$

$$C_{2|1}(v|u;\theta,\delta) = u^{-1}\overline{G}(x,y;\theta,\delta) \cdot t^{1/\theta-1}\{x^{\theta-1} - (x^{-\theta\delta} + y^{-\theta\delta})^{-1/\delta-1}x^{-\theta\delta-1}\},$$

$$c(u,v;\theta,\delta) = (uv)^{-1}\overline{G}(x,y;\theta,\delta) \cdot t^{1/\theta-2}\Big\{[t^{1/\theta} + (\theta-1)]z(x,y)\,z(y,x)$$

$$+ \theta(1+\delta)t(x^{-\theta\delta} + y^{-\theta\delta})^{-1/\delta-2}x^{-\theta\delta-1}y^{-\theta\delta-1}\Big\},$$

$$z(a,b) = a^{\theta-1} - (a^{-\theta\delta} + b^{-\theta\delta})^{-1/\delta-1}a^{-\theta\delta-1}.$$

For $v = C_{2|1}^{-1}(p|u;\theta,\delta)$, solve (4.67) via

$$h(y) = -t^{1/\theta} + (\theta^{-1} - 1)\log t + \log z(x,y) + x - \log p = 0$$

to get y_0 and set $v = e^{-y_0}$.

Some properties of the family of copulas (4.66) are the following.

- The Gumbel family is obtained when $\delta \to 0^+$ and the Galambos family obtains when $\theta = 1$.

- C^+ obtains as $\theta \to \infty$ or $\delta \to \infty$.

- The upper tail dependence parameter is $\lambda_U = 2 - (2 - 2^{-1/\delta})^{1/\theta}$. This is a 2-parameter extreme value copula family, with exponent $A(x,y;\theta,\delta) = [x^\theta + y^\theta - (x^{-\theta\delta} + y^{-\theta\delta})^{-1/\delta}]^{1/\theta}$. It is an extension of the Gumbel bivariate extreme value family. A similar 2-parameter extension of the Galambos family is given in (4.63).

- The lower tail order is $\kappa_L = A(1,1;\theta,\delta) = [2 - 2^{-1/\delta}]^{1/\theta}$.

- $B'(0; \theta, \delta) = -1$ and $B'(1; \theta, \delta) = 1$ from

$$B(w; \theta, \delta) = A(w, 1 - w; \theta, \delta) = [w^\theta + (1 - w)^\theta - \{w^{-\theta\delta} + (1 - w)^{-\theta\delta}\}^{-1/\delta}]^{1/\theta},$$

$$\frac{B'(w; \theta, \delta)}{[B(w; \theta, \delta)]^{1-\theta}} = w^{\theta-1} - (1 - w)^{\theta-1} + \frac{-w^{\theta\delta-1} + (1 - w)^{\theta\delta-1}}{\{w^{-\theta\delta} + (1 - w)^{-\theta\delta}\}^{1/\delta+1}}.$$

- With $\lambda = \lambda_U = 2 - (2 - 2^{-1/\delta})^{1/\theta}$, Blomqvist's $\beta = 2^\lambda - 1$. Note that $2 - \lambda = (2 - 2^{-1/\delta})^{1/\theta} \in [1, 2]$ so $(2 - \lambda)^\theta = 2 - 2^{-1/\delta} \in [1, 2]$ since $\delta > 0$. Given λ, there is a limited range for θ: $1 \le \theta \le (\log 2)/\log(2 - \lambda)$. From plots of the B function, with λ fixed, there seems to be little difference when $1 \le \theta \le (\log 2)/\log(2 - \lambda)$ and $\delta = (\log 2)/\{-\log[2 - (2 - \lambda)^\theta]\}$.

- There is not much flexibility for the extra parameter in a bivariate model, but the importance is that this model extends to a multivariate extreme value copula model via the construction in Section 3.5; an example is family MM3 in (4.18) in Section 4.8 on copulas based on the positive stable LT.

- Concordance increases as δ increases because the bivariate Galambos family used for K is increasing in $\prec_c$.

- Concordance increases as θ increases because $\partial A(x, y; \theta, \delta)/\partial\theta \le 0$ applying two majorization inequalities. The derivative with respect to θ uses A is the form $A(x, y; \theta, \delta) = [x^\theta + y^\theta - z^\theta(\theta)]^{1/\theta}$, with $z(\theta) = (x^{-\theta\delta} + y^{-\theta\delta})^{-1/(\theta\delta)}$.

- Boundary conditional cdfs: $C_{2|1}(\cdot|0; \theta, \delta)$ is degenerate at 0, $C_{2|1}(\cdot|1; \theta, \delta)$ is degenerate at 1.

4.22 Copulas based on the Sibuya stopped positive stable LT

Details and properties are mainly given for the bivariate copula based on the 2-parameter LT family.

4.22.1 Bivariate BB6: Joe and Hu, 1996

In (4.1), let K be the Gumbel family and let ψ be the Sibuya LT family. Then the two-parameter copula family of form (4.2) is

$$C(u, v; \theta, \delta) = 1 - \left(1 - \exp\left\{-[(-\log(1 - \overline{u}^\theta))^\delta + (-\log(1 - \overline{v}^\theta))^\delta]^{1/\delta}\right\}\right)^{1/\theta} \quad (4.68)$$

$$= \varphi(\varphi^{-1}(u) + \varphi^{-1}(v)), \quad 0 \le u, v \le 1, \ \theta \ge 1, \delta \ge 1,$$

with $\overline{u} = 1 - u$ and $\overline{v} = 1 - v$, where $\varphi(s) = \varphi(s; \theta, \delta) = 1 - [1 - \exp\{-s^{1/\delta}\}]^{1/\theta}$ (Sibuya stopped positive stable LT in the Appendix) and $\varphi^{-1}(t; \theta, \delta) = \{-\log[1 - (1 - t)^\theta]\}^\delta$ for $0 \le t \le 1$.

Let $x = -\log(1 - [1 - u]^\theta)$ and $y = -\log(1 - [1 - v]^\theta)$; these are monotone decreasing transforms taking 0 to ∞ and 1 to 0. The inverse transforms are $u = 1 - [1 - e^{-x}]^{1/\theta}$ and $v = 1 - [1 - e^{-y}]^{1/\theta}$. Then $C(u, v; \theta, \delta) = \overline{G}(x, y; \theta, \delta) = 1 - [1 - \exp\{-(x^\delta + y^\delta)^{1/\delta}\}]^{1/\theta}$, where $\overline{G}(x, y; \theta, \delta)$ is a bivariate survival function with univariate margins $\overline{G}_1 = \overline{G}_2$ and $\overline{G}_1(x) = 1 - [1 - e^{-x}]^{1/\theta}$, a survival function on $[0, \infty)$, $g_1(x) = -\partial\overline{G}_1(x)/\partial x = \theta^{-1}[1 - e^{-x}]^{1/\theta-1}e^{-x}$. The conditional cdf and copula density are:

$$C_{2|1}(v|u; \theta, \delta) = \frac{\partial\overline{G}}{\partial x} \cdot \frac{\partial x}{\partial u}, \qquad \frac{\partial x}{\partial u} = -\theta(1 - [1 - u]^\theta)^{-1}[1 - u]^{\theta-1},$$

$$c(u, v; \theta, \delta) = \frac{\partial^2\overline{G}}{\partial x\partial y} \cdot \frac{\partial x}{\partial u} \cdot \frac{\partial y}{\partial v}.$$

For the inverse of the conditional cdf, solve for y or equivalently v in the equation:

$$C_{2|1}(v|u;\theta,\delta) = \overline{G}_{2|1}(y|x;\theta,\delta) = p, \quad \overline{G}_{2|1}(y|x;\theta,\delta) = \frac{-\partial G(x,y;\theta,\delta)/\partial x}{g_1(x)}. \qquad (4.69)$$

Furthermore, with $w = \exp\{-(x^\delta + y^\delta)^{1/\delta}\}$,

$$\overline{G}_{2|1}(y|x;\theta,\delta) = \frac{\frac{1}{\theta}[1-w]^{1/\theta-1}w \cdot (x^\delta + y^\delta)^{1/\delta-1}x^{\delta-1}}{\frac{1}{\theta}[1-e^{-x}]^{1/\theta-1}e^{-x}},$$

$$C_{2|1}(v|u;\theta,\delta) = [1-w]^{1/\theta-1}w \cdot (x^\delta + y^\delta)^{1/\delta-1}x^{\delta-1}e^x[1-e^{-x}]^{1-1/\theta},$$

$$c(u,v;\theta,\delta) = [1-w]^{1/\theta-2}w \cdot (x^\delta + y^\delta)^{1/\delta-2}[(\theta-w)(x^\delta + y^\delta)^{1/\delta} + \theta(\delta-1)(1-w)]$$
$$\cdot (xy)^{\delta-1}(1-\overline{u}^\theta)^{-1}(1-\overline{v}^\theta)^{-1}(\overline{u}\cdot\overline{v})^{\theta-1}.$$

For $v = C_{2|1}^{-1}(p|u;\theta,\delta)$, solve (4.69) via

$$h(y) = (\theta^{-1}-1)\log(1-w) - (x^\delta + y^\delta)^{1/\delta} + (\delta^{-1}-1)\log(x^\delta + y^\delta)$$
$$+ (\delta-1)\log x - (\theta^{-1}-1)\log(1-e^{-x}) + x - \log p = 0$$

to get y_0 and set $v = 1 - (1-e^{-y_0})^{1/\theta}$.

Some properties of the family of copulas (4.68) are the following.

- The Gumbel family is obtained when $\theta = 1$, and the Joe/B5 family is obtained when $\delta = 1$. C^+ obtains as $\theta \to \infty$ or $\delta \to \infty$.

- Concordance increases as δ increases because the bivariate Gumbel family used for K is increasing in $\prec_c$.

- Concordance increases as θ increases because $\omega(s)$ is concave in $s > 0$, where $\omega(s) = \varphi^{-1}(\varphi(s;\theta_2,\delta);\theta_1,\delta) = [\sigma(s^{1/\delta})]^\delta$, $1 < \theta_1 < \theta_2$, and $\sigma(y) = -\log(1 - [1-e^{-y}]^\gamma)$, with $\gamma = \theta_1/\theta_2 < 1$.

- There is upper tail dependence with $\lambda_U = 2 - 2^{1/(\theta\delta)}$. The upper extreme value limit is the Gumbel family.

- Given $\lambda_U \in (0,1)$, then δ as a function of λ_U, θ is $\delta = \log(2)/[\theta\log(2 - \lambda_U)]$ provided this is ≥ 1. Hence the largest value of θ compatible with λ_U is $\log(2)/\log(2-\lambda_U)$.

- The lower tail order is $\kappa_L = 2^{1/\delta}$, by Theorem 8.37, because as $s \to \infty$, $\varphi(s) \sim T(s) = \theta^{-1}\exp\{-s^{1/\delta}\}$ and $\varphi'(s) = -(\theta\delta)^{-1}[1 - \exp\{-s^{1/\delta}\}]^{1/\theta-1}\exp\{-s^{1/\delta}\} \cdot s^{1/\delta-1} \sim T'(s)$.

- Blomqvist's $\beta = 4\beta^* - 1$, where

$$\beta^* = 1 - [1 - (1-2^{-\theta})^{2^{1/\delta}}]^{1/\theta} = 1 - [1 - (1-2^{-\theta})^{(2-\lambda_U)^\theta}]^{1/\theta}.$$

- Range of (β,λ_U): given $\lambda = \lambda_U \in (0,1)$, numerically β is decreasing in θ for fixed λ and the minimum and maximum values of β come from $4\beta^* - 1$ with $\theta \to \log(2)/\log(2-\lambda)$ and $\theta \to 1^+$. This leads to the range $[\beta_{\min}(\lambda), 2^\lambda - 1]$ for β, where

$$\beta_{\min}(\lambda) = 3 - 4[1 - (1-2^{-\theta^*})^{(2-\lambda)^{\theta^*}}]^{1/\theta^*}, \quad \theta^{\mathrm{w}} = \log(2)/\log(2-\lambda).$$

- Boundary conditional cdfs: $C_{2|1}(\cdot|0;\theta,\delta)$ is degenerate at 0, $C_{2|1}(\cdot|1;\theta,\delta)$ is degenerate at 1.

4.23 Copulas based on the Sibuya stopped gamma LT

Details and properties are mainly given for the bivariate copula based on the 2-parameter LT family.

4.23.1 *Bivariate BB7: Joe and Hu, 1996*

In (4.1), let K be the MTCJ family and let ψ be the Sibuya LT family. Then the resulting two-parameter copula family of the form (4.2) is

$$C(u, v; \theta, \delta) = 1 - \left(1 - \left[(1 - \bar{u}^\theta)^{-\delta} + (1 - \bar{v}^\theta)^{-\delta} - 1\right]^{-1/\delta}\right)^{1/\theta} \qquad (4.70)$$
$$= \varphi(\varphi^{-1}(u) + \varphi^{-1}(v)), \quad 0 \le u, v \le 1, \ \theta \ge 1, \ \delta > 0,$$

where $\varphi(s) = \varphi(s; \theta, \delta) = 1 - [1 - (1+s)^{-1/\delta}]^{1/\theta}$ (Sibuya stopped gamma LT in the Appendix) and $\varphi^{-1}(t; \theta, \delta) = [1 - (1 - t)^\theta]^{-\delta} - 1$ for $0 \le t \le 1$.

Let $x = (1 - [1 - u]^\theta)^{-\delta} - 1$ and $y = (1 - [1 - v]^\theta)^{-\delta} - 1$; these are monotone decreasing transforms taking 0 to ∞ and 1 to 0. The inverse transforms are $u = 1 - [1 - (x+1)^{-1/\delta}]^{1/\theta}$ and $v = 1 - [1 - (y+1)^{-1/\delta}]^{1/\theta}$. Then $C(u, v; \theta, \delta) = \overline{G}(x, y; \theta, \delta) = 1 - [1 - (x+y+1)^{-1/\delta}]^{1/\theta}$, where $\overline{G}(x, y; \theta, \delta)$ is a bivariate survival function with univariate margins $\overline{G}_1 = \overline{G}_2$ and $\overline{G}_1(x) = 1 - [1 - (x+1)^{-1/\delta}]^{1/\theta}$, a survival function on $[0, \infty)$, $g_1(x) = -\partial \overline{G}_1(x)/\partial x = (\delta\theta)^{-1}[1 - (x+1)^{-1/\delta}]^{1/\theta-1}(x+1)^{-1/\delta-1}$. The conditional cdf and copula density are:

$$C_{2|1}(v|u; \theta, \delta) = \frac{\partial \overline{G}}{\partial x} \cdot \frac{\partial x}{\partial u}, \qquad \frac{\partial x}{\partial u} = -\delta\theta(1 - [1 - u]^\theta)^{-\delta-1}[1 - u]^{\theta-1},$$

$$c(u, v; \theta, \delta) = \frac{\partial^2 \overline{G}}{\partial x \partial y} \cdot \frac{\partial x}{\partial u} \cdot \frac{\partial y}{\partial v}$$

For the inverse of the conditional cdf, solve for y or equivalently v in the equation:

$$C_{2|1}(v|u; \theta, \delta) = \overline{G}_{2|1}(y|x; \theta, \delta) = p, \quad \overline{G}_{2|1}(y|x; \theta, \delta) = \frac{-\partial G(x, y; \theta, \delta)/\partial x}{g_1(x)}. \qquad (4.71)$$

Furthermore,

$$\overline{G}_{2|1}(y|x; \theta, \delta) = \frac{\frac{1}{\delta\theta}[1 - (x+y+1)^{-1/\delta}]^{1/\theta-1}(x+y+1)^{-1/\delta-1}}{\frac{1}{\delta\theta}[1 - (x+1)^{-1/\delta}]^{1/\theta-1}(x+1)^{-1/\delta-1}},$$

$$C_{2|1}(v|u; \theta, \delta) = \{1 - (x+y+1)^{-1/\delta}\}^{1/\theta-1}(x+y+1)^{-1/\delta-1} \cdot (x+1)^{1+1/\delta}[1 - u]^{\theta-1},$$

$$c(u, v; \theta, \delta) = \{1 - (x+y+1)^{-1/\delta}\}^{1/\theta-2}(x+y+1)^{-1/\delta-2}[(x+1)(y+1)]^{1+1/\delta}$$
$$\cdot [\theta(\delta+1) - (\theta\delta+1)(x+y+1)^{-1/\delta}]([1 - u][1 - v])^{\theta-1}.$$

For a starting value y_0 to solve (4.71) via Newton-Raphson iterations, a possible choice is to set $y = x$ in the nondominant term and get an equation for y_0:

$$(x + y_0 + 1)^{-1/\delta-1}[1 - (2x+1)^{-1/\delta}]^{1/\theta-1} = p\delta\theta g_1(x)$$

or

$$y_0 = \{p\delta\theta g_1(x)[1 - (2x+1)^{-1/\delta}]^{1-1/\theta}\}^{-\delta/(1+\delta)} - 1 - x,$$

provided the above is > 1.

Some properties of the family of copulas (4.70) are the following.

- The MTCJ family is obtained when $\theta = 1$, and the Joe/B5 family obtains as $\delta \to 0^+$.
- C^+ obtains as $\theta \to \infty$ or $\delta \to \infty$.
- The lower tail dependence parameter is $\lambda_L = 2^{-1/\delta}$, independent of θ, and the upper tail dependence parameter is $\lambda_U = 2 - 2^{1/\theta}$, independent of δ. The range of (λ_L, λ_U) is $(0, 1)^2$,
- Given tail parameters λ_U, λ_L: $\delta = [\log 2]/(-\log \lambda_L)$ and $\theta = [\log 2]/\log(2 - \lambda_U)$.

- The extreme value limits from the lower and upper tails are, respectively, the Galambos and Gumbel families.

- Concordance increases as δ increases because the MTCJ family used for K is increasing in $\prec_c$.

- Concordance increases as θ increases when $\delta \leq 1$ by Theorem 8.22 because $\omega''(s) \leq 0$, where $\omega(s) = \varphi^{-1}(\varphi(s; \theta_2, \delta); \theta_1, \delta) = \{1 - [1 - (1 + s)^{-1/\delta}]^\gamma\}^{-\delta} - 1$, with $\theta_1 < \theta_2$, $\gamma = \theta_1/\theta_2 < 1$ and $0 < \delta \leq 1$.

- The concordance ordering in θ can fail for $\delta > 1$. In fact even Blomqvist's beta can fail to be increasing in θ for part of this region; see Joe (2011b). The numerical counterexamples are most easily found from

$$\frac{\partial C(u, v; \theta, \delta)}{\partial \theta}\bigg|_{\theta=1} < 0 \tag{4.72}$$

  for some choices of (u, v); (4.72) is negative for $u = v = \frac{1}{2}$ for $\delta \geq 2.23$.

- Kendall's $\tau = \int_0^\infty s[\varphi'(s; \theta, \delta)]^2 ds = 1 - 2\delta^{-1}(2 - \theta)^{-1} + 4\delta^{-1}\theta^{-2}B(\delta + 2, 2/\theta - 1)$ for $1 < \theta < 2$; see Schepsmeier (2010). For $\theta > 2$, $\tau = 1 - 4\theta^{-2}\sum_{i=0}^\infty B(2, 2/\theta + i)\prod_{k=1}^i \frac{(k-\delta)}{(k+1)}$.

- Blomqvist's $\beta = 4\beta^* - 1$ and

$$\beta^* = 1 - \left(1 - [2\{1 - 2^{-\theta}\}^{-\delta} - 1]^{-1/\delta}\right)^{1/\theta}.$$

  Given λ_L, λ_U, the β value can be quite different from that of the BB1 and BB4 copula families.

- Range of (β, λ_U): Given $\lambda_U \in (0, 1)$ and the increasing in concordance in δ, the minimum and maximum values of β come from $4\beta^* - 1$ with $\delta \to 0^+$ and $\delta \to \infty$. With

$$\beta^* = 1 - (1 - [2(1 - 2^{-\log 2/\log(2-\lambda_U)})^{-\delta} - 1]^{-1/\delta})^{\log(2-\lambda_U)/\log 2},$$

  this leads to the range $[\beta_{\text{Umin}}(\lambda_U), 1]$ given λ_U, where

$$\beta_{\text{Umin}}(\lambda) = 3 - 4\{1 - (1 - 2^{-\log 2/\log(2-\lambda_U)})^2\}^{\log(2-\lambda_U)/\log 2}.$$

  As $\delta \to 0^+$, $\beta^* \sim 1 - (1 - [1 - 2\delta\log(1 - 2^{-\log 2/\log(2-\lambda_U)})]^{-1/\delta})^{\log(2-\lambda_U)/\log 2}$
  $\sim 1 - (1 - \exp\{2\log(1 - 2^{-\log 2/\log(2-\lambda_U)})\})^{\log(2-\lambda_U)/\log 2}$.
  As $\delta \to \infty$, $\beta^* \sim 1 - \{1 - 2^{-1/\delta}(1 - 2^{-\log 2/\log(2-\lambda_U)})\}^{\log(2-\lambda_U)/\log 2} \sim \frac{1}{2}$.

- Range of (β, λ_L): Given $\lambda_L \in (0, 1)$ and the increasing in concordance in θ, the minimum and maximum values of β come from $4\beta^* - 1$ with $\theta \to 1^+$ and $\theta \to \infty$. With $\beta^* = 1 - (1 - [2(1 - 2^{-\theta})^{\log 2/\log \lambda_L} - 1]^{\log \lambda_L/\log 2})^{1/\theta}$, this leads to the range $[\beta_{\text{Lmin}}(\lambda_L), 1]$ given λ_L, where as $\theta \to 1^+$,

$$\beta_{\text{Lmin}}(\lambda) = 4\{2^{1-\log 2/\log \lambda} - 1\}^{\log \lambda/\log 2} - 1 \geq \beta_{\text{Umin}}(\lambda).$$

  As $\theta \to \infty$, $\beta^* \sim 1 - (1 - [1 - 2^{1-\theta}\log 2/\log \lambda_L]^{\log \lambda_L/\log 2})^{1/\theta} \sim 1 - (2^{1-\theta})^{1/\theta} \sim \frac{1}{2}$.

- BB7 has same range for (β, λ_L) as BB1 and more range for (β, λ_U) compared with BB1.

- Boundary conditional cdfs: $C_{2|1}(\cdot|0; \theta, \delta)$ is degenerate at 0, $C_{2|1}(\cdot|1; \theta, \delta)$ is degenerate at 1.

4.24   Copulas based on the generalized Sibuya LT

Details and properties are mainly given for the bivariate copula based on the 2-parameter LT family.

4.24.1 Bivariate BB8; Joe 1993

This is a two-parameter family of Archimedean copulas based on a two-parameter LT family that does not come from a mixture of max-id composition. The copula cdf is:

$$C(u,v;\vartheta,\delta) = \delta^{-1}\Big(1 - \{1 - \eta^{-1}[1-(1-\delta u)^{\vartheta}][1-(1-\delta v)^{\vartheta}]\}^{1/\vartheta}\Big), \quad \vartheta \geq 1, 0 < \delta \leq 1, \quad (4.73)$$

where $\eta = 1 - (1-\delta)^{\vartheta}$ and $0 \leq u, v \leq 1$.

Let $x = 1 - (1-\delta u)^{\vartheta}$ and $y = 1 - (1-\delta v)^{\vartheta}$; these are monotone increasing transforms from $[0,1]$ to $[0,\eta]$. The inverse transforms are $u = \delta^{-1}[1-(1-x)^{1/\vartheta}]$ and $v = \delta^{-1}[1-(1-y)^{1/\vartheta}]$. Then $C(u,v;\vartheta,\delta) = G(x,y;\vartheta,\delta) = \delta^{-1}[1-(1-\eta^{-1}xy)^{1/\vartheta}]$, where $G(x,y;\vartheta,\delta)$ is a bivariate cdf with univariate margins $G_1 = G_2$ and $G_1(x) = \delta^{-1}[1-(1-x)^{1/\vartheta}]$, $g_1(x) = \partial G_1(x)/\partial x = (\delta\vartheta)^{-1}(1-x)^{1/\vartheta-1}$. The conditional cdf and copula density are:

$$C_{2|1}(v|u;\vartheta,\delta) = \frac{\partial G}{\partial x} \cdot \frac{\partial x}{\partial u}, \qquad \frac{\partial x}{\partial u} = \delta\vartheta(1-\delta u)^{\vartheta-1},$$

$$c(u,v;\vartheta,\delta) = \frac{\partial^2 G}{\partial x \partial y} \cdot \frac{\partial x}{\partial u} \cdot \frac{\partial y}{\partial v}.$$

For the inverse of the conditional cdf, solve for y or equivalently v in the equation:

$$C_{2|1}(v|u;\vartheta,\delta) = G_{2|1}(y|x;\vartheta,\delta) = p, \quad G_{2|1}(y|x;\vartheta,\delta) = \frac{\partial G(x,y;\vartheta,\delta)/\partial x}{g_1(x)}.$$

Furthermore,

$$G_{2|1}(y|x;\vartheta,\delta) = \frac{(\eta\delta\vartheta)^{-1}y(1-\eta^{-1}xy)^{1/\vartheta-1}}{(\delta\vartheta)^{-1}(1-x)^{1/\vartheta-1}},$$

$$C_{2|1}(v|u;\vartheta,\delta) = \frac{\eta^{-1}y(1-\eta^{-1}xy)^{1/\vartheta-1}}{(1-x)^{1/\vartheta-1}},$$

$$c(u,v;\vartheta,\delta) = \eta^{-1}\delta(1-\eta^{-1}xy)^{1/\vartheta-2}(\vartheta-\eta^{-1}xy)(1-\delta u)^{\vartheta-1}(1-\delta v)^{\vartheta-1}.$$

Some properties of the family of copulas (4.73) are the following.

- $C^{\perp}$ obtains as $\delta \to 0^+$ or $\vartheta \to 1^+$. The Joe/B5 family is obtained when $\delta = 1$, and the Frank family is obtained as $\vartheta \to \infty$ with $\eta = 1 - (1-\delta)^{\vartheta}$ held constant (or with $\delta = 1 - (1-\eta)^{1/\vartheta}$).

- Archimedean with LT family $\zeta(s;\vartheta,\delta) = \delta^{-1}[1 - \{1-\eta e^{-s}\}^{1/\vartheta}]$, $\vartheta \geq 1$, $0 < \delta, \eta \leq 1$ (this is a LT family that includes the Sibuya LT). The functional inverse is $\zeta^{-1}(s;\vartheta,\delta) = -\log\{[1-(1-\delta t)^{\vartheta}]/\eta\}$ for $0 \leq t \leq 1$.

- There is no tail dependence except when $\delta = 1$.

- The lower tail order is $\kappa_L = 2$ by Theorem 8.37 because $\zeta(s;\vartheta,\delta) \sim [1-(1-\delta)^{\vartheta}](\delta\vartheta)^{-1}e^{-s}$ as $s \to \infty$.

- The upper tail order is $\kappa_U = 2$ for $0 < \delta < 1$ because $1 - \zeta(s;\vartheta,\delta)$ has a Taylor series expansion about $s = 0$.

- $c(0,0;\vartheta,\delta) = \delta\vartheta/[1-(1-\delta)^{\vartheta}]$, $c(1,1;\vartheta,\delta) = \delta(1-\delta)^{-1}[\vartheta-1+(1-\delta)^{\vartheta}]/[1-(1-\delta)^{\vartheta}]$, $c(0,0;\vartheta,\delta) \leq c(1,1;\vartheta,\delta)$ because $h(\delta;\vartheta) = \vartheta\delta-1+(1-\delta)^{\vartheta} \geq 0$ for $\vartheta \geq 1$, $0 < \delta \leq 1$. The preceding holds because $h(0;\vartheta) = 0$, $h(1;\vartheta) = \vartheta-1 \geq 0$, and $h'(\delta;\vartheta) = \vartheta[1-(1-\delta)^{\vartheta-1}] \geq 0$.

- Concordance increases as ϑ or δ increases. The proof of the concordance ordering is non-trivial.

- Boundary conditional cdfs: $C_{2|1}(v|0;\vartheta,\delta) = \eta^{-1}[1-(1-\delta v)^{\vartheta}]$ and $C_{2|1}(v|1;\vartheta,\delta) = \eta^{-1}(1-\delta)^{\vartheta-1}y(1-y)^{1/\vartheta-1}$, where $y = 1-(1-\delta v)^{\vartheta}$.

Origins and further notes

In Joe (1993), this copula family is the max-id family used to get the hierarchical Archimedean extension of the Frank copula. The LT for this copula has the LTs of the logarithmic series and Sibuya distributions as special cases. See the Appendix.

4.25   Copulas based on the tilted positive stable LT

Details and properties are mainly given for the bivariate copula based on the 2-parameter LT family.

4.25.1   Bivariate BB9 or Crowder

From the two-parameter family of LTs, $\zeta(s; \vartheta, \delta) = \exp\{-(\delta^{-\vartheta} + s)^{1/\vartheta} + \delta^{-1}\}$, $\vartheta \geq 1$ and $\delta > 0$ (exponentially tilted positive stable LT in the Appendix), the two-parameter family of Archimedean copulas is

$$C(u, v; \vartheta, \delta) = \exp\{-[(\delta^{-1} - \log u)^{\vartheta} + (\delta^{-1} - \log v)^{\vartheta} - \delta^{-\vartheta}]^{1/\vartheta} + \delta^{-1}\} \quad 0 \leq u, v \leq 1. \ (4.74)$$

Let $x = \delta^{-1} - \log u$ and $y = \delta^{-1} - \log v$; these are monotone decreasing transforms taking 0 to ∞ and 1 to δ^{-1}. The inverse transforms are $u = \exp\{-x + \delta^{-1}\}$ and $v = \exp\{-y + \delta^{-1}\}$. Then $C(u, v; \vartheta, \delta) = G(x, y; \vartheta, \delta) = \exp\{-(x^{\vartheta} + y^{\vartheta} - \delta^{-\vartheta})^{1/\vartheta} + \delta^{-1}\}$, where $\overline{G}(x, y; \vartheta, \delta)$ is a bivariate survival function with univariate margins $\overline{G}_1 = \overline{G}_2$ and $\overline{G}_1(x) = \exp\{-x + \delta^{-1}\}$, a shifted exponential survival function on $[\delta^{-1}, \infty)$, $g_1(x) = -\partial \overline{G}_1(x)/\partial x = \exp\{-x + \delta^{-1}\}$. The conditional cdf and copula density are:

$$C_{2|1}(v|u; \vartheta, \delta) = \frac{\partial \overline{G}}{\partial x} \cdot \frac{\partial x}{\partial u}, \qquad\qquad \frac{\partial x}{\partial u} = -u^{-1},$$

$$c(u, v; \vartheta, \delta) = \frac{\partial^2 \overline{G}}{\partial x \partial y} \cdot \frac{\partial x}{\partial u} \cdot \frac{\partial y}{\partial v}.$$

For the inverse of the conditional cdf, solve for y or equivalently v in the equation:

$$C_{2|1}(v|u; \vartheta, \delta) = \overline{G}_{2|1}(y|x; \vartheta, \delta) = p, \quad \overline{G}_{2|1}(y|x; \vartheta, \delta) = \frac{-\partial \overline{G}(x, y; \vartheta, \delta)/\partial x}{g_1(x)}. \quad (4.75)$$

Furthermore,

$$\overline{G}_{2|1}(y|x; \vartheta, \delta) = \frac{\overline{G}(x, y; \vartheta, \delta)(x^{\vartheta} + y^{\vartheta} - \delta^{-\vartheta})^{1/\vartheta - 1} x^{\vartheta - 1}}{\exp\{-x + \delta^{-1}\}}, \quad (4.76)$$

$$C_{2|1}(v|u; \vartheta, \delta) = \overline{G}(x, y; \vartheta, \delta)(x^{\vartheta} + y^{\vartheta} - \delta^{-\vartheta})^{1/\vartheta - 1} x^{\vartheta - 1} u^{-1},$$

$$c(u, v; \vartheta, \delta) = \overline{G} \cdot (x^{\vartheta} + y^{\vartheta} - \delta^{-\vartheta})^{1/\vartheta - 2} \{(x^{\vartheta} + y^{\vartheta} - \delta^{-\vartheta})^{1/\vartheta} + \vartheta - 1\} \cdot (xy)^{\vartheta - 1}(uv)^{-1}.$$

To solve (4.75) via Newton-Raphson iterations, it is simplest to take logarithms on both sides of (4.76) to get:

$$-(x^{\vartheta} + y^{\vartheta} - \delta^{-\vartheta})^{1/\vartheta} + (\vartheta^{-1} - 1) \log(x^{\vartheta} + y^{\vartheta} - \delta^{-\vartheta}) + (\vartheta - 1) \log x + x = \log p.$$

Let $z = (x^{\vartheta} + y^{\vartheta} - \delta^{-\vartheta})^{1/\vartheta} \geq x$. Then the above reduces to solving the equation

$$h(z) = z + (\vartheta - 1) \log z - [x + (\vartheta - 1) \log x - \log p] = 0$$

for z, given p, x, ϑ. This is the same nonlinear equation in (4.15), except the root z is transformed to y as $y = (z^{\vartheta} - x^{\vartheta} + \delta^{-\vartheta})^{1/\vartheta}$.

Some properties of the family of copulas (4.74) are the following.

- $C^\perp$ obtains as $\delta \to 0^+$ or for $\vartheta = 1$, and C^+ obtains as $\vartheta \to \infty$.
- Archimedean with LT family $\zeta(s; \vartheta, \delta)$ where $\zeta^{-1}(t; \vartheta, \delta) = (\delta^{-1} - \log t)^\vartheta - \delta^{-\vartheta}$.
- The Gumbel family B6 is a subfamily when $\delta \to \infty$ or $\delta^{-1} \to 0^+$.
- Concordance increases as either ϑ increases or δ increases by Theorem 8.22, because $\omega''(s) \le 0$, where

$$\omega(s) = \varphi^{-1}(\varphi(s; \vartheta_2, \delta_2); \vartheta_1, \delta_1) = [\delta_1^{-1} + (\delta_2^{-\vartheta_2} + s)^{1/\vartheta_2} - \delta_2^{-1}]^{\vartheta_1} - \delta_1^{-\vartheta_1},$$

  for $\vartheta_1 < \vartheta_2$, $\delta_1 = \delta_2$ or $\vartheta_1 = \vartheta_2$, $\delta_1 < \delta_2$. A stronger statement is that $\omega \in \mathcal{L}_\infty^*$ if $\vartheta_1 < \vartheta_2$, $\delta_1 = \delta_2$, and $\omega \in \mathcal{L}_2^*$ if $\vartheta_1 = \vartheta_2$, $\delta_1 < \delta_2$.
- The LT is that of the Archimedean copula of a conditional distribution of the trivariate version of (4.28)
- The lower tail order is $\kappa_L = 2^{1/\vartheta}$.
- The upper tail order is $\kappa_U = 2$.
- Based on contour plots, BB9 is mostly skewed to the lower tail (for most of parameter space with δ not large).
- Boundary conditional cdfs: $C_{2|1}(v|1; \vartheta, \delta) = v(1 - \delta \log v)^{1-\vartheta}$ and $C_{2|1}(\cdot|0; \vartheta, \delta)$ is degenerate at 0.
- This family extends to negative dependence with $0 < \vartheta < 1$ and $\delta^{-1} \ge 1 - \vartheta$ (see pages 161–162 of Joe (1997)). As $\vartheta \to 0^+$, this leads to $C(u, v; 0, \delta) = uv \exp\{-\delta(\log u)(\log v)\}$ for $0 < \delta \le 1$. This is the copula of the bivariate exponential distribution in Gumbel (1960a); it is a negatively dependent distribution, and included in Barnett (1980).

Origins and further notes

The multivariate Archimedean copula with the LT is the copula of the multivariate distribution with Weibull connections, studied in Crowder (1989).

In Joe (1993), this copula family is the max-id family needed to get the hierarchical Archimedean extension of the MTCJ copula. The parametrization given here is different and is appropriate to get increasing in concordance for both parameters.

4.26 Copulas based on the shifted negative binomial LT

Details and properties are mainly given for the bivariate copula based on the 2-parameter LT family.

4.26.1 Bivariate BB10

This is a 2-parameter family of copulas that is based on a 2-parameter family of LTs that does not come from a composition of the form (4.2); it is given as Example 4.4 in Marshall and Olkin (1988). The copula cdf is

$$C(u, v; \theta, \pi) = uv[1 - \pi(1 - u^\theta)(1 - v^\theta)]^{-1/\theta}, \quad 0 \le u, v \le 1,\ 0 \le \pi \le 1,\ \theta > 0. \quad (4.77)$$

The conditional cdf is

$$C_{2|1}(v|u; \theta, \pi) = [1 - \pi(1 - u^\theta)(1 - v^\theta)]^{-1/\theta - 1} \cdot v\{1 - \pi(1 - v^\theta)\},$$

and the density is:

$$c(u, v; \theta, \pi) = [1 - \pi(1 - u^\theta)(1 - v^\theta)]^{-1/\theta - 2} \cdot \{1 - \pi + \pi(1 + \theta)u^\theta v^\theta - \pi(1 - \pi)(1 - u^\theta)(1 - v^\theta)\},$$

for $0 \leq u, v \leq 1$.

Some properties of the family of copulas (4.77) are the following.

- $C^{\perp}$ obtains as $\theta \to 0^+$; C^+ obtains as $\theta \to \infty$ when $\pi = 1$, but $C^{\perp}$ obtains as $\theta \to \infty$ for $0 < \pi < 1$.

- For fixed θ, concordance increases as π increases; this can be verified by the derivative.

- For $\pi = 1$, concordance increases in θ. For $0 < \pi < 1$ fixed, and with fixed u, v, $C(u, v; \theta, \pi)$ increases and then decreases as θ increases.

- Archimedean based on the LT of the (shifted) negative binomial distribution is $\varphi(s; \theta, \pi) = [(1 - \pi)e^{-s}/(1 - \pi e^{-s})]^{1/\theta} = [(1 - \pi)/(e^s - \pi)]^{1/\theta}$, where $\theta > 0$ and $0 \leq \pi < 1$. The inverse is $\varphi^{-1}(t; \theta, \pi) = \log[(1 - \pi)t^{-\theta} + \pi]$.

- With $\theta = 1$, the LT is geometric and the copula family of the bivariate logistic distribution in Ali et al. (1978) (it can be extended into negative dependence with $-1 \leq \pi < 0$ and the copula with range $-1 \leq \pi \leq 1$ is referred to as the Ali-Mikhail-Haq copula). In this case, Genest and MacKay (1986) show that $\tau = [3\pi^2 - 2\pi - 2(1 - \pi)^2 \log(1 - \pi)]/(3\alpha^2)$ and the maximum value of Kendall's tau is $1/3$ when $\pi = 1$.

- For fixed θ, as $\pi \to 1$, the copula is

$$C(u, v; \theta, 1) = uv[1 - (1 - u^\theta)(1 - v^\theta)]^{-1/\theta}.$$

- The upper tail order is $\kappa_U = 2$ because all moments exist for the NB distribution.

- The lower tail order is $\kappa_L = 2$ because as $s \to \infty$, the LT behaves like $(1 - \pi)^{1/\theta} e^{-s/\theta}$.

- Blomqvist's β: for fixed θ, the maximum value of $C(\frac{1}{2}, \frac{1}{2}; \theta, 1)$ is $\frac{1}{4}[1 - (1 - 2^{-\theta})^2]^{-1/\theta}$ $= \frac{1}{2} 2^{-1/\theta}[1 - 2^{-1-\theta}]^{-1/\theta}$.

- With $\theta = 1$, Ali et al. (1978) obtain the copula as the solution of

$$\frac{1 - C(u, v; 1, \pi)}{C(u, v; 1, \pi)} = \frac{1 - u}{u} + \frac{1 - v}{v} + (1 - \pi)\frac{1 - u}{u} \cdot \frac{1 - v}{v}.$$

(4.77) satisfies

$$\frac{1 - C^\theta(u, v; \theta, \pi)}{C^\theta(u, v; \theta, \pi)} = \frac{1 - u^\theta}{u^\theta} + \frac{1 - v^\theta}{v^\theta} + (1 - \pi)\frac{1 - u^\theta}{u^\theta} \cdot \frac{1 - v^\theta}{v^\theta}.$$

The proof is as follows: $C^\theta(u, v; \theta, \pi) = \frac{u^\theta v^\theta}{[1 - \pi(1 - u^\theta)(1 - v^\theta)]}$ and

$$\frac{1 - C^\theta}{C^\theta} = \frac{1 - \pi + \pi u^\theta + \pi v^\theta - (1 + \pi)u^\theta v^\theta}{u^\theta v^\theta}$$

$$= \frac{(1 - u^\theta)v^\theta + (1 - v^\theta)u^\theta + (\pi - 1)u^\theta + (\pi - 1)v^\theta + (1 - \pi) + (1 - \pi)u^\theta v^\theta}{u^\theta v^\theta}.$$

- Boundary conditional cdfs: $C_{2|1}(v|0; \theta, \pi) = [1 - \pi(1 - v^\theta)]^{-1/\theta - 1} \cdot v\{1 - \pi(1 - v^\theta)\}$, and $C_{2|1}(v|1; \theta, \pi) = v\{1 - \pi(1 - v^\theta)\}$.

- Density at corners: $c(0, 0; \theta, \pi) = (1 - \pi)^{-1/\theta}$ and $c(1, 1; \theta, \pi) = 1 + \pi\theta$.

- The direction of tail asymmetry depends on (θ, π). For example, for $\theta = 1$, the direction of asymmetry is towards the joint lower tail. For fixed $0 < \pi < 1$, the direction of asymmetry is towards the joint lower tail with sufficiently small θ, and towards the joint upper tail with sufficiently large θ.

- The reflected BB10 copula is called Lomax copula in Balakrishnan and Lai (2009), because the bivariate Lomax survival function is $\overline{F}(x, y) = [1 + \alpha_1 x + \alpha_2 y + \alpha_{12} xy]^{-\eta}$ with $x, y > 0$, $0 \leq \alpha_{12} \leq (\eta + 1)\alpha_1 \alpha_2$, $\alpha_1 > 0$, $\alpha_2 > 0$, $\eta > 0$. The survival copula of this distribution has form (4.77).

4.27   Multivariate GB2 distribution and copula

The multivariate generalized beta type 2 (GB2) distribution is proposed as a copula in Yang et al. (2011) and some properties are proved. The copula is a generalization of the Archimedean copula based on the gamma LT in that it comes from a gamma mixture of gamma distributions. In this section, the derivation is given as well as additional properties.

Let $[Y_j|Q = q] \sim \text{Gamma}(\eta_j, \text{rate} = q)$ be conditionally independent for $j = 1, \ldots, d$, and let $Q \sim \text{Gamma}(\zeta, 1)$. The non-negative shape parameters are $\eta_1, \ldots, \eta_d$ and ζ; also let $\delta = \zeta^{-1}$. Different scale parameters could be added to the variables, but they would not affect the resulting copula. With $\eta_d^* = \eta_1 + \cdots + \eta_d$ and $y_d^* = y_1 + \cdots + y_d$, the joint density is:

$$f_{\boldsymbol{Y}}(\boldsymbol{y}) = \int_0^\infty \prod_{j=1}^d \frac{y_j^{\eta_j-1} e^{-qy_j} q^{\eta_j}}{\Gamma(\eta_j)} \cdot \frac{e^{-q} q^{\zeta-1}}{\Gamma(\zeta)} \, \mathrm{d}q = \frac{\Gamma(\eta_d^* + \zeta) \prod_{j=1}^d y_j^{\eta_j-1}}{\Gamma(\eta_1) \cdots \Gamma(\eta_d) \, \Gamma(\zeta) \, (1 + y_d^*)^{\eta_d^* + \zeta}}, \quad (4.78)$$

for $\boldsymbol{y} \in \mathbb{R}_+^d$. This is the inverted beta Liouville density (Chapter 6 of Fang et al. (1990)).

For $j = 1, \ldots, d$, the univariate margins are:

$$f_{Y_j}(y) = \frac{\Gamma(\eta_j + \zeta) \, y^{\eta_j-1}}{\Gamma(\eta_j) \, \Gamma(\zeta) \, (1+y)^{\eta_j+\zeta}}; \tag{4.79}$$

the density arises as a transform of a $\text{Beta}(\eta_j, \zeta)$ random variable. Let $W_j = Y_j/(1 + Y_j)$ with $w = y/(1+y)$, $y = w/(1-w)$, $\mathrm{d}w = \mathrm{d}y/(1+y)^2$. Then

$$f_{W_j}(w) = \frac{\Gamma(\eta_j + \zeta) \, w^{\eta_j-1} (1-w)^{\zeta-1}}{\Gamma(\eta_j) \, \Gamma(\zeta)},$$

so that, with F_B being the Beta cdf,

$$F_{Y_j}(y) = F_{W_j}(y/[1+y]) = F_B(y/[1+y]; \eta_j, \zeta),$$
$$F_{Y_j}^{-1}(u) = F_B^{-1}(u; \eta_j, \zeta)/[1 - F_B^{-1}(u; \eta_j, \zeta)].$$

The density (4.79) is the $\text{GB2}(\eta_j, \zeta, \sigma)$ density with scale parameter $\sigma = 1$.

The copula cdf is:

$$C(\boldsymbol{u}; \zeta, \eta_1, \ldots, \eta_d) = F_{\boldsymbol{Y}}\left(F_{Y_1}^{-1}(u_1), \ldots, F_{Y_d}^{-1}(u_d)\right), \quad \boldsymbol{u} \in [0,1]^d, \tag{4.80}$$

and the copula density is

$$c(\boldsymbol{u}; \zeta, \boldsymbol{\eta}) = \frac{f_{\boldsymbol{Y}}\left(F_{Y_1}^{-1}(u_1), \ldots, F_{Y_d}^{-1}(u_d)\right)}{\prod_{j=1}^d f_{Y_j}\left(F_{Y_j}^{-1}(u_j)\right)} = \frac{\Gamma(\eta_d^* + \zeta) \, \Gamma^{d-1}(\zeta)}{\prod_{j=1}^d \Gamma(\eta_j + \zeta)} \cdot \frac{\prod_{j=1}^d (1+y_j)^{\eta_j+\zeta}}{(1 + y_1 + \cdots + y_d)^{\eta_d^* + \zeta}},$$

with $y_j = F_{Y_j}^{-1}(u_j)$ and $0 < u_j < 1$. The cdf $F_{\boldsymbol{Y}}$ can be written as a 1-dimensional integral:

$$F_{\boldsymbol{Y}}(\boldsymbol{y}; \zeta, \eta_1, \ldots, \eta_d) = \int_0^\infty \left\{ \prod_{j=1}^d F_\Gamma(y_j q; \eta_j) \right\} [\Gamma(\zeta)]^{-1} q^{\zeta-1} e^{-q} \, \mathrm{d}q.$$

Next we derive some conditional distributions of the multivariate GB2 distributions. Consider $[Y_{m+1}, \ldots, Y_d | Y_1 = y_1, \ldots, Y_d = y_d]$ where $1 < m < d$. Because of the stochastic representation, with $\eta_m^* = \eta_1 + \cdots + \eta_m$ and $y_m^* = y_1 + \cdots + y_m$, the marginal density of $Y_1, \ldots, Y_m$ is

$$f_{Y_1, \ldots, Y_m}(\boldsymbol{y}_{1:m}; \zeta, \eta_1, \ldots, \eta_m) = \frac{\Gamma(\eta_m^* + \zeta) \prod_{j=1}^m y_j^{\eta_j-1}}{\Gamma(\eta_1) \cdots \Gamma(\eta_m) \, \Gamma(\zeta) \, (1 + y_m^*)^{\eta_m^* + \zeta}}.$$

With $z_j = y_j/(1 + y_m^*)$ for $j = m+1, \ldots, d$,

$$f_{\boldsymbol{Y}_{(m+1):d}|\boldsymbol{Y}_{1:m}}(\boldsymbol{y}_{(m+1):d}|\boldsymbol{y}_{1:m}; \zeta, \eta_1, \ldots, \eta_d)$$

$$= \frac{\Gamma(\eta_d^* + \zeta)(1 + y_m^*)^{\eta_m^* + \zeta} \prod_{j=m+1}^d y_j^{\eta_j - 1}}{\Gamma(\eta_{m+1}) \cdots \Gamma(\eta_d) \, \Gamma(\eta_m^* + \zeta) \left([1 + y_m^*] + y_{m+1} + \cdots + y_d \right)^{\eta_d^* + \zeta}}$$

$$= (1 + y_m^*)^{-(d-m)} \frac{\Gamma(\eta_{m+1} + \cdots + \eta_d + (\eta_m^* + \zeta)) \prod_{j=m+1}^d z_j^{\eta_j - 1}}{\Gamma(\eta_{m+1}) \cdots \Gamma(\eta_d) \, \Gamma(\eta_m^* + \zeta) \, (1 + z_{m_1} + \cdots + z_d)^{\eta_{m+1} + \cdots + \eta_d + (\eta_m^* + \zeta)}}.$$

That is, this is a $(d-m)$-variate GB2 density with shape parameters $\eta_{m+1}, \ldots, \eta_d$ and $p_m^* + \zeta$ and $(d-m)$ scale parameters equal to $1 + y_m^*$. Hence if $d - m \geq 2$, the copula of conditional distribution does not depend on $y_1, \ldots, y_m$. By permutation of indices, this property holds for conditioning on any m variables with $1 < m < d$.

In particular, in the bivariate case, the conditional densities $f_{Y_2|Y_1}(\cdot|y_1)$ and $f_{Y_1|Y_2}(\cdot|y_2)$ are $GB2(\eta_2, \zeta + \eta_1, 1 + y_1)$ and $GB2(\eta_1, \zeta + \eta_2, 1 + y_2)$ respectively.

Other properties are the following.

- *Tail dependence and order*: Using the conditional distributions, Yang et al. (2011) derive the upper tail dependence coefficient and prove that there is no lower tail dependence. Some approximate analysis suggests that the lower tail order is 2.
- *Covariance and correlation*: $\mathbb{E}(Y_j|Q) = \eta_j/Q$, $\mathrm{Var}(Y_j|Q) = \eta_j/Q^2$. Hence $\mathrm{Var}(Y_j) = \eta_j \mathbb{E}(Q^{-2}) + \eta_j^2 \mathrm{Var}(Q^{-1})$ when the inverse square moment exists; also for $j \neq k$, $\mathrm{Cov}(Y_j, Y_k) = \eta_j \eta_k \mathrm{Var}(Q^{-1})$ and the correlation is $\sqrt{\eta_j \eta_k / [(r + \eta_j)(r + \eta_k)]}$ with $r = \mathbb{E}(Q^{-2})/\mathrm{Var}(Q^{-1})$. This is the positively dependent 1-factor correlation structure. Because the model has conditional independence, only positive dependence is possible.
- *Independence and comonotonicity*: If $\delta = \zeta^{-1}$ is fixed (and $\mathbb{E}(Q^{-2})$ exists), then the correlation goes to 1 as $\eta_j, \eta_k \to \infty$ and it goes to 0 as $\eta_j, \eta_k \to 0^+$.
- *Concordance ordering for the bivariate case*: Some numerical checks suggest that $C(\cdot; \zeta, \eta_1, \eta_2)$ is increasing in concordance in η_1 with η_2 and ζ fixed. Also $C(\cdot; \zeta, \eta_1, \eta_2)$ decreasing in concordance in ζ (or increasing in concordance in δ) with η_1, η_2 fixed.
- *Multivariate Gaussian limit*: From the Appendix of Yang et al. (2011), the following leads to a bivariate Gaussian in the limit. Transform with $Y_j = (X_j/b_j)^{a_j}$, $X_j = b_j Y_j^{1/a_j}$, with $a_j > 0$, $b_j > 0$. For the bivariate symmetric case, let $\eta_1 = \eta_2 = \eta = \gamma \zeta$, $b_1 = b_2 = b$ with $-\log b = \zeta(\log \gamma)/(1 + \gamma^{-1})^{1/2}$, $a_1 = a_2 = a = \zeta^{-1}(1 + \gamma^{-1})^{1/2}$. A bivariate Gaussian density arises for $(\log X_1, \log X_2)$ as $\zeta \to \infty$, $\eta \to \infty$, $b \to 0^+$, $a \to 0^+$, and the correlation is $\rho = \gamma/(1 + \gamma) > 0$ (or $\gamma = \rho/(1 - \rho)$). The proof uses the cumulant generating function. Since a, b do not affect the copula, a bivariate Gaussian copula with correlation $\rho > 0$ arises with $\eta_1 = \eta_2 = \gamma \zeta$ as $\zeta \to \infty$ and $\gamma = \rho/(1 - \rho)$.
- There are copula families that bridge Archimedean and multivariate extreme value copulas. But (4.80) bridges an Archimedean copula family and a subfamily of Gaussian copulas, and could be considered as skewed-Gaussian copulas.

1-factor multivariate Gaussian as a limit: Technical details

The mixed moment of the multivariate density (4.78) is:

$$\mathbb{E}[Y_1^{n_1} \cdots Y_d^{n_d}] = \mathbb{E}(\mathbb{E}[Y_1^{n_1} \cdots Y_d^{n_d}|Q]) = \mathbb{E}\left(\prod_{j=1}^d \mathbb{E}[Y_j^{n_j}|Q]\right)$$

$$= \mathbb{E}\left(\prod_{j=1}^d Q^{-n_j}\right) \prod_{j=1}^d \frac{\Gamma(\eta_j + n_j)}{\Gamma(\eta_j)} = \frac{\Gamma(\zeta - n_\bullet)}{\Gamma(\zeta)} \prod_{j=1}^d \frac{\Gamma(\eta_j + n_j)}{\Gamma(\eta_j)},$$

where $n_\bullet = n_1 + \cdots + n_d$, provided all of the arguments of the Gamma function are positive. The moments exist in a neighborhood of $\mathbf{0}$.

Next let $X_j = b_j Y_j^{1/a_j}$ for $j = 1, \ldots, d$. Then

$$\mathbb{E}[X_1^{t_1} \cdots X_d^{t_d}] = \frac{\Gamma(\zeta - \sum_j t_j/a_j)}{\Gamma(\zeta)} \prod_{j=1}^{d} \frac{b_j^{t_j} \Gamma(\eta_j + t_j/a_j)}{\Gamma(\eta_j)}.$$

In a neigborhood of $\mathbf{0}$, the multivariate cumulant generating function $K(t_1, \ldots, t_d)$ of $(\log X_1, \ldots, \log X_d)$ is,

$$\log \mathbb{E}[X_1^{t_1} \cdots X_d^{t_d}] = \log \Gamma\Big(\zeta - \sum_j t_j/a_j\Big) - \log \Gamma(\zeta) + \sum_{j=1}^{d} \{t_j \log b_j + \log \Gamma(\eta_j + t_j/a_j) - \log \Gamma(\eta_j)\}.$$

For the case of $d = 2$, this becomes

$$\begin{aligned} K(t_1, t_2) = {}& t_1 \log b_1 + t_2 \log b_2 + \log \Gamma(\zeta - t_1/a_1 - t_2/a_2) - \log \Gamma(\zeta) \\ & + \log \Gamma(\eta_1 + t_1/a_1) - \log \Gamma(\eta_1) + \log \Gamma(\eta_2 + t_2/a_2) - \log \Gamma(\eta_1). \end{aligned}$$

Let $\eta_1 = \gamma_1 \zeta$, $\eta_2 = \gamma_2 \zeta$ with $\zeta \to \infty$. Apply Stirling's approximation, namely, $\log \Gamma(z) \sim (z - \frac{1}{2}) \log z - z + \frac{1}{2} \log(2\pi)$ as $z \to \infty$, one gets for ϵ near 0 that

$$\begin{aligned} \log \Gamma(z + \epsilon) - \log \Gamma(z) &\sim (z - \tfrac{1}{2} + \epsilon)(\log z + \epsilon z^{-1} - \tfrac{1}{2}\epsilon^2 z^{-2}) - (z - \tfrac{1}{2}) \log z - \epsilon \\ &\sim \epsilon \log z - \tfrac{1}{2}\epsilon z^{-1} + \tfrac{1}{2}\epsilon^2 z^{-1}(1 + \tfrac{1}{2}z^{-1}). \end{aligned}$$

As $\zeta \to \infty$, $K(t_1, t_2)$ is approximated by:

$$\begin{aligned} & t_1 \log b_1 + t_2 \log b_2 - (t_1/a_1 + t_2/a_2) \log \zeta + (t_1/a_1) \log(\zeta \gamma_1) + (t_2/a_2) \log(\zeta \gamma_2) \\ & + \tfrac{1}{2}(t_1/a_1 + t_2/a_2)\zeta^{-1} - \tfrac{1}{2}(t_1/a_1)\zeta^{-1}\gamma_1^{-1} - \tfrac{1}{2}(t_2/a_2)\zeta^{-1}\gamma_2^{-1} \\ & + \tfrac{1}{2}(t_1/a_1 + t_2/a_2)^2\zeta^{-1} + \tfrac{1}{2}(t_1/a_1)^2\zeta^{-1}\gamma_1^{-1} + \tfrac{1}{2}(t_2/a_2)^2\zeta^{-1}\gamma_2^{-1} \\ & + O(t_1^2\zeta^{-2}) + O(t_2^2\zeta^{-2}) + O(t_1 t_2\zeta^{-2}). \end{aligned}$$

- The coefficient of t_1 is: $\mu_1 = \log b_1 + a_1^{-1} \log \gamma_1 + O(\zeta^{-1})$.
- The coefficient of t_2 is: $\mu_2 = \log b_2 + a_2^{-1} \log \gamma_2 + O(\zeta^{-1})$.
- The coefficient of $\frac{1}{2}t_1^2$ is: $\sigma_1^2 = a_1^{-2}\zeta^{-1}(1 + \gamma_1^{-1}) + O(\zeta^{-2})$.
- The coefficient of $\frac{1}{2}t_2^2$ is: $\sigma_1^2 = a_2^{-2}\zeta^{-1}(1 + \gamma_2^{-1}) + O(\zeta^{-2})$.
- The coefficient of $t_1 t_2$ is: $\sigma_{12} = a_1^{-1} a_2^{-1}\zeta^{-1} + O(\zeta^{-2})$.

To get $\sigma_1^2 = \sigma_2^2 = 1$, let $a_1^2 = \zeta^{-1}(1 + \gamma_1^{-1}) \to 0$, $a_2^2 = \zeta^{-1}(1 + \gamma_2^{-1}) \to 0$. To get $\mu_1 = \mu_2 = 0$, let $-\log b_1 = a_1^{-1} \log \gamma_1 \to \infty$ $(b_1 \to 0^+)$, $-\log b_2 = a_2^{-1} \log \gamma_2 \to \infty$ $(b_2 \to 0^+)$. The correlation is then $\sigma_{12} = a_1^{-1} a_2^{-1} = (1 + \gamma_1^{-1})^{-1/2}(1 + \gamma_2^{-1})^{-1/2} = \{\gamma_1 \gamma_2/[(1 + \gamma_1)(1 + \gamma_2)]\}^{1/2}$.

For the multivariate extension, if $\eta_j = \gamma_j \zeta$ with $\gamma_j > 0$ for $j = 1, \ldots, d$, and $\zeta \to \infty$, then the above shows that the positively dependent 1-factor multivariate Gaussian copula is obtained. That is, the correlation matrix $(\rho_{jk})_{1 \leq j, k \leq d}$ has

$$\rho_{jk} = \sqrt{\frac{\gamma_j \gamma_k}{(1 + \gamma_j)(1 + \gamma_k)}}, \quad j \neq k.$$

4.28 Factor models based on convolution-closed families

Section 4.6 of Joe (1997) has a construction of multivariate distributions based on families of univariate distributions that are convolution-closed and infinitely divisible. Section 8.4 of

Joe (1997) uses this construction for transition probabilities of Markov time series models. In this section, we obtain some properties of the copulas for some continuous factor models based on this construction. For continuous random variables, convolution-closed families that are amenable to analysis are the Gamma and inverse Gaussian families.

The copulas of these distributions can have properties that are quite different from the Archimedean, extreme value and elliptical classes. We show some of these unusual properties for the case of the Gamma convolution 1-factor model.

A family $F(\cdot; \eta)$ is *convolution-closed* if $Y_i \sim F(\cdot; \eta_i)$, $i = 1, 2$, and Y_1, Y_2 independent implies the convolution $Y_1 + Y_2 \sim F(\cdot; \eta_1 + \eta_2)$, and η is said to be a convolution parameter. If $F(\cdot; \eta)$, $\eta > 0$, is a convolution-closed infinitely divisible parametric family such that $F(\cdot; \eta_1) * F(\cdot; \eta_2) = F(\cdot; \eta_1 + \eta_2)$, where $*$ is the convolution operator, then the n-fold convolution of $F(\cdot; \eta/n)$ is $F(\cdot; \eta)$. It is assumed that $F(\cdot; 0)$ corresponds to the degenerate distribution at 0.

Definition 4.1 Let $\mathcal{S}_d = \{S : S \subset \{1, \ldots, d\}, S \neq \emptyset\}$. Let $\{Z_S : S \in \mathcal{S}_d\}$, be a set of $2^d - 1$ independent random variables in the family $F(\cdot; \eta)$ such that Z_S has parameter $\theta_S \geq 0$ (if the parameter is zero, the random variable is also zero). The stochastic representation for a family of *multivariate distributions with univariate margins in a given convolution-closed infinitely divisible class*, parametrized by $\{\theta_S : S \in \mathcal{S}_d\}$, is

$$X_j = \sum_{S: j \in S} Z_S, \quad j = 1, \ldots, d.$$

X_j has distribution $F(\cdot; \eta_j)$, where $\eta_j = \sum_{S \in \mathcal{S}_d: j \in S} \theta_S$ for $j = 1, \ldots, d$.

The 1-factor submodel has non-zero parameters only when $S = \{1, \ldots, d\}$ and $S = \{j\}$ for $j = 1, \ldots, d$. In the next definition below, we replace $Z_{\{1, \ldots, d\}}$ with Z_0.

Definition 4.2 Let $\{Z_0, Z_1, \ldots, Z_d\}$, be a set of $d + 1$ independent random variables in the family $F(\cdot; \eta)$ such that Z_j has parameter $\theta_j \geq 0$ for $j = 0, 1, \ldots, d$. The stochastic representation for the *1-factor convolution-closed family* is

$$X_j = Z_0 + Z_j, \quad j = 1, \ldots, d. \tag{4.81}$$

X_j has distribution $F(\cdot; \eta_j)$, where $\eta_j = \theta_0 + \theta_j$ for $j = 1, \ldots, d$.

Next, we further assume that $F(\cdot; \eta)$ is absolutely continuous with density $f(\cdot; \eta)$ that is continuous in its support on the non-negative real line. We obtain the joint cdf and pdf of the 1-factor model in (4.81). With $\boldsymbol{\theta} = (\theta_0, \theta_1, \ldots, \theta_d)$, the joint cdf is

$$F_{1:d}(\boldsymbol{x}; \boldsymbol{\theta}) = \int_0^{\min\{x_1, \ldots, x_d\}} \left\{ \prod_{j=1}^d F(x_j - z; \theta_j) \right\} f(z; \theta_0) \mathrm{d}z, \quad \boldsymbol{x} \in \mathbb{R}_+^d, \tag{4.82}$$

and the joint pdf is

$$f_{1:d}(\boldsymbol{x}; \boldsymbol{\theta}) = \int_0^{\min\{x_1, \ldots, x_d\}} \left\{ \prod_{j=1}^d f(x_j - z; \theta_j) \right\} f(z; \theta_0) \mathrm{d}z. \tag{4.83}$$

The copula cdf for (4.82) is

$$C(\boldsymbol{u}; \boldsymbol{\theta}) = F_{1:d}\big(F^{-1}(u_1; \theta_0 + \theta_1), \ldots, F^{-1}(u_d; \theta_0 + \theta_d); \boldsymbol{\theta}\big), \quad \boldsymbol{u} \in [0, 1]^d, \tag{4.84}$$

and its density is

$$c(\boldsymbol{u}; \boldsymbol{\theta}) = \frac{f_{1:d}\big(F^{-1}(u_1; \theta_0 + \theta_1), \ldots, F^{-1}(u_d; \theta_0 + \theta_d); \boldsymbol{\theta}\big)}{\prod_{j=1}^{d} f\big(F^{-1}(u_j; \theta_0 + \theta_d); \theta_0 + \theta_d\big)}, \quad \boldsymbol{u} \in (0,1)^d.$$

If the random variables are non-negative and discrete, the joint cdf and pmf are similar, but the copula is not unique because of the discreteness.

The function (4.83) is bounded if $f(\cdot; \eta)$ is bounded, that is, if it cannot asymptote to ∞ at 0 for any $\eta > 0$. This is satisfied for example with the inverse Gaussian family. If $\lim_{z \to 0+} f(z; \eta)$ is infinite for part of the parameter space, then it is possible that $f_{1\ldots d}(x\boldsymbol{1}_d; \boldsymbol{\theta})$ is infinite for some parameter vectors $\boldsymbol{\theta}$. This is illustrated next for the Gamma family. If $x_1 = \cdots = x_d = x > 0$ in (4.83), then

$$f_{1:d}(x\boldsymbol{1}_d; \boldsymbol{\theta}) = \int_0^x \Big\{ \prod_{j=1}^d f(x - z; \theta_j) \Big\} f(z; \theta_0) \, \mathrm{d}z = \int_0^x \Big\{ \prod_{j=1}^d f(z; \theta_j) \Big\} f(x - z; \theta_0) \, \mathrm{d}z$$

$$= \prod_{j=0}^d [\Gamma(\theta_j)]^{-1} \int_0^x z^{\theta_1 + \cdots + \theta_d - d} e^{-dz} \cdot (x - z)^{\theta_0 - 1} e^{-(x-z)} \mathrm{d}z$$

and this is infinite if $\theta_1 + \cdots + \theta_d - d + 1 \le 0$ or $\theta_1 + \cdots + \theta_d \le d - 1$ or $d^{-1} \sum_j \theta_j \le 1 - d^{-1}$. If $\theta_1 = \cdots = \theta_d = \theta \le 1 - d^{-1}$, then the copula density is infinite for $c(u\boldsymbol{1}_d; \boldsymbol{\theta})$, and if $d^{-1} \sum_j \theta_j \le 1 - d^{-1}$ with unequal θ_j, the copula density is infinite on an increasing curve.

Next, we discuss some dependence and tail properties for (4.82) when $\eta_1, \ldots, \eta_d$ are fixed, $\theta_j = \eta_j - \theta_0$ and θ_0 varies from 0 to $\min\{\eta_1, \ldots, \eta_d\}$.

• The family is increasing in concordance as θ_0 increases. (4.84) becomes $C^\perp$ for $\theta_0 = 0$. If $\eta_1 = \ldots = \eta_d = \eta$, then (4.84) becomes C^+ for $\theta_0 = \eta$.

• The positive dependence property of positive orthant dependence holds. Also any variable is stochastically increasing in the others.

• If one or more θ_j is 0 for $1 \le j \le d$, then the support of copula (4.84) is a proper subset of $[0,1]^d$. For example, if $\theta_1 = 0$, then $X_1 \le \min\{X_2, \ldots, X_d\}$.

• Suppose $F(\cdot; \eta)$ has finite second moment. For $d \ge 3$, consider the conditional correlation

$$\rho_{12;3:d}(x_3, \ldots, x_d) = \mathrm{Cor}(X_1, X_2 | X_3 = x_3, \ldots, X_d = x_d).$$

Then

$$\lim_{x_3 \to 0^+, \ldots, x_d \to 0^+} \rho_{12;3:d}(x_3, \ldots, x_d) = 0 \quad \text{and} \quad \lim_{x_3 \to \infty, \ldots, x_d \to \infty} \rho_{12;3:d}(x_3, \ldots, x_d) = 1.$$

That is, the copula $C_{12;3:d}(\cdot; x_3, \ldots, x_d; \boldsymbol{\theta})$ based on $F_{1|3:d}$ and $F_{2|3:d}$ ranges from $C^\perp$ to C^+.

• Tail order/dependence properties. Consider the exchangeable bivariate case with $d = 2$ and $\eta = \eta_1 = \eta_2$. If $F(\cdot; \eta)$ is Gamma (with fixed scale parameter), then the upper tail order is 1 with upper tail dependence parameter of 0 (and slowly varying function in Definition 2.10 is $\ell^*(u) = (-\log u)^{-\theta_1}$) and the lower tail order is $\kappa_L = 1 + \theta_1/\eta$. See details in Example 8.7. If $F(\cdot; \eta)$ is inverse Gaussian (with fixed non-convolution parameter), then the upper tail order is 1 and the lower tail order is $\kappa_L = 1 + \theta_1/\eta$.

4.29 Morgenstern or FGM

The Morgenstern or Farlie-Gumbel-Morgenstern (FGM) family of copulas consists of small perturbations of $C^\perp$, so have limited range of dependence. This family, and others like it, are not useful for statistical modeling. However, because calculations of many things can be done in closed form, it is useful for demonstrating dependence concepts.

4.29.1   Bivariate FGM

For $-1 \leq \delta \leq 1$, the copula cdf is:

$$C(u, v; \delta) = uv[1 + \delta(1 - u)(1 - v)], \quad 0 \leq u, v \leq 1. \tag{4.85}$$

The conditional cdf is:

$$C_{2|1}(v|u; \delta) = v + \delta v(1 - v)(1 - 2u)$$

and the inverse is

$$C_{2|1}^{-1}(p|u; \delta) = \frac{[1 + \delta(1 - 2u)] - \sqrt{[1 + \delta(1 - 2u)]^2 - 4p(1 - 2u)}}{2\delta(1 - 2u)}$$

with $C_{2|1}^{-1}(p|u; \delta) = p$ if $u = 0.5$. The density is

$$c(u, v; \delta) = 1 + \delta(1 - 2u)(1 - 2v), \quad 0 \leq u, v \leq 1.$$

Some properties of (4.85) are the following.

- Increasing in $\prec_c$, increasing in $\prec_{SI}$.
- $C^\perp$ for $\delta = 0$, positive dependence for $\delta > 0$, negative dependence for $\delta < 0$.
- TP$_2$ density for $\delta \geq 0$, and RR$_2$ density for $\delta < 0$.
- Reflection symmetry.
- If $(U, V) \sim C(\cdot; \delta)$, then the copula of $(U, 1 - V)$ is $u - C(u, 1 - v; \delta) = C(u, v; -\delta)$.
- Kendall's $\tau = 2\delta/9$ with range from $-2/9$ to $2/9$.
- Blomqvist's $\beta = \delta/4$ with range from $-1/4$ to $1/4$.
- Spearman's $\rho_S = \delta/3$ with range from $-1/3$ to $1/3$.

4.29.2   Multivariate extensions of FGM

Several multivariate extensions (4.85) are possible; they are all perturbations of the d-variate $C^\perp$. One extension is

$$C_{1:d}\big(\boldsymbol{u}; \delta_{jk}, 1 \leq j < k \leq d\big) = u_1 \cdots u_d\Big[1 + \sum_{1 \leq j < k \leq d} \delta_{jk}(1 - u_j)(1 - u_k)\Big], \quad \boldsymbol{u} \in [0, 1]^d.$$

The parameter constraints include $|\delta_{jk}| \leq 1$ for all j, k and there are also other joint inequality constraints in the parameters to achieve a non-negative density (at the corner points).

Higher order products can also appear in the perturbation component; Johnson and Kotz (1975) and Shaked (1975) have:

$$C_{1:d}\big(\boldsymbol{u}; \delta_S, S \subset \{1, \ldots, d\}\big) = u_1 \cdots u_d\Big[1 + \sum_{S \subset \{1,\ldots,d\}, 2 < |S| < d} \delta_S \prod_{in S}(1 - u_j)\Big], \quad \boldsymbol{u} \in [0, 1]^d.$$

The constraints on $\{\delta_S, S \subset \{1, \ldots, d\}\}$ are that the mixed derivative is non-negative.

Origins and further notes

- The earliest reference of the bivariate copula is in Morgenstern (1956).
- Subsequent references are Gumbel (1958) and Farlie (1960); also Gumbel (1960a) uses this copula for a bivariate exponential distribution.

4.30 Fréchet's convex combination

Fréchet (1958) considered a convex combination of the independence, comonotonicity and countermonotonicity bivariate copulas; see also (4.3.9) of Mardia (1962):

$$C(u,v;\delta_1,\delta_2) = (1 - \delta_1 - \delta_2)C^\perp(u,v) + \delta_1 C^+(u,v) + \delta_2 C^-(u,v), \quad 0 \le u,v \le 1, \quad (4.86)$$

with $\delta_1 \ge 0$, $\delta_2 \ge 0$ and $\delta_1 + \delta_2 \le 1$. For $\delta = \delta_1$ and $\delta_2 = 0$, special case is:

$$C(u,v;\delta,0) = \delta \min\{u,v\} + (1-\delta)uv, \quad 0 \le \delta \le 1.$$

If $(U,V) \sim C$, then $\mathbb{P}(U = V) = \delta_1$ and $\mathbb{P}(U = 1-V) = \delta_2$, and the mass of the absolutely continuous component is $1 - \delta_1 - \delta_2$. The conditional cdf is:

$$C_{2|1}(v|u;\delta_1,\delta_2) = (1 - \delta_1 - \delta_2)v + \delta_1 I_{[u,1]}(v) + \delta_2 I_{[1-u,1]}(v).$$

Because of the singularity, this copula is not useful for two continuous variables, but it could be used for two discrete variables.

Some properties of (4.86) are the following.

- Increasing in $\prec_c$ and δ_1 increases and δ_2 decreases.
- $C^\perp$ for $\delta_1 = \delta_2 = 0$, C^+ for $\delta_1 = 1$, C^- for $\delta_2 = 1$.
- Reflection symmetry.
- Blomqvist's $\beta = \delta_1 - \delta_2$.
- Spearman's $\rho_S = \delta_1 - \delta_2$.
- Kendall's $\tau = \delta_1^2 - \delta_2^2 + 2\delta_1(1 - \delta_1 - \delta_2)/3 - 2\delta_2(1 - \delta_1 - \delta_2)/3 - 2\delta_1\delta_2/3 = (\delta_1^2 - \delta_2^2 + 2\delta_1 - 2\delta_2 - 2\delta_1\delta_2)/3$.

Multivariate extension

There is no countermonotonicity copula for dimensions $d \ge 3$, but there are other extremal copulas of form $C^-\big(C^+(u_1,\dots,u_m), C^+(u_{m+1},\dots,u_d)\big)$ if $U_1 \sim U(0,1)$ and $U_1 = \cdots = U_m = 1 - U_{m+1} = \cdots = 1 - U_d$; see the derivation in (2.39).

Let J_1, J_2, J_3 be a partition of $\{1,\dots,d\}$ where some of the subsets could be empty. Define

$$C^{(J_1,J_2,J_3)}(\boldsymbol{u}) = C^-\big(C^+(u_i, i \in J_1), C^+(u_i, i \in J_3)\big) \cdot C^\perp(u_i, i \in J_2),$$

where $C^+(u) = u$, $C^\perp(u) = u$, $C^+(\emptyset) = 1$ and $C^\perp(\emptyset) = 1$. Yang et al. (2009) consider multivariate copulas families that are convex combinations of the $C^{(J_1,J_2,J_3)}$ over all partitions J_1, J_2, J_3; each bivariate margin is in the family (4.86).

4.31 Additional parametric copula families

Many more parametric copula families can be constructed based on the approaches in Chapter 3, especially in the bivariate case. It is doubtful that the existing parametric bivariate copula families can approximate all bivariate copulas that have simple dependence structure such as stochastic increasing positive dependence. As the sample size increases, multi-parameter copula families become easier to distinguish.

One purpose of this section is to indicate the steps needed to show if a new family is similar to an existing family, using the approach in Section 5.7. A couple of the families are listed to show that even if copula cdfs have simple closed form with known dependence and tail properties, they might not be computable in double precision software because of nested exponentials.

4.31.1 Archimedean copula: LT is integral of Mittag-Leffler LT

The Mittag-Leffler LT is $\varphi(s;\theta,\delta) = (1+s^{1/\delta})^{-1/\theta}$, with $\theta > 0$ and $\delta > 1$. For this family, something similar to the Archimedean copula family based on the integrated positive stable LT can be obtained. This shows that one can get an Archimedean copula involving the incomplete beta function (compare Section 4.11 with the incomplete gamma function and Section 4.12 with the Bessel function of the second kind).

Let $\zeta = 1/\theta$ and $\vartheta^{-1} = \zeta - \delta > 0$, that is, $\vartheta = (\theta^{-1} - \delta)^{-1} > 0$. Then the integral of this LT is:

$$\int_s^\infty (1+t^{1/\delta})^{-\zeta}dt = \int_{s^{1/\delta}}^\infty (1+w)^{-\zeta}\delta w^{\delta-1}dw = \delta\int_{s^{1/\delta}/(1+s^{1/\delta})}^1 v^{\delta-1}(1-v)^{\zeta-\delta-1}dv$$

$$= \delta B(\delta,\zeta-\delta)\big[1 - F_B\big(s^{1/\delta}/(1+s^{1/\delta}),\delta,\zeta-\delta\big)\big],$$

where B is the Beta function and $F_B(\cdot,a,b)$ is the cdf of the Beta(a,b) random variable. By normalizing, one gets a LT family:

$$\psi(s;\vartheta,\delta) = 1 - F_B(s^{1/\delta}/(1+s^{1/\delta}),\delta,\vartheta^{-1}), \quad s \geq 0,$$

with

$$\psi^{-1}(t;\vartheta,\delta) = \Big\{\frac{1}{1 - F_B^{-1}(1-t,\delta,\vartheta^{-1})} - 1\Big\}^\delta, \quad 0 \leq t \leq 1.$$

Consider the Archimedean copula C_ψ with $\vartheta > 0$ and $\delta > 1$. The cdf, conditional cdf and density are:

$$C(u,v;\vartheta,\delta) = \psi\big(\psi^{-1}(u;\vartheta,\delta) + \psi^{-1}(v;\vartheta,\delta)\big), \quad 0 \leq u,v \leq 1,$$

$$C_{2|1}(v|u;\vartheta,\delta) = \frac{\psi'\big(\psi^{-1}(u;\vartheta,\delta) + \psi^{-1}(v;\vartheta,\delta)\big)}{\psi'\big(\psi^{-1}(u;\vartheta,\delta)\big)},$$

$$c(u,v;\vartheta,\delta) = \frac{\psi''\big(\psi^{-1}(u;\vartheta,\delta) + \psi^{-1}(v;\vartheta,\delta)\big)}{\psi'\big(\psi^{-1}(u;\vartheta,\delta)\big)\psi'\big(\psi^{-1}(u;\vartheta,\delta)\big)},$$

$$\psi'(s;\vartheta,\delta) = -(1+s^{1/\delta})^{-\zeta}/[\delta B(\delta,\vartheta^{-1})], \quad \zeta = \vartheta^{-1} + \delta,$$

$$\psi''(s;\vartheta,\delta) = \zeta s^{1/\delta-1}(1+s^{1/\delta})^{-\zeta-1}/[\delta^2 B(\delta,\vartheta^{-1})].$$

Some properties are the following.

- Numerically, C_ψ increases in concordance as ϑ increases with fixed δ, and also as δ increases with fixed ϑ. The limit is C^+ as $\delta \to \infty$ with ϑ fixed, or as $\vartheta \to \infty$ with δ fixed. As $\delta \to 1$ with $\vartheta > 0$ fixed, the limit is the MTCJ copula with parameter ϑ. As $\vartheta \to 0^+$ with δ fixed, the limit is the copula based on the integrated positive stable LT (Section 4.11).

- Using Theorem 8.26. Kendall's $\tau = 1 - 4B(2\delta,2\vartheta^{-1})/[\delta\, B^2(\delta,\vartheta^{-1})]$. Using Stirling's approximation, $\tau \to 1$ as $\delta \to \infty$ with ϑ fixed. Using $\Gamma(z) \sim z^{-1}$ as $z \to 0^+$, $\tau \to 1$ as $\vartheta \to \infty$ with δ fixed. Numerically, τ is increasing in ϑ,δ.

- The lower tail dependence parameter is

$$\lambda_L = 2 \lim_{s\to\infty} \frac{\psi'(2s;\vartheta,\delta)}{\psi'(s;\vartheta,\delta)} = 2 \times 2^{-\zeta/\delta} = 2^{-1/(\vartheta\delta)},$$

and the upper tail order is $\kappa_U = 1+\delta^{-1}$ because $\varphi(s)$ has maximal non-negative moment degree $M_\psi = \delta^{-1}$.

- If $\kappa = \kappa_U \in (1,2)$ and $\lambda = \lambda_L \in (0,1)$ are specified, then $\delta = (\kappa-1)^{-1} > 1$ and $\vartheta^{-1} = -\delta\log\lambda_L/\log 2 > 0$.

- Given $\lambda = \lambda_L \in (0,1)$ and $\delta > 1$, then $\vartheta^{-1} = -\delta \log \lambda / \log 2 > 0$ so that any combination is (λ, δ) is possible. Given (λ, δ), Blomqvist's beta is

$$\beta = 3 - 4F_B\left(\frac{s_0^{1/\delta}}{1 + s_0^{1/\delta}}; \delta, -\delta \log \lambda / \log 2\right), \quad s_0 = 2\left[\frac{1}{1 - F_B^{-1}(\frac{1}{2}; \delta, -\delta \log \lambda / \log 2)} - 1\right]^\delta.$$

  Numerically, this is increasing as δ increases. Hence the minimum and maximum values of β come from $\delta \to 1^+$ and $\delta \to \infty$. The maximum is 1 and the minimum is $4(2^{1+\vartheta_1} - 1)^{-1/\vartheta_1} - 1$ with $\vartheta_1 = -\log 2 / \log \lambda$. For the maximum, as $\delta \to \infty$, the Beta$(\delta, -\delta \log \lambda / \log 2)$ random variable converges in probability to a degenerate random variable at $m_0 = \log 2 / (\log 2 - \log \lambda)$. Therefore $1 - F_B^{-1}(\frac{1}{2}; \delta, -\delta \log \lambda / \log 2) \to (-\log \lambda) / (\log 2 - \log \lambda)$, $s_0^{1/\delta} \to \log 2 / (-\log \lambda)$, $\beta \sim 3 - 4F_B(m_0; \delta, -\delta \log \lambda / \log 2) \to 3 - 4 \times \frac{1}{2} = 1$. For the minimum, with $\delta \to 1^+$ and $\vartheta \to \vartheta_1$, $F_B(u; 1, \vartheta_1^{-1}) = 1 - (1-u)^{1/\vartheta_1}$, $F_B^{-1}(p; 1, \vartheta_1^{-1}) = 1 - (1-p)^{\vartheta_1}$, $s_0 = 2(2^{\vartheta_1} - 1)$ and $\beta \to 3 - 4F_B(s_0/(1 + s_0); 1, \vartheta_1^{-1}) = 4(1 + s_0)^{-1/\vartheta_1} - 1$.

- For bivariate copulas with tail dependence in one tail and intermediate tail dependence in the other, this family has more range of (β, λ) and (λ, κ) than the BB6 two-parameter family.

4.31.2 Archimedean copula based on positive stable stopped Sibuya LT

Consider the LT family $\varphi(s; \theta, \delta) = \exp\{-(-\log[1 - (1 - e^{-s})^{1/\delta}])^{1/\theta}\}$ for $\theta > 1$ and $\delta \geq 1$. Then $\varphi(s; \theta, \delta) = \psi(-\log \zeta(s; \delta); \theta)$ where ψ is positive stable and ζ is Sibuya, and $\varphi^{-1}(t; \theta, \delta) = -\log\{1 - [1 - e^{-(-\log t)^\theta}]^\delta\}$. The bivariate Archimedean copula cdf is

$$C(u, v; \theta, \delta) = \exp\{-(-\log[1 - (1 - xy)^{1/\delta}])^{1/\theta}\},$$
$$x = 1 - [1 - e^{-(-\log u)^\theta}]^\delta, \quad y = 1 - [1 - e^{-(-\log v)^\theta}]^\delta, \quad 0 \leq u, v \leq 1.$$

This copula is also a mixture of max-id copula with $K(\cdot; \delta)$ being the Joe/B5 copula and $\psi(\cdot; \theta)$ being the positive stable LT.

  Some properties are the following.

- Increasing in $\prec_c$ as δ increases.
- The Joe/B5 family obtains if $\theta \to 1^+$, and the Gumbel family obtains as $\delta \to 1^+$.
- $\varphi(s; \theta, \delta) \sim 1 - s^{1/(\delta\theta)}$ as $s \to 0^+$, and using Theorem 8.33 there is upper tail dependence with $\lambda_U = 2 - 2^{1/(\delta\theta)}$.
- $\varphi(s; \theta, \delta) \sim \exp\{-(\log \delta + s)^{1/\theta}\}$ as $s \to \infty$ and there is lower intermediate tail dependence with $\kappa_L = 2^{1/\theta}$.
- This copula involves the composition of two one-parameter LT families. Although dependence properties can be obtained, there are problems in computing the cdf and density as u, v get closer to 0. For example, for moderate dependence such as with $\kappa_L = 1.4$, $\lambda_U = 0.5$, $\theta = 2.060$, $\delta = 1.205$, $\beta = 0.572$, and $0 < u = v \leq 0.01 = u_0$, $x = y = 1 - [1 - e^{-(-\log u)^\theta}]^\delta$ cannot be computed in double precision. With more dependence, the boundary point u_0, where numerical problems occur, increases.

4.31.3 Archimedean copula based on gamma stopped Sibuya LT

Consider the LT family $\varphi(s; \theta, \delta) = \{1 - \log[1 - (1 - e^{-s})^{1/\delta}]\}^{-1/\theta}$ for $\theta > 0$ and $\delta \geq 1$. Then $\varphi(s; \theta, \delta) = \psi(-\log \zeta(s; \delta); \theta)$ where ψ is gamma and ζ is Sibuya, and $\varphi^{-1}(t; \theta, \delta) = -\log[1 - (1 - e^{1-t^{-\theta}})^\delta]$. The bivariate Archimedean copula is

$$C(u, v; \theta, \delta) = \{1 - \log[1 - (1 - xy)^{1/\delta}]\}^{-1/\theta}, \quad x = 1 - (1 - e^{1-u^{-\theta}})^\delta, \quad y = 1 - (1 - e^{1-v^{-\theta}})^\delta,$$

and $0 \le u, v \le 1$. This copula is also a mixture of max-id copula with $K(\cdot; \delta)$ being the Joe/B5 copula and $\psi(\cdot; \theta)$ being the gamma LT.

Some properties are the following.

- Increasing in $\prec_c$ as δ increases.
- The Joe/B5 family obtains if $\theta \to 0^+$ and the MTCJ family obtains as $\delta \to 1^+$.
- $\varphi(s; \theta, \delta) \sim 1 - \theta^{-1} s^{1/\delta}$ as $s \to 0^+$, and using Theorem 8.33 and there is upper tail dependence with $\lambda_U = 2 - 2^{1/\delta}$.
- $\varphi(s; \theta, \delta) \sim (1 + \log \delta + s)^{-1/\theta}$ as $s \to \infty$ and there is lower tail dependence with $\lambda_L = 2^{-1/\theta}$.
- This copula involves the composition of two one-parameter LT families. Although dependence properties can be obtained, there are problems in computing the cdf and density as u, v get closer to 0. For example, for moderate dependence such as with $\lambda_L = 0.5$, $\lambda_U = 0.5$, $\theta = 1.0$, $\delta = 1.710$, $\beta = 0.489$, and $0 < u = v \le 0.01 = u_0$, $x = y = 1 - (1 - e^{1 - u^{-\theta}})^\delta$ cannot be computed in double precision. With more dependence, the boundary point u_0, where numerical problems occur, increases.

4.31.4 3-parameter families with a power parameter

We list two 3-parameter families that are considered interesting from the theory in Section 3.6. That is, with $\eta > 0$, $C_{\psi,P}(u, v; \eta) = K^{1/\eta}(u^\eta, v^\eta)$ if $K(u, v) = \psi(\psi^{-1}(u) + \psi^{-1}(v))$ or $C_{\psi,R}(u, v; \eta) = K^{1/\eta}(u^\eta, v^\eta)$ if $K(u, v) = u + v - 1 + \psi(\psi^{-1}(1 - u) + \psi^{-1}(1 - v))$, where K is based on a 2-parameter Archimedean copula family.

The extension of BB1 with a power parameter only works for the reflected copula because the BB1 copula cdf already has powers of u, v. Similarly, BB7 can only be extended for $C_{\psi,P}$ because the BB7 cdf already has powers of $1 - u$ and $1 - v$. The idea of adding a third parameter is to see if these 3-parameter families have flexibility in separately affecting the joint upper tail, the middle, and the joint lower tail.

Case 1. *Power of reflected BB1*

By adding a power parameter to the reflected BB1 copula in (4.53), $C_{\psi,R}$ leads to

$$C(u, v; \theta, \delta, \eta) = \left[u^\eta + v^\eta - 1 + \left\{ 1 + \left[([1 - u^\eta]^{-\theta} - 1)^\delta + ([1 - v^\eta]^{-\theta} - 1)^\delta \right]^{1/\delta} \right\}^{-1/\theta} \right]^{1/\eta}, \quad (4.87)$$

with $0 \le u, v \le 1$.

Properties include the following.

- With $\psi(s) = (1 + s^{1/\delta})^{-1/\theta}$ for Table 3.2, this is case (c)(iii) in the rightmost columns with $\zeta = (\delta\theta)^{-1}$, so that the limit is the Galambos($\delta\theta$) copula as $\eta \to 0^+$ and C^+ as $\eta \to \infty$. The independence copula $C^\perp$ obtains as $\delta\theta \to 0^+$.
- Upper tail dependence: $\lambda_U = 2^{-1/(\delta\theta)}$.
- Lower tail dependence: $\lambda_L = (2 - 2^{1/\delta})^{1/\eta}$.
- Blomqvist's $\beta = 4\beta^* - 1$, where

$$\beta^* = C(\tfrac{1}{2}, \tfrac{1}{2}; \theta, \delta, \eta) = \left[2^{1-\eta} - 1 + \left\{ 1 + 2^{1/\delta}([1 - 2^{-\eta}]^{-\theta} - 1) \right\}^{-1/\theta} \right]^{1/\eta}.$$

For fixed θ, δ, Blomqvist's β is decreasing and then increasing in η. For fixed η, because the BB1 copula is increasing in concordance as θ, δ increase, then Blomqvist's β is increasing in θ, δ.

- With $\theta \to 0^+$, the result is a 2-parameter copula family that is the reflected Gumbel copula with a power parameter. For the positive stable LT, which is in case (b)(iii) of Table 3.2 so that $C^\perp$ obtains as $\eta \to 0^+$ and C^+ obtains as $\eta \to \infty$. The copula is

$$\left[u^\eta + v^\eta - 1 + \exp\left\{-\left([-\log(1-u^\eta)]^\delta + [-\log(1-v^\alpha)]^\delta\right)^{1/\delta}\right\}\right]^{1/\eta}, \quad \delta \geq 1, \eta > 0.$$

and Blomqvist's beta is

$$4\left[2^{1-\eta} - 1 + (1 - 2^{-\eta})^{2^{1/\delta}}\right]^{1/\eta} - 1.$$

The lower tail dependence is $\lambda_L = (2 - 2^{1/\delta})^{1/\eta}$. The upper tail order is $\kappa_L = 2^{1/\delta}$.

- For the 3-parameter family, given $\eta, \lambda_L, \lambda_U$, one can solve for δ, θ and get $\delta = (\log 2)/\log(2 - \lambda_L^\eta)$ and $\theta = (\log 2)/(-\delta \log \lambda_U) = \log(2 - \lambda_L^\eta)/(-\log \lambda_U)$. There is a wide range of β with given λ_L, λ_U. In this parametrization, as η increases with λ_L, λ_U fixed, then β decreases. It can be shown that $\beta \to 1$ as $\eta \to 0^+$ and $\beta \to \max\{0, 2\lambda_L - 1\}$ as $\eta \to \infty$.

Proof of the limits for β:

Note that $2^{1/\delta} = (2 - \lambda_L^\eta)$. Write as the following in order to make approximations in steps:

$$t_1 = (1 - 2^{-\eta})^{\log(2-\lambda_L^\eta)/\log \lambda_U} - 1, \quad t_2 = \{1 + (2 - \lambda_L^\eta)t_1\}^{(\log \lambda_U)/\log(2-\lambda_L^\eta)},$$
$$\beta^* = [2^{1-\eta} - 1 + t_2]^{1/\eta}.$$

As $\eta \to 0^+$ with $x > 0$, an approximation is $x^\eta \sim 1 + \eta \log x$ or $1 - x^\eta \sim -\eta \log x$.

- $\eta \to 0^+$: let $p = -\log \lambda_U > 0$, $q = -\log \lambda_L > 0$;

$$t_1 \sim (\eta \log 2)^{-p^{-1}\log(1-\eta \log \lambda_L)} - 1 \sim (\eta \log 2)^{-\eta p^{-1} q} - 1,$$
$$t_2 \sim \{1 + (1 + \eta q)[(\eta \log 2)^{-\eta p^{-1} q} - 1]\}^{-p/(\eta q)}$$
$$\sim \{1 + [(\eta \log 2)^{-\eta p^{-1} q} - 1] + (\eta q)[(\eta \log 2)^{-\eta p^{-1} q} - 1]\}^{-p/(\eta q)}$$
$$\sim \{(\eta \log 2)^{-\eta p^{-1} q}\}^{-p/(\eta q)} \sim \eta \log 2 \quad \text{(above third term is order of magnitude smaller)},$$
$$\beta^* \sim [2(1 - \eta \log 2) - 1 + \eta \log 2]^{1/\eta} \sim [1 - \eta \log 2]^{1/\eta} \to e^{-\log 2} = \tfrac{1}{2}.$$

Hence $\beta \to 4 \cdot \tfrac{1}{2} - 1 = 1$.

- $\eta \to \infty$:

$$t_1 \sim \{1 + 2^{-\eta}p^{-1}\log(2 - \lambda_L^\eta) + \tfrac{1}{2}2^{-2\eta}[-p^{-1}\log(2-\lambda_L^\eta)][-p^{-1}\log(2-\lambda_L^\eta) - 1]\} - 1$$
$$\sim 2^{-\eta}p^{-1}\log(2-\lambda_L^\eta) + \tfrac{1}{2}2^{-2\eta}p^{-1}\log(2-\lambda_L^\eta)[1 + p^{-1}\log(2-\lambda_L^\eta)],$$
$$t_2 \sim \{1 + (2 - \lambda_L^\eta)2^{-\eta}p^{-1}\log(2-\lambda_L^\eta)$$
$$\qquad + (2 - \lambda_L^\eta) \cdot \tfrac{1}{2}2^{-2\eta}[p^{-1}\log(2-\lambda_L^\eta)][1 + p^{-1}\log(2-\lambda_L^\eta)]\}^{-p/\log(2-\lambda_L^\eta)}$$
$$\sim 1 - (2 - \lambda_L^\eta)2^{-\eta} - \tfrac{1}{2}(2 - \lambda_L^\eta)2^{-2\eta}[1 + p^{-1}\log(2-\lambda_L^\eta)]$$
$$\qquad + \tfrac{1}{2}[p^{-1}\log(2-\lambda_L^\eta)][1 + p/\log(2-\lambda_L^\eta)](2 - \lambda_L^\eta)^2 2^{-2\eta}$$
$$\sim 1 - (2 - \lambda_L^\eta)2^{-\eta} + 2^{-2\eta}(1 + p^{-1}\log 2),$$
$$\beta^* \sim [\lambda_L^\eta 2^{-\eta} + 2^{-2\eta}(1 + p^{-1}\log 2)]^{1/\eta} \to \max\{\tfrac{1}{2}\lambda_L, \tfrac{1}{4}\}.$$

Hence $\beta \to \max\{2\lambda_L - 1, 0\}$.

Case 2. *Power of BB7*

By adding a power parameter to the BB7 copula (4.70), $C_{\psi,P}$ leads to

$$C(u, v; \theta, \delta, \eta) = \left\{ 1 - \left(1 - \left[(1 - \{1 - u^\eta\}^\theta)^{-\delta} + (1 - \{1 - v^\eta\}^\theta)^{-\delta} - 1 \right]^{-1/\delta} \right)^{1/\theta} \right\}^{1/\eta}, \quad (4.88)$$

with $0 \le u, v \le 1$.

Properties include the following.

- With $\psi(s) = 1 - [1 - (1 + s)^{-1/\delta}]^{1/\theta}$ for Table 3.2, this is case (c)(iii) in the middle columns with $\xi = \theta^{-1}$, so that the limit is the Gumbel(θ) copula as $\eta \to 0^+$ and C^+ as $\eta \to \infty$. The independence copula $C^\perp$ obtains as $\theta \to 1$.

- Upper tail dependence: $\lambda_U = 2 - 2^{1/\theta}$.

- Lower tail dependence: $\lambda_L = 2^{-1/(\delta\eta)}$.

- Blomqvist's $\beta = 4\beta^* - 1$, where

$$\beta^* = C(\tfrac{1}{2}, \tfrac{1}{2}; \theta, \delta, \eta) = \left\{ 1 - \left(1 - [2(1 - \{1 - 2^{-\eta}\}^\theta)^{-\delta} - 1]^{-1/\delta} \right)^{1/\theta} \right\}^{1/\eta}.$$

For fixed θ, δ, Blomqvist's β is decreasing and then increasing in η. For fixed η, because the BB7 copula is increasing in concordance as δ increases with fixed θ and increasing in concordance as θ increases with fixed $0 < \delta \le 1$, then Blomqvist's β is increasing for much of the range of θ, δ.

- Given $\eta, \lambda_L, \lambda_U$, one can solve for δ, θ to get $\theta = (\log 2)/\log(2 - \lambda_U)$ and $\delta = (\log 2)/(-\eta \log \lambda_L)$ (latter $\to \infty$ as $\eta \to 0^+$). In this parametrization, as η increases with λ_L, λ_U fixed, then β decreases for part of the range; the range of β's is not as wide as for (4.87).

- It can be shown that $\beta \to 4 \cdot 2^{\,2^{1/\theta(\lambda_U)}} - 1$ as $\eta \to 0^+$, and $\beta \to 4 \cdot (2^{1+\Delta(\lambda_L)} \ 1)^{-1/\Delta(\lambda_L)} - 1$ as $\eta \to \infty$, where $\Delta = (\log 2)/(-\log \lambda_L)$, $\theta = (\log 2)/\log(2 - \lambda_U)$. With fixed λ_L, λ_U, sometimes the limiting β as $\eta \to 0^+$ is larger and sometimes the limiting β as $\eta \to \infty$ is larger. This means that the largest possible β might be at a non-boundary η value when λ_L, λ_U are fixed.

Proof of the limits for β:

Let $\delta = \Delta/\eta$ where $\Delta = (\log 2)/(-\log \lambda_L)$; θ depends on λ_U but not η. Write as the following in order to make approximations in steps:

$$t_1 = (1 - \{1 - 2^{-\eta}\}^\theta)^{-\Delta/\eta}, \quad t_2 = [2t_1 - 1]^{-\eta/\Delta},$$
$$t_3 = (1 - t_2)^{1/\theta}, \quad \beta^* = (1 - t_3)^{1/\eta}.$$

As $\eta \to 0^+$ with $x > 0$, an approximation is $x^\eta \sim 1 + \eta \log x$ or $1 - x^\eta \sim -\eta \log x$.

- $\eta \to 0^+$: let $p = (\log 2)^\theta$, $\beta \to 2^{2 - 2^{1/\theta}} - 1$:

$$t_1 \sim (1 - \{\eta \log 2\}^\theta)^{-\Delta/\eta} = (1 - \eta^\theta p)^{-\Delta/\eta} \sim (e^{p\Delta})^{\eta^{\theta-1}} \sim 1 + \eta^{\theta-1} p\Delta,$$
$$t_2 \sim [1 + 2\eta^{\theta-1} p\Delta]^{-\eta/\Delta} \sim (e^{-2p})^{\eta^\theta} \sim 1 - 2p\eta^\theta,$$
$$t_3 \sim (2p\eta^\theta)^{1/\theta} = (2p)^{1/\theta}\eta,$$
$$\beta^* \sim (1 - \eta(2p)^{1/\theta})^{1/\eta} \to \exp\{-(2p)^{1/\theta}\} = \exp\{-2^{1/\theta}(\log 2)\} = 2^{-2^{1/\theta}}.$$

- $\eta \to \infty$: $\beta \to 4[2^{1+\Delta} - 1]^{-1/\Delta} - 1$:

$$t_1 \sim (\theta 2^{-\eta})^{-\Delta/\eta} = 2^\Delta \theta^{-\Delta/\eta},$$
$$t_2 \sim [2^{1+\Delta}\theta^{-\Delta/\eta} - 1]^{-\eta/\Delta} \to 0,$$
$$t_3 \sim 1 - \theta^{-1}t_2,$$
$$\beta^* \sim (\theta^{-1}t_2)^{1/\eta} \sim [2^{1+\Delta}\theta^{-\Delta/\eta} - 1]^{-1/\Delta} \to [2^{1+\Delta} - 1]^{-1/\Delta}.$$

Comparison of the two 3-parameter copula families (4.87) **and** (4.88)

With (λ_L, λ_U) fixed, when λ_U is small and λ_L is large, then the power of BB7 has a wider range of β's than the power of reflected BB1, but there are computational roundoffs from the small η's needed to get a smaller β (than is achievable with power of reflected BB1).

4.32 Dependence comparisons

In this section, some dependence comparisons are made for some bivariate copula families that (a) have both lower and upper tail dependence; (b) have tail dependence in one tail and intermediate tail dependence in the other tail.

Table 4.3 is expanded from the Appendix of Nikoloulopoulos et al. (2012), comparing copula families with both lower and upper tail dependence.

$\lambda_L = \lambda_U$	BB1/BB4	BB7	gaSSiA	t_2	t_5	t_{10}
0.1	0.181	0.156	0.174	-0.192	0.146	0.368
0.2	0.274	0.226	0.259	0.036	0.323	0.503
0.3	0.363	0.294	0.338	0.204	0.448	0.597
0.4	0.450	0.363	0.415	0.345	0.549	0.671
0.5	0.537	0.433	0.489	0.471	0.637	0.736
0.6	0.625	0.508	0.558	0.586	0.717	0.795
0.7	0.715	0.588	0.617	0.694	0.792	0.849
0.8	0.807	0.680	0.677	0.798	0.863	0.901

Table 4.3 *Blomqvist's β for fixed $\lambda_L = \lambda_U$ for BB1, BB4, BB7 and the t-copula with $\nu = 2, 5, 10$. gaSSiA is the 2-parameter Archimedean copula family based on the gamma stopped Sibuya LT in Section 4.31.3.*

Further comparisons can be made with t_ν copulas when $\lambda_L = \lambda_U$. Numerically β is smaller for BB7 than BB1/BB4, when λ_L, λ_U are fixed. The difference of the β values can be over 0.1 for some (λ_L, λ_U) pairs. For the 2-parameter Archimedean copula family based on the gamma stopped Sibuya LT, the β values are in between those of BB1/BB4 and BB7. Hence, for a fixed β, BB7 has heavier tail dependence. Table 4.3 shows that BB1/BB4 are closer to the t_5 copula than BB7. For the t_ν copula, as ν increases, fixed $\lambda_U = \lambda_L$ means that β increases as ν increases. Equivalently for fixed β, tail dependence decreases as ν increases.

With λ_U, λ_L fixed in order to determine θ, δ and θ_R, δ_R, the L_∞ distance of $C(\cdot; \theta, \delta)$ and the reflected copula $\widehat{C}(\cdot; \theta_R, \delta_R)$ were compared for BB1 and BB7. In this comparison, the BB7 copula had smaller L_∞ distance than the BB1 copula. That is, for BB7, $C(\cdot; \theta, \delta)$ and $\widehat{C}(\cdot; \theta_R, \delta_R)$ are closer to each other.

Because of the above properties, for the modeling with vine and factor copulas, it might be sufficient to consider the BB1, reflected BB1 and BB7 for the bivariate copula familes with asymmetric lower and upper tail dependence.

There are several bivariate 2-parameter copula families which have tail dependence in one tail and intermediate tail dependence in the other tail. BB5 is a 2-parameter extreme value copula family so that $\kappa_L = 2 - \lambda_U$; that is, the lower tail order is determined by the upper tail dependence parameter. The BB6 family has more flexibility in that $\kappa_L = (2 - \lambda_U)^\theta$ with $\theta \geq 1$ so that $2 - \lambda_U \leq \kappa_L$. For the 2-parameter Archimedean family based on the positive stable stopped Sibuya LT, $\kappa_L = (2 - \lambda_U)^\delta$ with $\delta \geq 1$ so that also $2 - \lambda_U \leq \kappa_L$. The 2-parameter Archimedean family based on the integrated Mittag-Leffler LT has more flexibility in that $(\kappa_U - 1, \lambda_U) \in (0, 1)^2$. Table 4.4 has values of Blomqvist's β for values of

(κ, λ) for BB6, the Archimedean copula family based on the positive stable stopped Sibuya LT and the Archimedean copula family based on the integrated Mittag-Leffler LT; it shows that the latter has different behavior than the BB6 copula.

κ	λ	β_{BB6}	β_{psSSiA}	β_{iMitLefA}
1.9	0.2	0.116	0.116	0.208
1.9	0.4	0.226	0.226	0.300
1.9	0.6	0.383	0.384	0.428
1.9	0.9	0.801	0.803	0.802
1.7	0.4	0.283	0.284	0.366
1.6	0.5	0.377	0.378	0.451
1.5	0.2	NA	NA	0.383
1.5	0.3	NA	NA	0.415
1.5	0.4	NA	NA	0.450
1.5	0.5	0.414	0.414	0.490
1.5	0.6	0.477	0.479	0.539
1.5	0.8	0.661	0.671	0.687
1.4	0.7	0.585	0.590	0.634
1.3	0.8	0.703	0.712	0.737

Table 4.4 *Blomqvist's β for fixed κ, λ (tail order at one tail and tail dependence parameter at the other) for BB6, psSSiA (the Archimedean copula family in Section 4.31.2 based on the positive stable stopped Sibuya LT) and iMitLefA (the Archimedean copula family in Section 4.31.1 based on the integrated Mittag-Leffler LT). NA means that a combination of (κ, λ) is not available or possible.*

For the BB6 copula, given $\lambda = \lambda_U$, the largest possible value of Blomqvist's β is $2^\lambda - 1$, and then one can solve for θ in the equation

$$\beta = 3 - 4\left[1 - (1 - 2^{-\theta})^{(2-\lambda)^\theta}\right]^{1/\theta}$$

where $\delta = \theta^{-1}(\log 2)/\log(2-\lambda)$ and $\kappa = 2^{1/\delta} = (2-\lambda)^\theta$. For the Archimedean family based on the integrated Mittag-Leffler LT, given λ, β, one can solve for (ϑ, δ) from the equations:

$$\lambda = 2^{-(\vartheta\delta)}, \quad \beta = 3 - 4F_B(s_0^{1/\delta}/(1 + s_0^{1/\delta}); \delta, \vartheta^{-1}), \quad s_0 = 2\left[\frac{1}{1 - F_B^{-1}(\tfrac{1}{2}, \delta, \vartheta^{-1})} - 1\right]^\delta.$$

and then $\kappa = 1 + \delta^{-1}$.

λ	$\beta_{\text{BB6}}(\text{LB})$	$\beta_{\text{BB6}}(\text{UB})$	$\beta_{\text{iMitLefA}}(\text{LB})$	$\beta_{\text{iMitLefA}}(\text{UB})$
0.1	0.040	0.072	0.127	1
0.2	0.086	0.149	0.174	1
0.3	0.140	0.231	0.220	1
0.4	0.203	0.320	0.272	1
0.5	0.277	0.414	0.333	1
0.6	0.368	0.516	0.408	1
0.7	0.481	0.625	0.504	1
0.8	0.623	0.741	0.631	1
0.9	0.801	0.866	0.801	1

Table 4.5 *Lower bound (LB) and upper bound (UB) of Blomqvist's β for fixed λ (tail dependence at one tail) for the BB6 copula and the Archimedean copula based on the integrated Mittag-Leffler LT (Section 4.31.1).*

Table 4.5 shows the range of β for different λ values for the BB6 copula and the Archimedean copula based on the integrated Mittag-Leffler LT. The latter has more flexibility which may explain why it seems to lead to larger log-likelihoods when both are used for some data sets in Chapter 7.

4.33 Summary for parametric copula families

The multivariate copulas arising from the vine pair-copula construction and various factor copula models make use of a sequence of bivariate parametric copula families; the bivariate copulas that can be used for different edges of the vine are algebraically independent. Hence these copula models are flexible and can approximate many multivariate distributions. It is important to understand the differences among parametric copula families, so this chapter has summarized dependence and tail properties of bivariate copula families.

Different parametric copula families can be distinguished more easily when there is stronger dependence. With weaker dependence, parametric copula families become less distinguishable. This is based on Kullback-Leibler divergences; see Section 5.7. As the sample size increases, parametric copula models with more parameters can be considered.

In particular, for the vine pair-copula construction, where typically $d-1$ pairs of variables with stronger dependence are assigned to the edges of tree $\mathcal{T}_1$, one could consider bivariate parametric copula families with three or more parameters. For example, one of the parameters could account for permutation asymmetry. For higher-level trees of the vine, including factor copula models, 1-parameter bivariate copula families may be adequate, especially when the conditional dependence is relatively weaker.

Some of the techniques in Sections 3.17 and 3.18 could be used to get bivariate copula families with additional parameters.

Chapter 5

Inference, diagnostics and model selection

This chapter mainly emphasizes likelihood-based inference methods for estimation in parametric models and the use of diagnostics as a guide to choosing models and assessing their fits. Dependence models based on copulas are generally approximations and some results are stated with this point of view. Parametric families that can be easily constructed tend to have the property of monotone positive (or negative) dependence for bivariate margins. When bivariate margins do not have simple relationships then non-simple parametric families or non-parametric approaches can be considered; there is some discussion of the latter in Section 5.10.3.

Diagnostics include assessment of univariate and multivariate tail asymmetry, strength of dependence in the joint tails relative to Gaussian, conditional dependence, and adequacy of fit for tail inference.

Section 5.1 discusses parametric inference for copula models. Section 5.2 summarizes some main results on likelihood inference with maximum likelihood. Section 5.3 has the log-likelihood for copula models for continuous variables, discrete variables, and mixed continuous-discrete. Section 5.4 has main results for asymptotic theory of maximum likelihood, including the case that the true model is not necessarily the model being fitted. Section 5.5 discusses inference functions and estimating equations which are used for sequential estimation of parameters. Section 5.6 discusses composite likelihood which might be suitable when there are common parameters in different margins. Section 5.7 has results on Kullback-Leibler divergence and the Kullback-Leibler sample size needed to differentiate parametric copula families with high probability; this type of analysis is relevant for model comparisons.

The second part of the chapter is on initial data analysis and diagnostics for assessing fit and deciding among copula models. Section 5.8 discusses initial data analyses to decide on models for univariate margins and the dependence structure. Section 5.9 introduces the pseudo likelihood for copulas with empirical univariate margins; this can be used for sensitivity analysis on the suitability of univariate and dependence models. Section 5.10 discusses non-parametric methods for copulas that can be used for further models or diagnostics, and includes some results on the empirical copula. Section 5.11 has some diagnostics for conditional dependence. Section 5.12 discusses diagnostic methods for assessing the fit of copula models for continuous and discrete response variables. Section 5.13 discusses the comparison of parametric models based on Vuong's log-likelihood ratio method.

Section 5.14 contains a chapter summary.

5.1 Parametric inference for copulas

For dependence modeling with copulas, most common is the use of parametric copula families. Advantages of parametric inference methods for copula models are the following.

- They are easier to numerically implement than non-parametric approaches.
- They can be used in high dimensions.

- With the use of the likelihood, it is the same theory for continuous, discrete or mixed response variables, and censored or missing data can be accommodated.
- They can be adapted to univariate margins that are regression or time series models, or to time-varying dependence. Covariates can be included in univariate or dependence parameters.
- They can easily be used to compare competing models.

Chapter 3 has copula construction methods for different dependence structures such as exchangeable, 1-factor, truncated vine, multiple-factor, structured-factor, hierarchical, etc. For likelihood inference with copula models, two key aspects to capture are (i) dependence structure and (ii) tail behavior. Kullback-Leibler divergence is linked to likelihood inference with parametric models even when models under consideration are "misspecified." Kullback-Leibler divergence can help in understanding what types of copula models are similar or can be easily distinguished. In this chapter, it will be shown through Kullback-Leibler divergence that the concept of tail order (Section 2.16) is relevant and can distinguish copula models. The dependence structure is important for copula models with continuous and discrete response variables. Tail behavior is important for continuous variables and less important for discrete response variables with few categories; in the latter case, one can not readily see extra dependence in the joint tails, so copula models can be less easy to distinguish.

Based on empirical experience in fitting parametric vine and factor copula models to a moderate to large number of response variables, we note the following comparison summarized in Table 5.1. In this table, the first column might be Gaussian and the second column might be something with tail dependence, or tail asymmetry, or both. If the tail properties of the bivariate copula families used within the vine/factor models are quite different, then generally the rows and columns of the table will be similarly ordered; this applies also when the numbers of dependence structures and tail behaviors are each greater than two. This motivates the use of Gaussian models as baseline models from which to find copula models with improved fits to the data.

	tail behavior 1	tail behavior 2
dependence structure 1	L_{11}	L_{12}
dependence structure 2	L_{21}	L_{22}

Table 5.1 *Cross-classification to parametric vine/factor copula models with two different dependence structures and two families of bivariate copulas with different tail behaviors: log-likelihoods are usually similarly ordered over rows or columns, or approximately similarly ordered when the number of rows/columns is more than two.*

Because Gaussian copula models correspond to classical multivariate normal methods after variables have been transformed to $N(0, 1)$, the methodology has been more developed and it is computationally easier to compare many different dependence structures. After some candidate dependence structures are found that provide good fits and that are reasonable based on the context of the application, one can try for improved model fits with copula versions having the same dependence structures but allowing for different tail properties.

The above heuristics for copula selection are summarized in the following remark.

Remark 5.1 Empirical experience with many data sets suggests the following when all variables are monotonically related to each other.

- Use the Gaussian copula to determine suitable dependence structures among exchangeable, conditionally independent (p-factor), hierarchical, autoregressive, Markov tree, truncated vine, etc.; these are discussed in Chapter 3.

- For the best few models found among Gaussian copulas with special dependence structures, find corresponding parametric copula models that better match the tails of the data. The better fits lead to smaller values of AIC or BIC as defined in (1.11). Initial data analysis can be a guide to tail asymmetry in the bivariate margins or more dependence in the joint tails than Gaussian.

5.2 Likelihood inference

The most general methods of estimation for parametric models are based on the likelihood. Maximum likelihood inference is straightforward to do in statistical software if the likelihood is computationally feasible. If (i) the log-likelihood of the multivariate model is computationally demanding or infeasible but log-likelihoods of low-dimensional margins are computationally feasible, and (ii) parameters are identifiable from the low-dimensional margins, then composite likelihood methods are good options. If univariate parameters are different for different univariate margins, then a sequential method (inference functions for margins) of estimating univariate parameters followed by dependence parameters is a computational method to avoid a likelihood optimization over a large parameter vector (consisting of the union of d univariate parameter vectors and a vector of dependence/copula parameters). The estimator from the method of inference functions for margins can be a reported summary from the data analysis or it can be used as a good starting point for full maximum likelihood. See Section 5.5 for more details.

For full maximum likelihood, composite likelihood and inference functions for margins, computational implementation minimally requires the coding of the log-likelihood of the multivariate density, or the log-likelihoods for some of the low-dimensional marginal densities. The corresponding negative log-likelihoods or composite log-likelihoods can be input into a numerical minimizer and the output includes the point of minimum $\hat{\boldsymbol{\theta}}$ of the objective function and the Hessian matrix $\boldsymbol{H}(\hat{\boldsymbol{\theta}})$ of second order derivatives at the point of minimum.

Bayesian inference can be used if there is historical information that can be included in a prior; this can be an advantage if the sample size is small. For a large sample size, typically Bayesian inference proceeds without much thought on the nearly flat prior since the log-likelihood would dominate the prior. For models based on copulas, Min and Czado (2010), Min and Czado (2011), and Smith and Khaled (2012) and others have implemented Markov chain Monte Carlo (MCMC) methods, including reversible jump MCMC methods, for comparisons of different pair-copula constructions with continuous and discrete response data. There is no special Bayesian or MCMC theory specific for copula models; for multivariate copula models outside of multivariate Gaussian, copula densities are not in the exponential family, and there are no convenient conjugate families. MCMC methods take much more computational running time than those using likelihood-based estimating/inference equations.

With a flat prior in the neighborhood of the maximum likelihood estimate and a large enough sample size, the Bayesian posterior is approximately multivariate normal with center at the maximum likelihood estimate $\hat{\boldsymbol{\theta}}_{\text{obs}}$ and covariance matrix equal to the inverse Hessian $\boldsymbol{H}^{-1}(\hat{\boldsymbol{\theta}}_{\text{obs}})$. That is, it is approximately the same as the plug-in sampling distribution $\text{N}(\hat{\boldsymbol{\theta}}_{\text{obs}}, \boldsymbol{H}^{-1}(\hat{\boldsymbol{\theta}}_{\text{obs}}))$ of the maximum likelihood estimator from the observed sample. Interval estimates will then be similar.

If $\boldsymbol{\theta}$ is the parameter vector in the original parametrization and $\boldsymbol{\vartheta}$ is the parameter vector in alternative parametrization (so that there is a 1-1 mapping from $\boldsymbol{\theta}$ to $\boldsymbol{\vartheta}$), then asymptotic maximum likelihood theory (see Section 5.4 for a summary) says that with sample size n,

$$n^{1/2}(\hat{\boldsymbol{\theta}} - \boldsymbol{\theta}) \rightarrow_d \text{N}(\boldsymbol{0}, \mathcal{I}^{-1}(\boldsymbol{\theta})),$$

$$n^{1/2}(\hat{\boldsymbol{\vartheta}} - \boldsymbol{\vartheta}) \rightarrow_d \text{N}(\boldsymbol{0}, \mathcal{I}_t^{-1}(\boldsymbol{\vartheta})),$$

as $n \to \infty$, where $\mathcal{I}_t = \mathcal{J} \mathcal{I} \mathcal{J}^\top$ and $\mathcal{J}$ is the Jacobian matrix of the transformation. An advantage of Bayesian MCMC is that the posterior sample from MCMC can be inspected for closeness to multivariate normality and if needed, a transformation of parameters can be found that makes the multivariate normality a better approximation. That is, the posterior sample can suggest the transform $\boldsymbol{h}$ to use to get $\boldsymbol{\vartheta} = \boldsymbol{h}(\boldsymbol{\theta})$ that has a posterior distribution that is closer to multivariate normal.

5.3 Log-likelihood for copula models

In this section, the copula log-likelihoods are given in different situations, such as response variables types, censoring and missingness. For ease of notation, the assumption is that the log-likelihoods are based on multivariate i.i.d. data $(y_{i1} \ldots, y_{id})$, $i = 1, \ldots, d$, from the model

$$F(\cdot; \boldsymbol{\eta}_1, \ldots, \boldsymbol{\eta}_d, \boldsymbol{\delta}) = C\big(F_1(\cdot; \boldsymbol{\eta}_1), \ldots, F_d(\cdot; \boldsymbol{\eta}_d); \boldsymbol{\delta}\big).$$

If there are covariates, then any parameter (univariate or dependence) could be considered a function of the covariates.

Let the log-likelihood be denoted as $L = L(\boldsymbol{\eta}_1, \ldots, \boldsymbol{\eta}_d, \boldsymbol{\delta})$,

- If all variables are continuous and $F_1, \ldots, F_d, C$ are absolutely continuous with respective densities $f_1, \ldots, f_d, c$, then the log-likelihood is

$$L = \sum_{i=1}^{n} \Big\{ \log c\big(F_1(y_{i1}; \boldsymbol{\eta}_1), \ldots, F_d(y_{id}; \boldsymbol{\eta}_d); \boldsymbol{\delta}\big) + \sum_{j=1}^{d} \log f_j(y_{ij}; \boldsymbol{\eta}_j) \Big\}.$$

- If all variables are discrete, then the log-likelihood is:

$$L = \sum_{i=1}^{n} \log \Big\{ \sum_{(\diamond_1 \cdots \diamond_d) \in \{-,+\}^d} (-1)^{\#\{\diamond_j = -\}} C\big(F_1(y_{i1}^{\diamond_1}; \boldsymbol{\eta}_1), \ldots, F_d(y_{id}^{\diamond_d}; \boldsymbol{\eta}_d); \boldsymbol{\delta}\big) \Big\},$$

where y_{ij}^+ (with $\diamond_j = +$) and y_{ij}^- (with $\diamond_j = -$) in the cdf refer to evaluation as right limits and left limits respectively.

- In case of mixed variable types with S_1 being the set of indices of the continuous variables and S_2 being the set of indices of the discrete variables, the log-likelihood is:

$$L = \sum_{i=1}^{n} \Big\{ \log c_{S_1}\big(F_j(y_{ij}; \boldsymbol{\eta}_j) : j \in S_1; \boldsymbol{\delta}\big) + \sum_{j \in S_1} \log f_j(y_{ij}; \boldsymbol{\eta}_j)$$

$$+ \log \sum_{\diamond_k \in \{-,+\} : k \in S_2} (-1)^{\#\{\diamond_k = -\}} C_{S_2|S_1}\big(F_k(y_{ik}^{\diamond_k}; \boldsymbol{\eta}_k) : k \in S_2 \mid F_j(y_{ij}; \boldsymbol{\eta}_1) : j \in S_1; \boldsymbol{\delta}\big) \Big\},$$

- If all variables are continuous and variables indexed in S_{2i} are right-censored at $(y_{ik}, k \in S_{2i})$ for the ith observation, then the contribution to the log-likelihood is:

$$\log c_{S_{1i}}\big(F_j(y_{ij}; \boldsymbol{\eta}_j) : j \in S_{1i}; \boldsymbol{\delta}\big) + \sum_{j \in S_{1i}} \log f_j(y_{ij}; \boldsymbol{\eta}_j)$$

$$+ \log \overline{C}_{S_{2i}|S_{1i}}\big(F_k(y_{ik}; \boldsymbol{\eta}_k) : k \in S_{2i} \mid F_j(y_{ij}; \boldsymbol{\eta}_j) : j \in S_{1i}; \boldsymbol{\delta}\big);$$

where S_{1i} is the set of indices of the uncensored observations.

- If variables indexed in S_{2i} are missing at random for the ith observation, then the contribution to the log-likelihood is:

$$\log c_{S_{1i}}\big(F_j(y_{ij}; \boldsymbol{\eta}_j) : j \in S_{1i}; \boldsymbol{\delta}\big) + \sum_{j \in S_{1i}} \log f_j(y_{ij}; \boldsymbol{\eta}_j),$$

where S_{1i} is the set of indices of the non-missing observations. The maximum likelihood estimator is consistent if the pattern of missingness over all observations is such that all parameters are identifiable in the combined log-likelihood. If the missingness is non-ignorable and there is a probability model for the non-missingness, the log probability of missingness can be added to the above.

For non-Gaussian stationary time series models for $\{Y_t : 1 \leq t \leq n\}$, where a copula is used for an Markov transition function of order $(d - 1)$, the joint cdf of $Y_t, Y_{t+1}, \ldots, Y_{t+d-1}$ has form $C_{1:d}(F_Y(z_1), \ldots, F_Y(z_d); \boldsymbol{\delta})$. The transition cdf is $C_{d|1:(d-1)}(F_Y(z_d)|F_Y(z_1), \ldots, F_Y(z_{d-1}); \boldsymbol{\delta})$, and the transition density is

$$c_{d|1:(d-1)}\big(F_Y(z_d)|F_Y(z_1), \ldots, F_Y(z_{d-1}); \boldsymbol{\delta}\big) f_Y(z_d)$$

for absolutely continuous F_Y and

$$C_{d|1:(d-1)}\big(F_Y(z_d)|F_Y(z_1), \ldots, F_Y(z_{d-1}); \boldsymbol{\delta}\big) - C_{d|1:(d-1)}\big(F_Y(z_d^-)|F_Y(z_1), \ldots, F_Y(z_{d-1}); \boldsymbol{\delta}\big)$$

for discrete F_Y. The conditional log-likelihood is the sum of logarithms of these transition densities applied to $(y_{t-d}, \ldots, y_{t-1}, y_t)$ for $t = d, \ldots, n$.

5.4   Maximum likelihood: asymptotic theory

The presentation here is for a random sample (i.i.d. observations), but the theory is similar if there are covariates. The asymptotic result is given for the case where the parametric density $f(\cdot; \boldsymbol{\theta})$ used for inference is not necessarily the true density that generated the data.

Let $\boldsymbol{Y}_1, \ldots, \boldsymbol{Y}_n$ be a random sample from a density g with respect to ν (Lebesgue measure or counting measure), and let the realizations be $\boldsymbol{y}_1, \ldots, \boldsymbol{y}_n$. Let $f(\cdot; \boldsymbol{\theta})$ be a parametric model with parameter $\boldsymbol{\theta}$ which is a subset of a Euclidean space. Let $\ell(\boldsymbol{\theta}; \boldsymbol{y}) = \log f(\boldsymbol{y}; \boldsymbol{\theta})$. Consider maximum likelihood based on this model where the log-likelihood is

$$L(\boldsymbol{\theta}) = L(\boldsymbol{\theta}; \boldsymbol{y}_1, \ldots, \boldsymbol{y}_n) = \sum_{i=1}^{n} \ell(\boldsymbol{\theta}; \boldsymbol{y}_i)$$

and the maximum likelihood estimator (MLE) is

$$\hat{\boldsymbol{\theta}} = \hat{\boldsymbol{\theta}}(\boldsymbol{y}_1, \ldots, \boldsymbol{y}_n) = \operatorname{argmax} L(\boldsymbol{\theta}; \boldsymbol{y}_1, \ldots, \boldsymbol{y}_n).$$

Let $\boldsymbol{\Psi} = \frac{\partial \ell}{\partial \boldsymbol{\theta}}$ be considered as a vector of inference functions.

Under some regularity conditions, such as in White (2004) and Vuong (1989), including the MLE asymptotically being in the interior of the parameter space, the following hold.

- $\hat{\boldsymbol{\theta}}$ is the root of

$$\sum_{i=1}^{n} \frac{\partial \ell(\boldsymbol{\theta}; \boldsymbol{y}_i)}{\partial \boldsymbol{\theta}} = \sum_{i=1}^{n} \boldsymbol{\Psi}(\boldsymbol{\theta}; \boldsymbol{y}_i) = \boldsymbol{0}.$$

- $\hat{\boldsymbol{\theta}} \to_p \boldsymbol{\theta}^*$, where $\boldsymbol{\theta}^*$ maximizes

$$\int g(\boldsymbol{y}) \log f(\boldsymbol{y}; \boldsymbol{\theta}) \, d\nu(\boldsymbol{y}),$$

or minimizes the divergence

$$\int g(\boldsymbol{y}) \log[g(\boldsymbol{y})/f(\boldsymbol{y}; \boldsymbol{\theta})] \, d\nu(\boldsymbol{y}),$$

and is the root of

$$\int g(\boldsymbol{y})\frac{\partial \ell(\boldsymbol{\theta};\boldsymbol{y})}{\partial \boldsymbol{\theta}}\,\mathrm{d}\nu(\boldsymbol{y}) = \mathbb{E}_g[\boldsymbol{\Psi}(\boldsymbol{\theta};\boldsymbol{Y})] = \boldsymbol{0},$$

with $\boldsymbol{Y} \sim g$. Here $\mathbb{E}_g$ is used to indicate that expectation is taken assuming the true density is g. $\boldsymbol{\Psi}$ is called the inference function vector in Section 5.5.

- $n^{1/2}(\hat{\boldsymbol{\theta}} - \boldsymbol{\theta}^*) \to_d \mathrm{N}(\boldsymbol{0}, \boldsymbol{H}^{-1}\boldsymbol{J}\boldsymbol{H}^{-1})$, where

$$\boldsymbol{H} = -\mathbb{E}_g\left[\frac{\partial^2 \ell}{\partial \boldsymbol{\theta}\partial \boldsymbol{\theta}^\top}\right] = -\mathbb{E}_g\left[\frac{\partial \boldsymbol{\Psi}(\boldsymbol{\theta},\boldsymbol{Y})}{\partial \boldsymbol{\theta}^\top}\right] \tag{5.1}$$

is the Hessian matrix of the negative log density and the negative expected value of the derivative of the inference function vector, and

$$\boldsymbol{J} = \mathrm{Cov}_g(\boldsymbol{\Psi}(\boldsymbol{\theta};\boldsymbol{Y})) \tag{5.2}$$

is the covariance matrix of the inference function vector.

- When $g = f(\cdot; \boldsymbol{\theta}^*)$ and the standard regularity conditions of maximum likelihood hold (see Serfling (1980)), then $\boldsymbol{H} = \boldsymbol{J}$ is the Fisher information matrix evaluated at $\boldsymbol{\theta}^*$.

With a correctly specified model, for further inference following maximum likelihood, with the plug-in method, the asymptotic approximation to the sampling distribution of the MLE as a random vector is

$$\mathrm{N}(\hat{\boldsymbol{\theta}}_{\mathrm{obs}}, n^{-1}\hat{\mathcal{I}}^{-1}),$$

where $n\hat{\mathcal{I}}$ is the Hessian of the negative log-likelihood at the MLE $\hat{\boldsymbol{\theta}}_{\mathrm{obs}}$. This can be used for confidence intervals with the delta method. With the expected information matrix, one can use $\mathrm{N}(\hat{\boldsymbol{\theta}}_{\mathrm{obs}}, n^{-1}\mathcal{I}^{-1}(\hat{\boldsymbol{\theta}}_{\mathrm{obs}}))$. For the delta method, one has asymptotically

$$\boldsymbol{k}(\hat{\boldsymbol{\theta}}) \stackrel{\cdot}{\sim} \mathrm{N}\big(\boldsymbol{k}(\boldsymbol{\theta}^*), n^{-1}(\nabla \boldsymbol{k})^\top \mathcal{I}^{-1}(\boldsymbol{\theta}^*)\nabla \boldsymbol{k}\big),$$

where $\boldsymbol{k}$ can have any dimension. For interval estimates or confidence intervals etc., with plug-in, one can use the approximation $\mathrm{N}(\boldsymbol{k}(\hat{\boldsymbol{\theta}}_{\mathrm{obs}}), n^{-1}(\nabla \boldsymbol{k})^\top(\hat{\boldsymbol{\theta}}_{\mathrm{obs}})\hat{\mathcal{I}}^{-1}\nabla \boldsymbol{k}(\hat{\boldsymbol{\theta}}_{\mathrm{obs}}))$. If there is concern that the model is misspecified, replace the inverse Fisher information matrix by an estimate of the inverse Godambe matrix in Section 5.5.

In practice, the *rate of convergence to normality* and the quality of the asymptotic approximation depends on the parametrization. For an alternative to the delta method for interval estimates, obtain a good transformation function and reparametrize with $\boldsymbol{\vartheta} = \boldsymbol{h}(\boldsymbol{\theta})$. Get the observed Fisher information for the $\boldsymbol{\vartheta}$ parametrization. Simulate from the plug-in sampling distribution, $\mathrm{N}(\hat{\boldsymbol{\vartheta}}_{\mathrm{obs}}, \mathcal{I}_t^{-1}(\hat{\boldsymbol{\vartheta}}_{\mathrm{obs}}))$. From this interval estimates can be obtained for $k(\boldsymbol{\vartheta})$ for any real-valued function k.

5.5 Inference functions and estimating equations

There are a number of likelihood-based methods where parameter estimates are based on solving score equations of (weighted sums of) low-dimensional log-likelihoods, provided the parameters are identifiable from lower-dimensional margins. With multivariate models, inference based on lower-dimensional margins is often chosen for numerical efficiency and feasibility, or as a preliminary step for full maximum likelihood estimation. Numerical considerations are the following.

- A numerical optimization with many parameters is replaced by a sequence of smaller numerical optimization problems. This sometimes is a practical need because numerical optimization methods might behave worse as the dimension of the parameter vector increase.

- The full likelihood might involve multi-dimensional integrals which are numerically too time-consuming to evaluate.

Also the estimates of parameters of low-dimensional margins will be less sensitive to the correct specification of the higher-dimensional margins.

Inference based on low-dimension margins include the following.

- The method called *inference functions for margins* (IFM) in Joe and Xu (1996) and Joe (1997), for which univariate parameters are first estimated based on individual univariate log-likelihoods and then multivariate parameters are estimated from the multivariate log-likelihood or lower-dimensional log-likelihoods with univariate parameter estimates held fixed. Some analysis of asymptotic efficiency of IFM is given in Joe (2005).
- The method of *composite marginal likelihood estimation*, with typical examples of maximizing the (weighted) sum of log-likelihoods of bivariate margins or of trivariate margins. This method is used when there are parameters common to different margins. A survey of composite likelihood methodology is given in Varin et al. (2011).

For multivariate models, where there are independent replicates and the sample size n can be large, the analysis of these methods is based on the theory of inference functions or estimating equations.

A brief summary of the theory is given for the i.i.d. case with data vector $\boldsymbol{y}_1, \ldots, \boldsymbol{y}_n$ but this extends to independent observations with covariates.

Let $\boldsymbol{\Psi}$ be a vector of functions with the same dimension as parameter vector $\boldsymbol{\theta} = (\theta_1, \ldots, \theta_q)^\top$. The vector of *inference functions* is

$$\sum_{i=1}^{n} \boldsymbol{\Psi}(\boldsymbol{\theta}, \boldsymbol{y}_i). \tag{5.3}$$

Let $\boldsymbol{Y}, \boldsymbol{Y}_1, \ldots, \boldsymbol{Y}_n$ be i.i.d. with the density g. The inference functions are associated with a parametric model $f(\cdot; \boldsymbol{\theta})$, so that $\mathbb{E}_g[\boldsymbol{\Psi}(\boldsymbol{\theta}^*, \boldsymbol{Y})] = \boldsymbol{0}$ when $g = f(\cdot; \boldsymbol{\theta}^*)$.

For a typical use of IFM with the copula model with continuous variables, let $\boldsymbol{Y}_1, \ldots, \boldsymbol{Y}_n$ be i.i.d. from the cdf $F(\cdot; \boldsymbol{\theta}) = C(F_1(\cdot; \boldsymbol{\eta}_1), \ldots, F_d(\cdot; \boldsymbol{\eta}_d); \boldsymbol{\delta})$ where $\boldsymbol{\theta} = (\boldsymbol{\eta}_1^\top, \ldots, \boldsymbol{\eta}_d^\top, \boldsymbol{\delta}^\top)^\top$. Let the density functions be $f_1, \ldots, f_d, c$. For the terms of the log-likelihoods, let $\ell_j = \log f_j$ for $j = 1, \ldots, d$ and let $\ell_{\text{mult}} = \log c(F_1, \ldots, F_d)$. Then the IFM estimator is the root of (5.3) with

$$\boldsymbol{\Psi}(\boldsymbol{\theta}, \boldsymbol{y}) = \begin{pmatrix} \partial \ell_1(\boldsymbol{\eta}_1, y_1)/\partial \boldsymbol{\eta}_1 \\ \vdots \\ \partial \ell_d(\boldsymbol{\eta}_d, y_d)/\partial \boldsymbol{\eta}_d \\ \partial \ell_{\text{mult}}(\boldsymbol{\delta}, \boldsymbol{\eta}_1, \ldots, \boldsymbol{\eta}_d, \boldsymbol{y})/\partial \boldsymbol{\delta} \end{pmatrix}.$$

Let $f(\cdot; \boldsymbol{\theta}) = c(F_1(\cdot; \boldsymbol{\eta}_1), \ldots, F_d(\cdot; \boldsymbol{\eta}_d); \boldsymbol{\delta}) \cdot \prod_{j=1}^{d} f_j(\cdot; \boldsymbol{\eta}_j)$. If the true density is $g = f(\cdot; \boldsymbol{\theta}^*)$ where $\boldsymbol{\theta}^*$ is in the interior of the parameter space, then $\mathbb{E}_g[\boldsymbol{\Psi}(\boldsymbol{\theta}^*; \boldsymbol{Y}] = \boldsymbol{0}$ for $\boldsymbol{Y} \sim g$. If the true g with univariate marginal densities $g_1, \ldots, g_d$ is not a member of the candidate model, then $\mathbb{E}_g[\boldsymbol{\Psi}(\boldsymbol{\theta}^*; \boldsymbol{Y}] = \boldsymbol{0}$ under some regularity conditions, where

$$\boldsymbol{\eta}_j^* = \text{argmax} \int g_j(y_j) \log f_j(y_j; \boldsymbol{\eta}_j) \, dy_j, \quad j = 1, \ldots, d,$$

and

$$\boldsymbol{\delta}^* = \text{argmax} \int g(\boldsymbol{y}) \log \Big\{ c(F_1(y_1; \boldsymbol{\eta}_1^*), \ldots, F_d(y_d; \boldsymbol{\eta}_d^*); \boldsymbol{\delta}) \prod_{j=1}^{d} f_j(y_j; \boldsymbol{\eta}_j^*) \Big\} d\boldsymbol{y}.$$

Note that if maximum likelihood with the full multivariate likelihood is feasible, that is, simultaneous estimation of $\boldsymbol{\eta}_1, \ldots, \boldsymbol{\eta}_d, \boldsymbol{\delta}$, then the IFM estimator can be considered as a good starting point for the numerical maximization of the log-likelihood.

Example 5.1 (Sequential estimation).

When there is more structure to the dependence parameters in the copula, the copula parameters need not be estimated simultaneously after the first step of separate univariate likelihoods. This is illustrated for a 2-truncated vine copula for $d \geq 3$ continuous variables with vine array A with diagonal of $1, \ldots, d$. Let $\{a_{1j} : j = 1, \ldots, d\}$ and $\{a_{2j} : j = 2, \ldots, d\}$ be the elements of the first and second row of A.

Let $f_1(\cdot; \boldsymbol{\eta}_1), \ldots, f_d(\cdot; \boldsymbol{\eta}_d)$ be the univariate densities, and let $\ell_j = \log f_j$. For $j = 2, \ldots, d$, let $c_{a_{1j}j}(\cdot; \boldsymbol{\delta}_{a_{1j}j})$ be a copula density for tree $\mathcal{T}_1$ and let $\ell_{a_{1j}j} = \log c_{a_{1j}j}(F_{a_{1j}}, F_j; \boldsymbol{\delta}_{a_{1j}j})$. For $j = 3, \ldots, d$, let $c_{a_{2j}j;a_{1j}}(\cdot; \boldsymbol{\delta}_{a_{2j}j})$ be a copula density for tree $\mathcal{T}_2$ and let $\ell_{a_{2j}j} = \log c_{a_{2j}j;a_{1j}}(F_{a_{2j}|a_{1j}}, F_{j|a_{1j}}; \boldsymbol{\delta}_{a_{2j}j})$.

In Hobæk Haff (2012) and Dissmann et al. (2013), the parameters are estimated sequentially; firstly $\boldsymbol{\eta}_1, \ldots, \boldsymbol{\eta}_d$ from the univariate log-likelihoods, secondly $\boldsymbol{\delta}_{12}, \ldots; \boldsymbol{\delta}_{a_{1d}d}$ from the log-likelihoods for the copula densities in $\mathcal{T}_1$, and then $\boldsymbol{\delta}_{a_{23}3}, \ldots, \boldsymbol{\delta}_{a_{2d}d}$ in $\mathcal{T}_2$ etc.

Then the IFM estimator is the root of (5.3) with

$$
\boldsymbol{\Psi}(\boldsymbol{\theta}, \boldsymbol{y}) =
\begin{pmatrix}
\partial\ell_1(\boldsymbol{\eta}_1, y_1)/\partial\boldsymbol{\eta}_1 \\
\vdots \\
\partial\ell_d(\boldsymbol{\eta}_d, y_d)/\partial\boldsymbol{\eta}_d \\
\partial\ell_{12}(\boldsymbol{\delta}_{12}, \boldsymbol{\eta}_2, \boldsymbol{\eta}_2, y_1, y_2)/\partial\boldsymbol{\delta}_{12} \\
\vdots \\
\partial\ell_{a_{1d}d}(\boldsymbol{\delta}_{a_{1d}d}, \boldsymbol{\eta}_{a_{1d}}, \boldsymbol{\eta}_d, y_{a_{1d}}, y_d)/\partial\boldsymbol{\delta}_{a_{1d}d} \\
\partial\ell_{a_{23}3}(\boldsymbol{\delta}_{a_{23}3}, \boldsymbol{\delta}_{a_{13}3}, \boldsymbol{\delta}_{a_{13}a_{23}}, \boldsymbol{\eta}_{a_{13}}, \boldsymbol{\eta}_{a_{23}}, \boldsymbol{\eta}_3, y_{a_{13}}, y_{a_{23}}, y_3)/\partial\boldsymbol{\delta}_{a_{23}3} \\
\vdots \\
\partial\ell_{a_{2d}d}(\boldsymbol{\delta}_{a_{2d}d}, \boldsymbol{\delta}_{a_{1d}d}, \boldsymbol{\delta}_{a_{1d}a_{2d}}, \boldsymbol{\eta}_{a_{1d}}, \boldsymbol{\eta}_{a_{2d}}, \boldsymbol{\eta}_d, y_{a_{1d}}, y_{a_{2d}}, y_d)/\partial\boldsymbol{\delta}_{a_{2d}d}
\end{pmatrix}.
$$

More detailed comparisons of related stepwise semi-parametric estimation methods for vine copulas are given in Hobæk Haff (2013). □

Next, let's continue with theory on estimating equations. An outline of the asymptotic distribution is given for the root of (5.3), when the true density is g. Let $\partial\boldsymbol{\Psi}/\partial\boldsymbol{\theta}^\top$ be the matrix with (r, s) component $\partial\Psi_r(\boldsymbol{y}, \boldsymbol{\theta})/\partial\theta_s$, where Ψ_r is the rth component of $\boldsymbol{\Psi}$ and θ_s is the sth component of $\boldsymbol{\theta}$. Suppose $\tilde{\boldsymbol{\theta}} = \tilde{\boldsymbol{\theta}}(\boldsymbol{y}_1, \ldots, \boldsymbol{y}_n)$ is the only root satisfying

$$
\sum_{i=1}^n \boldsymbol{\Psi}(\tilde{\boldsymbol{\theta}}; \boldsymbol{y}_i) = \mathbf{0}, \tag{5.4}
$$

and suppose $\boldsymbol{\theta}^*$ is the only root of $\mathbb{E}_g[\boldsymbol{\Psi}(\boldsymbol{\theta}; \boldsymbol{Y})] = \mathbf{0}$. Assume the regularity conditions of score equations in asymptotic likelihood theory hold for $\boldsymbol{\Psi}$. From a first order Taylor expansion of (5.4),

$$
n^{1/2}(\tilde{\boldsymbol{\theta}} - \boldsymbol{\theta}^*) \approx \left\{ -\mathbb{E}_g\left[\frac{\partial\boldsymbol{\Psi}(\boldsymbol{\theta}^*, \boldsymbol{Y})}{\partial\boldsymbol{\theta}^\top}\right] \right\}^{-1} n^{-1/2} \sum_{i=1}^n \boldsymbol{\Psi}(\boldsymbol{\theta}^*, \boldsymbol{Y}_i) + O_p(n^{-1/2}). \tag{5.5}
$$

Hence the asymptotic distribution of $n^{1/2}(\tilde{\boldsymbol{\theta}} - \boldsymbol{\theta}^*)$ is equivalent to that of

$$
\left\{ -\mathbb{E}_g\left[\frac{\partial\boldsymbol{\Psi}(\boldsymbol{\theta}^*, \boldsymbol{Y})}{\partial\boldsymbol{\theta}^\top}\right] \right\}^{-1} \boldsymbol{Z},
$$

where $\boldsymbol{Z} \sim N(\mathbf{0}, \mathrm{Cov}_g[\boldsymbol{\Psi}(\boldsymbol{\theta}^*, \boldsymbol{Y})])$. That is, the asymptotic covariance matrix of $n^{1/2}(\tilde{\boldsymbol{\theta}} - \boldsymbol{\theta}^*)$, called the inverse Godambe information matrix, is

$$
\boldsymbol{V} = \boldsymbol{H}^{-1} \boldsymbol{J} (\boldsymbol{H}^{-1})^\top,
$$

where

$$H = -\mathbb{E}_g\left[\frac{\partial \boldsymbol{\Psi}(\boldsymbol{\theta}^*, \boldsymbol{Y})}{\partial \boldsymbol{\theta}^\top}\right], \quad J = \mathbb{E}_g\left[\boldsymbol{\Psi}(\boldsymbol{\theta}^*, \boldsymbol{Y})\boldsymbol{\Psi}^\top(\boldsymbol{\theta}^*, \boldsymbol{Y})\right]. \tag{5.6}$$

Estimation of the asymptotic covariance matrix $n^{-1}\boldsymbol{V}$ of $\tilde{\boldsymbol{\theta}}$ involves $-\mathbb{E}_g[\partial\boldsymbol{\Psi}(\boldsymbol{\theta}^*, \boldsymbol{Y})/\partial\boldsymbol{\theta}^\top]$ and can be obtained if the derivatives of $\boldsymbol{\Psi}$ can be analytically calculated. Note that if $\boldsymbol{\Psi}$ are the partial derivative equations (score equations of univariate, bivariate or multivariate density) and numerical minimization with a quasi-Newton numerical method is used, then $\boldsymbol{\Psi}$ is not explicitly needed.

5.5.1 Resampling methods for interval estimates

In this subsection, we summarize some resampling methods to get interval estimates of quantities which are computed based on inference equations. More details are given in Shao and Tu (1995).

If data are i.i.d., a general estimation approach of $n^{-1}\boldsymbol{V}$ is via the jackknife. Let $\tilde{\boldsymbol{\theta}}^{(i)}$ be the estimator of $\boldsymbol{\theta}$ with the ith observation $\boldsymbol{Y}_i$ deleted, $i = 1, \ldots, n$. The jackknife estimator of $n^{-1}\boldsymbol{V}$ is

$$\sum_{i=1}^{n}(\tilde{\boldsymbol{\theta}}^{(i)} - \tilde{\boldsymbol{\theta}})(\tilde{\boldsymbol{\theta}}^{(i)} - \tilde{\boldsymbol{\theta}})^\top.$$

Proof: An outline of the derivation is given for the case where the true density is g, which is not necessarily a member of $f(\cdot; \boldsymbol{\theta})$. Let ϵ be the $O_p(n^{-1/2})$ term in (5.5), and suppose the next order term after this is $O_p(n^{-1})$. Substitute an analogy of (5.5) for $\tilde{\boldsymbol{\theta}}^{(i)}$ and use (5.6) to get

$$(\tilde{\boldsymbol{\theta}}^{(i)} - \boldsymbol{\theta}^*) \approx \boldsymbol{H}^{-1}(n-1)^{-1}\sum_{k \neq i}\boldsymbol{\Psi}(\boldsymbol{\theta}^*, \boldsymbol{Y}_k) + n^{-1/2}\epsilon$$

$$\approx \boldsymbol{H}^{-1}(n-1)^{-1}\cdot\left[n\boldsymbol{H}\cdot(\tilde{\boldsymbol{\theta}} - \boldsymbol{\theta}^*) - \boldsymbol{\Psi}(\boldsymbol{\theta}, \boldsymbol{Y}_i) - n^{1/2}\boldsymbol{H}\epsilon\right] + n^{-1/2}\epsilon$$

$$= n(n-1)^{-1}(\tilde{\boldsymbol{\theta}} - \boldsymbol{\theta}^*) - (n-1)^{-1}\boldsymbol{H}^{-1}\boldsymbol{\Psi}(\boldsymbol{\theta}^*, \boldsymbol{Y}_i) + O_p(n^{-3/2}).$$

After eliminating $\boldsymbol{\theta}^*$ from both sides,

$$(\tilde{\boldsymbol{\theta}}^{(i)} - \tilde{\boldsymbol{\theta}}) \approx -n^{-1}\boldsymbol{H}^{-1}\boldsymbol{\Psi}(\boldsymbol{\theta}^*, \boldsymbol{Y}_i) + O_p(n^{-3/2}).$$

Then,

$$\sum_{i=1}^{n}(\tilde{\boldsymbol{\theta}}^{(i)} - \tilde{\boldsymbol{\theta}})(\tilde{\boldsymbol{\theta}}^{(i)} - \tilde{\boldsymbol{\theta}})^\top \approx n^{-2}\boldsymbol{H}^{-1}\cdot\left[\sum_{i=1}^{n}\boldsymbol{\Psi}(\boldsymbol{\theta}^*, \boldsymbol{Y}_i)\,\boldsymbol{\Psi}(\boldsymbol{\theta}^*, \boldsymbol{Y}_i)\right]\cdot\boldsymbol{H}^{-1} + O_p(n^{-3/2})$$

$$\approx n^{-1}\boldsymbol{H}^{-1}\cdot\mathrm{Cov}(\boldsymbol{\Psi}(\boldsymbol{\theta}^*, \boldsymbol{Y}))\cdot\boldsymbol{H}^{-1} + O_p(n^{-3/2})$$

$$= n^{-1}\boldsymbol{V} + O_p(n^{-3/2}).$$

$\square$

For large samples, the jackknife can be modified into estimates from deletions of more than one observation at a time in order to reduce the total amount of computation time. Suppose $n = n_w n_g$, with n_g groups or blocks of size n_w; n_g estimators can be obtained with the kth estimate based on $n - n_w$ observations after deleting the n_w observations in the kth block. (It is probably best to randomize the n observations into the n_g blocks. The jackknife estimates will depend a little on the randomization if the block size is more than 1.) Note that the simple (delete-one) jackknife has $n_w = 1$.

Let $\tilde{\boldsymbol{\theta}}^{(k)}$ be the estimator of $\boldsymbol{\theta}$ with the kth block deleted, $k = 1, \ldots, n_g$. For the asymptotic approximation, consider n_w as fixed with $n_g \to \infty$. The jackknife estimator of $n^{-1}\boldsymbol{V}$ is

$$\sum_{k=1}^{n_g} (\tilde{\boldsymbol{\theta}}^{(k)} - \tilde{\boldsymbol{\theta}})(\tilde{\boldsymbol{\theta}}^{(k)} - \tilde{\boldsymbol{\theta}})^\top.$$

The jackknife method can also be used for estimates of functions of parameters (such as probabilities of being in some category or probabilities of exceedances). The delta or Taylor method requires partial derivatives (of the function with respect to the parameters) and the jackknife method eliminates the need for this. Let $s(\boldsymbol{\theta})$ be a (real-valued) quantity of interest. In addition to $s(\tilde{\boldsymbol{\theta}})$, the estimates $s(\tilde{\boldsymbol{\theta}}^{(k)})$, $k = 1, \ldots, n_g$, from the subsamples can be obtained. The jackknife estimate of the SE of $s(\tilde{\boldsymbol{\theta}})$ is

$$\left\{ \sum_{k=1}^{n_g} \left[s(\tilde{\boldsymbol{\theta}}^{(k)}) - s(\tilde{\boldsymbol{\theta}}) \right]^2 \right\}^{1/2}.$$

Quantities of interest include tail probability such as simultaneously exceeding the 80th percentile for all variables, quantiles of linear combinations, measures of dependence such as Kendall's τ and Spearman's ρ_S; the above would lead to the model-based estimates which could be compared with the empirical estimates as an assessment of the adequacy of the model.

If the copula model has upper (lower) tail dependence, the model-based estimate of the tail dependence parameter may not be very good, as the log-likelihood is dominated by data in the middle. However, as noted in Section 2.13, there is no true empirical version of the tail dependence parameter.

If data are obtained sequentially over a short period of time, and an assumption of stationarity with weak dependence is more appropriate than i.i.d.; then the jackknife resampling method is not appropriate, but stationary bootstrap methods can be combined with estimators from inference equations or averages. One stationary bootstrap method is given in Politis and Romano (1994), and it is described below.

Suppose the stationary data values are $\boldsymbol{Y}_1, \ldots, \boldsymbol{Y}_n$. For a bootstrap sample, $\boldsymbol{Y}_1^* = \boldsymbol{Y}_{I_1}$, where I_1 is uniform on the integers 1 to n. Next let $\boldsymbol{Y}_2^* = \boldsymbol{Y}_{I_2}$ with probability p where I_2 is uniform on the integers 1 to n, and let $\boldsymbol{Y}_2^* = \boldsymbol{Y}_{(I_1+1) \bmod n}$ with probability $1-p$. Continue in this way for $i = 3, \ldots, n$, so that if $\boldsymbol{Y}_i^* = \boldsymbol{Y}_J$ for some $1 \leq J \leq n$, then $\boldsymbol{Y}_{i+1}^* = \boldsymbol{Y}_{(J+1) \bmod n}$ with probability $1 - p$ and picked at random from the original observations with probability p. This is the same as having random blocks of Geometric(p) length from the original data series with wrapping of indices so that $\boldsymbol{Y}_1$ follows $\boldsymbol{Y}_n$. Politis and Romano (1994) suggest $p \propto n^{-1/3}$.

5.6   Composite likelihood

Another version of sets of estimating equations occurs with composite likelihood methods. Composite marginal log-likelihoods are weighted sums of logarithms of low-dimensional marginal densities, such as univariate, bivariate and trivariate. Composite conditional log-likelihoods are weighted sums of logarithms of low-dimensional conditional densities based on subsets of the variables. Composite marginal log-likelihoods are useful when (a) the full multivariate log-likelihood is computationally difficult and infeasible, and log-likelihood of low-dimensional margins are easily computationally feasible, and (b) there are parameters that are common to more than one univariate and bivariate margin. Examples of such parameters are univariate regression parameters common to several margins (in multivariate models for repeated measures data), and dependence parameters common to several bivariate margins because of a partial exchangeable dependence structure. There are many applications of composite likelihood, also called pseudo-likelihood; see Varin (2008) and Varin et al. (2011) for a review. The term "composite likelihood" comes from Lindsay (1988).

For copula models, composite marginal log-likelihoods can be useful for multivariate discrete data when univariate margins are regressions with common parameters and the multivariate copula would require multidimensional integrations to get rectangle probabilities for observed discrete vectors.

Some additional terminology and notation are introduced below.

- The composite log-density based on K different margins or conditional distributions has the form

$$cl(\boldsymbol{\theta}, \boldsymbol{y}) = \sum_{k=1}^{K} w_k l_{S_k}(\boldsymbol{\theta}, \boldsymbol{y}), \quad l_{S_k}(\boldsymbol{\theta}, \boldsymbol{y}) = \log f_{S_k}(y_j, j \in S_k; \boldsymbol{\theta}) \text{ for margin } S_k.$$

- For a sample of size n, the maximum composite log-likelihood estimator $\hat{\boldsymbol{\theta}}$ maximizes $\sum_{i=1}^{n} cl(\boldsymbol{\theta}, \boldsymbol{y}_i)$, where $\boldsymbol{y}_i = (y_{i1}, \ldots, y_{id})$.

- The composite score function is the partial derivative of the composite log-density with respect to the parameter vector:

$$\boldsymbol{\Psi}(\boldsymbol{\theta}, \boldsymbol{y}) = \sum_{k=1}^{K} w_k \nabla_{\boldsymbol{\theta}} l_{S_k}(\boldsymbol{\theta}, \boldsymbol{y}).$$

- Sensitivity or Hessian matrix: $\boldsymbol{H}(\boldsymbol{\theta}) = \mathbb{E}\{-\nabla_{\boldsymbol{\theta}} \boldsymbol{\Psi}^{\top}(\boldsymbol{\theta}, \boldsymbol{Y})\}$.
- Variability matrix: $\boldsymbol{J}(\boldsymbol{\theta}) = \text{Cov}\{\boldsymbol{\Psi}(\boldsymbol{\theta}, \boldsymbol{Y})\}$.
- Godambe information matrix: $\boldsymbol{G}(\boldsymbol{\theta}) = \boldsymbol{H}(\boldsymbol{\theta})\boldsymbol{J}^{-1}(\boldsymbol{\theta})\boldsymbol{H}(\boldsymbol{\theta})$. As $n \to \infty$, $\boldsymbol{V} = \boldsymbol{G}^{-1}(\boldsymbol{\theta})$ is the asymptotic covariance matrix of $n^{1/2}(\hat{\boldsymbol{\theta}} - \boldsymbol{\theta})$ under some regularity conditions. In the case of observation of a single time series or random field, the asymptotics depend on ergodicity conditions and the above n is replaced by the observation length d.
- The composite likelihood version of AIC, as given in Varin and Vidoni (2005), is

$$\text{CLAIC} = -2 \sum_i cl(\hat{\boldsymbol{\theta}}, \boldsymbol{y}_i) + 2 \,\text{tr}[\boldsymbol{J}(\hat{\boldsymbol{\theta}})\boldsymbol{H}^{-1}(\hat{\boldsymbol{\theta}})].$$

- The composite likelihood version of BIC, as given in Gao and Song (2010), is

$$\text{CLBIC} = -2 \sum_i cl(\hat{\boldsymbol{\theta}}, \boldsymbol{y}_i) + (\log n) \,\text{tr}[\boldsymbol{J}(\hat{\boldsymbol{\theta}})\boldsymbol{H}^{-1}(\hat{\boldsymbol{\theta}})].$$

The above could be modified to allow for covariates $\boldsymbol{x}_1, \ldots, \boldsymbol{x}_n$, which are assumed to be well-behaved in order that asymptotic normality holds. Then, as $n \to \infty$, $\boldsymbol{H}(\boldsymbol{\theta}) = \lim_{n\to\infty} n^{-1} \sum_{i=1}^{n} \mathbb{E}\{-\nabla_{\boldsymbol{\theta}} \boldsymbol{\Psi}(\boldsymbol{\theta}, \boldsymbol{Y}_i; \boldsymbol{x}_i)\}$ and $\boldsymbol{J}(\boldsymbol{\theta}) = \lim_{n\to\infty} n^{-1} \sum_{i=1}^{n} \text{Cov}\{\boldsymbol{\Psi}(\boldsymbol{\theta}, \boldsymbol{Y}_i; \boldsymbol{x}_i)\}$.

For numerical estimation, note that $\hat{\boldsymbol{\theta}}$ is straightforward to obtain by inputting $-\sum_i cl(\boldsymbol{\theta}, \boldsymbol{y}_i)$ into a numerical minimizer, and this can also yield an estimate $\widehat{\boldsymbol{H}}$ at $\hat{\boldsymbol{\theta}}$. But estimation of the variability matrix $\boldsymbol{J}$ needed for $\boldsymbol{G}^{-1}$ and standard errors of components of $\hat{\boldsymbol{\theta}}$ can be computationally challenging. With $\boldsymbol{V} = \boldsymbol{H}^{-1}\boldsymbol{J}(\boldsymbol{H}^{-1})^{\top}$, one approach is that $n^{-1}\boldsymbol{V}$ at $\hat{\boldsymbol{\theta}}$ is estimated via jackknife, leading to $\widehat{\boldsymbol{V}}$; this was suggested and used in Zhao and Joe (2005) and is a special case of its use for estimating equations (Section 5.5). Then $\widehat{\boldsymbol{J}} = \widehat{\boldsymbol{H}}\widehat{\boldsymbol{V}}\widehat{\boldsymbol{H}}^{\top}$. The penalty term of CLAIC and CLBIC can be written as $\text{tr}(\boldsymbol{J}(\boldsymbol{H}^{-1})^{\top}) = \text{tr}(\boldsymbol{H}\boldsymbol{V})$, and can be estimated with $\widehat{\boldsymbol{H}}$ and $\widehat{\boldsymbol{V}}$.

An example of the use of composite likelihood is given in Chapter 7.

5.7 Kullback-Leibler divergence

For inferences based on likelihood, the Kullback-Leibler divergence is relevant, especially as the parametric model used in the likelihood could be misspecified; that is, the model might not be the true distribution generating the data, but just a means to make inferences on joint probabilities. Typically, one considers several different models when analyzing data, and from a theoretical point of view, the Kullback-Leibler divergence of pairs of competing models provides information on the sample size need to discriminate them.

We will define Kullback-Leibler divergence for two densities and the expected log-likelihood ratio. Because Kullback-Leibler divergence is a non-negative quantity that is not bounded above, in Section 5.7.1 we use also the expected value of the square of the log-likelihood ratio in order to get a sample size value that is an indication of how different two densities are.

Consider two densities (competing models) f and g with respect to measure ν (usually Lebesgue or counting measure) in $\mathbb{R}^d$. The Kullback-Leibler divergence of f relative to g is

$$KL(f|g) = \int g \log[g/f] \, \mathrm{d}\nu \geq 0; \tag{5.7}$$

this is asymmetric and permuting f, g leads to

$$KL(g|f) = \int f \log[f/g] \, \mathrm{d}\nu = -\int f \log[g/f] \, \mathrm{d}\nu \geq 0. \tag{5.8}$$

Note that $KL(f; g)$ is the expected log-likelihood ratio of g to f when g is the true density — one would expect $\log[g(\boldsymbol{Y})/f(\boldsymbol{Y})]$ to be more often positive, when $\boldsymbol{Y} \sim g$ (and by symmetry, one would expect it to be more often negative when $\boldsymbol{Y} \sim f$). Jeffreys' divergence is the symmetrized version of Kullback-Leibler divergence; it is defined as

$$J(f, g) = J(g, f) := KL(f|g) + KL(g|f) = \int (g - f) \log[g/f] \, \mathrm{d}\nu.$$

The Kullback-Leibler and Jeffreys divergences are applied to copula models in Sections 5.7.2 to 5.7.4. With illustration through the examples, the main results are the following.

1. If two copula densities have a positive dependence property such as stochastically increasing and the same strength of dependence as given by a measure of monotone association (Kendall's tau, Spearman's rho or Blomqvist's beta), then the magnitude of the Kullback-Leibler divergence is related to the closeness of the strength of dependence in the tails. This is because copula densities can asymptote to infinity in a joint tail at different rates (tail order less than dimension d) or converge to a constant in the joint tail (if tail order is the dimension d or larger).

2. Copula models with stronger dependence have larger Kullback-Leibler divergence than those with weaker dependence when strength of dependence in the tails are different based on the tail orders.

3. If two copula models are applied to discrete variables and have the same strength of dependence as given by a measure of monotone association, then the Kullback-Leibler divergence gets smaller with more discretization. This is because the different asymptotic rates in the joint tails of the copula density do not affect rectangle probabilities for the log-likelihood with discrete response.

4. For discrete response, because copula models are less differentiated, the discretized multivariate Gaussian model is a good choice with flexible dependence structure. However, numerical integration is needed for high-dimensional rectangle probabilities. The Kullback-Leibler divergence with discrete-response vine models can be quite small. These vine models with discrete variables also have flexible dependence and their simpler computational effort (see Algorithm 7 in Chapter 6) make them good candidates for modeling.

The examples are also an illustration of a methodology to apply as new parametric copula models are derived and compared in the future.

5.7.1 Sample size to distinguish two densities

We use the log-likelihood ratio to get a sample size n_{fg} which gives an indication of the sample size needed to distinguish f and g with probability at least 0.95. If f, g are similar, then n_{fg} will be larger, and if f, g are far apart, then n_{fg} will be smaller. The calculation is based on an approximation from the Central Limit Theorem and assumes that the square of the log-likelihood ratio has finite variance when computed with f or g being the true density.

Let variances of the log density ratios be:

$$\sigma_g^2 = \int g(\boldsymbol{y})\Big[\log\frac{g(\boldsymbol{y})}{f(\boldsymbol{y})}\Big]^2 \, \mathrm{d}\nu(\boldsymbol{y}) - [KL(f|g)]^2, \tag{5.9}$$

$$\sigma_f^2 = \int f(\boldsymbol{y})\Big[\log\frac{f(\boldsymbol{y})}{g(\boldsymbol{y})}\Big]^2 \, \mathrm{d}\nu(\boldsymbol{y}) - [KL(g|f)]^2. \tag{5.10}$$

For a random sample $\boldsymbol{Y}_1, \ldots, \boldsymbol{Y}_n$ from one of f or g, let

$$\Delta = n^{-1}\sum \log[f(\boldsymbol{Y}_i)/g(\boldsymbol{Y}_i)]$$

be the averaged log-likelihood ratio. If the true model is f, then as $n \to \infty$,

$$\mathbb{P}(\Delta > 0; f \text{ is true}) = \mathbb{P}\Big(\frac{\Delta - KL(g|f)}{n^{-1/2}\sigma_f} > \frac{-KL(g|f)}{n^{-1/2}\sigma_f}; f \text{ is true}\Big) \approx \Phi\big(n^{1/2}KL(g|f)/\sigma_f\big).$$

For small $\alpha > 0$, the above probability asymptotically exceeds $1 - \alpha$ if $n^{1/2}KL(g|f)/\sigma_f > \Phi^{-1}(1 - \alpha)$ or $n^{1/2} > \sigma_f\Phi^{-1}(1 - \alpha)/KL(g|f)$. Similarly, if the true model is g, then one gets $n^{1/2} > \sigma_g\Phi^{-1}(1 - \alpha)/KL(f|g)$.

For comparison of copula families, we will take $\alpha = 0.05$ and report

$$n_{fg} = \Big[\Phi^{-1}(1 - \alpha) \cdot \max\Big\{\frac{\sigma_f}{KL(g|f)}, \ \frac{\sigma_g}{KL(f|g)}\Big\}\Big]^2, \tag{5.11}$$

which we call a *KL sample size*. This is larger when one of $KL(g|f), KL(f|g)$ is small or if one of the variances σ_f^2, σ_g^2 is large.

Next we indicate a modification of KL divergence with covariates. In this case, we write densities as $f(\cdot; \boldsymbol{x})$ and $g(\cdot; \boldsymbol{x})$ where $\boldsymbol{x}$ is a vector of covariates. To get a KL divergence of competing models, we will assume $\boldsymbol{x}$ is the realization of a random vector $\boldsymbol{X}$ with distribution $F_{\boldsymbol{X}}$. Then the KL divergence is like in (5.7), with an extra integral/summation with respect to $F_{\boldsymbol{X}}(\mathrm{d}\boldsymbol{x})$, that is,

$$KL(f|g) = \int\int g(\boldsymbol{y}; \boldsymbol{x}) \log[g(\boldsymbol{y}; \boldsymbol{x})/f(\boldsymbol{y}; \boldsymbol{x})] \, \mathrm{d}\nu(\boldsymbol{y}) \, \mathrm{d}F_{\boldsymbol{X}}(\boldsymbol{x}),$$

where the integral over $\boldsymbol{x}$ is a summation if $\boldsymbol{X}$ is assumed discrete.

To assess the closeness of copula models combined with univariate regression models for discrete $\boldsymbol{Y}$, it is convenient to also take $\boldsymbol{X}$ to be discrete with a finite number of possible values in order to compute KL sample sizes.

The next three subsections have examples for continuous and discrete variables.

5.7.2   Jeffreys' divergence and KL sample size

In this section, we tabulate some KL sample sizes to show the similarity or dissimilarity of different copula families that all have positive dependence properties like stochastically increasing. Because the positive dependence properties are similar, the differences are due to tail properties and the magnitudes of the KL sample sizes can be explained by the similarity or dissimilarity of lower and upper tail orders. Hence the lower and upper tail orders of copula families are relevant for likelihood inference. A small sample size is sufficient to distinguish copula families with quite different tail behaviors.

If $f = c_1$ and $g = c_2$ are two parametric families of copulas, then for (5.7) to (5.10), the following integrals must be numerically computed:

$$\Delta_1 = \int_{[0,1]^2} c_1 \log[c_1/c_2] \, \mathrm{d}u \, \mathrm{d}v \approx \int_{[-B,B]^2} c_1(\Phi(z_1), \Phi(z_2)) \, \phi(z_1) \, \phi(z_2) \, \mathrm{d}z_1 \mathrm{d}z_2, \quad \text{large } B,$$

$$\sigma_1^2 = \int c_1 \{\log[c_1/c_2]\}^2 \, \mathrm{d}u \, \mathrm{d}v - \Delta_1^2,$$

$$\Delta_2 = \int c_1 [c_2/c_1] \log[c_2/c_1] \, \mathrm{d}u \, \mathrm{d}v,$$

$$\sigma_2^2 = \int c_1 [c_2/c_1] \{\log[c_2/c_1]\}^2 \, \mathrm{d}u \, \mathrm{d}v - \Delta_2^2.$$

Tables 5.2 and 5.3 show Jeffreys' divergences and KL sample sizes for comparing pairs among eight bivariate one-parameter copula families when one of Kendall's τ or Spearman's ρ_S is 0.5. The numerical summaries with fixed Blomqvist's β are close to those with fixed Kendall's τ. There is no monotone dependence measure that matches exactly the divergence of pairs of copula families based on likelihood. Table 5.4 has KL sample sizes as ρ_S varies from 0.1 to 0.9 for comparing some copula families versus bivariate Gaussian.

From the Tables 5.2 and 5.3, note that for MTCJ and reflected MTCJ, tail dependence in one tail and tail quadrant independence in the other tail, are far apart and require a sample size of 22–36 to distinguish when the Kendall τ or Spearman ρ_S values are 0.5. Copula families that are very close to each other are (a) Gumbel and Galambos (KL sample size 5900–7400); (b) Galambos and Hüsler-Reiss (KL sample size 2800–5800); (c) reflected MTCJ and Joe/B5 (KL sample size 2200–4500). Of the reflection symmetric copula families (B1–B3), the Plackett and Frank families are the closest (KL sample size 400–1600); they are closer to each other than with the bivariate Gaussian copula family because they both have tail quadrant independence whereas with a positive correlation parameter, the bivariate Gaussian copula has intermediate tail dependence. Hence the log-likelihood ratio which is part of KL divergence can distinguish the strength of dependence in the tails. One of the first papers to have Jeffreys' divergence is Nikoloulopoulos and Karlis (2008a) but this was without the concept of tail order to explain all of the similarities and differences of these one-parameter copula families that interpolate independence and comonotonicity.

Note that a common $\tau = 0.5$ has more dependence than a common $\rho_S = 0.5$. Under mild positive dependence properties for the copula family, $\rho_S \geq \tau$ for a fixed dependence parameter; see Capéraà and Genest (1993) and Fredricks and Nelsen (2007)). For the eight copula families in Tables 5.2 and 5.3, Blomqvist's β is quite close to Kendall's τ, and Spearman's ρ_S is larger for a fixed copula parameter. This combined with Table 5.4 explains why the KL sample sizes are mostly larger (copulas less different) in Table 5.3.

	B1	B2	B3	B4	B5	B6	B7	B8
	Jeffreys' divergence							
B1		0.087	0.068	0.260	0.293	0.076	0.072	0.078
B2	0.087		0.027	0.290	0.307	0.110	0.126	0.187
B3	0.068	0.027		0.319	0.345	0.135	0.144	0.191
B4	0.260	0.290	0.319		0.002	0.059	0.060	0.068
B5	0.293	0.307	0.345	0.002		0.074	0.078	0.090
B6	0.076	0.110	0.135	0.059	0.074		0.002	0.012
B7	0.072	0.126	0.144	0.060	0.078	0.002		0.005
B8	0.078	0.187	0.191	0.068	0.090	0.012	0.005	
	KL sample sizes							
B1		199	212	58	50	159	176	218
B2	199		414	66	60	133	126	150
B3	212	414		62	55	112	115	148
B4	58	66	62		4486	210	198	171
B5	50	60	55	4486		163	150	128
B6	159	133	112	210	163		7324	1220
B7	176	126	115	198	150	7324		2831
B8	218	150	148	171	128	1220	2831	

Table 5.2 *Jeffreys' divergence and Kullback-Leibler sample sizes to show "distances" of copula families B1–B8 when Kendall's τ is 0.5; B1=bivariate Gaussian, B2=Plackett, B3=Frank, B4=reflected MTCJ, B5=Joe, B6=Gumbel, B7=Galambos, B8=Hüsler-Reiss. For comparison, for reflected MTCJ versus MTCJ (upper versus lower tail dependence), the Jeffreys divergence value is 0.903 and the Kullback-Leibler sample size is 22.*

	B1	B2	B3	B4	B5	B6	B7	B8
	Jeffreys' divergence							
B1		0.034	0.028	0.119	0.157	0.047	0.043	0.046
B2	0.034		0.007	0.139	0.169	0.063	0.072	0.092
B3	0.028	0.007		0.148	0.185	0.074	0.079	0.094
B4	0.119	0.139	0.148		0.005	0.022	0.021	0.024
B5	0.157	0.169	0.185	0.005		0.036	0.039	0.046
B6	0.047	0.063	0.074	0.022	0.036		0.002	0.008
B7	0.043	0.072	0.079	0.021	0.039	0.002		0.002
B8	0.046	0.092	0.094	0.024	0.046	0.008	0.002	
	KL sample sizes							
B1		394	440	110	83	249	283	323
B2	394		1579	107	86	210	192	207
B3	440	1579		103	81	184	178	203
B4	110	107	103		2252	544	537	464
B5	83	86	81	2252		327	290	247
B6	249	210	184	544	327		5901	1645
B7	283	192	178	537	290	5901		5797
B8	323	207	203	464	247	1645	5797	

Table 5.3 *Jeffreys' divergence and Kullback-Leibler sample sizes to show "distances" of copula families B1–B8 when Spearman's ρ_S is 0.5; B1=bivariate Gaussian, B2=Plackett, B3=Frank, B4=reflected MTCJ, B5=Joe, B6=Gumbel, B7=Galambos, B8=Hüsler-Reiss. For comparison, for reflected MTCJ versus MTCJ (upper versus lower tail dependence), the Jeffreys divergence value is 0.414 and the Kullback-Leibler sample size is 36.*

As the dependence increases through one of these measures, the families become farther apart (smaller KL sample sizes). With weak positive dependence, all of these families are harder to distinguish as seen in Table 5.4.

ρ_S	0.1	0.2	0.3	0.4	0.5	0.6	0.7	0.8	0.9
Gaussian vs. Frank	11200	2802	1244	697	440	297	206	143	99
Gaussian vs. Gumbel	2743	972	531	345	249	192	152	121	102
Gaussian vs. MTCJ	2309	636	290	170	110	76	54	40	34

Table 5.4 *Kullback-Leibler sample sizes for bivariate Gaussian copulas versus Frank, Gumbel and MTCJ as Spearman's ρ_S varies from 0.1 to 0.9.*

Although some of these bivariate copula families are quite close to each other, they are still useful, as they have different types of multivariate extensions. But there are implications for vine and factor copulas which are built from bivariate linking copulas; the bivariate Gumbel and reflected Gumbel copulas is sufficient for 1-parameter bivariate extreme value copulas (as substitutes for the Galambos and Hüsler-Reiss copulas), and MTCJ or reflected MTCJ is an adequate subsitute for the Joe/B5 copula (MTCJ is easier to simulate from).

Next, we show that if the same copula families are used with discrete response variables Y_1, Y_2 with regressions on $\boldsymbol{x}$, then the KL sample sizes are (much) larger. That is, with discrete response variables, it takes larger sample sizes to distinguish copula models (because tails of the copula densities would not be "observed"). The KL sample sizes get smaller with less discretization.

The setup for the summary on Table 5.5 is as follows. The bivariate models combine a one-parameter bivariate copula family with ordinal logistic regressions for the two univariate margins. The scalar covariate x is assume to take n_x values equally spaced in $[-1, 1]$. The number of ordinal categories for variables y_1, y_2 are n_{1c} and n_{2c} respectively. The univariate marginal probabilities are:

$$\mathbb{P}(Y_j = k; x) = \frac{e^{\gamma_{k+1} + \beta_{j1} x}}{1 + e^{\gamma_{k+1} + \beta_{j1} x}}, \quad k = 0, \ldots, n_{jc} - 1, \; j = 1, 2,$$

where the slopes are β_{11}, β_{21} and the cutpoints are $\gamma_{k+1} = \log(k/(n_{jc} - k))$ for $k = 0, \ldots, n_{jc}$ (so that $\gamma_0 = -\infty$ and $\gamma_{n_{jc}} = \infty$). The cutpoints are chosen so that for the median x value of 0, the ordinal categories have an equal probability of occurring. With copula families C_1, C_2, bivariate probabilities $f(y_1, y_2; \boldsymbol{x})$ and $g(y_1, y_2; \boldsymbol{x})$ can be obtained based on rectangle probabilities (as explained in Section 1.3).

copula1	copula2	$n_c = 2$	$n_c = 3$	$n_c = 4$	$n_c = 5$	∞
Gaussian	Plackett	11600	1490	716	499	199
Gaussian	Frank	6350	2400	1330	908	212
Gaussian	Gumbel	12800	1850	947	656	159
Gaussian	MTCJ	2840	470	254	182	58

Table 5.5 *Kullback-Leibler sample sizes for ordinal logistic regression with bivariate Gaussian copula versus other copulas with different tail properties; $n_c = n_{1c} = n_{2c}$ is the number of ordinal categories and is varying from 2 to 5; $n_x = 5$ is the number of values of the covariate x and is uniformly spaced in $[-1, 1]$; $\beta_{11} = 1$, $\beta_{21} = 2$; Kendall's τ for the copula families is 0.5. The KL sample sizes for the ∞ case are taken from Table 5.2 for comparisons.*

The results in Table 5.5 change somewhat (mostly for n_{1c}, n_{2c} being 2 or 3) as n_x increases or the slopes β_{11}, β_{21} change, but the pattern does not change. Univariate ordinal probit regression lead to similar results since logit and probit regressions are similar. The multivariate Gaussian copula with univariate ordinal probit regression is just the commonly used multivariate probit model or discretized multivariate Gaussian model.

Next, we show how some of the 2-parameter and 3-parameter families in Chapter 4, with both lower and upper tail dependence, can be compared with KL divergence.

Table 5.6 compares the BB1, BB4, BB7 copulas and their reflected versions; two cases of fixed (λ_L, λ_U) are summarized. The patterns are similar for other choices of (λ_L, λ_U). The BB1 and BB4 families are very close (this is related to the closeness of the Gumbel and Galambos copula families), even though only one of them is an Archimedean copula. The BB7 family is a bit different. When $\lambda_L \neq \lambda_U$, BB7 is closer to its reflected copula than BB1 (or BB4). So for vine and factor copulas, it is adequate to consider BB1, reflected BB1 and BB7 as options for asymmetric tail dependence.

copula1	copula2	Jeffreys	KLss	Jeffreys	KLss
		$\lambda_L = \lambda_U = 0.4$		$\lambda_L = 0.3, \lambda_U = 0.7$	
BB1	BB4	0.0018	6360	0.0010	10800
BB1	BB7	0.0231	490	0.1316	94
BB4	BB7	0.0270	435	0.1393	92
BB1	BB1r	0.0075	1480	0.0980	127
BB4	BB4r	0.0055	1970	0.1023	121
BB7	BB7r	0.0083	1320	0.0061	1860
BB1	BB7r	0.0307	380	0.1711	75
BB7	BB1r	0.0307	380	0.0072	1570

Table 5.6 *Comparisons of BB1, BB4, BB7, BB1r, BB4r, BB7r (the latter three are reflected or survival versions of the copulas) using Jeffreys' divergence and Kullback-Leibler sample size*

For the 3-parameter copula family in Section 4.31.4 based on reflected BB1 with a power parameter, Table 5.7 has some Kullback-Leibler sample sizes to distinguish two extreme Blomqvist's β with λ_L, λ_U fixed. This table shows that, for a model with flexible tail dependence parameters and Blomqvist's β, extreme beta values for the same tail dependence parameters can be distinguished with a sample of around 100 or less.

λ_L	λ_U	η_1 (β_1)	η_2 (β_2)	Jeffreys	KLss
0.4	0.4	0.2 (0.660)	1.0 (0.450)	0.301	44
0.6	0.6	0.2 (0.786)	1.0 (0.625)	0.386	36
0.4	0.6	0.2 (0.696)	1.0 (0.526)	0.256	52
0.6	0.4	0.2 (0.769)	1.0 (0.582)	0.457	31
0.3	0.7	0.2 (0.695)	1.0 (0.558)	0.199	67
0.7	0.3	0.2 (0.822)	1.0 (0.655)	0.592	25
0.4	0.4	3.0 (0.249)	1.0 (0.450)	0.112	111
0.6	0.6	3.0 (0.453)	1.0 (0.537)	0.129	101
0.4	0.6	3.0 (0.323)	1.0 (0.526)	0.136	93
0.6	0.4	3.0 (0.410)	1.0 (0.582)	0.117	110
0.3	0.7	3.0 (0.354)	1.0 (0.558)	0.162	77
0.7	0.3	3.0 (0.518)	1.0 (0.655)	0.119	109

Table 5.7 *Comparisons of reflected BB1 with different power parameters η (or different Blomqvist β values) with fixed tail dependence parameters using Jeffreys' divergence and Kullback-Leibler sample size; note that numerical problems occur with more extreme η values than given here. The pattern is similar for other choices of (λ_L, λ_U).*

5.7.3 Kullback-Leibler divergence and maximum likelihood

The Kullback-Leibler divergence is relevant for maximum likelihood from a parametric family. We give an explanation for the case of i.i.d. observations but the ideas extend to independent observations with covariates. Suppose the true density is g and a parameter model of densities is $f(\cdot; \boldsymbol{\theta})$. Let $\{\boldsymbol{Y}_i, \ i = 1, \ldots, n\}$ be a random sample. The averaged log-likelihood is:

$$n^{-1} L(\boldsymbol{\theta}) = n^{-1} L(\boldsymbol{\theta}; \boldsymbol{Y}_1, \ldots, \boldsymbol{Y}_n) = n^{-1} \sum_{i=1}^{n} \log f(\boldsymbol{Y}_i; \boldsymbol{\theta})$$

and the maximum likelihood estimator is

$$\hat{\boldsymbol{\theta}}_n = \operatorname{argmax} L(\boldsymbol{\theta}; \boldsymbol{Y}_1, \ldots, \boldsymbol{Y}_n).$$

As $n \to \infty$, $\hat{\boldsymbol{\theta}}_n \to_p \boldsymbol{\theta}^*$, where

$$\boldsymbol{\theta}^* = \operatorname*{argmax}_{\boldsymbol{\theta}} \int g(\boldsymbol{y}) \log f(\boldsymbol{y}; \boldsymbol{\theta}) \, d\nu = \operatorname*{argmin}_{\boldsymbol{\theta}} \int g(\boldsymbol{y}) \log[g(\boldsymbol{y})/f(\boldsymbol{y}; \boldsymbol{\theta})] \, d\nu,$$

that is, $\boldsymbol{\theta}^*$ is the parameter such that $f(\cdot; \boldsymbol{\theta}^*)$ is closest to g among $\{f(\cdot; \boldsymbol{\theta})\}$ in the Kullback-Leibler divergence.

For two parametric families of densities $f_1(\cdot; \theta_1)$ and $f_2(\cdot; \theta_2)$. Let $\boldsymbol{Y}_1, \ldots, \boldsymbol{Y}_n$ be a random sample from one of these densities; θ_1, θ_2 and $\boldsymbol{Y}_i$ could be vectors in general.

If data are from f_1 (model 1) with parameter θ_1 and f_2 (model 2) is used for likelihood inference, then asymptotically the quasi-MLE is $\theta_2^*(\theta_1)$ which satisfies

$$\theta_2^*(\theta_1) = \operatorname*{argmax}_{\theta_2} \int f_1(\boldsymbol{y}; \theta_1) \{\log f_2(\boldsymbol{y}; \theta_2)\} \, d\boldsymbol{y}.$$

Similarly, let

$$\theta_1^*(\theta_2) = \operatorname*{argmax}_{\theta_1} \int f_2(\boldsymbol{y}; \theta_1) \{\log f_1(\boldsymbol{y}; \theta_1)\} \, d\boldsymbol{y}.$$

be the asymptotic quasi-MLE when model 2 is the true model and model 1 is used for likelihood inference. Note that there is no symmetry of θ_1^* and θ_2^*, that is, $\theta_1^*[\theta_2^*(\theta_1)]$ is not θ_1 in general.

If $c_1 = f_1$ and $c_2 = f_2$ are two different parametric copula families, and δ_1, δ_2 are values of a dependence measure for $c_1(\cdot; \theta_1)$ and $c_2(\cdot; \theta_2^*(\theta_1))$, then often δ_1, δ_2 are close to each other. That is, if the copula model is not misspecified badly, then the model-based estimates of dependence measures such as Blomqvist's β, Kendall's τ and Spearman's ρ_S are usually quite good. This is indicated with some numerical results in Table 5.8. The biggest differences are when the true family c_1 is reflection symmetric and the quasi family c_2 is quite tail asymmetric.

Table 5.8 is based on the copula parameter for the true family corresponding to a Spearman ρ_S value of 0.5. The pattern is similar for other ρ_S as shown in Table 5.9 for $\rho_S = 0.3, 0.7$ with the Gaussian and Gumbel copulas.

copula	t or q	β	τ	ρ_S	$C(0.2, 0.2)$	$\overline{C}(0.8, 0.8)$
Gaussian	t	0.346	0.346	0.500	0.089	0.089
Plackett	q	0.367	0.332	0.476	0.087	0.087
Frank	q	0.386	0.345	0.500	0.086	0.086
MTCJ	q	0.257	0.258	0.380	0.060	0.096
Gumbel	q	0.310	0.312	0.449	0.075	0.098
Gaussian	q	0.321	0.321	0.466	0.085	0.085
Plackett	t	0.387	0.350	0.500	0.090	0.090
Frank	q	0.396	0.354	0.512	0.087	0.087
MTCJ	q	0.246	0.248	0.365	0.059	0.093
Gumbel	q	0.303	0.305	0.440	0.074	0.097
Gaussian	q	0.317	0.317	0.460	0.085	0.085
Plackett	q	0.366	0.330	0.474	0.087	0.087
Frank	t	0.386	0.345	0.500	0.086	0.086
MTCJ	q	0.236	0.238	0.352	0.058	0.091
Gumbel	q	0.290	0.293	0.423	0.072	0.095
Gaussian	q	0.348	0.348	0.502	0.089	0.089
Plackett	q	0.386	0.349	0.499	0.090	0.090
Frank	q	0.393	0.352	0.509	0.087	0.087
MTCJ	t	0.350	0.349	0.500	0.068	0.114
Gumbel	q	0.362	0.364	0.517	0.082	0.107
Gaussian	q	0.352	0.352	0.507	0.090	0.090
Plackett	q	0.390	0.353	0.504	0.090	0.090
Frank	q	0.396	0.354	0.513	0.087	0.087
MTCJ	q	0.320	0.320	0.462	0.065	0.108
Gumbel	t	0.349	0.351	0.500	0.080	0.105

Table 5.8 *Comparison of five 1-parameter copula families in 25 pairs when one is the true (t) family and the other is the quasi (q) family used for likelihood inference. In each block of five lines, one is the true family and the other four are quasi families. The copula parameter of the true family is chosen to have a Spearman ρ_S value of 0.5, and the copula parameter of the quasi family is found from minimizing the Kullback-Leibler divergence. Then Blomqvist's β, Kendall's τ, Spearman ρ_S, $C(0.2, 0.2)$ and $\overline{C}(0.8, 0.8)$ are compared. The absolute differences of these summaries are smaller when the true family and the quasi family are closer to each other; see Table 5.3. The bigger differences occur when the true family is one of Gaussian, Plackett or Frank, and the quasi family is MTCJ (the most tail asymmetric among the five copula families in this table).*

copula	t or q	β	τ	ρ_S	$C(0.2, 0.2)$	$\overline{C}(0.8, 0.8)$
		$\rho_S = 0.3$				
Gaussian	t	0.203	0.203	0.300	0.067	0.067
Gumbel	q	0.172	0.175	0.258	0.058	0.073
		$\rho_S = 0.5$				
Gaussian	t	0.346	0.346	0.500	0.089	0.089
Gumbel	q	0.310	0.312	0.449	0.075	0.098
		$\rho_S = 0.7$				
Gaussian	t	0.509	0.509	0.700	0.115	0.115
Gumbel	q	0.472	0.472	0.650	0.098	0.125

Table 5.9: *Continuation of Table 5.8 with varying ρ_S for the Gaussian copula.*

5.7.4 Discretized multivariate Gaussian

Table 5.5 shows that when the univariate margins are regression models for discrete response, then copula models can be more difficult to discriminate. Hence the multivariate Gaussian copula or discretized multivariate Gaussian model is a good choice because of its flexible dependence structure. It is commonly used in psychometrics but estimation might be with a limited information method based on low-dimensional margins in order to avoid the high-dimensional integration. That is, as the dimension d increases, the computations for multivariate Gaussian rectangle probabilities take more time to achieve accuracy of a few decimal places, unless the correlation matrix is positive exchangeable or has a 1-factor structure (in these cases, probabilities can be computed as 1-dimensional integrals for any $d > 2$).

The discrete vine pair-copula construction in Section 3.9.5 requires only specification of a sequence of $d(d-1)/2$ bivariate copulas. In this section, we show some calculations of Kullback-Leibler divergences and sample sizes, assuming discretized multivariate Gaussian is the true model, to show the quality of the approximation of the discrete vine pair-copula construction with bivariate Gaussian copulas. Because of the very good approximation, this discrete vine construction can be considered as a "replacement" model for discretized multivariate Gaussian when the dependence structure is not positive exchangeable or 1-factor. That is, maximum likelihood inference is possible; examples of this are given in Chapter 7.

With dimensions $d = 3$ to 5 for which multivariate Gaussian rectangle probabilities can be computed reasonably accurately and quickly, using the randomized quasi-Monte Carlo method of Genz and Bretz (2009) or the numerical integration method of Schervish (1984); the latter can fail on some rare combinations of correlation matrices and rectangles for $d > 3$. Comparisons were made of the discretized multivariate Gaussian model and the discrete vine construction with bivariate Gaussian copulas. Ordinal response variables with a few categories were considered and latent correlation matrices having autoregressive dependence and unstructured dependence were used. Examples of correlation matrices with autoregressive dependence are the following.

$$\boldsymbol{R}_1 = \begin{pmatrix} 1.000 & 0.600 & 0.360 & 0.216 & 0.130 \\ 0.600 & 1.000 & 0.600 & 0.360 & 0.216 \\ 0.360 & 0.600 & 1.000 & 0.600 & 0.360 \\ 0.216 & 0.360 & 0.600 & 1.000 & 0.600 \\ 0.130 & 0.216 & 0.360 & 0.600 & 1.000 \end{pmatrix}, \quad \boldsymbol{R}_2 = \begin{pmatrix} 1.000 & 0.500 & 0.600 & 0.393 & 0.385 \\ 0.500 & 1.000 & 0.500 & 0.600 & 0.393 \\ 0.600 & 0.500 & 1.000 & 0.500 & 0.600 \\ 0.393 & 0.600 & 0.500 & 1.000 & 0.500 \\ 0.385 & 0.393 & 0.600 & 0.500 & 1.000 \end{pmatrix},$$

$$\boldsymbol{R}_3 = \begin{pmatrix} 1.0 & 0.6 & 0.2 & 0.2 \\ 0.6 & 1.0 & 0.2 & 0.1 \\ 0.2 & 0.2 & 1.0 & 0.8 \\ 0.2 & 0.1 & 0.8 & 1.0 \end{pmatrix}, \quad \boldsymbol{R}_4 = \begin{pmatrix} 1.0 & 0.3 & 0.6 & 0.4 \\ 0.3 & 1.0 & 0.4 & 0.5 \\ 0.6 & 0.4 & 1.0 & 0.0 \\ 0.4 & 0.5 & 0.0 & 1.0 \end{pmatrix}.$$

$\boldsymbol{R}_1$ is an AR(1) matrix with lag-1 correlation of 0.6, and $\boldsymbol{R}_2$ is an AR(2) matrix with first two lags of $0.5, 0.6$ and these structured correlation matrices exist in any dimension $d \geq 3$. $\boldsymbol{R}_3, \boldsymbol{R}_4$ are randomly generated correlation matrices with correlations made non-negative and rounded to the nearest tenth.

For the correlation matrices with autoregressive structure, the D-vine is reasonable for discrete pair-copula construction. For the correlation matrices with unstructured dependence, we compare the approximations with all of the possible vines (3 for $d = 3$, 24 for $d = 4$, 480 for $d = 5$). To get multivariate ordinal probit models, three ordinal categories were chosen and then cutpoints on the $U(0, 1)$ scale for each variable were taken as random multiples of 0.1 and converted with Φ^{-1} to cutpoints on the $N(0, 1)$ scale. Summary results are given in Table 5.10.

case	d	lag1/lag2	L_∞ distance	KLdiv	KLss
AR1	3	0.6	0.001–0.006	0.000–0.002	1100–4100
AR1	3	0.7	0.001–0.008	0.000–0.004	1200–4200
AR1	3	0.8	0.001–0.013	0.000–0.008	600–5300
AR2	3	0.5,0.6	0.000–0.002	0.000–0.001	100–5300
AR2	3	0.6,0.5	0.000–0.005	0.000–0.002	800–4700
AR2	3	0.7,0.6	0.001–0.008	0.000–0.003	1300–4700
AR1	4	0.6	0.001–0.007	0.001–0.003	1500–3000
AR1	4	0.7	0.002–0.011	0.001–0.006	900–3800
AR1	4	0.8	0.002–0.015	0.001–0.011	500–2000
AR2	4	0.5,0.6	0.001–0.007	0.001–0.002	1700–3300
AR2	4	0.6,0.5	0.001–0.009	0.001–0.003	1600–3100
AR2	4	0.7,0.6	0.001–0.012	0.001–0.005	900–3200
AR1	5	0.6	0.002–0.007	0.001–0.004	1200–2100
AR1	5	0.7	0.003–0.010	0.003–0.008	600–1500
AR1	5	0.8	0.006–0.016	0.005–0.015	300–1000
AR2	5	0.5,0.6	0.002–0.009	0.001–0.003	1300–2300
AR2	5	0.6,0.5	0.002–0.009	0.001–0.004	1100–2200
AR2	5	0.7,0.6	0.004–0.013	0.002–0.008	600–1600

Table 5.10 *D-vine approximation of multivariate probit with AR correlation structure. In each row, the range of (a) L_∞ distance of multivariate probit probability distributions and those based of the D-vine approximation (b) Kullback-Leibler divergences and (c) Kullback-Leibler sample sizes for 0.95 probability of discrimination, are based on centered 95% intervals from 100 sets of cutpoint parameters. The form of (5.11) that is used is $n_{fg} = [\Phi^{-1}(0.95)\sigma_g/KL(f|g)]^2$ with g being the pmf for discretized multivariate Gaussian and f being the pmf for the approximation. The D-vine approximation tends to get slowly worse with more dependence, so the Kullback-Leibler sample sizes tend to decrease with more dependence.*

Given a partial correlation vine representation $\boldsymbol{\theta}$ with vine array A of the latent correlation matrix $\boldsymbol{R}$, the discrete R-vine with array A that is closest in Kullback-Leibler divergence (usually) has a parameter vector that is close to the correlations and partial correlations in $\boldsymbol{\theta}$. For unstructured correlation matrices, the performance of the vine pair-copula approximation depends a lot on the specific correlation matrix $\boldsymbol{R}$ and not so much on the cutpoint parameters. We give an indication of the variation with the above $\boldsymbol{R}_3$ and $\boldsymbol{R}_4$ with dimension $d = 4$. For both $\boldsymbol{R}_3$ and $\boldsymbol{R}_4$, over 24 possible vines and five sets of cutpoint parameters, the 95% center range of L_∞ distances of multivariate probit probabilities and the R-vine approximation is 0.000 to 0.005 and the 95% center range of Kullback-Leibler divergences is 0.000 to 0.002. For $\boldsymbol{R}_3$, the 95% center range of Kullback-Leibler sample sizes was 6000 to 300000 with the enumeration over R-vines; and for $\boldsymbol{R}_4$, the 95% center range was 2000 to 40000. So there are plenty of R-vines that provide good approximations on the three measures used in Table 5.10, but the approximation need not be equally good for all R-vines. The big variation in Kullback Leibler sample sizes is due to the value of σ_g^2 in (5.9).

In Section 6.17, there is discussion of methods to find good truncated R-vine approximations with small partial correlations in the higher tree levels.

5.8   Initial data analysis for copula models

Section 5.7 on Kullback-Leibler divergence provides information on when different copula models are similar for continuous and discrete responses. With that background, in this section, we discuss some of the steps of initial data analysis that are helpful in deciding on candidate copula models for data sets with continuous or discrete response variables.

The first step consists of univariate and bivariate plots and even high-dimensional plots to see the structure of the data. The detailed guidelines below are aimed at the situation where bivariate and multi-dimensional plots suggest unimodal densities. This is the common situation where parametric copula models are used. If plots show multimodality, then multivariate techniques of mixture models, cluster analysis and non-parametric functional data analysis might be more appropriate.

For dependence models with copulas, the basic steps are (i) candidates for univariate models, (ii) determination of appropriate dependence structure, (iii) determination of appropriate joint tail behavior, and (iv) candidate copula models that meet the requirements of (ii) and (iii).

Initial data analysis for univariate models is easier as there is lots of literature on this. Section 5.8.1 has some details, and references to parametric families of univariate densities with a variety of tail behavior. For dependence structure, in Section 5.8.2, we discuss the use of the correlation matrix of normal scores (in the case continuous response variables) or the latent correlation matrix (in the case of discrete or mixed response variables). For tail behavior and possible departures from the Gaussian copula, we recommend the use of bivariate normal score plots and semi-correlations in Section 5.8.3. Then the copula construction theory in Chapter 3 and the properties of parametric families in Chapter 4 can be use to obtain candidate copula models. For flexible dependence, the required models might be within the classes of vine and factor copulas. The comparisons in Section 5.7 help in the choice of bivariate linking copulas, as it is only needed to choose among families with the appropriate tail properties and to avoid "duplication" among families that are difficult to differentiate. Some of the steps for initial data analysis and model choice are illustrated in Chapter 1 and more detailed examples are given in Chapter 7.

5.8.1   Univariate models

For dependence models with copulas, the first step is to choose appropriate models for each response variable. In many contexts, there is much past experience that can suggest parametric models for individual variables. Asymmetry and tail heaviness are also considerations in the univariate candidate models. If tail inference is important, one can apply extreme value inference methods such as peaks over thresholds or generalized Pareto tails (Pickands (1975), Coles (2001)) to get a better idea if unbounded upper or lower tails are decreasing exponentially or at a rate of an inverse power.

There are many univariate unimodal models for support of the entire real line, the positive real line and the non-negative integers. These can have a variety of asymmetry and tail behavior. A list of selected references is given below.

- For insurance applications, an appendix of Klugman et al. (2010) has a variety of 1- to 4-parameter families; this includes Pareto, Weibull, gamma, generalized gamma, inverse Gaussian, generalized beta of the second kind, Burr, inverse Burr (or Dagum), negative binomial, and compound Poisson.
- The skew-normal distribution in Azzalini (1985) and skew-t distributions in Azzalini and Capitanio (2003), Hansen (1994), Jones and Faddy (2003).
- Generalized hyperbolic, normal inverse Gaussian and extensions with support on the real line in Barndorff-Nielsen (1977, 1997), McNeil et al. (2005), Aas and Hobæk Haff (2006).
- Families involving transform families to achieve different amounts of skewness and tail-weights, such as in Jones and Pewsey (2009), Rosco et al. (2011).
- For count distributions with support on the non-negative integers, see Johnson et al. (2005), Kokonendji et al. (2004), El-Shaarawi et al. (2011) and references therein for models with different amounts of skewness and tailweights.

As the sample size increases, it becomes easier to detect that simple parametric families with a few parameters are not adequate fits. To stay within the parametric framework, one possibility for continuous variables is to use a smoothed histogram in the middle and generalized Pareto densities for unbounded tails; assuming there is one parameter for each bin in the histogram, this would be parametric with the number of parameters being equal to two times the number of unbounded tails plus the number of bins.

If a variable has support on the whole real line, then the univariate density would usually include a location parameter and a scale parameter. If a variable has support on the positive real line, then the univariate density would usually include a scale parameter. In either case, an assessment of the fit of a scale or location-scale family can be made with a quantile-quantile plot relative to the base member of the family, after the parameters are estimated.

5.8.2 Dependence structure

If variables are continuous and there are no covariates, then one can estimate univariate distributions and then transform to normal scores via $\Phi^{-1}\circ F_j$ for the jth variable ($j = 1, \ldots, d$) or one can empirically transform based on $\Phi^{-1}([r_{ij} - \frac{1}{2}]/n)$ where r_{ij} is the (increasing) rank of the jth variable for the ith observation $(y_{i1}, \ldots, y_{id})$ and n is the sample size. Let $u_i = (i - \frac{1}{2})/n$ and $z_i = \Phi^{-1}(u_i)$; these conversions to normal scores are better than the use of $i/(n+1)$ because $n^{-1}\sum_{i=1}^n z_i^2$ is closer to 1. Let the resulting correlation matrix of normal scores be denoted as $\boldsymbol{R}_{\text{observed}}$.

With plots of normal scores for two variables at a time, one can check if the dependence structure is simple positive (negative) dependence such as stochastically increasing (decreasing) with regression, quantile and standard deviation smoothers. The common parametric bivariate copula families satisfy the dependence criterion of stochastically increasing or decreasing, and hence these families would be inappropriate if the regression or conditional mean smoother through a bivariate normal scores plot is not monotone. Note also that measures of monotone association such as Kendall's τ and Spearman's ρ_S are not good summaries if the regression relationship is not monotone.

If some variables are discrete or if there are covariates, then a latent correlation matrix can be obtained based on each pair of variables. This corresponds to first fitting a univariate model (distribution or regression) separately to each variable, and then to getting a dependence measure for each pair of variables by fitting a bivariate Gaussian copula (see Section 2.12.7). Let the resulting latent correlation matrix be denoted as $\boldsymbol{R}_{\text{latent}}$. This latent correlation matrix will generally be positive definite if the sample size is large enough. Assuming it is positive definite, then methods in the psychometrics and structural covariance literature (Browne and Arminger (1995) and Mulaik (2009)) can be used for check if some dependence structures are good approximations. These methods also depend on the variables being monotonically related so bivariate marginal distributions of the data should be checked for this property.

Assuming the variables are monotonically related, one could use $\boldsymbol{R}_{\text{observed}}$ or $\boldsymbol{R}_{\text{latent}}$ to fit dependence structures such as exchangeable, 1-factor, truncated vine, multiple-factor, structured-factor, hierarchical, etc. The truncated vine structure includes the autoregressive dependence structure when there is a time or sequence to the variable indexing. For dimension d, consider models than lead to a "parsimonious" correlation structure with fewer than the saturated model with $\binom{d}{2}$ correlation parameters.

For a particular dependence structure, let its parameter vector be $\boldsymbol{\delta}$, with $1 \leq \dim(\boldsymbol{\delta}) < \binom{d}{2}$. Let $\boldsymbol{R}_{\text{model}}(\tilde{\boldsymbol{\delta}})$ be the best fit where

$$\tilde{\boldsymbol{\delta}} = \underset{\boldsymbol{\delta}}{\text{argmin}}\big\{\log(|\boldsymbol{R}_{\text{model}}(\boldsymbol{\delta})|) - \log(|\boldsymbol{R}_{\text{data}}|) + \text{tr}[\boldsymbol{R}_{\text{model}}^{-1}(\boldsymbol{\delta})\boldsymbol{R}_{\text{data}}] - d\big\},$$

and $\boldsymbol{R}_{\text{data}}$ is either $\boldsymbol{R}_{\text{observed}}$ or $\boldsymbol{R}_{\text{latent}}$, and $|\boldsymbol{R}| = \det(\boldsymbol{R})$. The model discrepancy measure is

$$D_{\text{model}} = \log(|\boldsymbol{R}_{\text{model}}(\hat{\boldsymbol{\delta}})|) - \log(|\boldsymbol{R}_{\text{data}}|) + \text{tr}[\boldsymbol{R}_{\text{model}}^{-1}(\hat{\boldsymbol{\delta}})\boldsymbol{R}_{\text{data}}] - d, \qquad (5.12)$$

where $\hat{\boldsymbol{\delta}}$ is $\tilde{\boldsymbol{\delta}}$ or another estimator.

The discrepancy measure is a log-likelihood ratio multiplied by 2. For realizations of Gaussian random variables $\boldsymbol{z}_1, \ldots, \boldsymbol{z}_n$ that have been standardized to have sample means of 0 and sample standard deviations of 1, n^{-1} times the negative log-likelihood of the model at the MLE $\tilde{\boldsymbol{\delta}}$ is

$$\tfrac{1}{2}\log(|\boldsymbol{R}_{\text{model}}(\tilde{\boldsymbol{\delta}})|) + \tfrac{1}{2}\text{tr}[\boldsymbol{R}_{\text{model}}^{-1}(\tilde{\boldsymbol{\delta}})\boldsymbol{R}_{\text{data}}] + \tfrac{1}{2}d\log(2\pi),$$

and this is bounded below by

$$\tfrac{1}{2}\log(|\boldsymbol{R}_{\text{data}}|) + \tfrac{1}{2}d + \tfrac{1}{2}d\log(2\pi)$$

for the saturated model with $\binom{d}{2}$ correlation parameters.

Note that the above is a diagnostic method to quickly compare dependence structures. It is valid for a Gaussian copula model and is reasonable if the normal scores correlation ρ_N in Section 2.17 is a good summary of bivariate dependence,

For multivariate extreme value data, the bivariate normal scores plots can be used to check that there is skewness to the upper tail for maxima and skewness to the lower tail for minima. If the extremes are maxima or minima of only a few independent replications, then the skewness might not be obvious and a multivariate extreme value copula might be inappropriate. Assuming the number of independent replications is large enough, there are further diagnostic checks for the appropriateness of a multivariate extreme value copula. A simple check for max-stability (after conversion to Fréchet marginal distributions) is to check some weighted maxima with a Fréchet quantile-quantile plot, or equivalently, a check for min-stability (after conversion to exponential survival margins) is to check some weighted minima with an exponential quantile-quantile plot. Using many random weight vectors will give an idea about max/min-stability assumption.

5.8.3 Joint tails

The results in Section 2.17 show that semi-correlations can distinguish tail behavior as measured by the tail order. In addition to multivariate data displayed in bivariate normal scores plots, the lower and upper semi-correlations, for the lower and upper quadrants respectively, can be used to assess tail asymmetry.

In bivariate normal scores plots, one looks for deviations from elliptical-shape scatter to suggest non-Gaussian copula families. In particular, (a) tail dependence is suggested with sharper corners than that in an ellipse; (b) asymmetry is suggested if the scatter is quite different across one of the diagonals connecting $(0,0)$ to $(1,1)$ or $(0,1)$ to $(1,0)$.

5.9   Copula pseudo likelihood, sensitivity analysis

In dependence modeling with copulas, with a parametric approach, there are many possible choices for the univariate margins and copula family. Hence it is important to do some sensitivity analysis of inferences to the choices.

Section 1.5 refers to estimation consistency, where parameter estimates from different estimation methods such as IFM and full likelihood can be compared. If they yield quite different estimates (relative to standard errors) of univariate or copula parameters, then this is an indication that part of the model may need improvement.

For continuous variables, the estimated copula parameters can also be compared using the maximum pseudo likelihood approach of Genest et al. (1995a). This is a semi-parametric method and uses empirical distribution functions $\hat{F}_1, \ldots, \hat{F}_d$ and maximizes

$$L_{\text{pseudo}}(\boldsymbol{\delta}) = \sum_{i=1}^{n} \log c(\hat{F}_1(y_{i1}), \ldots, \hat{F}_d(y_{id}); \boldsymbol{\delta})$$

in the copula parameter $\boldsymbol{\delta}$; here, $\hat{F}_j(y) = n^{-1} \left[\sum_{i=1}^{n} I(y_{ij} \leq y) - \frac{1}{2} \right]$. Let the resulting estimator be denoted as $\tilde{\boldsymbol{\delta}}_{\text{pseudo}}$. Note that the pseudo likelihood with univariate empirical distributions can only be implemented if the variables are continuous and there are no covariates, or censored observations.

If the copula model is specified correctly, then Genest et al. (1995a) show that $\tilde{\boldsymbol{\delta}}_{\text{pseudo}}$ is consistent and asymptotically normal. Chen and Fan (2006) prove further asymptotic properties when the pseudo likelihood is used in a time series context, and develop estimation methods for functions of the copula parameter and univariate margins. If the copula model is not correctly specified and the true copula density is c^*, then $\tilde{\boldsymbol{\delta}}_{\text{pseudo}}$ converges to $\boldsymbol{\delta}^*$ where $c(\cdot; \boldsymbol{\delta}^*)$ is closest in the Kullback-Leibler divergence to c^*.

Wang et al. (2013) propose a jackknife empirical likelihood method to construct confidence intervals of (functions of) parameters without the need of estimating the asymptotic covariance matrix.

When applicable, the pseudo likelihood approach is useful for copula selection/comparison; with different models and copula densities, $c^{(1)}, \ldots, c^{(M)}$ can be informally compared with AIC or BIC based on the corresponding maximized pseudo log-likelihoods: $L_{\text{pseudo}}^{(m)}(\tilde{\boldsymbol{\delta}}^{(m)})$ for $m = 1, \ldots, M$. See Chen and Fan (2006) for a more formal pseudo likelihood ratio procedure.

However, for inferences on tail probabilities and quantiles of linear combinations, the parametric approach for univariate models is more flexible; as indicated in Section 5.8.1, one can use nearly non-parametric univariate margins with "many" parameters.

5.10 Non-parametric inference

When bivariate normal scores plots suggest the bivariate margins are not well fitted by copula families with a few parameters, non-parametric estimation methods in Section 5.10.1 and 5.10.3 might be considered.

Even if parametric copula models are used, non-parametric inferential methods can be used to assess the adequacy of fit by, for example, comparison of parametric and non-parametric summary statistics of relevance; see Section 5.10.2 for summary statistics that are functionals of the copula. This has analogies to some limited information methods in psychometrics.

5.10.1 Empirical copula

Similar to the empirical distribution, the empirical copula can be defined for multivariate i.i.d. data after a transform to ranks.

Suppose data $(y_{i1}, \ldots, y_{id})$, $i = 1, \ldots, n$, are realizations of a random sample from a continuous cdf F and there are no ties. Let $r_{1j}, \ldots, r_{nj}$ be the (increasing) ranks for the jth variable, The *empirical copula* can be defined as

$$\widehat{C}_n(\boldsymbol{u}) = n^{-1} \sum_{i=1}^{n} I\big([r_{ij} - \tfrac{1}{2}]/n \leq u_j, \, j = 1, \ldots, d\big). \tag{5.13}$$

The adjustment of the uniform score $[r_{ij} - \frac{1}{2}]/n$ could be done in an alternative form, but there is asymptotically equivalence.[1] This empirical copula puts mass of n^{-1} at the tuples $([r_{i1} - \frac{1}{2}]/n, \ldots, [r_{id} - \frac{1}{2}]/n)$ for $i = 1, \ldots, n$.

Note that the empirical copula (5.13) is not a copula but it can be converted to a copula $\widetilde{C}$ with density:

$$\widetilde{c}(\boldsymbol{u}) = \begin{cases} n^{d-1}, & \dfrac{\boldsymbol{u} \in \prod_{j=1}^{d} \left[\frac{(r_{ij}-1)}{n}, r_{ij} \right]}{n}, \quad i = 1, \ldots, n, \\ 0 & \text{otherwise.} \end{cases} \tag{5.14}$$

That is, $\widetilde{C}$ has density n^{d-1} on the rectangle with width n^{-1} in each dimension, centered at $([r_{i1} - \frac{1}{2}]/n, \ldots, [r_{id} - \frac{1}{2}]/n)$. Hence $\widetilde{c}$ has positive density only for n rectangles, each with volume n^{-d}. By definition, each of its univariate margin density is $U(0,1)$. As an alternative, Malevergne and Sornette (2006)) linearly interpolate the "empirical copula" to get a copula.

Let $\mathcal{G}_C$ be the limiting (Gaussian) process of $n^{1/2}[\widehat{C}_n - C]$. This has been well studied and the conditions for the derivation have been weakened over time. References for empirical copulas and the limiting Gaussian process theory are Rüschendorf (1976), Gaenssler and Stute (1987), van der Vaart and Wellner (1996), Fermanian et al. (2004), Tsukahara (2005), and Segers (2012).

Suppose C has continuous partial derivatives (of first order). Under some regularity conditions, given in the above references, the limiting Gaussian process is

$$\mathcal{G}_C(\boldsymbol{u}) = \mathcal{B}_C(\boldsymbol{u}) - \sum_{j=1}^{d} (D_j C)(\boldsymbol{u}) \, \mathcal{B}_j(u_j),$$

where $\mathcal{B}_C$ is Brownian bridge on $[0,1]^d$ with covariance function

$$\mathbb{E}[\mathcal{B}_C(\boldsymbol{u}) \mathcal{B}_C \boldsymbol{u}')] = C(u_1 \wedge u_1', \ldots, u_d \wedge u_d') - C(\boldsymbol{u}) \, C(\boldsymbol{u}')$$

and $\mathcal{B}_j(u_j) = \mathcal{B}_C(1, \ldots, 1, u_j, 1, \ldots, 1)$. The second term in $\mathcal{G}_C$ is due to not assuming known univariate margins in the original pre-ranked data.

The asymptotic theory is the same if $\widehat{C}_n$ is replaced by $\widetilde{C}_n$.

Rüschendorf (1976) introduced the empirical copula to develop multivariate rank order statistics. Functionals of empirical copulas have been used for tests of asymmetry (e.g., Genest et al. (2012)), measures of dependence (e.g., Schmid and Schmidt (2007a,b)) and copula goodness-of-fit procedures (reviews are in Genest et al. (2009b), Berg (2009)).

5.10.2 Estimation of functionals of a copula

Summary statistics for comparison of empirical versus model-based are useful for low-dimensional margins such as bivariate and trivariate. Functionals of copulas in the literature have mainly been for bivariate copulas, so for ease of notation, in this section, we provide details for bivariate copulas.

Consider a functional (mapping of the copula space to real numbers) of the form:

$$\alpha(C) = \int_0^1 \int_0^1 a(u, v) \, \mathrm{d}C(u, v), \tag{5.15}$$

where $a : [0,1]^2 \to \mathbb{R} \cup \{\infty\}$ is square integrable with possible infinite values on the boundary

---

[1]The conversion $r_{ij}/(n+1)$ is commonly used for the empirical copula; the choice here is to match the transform for normal scores where the use of $z_{ij} = \Phi^{-1}([r_{ij} - \frac{1}{2}]/n)$ is better than $z_{ij}^* = \Phi^{-1}(r_{ij}/[n+1])$ in that $\sum_{i=1}^{n} z_{ij}^2$ is closer to 1 than $\sum_{i=1}^{n} (z_{ij}^*)^2$ for large n.

of the unit square, and is such that $\alpha(C)$ is well defined for any bivariate copula C. Examples are measures of monotone association such as Spearman's ρ_S and the correlation of normal scores ρ_N. The theory below can be adapted to other types of functionals.

Assume that a is twice continuously differentiable in the region where it is non-zero, and that $a_{12} = \frac{\partial^2 a}{\partial u \partial v}$ is integrable on $(0, 1)^2$. With (u_0, v_0) and (u, v) in the interior of the unit square (or on the boundary if $a(\cdot)$ is finite everywhere),

$$a(u, v) - a(u_0, v) - a(u, v_0) + a(u_0, v_0) = \int_{u_0}^{u} \int_{v_0}^{v} a_{12}(s, t) \, ds dt.$$

After substituting for $a(u, v)$ and changing the order of integration, (5.15) is the same as:

$$\alpha^*(C) = \int_0^1 \int_0^1 a_{12}(u, v) \left[C(u, v) - C(u, v_0) - C(u_0, v) + C(u_0, v_0) \right] du dv$$

$$+ \int_0^1 a(u_0, v) \, dv + \int_0^1 a(u, v_0) \, du - a(u_0, v_0). \tag{5.16}$$

The sample version of (5.15) is

$$\hat{\alpha} = \alpha(\widehat{C}_n) = \int_0^1 \int_0^1 a(u, v) \, d\widehat{C}_n(u, v) = n^{-1} \sum_{i=1}^n a\left(\frac{r_{i1} - \frac{1}{2}}{n}, \frac{r_{i2} - \frac{1}{2}}{n} \right)$$

$$= \alpha^*(\widehat{C}_n) + O_p(n^{-1})$$

$$= \int_0^1 \int_0^1 a(u, v) \, d\widetilde{C}_n(u, v) + O_p(n^{-1}).$$

Note that with $S_{in} = [(r_{i1} - 1)/n, r_{i1}/n] \times [(r_{i2} - 1)/n, r_{i2}/n]$,

$$\int_{(r_{i1}-1)/n}^{r_{i1}/n} \int_{(r_{i2}-1)/n}^{r_{i2}/n} a(u, v) \, d\widehat{C}_n(u, v) = n^{-1} a\left(\frac{r_{i1} - \frac{1}{2}}{n}, \frac{r_{i2} - \frac{1}{2}}{n} \right),$$

and from a Taylor expansion to first order,

$$\left| \int_{S_{in}} a(u, v) \, \widetilde{c}_n(u, v) \, du dv - n^{-1} a\left(\frac{r_{i1} - \frac{1}{2}}{n}, \frac{r_{i2} - \frac{1}{2}}{n} \right) \right| \le n^{-2} \max_{(u,v) \in S_{in}} \left\{ \frac{\partial a}{\partial u}, \frac{\partial a}{\partial v} \right\}.$$

For asymptotics, with the continuous mapping theorem,

$$n^{1/2}[\alpha(\widehat{C}_n) - \alpha(C)] = \int_{[0,1]^2} a_{12}(u, v) \, n^{1/2} \Big\{ [\widehat{C}_n(u, v) - \widehat{C}_n(u, v_0) - \widehat{C}_n(u_0, v) + \widehat{C}_n(u_0, v_0)]$$

$$- [C(u, v) - C(u, v_0) - C(u_0, v) + C(u_0, v_0)] \Big\} du dv + O_p(n^{-1/2})$$

$$\to_d \int_{[0,1]^2} a_{12}(u, v) \left[\mathcal{G}_C(u, v) - \mathcal{G}_C(u, v_0) - \mathcal{G}_C(u_0, v) + \mathcal{G}_C(u_0, v_0) \right] du dv.$$

Hence $\alpha(\widehat{C}_n)$ is asymptotically normal. If (5.16) doesn't hold, asymptotic normality holds under some conditions and the asymptotic representation has a different form.

An example of a tail-weighted measure of dependence is $\text{Cor}[b(U), b(V)|U > \frac{1}{2}, V > \frac{1}{2}]$ if b is increasing on $[\frac{1}{2}, 1]$ with $b(\frac{1}{2}) = 0$. For the upper semi-correlation of normal scores, $b(u) = \Phi^{-1}(u)$ for $\frac{1}{2} \le u < 1$ (but the second version α^* after integration by parts might not be valid). This conditional correlation involves $\mathbb{E}[b^{m_1}(U) b^{m_2}(V)|U > \frac{1}{2}, V > \frac{1}{2})$. with $(m_1, m_2) \in \{(1, 1), (2, 0), (0, 2), (1, 0), (0, 1)\}$, that is, (5.15) for five different $a(\cdot)$ functions

$b^{m_1}(u)b^{m_2}(v)I_{(1/2,1]^2}(u,v)$. The resulting vector of five $n^{1/2}[\hat{\alpha}^{(m_1.m_2)} - \alpha^{(m_1.m_2)}]$ based on the a function is asymptotically multivariate normal. By the Cramer-Wold device,

$$n^{-1/2}\{\hat{\rho}_b - \text{Cor}[b(U), b(V)|U > \tfrac{1}{2}, V > \tfrac{1}{2}]\}$$

is asymptotical normal, where

$$\hat{\rho}_b = \frac{\hat{\mu}^{(1,1)} - \hat{\mu}^{(1,0)}\hat{\mu}^{(0,1)}}{[\hat{\mu}^{(2,0)} - (\hat{\mu}^{(1,0)})^2]^{1/2}[\hat{\mu}^{(0,2)} - (\hat{\mu}^{(0,1)})^2]^{1/2}},$$

and

$$\hat{\mu}^{(m_1,m_2)} = n^{-1}\sum_{i=1}^{n} b^{m_1}\left(\frac{r_{i1} - \tfrac{1}{2}}{n}\right) b^{m_2}\left(\frac{r_{i1} - \tfrac{1}{2}}{n}\right) I\left(r_{i1} - \tfrac{1}{2} > \tfrac{1}{2}n,\, r_{i2} - \tfrac{1}{2} > \tfrac{1}{2}n\right).$$

5.10.3 Non-parametric estimation of low-dimensional copula

Non-parametric estimation of a copula is desirable if univariate margins are well-behaved and can be fit with parametric families, and the multivariate dependence structure is more complex than monotone relations among variables. If the univariate margins are also not simple, then multivariate non-parametric estimation approaches can be applied to the multivariate distribution directly, rather than estimation of non-parametric univariate margins with a non-parametric copula.

If response variables are continuous and there are no covariates, then non-parametric estimation of the copula cdf and density can be considered if the dimension is low (otherwise there is the curse of dimensionality). but possibily a bivariate approach can work with vines.

As an estimate of the copula cdf and density, $\widetilde{C}$ with density $\widetilde{c}$ in (5.14) is not completely satisfactory. If non-parametric estimation of the copula cdf and density is really desired, some local smoothing can be used.

Note that most commonly-used parametric copula densities asymptote to infinity for at least one corner, so the non-parametric density estimation might be better done on a transformed scale. Below is a summary of a few non-parametric methods proposed in the literature for copula estimation.

- Kernel estimation of the copula density or cdf (Gijbels and Mielniczuk (1990) and Fermanian and Scaillet (2003)): The usual kernel-based method can be applied for a multivariate copula cdf or density and, for the bivariate case, Omelka et al. (2009) adjust for the corners where the copula density might be infinite. For kernel density estimation applied to a copula, there are several things that would have to be considered: (a) how to adjust for the fixed univariate margins; (b) should the kernel be applied U(0, 1), or N(0, 1) or some other margins and transformed to U(0, 1); (c) for the multivariate cdf, should the multivariate kernel be the product of independent univariate kernels or a kernel with dependence that better match the data. The adequacy of this approach is unclear, and different choices can have an influence on tail inferences.
- Approximations with Bernstein polynomials (Sancetta and Satchell (2004)): this is an approximation approach of a copula based on d-dimensional Bernstein polynomials with a grid points i/m for $i = 0, 1, \ldots, m$ for each dimension. Then differentiation leads to an estimate of the copula density. Some asymptotic results, as m increases with an appropriate rate relative to the sample size n, are given in Bouezmarni et al. (2010) and Janssen et al. (2012).
- Wavelets (Genest et al. (2009a) and Morettin et al. (2010)): this has been studied for the bivariate case.

- Total variation penalized likelihood (Qu et al. (2009) and Qu and Yin (2012)): this has been studied for the bivariate case, and handles asymptotes to infinity in the corners and preserves the margins.
- Copula density estimation with B-splines (Shen et al. (2008), Kauermann et al. (2013)). Shen et al. (2008) use bivariate linear B-spline copulas as approximations. Kauermann et al. (2013) use penalized hierarchical B-splines with theory that is multivariate, and handle asymptotes to infinity in the corners and preserves $U(0,1)$ margins. Their comparisons show that the method is better than Bernstein polynomials and the kernel approach; curse-of-dimensionality is mentioned for higher dimensions.

5.11 Diagnostics for conditional dependence

In this section we mention two topics. The first topic concerns some diagnostics for conditional dependence as a function of values of conditioning variables, or a check of the simplifying assumption for vine copulas. The second topic concerns the use of diagnostics for pseudo-observations after sequentially fitting trees of a vine copula.

For conditional dependence for a multivariate distribution, one can numerically or analytically evaluate a measure of monotone association, such as Spearman's ρ_S, for $C_{jk;S}(\cdot|\boldsymbol{u}_S)$ over values of $\boldsymbol{u}_S$ for a conditional distribution $F_{jk|S}$ with $j \neq k$ and $j, k \notin S$.

For the case of two response variables Y_1, Y_2 of interest plus a third variable X considered as a covariate, Gijbels et al. (2011) and Acar et al. (2011) use local smoothing methods to estimate the copula of $F_{Y_1,Y_2|X}(\cdot|x)$. Gijbels et al. (2011) also estimate Spearman's $\rho_S(x)$ and Kendall's $\tau(x)$ for the conditional distributions as a function of x. The non-parametric functions of conditional monotone dependence can be assessed for trends.

If the data are observational and have the form (y_{i1}, y_{i2}, x_i) (or a multivariate extension) for $i = 1, \ldots, n$, then the data vectors with the covariates can be considered as a random sample. If inference on the conditional dependence is important, one could fit a parametric copula model to (y_{i1}, y_{i2}, x_i) and compare the model-based and non-parametric estimates of Spearman's ρ_S for $F_{Y_1,Y_2|X}(\cdot|x)$ as a function of x.

If all variables are considered as response variables, and one wants to check on the reasonableness of the simplifying assumption (3.35), one could compute some non-parametric estimates of Spearman's ρ_S for $C_{j_1j_2;k}(\cdot; u_k)$ for subsets of three variables. Alternatively, in the vine copula, one could in tree $\mathcal{T}_2$ with bivariate copula $C_{a_2j;a_1j}(\cdot; \delta_{a_2j;a_1j})$ have a parameter $\delta_{a_2j;a_1j}(u_{a_1j})$ that depends on the value of the conditioning variable in a functional relationship. For example, one can have a log-likelihood where a 1-1 transform of δ is linear or quadratic in u_{a_1j} to check if there is a significant monotone or non-monotone trend in the conditional dependence. The approach of Acar et al. (2012) is to use methods from non-parametric regression to do local smoothing.

Now, consider three variables Y_1, Y_2, X where X may be a covariate or a third response variable. The conditional distributions $F_{Y_1|X}, F_{Y_2|X}$ can be estimated non-parametrically, or parametrically after F_{Y_1X} and F_{Y_2X} are obtained after fitting a parametric copula to (Y_1, Y_2, X). We next summarize the non-parametric smoothing method given in Gijbels et al. (2011).

Suppose data are (y_{i1}, y_{i2}, x_i) for $i = 1, \ldots, n$. For $j = 1, 2$, the conditional cdf $F_{Y_j|X}(y|x)$ is estimated by

$$\tilde{F}_{Y_j|X}(y|x) = \sum_{i'=1}^{n} w_{i'j}(x)I(y_{i'j} \leq y),$$

for appropriately chosen weights $w_{i'j}(x)$ that depend on $x_{i'}$. For the conditional cdfs given $X = x$, these estimates are based on a weighted empirical cdf, with more weight given to the x_i values that are closer to x. An example consists of kernel-based weights (Nadaraya

(1964), Watson (1964)) with

$$w_{mj}(x) = \frac{k((x_m - x)/h_{jn})}{\sum_{m'=1}^n k((x_{m'} - x)/h_{jn})};$$

h_{jn} is a bandwidth which should be $O(n^{-1/5})$. The function k is a kernel density that integrates to 1, such as $k(z) = \frac{35}{32}(1 - z^2)^3 I_{[-1,1]}(z)$. Other examples of weights are given in Gijbels et al. (2011).

The transformed values of (y_{i1}, y_{i2}) based on the conditional cdfs at x_i are $\tilde{u}_{i1} = \tilde{F}_{Y_1|X}(y_{i1}|x_i)$ and $\tilde{u}_{i2} = \tilde{F}_{Y_2|X}(y_{i2}|x_i)$. These are now fixed for estimating $\rho_S(x)$ for any x in the range of the space for this variable.

Let $C_{12;X}(\cdot; x)$ be the copula for $F_{1|X}(\cdot|x)$ and $F_{2|X}(\cdot|x)$. The estimate of $C_{12;X}(u_1, u_2; x)$ is as follows. Let $w_m(x) = k((x_m - x)/h_n)/\sum_{m'=1}^n k((x_{m'} - x)/h_n)$ where h_n is $O(n^{-1/5})$. Let

$$\tilde{G}_{12;X}(v_1, v_2; x) = \sum_{i'=1}^n w_{i'}(x)I(\tilde{u}_{i'1} \le v_1, \tilde{u}_{i'2} \le v_2)$$

be an estimate of $G_{12;X}(v_1, v_2; x) = \mathbb{P}(F_{Y_1|X}(Y_1|x) \le v_1, F_{Y_2|X}(Y_2|x) \le v_2)$. Because of the smoothing for $\tilde{F}_{Y_1|X}$ and $\tilde{F}_{Y_2|X}$, $\tilde{G}_{12;X}(\cdot; x)$ does not have $U(0,1)$ margins, so one can obtain its margins $\tilde{G}_{1;X}(v_1; x)$ and $\tilde{G}_{2;X}(v_2; x)$. Note that

$$\tilde{G}_{j;X}(v_j; x) = \sum_{i'=1}^n w_{i'}(x)I(\tilde{u}_{i'j} \le v_j)$$

has jumps. Then an estimate $C_{12;X}(u_1, u_2; x)$ is

$$\tilde{C}_{12;X}(u_1, u_2; x) = \tilde{G}_{12;X}\big(\tilde{G}_{1;X}^{-1}(u_1; x), \tilde{G}_{2;X}^{-1}(u_2; x); x\big),$$

this is also a function with jumps. Spearman's rho for $C_{12;X}(u_1, u_2; x)$ is $\rho_S(x) = 12 \int \int C_{12;X}(u_1, u_2; x)\, du_1 du_2 - 3$, and this is estimated by

$$
\begin{aligned}
\tilde{\rho}_S(x) &= 12 \int \int \tilde{C}_{12;X}(u_1, u_2; x)\, du_1 du_2 - 3 \\
&= 12 \sum_{i=1}^n w_i(x) \int \int I\big(\tilde{u}_{i1} \le \tilde{G}_{1;X}^{-1}(u_1; x), \tilde{u}_{i2} \le \tilde{G}_{2;X}^{-1}(u_2; x)\big)\, du_1 du_2 - 3 \\
&= 12 \sum_{i=1}^n w_i(x) \int \int I\big(\tilde{G}_{1;X}(\tilde{u}_{i1}; x) \le u_1\big), \tilde{G}_{2;X}(\tilde{u}_{i2}; x) \le u_2\big)\, du_1 du_2 - 3 \\
&= 12 \sum_{i=1}^n w_i(x)[1 - \tilde{G}_{1;X}(\tilde{u}_{i1}; x)][1 - \tilde{G}_{2;X}(\tilde{u}_{i2}; x)] - 3 \\
&:= 12 \sum_{i=1}^n w_i(x)(1 - \hat{u}_{i1})(1 - \hat{u}_{i2}) - 3,
\end{aligned}
$$

where for $j = 1, 2$, $\hat{u}_{ij} := \tilde{G}_{j;X}(\tilde{u}_{ij}; x) = \sum_{i'=1}^n w_{i'}(x)I(\tilde{u}_{i'j} \le \tilde{u}_{ij})$. The Kendall tau estimate as a function of x in Gijbels et al. (2011) involves a double sum so that $\tilde{\rho}_S(x)$ is computationally faster.

The above $\tilde{\rho}_S(x)$ is not invariant to transforms of x. If $z = g(x)$ for a monotone function and one gets a function $\tilde{\rho}_S^*(z)$ as a function of z, then $\tilde{\rho}_S(x_0)$ and $\tilde{\rho}_S^*(z_0)$ are in general different when $z_0 = g(x_0)$. The use of non-parametric smoothing is not simple, as there are several choices to make: (i) smoothing method; (ii) choice of kernel and bandwidths for kernel-based methods; (iii) choice of transformation for covariate or conditioning variable. Also needed would be estimates of standard errors.

Next is an example to evaluate $\tilde{\rho}_S(x)$, using the trivariate case of the gamma factor model in Section 4.28 and Stöber et al. (2013b).

Example 5.2 (Gamma factor model). Suppose $Y_j = Z_0 + Z_j$, $j = 1, 2, 3$ where Z_0, Z_1, Z_2, Z_3 are independent Gamma random variables with respective shape parameter θ_j (and scale=1). Marginally Y_j is Gamma$(\eta_j, 1)$ with $\eta_j = \theta_0 + \theta_j$. Given $Y_3 = x$, a stochastic representation for (Y_1, Y_2) is $(Z_0^* + Z_1, Z_0^* + Z_2)$, where Z_0^* is a x times a Beta(θ_0, θ_3) random variable. The conditional correlation increases from 0 to 1 as x increases. The copula of Y_1, Y_2 given $Y_3 = x$ is: $C_{12;3}(u_1, u_2; x) := F_{Y_1 Y_2 | Y_3}\big(F_{Y_1 | Y_3}^{-1}(u_1 | x), F_{Y_2 | Y_3}^{-1}(u_2 | x) | x\big)$, where

$$F_{Y_j | Y_3}(y_j | x) = \int_0^{(y_j / x) \wedge 1} F_\Gamma(y_j - xb; \theta_j) \, f_B(b; \theta_0, \theta_3) \, db, \quad j = 1, 2,$$

$$F_{Y_1 Y_2 | Y_3}(y_1, y_2 | x) = \int_0^{(y_1 / x) \wedge (y_2 / x) \wedge 1} F_\Gamma(y_1 - xb; \theta_1) \, F_\Gamma(y_2 - xb; \theta_2) \, f_B(b; \theta_0, \theta_3) \, db,$$

and f_B is the Beta density function and F_Γ is the Gamma cdf.

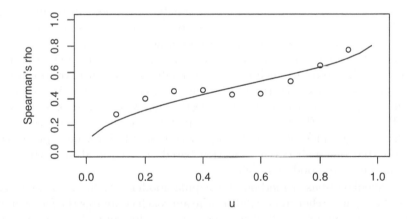

Figure 5.1 *Trivariate gamma factor model with $(\theta_0, \theta_1, \theta_2, \theta_3) = (3, 1, 1.5, 2)$: simulated data with sample size $n = 1000$; conditional Spearman rho estimates are shown with $X = Y_3$ transformed to $U(0, 1)$. The solid line is the Spearman rho of the conditional distribution of $[X_1, X_2 | F_3(X_3) = u]$ computed by numerical integration, and the circles show the estimates at a grid of u values.*

Figure 5.1 has plot of $\tilde{\rho}(\cdot)$ for a simulated data set of size $n = 1000$ for the case of $\theta_0 = 3$ and $(\theta_1, \theta_2, \theta_3) = (1, 1.5, 2)$ (all correlations are above $2/3$). Also shown is the exact $\rho_S(\cdot)$ for $C_{12;3}(\cdot; u)$ computed via numerical integration. The weights are based on the kernel method. For this example, transforming Y_3 to $U(0, 1)$ before computing $\tilde{\rho}(\cdot)$ produces better estimates with a bandwidth h_n that doesn't depend on value of the conditioning variable.

This example suggests that a large sample size is needed to detect trends on $\rho_S(x)$. With financial return data with subsets of three assets and a sample size of the order of a few hundred (for 2–3 years), the $\tilde{\rho}(x)$ values do not show a pattern like Figure 5.1. That is, the pattern looks more like roughly constant strength of conditional dependence plus some sampling variation. □

Next we discuss diagnostics assuming the simplifying assumption is acceptable as an approximation. Suppose the data have been converted to $U(0,1)$ and a vine copula is sequentially estimated as in Example 5.1. After tree $\mathcal{T}_1$, one can create pseudo-observations of the form $(v_{ij_1}, v_{ij_2}) = (C_{j_1|k}(u_{ij_1}|u_{ik}; \tilde{\boldsymbol{\theta}}_{j_1k}), C_{j_2|k}(u_{ij_2}|u_{ik}; \tilde{\boldsymbol{\theta}}_{j_2k}))$ for $i = 1, \ldots, n$, assuming (j_1, k) and (j_2, k) are edges in tree $\mathcal{T}_1$ and $\tilde{\boldsymbol{\theta}}_{j_1k}, \tilde{\boldsymbol{\theta}}_{j_2k}$ are estimates based on maximum likelihood with only tree $\mathcal{T}_1$. For any such pair of such pseudo-variables, diagnostics such as bivariate normal scores plots and semi-correlations can be used to assess tail asymmetry and strength of dependence in the tails. This can continued to higher-order trees. For example, after tree $\mathcal{T}_\ell$, one can get pseudo-observations of the form $(v_{ij_1}, v_{ij_2}) = (C_{j_1|S}(u_{ij_1}|\boldsymbol{u}_{iS}; \tilde{\boldsymbol{\theta}}_{j_1S}), C_{j_2|S}(u_{ij_2}|\boldsymbol{u}_{iS}; \tilde{\boldsymbol{\theta}}_{j_2S}))$ for $i = 1, \ldots, n$ assuming that $\{j_1 \cup S\}$ and $\{j_2 \cup S\}$ are margins that are available after tree $\mathcal{T}_\ell$ and $\tilde{\boldsymbol{\theta}}_{j_1S}, \tilde{\boldsymbol{\theta}}_{j_1S}$ are vectors of parameters that have been estimated sequentially up to tree $\mathcal{T}_\ell$.

5.12   Diagnostics for adequacy of fit

In this section, some methods are given for assessing copula adequacy of fit from bivariate margins. If the data set is sufficiently large, similar methods can also be applied to trivariate and higher-dimensional margins. As part of model building, diagnostics for adequacy of fit can be more informative than formal goodness-of-fit procedures.

There is literature of copula goodness-of-fit procedures, see for example, Genest et al. (2009b), Berg (2009). Kojadinovic and Yan (2011) and Kojadinovic et al. (2011), but the implementation and discussion is mainly for 1-parameter Archimedean copula families and multivariate elliptical copulas. The goodness-of-fit procedures are based on the Rosenblatt transform, the Cramer-von Mises or Kolmogorov-Smirnov functionals and other functions of the empirical cdf, so that they involve a global distance measure between the model-based and empirical distribution. Such procedures might not be sensitive to tail behaviors. But possibly bigger problems with these procedures, based on the empirical cdf, are that (a) they are not diagnostic in the sense of suggesting improved parametric models in the case of small P-values and (b) they have less capability to discriminate copula families compared with diagnostic methods if the sample size is small. For vine copulas, Dissmann et al. (2013) reported that pair-copula selection based on likelihood and AIC seem to be better than using bivariate goodness-of-fit tests.

More recent likelihood-based methods for copula goodness-of-fit are studied in Huang and Prokhorov (2014) and Schepsmeier (2014); the approach compares the Hessian matrix in (5.1) and the covariance matrix (5.2) of the score function for equivalence, and requires the computation of the gradient and Hessian of the negative log-likelihood. Schepsmeier (2014) applies these methods to vine copulas. This procedure also is not diagnostic in suggesting improved parametric models.

In the psychometrics literature (e.g., Maydeu-Olivares and Joe (2006) and references therein), there are limited information procedures based on a combining goodness-of-fit statistics applied to low-dimensional (such as bivariate and trivariate) margins. The approach was developed because the data are sparse in high dimensions but not in low dimensions. For high-dimensional copula models, sparsity issues are also relevant so that a copula goodness-of-fit approach that combines low-dimensional margins might be useful.

In Section 5.1, it is mentioned that models with smaller values of AIC or BIC are better in the sense that they provide better fits to the data. If several parametric models are fitted, the AIC and BIC values do not indicate whether any of the models is adequate. In practice, one needs to decide whether or not to continue trying to find parametric models that can yield smaller AIC/BIC values. The diagnostic approaches presented in the subsections below can detect model inadequacy in specified directions or provide indications of whether the best models based on AIC/BIC are adequate for relevant inferences. For high-dimensional parsimonious copula models, minimally one diagnostic is to check if the dependence structure is adequately approximated. The diagnostics could be converted into more formal statistical methods.

5.12.1 Continuous variables

For comparing the fits of different parametric copula models with continuous response variables, experience from real data sets and simulated data sets from multivariate non-copula-based constructions indicate that the log-likelihood is larger or AIC/BIC values are smaller when the copula model (i) firstly, better approximates the dependence structure of the data, and (ii) secondly, better approximates strength of dependence in the tails.

To assess whether a parametric copula model well approximates the dependence structure, one simple method is to compare empirical versus model-based for measures of association such as Kendall's τ or Spearman's ρ_S. This allows a check if copula models with parsimonious dependence adequately approximate the dependence structure of the multivariate data. Note that Table 2.3 has asymptotic variances of the empirical Kendall's τ and Spearman's ρ_S for some bivariate copula families, from which one can convert to typical standard errors for different sample sizes.

Checking the dependence structure is only a first step because parametric copula models could adequately match model-based Kendall's τ with empirical counterparts without being a good fit in the tail. For several competing parametric copula models, the model with the smallest value of AIC or BIC might fit fine for central measures of dependence but not necessarily as well for strength of dependence in the tails. This is because the log-likelihood is dominated by data in the "middle."

To assess whether a copula model adequately approximates the strength of dependence in the tails, empirical lower and upper semi-correlations of normal scores can be compared with model-based counterparts for each pair of variables. Also other tail-weighted dependence measures can be used (and these might be depend in the application and inferences of interest). The model-based bivariate tail-weighted measures typically involved two-dimensional numerical integration. Because of this, it is desirable to consider measures that are fast to compute; examples are correlation of powers after truncation to the lower or upper quadrant for each pair of variables.

If C_{jk} is the bivariate copula margin of a fitted d-variate copula, consider:

$$
\begin{aligned}
&\mathrm{Cor}[(1 - 2U_j)^r, (1 - 2U_2)^r | U_j < \tfrac{1}{2}, U_k < \tfrac{1}{2}], \\
&\mathrm{Cor}[(2U_j - 1)^r, (2U_2 - 1)^r | U_j > \tfrac{1}{2}, U_k > \tfrac{1}{2}], \quad (U_j, U_k) \sim C_{jk},
\end{aligned}
\tag{5.17}
$$

for a suitably chosen positive power r. These are faster to compute than semi-correlations for the model-based versions, especially for a d-dimensional copula with large d and $d(d-1)/2$ bivariate margins. These types of measures are studied in Krupskii (2014), and some analysis suggest a power r of around 5 or 6 to be a good choice that balances (i) variability and (ii) capability to discriminate tail dependence from intermediate tail dependence (for example, joint tails of the bivariate t_4 versus Gaussian copula). Some theory in Section 5.10.3 can be used to analyze some asymptotic properties of these tail-weighted dependence measures.

Examples of the use of (5.17), in its model-based and empirical versions, are given in Chapter 7. The empirical versions are sample correlations of transformed data, after truncation. Note that assessment of comparative adequacy of models based on tail-based diagnostics could provide a different comparison than AIC/BIC values, because the latter put more weight on the tails than the log-likelihood,

5.12.2 *Multivariate discrete and ordinal categorical*

For discrete response, methods for assessing copula model adequacy are not the same as for continuous response. If a copula model is used as a latent variable model for multivariate discrete response, any strong dependence in the tails would be somewhat attenuated by the discreteness.

Also for multivariate discrete response data, there are typically covariates. Without covariates, bivariate and higher-order margins (depending on sparsity) can be compared in marginal empirical versus model-based tables. A similar comparison can even be done when there are covariates.

Let the discrete response variables be denoted as $y_1, \ldots, y_d$. Suppose the observation vectors are $(y_{i1}, \ldots, y_{id})$ with corresponding covariates $(\boldsymbol{x}_{i1}, \ldots, \boldsymbol{x}_{id})$, $i = 1, \ldots, n$. Let

$$F(\cdot; \boldsymbol{\theta}; \boldsymbol{x}_1, \ldots, \boldsymbol{x}_d) = C\big(F_1(\cdot; \boldsymbol{\eta}_1; \boldsymbol{x}_1), \ldots, F_d(\cdot; \boldsymbol{\eta}_d; \boldsymbol{x}_d); \boldsymbol{\delta}\big)$$

be the model, and let $f_1, \ldots, f_d$ be the univariate pmfs. Let $\tilde{\boldsymbol{\theta}} = (\tilde{\boldsymbol{\eta}}_1, \ldots, \tilde{\boldsymbol{\eta}}_d, \tilde{\boldsymbol{\delta}})$ be estimates.

No covariates

The case of no covariates refers to empty sets for $(\boldsymbol{x}_{i1}, \ldots, \boldsymbol{x}_{id})$, $i = 1, \ldots, n$. The copula model is

$$F(\cdot; \boldsymbol{\theta}) = C\big(F_1(\cdot; \boldsymbol{\eta}_1), \ldots, F_d(\cdot; \boldsymbol{\eta}_d); \boldsymbol{\delta}\big).$$

If there are no covariates, after fitting a multivariate model, one can compare model-based frequencies versus observed frequencies for all univariate and bivariate margins, and up to the d-dimensional table if the d-dimensional data are not sparse with lots of zero counts for d-vectors. To be formal, one could combine all of the univariate and bivariate tables, and use the goodness-of-fit statistic M_2 of Maydeu-Olivares and Joe (2006) if parameters are all identifiable from bivariate margins. The corners of tables of the model-based versus observed bivariate frequencies could be inspected for possible patterns such as over-estimation/underestimation; formal test statistics might not be sensitive to tail behavior.

Covariates

If there are covariates, we will only suggest an informal comparison of model-based frequencies versus observed frequencies for all univariate and bivariate margins, based on latent or polychoric correlations. Because of its wide range of dependence, the multivariate Gaussian copula would usually be the first choice to consider as a model. If the multivariate Gaussian copula is used as the initial tentative model, this will be sufficient to determine if there are large deviations suggesting an alternative copula model. The estimated correlation parameters of the multivariate Gaussian copula are called polychoric correlations for ordinal response and tetrachoric correlations for binary response (Section 2.12.7).

For fitting the multivariate Gaussian copula to discrete data, full maximum likelihood, IFM or composite likelihood can be used for parameter estimation. For two-stage IFM with different parameters for different univariate margins, univariate parameters are estimated first and then latent correlation parameters are estimated from bivariate margins, separately or combined. If separate bivariate margins are used in the second stage, the correlation matrix may need adjustment in order to be non-negative definite. Composite likelihood can be used with the sum of univariate log-likelihoods to estimate univariate parameters if they are common to all univariate margins, followed by a second stage estimation of the latent correlation parameters.

Below is a summary of the algorithm for Observed versus model-based Expected frequencies for a multivariate pmf $f_{1:d}(\cdot; \boldsymbol{\eta}_1, \ldots, \boldsymbol{\eta}_d, \boldsymbol{\delta}; \boldsymbol{x}_1, \ldots, \boldsymbol{x}_d)$.

- For variable y_j $(j = 1, \ldots, d)$, possible collapse some categories to avoid zero or small counts. For example, m_j categories might be formed from $0, 1, \ldots, m_j - 1$, where $m_j - 1$ refers to the category formed from $\{y_{ij} : y_{ij} \geq m_j - 1\}$. Let the resulting variables be y_j^*.

- For the univariate summary for y_j^*, get a one-way table $\boldsymbol{O}_j$ with say m_j observed counts. Compare with a table $\boldsymbol{E}_j$ of model-based expected frequencies with

$$E_j(a) = \sum_{i=1}^{n} f_j^*(a; \tilde{\boldsymbol{\eta}}_j; \boldsymbol{x}_{ij}).$$

where f^* is the modification of the univariate marginal pmf f_j to account for combination of categories. Large discrepancies of $\boldsymbol{O}_j$ and $\boldsymbol{E}_j$ for at least one j would indicate a need for an improved univariate model.

- Consider a pair of variables y_j^*, y_k^* with $1 \le j < k \le d$. Get a two-way table $\boldsymbol{O}_{jk}$ with say, dimension $m_j \times m_k$, of observed counts. Let f_{jk}^* be a bivariate marginal pmf that accounts for the combination of categories.

- Compare with a table of model-based expected frequencies obtained as follows. Compute the $m_j \times m_k$ table of expected counts $\boldsymbol{E}_{jk}$ with

$$E_{jk}(a, b) = \sum_{i=1}^{n} f_{jk}^*(a, b; \tilde{\boldsymbol{\eta}}_j, \tilde{\boldsymbol{\eta}}_k, \tilde{\boldsymbol{\delta}}; \boldsymbol{x}_{ij}, \boldsymbol{x}_{ik}).$$

- Compare $\boldsymbol{O}_{jk}$ with $\boldsymbol{E}_{jk}$ for possible discrepancies in the middle or in one of the corners.

The above could also be done on different non-overlapping subsets of the covariates vectors $\{\boldsymbol{x}_{i1}, \ldots, \boldsymbol{x}_{id}\}$ if there is concern that the copula might not be constant over the varying covariate vectors.

For univariate count regression models, inference based on the aggregated model-based expected counts are studied in Cameron and Trivedi (1998) and Andrews (1988).

5.13 Vuong's procedure for parametric model comparisons

Vuong's procedure (from Vuong (1989)) is the sample version of the calculations for Kullback-Leibler divergence and sample size to differentiate two models which could be non-nested. In the copula literature, see for example, Brechmann et al. (2012), it has been used to compare copula models. Chapter 4 of Belgorodski (2010) has detailed comparisons of its use for copula models. It is also used within the sequential approach in Brechmann et al. (2012) to fit truncated R-vine copulas with a suitable truncated level; the sequential procedure of adding extra levels or trees to the vine is terminated when the gain is determined to be negligible based on Vuong's procedure.

Some details in a general context are given below. With a sample size n, suppose the observed responses are the vectors $\boldsymbol{y}_1, \ldots, \boldsymbol{y}_n$ which are realizations of $\boldsymbol{Y}_1, \ldots, \boldsymbol{Y}_n$; there could be corresponding covariate vectors $\boldsymbol{x}_1, \ldots, \boldsymbol{x}_n$. Let $\mathbb{E}_g$ denote expectation with respect to the true density g. With two models M_1 and M_2 with parametric densities $f^{(1)}$ and $f^{(2)}$ respectively, we can compare

$$\Delta_{1g} = n^{-1} \left[\sum_i \{ \mathbb{E}_g[\log g(\boldsymbol{Y}_i; \boldsymbol{x}_i)] - \mathbb{E}_q[\log f^{(1)}(\boldsymbol{Y}_i; \boldsymbol{x}_i, \boldsymbol{\theta}^{(1)})] \} \right], \tag{5.18}$$

and

$$\Delta_{2g} = n^{-1} \left[\sum_i \{ \mathbb{E}_g[\log g(\boldsymbol{Y}_i; \boldsymbol{x}_i)] - \mathbb{E}_g[\log f^{(2)}(\boldsymbol{Y}_i; \boldsymbol{x}_i, \boldsymbol{\theta}^{(2)})] \}, \right]. \tag{5.19}$$

where $\boldsymbol{\theta}^{(1)}, \boldsymbol{\theta}^{(2)}$ are the parameters in models M_1 and M_2 that lead to the closest Kullback-Leibler divergence to g; equivalently, they are the limits in probability of the maximum likelihood estimators based on models M_1 and M_2 respectively. Model M_1 is closer to the true g if the divergence (5.18) is smaller than (5.19). Let $\Delta_{12} = \Delta_{1g} - \Delta_{2g}$ be the difference in divergence from models M_1 and M_2 when the true density is g. That is,

$$\Delta_{12} = n^{-1}\left[\sum_i \{\mathbb{E}_g[\log f^{(2)}(\boldsymbol{Y}_i; \boldsymbol{x}_i, \boldsymbol{\theta}^{(1)})] - \mathbb{E}_g[\log f^{(1)}(\boldsymbol{Y}_i; \boldsymbol{x}_i, \boldsymbol{\theta}^{(2)})]\}\right]. \qquad (5.20)$$

Hence $\Delta_{12} > 0$ if model M_2 is the better fitting model, and $\Delta_{12} < 0$ if model M_1 is the better fitting model. The sample version of (5.20) with maximum likelihood estimators $\hat{\boldsymbol{\theta}}^{(1)}, \hat{\boldsymbol{\theta}}^{(2)}$ is:

$$\widehat{D}_{12} = n^{-1}\sum_i D_i = n^{-1}(LLR),$$

where LLR is the log-likelihood ratio and

$$D_i = \log\left\{\frac{f^{(2)}(\boldsymbol{y}_i; \boldsymbol{x}_i, \hat{\boldsymbol{\theta}}^{(2)})}{f^{(1)}(\boldsymbol{y}_i; \boldsymbol{x}_i, \hat{\boldsymbol{\theta}}^{(1)})}\right\}.$$

For non-nested or nested models where $f^{(1)}(\cdot; \boldsymbol{x}, \boldsymbol{\theta}^{(1)})$ and $f^{(2)}(\cdot; \boldsymbol{x}, \boldsymbol{\theta}^{(2)})$ are not the same density, a large sample 95% confidence interval for the parameter Δ_{12} is

$$\widehat{D}_{12} \pm 1.96 \times n^{-1/2}\hat{\sigma}_{12}, \qquad (5.21)$$

where

$$\hat{\sigma}_{12}^2 = (n-1)^{-1}\sum_{i=1}^n (D_i - \widehat{D}_{12})^2.$$

Vuong (1989) also has versions with adjusted log-likelihood ratios based on the Akaike or AIC correction and based on the Schwarz or BIC correction. These are respectively

$$\begin{gathered}
\widehat{D}_{12} - n^{-1}[\dim(\boldsymbol{\theta}^{(2)}) - \dim(\boldsymbol{\theta}^{(1)})] \pm 1.96 \times n^{-1/2}\hat{\sigma}_{12}, \\
\widehat{D}_{12} - \tfrac{1}{2}n^{-1}\log(n)[\dim(\boldsymbol{\theta}^{(2)}) - \dim(\boldsymbol{\theta}^{(1)})] \pm 1.96 \times n^{-1/2}\hat{\sigma}_{12}.
\end{gathered} \qquad (5.22)$$

The above can be used as a diagnostic as follows. If the interval in (5.21) or (5.22) contains 0, models M_1 and M_2 would not be considered significantly different. If the interval does not contain 0, then model M_1 or M_2 is the better fit depending on whether the interval is completely below 0 or above 0 respectively. The AIC or BIC correction means that a model with fewer parameters will compare more favorably with a model with more parameters.

If two models are such that one is nested within the other, then the log-likelihood ratio procedure with chi-square reference distribution can also be used for a comparison.

The Vuong procedure compares different models and assesses whether they provide similar fits to the data. It does not assess whether any of the models is a good enough fit. Hence this procedure should be combined with methods in Section 5.12.

5.14 Summary for inference

Some of the inference and diagnostics methods in this chapter were used in Chapter 1. All of the methods are covered with the variety of data examples in Chapter 7.

Relating to Sections 5.7 and 2.16, copulas with tail dependence have densities that asymptote to infinity in the relevant corners at a faster rate copulas with intermediate tail dependence. For multivariate data that seem to have strong dependence in the tails based on initial data analysis, the best fitting copula models based on AIC and BIC have tail dependence because the "concentration" of data in the tail leads to larger log-likelihoods. But this doesn't mean that that these best fitting copula models can provide adequate tail inference. This is one reason for the diagnostic methods in Section 5.12. Further research might be needed on some form of weighted likelihood and some use of multivariate extreme value inference to get better inferences for tail quantities.

Chapter 6

Computing and algorithms

For estimation and simulation from high-dimensional copulas, such as vines and factor models, and for converting from one parametrization to another for Gaussian dependence models, there are sequential procedures that are best summarized as algorithms. In this chapter, the algorithms are presented as pseudo-code[1]. Algorithms include mappings from partial correlations to correlations and/or regression coefficients for Gaussian truncated vines and factor models. This is important because Section 5.1 suggests to initially fit and compare different dependence structures based on an assumption of multivariate Gaussian copulas.

For computations of functionals of copulas such as dependence measures, and estimation of copula parameters via likelihood methods, quantities of interest often do not have closed form. In working with copula models, familiarity with basic numerical methods can help in producing faster-running and more reliable code. Numerical methods include finding roots of nonlinear equations in Section 6.1, numerical optimization in Section 6.2, numerical integration and quadrature in Section 6.3, interpolation in Section 6.4, and some numerical methods involving matrices in Section 6.5. Section 6.6 discusses some other numerical algorithms for graphs and spanning trees.

The numerical methods that are summarized in the first part of this chapter are then applied in the context of copulas. Particular topics include numerically stable computation of Kendall's τ and Spearman's ρ_S and related quantities in Sections 6.7 and 6.8, and simulation from multivariate distributions and copulas in general and for special classes in Section 6.9.

For vine and factor copulas, algorithms are presented in different sections: likelihoods of vines for continuous and discrete variables in Section 6.10; likelihoods of factor copula models in Section 6.11. Section 6.12 has a summary of the copula derivatives that are needed in order to optimize vine copula and factor copula log-likelihoods with a modified Newton-Raphson iterative method. Section 6.13 has algorithms for vine arrays and enumeration or for random selection of vines. Section 6.14 has algorithms for simulation of vines and factor copulas. Section 6.15 has algorithms for converting among different parametrizations of truncated Gaussian vines: mappings among sets of correlations, sets of partial correlations, sets of regression coefficients of linear representations. Section 6.16 has algorithms in converting among different parametrizations of loadings and partial correlations for multivariate t or Gaussian factor models. Section 6.17 discusses some strategies for search through the space of (truncated) R-vines combined with choices of pair-copulas on edges of the vines.

Section 6.18 contains a chapter summary.

---

[1]No computer code is presented in this book; see the author's publication web page for the information on code/software that implements all of the algorithms in this chapter.

6.1 Roots of nonlinear equations

The solving of nonlinear equations is implicit in univariate quantiles of conditional distributions and in estimation with maximum likelihood and other likelihood-based methods in Sections 5.5 and 5.6.

For solving the quantile or inverse cdf, a nonlinear equation must be numerically solved if the inverse cdf does not have closed form. Solving for a maximum likelihood estimate usually corresponds to finding a root of the nonlinear system of score equations (first order partial derivatives of log-likelihood with respect to the parameters).

If possible, the Newton-Raphson method or a modification of it is most efficient in the sense of a quadratic convergence rate (or fewest iterations) to the root. For a single nonlinear equation, the bisection method is always a possibility if one can easily bracket the root with a lower bound and an upper bound.

The Newton-Raphson method is summarized first for solving a single nonlinear differentiable equation in one unknown; that is, solving $g(z) = 0$ for a root z. Suppose the current guess is $z^{(k)}$ and a tangent line $g(z^{(k)}) + [z - z^{(k)}] g'(z^{(k)})$ is drawn at $(z^{(k)}, g(z^{(k)}))$. Solving the root of linear tangent line leads to the next iteration:

$$g(z^{(k)}) + [z^{(k+1)} - z^{(k)}] g'(z^{(k)}) = 0 \quad \text{or}$$

$$z^{(k+1)} \leftarrow z^{(k)} - \frac{g(z^{(k)})}{g'(z^{(k)})}.$$

Iterations continue until $|z^{(k+1)} - z^{(k)}| < \epsilon$, where the tolerance ϵ is typically something like 10^{-6}. Sometimes it is better to solve $g(z) = g_0(z) - p = 0$ as $h(g_0[z(y)]) - h(p) = 0$ if h has range and y has domain in $(-\infty, \infty)$. This is especially true if $g(z) = C_{2|1}(z|u) - p$ $(0 < z < 1)$ where C is a bivariate copula, and $0 < u < 1$, $0 < p < 1$ are given. Solving $C_{2|1}(z|u) - p$ can be difficult with the Newton-Raphson method when the copula and value of u are such that $C_{2|1}(\cdot|u)$ has a steep rise from 0 to 1.

Because the Newton-Raphson method can be very sensitive to the starting point, implementation usually involves modifications that limits the step size $g(z^{(k)})/g'(z^{(k)})$ and decreases the step in order that $z^{(k+1)}$ is within the boundaries.

The multidimensional nonlinear system follows the same idea. Consider m equations in m variables. We use $\boldsymbol{\theta}$ for the argument of the nonlinear equations as in statistics the nonlinear system often comes from a set of estimating or inference equations for parameters of interest.

Suppose we want to solve the system:

$$g_r(\theta_1, \ldots, \theta_m) = 0, \quad r = 1, \ldots, m,$$

where g_r are differentiable. The generalization of the above is as follows. The tangent planes are:

$$g_r(t_1, \ldots, t_m) = g_r(t_1^*, \ldots, t_m^*) + \sum_{s=1}^{m} \frac{\partial g_r}{\partial \theta_s}(t_1^*, \ldots, t_m^*) \cdot (t_s - t_s^*), \quad r = 1, \ldots, m.$$

Using these hyperplanes for a linear approximation, if the current guess of the root is $(\theta_1^{(k)}, \ldots, \theta_m^{(k)})$, the next guess $(\theta_1^{(k+1)}, \ldots, \theta_m^{(k+1)})$ satisfies

$$0 = g_r(\theta_1^{(k)}, \ldots, \theta_m^{(k)}) + \sum_{s=1}^{m} \frac{\partial g_i}{\partial \theta_s}(\theta_1^{(k)}, \ldots, \theta_m^{(k)}) \cdot (\theta_s^{(k+1)} - \theta_s^{(k)}), \quad r = 1, \ldots, m,$$

or

$$\begin{pmatrix} \theta_1^{(k+1)} \\ \vdots \\ \theta_m^{(k+1)} \end{pmatrix} = \begin{pmatrix} \theta_1^{(k)} \\ \vdots \\ \theta_m^{(k)} \end{pmatrix} - \begin{pmatrix} \frac{\partial g_1}{\partial \theta_1} & \cdots & \frac{\partial g_1}{\partial \theta_m} \\ \vdots & \ddots & \vdots \\ \frac{\partial g_m}{\partial \theta_1} & \cdots & \frac{\partial g_m}{\partial \theta_m} \end{pmatrix}^{-1} \Bigg|_{(\theta_1^{(k)}, \ldots, \theta_m^{(k)})} \cdot \begin{pmatrix} g_1(\theta_1^{(k)}, \ldots, \theta_m^{(k)}) \\ \vdots \\ g_m(\theta_1^{(k)}, \ldots, \theta_m^{(k)}) \end{pmatrix}.$$

Numerically the last term of the above is implemented as solving a linear system, that is, one solves something of the form $\boldsymbol{Ax} = \boldsymbol{b}$ instead of inverting the matrix $\boldsymbol{A}$ and then multiplying by $\boldsymbol{b}$.

A good starting point is important, especially if there is more than one root. If there is more than one root, the statistical or probabilistic context can be used to get the desired root as well as a good starting point for the numerical iterations.

Pseudo-code for modified Newton-Raphson method for solving $\boldsymbol{g}(\boldsymbol{\theta}) = 0$.

Inputs: functions $\boldsymbol{g}$ (vector) and $\boldsymbol{H} = \partial \boldsymbol{g}/\partial \boldsymbol{\theta}^{\top}$ (matrix); maxstep = maximum step size; bounds on parameters $\boldsymbol{\theta}$; maxitn = maximum number of iterations; tolerance ϵ; starting point $\boldsymbol{\theta}^{(0)}$, and a flag for printing iterations.

1. $k \leftarrow 0; \boldsymbol{\theta} \leftarrow \boldsymbol{\theta}^{(0)}$.

2. Compute $\boldsymbol{g}_k = \boldsymbol{g}(\boldsymbol{\theta})$ and $\boldsymbol{H}_k = \boldsymbol{H}(\boldsymbol{\theta})$.

3. Set $\boldsymbol{\Delta} \leftarrow \boldsymbol{H}_k^{-1}\boldsymbol{g}_k$. Update $\boldsymbol{\theta} \leftarrow \boldsymbol{\theta} - \boldsymbol{\Delta}; k \leftarrow k + 1$.

4. [Modification step]. While some component of $\boldsymbol{\theta}$ is outside boundary or $||\boldsymbol{\Delta}||_{\infty} >$maxstep, reset $\boldsymbol{\Delta} \leftarrow \boldsymbol{\Delta}/2$ and $\boldsymbol{\theta} \leftarrow \boldsymbol{\theta} + \boldsymbol{\Delta}$.

5. If print flag is on, print $k, \boldsymbol{\theta}, \boldsymbol{\Delta}$.

6. If $||\boldsymbol{\Delta}||_{\infty} < \epsilon$, declare convergence and return $\boldsymbol{\theta}$. Also return $\boldsymbol{H}_k$ if $\boldsymbol{g}(\boldsymbol{\theta})$ is a vector of score equations.

7. If $k \geq$ maxitn, declare non-convergence and return an error code.

8. Otherwise, go to step 2.

6.2   Numerical optimization and maximum likelihood

In this section, we present numerical optimization in the context of maximizing the log-likelihood or minimizing the negative log-likelihood. The theory applies more generally, but what is most relevant to inference with parametric copula models is the numerical optimization of the log-likelihood.

Suppose the model consists of a density $f(\cdot; \boldsymbol{\theta})$, where $\boldsymbol{\theta}$ is parameter vector. For simpler notation, we assume that data are $y_1, \ldots, y_n$ considered as a random sample of size n. The details are similar for independent data that are not identically distributed because of covariates.

If one is using the Newton-Raphson method for optimization, say maximizing log-likelihood $L(\theta_1, \ldots, \theta_m)$ or minimizing the negative log-likelihood $-L$, then under regularity conditions, the maximum likelihood estimator is the root of the above equations with $g_r = -\partial L/\partial\theta_r, r = 1, \ldots, m$, and

$$\frac{\partial g_r}{\partial \theta_s} = -\frac{\partial^2 L}{\partial \theta_r \theta_s}, \quad r, s = 1, \ldots, m.$$

The iterative equation has the Hessian matrix $(-\partial^2 L/\partial\theta_r\partial\theta_s)$ of second order derivatives.

For the quasi-Newton method, the partial derivatives g_r are obtained numerically as Newton quotients: $g_r(\boldsymbol{\theta}_r) \approx [-L(\boldsymbol{\theta} + \epsilon \boldsymbol{e}_r) + L(\boldsymbol{\theta})]/\epsilon$, where ϵ is small, e.g., 10^{-6}, and $\boldsymbol{e}_r$ is a vector with 1 in the rth position and 0s elsewhere. The Hessian is obtained numerically through an updating method, such as the Broyden, Fletcher, Goldfarb and Shanno (BFGS) method. A detailed reference for numerical optimization methods is Nash (1990).

In terms of the combination of programming time and computer run time, the quasi-Newton method is a good choice when the optimization is not repeated many times and if the dimension m of $\boldsymbol{\theta}$ is not too large. For quasi-Newton minimization methods that are implemented in mathematical and statistical software, as m increases over say 30, the numerical method might get to a local minimum quite quickly, but take a long time to get a Hessian estimate, or fail to get a positive definite Hessian at the local minimum.

For models, such as factor copula models, for which the number of parameters increases as the number of response variables increases, a modified Newton-Raphson implementation is desirable for faster numerical maximum likelihood.

The pseudo-code in Section 6.1 is for solving a nonlinear system of equations. Further modifications are the following when the equations for gradient equations are associated with numerical minimization.

(i) If the Newton-Raphson step leads to $\boldsymbol{\theta}^{(k+1)}$, check that the negative log-likelihood is decreasing, that is, $-L(\boldsymbol{\theta}^{(k+1)}) < -L(\boldsymbol{\theta}^{(k)})$; otherwise do not accept $\boldsymbol{\theta}^{(k+1)}$ and take a smaller step from $\boldsymbol{\theta}^{(k)}$ in the direction of $\boldsymbol{\theta}^{(k+1)}$.

(ii) If the negative log-likelihood is multimodal with many local maxima and minima, the Hessian of $-L$ may not be positive definite at the iteration $\boldsymbol{\theta}^{(k)}$ and then the Newton-Raphson may lead to a direction where $-L$ is not decreasing. If the Hessian $\boldsymbol{H}_k = (-\partial^2 L(\boldsymbol{\theta}^{(k)})/\partial\theta_r\partial\theta_s)$ is not positive definite, then replace $\boldsymbol{H}_k$ by a positive definite modification $\widetilde{\boldsymbol{H}}_k$ before evaluating the new direction $\boldsymbol{\Delta} \leftarrow \widetilde{\boldsymbol{H}}_k^{-1}\boldsymbol{g}_k$. An example of a positive definite modification is to replace negative eigenvalues in the spectral or eigen decomposition of $\boldsymbol{H}_k$ by small positive values.

The above modifications were implemented by Krupskii (2014) for maximum likelihood with factor copula models with a large number of variables and parameters, and works when the number of parameters is in the hundreds.

6.3   Numerical integration and quadrature

Log-likelihoods which involve 1-dimensional integrals are typically manageable for numerical optimizers. Even log-likelihoods with 2-dimensional and 3-dimensional integrals can be handled if the integrals are computed with Gaussian quadrature methods. For example, the densities of factor copula models with either continuous or discrete response variables have dimension of integrals equal to the number of latent variables; see Section 3.10.

In this section, we briefly summarize numerical integration methods that are useful for computing log-likelihoods factor copula models and for computing bivariate measures of dependence.

One-dimensional numerical integration has been well studied by numerical analysts. A method with a brief algorithm and implementation into code is the Romberg method (Ralston and Rabinowitz (1978)) which is an extrapolation of the trapezoidal rule. However, mathematical and statistical software has more robust methods of adaptive integration with Gauss-Kronrod quadrature rules (Kahaner et al. (1989)). The latter can handle integrals that have unbounded limits.

Gauss-Kronrod quadrature and adaptive integration have been extended for multidimensional numerical integration or cubature; a reference is Berntsen et al. (1991). These methods assume integration over a bounded hyperrectangle. For two-dimensional integrals, cubature is fast when integrands are smooth (without steep peaks).

In numerical minimization of objective functions, such as negative log-likelihoods, that require numerical integration in their evaluations, there is an advantage to using Gaussian quadrature methods (Stroud and Secrest (1966)). Useful versions for statistical applications are Gauss-Legendre, Gauss-Hermite, Gauss-Laguerre, Gauss-Jacobi for integrals of the form

$$\int h(x)\, f(x)\, \mathrm{d}x,$$

f is the U$(0, 1)$ density for Gauss-Legendre, the N$(0, 1)$ density for Gauss-Hermite, the

Gamma$(\alpha, 1)$ density for Gauss-Laguerre and the Beta(α, β) density for Gauss-Jacobi. That is, these methods are used for integrals that are expectations with respect to Uniform, Gaussian, Gamma and Beta random variables. The integral is approximated as

$$\int h(x)\, f(x)\, \mathrm{d}x \approx \sum_{k=1}^{n_q} w_k h(x_k),$$

where n_q is the number of quadrature points, $\{x_k\}$ are the quadrature points or nodes and $\{w_k\}$ are the quadrature weights. The approximation tends to get better as n_q increases. For n_q points, the approximation becomes an equality for functions h that are polynomials of degrees n_q or less. With $h \equiv 1$, one gets that $\sum_{k=1}^{n_q} w_k = 1$, and the approximation in the quadrature can be viewed as expectation with respect to a discrete random variable that has mass w_k at x_k for $k = 1, \ldots, n_q$; the discrete random variable is a discretization of the continuous random variable.

Gaussian quadrature extends to higher-dimensional integrals in that

$$\int_{\mathbb{R}^m} h(x_1, \ldots, x_m) f(x_1) \cdots f(x_m)\, \mathrm{d}\boldsymbol{x}$$

is approximated by

$$\sum_{k_1=1}^{n_q} \cdots \sum_{k_m=1}^{n_q} w_{k_1} \cdots w_{k_m} h(x_{k_1}, \ldots, x_{k_m}).$$

Often the choice of n_q for 2- or 3-dimensional integrals can be taken as smaller than that for 1-dimensional integrals.

For objective function that require numerical integration, an advantage of Gaussian quadrature over adaptive integration methods in numerical minimization is that n_q is fixed so that the approximated objective function is smooth as the parameter $\boldsymbol{\theta}$ varies, and numerical derivatives in quasi-Newton methods are reliable. For adaptive integration methods, it is possible that the subregions change with small changes to the parameter $\boldsymbol{\theta}$ and then numerical derivatives would not be reliable. With the modified Newton-Raphson method, the gradient vector and Hessian can be evaluated at the same time as the objective function using Gaussian quadrature (assuming the derivatives of the integral are the same as the integrals of the derivatives).

Krupskii and Joe (2013) reported implementations of numerical maximum likelihood for 1-factor and 2-factor copulas models with Gauss-Legendre quadature within a modifed Newton-Raphson method. The experience is that n_q between 21 and 31 is usually adequate for 1-dimensional integrals and n_q between 15 and 25 per dimension is adequate for 2-dimensional integrals.

The computations of Kullback-Leibler divergence between copula families in Section 5.7 were done with a combination of adaptive quadrature and Gauss-Legendre quadature. Gauss-Legendre quadature was use to find the parameter in a parametric copula family that leads to the minimum Kullback-Leibler divergence with a given copula. With two fixed copulas, adaptive quadrature was used in determining the Kullback-Leibler sample size (for differentiating the two families with 95% probability).

Besides integrals for likelihoods, numerical integration also are needed for copula models in computation of model-based values of measures of monotone association such as Kendall's τ and Spearman's ρ_S. Also they are needed for other functions of copulas; Section 5.10.2 mentions the comparison of empirical and model-based quantities in order to assess the adequacy of fit of a copula model.

For functionals of a copula, the general recommendation is that it is usually better to convert the integrand to a form where it is bounded; this speeds up the computational time and increases the accuracy. This is illustrated below for first and second order moments (assuming they exist); see also Section 6.7 for the forms of Kendall's τ and Spearman's ρ_S that are most suitable for numerical integrals.

When densities are unbounded, it is better to use a form with (conditional) cdfs and survival functions. Let $X = X^+ - X^-$ be a random variable with finite mean and $X^+ = \max\{0, X\}$ and $X^- = \max\{0, -X\}$. Then

$$\mathbb{E}(X) = \int_0^\infty \overline{F}_X(x)\,\mathrm{d}x - \int_0^\infty F(-x)\,\mathrm{d}x.$$

If $X_j = X_j^+ - X_j^-$ with $X_j^+ = \max\{0, X_j\}$ and $X_j^- = \max\{0, -X_j\}$ for $j = 1, 2$ and $\mathbb{E}(X_1 X_2)$ exists, then

$$\mathbb{E}(X_1 X_2) = \mathbb{E}(X_1^+ X_2^+) + \mathbb{E}(X_1^- X_2^-) - \mathbb{E}(X_1^+ X_2^-) - \mathbb{E}(X_1^- X_2^+)$$

$$= \int_0^\infty \int_0^\infty x_1 f_1(x_1)\{\overline{F}_{2|1}(x_2|x_1) + F_{2|1}(-x_2|-x_1) - F_{2|1}(-x_2|x_1) - \overline{F}_{2|1}(x_2|-x_1)\}\mathrm{d}x_2\mathrm{d}x_1$$

where $F_{2|1} = F_{X_2|X_1}$ and $f_1 = F_{X_1}$. An alternative form using Hoeffding's identity is:

$$\mathrm{Cov}(X_1, X_2) = \int_{-\infty}^\infty \int_{-\infty}^\infty \{\overline{F}_{X_1 X_2}(x_1, x_2) - \overline{F}_1(x_1)\,\overline{F}_2(x_2)\}\,\mathrm{d}x_1 \mathrm{d}x_2$$

$$= \int_{-\infty}^\infty \int_{-\infty}^\infty \{F_{X_1 X_2}(x_1, x_2) - F_1(x_1)\,F_2(x_2)\}\,\mathrm{d}x_1 \mathrm{d}x_2.$$

For numerical implementation, the integrands in four quadrants can be combined to get an integral over $[0, \infty)^2$. If $C(F_1, F_2) \in \mathcal{F}(F_1, F_2)$ and F_1, F_2 have densities, then $F_{2|1}(x_2|x_1) = C_{2|1}(F_2(x_2)|F_1(x_1))$. These two forms for the second mixed moment are useful for the numerical computation for moments of $C(\Phi, \Phi)$ such as in (2.57).

6.4   Interpolation

Interpolation preserving monotonicity is useful in a few situations for computations with copulas. The generic setting is that interpolation is needed for two columns of numbers that are monotonically related when it can be time consuming to evaluate one of these two columns.

Examples are the following.

(a) Interpolating between a copula parameter and a dependence measure such as Kendall's τ, Spearman's ρ_S for a 1-parameter copula family that is increasing in concordance as the copula parameter δ increases.

(b) Solving for a conditional quantile $C_{2|1}(v|u) = p$ near the boundary (u or p near 0 or 1) and $C_{2|1}(v|u)$ is susceptible to numerical roundoff. In this case, with fixed u, evaluate $C_{2|1}(\cdot|u)$ for a grid of values $\{v_k\}$ near the boundary to get corresponding values $\{p_k\}$; then add the pair $(v, p) = (0, 0)$ or $(1, 1)$ and apply monotone interpolation for other (v, p) values near the boundary.

(c) Numerical maximum likelihood for a copula model with density $c_{1:d}(\boldsymbol{u}) = f_{1:d}(F_1^{-1}(u_1; \boldsymbol{\eta}_1), \ldots, F_d^{-1}(u_d; \boldsymbol{\eta}_d); \boldsymbol{\delta}) / \prod_{j=1}^d f_j[F_j^{-1}(u_j; \boldsymbol{\eta}_j); \boldsymbol{\eta}_j]$, where F_j have closed forms but not F_j^{-1}. If $F_j^{-1}(u; \boldsymbol{\eta}_j)$ have to be obtained by solving equations, and the density $c_{1:d}(\boldsymbol{u}_i)$ has to be evaluated for n vectors $\boldsymbol{u}_i$ with n large in each step of numerical optimization, then it is numerically faster to compute $F_j(x, \boldsymbol{\eta}_j) = p$ on a grid (x_k, p_k) and use monotone interpolation to approximate $F_j^{-1}(u; \boldsymbol{\eta}_j)$. This was done for the simulation with the bivariate skew-normal copula in Section 7.1.

(d) A parametric Archimedean copula model C_ψ with numerical maximum likelihood where $\psi(\cdot; \delta)$ can be evaluated in closed form but not ψ^{-1}. Then monotone interpolation can be used for computing ψ^{-1}. An example is the copula in Section 4.12, where ψ is the LT of an inverse gamma random variable and is expressed in terms of a modified Bessel function of the second kind.

Chapter 4 of Kahaner et al. (1989) describes a method using piecewise cubic interpolation to preserve monotonicity; it is based on Fritsch and Carlson (1980). If the derivatives at the evaluated points can be computed, these can be used instead of being estimated as part of the algorithm for piecewise cubic interpolation.

For a truncated R-vine with some pair-copulas that are t_ν in the lower trees, or for 2-factor copula models with t_ν linking copulas to the first latent variable, many evaluations of the $T_{\nu+1}$ univariate (conditional) cdf are needed for the log-likelihood. When this is numerically too time-consuming, the use of the piecewise cubic interpolation leads to a tremendous reduction in total computational time with almost no loss of accuracy. The same situation occurs for simulation of a copula-GARCH time series model (see Section 7.8), where the vector of innovations is distributed with a multivariate t_ν copula.

In summary, when a smooth monotone real-valued function g is numerically non-trivial and the number of evaluations of it is sufficiently large ($> N$), then interpolation from a table of $g(x_k)$ at N well chosen points $\{x_k\}$ can decrease the total computational time.

6.5   Numerical methods involving matrices

Numerical methods involving matrices are covered in many books on numerical analysis; see, for example, Stewart (1973) and Nash (1990). In this section, we just mention a few things relevant for numerical optimization with copula models.

Linear system solving is needed within Newton-Raphson iterations and solving for partial correlations from sub-correlation matrices. The notation $\boldsymbol{H}^{-1}\boldsymbol{g}$ numerically means to solve $\boldsymbol{H}\boldsymbol{b} = \boldsymbol{g}$ for $\boldsymbol{b}$, that is, input $\boldsymbol{H}$ as the left-hand side and $\boldsymbol{g}$ as the right-hand side into a numerical linear system solver; unless there is a closed form for $\boldsymbol{H}^{-1}$, this is generally more numerically efficient than evaluating $\boldsymbol{H}^{-1}$ numerically and then multiplying by $\boldsymbol{g}$.

The triangular matrix from the Cholesky decomposition of a positive definite covariance matrix is used for simulation from multivariate Gaussian and elliptical distributions. Also the Cholesky decomposition algorithm can be used to check if a symmetric matrix is positive definite. The Cholesky triangular matrix can also be used as a parametrization of a covariance matrix in numerical optimizations to avoid the checking of the constraint positive definiteness. For a positive definite (or non-negative definite) symmetric $d \times d$ matrix $\boldsymbol{\Sigma} = (\sigma_{ij})$, the Cholesky decomposition is $\boldsymbol{\Sigma} = \boldsymbol{A}\boldsymbol{A}^T$ where $\boldsymbol{A} = (a_{ij})$ is lower (or upper) triangular. Note that the Cholesky decomposition is unique among lower triangular $\boldsymbol{A}$ if the diagonal of $\boldsymbol{A}$ is non-negative. Actual implementation of the decomposition may require some tolerances for numerical roundoffs. Brief pseudo-code is as follows.

- If $\sigma_{11} \leq 0$ return flag; else $a_{11} \leftarrow \sqrt{\sigma_{11}}$
- for $k = 2$ to d:
    - for $i = 1$ to $k - 1$:
    - $s \leftarrow \sum_{j=1}^{i-1} a_{ij} a_{kj}$;
    - $a_{ki} \leftarrow (\sigma_{ki} - s)/a_{ii}$.
    - $s \leftarrow \sum_{j=1}^{k-1} a_{kj}^2$
    - if $\sigma_{kk} - s < 0$, return flag indicating the matrix is not non-negative definite.
    - $a_{kk} \leftarrow \sqrt{\sigma_{kk} - s}$.
- return $\boldsymbol{A}$.

Note that the Cholesky decomposition algorithm is obtained sequentially by solving for the lower triangular matrix A one element at a time:

$$a_{11}, a_{21}, a_{22}, a_{31}, a_{32}, a_{33}, a_{41}, \ldots.$$

To simulate $X \sim N_d(\mu, \Sigma)$, first simulate $Z_1, \ldots, Z_d$ i.i.d. $N(0,1)$ and then set $X = \mu + AZ$, where $\Sigma = AA^T$. If AZ is computed using *for* loops, then there are fewer multiplications/additions if A is triangular.

6.6   Graphs and spanning trees

In this section, we discuss some numerical techniques from graph theory for working with the trees within vines.

A vine consists of a sequence of trees so that optimization over vines can involve algorithms for graphs and spanning trees. The number of vines increases exponentially with the number of variables d so that optimization via enumeration becomes impossible beyond $d = 7$ or 8. An alternative is a sequential approach by tree $T_1, T_2, \ldots$. For example, one might want to choose tree T_1 to have pairs of variables with the strongest dependence; this corresponds to maximizing or minimizing a sum of transforms of a dependence measure such as $|\tau|$ or $|\rho_N|$, After the first tree is "optimized" at T_1^*, one might want to choose a second tree T_2 so that there is strong conditional dependence given the pairs in T_1^*. One could maximize or minimize a sum of transforms of a conditional dependence measure, such as partial correlation or partial tau, over trees T_2 that are compatible with T_1^*. One could continue in this way with a greedy stepwise algorithm. The result need not be a global optimum over vines but one would hope that the result is a good sub-optimal vine.

References for an efficient minimum spanning tree algorithm are Prim (1957) and Papadimitriou and Steiglitz (1982). These make use of ideas of graph theory, where a *graph* is a set of vertices (nodes) with edges over some pairs of the vertices. There are software libraries with lots of graph algorithms for creating and manipulating graphs; see for example Csárdi and Nepusz (2012).

A complete graph on d vertices has $\binom{d}{2}$ edges. Weights are assigned to each edge of a graph. For first the optimal T_1 in a greedy approach, the weight for an edge for two variables could be a transform of the absolute value of a dependence measure for the two variables, such as $|\tau|$ or $|\rho_N|$; for the minimum spanning tree algorithm, the weight is smaller for stronger dependence. Section 6.17.1 has more details of a greedy approach for partial correlation vines that makes use of the minimum spanning tree algorithm.

To go beyond the greedy approach, one needs to consider "neighbors" of the minimum spanning tree or K best spanning trees at each step. A 1-neighbor of a tree T can be defined as follows.

1. Choose an edge $e \notin T$ by joining two previously unconnected vertices, and set $T' = T \cup \{e\}$. Then the graph T' is no longer a tree but has a cycle C of edges including e.

2. Remove an edge $e' \in C \setminus \{e\}$ from T' to obtain $T'' = T' \setminus \{e'\}$. T'' is a tree and called a 1-neighbor of T.

In a similar way, we can define n-neighbors, where n edges are changed in the tree.

A straightforward extension to go beyond the minimum spanning tree is to take into account a range of K best spanning trees, that is, the K spanning trees with smallest sums of weights. The problem of identifying these K best spanning trees is NP-hard. Several algorithms have been described in the literature to have efficient running time; see for example Gabow (1977). Note that there may be 1-neighbors of the minimum spanning tree among the K best spanning trees but this is not necessarily so.

If an ℓ-truncated vine structure is built, only the ℓth tree is selected as a minimum spanning tree given the first $\ell - 1$ trees and the sequential one-tree-at-a-time selection of minimum spanning trees is very likely to not lead to the optimum. The search space consists of all $(\ell - 1)$-truncated regular vines and can not be explored entirely. Combinations of the n-neighbors and best spanning trees may be used to find good suboptimal solutions. These non-greedy algorithms are included in Brechmann and Joe (2014a).

6.7 Computation of τ, ρ_S and ρ_N for copulas

In this section, we discuss computation of dependence measures for bivariate copulas.

The general mathematical form of Kendall's τ, Spearman's ρ_S and the normal scores correlation ρ_N involve two-dimensional integrals. Sometimes there is a reduction to a one-dimensional integral, and in some cases, there is a closed form expression as a function of the copula parameter.

For most one-parameter bivariate copula families, Blomqvist's β is easiest to compute as a function of the copula parameter δ. Given a specified β, the parameter δ is either a simple function of β or it can be computed (to desired accuracy of at least 10^{-6}) by solve a nonlinear equation. The family in Section 4.11 is an exception where this nonlinear equation is harder to solve than computing Kendall's τ. Also for one-parameter copula families with tail dependence in one of the tails (or reflection symmetric copulas with the same tail dependence parameter in the two tails), the mapping from δ to λ and its inverse is simple for those in Chapter 4.

Kendall's τ for an Archimedean copula with generator $\psi \in \mathcal{L}_2$ can be expressed as a one-dimensional integral: $\tau = 1 - 4 \int_0^\infty s[\psi'(s)]^2 \, ds$ is numerically better than the formula in Genest and MacKay (1986) because the latter has an unbounded integrand; see Theorem 8.26. Also Kendall's τ and Spearman's ρ_S can be expressed as one-dimensional integrals for bivariate extreme value copulas; see Theorem 8.44.

Let $C(u, v)$ be a bivariate copula cdf with density $c(u, v)$. Spearman's ρ_S can be computed as a 2-dimensional numerical integral using (2.47) or (2.48), i.e.,

$$\rho_S(C) = 12 \int_0^1 \int_0^1 C(u, v) \, dudv - 3 = 3 - 12 \int_0^1 \int_0^1 u \, C_{2|1}(v|u) \, dudv,$$

so that the integrand is always bounded by 1. The second form with a conditional cdf is better for something like a t_ν copula. Kendall's τ can be computed as 2-dimensional numerical integral using (2.42), i.e.,

$$\tau(C) = 1 - 4 \int_0^1 \int_0^1 C_{1|2}(u|v) \, C_{2|1}(v|u) \, dudv.$$

As alternatives to the original definitions in (2.45) and (2.41), these formulas avoid the use of the copula density, which can be infinite at corners when the tail order is strictly less than 2.

ρ_N as a two-dimensional numerical integral can be numerically evaluated with one of the following, with $u = \Phi(z_1)$, $v = \Phi(z_2)$ in the integrand:

$$\rho_N(C) = \int_{-\infty}^\infty \int_{-\infty}^\infty \{C(u, v) - uv\} \, dz_2 dz_1$$
$$= \int_0^\infty z_1 \phi(z_1) \int_0^\infty \{\overline{C}_{2|1}(v|u) + C_{2|1}(1 - v|1 - u) - C_{2|1}(1 - v|u) - \overline{C}_{2|1}(v|1 - u)\} dz_2 dz_1,$$

where ∞ in the integral limits is replaced with a value around 6 or 7, so that all tails of the integrals are negligible to around 10^{-4}.

Computations of (2.44) and (2.50) are also numerically more reliable with an alternative form after integration by parts. To avoid an unbounded copula density c, the general integration by parts formula that is relevant is:

$$\iint H\mathrm{d}C = \int_0^1 H(u,1)\,\mathrm{d}u - \int_0^1\int_0^1 \frac{\partial H(u,v)}{\partial v} C_{2|1}(v|u)\,\mathrm{d}v\,\mathrm{d}u.$$

For the asymptotic variance of $\hat{\tau}$, as $n \to \infty$,

$$n\,\mathrm{Var}(\hat{\tau}) \to 16\int_0^1\int_0^1 [C(u,v) + \overline{C}(u,v)]^2 \mathrm{d}C(u,v) - 4(\tau+1)^2.$$

With $H = [C + \overline{C}]^2$, $H(u,1) = u^2$, and

$$\iint H\mathrm{d}C = \tfrac{1}{3} - 2\int_0^1\int_0^1 [C(u,v) + \overline{C}(u,v)][2C_{1|2}(u|v) - 1]C_{2|1}(v|u)\,\mathrm{d}v\,\mathrm{d}u.$$

For the asymptotic variance of $\hat{\rho}_S$, as $n \to \infty$,

$$n\,\mathrm{Var}(\tilde{\rho}_S) = 144\int_0^1\int_0^1 \{u^2v^2 + 2uvg_1(u) + 2uvg_2(v) + g_1^2(u) + g_2^2(v) + 2g_1(u)g_2(v)\}\mathrm{d}C(u,v)$$
$$-9(\rho_S + 3)^2,$$

where $g_1(u) = \int_0^1 \overline{C}(u,w)\mathrm{d}w = \tfrac{1}{2} - u + \int_0^1 C(u,w)\mathrm{d}w$ and $g_2(v) = \int_0^1 \overline{C}(x,v)\mathrm{d}x = \tfrac{1}{2} - v + \int_0^1 C(x,v)\mathrm{d}x$. Let $H(u,v) = u^2v^2 + 2uvg_1(u) + 2uvg_2(v) + g_1^2(u) + g_2^2(v) + 2g_1(u)g_2(v)$. Then

$$\iint H\mathrm{d}C = \int_0^1 \{u^2 + 2ug_1(u) + g_1^2(u)\}\mathrm{d}u$$
$$- \int_0^1\int_0^1 2\{u^2v + ug_1(u) + ug_2(v) + uvg_2'(v) + g_2(v)g_2'(v) + g_1(u)g_2'(v)\}C_{2|1}(v|u)\,\mathrm{d}v\mathrm{d}u.$$

with $g_2'(v) = -\int_0^1 \overline{C}_{1|2}(x|v)\mathrm{d}x = \int_0^1 C_{1|2}(x|v)\mathrm{d}x - 1$. Since $g_2(1) = 0$, then $H(u,1) = u^2 + 2ug_1(u) + g_1^2(u)$.

For comparing density contours and dependence properties of different bivariate one-parameter copula families, it is useful to be able to compute the copula parameter δ given a fixed value of β, τ, ρ_S or ρ_N. This is done in many examples and tables in this book.

The following is a general way to do this in a computer implementation for a one-parameter copula family $C(u,v;\delta)$ with density $c(u,v;\delta)$. The family is assumed to include $C^\perp$ and C^+, and assumed to be increasing in concordance as δ increases, so that the dependence measures are all increasing in δ.

- Select a grid of β values that include 0 and 1, for example $0, 0.02, \ldots, 0.96, 0.98, 1$.
- Determine the corresponding δ values for the copula family.
- For these δ values, compute τ, ρ_S, ρ_N either in closed form or using numerical integration.
- Now there is a table with columns $\delta, \beta, \tau, \rho_S, \rho_N$, which can be stored (say, in a statistical software).
- For a given fixed value of τ, ρ_S or ρ_N, apply a monotone interpolator (see Section 6.4) to get β; then solve a nonlinear equation or apply an inverse transform to get δ. The interpolation through β should be better than through δ, as all of the dependence measures take values in $[-1, 1]$, and the copula parameter δ is often infinite for C^+.

An alternative approach using penalized smoothing splines is in the Appendix of Kojadinovic and Yan (2010) for inverting for τ and ρ_S to the copula parameter.

6.8   Computation of empirical Kendall's τ

In this section, we indicate methods for efficient computation of Kendall's τ for a sample of n pairs. For Spearman's ρ_S, the two variables can be converted to ranks via sort algorithms that are $O(n \log n)$ in computational complexity. If Kendall's τ is computed as a double sum (as in (2.43)), the computational complexity is $O(n^2)$. However, Knight (1966) has an algorithm that is $O(n \log n)$ based on separate sorts of the two variables in blocks of 2^k, $k = 1, 2, \ldots$, and counting the number of exchanges to sort the second variable when the records are ordered in the first variable; this is one of the methods given in Kendall (1948).

Algorithm 1 $O(n \log n)$ algorithm for computing Kendall's tau, from Knight (1966).

1: input x_i, y_i, $i = 1, \ldots, n$ as vectors $\boldsymbol{x}, \boldsymbol{y}$.

2: Phase 1: use $O(n \log n)$ algorithm and reindex so that $\boldsymbol{x}$ is a sorted vector (in ascending order) and $\boldsymbol{y}$ is sorted in the blocks of ties in $\boldsymbol{x}$.

3: Adjustments for ties on $\boldsymbol{x}$ and joint ties; set $t_x \leftarrow 0$; $t_{xy} \leftarrow 0$, $i_x \leftarrow 1$, $i_{xy} \leftarrow 1$.

4: **for** $i = 2, \ldots, n$: **do**

5:     if $(x_i = x_{i-1})$ then $\{\ i_x \leftarrow i_x + 1$;
        if $(y_i = y_{i-1}$) then $i_{xy} \leftarrow i_{xy} + 1$ else $\{\ t_{xy} + i_{xy}(i_{xy} - 1)/2$; $i_{xy} \leftarrow 1\ \}\ \}$

6:     else $\{\ t_x \leftarrow t_x + i_x(i_x - 1)/2$; $t_{xy} \leftarrow t_{xy} + i_{xy}(i_{xy} - 1)/2$; $i_x \leftarrow 1$; $i_{xy} \leftarrow 1\ \}$.

7: **end for**

8: $t_x \leftarrow t_x + i_x(i_x - 1)/2$; $t_{xy} \leftarrow t_{xy} + i_{xy}(i_{xy} - 1)/2$.

9: Phase 2: sort $\boldsymbol{y}$ vector in blocks of $2, 2^2, 2^3 \ldots$, with same switches in $\boldsymbol{x}$ vector, using the steps below.

10: Allocate temporary vectors $\boldsymbol{x}', \boldsymbol{y}'$ of length n.

11: Initialize $k \leftarrow 1$, $n_{\text{exch}} \leftarrow 0$.

12: **repeat**

13:     $\ell \leftarrow 1$;

14:     **repeat**

15:        $i \leftarrow \ell$; $j \leftarrow \min\{i + k, n + 1\}$; $i_{\text{end}} \leftarrow j$; $j_{\text{end}} \leftarrow \min\{j + k, n + 1\}$.

16:        **repeat**

17:           iflag$\leftarrow I(i < i_{\text{end}})$; jflag$\leftarrow I(j < j_{\text{end}})$; yflag$\leftarrow I(y_i > y_j)$;

18:           if [iflag & ((jflag & !yflag) or !jflag)] then $\{x'_\ell \leftarrow x_i$; $y'_\ell \leftarrow y_i$; $i \leftarrow i + 1$; $\ell \leftarrow \ell + 1\}$.

19:           else if [(!iflag & jflag) or (iflag & jflag & yflag)] then $\{x'_\ell \leftarrow x_j$; $y'_\ell \leftarrow y_j$; $j \leftarrow j + 1$;
          $\ell \leftarrow \ell + 1$; $n_{\text{exch}} \leftarrow n_{\text{exch}} + (i_{\text{end}} - i)\}$.

20:        **until** (!iflag & !jflag)

21:     **until** $(\ell > n)$

22:     swap $\boldsymbol{y}, \boldsymbol{y}'$; swap $\boldsymbol{x}, \boldsymbol{x}'$;

23:     $k \leftarrow 2k$;

24: **until** $(k >= n)$

25: End of phase 2.

26: Phase 2': repeat the steps in Phase 2 with the roles or $\boldsymbol{x}, \boldsymbol{y}$ interchanged,
    xflag$\leftarrow I((x_i > x_j)$ or $(x_i = x_j)\&(y_i > y_j))$ replacing yflag, and n_{exch} not used.

27: Adjustments for ties on $\boldsymbol{y}$; set $t_y \leftarrow 0$ and $i_y \leftarrow 0$;

28: **for** $i = 2, \ldots, n$: **do**

29:     if $(y_i = y_{i-1})$ then $i_y \leftarrow i_y + 1$;

30:     else $\{\ t_y \leftarrow t_y + i_y(i_y - 1)/2$; $i_y \leftarrow 1\ \}$.

31: **end for**

32: $t_y \leftarrow t_y + i_y(i_y - 1)/2$.

33: Numerator of τ is $\frac{1}{2}n(n - 1) - (2n_{\text{exch}} + t_x + t_y - t_{xy}))$.

34: Denominator of τ is $[(\frac{1}{2}n(n - 1) - t_x)(\frac{1}{2}n(n - 1) - t_y)]^{1/2}$.

35: Return the ratio τ.

6.9 Simulation from multivariate distributions and copulas

This section summarizes methods for simulation of random vectors from some classes of multivariate distributions and copulas. Section 6.9.1 outlines the general sequential conditional method for simulation from a multivariate distribution. Section 6.9.2 has the conditional distributions when some uniform random variables are reflected, as this is useful for the simulation from conditional method with reflected copulas. Section 6.9.3 summarizes the simulation via componentwise maxima or minima for multivariate cdfs or survival functions that are products of others. Section 6.9.4 outlines the simulation for Archimedean copula based on LT from the mixture representation, and Section 6.9.5 has similar results for mixture of max-id copulas. Section 6.9.6 mentions the subclasses of multivariate extreme value copulas for which simulation from stochastic representations can be used.

As an example, random vectors from elliptical distribution can be simulated based on a multivariate Gaussian random vector and the radial random variable. A multivariate Gaussian random vector with mean vector $\mathbf{0}$ and correlation matrix $\mathbf{R}$ is most easily simulated based on the lower triangular Cholesky matrix $\mathbf{A}$ such that $\mathbf{A}\mathbf{A}^\top = \mathbf{R}$, with $\mathbf{A}\mathbf{Z}^\perp$ where $\mathbf{Z}^\perp$ is a vector of i.i.d. $N(0,1)$ random variables. To simulate from the corresponding elliptical copula, one needs the inverse of the univariate marginal cdf. The relation of the univariate marginal cdf and cdf of the radial random variable is given in Section 2.7.

For multivariate $t_{d,\nu}(\cdot; \mathbf{R})$ where $\mathbf{R}$ is a correlation matrix, generate $\mathbf{Z} = (Z_1, \ldots, Z_d)^\top \sim N_d(\mathbf{0}, \mathbf{R})$ using the Cholesky decomposition $\mathbf{R} = \mathbf{A}\mathbf{A}^\top$ to get $\mathbf{Z} = \mathbf{A}\mathbf{Z}^\perp$. Then generate $W \sim \text{Gamma}(\nu/2, \text{scale} = 2/\nu)$ and $\mathbf{Y} = (Y_1, \ldots, Y_d)^\top = \mathbf{Z}/\sqrt{W}$ is multivariate $t_{d,\nu}$. Finally, let $U_j = T_{1,\nu}(Y_j)$ for $j = 1, \ldots, d$ to get a multivariate vector from the $t_{d,\nu}$ copula with parameter $\mathbf{R}$.

There are standard and specialized techniques for generating random deviates from the common univariate distributions; see, for example, Devroye (1986). They can be used within the conditional approach when conditional distributions match known univariate distributions, and they can also be used when there are stochastic representations for the multivariate distribution. The basic results for simulation of a random variable with cdf F are given in Section 2.5.

6.9.1 Conditional method or Rosenblatt transform

The most general approach is based on a sequence of conditional distributions. This is called the conditional approach in Devroye (1986); because of Rosenblatt (1952), it is also known as the Rosenblatt transform for a continuous random vector to/from a vector of $U(0,1)$ random variables. Let $F_{1:d}(x_1, \ldots, x_d)$ be a multivariate distribution with univariate margins $F_1, \ldots, F_d$; the corresponding random variables can be continuous, discrete or mixed. With the given indexing, conditional distributions are

$$F_{2|1}(x_2|x_1),\ F_{3|12}(x_3|x_1, x_2), \ldots,\ F_{d|1\cdots d-1}(x_d|x_1, \ldots, x_{d-1}),$$

with functional inverses

$$F_{2|1}^{-1}(p_2|x_1),\ F_{3|12}^{-1}(p_3|x_1, x_2), \ldots,\ F_{d|1\cdots d-1}^{-1}(p_d|x_1, \ldots, x_{d-1}),$$

where $0 < p_j < 1$ for all j.

To generate a random vector $(X_1, \ldots, X_d) \sim F_{1:d}$, let $P_1, \ldots, P_d$ be i.i.d. $U(0,1)$ and let

- $X_1 \leftarrow F_1^{-1}(P_1)$,
- $X_2 \leftarrow F_{2|1}^{-1}(P_2|X_1)$,
- $X_3 \leftarrow F_{3|12}^{-1}(P_3|X_1, X_2), \ldots,$
- $X_d \leftarrow F_{d|1\cdots d-1}^{-1}(P_d|X_1, \ldots, X_{d-1})$.

If $F_{1:d}$ is a copula, the resulting $(X_1, \ldots, X_d)$ is a vector of dependent $U(0,1)$ random variables. A permutation of the order $1, 2, \ldots, d$ can also be used. In practice, one would choose a permutation where the computations are simplest.

We indicate an outline of the proof of this conditional approach for $d = 2$ and 3. For $d = 2$, with F_1 absolutely continuous and $F_{2|1}$ continuous, and P_1, P_2 independent $U(0,1)$, then $X_1 = F_1^{-1}(P_1) \sim F_1$ and

$$\mathbb{P}\big(X_1 = F_1^{-1}(P_1) \leq x_1, \, X_2 = F_{2|1}^{-1}(P_2|X_1) \leq x_2\big) = \mathbb{P}\big(X_1 \leq x_1, \, P_2 \leq F_{2|1}(x_2|X_1)\big)$$

$$= \int_{-\infty}^{x_1} \mathbb{P}\big(P_2 \leq F_{2|1}(x_2|y)|X_1 = y\big) \, dF_1(y) = \int_{-\infty}^{x_1} F_{2|1}(x_2|y) \, dF_1(y) = F_{12}(x_1, x_2).$$

Next, for $d = 3$ with F_{12} absolutely continuous and $F_{3|12}$ continuous, there is an additional step of $X_3 = F_{3|12}^{-1}(P_3|X_1, X_2)$. and

$$\mathbb{P}\big(X_1 \leq x_1, X_2 \leq x_2, F_{3|12}^{-1}(P_3|X_1, X_2) \leq x_3\big)$$

$$= \int_{-\infty}^{x_1} \int_{-\infty}^{x_2} \mathbb{P}\big(P_3 \leq F_{3|12}(x_3|y_1, y_2)|X_1 = y_1, X_2 = y_2\big) \, dF_{12}(y_1, y_2)$$

$$= \int_{-\infty}^{x_1} \int_{-\infty}^{x_2} F_{3|12}(x_3|y_1, y_2) \, dF_{12}(y_1, y_2) = F_{123}(x_1, x_2, x_3).$$

Assuming $F_{1:d}$ has simple form, the conditional method is easiest to implement when the conditional inverse cdfs have closed form. Otherwise,

$$F_{j|1\cdots j-1}^{-1}(p|x_1, \ldots, x_{j-1}), \quad 0 < p < 1,$$

is obtained numerically by solving

$$F_{j|1\cdots j-1}(z|x_1, \ldots, x_{j-1}) - p = 0$$

for z. See Section 6.1 for discussion of solving nonlinear equations.

Vine copulas and factor copulas just depend on all of the conditional cdfs of the bivariate distributions in the vine to generate random vector via the conditional method. The other details on the order of processing the vine array, etc., are given as algorithms in Sections 6.14.

6.9.2   Simulation with reflected uniform random variables

For the parametric copula families in Chapter 4, the directions of tail asymmetry are summarized in Table 4.1. For particular applications, one would consider copula families that can match the asymmetries and dependence seen in the data. Sometimes a reflection of one of the families will lead to a better match of the tail asymmetry and sometimes a partial reflection (of subset of variables) might be considered to match the pattern of negative dependence.

We show some details for the bivariate case how simulation is affected if the conditional method is used. For bivariate, suppose $(U, V) \sim C(u, v)$. The copula of $(1 - U, 1 - V)$ is denoted in Chapter 2 as

$$\widehat{C}(u, v) = u + v - 1 + C(1 - u, 1 - v).$$

Also let $\acute{C}$ be the copula of $(1 - U, V)$ and $\grave{C}$ be the copula of $(U, 1 - V)$. Then

$$\acute{C}(u, v) = v - C(1 - u, v), \quad \grave{C}(u, v) = u - C(u, 1 - v).$$

The conditional distributions and quantile functions of these can be written in terms of $C_{2|1}$:

$$\widehat{C}_{2|1}(v|u) = 1 - C_{2|1}(1-v|1-u), \quad \widehat{C}_{2|1}^{-1}(p|u) = 1 - C_{2|1}^{-1}(1-p|1-u);$$

$$\acute{C}_{2|1}(v|u) = C_{2|1}(v|1-u), \quad \acute{C}_{2|1}^{-1}(p|u) = C_{2|1}^{-1}(p|1-u);$$

$$\grave{C}_{2|1}(v|u) = 1 - C_{2|1}(1-v|u), \quad \grave{C}_{2|1}^{-1}(p|u) = 1 - C_{2|1}^{-1}(1-p|u).$$

Similarly, $\widehat{C}_{1|2}, \acute{C}_{1|2}, \grave{C}_{1|2}$ are defined from $C_{1|2}$. These formulas are useful for computations and simulations when code has been written for C, $C_{2|1}, C_{2|1}^{-1}, C_{1|2}, C_{1|2}^{-1}$.

The ideas here extend to reflection of some or all of $U_1, \ldots, U_d$ when the joint copula is $C_{1:d}$.

6.9.3 Simulation from product of cdfs

When products of cdfs (survival functions) are used to obtain a multivariate cdf (survival function), then there is a stochastic representation involving maxima (minima). Hence if the terms in the product can be simulated from, the stochastic representation can be used to the product. As an example, the construction of mixture of max-id copulas can involve products of bivariate distributions.

Suppose G, H are d-variate cdfs with $d \geq 2$ and let $F = GH$. The simulation method based on the stochastic representation is to generate $\boldsymbol{Y} \sim G$ and $\boldsymbol{Z} \sim H$, followed by $\boldsymbol{X} = (Y_1 \vee Z_1, \ldots, Y_d \vee Z_d) \sim F$.

Similarly if G, H are d-variate survival functions with $d \geq 2$ and $\overline{F} = \overline{GH}$. The simulation method based on the stochastic representation is to generate $\boldsymbol{Y} \sim G$ and $\boldsymbol{Z} \sim H$, followed by $\boldsymbol{X} = (Y_1 \wedge Z_1, \ldots, Y_d \wedge Z_d) \sim F$.

6.9.4 Simulation from Archimedean copulas

Archimedean copulas C_ψ are examples of multivariate distributions with a mixture representation (and conditional independence when the generator is a LT in $\mathcal{L}_\infty$). The approach based on the stochastic representation of Marshall and Olkin (1988) when $\psi \in \mathcal{L}_\infty$ is the following (see Section 3.17.1). If Q is a random variable with cdf F_Q and LT $\psi_Q = \psi$, then generate $Q \sim F_Q$. Given $Q = q$, generate $Y_1, \ldots, Y_d$ conditionally independent as Exponential random variables with rate q. Then $Y_1, \ldots, Y_d$ are unconditionally dependent with survival function $\overline{F}_Y(y) = \psi(y)$. Hence $(U_1, \ldots, U_d) = (\psi(Y_1), \ldots, \psi(Y_d))$ is a random vector distributed with the Archimedean copula C_ψ.

The pseudo-code is as follows.

- Generate $Q = q$ with cdf F_Q having LT ψ.
- Generate $Y_1, \ldots, Y_d$ as Exponential(rate=q), for example, $Y_j \leftarrow q^{-1}(-\log V_j)$, where $V_1, \ldots, V_d$ are independent U(0, 1).
- Set $U_j \leftarrow \psi(Y_j)$ for $j = 1, \ldots, d$ and return $(U_1, \ldots, U_d)$.

For the parametric Archimedean copula families listed in Chapter 4, this algorithm depends on the capability of generating a random variable with the corresponding LT families. For most of them, an algorithm is available; if so, these are included with the summaries of the LT families in the Appendix.

If $C_\psi(\boldsymbol{u}_{1:d})$ is an Archimedean copula with $\psi \in \mathcal{L}_d \backslash \mathcal{L}_\infty$, then simulation can be based on the stochastic representation in (3.12). Random vectors from these Archimedean copulas are easy to simulate if the distribution F_R of the radial random variable can be simulated; in this case, ψ^{-1} need not have a simple form.

Alternatively, the conditional or Rosenblatt's method can be used by for $\psi \notin \mathcal{L}_\infty$. Let

$\xi = \psi^{-1}$, the functional inverse of ψ. For d-variate $C_\psi(\boldsymbol{u}) = C(\boldsymbol{u}) = \psi(\xi(u_1) + \cdots + \xi(u_d))$ with conditional cdf

$$C_{d|1\cdots d-1}(u_d|u_1,\ldots,u_{d-1}) = \frac{\psi^{(d-1)}(\xi(u_1) + \cdots + \xi(u_d)) \prod_{j=1}^{d-1} \xi'(u_j)}{c_{1\cdots d-1}(u_1,\ldots,u_{d-1}).}$$

$(-1)^{d-1}\psi^{(d-1)}$ is monotone increasing, so the conditional inverse is

$$C_{d|1\cdots d-1}^{-1}(p|u_1,\ldots,u_{d-1}) = \psi\big[(\psi^{(d-1)})^{-1}(p^*) - \xi(u_1) - \cdots - \xi(u_{d-1})\big],$$
$$p^* = p\, c_{1\cdots d-1}(u_1,\ldots,u_{d-1})/[\xi'(u_1)\cdots\xi'(u_{d-1})].$$

Note that p^* is negative if $d-1$ is an odd integer. Generally, this involves solving nonlinear equations.

For simulation from hierarchical Archimedean copulas, see the algorithms and details in Hofert (2008), Hofert (2011) and Hofert and Maechler (2011). For the density of hierarchical Archimedean copulas, see Hofert and Pham (2013).

6.9.5   Simulation from mixture of max-id

Mixtures of max-id distributions lead to a large class of copulas and we indicate the subclass based on a product of max-id bivariate copulas where simulation based on the stochastic representation is possible. Let

$$C(\boldsymbol{u}) = \int_0^\infty K^q\big(e^{-\psi^{-1}(u_1)},\ldots,e^{-\psi^{-1}(u_d)}\big)\mathrm{d}F_Q(q) = \psi\Big[-\log K\big(e^{-\psi^{-1}(u_1)},\ldots,e^{-\psi^{-1}(u_d)}\big)\Big],$$

where Q has LT $\psi \in \mathcal{L}_\infty$ and K is a max-id copula. A simulation algorithm based on the stochastic representation is the following.

- Generate $Q = q$ from F_Q.
- Generate $(V_1,\ldots,V_d) \sim K^q(v_1,\ldots,v_d)$ with Beta$(q,1)$ margins.
- Let $U_j = \psi(-\log V_j)$ for $j = 1,\ldots,d$.

If K is an extreme value copula, then in the above $(V_1,\ldots,V_d)$ can be generated as $(X_1^{1/q},\ldots,X_d^{1/q})$, where $(X_1,\ldots,X_d) \sim K$. So an approach for the BB1 copula is the following.

- Generate $Q = q$ from Gamma$(\theta^{-1},1)$.
- Generate $(X_1,X_2) \sim$ Gumbel with parameter δ and let $(V_1,V_2) = (X_1^{1/q}, X_2^{1/q})$.
- Let $U_j = [1 - \log V_j]^{-1/\theta}$ for $j = 1,2$.

Similarly for the BB4 copula, the Galambos copula replaces the Gumbel copula.

If K is the MTCJ copula with parameter δ, then K^q is a multivariate Beta$(q,1)$ distribution with MTCJ copula having parameter δ/q. So an approach for BB2 is the following.

- Generate $Q = q$ from Gamma$(\theta^{-1},1)$.
- Generate $(X_1,X_2) \sim$ MTCJ with parameter δ/q and let $(V_1,V_2) = (X_1^{1/q}, X_2^{1/q})$.
- Let $U_j = [1 - \log V_j]^{-1/\theta}$ for $j = 1,2$.

Similarly for the BB3 copula, use $Q \sim$ Postable(θ^{-1}) in step 1 and $U_j = \exp\{-(-\log V_j)^{1/\theta}\}$ in step 3; and for the BB7 copula, use $Q \sim$ Sibuya(θ^{-1}) in step 1 and $U_j = 1 - (1 - V_j)^{1/\theta}\}$ in step 3.

If $d = 2$ and there is no direct way to simulate from K^q, then use the conditional method applied to $G(v_1,v_2) = K^q(v_1,v_2)$. The conditional distribution $G_{2|1}(v_2|v_1)$ is

$$G_{2|1}(v_2|v_1) = \frac{\partial G/\partial v_1}{q v_1^{q-1}} = \frac{K^{q-1}(v_1,v_2)\, K_{2|1}(v_2|v_1)}{v_1^{q-1}},$$

and the partial derivative of $G_{2|1}(v_2|v_1)$ with respect to v_2 is

$$\frac{\partial^2 G/\partial v_1 \partial v_2}{q v_1^{q-1}} = \frac{K^{q-2}\{(q-1)K_{2|1} K_{1|2} + K\, k_{12}\}}{v_1^{q-1}}.$$

These two equations can be used for an iterative method to solve for $G_{2|1}(v_2|v_1) = p$ in v_2 given p is generated from $U(0,1)$ and v_1 is generated from $\text{Beta}(q,1)$.

Next consider a trivariate K that is the product of bivariate max-id copulas, for example,

$$K(x_1, x_2, x_3) = H_{12}(x_1^{1/2}, x_2^{1/2})\, H_{13}(x_1^{1/2}, x_3^{1/2})\, H_{23}(x_2^{1/2}, x_3^{1/2}),$$

$$K^q(v_1, v_2, v_3) = H_{12}^q(v_1^{1/2}, v_2^{1/2})\, H_{13}^q(v_1^{1/2}, v_3^{1/2})\, H_{23}^q(v_2^{1/2}, v_3^{1/2}).$$

For simulation, given $Q = q$, independently generate $(V_{1a}, V_{2a}) \sim H_{12}^q(v_1^{1/2}, v_2^{1/2})$, $(V_{1b}, V_{3a}) \sim H_{13}^q(v_1^{1/2}, v_3^{1/2})$, $(V_{2b}, V_{3b}) \sim H_{23}^q(v_2^{1/2}, v_3^{1/2})$, to get $(V_1 = V_{1a} \vee V_{1b}, V_2 = V_{2a} \vee V_{2b}, V_3 = V_{3a} \vee V_{3b}) \sim K^q(v_1, v_2, v_3)$. These steps can extended for the d-variate copula with K being a product of bivariate max-id copulas,

An example of a multivariate mixture of max-id copula is the MM1 copula in (4.17) with cdf

$$C_{\text{MM1}}(\boldsymbol{u}) = \exp\Big\{-\Big[\sum_{i<j}\Big\{\Big(\frac{x_i^\theta}{d-1}\Big)^{\delta_{ij}} + \Big(\frac{x_j^\theta}{d-1}\Big)^{\delta_{ij}}\Big\}^{1/\delta_{ij}}\Big]^{1/\theta}\Big\}$$

$$= \int_0^\infty \Big\{ \prod_{1 \le i < j \le d} H_{ij}^q\big(u_i^{1/(d-1)}, u_j^{1/(d-1)}; \delta_{ij}\big)\Big\}\, dF_Q(q),$$

where F_Q is positive stable with parameter θ^{-1} and H_{ij} is the Gumbel(δ_{ij}) copula. The steps for simulation from the MM1 copula are the following.

- Generate $Q = q \sim \text{Postable}(\theta^{-1})$.
- Generate $(X_i^{\{i,j\}}, X_j^{\{i,j\}})$ from the Gumbel(δ_{ij}) copula.
- For $j = 1, \ldots, d$, let $V_j = [\max_{i \ne j} X_i^{\{i,j\}})]^{(d-1)/q}$
- For $j = 1, \ldots, d$, let $U_i = \exp\{-(-\log V_j)^{1/\theta}\}$.
- Return the multivariate vector $(U_1, \ldots, U_d)$.

6.9.6 Simulation from multivariate extreme value copulas

In the preceding subsection, the MM1 family is an extreme value copula family. Simulation from extreme value copulas that are multivariate Gumbel extensions are possible (see Stephenson (2003)), even for asymmetric versions, based on stochastic representations. Also simulations of classes of Marshall-Olkin multivariate extreme value distributions can be done from the stochastic representation; see Chapter 3 of Mai and Scherer (2012b).

More generally, for other multivariate extreme value copulas one must obtain the derivatives of the copula and use the sequential conditional method.

6.10 Likelihood for vine copula

In this section, we summarize the algorithms for computing the likelihood of a vine copula density for continuous variables and the likelihood for a multivariate vine distribution for discrete variables. The recursions in the continuous case are simpler so we cover them before the discrete case.

For the continuous case, we assume data have been converted to $\boldsymbol{u}_i = (u_{i1}, \ldots, u_{id})$ which are considered as i.i.d. realizations of a d-dimensional $U(0,1)$ random vector.

The algorithm for the C-vine is simplest so we start with the C-vine in Algorithm 2, then the D-vine in Algorithm 3 and the general regular vines in Algorithm 4. For truncated vines, the loop on the trees stop at the truncation level instead of continuing to tree $\mathcal{T}_{d-1}$ for d-dimensional vines.

Algorithm 2 Log-likelihood evaluation for the C-vine density: continuous variables; modification of Aas et al. (2009).

1: The parameter of $c_{\ell j;1:(\ell-1)}$ is denoted as $\theta_{\ell j}$.
2: $(\mathcal{T}_1)$ loglik $\leftarrow \sum_{i=1}^{n}\sum_{j=2}^{d}\log c_{1j}(u_{i1},u_{ij};\theta_{1j})$;
3:    let $v_{ij} \leftarrow u_{ij}$ for $i=1,\ldots,n$ and $j=1,\ldots,d$;
4: **for** $\ell = 2,\ldots,d-1$: **do**
5:    let $v_{ij} \leftarrow C_{j|\ell-1;1:(\ell-2)}(v_{ij}|v_{i,\ell-1};\theta_{\ell-1,j})$ for $i=1,\ldots,n$ and $j=\ell,\ldots,d$;
6:    update loglik $\leftarrow$ loglik $+ \sum_{i=1}^{n}\sum_{j=\ell+1}^{d}\log c_{\ell j;1:(\ell-1)}(v_{i,\ell},v_{i,j};\theta_{\ell j})$.
7: **end for**
8: Return loglik.

Algorithm 3 Log-likelihood evaluation for the D-vine density: continuous variables; modification of Aas et al. (2009).

1: The parameter of $c_{j-\ell,j;(j-\ell+1):j}$ is denoted as $\theta_{j-\ell,j}$.
2: $(\mathcal{T}_1)$ loglik $\leftarrow \sum_{i=1}^{n}\sum_{j=2}^{d}\log c_{j-1,j}(u_{i,j-1},u_{i,j};\theta_{j-1,j})$;
3:    let $w'_{ij} \leftarrow u_{ij}$ and $w_{ij} \leftarrow u_{ij}$ for $i=1,\ldots,n$ and $j=1,\ldots,d$;
4: **for** $\ell = 2,\ldots,d-1$: **do**
5:    let $v'_{ij} \leftarrow C_{j-\ell+1|j;(j-\ell+2):(j-1)}(w'_{i,j-1}|w_{ij};\theta_{j-\ell+1,j})$ for $j=\ell,\ldots,d-1$ and
6:    let $v_{ij} \leftarrow C_{j|j-\ell+1;(j-\ell+2):(j-1)}(w_{ij}|w'_{i,j-1};\theta_{j-\ell+1,j})$ for $j=\ell+1,\ldots,d$;
7:    update loglik $\leftarrow$ loglik $+ \sum_{i=1}^{n}\sum_{j=\ell+1}^{d}\log c_{j-\ell,j;(j-\ell+1):(j-1)}(v'_{i,j-1},v_{ij};\theta_{j-\ell,j})$;
8:    reset $w'_{ij} \leftarrow v'_{ij}$, $w_{ij} \leftarrow v_{ij}$ for $j=\ell+1,\ldots,d$.
9: **end for**
10: Return loglik.

The algorithm for the log-likelihood for the regular vine is in Algorithm 4. The vine array $A = (a_{kj} : k = 1,\ldots,j; j = 1,\ldots,d)$ in Section 3.9.4 is used, assuming the indices have been permuted so that $a_{jj} = j$ on the diagonal of A. The upper triangular matrix $M = (m_{kj})$ is defined as $m_{kj} = \max\{a_{1j},\ldots,a_{kj}\}$ for $k = 1,\ldots,j-1$, $j = 2,\ldots,d$; this keeps track of the largest variable index paired to variable j in an edge in trees $\mathcal{T}_1,\ldots,\mathcal{T}_k$.

We give an illustrative example with $d = 5$, where

$$A = \begin{bmatrix} 1 & 1 & 2 & 2 & 4 \\ & 2 & 1 & 1 & 2 \\ & & 3 & 3 & 1 \\ & & & 4 & 3 \\ & & & & 5 \end{bmatrix}, \quad M = \begin{bmatrix} 1 & 1 & 2 & 2 & 4 \\ & 2 & 2 & 2 & 4 \\ & & 3 & 3 & 4 \\ & & & 4 & 4 \\ & & & & 5 \end{bmatrix}.$$

For this example in the upper triangle, the only case of $a_{kj} = m_{kj}$ is when $j = 4$, $k = 3$. This is the only case which has v_k in the first argument of the copula $C_{a_{kj}j;a_{1j}\cdots a_{k-1,j}}$ instead of $v'_{m_{kj}}$. The calculations in the sequence are:

- $c_{12}(u_{i1},u_{i2})$, $c_{23}(u_{i2},u_{i3})$, $c_{24}(u_{i2},u_{i4})$, $c_{45}(u_{i4},u_{i5})$;
- $v'_{i2} = C_{1|2}(u_{i1}|u_{i2})$, $v'_{i3} = C_{2|3}(u_{i2}|u_{i3})$, $v'_{i4} = C_{2|4}(u_{i2}|u_{i4})$, $v'_{i5} = C_{4|5}(u_{i4}|u_{i5})$;
- $v_{i2} = C_{2|1}(u_{i2}|u_{i1})$, $v_{i3} = C_{3|2}(u_{i3}|u_{i2})$, $v_{i4} = C_{4|2}(u_{i4}|u_{i2})$, $v_{i5} = C_{5|4}(u_{i5}|u_{i4})$;
- $c_{13;2}(v'_{i2},v_{i3})$, $c_{14;2}(v'_{i2},v_{i4})$, $c_{25;4}(v'_{i4},v_{i5})$;

- $w'_{ij} \leftarrow v'_{ij}$, $w_{ij} \leftarrow v_{ij}$ for $j = 3, 4, 5$;
- $v'_{i3} = C_{1|3;2}(w'_{i2}|w_{i3})$, $v'_{i4} = C_{1|4;2}(w'_{i2}|w_{i4})$, $v'_{i5} = C_{2|5;4}(w'_{i4}|w_{i5})$;
- $v_{i3} = C_{3|1;2}(w_{i3}|w'_{i2})$, $v_{i4} = C_{4|1;2}(w_{i4}|w'_{i2})$, $v_{i5} = C_{5|2;4}(w_{i5}|w'_{i4})$;
- $c_{34;12}(v_{i3}, v_{i4})$, $c_{15;24}(v'_{i4}, v_{i5})$;
- $w'_{ij} \leftarrow v'_{ij}$, $w_{ij} \leftarrow v_{ij}$ for $j = 4, 5$;
- $v'_{i4} = C_{3|4;12}(w_{i3}|w_{i4})$, $v'_{i5} = C_{1|5;24}(w'_{i4}|w_{i5})$;
- $v_{i4} = C_{4|3;12}(w_{i4}|w_{i3})$, $v_{i5} = C_{5|1;24}(w_{i5}|w'_{i4})$;
- $c_{35;124}(v'_{i4}, v_{i5})$.

Algorithm 4 Log-likelihood evaluation for the R-vine density: continuous variables; modification of Dissmann et al. (2013) (for the latter, the vine array A^* is the reflected form of A with $a^*_{kj} = a_{d+1-k, d+1-j}$).

1: The parameter of the copula density c_e is θ_e but θ_e will not be shown.
2: Input vine array $A = (a_{kj})$ with $a_{jj} = j$ for $j = 1, \ldots, d$ on the diagonal.
3: Compute $M = (m_{kj})$ in the upper triangle, where $m_{kj} = \max\{a_{1j}, \ldots, a_{kj}\}$ for $k = 1, \ldots, j-1$, $j = 2, \ldots, d$.
4: Compute the $I = (I_{kj})$ indicator array as in Algorithm 5.
5: $(\mathcal{T}_1)$ loglik $\leftarrow \sum_{i=1}^n \sum_{j=2}^d \log c_{a_{1j}j}(u_{ia_{1j}}, u_{ij})$;
6: **for** $j = 2, \ldots, d$: **do**
7:    if $(I_{1j} = 1)$ then let $v'_{ij} \leftarrow C_{a_{1j}|j}(u_{ia_{1j}}|u_{ij})$,
8:    set $v_{ij} \leftarrow C_{j|a_{1j}}(u_{ij}|u_{ia_{1j}})$;
9: **end for**
10: update loglik $\leftarrow$ loglik $+ \sum_{i=1}^n \sum_{j=3}^d \log c_{a_{2j},j;a_{1j}}(s_{ij}, v_{ij})$, where $s_{ij} \leftarrow v_{im_{2j}}$ if $a_{2j} = m_{2j}$ and $s_{ij} \leftarrow v'_{im_{2j}}$ if $a_{2j} < m_{2j}$;
11: let $w'_{ij} \leftarrow v'_{ij}$, $w_{ij} \leftarrow v_{ij}$, $j = 3, \ldots, d$;
12: **for** $\ell = 3, \ldots, d-1$: **do**
13:    **for** $j = \ell, \ldots, d$: **do**
14:       if $(I_{\ell-1,j} = 1)$ then set $v'_{ij} \leftarrow C_{a_{\ell-1,j}|j;a_{1j}\cdots a_{\ell-2,j}}(s_{ij}|w_{ij})$,
15:       set $v_{ij} \leftarrow C_{j|a_{\ell-1,j};a_{1j}\cdots a_{\ell-2,j}}(w_{ij}|s_{ij})$;
16:    **end for**
17:    update loglik $\leftarrow$ loglik $+ \sum_{i=1}^n \sum_{j=\ell+1}^d \log c_{a_{\ell j},j}; a_{1j} \cdots a_{\ell-1,j}(s_{ij}, v_{ij})$,
18:    where $s_{ij} \leftarrow v_{im_{\ell,j}}$ if $a_{\ell,j} = m_{\ell,j}$ and $s_{ij} \leftarrow v'_{im_{\ell,j}}$ if $a_{\ell,j} < m_{\ell,j}$;
19:    reset $w'_{ij} \leftarrow v'_{ij}$, $w_{ij} \leftarrow v_{ij}$, $j = \ell+1, \ldots, d$.
20: **end for**
21: Return loglik.

In Algorithm 4, note that not all of the v'_{im} need to be evaluated, as only some of them are needed in later steps as s_{ij} in the first argument of a bivariate copula density. For the C-vine (Algorithm 2), s_{ij} is set to $v_{i\ell}$ in the ℓth tree and v'_{im} is not needed. For the D-vine (Algorithm 3), s_{ij} is always set to $v'_{i,j-1}$ so that v'_{ik} is needed in the ℓth tree for $k = \ell, \ldots, d-1$ (that is, v'_{id} is not needed). For a vine between the C-vine and D-vine, the number of v'_{im} needed is less if the vine is closer to the C-vine. Also for some vines, such as the D-vine, $v_{i\ell}$ for $\mathcal{T}_\ell$ is not needed.

Algorithm 4 involves less computations if a second array I of indicator variables is computed after the M array; the details of this second array I are given in Algorithm 5. That is, Algorithm 4 in $\mathcal{T}_{\ell-1}$ evaluates v'_{ij} (for all i) only if $I_{\ell-1,j} = 1$.

Algorithm 5 Indicator I array derived from an R-vine array A: it is used to decide on necessary computations.

1: Input vine array $A = (a_{kj})$.
2: Compute $M = (m_{kj})$ in the upper triangle, where $m_{kj} = \max\{a_{1j}, \ldots, a_{kj}\}$ for $k = 1, \ldots, j - 1$, $j = 2, \ldots, d$.
3: I is initialized to a $d \times d$ array of 0s.
4: **for** $k = 2, \ldots, d - 1$: **do**
5: $\quad$ **for** $j = k + 1, \ldots, d$: **do**
6: $\qquad$ if $a_{kj} < m_{kj}$ then $I_{k-1,m_{k,j}} \leftarrow 1$.
7: $\quad$ **end for**
8: **end for**
9: Return $I = (I_{kj})$.

Next we go to the discrete case with the D-vine and then the general R-vine.

Let $\boldsymbol{Y} = (Y_1, Y_2, \ldots, Y_d)$ be a discrete-valued random d-vector. For a vector of integers $\boldsymbol{i}$, let $\boldsymbol{Y}_{\boldsymbol{i}} = (Y_i : i \in \boldsymbol{i})$. For notation, with $\boldsymbol{y}_{\boldsymbol{i}}$ a mass point of $\boldsymbol{Y}_{\boldsymbol{i}}$ and y_g a mass point of Y_g, let $F_{g|\boldsymbol{i}}^+ := \mathbb{P}(Y_g \leq y_g | \boldsymbol{Y}_{\boldsymbol{i}} = \boldsymbol{y}_{\boldsymbol{i}})$, $F_{g|\boldsymbol{i}}^- := \mathbb{P}(Y_g < y_g | \boldsymbol{Y}_{\boldsymbol{i}} = \boldsymbol{y}_{\boldsymbol{i}})$, $f_{g|\boldsymbol{i}} := \mathbb{P}(Y_g = y_g | \boldsymbol{Y}_{\boldsymbol{i}} = \boldsymbol{y}_{\boldsymbol{i}})$. The case of $\boldsymbol{i}$ empty corresponds to marginal probabilities. Let $C_{gh;\boldsymbol{i}}$ be a bivariate copula that applies to conditional cdfs $F_{g|\boldsymbol{i}}$ and $F_{h|\boldsymbol{i}}$. Denote

$$C_{gh;\boldsymbol{i}}^{++} := C_{gh;\boldsymbol{i}}\big(F_{g|\boldsymbol{i}}^+, F_{h|\boldsymbol{i}}^+\big), \quad C_{gh;\boldsymbol{i}}^{+-} := C_{gh;\boldsymbol{i}}\big(F_{g|\boldsymbol{i}}^+, F_{h|\boldsymbol{i}}^-\big),$$
$$C_{gh;\boldsymbol{i}}^{-+} := C_{gh;\boldsymbol{i}}\big(F_{g|\boldsymbol{i}}^-, F_{h|\boldsymbol{i}}^+\big), \quad C_{gh;\boldsymbol{i}}^{--} := C_{gh;\boldsymbol{i}}\big(F_{g|\boldsymbol{i}}^-, F_{h|\boldsymbol{i}}^-\big).$$

The keys to the algorithm for the joint pmf for D-vine for discrete variables are the recursions:

(i) $F_{j-t|(j-t+1):(j-1)}^+ = [C_{j-t,j-1;(j-t+1):(j-2)}^{++} - C_{j-t,j-1;(j-t+1):(j-2)}^{+-}]/f_{j-1|(j-t+1):(j-2)}$;

(ii) $F_{j-t|(j-t+1):(j-1)}^- = [C_{j-t,j-1;(j-t+1):(j-2)}^{-+} - C_{j-t,j-1;(j-t+1):(j-2)}^{--}]/f_{j-1|(j-t+1):(j-2)}$;

(iii) $f_{j-t|(j-t+1):(j-1)} = F_{j-t|(j-t+1):(j-1)}^+ - F_{j-t|(j-t+1):(j-1)}^-$;

(iv) $F_{j|(j-t+1):(j-1)}^+ = [C_{j-t+1,j;(j-t+2):(j-1)}^{++} - C_{j-t+1,j;(j-t+2):(j-1)}^{-+}]/f_{j-t+1|(j-t+2):(j-1)}$;

(v) $F_{j|(j-t+1):(j-1)}^- = [C_{j-t+1,j;(j-t+2):(j-1)}^{+-} - C_{j-t+1,j;(j-t+2):(j-1)}^{--}]/f_{j-t+1|(j-t+2):(j-1)}$;

(vi) $f_{j|(j-t+1):(j-1)} = F_{j|(j-t+1):(j-1)}^+ - F_{j|(j-t+1):(j-1)}^-$;

(vii) values based on $C_{j-t,j;(j-t+1):(j-1)}$ can be computed, and then back to (i) with t incremented by 1.

The key identity being used is

$$\mathbb{P}(Y_g \leq y_g | Y_h = y_h, \boldsymbol{Y}_{\boldsymbol{i}} = \boldsymbol{y}_{\boldsymbol{i}}) = \frac{\mathbb{P}(Y_g \leq y_g, Y_h \leq y_h | \boldsymbol{Y}_{\boldsymbol{i}} = \boldsymbol{y}_{\boldsymbol{i}}) - \mathbb{P}(Y_g \leq y_g, Y_h < y_h | \boldsymbol{Y}_{\boldsymbol{i}} = \boldsymbol{y}_{\boldsymbol{i}})}{\mathbb{P}(Y_h = y_h | \boldsymbol{Y}_{\boldsymbol{i}} = \boldsymbol{y}_{\boldsymbol{i}})}.$$

Algorithm 6 Joint pmf for D-vine for discrete variables (Panagiotelis et al. (2012)); superscript $+$ $(-)$ indicates right (left) limit of univariate or bivariate cdf and the subscript indicates which univariate or bivariate cdf.

1: Input $\boldsymbol{y} = (y_1, \ldots, y_d)$.
2: Allocate $d \times d$ matrix P where $P_{\ell j} = f_{(j-\ell+1):j}$ for $\ell = 1, \ldots, d$ and $j = \ell+1, \ldots, d$ and the joint probability $f_{1:d}$ will appear as P_{dd}.
3: Allocate $C^{++}, C^{+-}, C^{-+}, C^{--}, V'^{+}, V'^{-}, V^{+}, V^{-}, v', v, w', w$ as vectors of length d.
4: Evaluate F_j^+, F_j^-, $f_j = F_j^+ - F_j^-$ and let $P_{1j} \leftarrow f_j$ for $j = 1, \ldots, d$;
5: let $C_j^{++} \leftarrow C_{j-1,j}(F_{j-1}^+, F_j^+)$, $C_j^{+-} \leftarrow C_{j-1,j}(F_{j-1}^+, F_j^-)$, $C_j^{-+} \leftarrow C_{j-1,j}(F_{j-1}^-, F_j^+)$ and $C_j^{--} \leftarrow C_{j-1,j}(F_{j-1}^-, F_j^-)$ for $j = 2, \ldots, d$;
6: set $P_{2j} \leftarrow C_j^{++} - C_j^{+-} - C_j^{-+} + C_j^{--}$ for $j = 2, \ldots, d$;
7: **for** $j = 2, \ldots, d$: $(\mathcal{T}_1)$ **do**
8:   (i) $V'^{+}_j \leftarrow F^+_{j-1|j} = \frac{C_j^{++} - C_j^{+-}}{f_j}$, $V'^{-}_j \leftarrow F^-_{j-1|j} = \frac{C_j^{-+} - C_j^{--}}{f_j}$ and $v'_j \leftarrow f_{j-1|j} = F^+_{j-1|j} - F^-_{j-1|j}$;
9:   (ii) $V^+_j \leftarrow F^+_{j|j-1} = \frac{C_j^{++} - C_j^{-+}}{f_{j-1}}$, $V^-_j \leftarrow F^-_{j|j-1} = \frac{C_j^{+-} - C_j^{--}}{f_{j-1}}$ and $v_j \leftarrow f_{j|j-1} = F^+_{j|j-1} - F^-_{j|j-1}$;
10: **end for**
11: **for** $\ell = 2, \ldots, d-1$: $(\mathcal{T}_2, \ldots, \mathcal{T}_{d-1})$ **do**
12:   let $C_j^{\alpha\beta} \leftarrow C_{j-\ell,j;(j-\ell+1):(j-1)}(V'^{\alpha}_{j-1}, V^{\beta}_j)$, for $j = \ell+1, \ldots, d$ and $\alpha, \beta \in \{+, -\}$;
13:   let $w'_j \leftarrow v'_j$, $w_j \leftarrow v_j$ for $j = \ell, \ldots, d$;
14:   **for** $j = \ell+1, \ldots, d$: **do**
15:     (i) $V'^{+}_j \leftarrow \frac{C_j^{++} - C_j^{+-}}{w_j}$, $V'^{-}_j \leftarrow \frac{C_j^{-+} - C_j^{--}}{w_j}$ and $v'_j \leftarrow V'^{+}_j - V'^{-}_j$;
16:     (ii) $V^+_j \leftarrow \frac{C_j^{++} - C_j^{-+}}{w'_{j-1}}$, $V^-_j \leftarrow \frac{C_j^{+-} - C_j^{--}}{w'_{j-1}}$ and $v_j \leftarrow V^+_j - V^-_j$;
17:   **end for**
18:   let $P_{\ell+1,j} \leftarrow P_{\ell,j-1} \times v_j$ for $j = \ell+1, \ldots, d$.
19: **end for**
20: Return joint pmf P_{dd}, or all of P which has probabilities for all consecutive subsequences $f_{(j-k+1):j}$.

The algorithm for the R-vine with discrete variables is a combination of the R-vine for continuous variables and the D-vine for discrete variables.

In tree $\mathcal{T}_\ell$, with $S = \{a_{1j}, \ldots a_{\ell-1,j}\}$,

$$V'^{+}_j = F^+_{a_{\ell j}|j \cup S} = \frac{C_{a_{\ell j}j;S}(F^+_{a_{\ell j}|S}, F^+_{j|S}) - C_{a_{\ell j}j;S}(F^+_{a_{\ell j}|S}, F^-_{j|S})}{f_{j|S}},$$

$$V^+_j = F^+_{j|a_{\ell j} \cup S} = \frac{C_{a_{\ell j}j;S}(F^+_{a_{\ell j}|S}, F^+_{j|S}) - C_{a_{\ell j}j;S}(F^-_{a_{\ell j}|S}, F^+_{j|S})}{f_{a_{\ell j}|S}}.$$

With superscript of $+$ or $-$, the first argument $s_j := F_{a_{\ell j}|S}$ is the previous $V_{m_{\ell j}}$ if $a_{\ell j} = m_{\ell j}$ and it is the previous $V'_{m_{\ell j}}$ if $a_{\ell j} < m_{\ell j}$. Similarly, for the denominator, $t_j := f_{a_{\ell j}|S}$ is the previous $w_{m_{\ell j}}$ if $a_{\ell j} = m_{\ell j}$ and it is the previous $w'_{m_{\ell j}}$ if $a_{\ell j} < m_{\ell j}$.

For some vines such as C-vines, not all of the V' calculations are needed. But the decision is not as straightforward as the indicator matrix in Algorithm 5 because the denominator of the conditional probability of form V might require $V'^{+} - V'^{-}$ from the previous tree. One could just make one pass with an input vector $\boldsymbol{y}$ and then flag the case where V's are needed. Then for future $\boldsymbol{y}$, there is an indicator array.

Algorithm 7 Joint pmf for R-vine for discrete variables; the superscript $+$ ($-$) indicates right (left) limit of univariate or bivariate cdf and the subscript indicates which univariate or bivariate cdf.

1: Input $\boldsymbol{y} = (y_1, \ldots, y_d)$.
2: Input vine array $A = (a_{kj})$.
3: Compute $M = (m_{kj})$ in the upper triangle, where $m_{kj} = \max\{a_{1j}, \ldots, a_{kj}\}$ for $k = 1, \ldots, j-1$, $j = 2, \ldots, d$.
4: Allocate $d \times d$ matrix P where the joint probability $f_{1:d}$ will appear as P_{dd}.
5: Allocate $C^{++}, C^{+-}, C^{-+}, C^{--}, V'^{+}, V'^{-}, V^{+}, V^{-}, v', v, w', w$ as vectors of length d.
6: Evaluate $F_j^+, F_j^-, f_j = F_j^+ - F_j^-$ and let $P_{1j} \leftarrow f_j$ for $j = 1, \ldots, d$;
7: let $C_j^{++} \leftarrow C_{a_{1j},j}(F_{a_{1j}}^+, F_j^+)$, $C_j^{+-} \leftarrow C_{a_{1j},j}(F_{a_{1j}}^+, F_j^-)$, $C_j^{-+} \leftarrow C_{a_{1j},j}(F_{a_{1j}}^-, F_j^+)$ and $C_j^{--} \leftarrow C_{a_{1j},j}(F_{a_{1j}}^-, F_j^-)$ for $j = 2, \ldots, d$;
8: set $P_{2j} \leftarrow C_j^{++} - C_j^{+-} - C_j^{-+} + C_j^{--}$ for $j = 2, \ldots, d$;
9: **for** $j = 2, \ldots, d$: **do**
10:    (i) $V'^{+}_j \leftarrow F_{a_{1j}|j}^+ = \dfrac{C_j^{++} - C_j^{+-}}{f_j}$, $V'^{-}_j \leftarrow F_{a_{1j}|j}^- = \dfrac{C_j^{-+} - C_j^{--}}{f_j}$, $v'_j \leftarrow f_{a_{1j}|j} = F_{a_{1j}|j}^+ - F_{a_{1j}|j}^-$;
11:    (ii) $V_j^+ \leftarrow F_{j|a_{1j}}^+ = \dfrac{C_j^{++} - C_j^{-+}}{f_{a_{1j}}}$, $V_j^- \leftarrow F_{j|a_{1j}}^- = \dfrac{C_j^{+-} - C_j^{--}}{f_{a_{1j}}}$, $v_j \leftarrow f_{j|a_{1j}} = F_{j|a_{1j}}^+ - F_{j|a_{1j}}^-$;
12: **end for**
13: **for** $\ell = 2, \ldots, d-1$: $(\mathcal{T}_2, \ldots, \mathcal{T}_{d-1})$ **do**
14:    let $C_j^{\alpha\beta} \leftarrow C_{a_{\ell j},j;a_{1j}\cdots a_{\ell-1,j}}(s_j^{\alpha}, V_j^{\beta})$, where $s_j^{\alpha} \leftarrow V_{m_{\ell j}}^{\alpha}$ if $a_{\ell j} = m_{\ell j}$ and $s_j^{\alpha} \leftarrow V'^{\alpha}_{m_{\ell j}}$ if $a_{\ell j} < m_{\ell j}$ for $j = \ell + 1, \ldots, d$ and $\alpha, \beta \in \{+, -\}$;
15:    let $w'_j \leftarrow v'_j$, $w_j \leftarrow v_j$ for $j = \ell, \ldots, d$;
16:    **for** $j = \ell + 1, \ldots, d$: **do**
17:       (i) $V'^{+}_j \leftarrow \dfrac{C_j^{++} - C_j^{+-}}{w_j}$, $V'^{-}_j \leftarrow \dfrac{C_j^{-+} - C_j^{--}}{w_j}$ and $v'_j \leftarrow V'^{+}_j - V'^{-}_j$;
18:       (ii) $t_j \leftarrow w_{m_{\ell j}}$ if $a_{\ell j} = m_{\ell j}$ and $t_j \leftarrow w'_{m_{\ell j}}$ if $a_{\ell j} < m_{\ell j}$;
19:       (iii) $V_j^+ \leftarrow \dfrac{C_j^{++} - C_j^{-+}}{t_j}$, $V_j^- \leftarrow \dfrac{C_j^{+-} - C_j^{--}}{t_j}$ and $v_j \leftarrow V_j^+ - V_j^-$;
20:    **end for**
21:    let $P_{\ell+1,j} \leftarrow P_{\ell,m_{\ell j}} \times v_j$ for $j = \ell + 1, \ldots, d$.
22: **end for**
23: Return joint pmf P_{dd},

6.11 Likelihood for factor copula

For the factor copula model for continuous variables, the integrand is straightforward to compute, and then the integral can be evaluated with Gauss-Legendre quadrature. Typically, the number of quadrature points n_q per dimension around 20–30 is adequate; in practice, one can use some test data sets and slowly increase n_q until the maximum likelihood estimate is numerically stable.

For factor copula models for ordinal response and no covariates, the computation is faster by first storing in arrays all of the possible conditional probability values given latent variables at the quadrature points, and then computing the integrals by accessing the arrays. This is indicated in Algorithms 8 and 9.

If only the log-likelihood is evaluated, it can be inputted into a numerical quasi-Newton routine. This is fine if the number of variables d and the total number n_p of copula parameters is not too large (say, $n_p < 30$). As d and n_p get larger, there is much faster convergence to the MLE when a modified Newton-Raphson method is used with the evaluation of the gradient and Hessian of the log-likelihood. The steps for obtaining the gradient and Hessian are outlined in the Section 6.12.

Algorithm 8 Likelihood for 1-factor copula with ordinal response and no covariates, with cutpoints based on univariate margins (Nikoloulopoulos and Joe (2014)). There are d ordinal variables, category labels $0, \ldots, K-1$ for each variable. V is the latent variable, assumed to be U(0,1).

1: Input Q = number of quadrature points for Gauss-Legendre, the nodes $\{x_q\}$ and the weights $\{w_q\}$, d = number of variables, K = number of categories for each variable, n=sample size, d-vectors $\boldsymbol{y}_1, \ldots, \boldsymbol{y}_n$ with elements in $\{0, \ldots, K-1\}^d$, copula C_{Vj} (linking jth observed variable Y_j to the latent variable V) for $j = 1, \ldots, d$ and their parameter values.

2: Set the cutpoints on the uniform scale: $a_{11}, \ldots, a_{1,K-1}, \ldots, a_{d1}, \ldots, a_{d,K-1}$ based on the univariate margins pmfs. Also set boundary cutpoints $a_{j0} = 0$ and $a_{jK} = 1$ for each variable.

3: For $j = 1, \ldots, d$, compute/store $C_{j|V}(a_{jk}|x_q)$ for $k = 0, \ldots, K$ and $q = 1, \ldots, Q$.

4: For $j = 1, \ldots, d$, compute the probability $f_{j|V}(k-1|x_q) = C_{j|V}(a_{jk}|x_q) - C_{j|V}(a_{j,k-1}|x_q)$ for $k = 1, \ldots, K$ and $q = 1, \ldots, Q$. After this step, assume that these conditional pmfs are stored in a $Q \times K \times d$ array.

5: loglik $\leftarrow 0$.

6: **for** $i = 1, \ldots, n$: (data loop) **do**

7:     for $\boldsymbol{y}_i$, let $p_{iq} \leftarrow \prod_{j=1}^{d} f_{j|V}(y_{ij}|x_q)$ by extracting from the 3-dimensional array;

8:     update the log-likelihood with loglik $\leftarrow$ loglik $+ \log(\sum_{q=1}^{Q} p_{iq} w_q)$.

9: **end for**

10: Return loglik.

Algorithm 9 Likelihood for 2-factor copula with ordinal response and no covariates, with cutpoints based on univariate margins (Nikoloulopoulos and Joe (2014)). There are d ordinal variables, category labels $0, \ldots, K-1$ for each variable. V_1, V_2 are latent variables, assumed to be independent U(0,1).

1: Input Q = number of quadrature points for Gauss-Legendre, the nodes $\{x_q\}$ and the weights $\{w_q\}$, d = number of variables, K = number of categories for each variable, n=sample size, d-vectors $\boldsymbol{y}_1, \ldots, \boldsymbol{y}_n$ with elements in $\{0, \ldots, K-1\}^d$, copula C_{V_1j} (linking jth observed variable Y_j to the latent variable V_1) and $C_{V_2j;V_1}$ (linking jth variable to latent variable V_2 given V_1) for $j = 1, \ldots, d$ and their parameter values.

2: Set the cutpoints on the uniform scale: $a_{11}, \ldots, a_{1,K-1}, \ldots, a_{d1}, \ldots, a_{d,K-1}$ based on the univariate margins pmfs. Also set boundary cutpoints $a_{j0} = 0$ and $a_{jK} = 1$ for each variable.

3: For $j = 1, \ldots, d$, compute/store $s_{q_1,k,j} = C_{j|V}(a_{jk}|x_{q_1})$ for $k = 0, \ldots, K$, $q_1 = 1, \ldots, Q$,

4: For $j = 1, \ldots, d$, compute/store $t_{q_1,q_2,k,j} = C_{j|V_2;V_1}(C_{j|V_1}(a_{jk}|x_{q_1})|x_{q_2}) = C_{j|V_2;V_1}(s_{q_1,k,j}|x_{q_2})$ for $k = 0, \ldots, K$, $q_1, q_2 = 1, \ldots, Q$,

5: For $j = 1, \ldots, d$, compute the probability $f_{j|V_1,V_2}(k-1|x_{q_1}, x_{q_2}) = t_{q_1,q_2,k,j} - t_{q_1,q_2,k-1,j}$ $k = 1, \ldots, K$ and $q_1, q_2 = 1, \ldots, Q$, After this step, assume that these conditional pmfs are stored in a $Q \times Q \times K \times d$ array.

6: loglik $\leftarrow 0$.

7: **for** $i = 1, \ldots, n$: (data loop) **do**

8:     for $\boldsymbol{y}_i$, let $p_{iq_1q_2} \leftarrow \prod_{j=1}^{d} f_{j|V_1,V_2}(y_{ij}|x_{q_1}, x_{q_2})$ by extracting from the 4-dimensional array;

9:     update the log-likelihood with loglik $\leftarrow$ loglik $+ \log(\sum_{q_1=1}^{Q} \sum_{q_2=1}^{Q} p_{iq_1q_2} w_{q_1} w_{q_2})$.

10: **end for**

11: Return loglik.

6.12 Copula derivatives for factor and vine copulas

A parametric factor copula density involve integrals of a C-vine density rooted at latent variables. However, there is enough structure that derivatives of the factor copula density with respect to dependence parameters can be obtained as an integral of derivatives of the C-vine density. This is outlined below for the 1-factor and 2-factor copula density for continuous and ordinal responses. Then recursions for derivatives of D-vine and R-vine copulas are given. Schepsmeier and Stöber (2014) have derived derivatives for some bivariate copulas and Stöber and Schepsmeier (2013) use them for the gradient and Hessian of the log-likelihood of vine copulas. The implementation is in Schepsmeier et al. (2013).

1-factor with parameter θ_j for the bivariate copula linking the jth observed variable with the latent variable V. From (3.53), let the integrand be e^I, where $I(\boldsymbol{u}; v) = \sum_{i=1}^d \log c_{Vj}(v, u_j; \theta_j)$. Let $\ell_{j1}(\theta_j; \cdot) = \log c_{Vj}(\cdot; \theta_j)$ for $j = 1, \ldots, d$. Then, with $k \neq j$,

$$\frac{\partial e^I}{\partial \theta_j} = e^I \frac{\partial \ell_{j1}}{\partial \theta_j}, \quad \frac{\partial^2 e^I}{\partial \theta_j \partial \theta_j^\top} = e^I \Big[\frac{\partial \ell_{j1}}{\partial \theta_j} \frac{\partial \ell_{j1}}{\partial \theta_j^\top} + \frac{\partial^2 \ell_{j1}}{\partial \theta_j \partial \theta_j^\top} \Big], \quad \frac{\partial^2 e^I}{\partial \theta_j \partial \theta_k^\top} = e^I \frac{\partial \ell_{j1}}{\partial \theta_j} \frac{\partial \ell_{k1}}{\partial \theta_k^\top}.$$

For an observation $\boldsymbol{u} = (u_{i1}, \ldots, u_{id})$,

- the contribution to the log-likelihood is $L := \log \int_0^1 \exp\{I(\boldsymbol{u}; v)\} \, dv$;
- the gradient of the integrand with respect to θ_j is $g_j(\boldsymbol{u}, v) := e^I \frac{\partial c_{Vj}}{\partial \theta_j} / c_{Vj} = e^I \frac{\partial \ell_{j1}}{\partial \theta_j}$ and the contribution to the gradient of the log-likelihood is $L_j := \int_0^1 g_j(\boldsymbol{u}, v) \, dv / e^L$;
- the (j, j) term of the Hessian of the integrand is $h_{jj}(\boldsymbol{u}, v) := e^I \Big[\frac{\partial \ell_{j1}}{\partial \theta_j} \frac{\partial \ell_{j1}}{\partial \theta_j^\top} + \frac{\partial^2 \ell_{j1}}{\partial \theta_j \partial \theta_j^\top} \Big]$;
- for $j \neq k$, the (j, k) term of the Hessian of the integrand is $h_{jk}(\boldsymbol{u}, v) := e^I \frac{\partial \ell_{j1}}{\partial \theta_j} \frac{\partial \ell_{k1}}{\partial \theta_k^\top}$;
- for (j, k) with $j = k$ or $j \neq k$, the contribution to Hessian of the log-likelihood is $\int_0^1 h_{jk}(\boldsymbol{u}, v) \, dv / e^L - L_j L_k$;
- all integrals can be numerically evaluated via Gauss-Legendre quadrature.

With code for the first and second order partial derivatives of the log density of each bivariate linking copula, it is straightforward to code the gradient and Hessian of the log-likelihood of a factor copula model. The integrals for the gradient and Hessian can be computed at the same time as integrals for the log-likelihood. Then numerical maximum likelihood can proceed with a modified Newton-Raphson algorithm (see Section 6.2).

2-factor with parameters θ_j and γ_j for the bivariate copula linking the jth observed variable with the latent variable V_1 and V_2 respectively. From (3.55), let the integrand be e^I, where

$$I(\boldsymbol{u}; v_1, v_2) = \sum_j \log c_{V_1 j}(v_1, u_j; \theta_j) + \sum_j \log c_{V_2 j; V_1}[v_2, C_{j|V_1}(u_j|v_1; \theta_j)); \gamma_j].$$

Let ℓ_{j1} be defined as above for the 1-factor copula model, and let $\ell_{j2}(\gamma; x_j, v_2) = \log c_{V_2 j; V_1}(v_2, x_j; \gamma)$ with $x_j = C_{j|V_1}(u_j|v_1; \theta_j)$.

Details are similar to the 1-factor copula, with two-dimensional Gauss-Legendre quadature, and more partial derivatives of the integrand.

For a bivariate copula family $c_{V_1 j}$ for the first factor, required derivatives are the first and second order derivatives of $\log c_{V_1 j}$ with respect to θ_j. Additional required derivatives are $\frac{\partial C_{j|V_1}(u|v;\theta)}{\partial \theta_j}$, $\frac{\partial^2 C_{1|V_1}(u|v;\theta)}{\partial \theta_j \partial \theta_j^\top}$. For a bivariate copula family $c_{V_2 j}(v_2, x) = c_{V_2 j; V_1}(v_2, x)$ for the second factor, the required derivatives are $\ell_{j2}(\gamma_j; x, v) = \log c_{V_2 j}$, $\frac{\partial \ell_{j2}}{\partial \gamma_j}$, $\frac{\partial^2 \ell_{j2}}{\partial \gamma_j \partial \gamma_j^\top}$, $\frac{\partial \ell_{j2}}{\partial x}$, $\frac{\partial^2 \ell_{j2}}{\partial x^2}$, $\frac{\partial^2 \ell_{j2}}{\partial \gamma_j \partial x}$.

The contribution to integrand product for the jth variable is

$$P_j := \exp\{\ell_{j1}(\theta_j; u_j, v_1) + \ell_{j2}(\gamma_j; x_j(\theta_j), v_2)\}.$$

Partial derivatives of P_j (with $x = x_j$) are:

- $\frac{\partial}{\partial \theta_j}$: $P_j[\frac{\partial \ell_{j1}}{\partial \theta_j} + \frac{\partial \ell_{j2}}{\partial x}\frac{\partial x}{\partial \theta_j}]$;

- $\frac{\partial}{\partial \gamma_j}$: $P_j \frac{\partial \ell_{j2}}{\partial \gamma_j}$;

- $\frac{\partial^2}{\partial \theta_j \partial \theta_j^\top}$: $P_j\{[\frac{\partial \ell_{j1}}{\partial \theta_j} + \frac{\partial \ell_{j2}}{\partial x}\frac{\partial x}{\partial \theta_j}][\frac{\partial \ell_{j1}}{\partial \theta_j^\top} + \frac{\partial \ell_{j2}}{\partial x}\frac{\partial x}{\partial \theta_j^\top}] + \frac{\partial^2 \ell_{j1}}{\partial \theta_j \partial \theta_j^\top} + \frac{\partial^2 \ell_{j2}}{\partial x^2}\frac{\partial x}{\partial \theta_j}\frac{\partial x}{\partial \theta_j^\top} + \frac{\partial \ell_{j2}}{\partial x}\frac{\partial^2 x}{\partial \theta_j \partial \theta_j^\top}\}$;

- $\frac{\partial^2}{\partial \gamma_j \gamma_j^\top}$: $P_j[\frac{\partial^2 \ell_{j2}}{\partial \gamma_j \gamma_j^\top} + \frac{\partial \ell_{j2}}{\partial \gamma_j}\frac{\partial \ell_{j2}}{\partial \gamma_j^\top}]$;

- $\frac{\partial^2}{\partial \theta_j \partial \gamma_j^\top}$: $P_j\{[\frac{\partial \ell_{j1}}{\partial \theta_j} + \frac{\partial \ell_{j2}}{\partial x}\frac{\partial x}{\partial \theta_j}]\frac{\partial \ell_{j2}}{\partial \gamma_j^\top} + \frac{\partial x}{\partial \theta_j}\frac{\partial^2 \ell_{j2}}{\partial x \partial \gamma_j^\top}\}$.

These are incorporated into the gradient and Hessian of the integrand and of the ith term of the log-likelihood in a similar way to the 1-factor model.

There are similar gradients and Hessians for the log-likelihood of the discrete (item response) factor copula model. Let the cutpoints in the $U(0,1)$ scale for the jth item/variable be $a_{j,k}$, $k = 1, \ldots, K-1$, with $a_{j,0} = 0$ and $a_{j,K} = 1$. These correspond to $a_{j,k} = \Phi(\zeta_{j,k})$ where $\zeta_{j,k}$ are cutpoints in the $N(0,1)$ scale.

1-factor ordinal response:

From (3.60), the pmf for the 1-factor model is

$$\pi_d(\mathbf{y}) = \int_0^1 \prod_{j=1}^d f_{V_1 j}(v, y_j)\, \mathrm{d}v,$$

where $f_{V_1 j}(v, y) = f_{V_1 j}(v, y; \mathbf{a}_j) = C_{j|V_1}(a_{j,y+1}|v) - C_{j|V_1}(a_{j,y}|v)$ is the joint density of V_1 and Y_j. If θ_j is parameter of $C_{j|V_1}$, then for the gradient and Hessian of the integrand, one needs $\partial C_{j|V_1}/\partial \theta_j$ and $\partial^2 C_{j|V_1}/\partial \theta_j \partial \theta_j^\top$.

2-factor ordinal response:

Let θ_j and γ_j be the parameters for linking the jth observed variable to the first and second latent variable respectively. From (3.61), the pmf for the 2-factor model is

$$\pi_d(\boldsymbol{y}) = \int_0^1 \int_0^1 \prod_{j=1}^d f_{V_2 j|V_1}(v_2, y_j|v_1)\, \mathrm{d}v_1 \mathrm{d}v_2,$$

where

$$f_{V_2 j|V_1}(v_2, y|v_1) = C_{j|V_2;V_1}\big(C_{j|V_1}(a_{j,y+1}|v_1; \theta_j)|v_2; \gamma_j\big) - C_{j|V_2;V_1}\big(C_{j|V_1}(a_{j,y}|v_1; \theta_j)|v_2; \gamma_j\big).$$

With $\theta = \theta_j$ and $\gamma = \gamma_j$, let

$$f_j^*(\theta, \gamma, p_1, p_2; v_1, v_2) = C_{j|V_2;V_1}\big(p_2(v_1; \theta)|v_2; \gamma\big) - C_{j|V_2;V_1}\big(p_1(v_1; \theta)|v_2; \gamma\big),$$

where $p_2(v_1; \theta) = C_{j|V_1}(a_{j,y+1}|v_1; \theta)$ and $p_1(v_1; \theta) = C_{j|V_1}(a_{j,y}|v_1; \theta)$. With u as an argument of $C_{V_1 j}(v_1, u)$,

$$\frac{\partial f_j^*}{\partial \gamma} = \frac{\partial C_{j|V_2;V_1}(p_2(v_1; \theta)|v_2; \gamma)}{\partial \gamma} - \frac{\partial C_{j|V_2;V_1}(p_1(v_1; \theta)|v_2; \gamma)}{\partial \gamma},$$

$$\frac{\partial^2 f_j^*}{\partial \gamma \partial \gamma^\top} = \frac{\partial^2 C_{j|V_2;V_1}(p_2(v_1; \theta)|v_2; \gamma)}{\partial \gamma \partial \gamma^\top} - \frac{\partial^2 C_{j|V_2;V_1}(p_1(v_1; \theta)|v_2; \gamma)}{\partial \gamma \partial \gamma^\top},$$

$$\frac{\partial f_j^*}{\partial \theta} = \frac{\partial C_{j|V_2;V_1}(p_2(v_1;\theta)|v_2;\gamma)}{\partial u}\frac{\partial p_2}{\partial \theta} - \frac{\partial C_{j|V_2;V_1}(p_1(v_1;\theta)|v_2;\gamma)}{\partial u}\frac{\partial p_1}{\partial \theta}$$

$$= c_{V_2j;V_1}(v_2,p_2;\gamma)\frac{\partial C_{j|V_1}(a_{j,y+1}|v_1;\theta)}{\partial \theta} - c_{V_2j;V_1}(v_2,p_1;\gamma)\frac{\partial C_{j|V_1}(a_{j,y}|v_1;\theta)}{\partial \theta},$$

$$\frac{\partial^2 f_j^*}{\partial \theta \partial \theta^\top} = \frac{\partial c_{V_2j;V_1}(v_2,p_2;\gamma)}{\partial u}\left[\frac{\partial C_{j|V_1}(a_{j,y+1}|v_1;\theta)}{\partial \theta}\right]\left[\frac{\partial C_{j|V_1}(a_{j,y+1}|v_1;\theta)}{\partial \theta^\top}\right]$$

$$- \frac{\partial c_{V_2j;V_1}(v_2,p_1;\gamma)}{\partial u}\left[\frac{\partial C_{j|V_1}(a_{j,y}|v_1;\theta)}{\partial \theta}\right]\left[\frac{\partial C_{j|V_1}(a_{j,y}|v_1;\theta)}{\partial \theta^\top}\right]$$

$$+ c_{V_2j;V_1}(v_2,p_2;\gamma)\frac{\partial^2 C_{j|V_1}(a_{j,y+1}|v_1;\theta)}{\partial \theta \partial \theta^\top} - c_{V_2j;V_1}(v_2,p_1;\gamma)\frac{\partial^2 C_{j|V_1}(a_{j,y}|v_1;\theta)}{\partial \theta \partial \theta^\top},$$

$$\frac{\partial^2 f_j^*}{\partial \theta \partial \gamma^\top} = \frac{\partial C_{j|V_1}(a_{j,y+1}|v_1;\theta)}{\partial \theta}\frac{\partial c_{V_2j;V_1}(v_2,p_2;\gamma)}{\partial \gamma^\top} - \frac{\partial C_{j|V_1}(a_{j,y}|v_1;\theta)}{\partial \theta}\frac{\partial c_{V_2j;V_1}(v_2,p_1;\gamma)}{\partial \gamma^\top}.$$

For the gradient and Hessian of the integrand, one needs $\partial C_{j|V_1}/\partial \theta_j$ and $\partial^2 C_{j|V_1}/\partial \theta_j \partial \theta_j^\top$, $\partial C_{j|V_2;V_1}/\partial u = c_{V_2j;V_1}$, $\partial c_{V_2j;V_1}/\partial u$, $\partial c_{V_2j;V_1}/\partial \gamma_j$, $\partial C_{j|V_2;V_1}/\partial \gamma_j$ and $\partial^2 C_{j|V_2;V_1}/\partial \gamma_j \partial \gamma_j^\top$.

D-vine and R-vine

For the gradient and Hessian of a parametric R-vine log-likelihood, recursion equations can be obtained from differentiation of (3.38). Algorithms 10 and 11 show how the gradients can be computed and stored at the same time as computing the log-likelihood. To avoid too many details, the Hessian is not included in the algorithm but the recursion is summarized later in this section.

For convenience of notation for partial derivatives, the generic arguments of a bivariate copula density $c_{kk'}$ are u, v and the generic arguments of a conditional cdf $C_{k|k'}$ are y for the conditioned variable and x for the conditioning variable.

For the gradient vector, consider first the D-vine. Let $c_{j-\ell,j;(j-\ell+1)} : (j-1)(u,v;\theta_{j-l,j})$ be a copula density in $\mathcal{T}_\ell$ and let $C_{j|j-\ell}(y|x;\theta_{j-l,j})$ and $C_{j-\ell|j}(y|x;\theta_{j-l,j})$ be its conditional cdfs.

For tree $\mathcal{T}_2$, one term of the log-likelihood is:

$$\log c_{j-2,j;j-1}\big(C_{j-2|j-1}(u_{j-2}|u_{j-1};\theta_{j-2,j-1}), C_{j|j-1}(u_j|u_{j-1};\theta_{j-1,j});\theta_{j-2,j}\big);$$

it has partial derivatives (with respect to parameters):

$$\frac{\partial \log c_{j-2,j;j-1}}{\partial \theta_{j-2,j}}, \quad \frac{\partial \log c_{j-2,j;j-1}}{\partial u}\cdot\frac{\partial C_{j-2|j-1}}{\partial \theta_{j-2,j-1}}, \quad \frac{\partial \log c_{j-2,j;j-1}}{\partial v}\cdot\frac{\partial C_{j|j-1}}{\partial \theta_{j-1,j}}.$$

For tree $\mathcal{T}_\ell$, with $S = (j-\ell+2) : (j-1)$, there are recursions (for forward and backward conditioning) from:

$$C_{j|(j-\ell+1):(j-1)}(u_j|\boldsymbol{u}_{(j-\ell+1):(j-1)};\boldsymbol{\theta}) = C_{j|j-\ell+1;S}(C_{j|S}|C_{j-\ell+1|S}),$$

$$C_{j-\ell+1|(j-\ell+2):j}(u_{j-\ell+1}|\boldsymbol{u}_{(j-\ell+2):j};\boldsymbol{\theta}) = C_{j-\ell+1|j;S}(C_{j-\ell+1|S}|C_{j|S}).$$

If θ is any parameter in $\boldsymbol{\theta}$ from previous conditionings (that is, except for $\theta_{j-\ell+1,j}$), then the recursions on the derivatives are:

$$\frac{\partial C_{j|(j-\ell+1):(j-1)}}{\partial \theta} = \frac{\partial C_{j|j-\ell+1;(j-\ell+2):(j-1)}}{\partial x}\cdot\frac{\partial C_{j-\ell+1|(j-\ell+2):(j-1)}}{\partial \theta}$$

$$+ \frac{\partial C_{j|j-\ell+1;(j-\ell+2):(j-1)}}{\partial y}\cdot\frac{\partial C_{j|(j-\ell+2):(j-1)}}{\partial \theta},$$

$$\frac{\partial C_{j-\ell+1|(j-\ell+2):j}}{\partial \theta} = \frac{\partial C_{j-\ell+1|j;(j-\ell+2):(j-1)}}{\partial x}\cdot\frac{\partial C_{j|(j-\ell+2):(j-1)}}{\partial \theta}$$

$$+ \frac{\partial C_{j-\ell+1|j;(j-\ell+2):(j-1)}}{\partial y}\cdot\frac{\partial C_{j-\ell+1|(j-\ell+2):(j-1)}}{\partial \theta}.$$

Algorithm 10 Log-density and gradient evaluation for the D-vine at one $\boldsymbol{u}$.

1: The parameter of $c_{j-\ell,j;j-\ell+1:j}$ is denoted as $\theta_{j-\ell,j}$. A generic bivariate copula density is $c_{kk'}(u,v)$ and its conditional cdfs are written as $C_{k|k'}(y|x)$ and $C_{k'|k}(y|x)$ for partial derivatives.

2: Initialize gradient vector ∇ as a vector of length n_p equal to the sum of sizes of the $\theta_{j-\ell,j}$.

3: Let $q_{j-\ell,j}$ be the position(s) of ∇ where $\theta_{j-\ell,j}$ is stored.

4: Initialize $n_p \times d$ matrices $\mathrm{fder}[\cdot,\cdot]$, $\mathrm{bder}[\cdot,\cdot]$, $\mathrm{fderprev}[\cdot,\cdot]$, $\mathrm{bderprev}[\cdot,\cdot]$ to 0.

5: $(\mathcal{T}_1)$ $\mathrm{logden} \leftarrow \sum_{j=2}^{d} \log c_{j-1,j}(u_{j-1}, u_j; \theta_{j-1,j})$;

6: update ∇ at position $q_{j-1,j}$ by adding $\partial \log c_{j-1,j}/\partial\theta(u_{j-1}, u_j; \theta_{j-1,j})$ for $j = 2, \dots, d$;

7: let $w'_j \leftarrow u_j$ and $w_j \leftarrow u_j$ for $j = 1, \dots, d$;

8: **for** $\ell = 2, \dots, d-1$: **do**

9:      let $v'_j \leftarrow C_{j-\ell+1|j;(j-\ell+2):(j-1)}(w'_{j-1}|w_j; \theta_{j-\ell+1,j})$, $v'_{j\theta} \leftarrow \partial C_{j-\ell+1|j;(j-\ell+2):(j-1)}/\partial\theta$, $v'_{jx} \leftarrow \partial C_{j-\ell+1|j;(j-\ell+2):(j-1)}/\partial x$, $v'_{jy} \leftarrow \partial C_{j-\ell+1|j;(j-\ell+2):(j-1)}/\partial y$, $\mathrm{bder}[,j] \leftarrow v'_{jx}\mathrm{fderprev}[,j] + v'_{jy}\mathrm{bderprev}[,j-1]$, for $j = \ell, \dots, d-1$;

10:      for position $q_{j-\ell+1,j}$, set $\mathrm{bder}[q_{j-\ell+1,j},j] \leftarrow v'_{j\theta}$ for $j = \ell, \dots, d-1$;

11:      let $v_j \leftarrow C_{j|j-\ell+1;(j-\ell+2):(j-1)}(w_j|w'_{j-1}; \theta_{j-\ell+1,j})$, $v_{j\theta} \leftarrow \partial C_{j|j-\ell+1;(j-\ell+2):(j-1)}/\partial\theta$, $v_{jx} \leftarrow \partial C_{j|j-\ell+1;(j-\ell+2):(j-1)}/\partial x$, $v_{jy} \leftarrow \partial C_{j|j-\ell+1;(j-\ell+2):(j-1)}/\partial y$, $\mathrm{fder}[,j] \leftarrow v_{jx}\mathrm{bderprev}[,j-1] + v_{jy}\mathrm{fderprev}[,j]$, for $j = \ell+1, \dots, d$;

12:      for position $q_{j-\ell+1,j}$, set $\mathrm{fder}[q_{j-\ell+1,j},j] \leftarrow v_{j\theta}$ for $j = \ell+1, \dots, d$;

13:      update $\mathrm{logden} \leftarrow \mathrm{logden} + \sum_{j=\ell+1}^{d} \log c_{j-\ell,j;(j-\ell+1):(j-1)}(v'_{j-1}, v_j; \theta_{j-\ell.j})$;

14:      update $\nabla \leftarrow \nabla + \sum_{j=\ell+1}^{d} \partial \log c_{j-\ell,j;(j-\ell+1):(j-1)}/\partial u \cdot \mathrm{bder}[,j-1] + \partial \log c_{j-\ell,j;(j-\ell+1):(j-1)}/\partial v \cdot \mathrm{fder}[,j]$;

15:      reset $w'_j \leftarrow v'_j$, $w_j \leftarrow v_j$, for $j = \ell+1, \dots, d$;

16:      reset $\mathrm{fderprev} \leftarrow \mathrm{fder}$ and $\mathrm{bderprev} \leftarrow \mathrm{bder}$.

17: **end for**

18: Return logden and ∇.

The extension of the algorithm to the general R-vine is similar except the notation involves the vine array A.

Algorithm 11 Log-density and gradient evaluation for the R-vine at one $\boldsymbol{u}$.

1: A generic copula density is denoted $c_{kk'}(u, v)$ and its conditional cdfs are denoted as $C_{k|k'}(y|x)$ and $C_{k'|k}(y|x)$.

2: Input vine array $A = (a_{kj})$.

3: Compute $M = (m_{kj})$ in the upper triangle, where $m_{kj} = \max\{a_{1j}, \ldots, a_{kj}\}$ for $k = 1, \ldots, j-1$, $j = 2, \ldots, d$. Compute the $I = (I_{kj})$ indicator array as in Algorithm 5.

4: The parameter of $c_{a_{\ell,j}|j;a_{1j}\cdots a_{\ell-1,j}}$ is denoted as $\theta_{\ell,j}$.

5: Initialize gradient vector ∇ as a vector of length n_p equal to the sum of sizes of the $\theta_{\ell,j}$.

6: Let $q_{\ell,j}$ be the position(s) of ∇ where $\theta_{\ell,j}$ is stored.

7: Initialize $n_p \times d$ matrices fder$[\cdot, \cdot]$, bder$[\cdot, \cdot]$, fderprev$[\cdot, \cdot]$, bderprev$[\cdot, \cdot]$, sder$[\cdot, \cdot]$ to 0.

8: $(\mathcal{T}_1)$ logden $\leftarrow \sum_{j=2}^{d} \log c_{a_{1j}j}(u_{a_{1j}}, u_j; \theta_{1j})$;

9:     update ∇ at position $q_{1,j}$ by adding $\partial \log c_{a_{1j}j}/(u_{a_{1j}}, u_j; \theta_{1j})\partial\theta$ for $j = 2, \ldots, d$;

10: **for** $j = 2, \ldots, d$: **do**

11:     if $(I_{1j} = 1)$ then let $v'_j \leftarrow C_{a_{1j}|j}(u_{a_{1j}}|u_j; \theta_{1j})$, $v'_{j\theta} \leftarrow \partial C_{a_{1j}|j}/\partial\theta$, $v'_{jx} \leftarrow \partial C_{a_{1j}|j}/\partial x$, $v'_{jy} \leftarrow \partial C_{a_{1j}|j}/\partial y$;

12:     for position $q_{1,j}$, set bder$[q_{1,j}, j] \leftarrow v'_{j\theta}$;

13:     let $v_j \leftarrow C_{j|a_{1j}}(u_j|u_{a_{1j}}; \theta_{1j})$, $v_{j\theta} \leftarrow \partial C_{j|a_{1j}}/\partial\theta$, $v_{jx} \leftarrow \partial C_{j|a_{1j}}/\partial x$, $v_{jy} \leftarrow \partial C_{j|a_{1j}}/\partial y$;

14:     for position $q_{1,j}$, set fder$[q_{1,j}, j] \leftarrow v_{j\theta}$;

15: **end for**

16: let fderprev $\leftarrow$ fder and bderprev $\leftarrow$ bder;

17: **for** $j = 3, \ldots, d$: **do**

18:     if $a_{2j} = m_{2j}$, then set $s_j \leftarrow v_{m_{2j}}$, sder$[\cdot, j] \leftarrow$ fder$[\cdot, m_{2j}]$;

19:     if $a_{2j} < m_{2j}$, then set $s_j \leftarrow v'_{m_{2j}}$, sder$[\cdot, j] \leftarrow$ bder$[\cdot, m_{2j}]$;

20: **end for**

21: $(\mathcal{T}_2)$ update logden $\leftarrow$ logden $+ \sum_{j=3}^{d} \log c_{a_{2j},j;a_{1j}}(s_j, v_j; \theta_{2,j})$,

22: update $\nabla \leftarrow \nabla + \sum_{j=3}^{d} \partial \log c_{a_{2j},j;a_{1j}}/\partial u \cdot$ sder$[\cdot, j] + \partial \log c_{a_{2j},j;a_{1j}}/\partial v \cdot$ fder$[\cdot, j]$, with the arguments $s_j, v_j, \theta_{2,j}$ in the jth term;

23: update ∇ at position $q_{2,j}$ by adding $\partial \log c_{a_{2j},j;a_{1j}}(s_j, v_j; \theta_{2,j})/\partial\theta$

24: let $w'_j \leftarrow v'_j$, $w_j \leftarrow v_j$, $j = 3, \ldots, d$;

25: set fderprev $\leftarrow$ fder and bderprev $\leftarrow$ bder;

26: **for** $\ell = 3, \ldots, d-1$: $(\mathcal{T}_3, \ldots, \mathcal{T}_{d-1})$ **do**

27:     **for** $j = \ell, \ldots, d$: **do**

28:         if $(I_{\ell-1,j} = 1)$ then set $v'_j \leftarrow C_{a_{\ell-1,j}|j;a_{1j}\cdots a_{\ell-2,j}}(s_j|w_j; \theta_{\ell-1,j})$, $v'_{j\theta} \leftarrow \partial C_{a_{\ell-1,j}|j;a_{1j}\cdots a_{\ell-2,j}}/\partial\theta$, $v'_{jx} \leftarrow \partial C_{a_{\ell-1,j}|j;a_{1j}\cdots a_{\ell-2,j}}/\partial x$, $v'_{jy} \leftarrow \partial C_{a_{\ell-1,j}|j;a_{1j}\cdots a_{\ell-2,j}}/\partial y$, bder$[\cdot, j] \leftarrow v'_{jx}$fderprev$[\cdot, j] + v'_{jy}$sder$[\cdot, j]$ and for position $q_{\ell-1,j}$, set bder$[q_{\ell-1,j}, j] \leftarrow v'_{j\theta}$;

29:         set $v_j \leftarrow C_{j|a_{\ell-1,j};a_{1j}\cdots a_{\ell-2,j}}(w_j|s_j; \theta_{\ell-1,j})$, $v_{j\theta} \leftarrow \partial C_{j|a_{\ell-1,j};a_{1j}\cdots a_{\ell-2,j}}/\partial\theta$, $v_{jx} \leftarrow \partial C_{j|a_{\ell-1,j};a_{1j}\cdots a_{\ell-2,j}}/\partial x$, $v_{jy} \leftarrow \partial C_{j|a_{\ell-1,j};a_{1j}\cdots a_{\ell-2,j}}/\partial y$, fder$[\cdot, j] \leftarrow v_{jx}$sder$[\cdot, j] + v_{jy}$fderprev$[\cdot, j]$, and for position $q_{\ell-1,j}$, set fder$[q_{\ell-1,j}, j] \leftarrow v_{j\theta}$;

30:     **end for**

31:     **for** $j = \ell+1, \ldots, d$: **do**

32:         if $a_{\ell,j} = m_{\ell,j}$, then $s_j \leftarrow v_{m_{\ell,j}}$, sder$[\cdot, j] \leftarrow$ fder$[\cdot, m_{\ell,j}]$;

33:         if $a_{\ell,j} < m_{\ell,j}$, then $s_j \leftarrow v'_{m_{\ell,j}}$, sder$[\cdot, j] \leftarrow$ bder$[\cdot, m_{\ell,j}]$;

34:     **end for**

35:     update logden $\leftarrow$ logden $+ \sum_{j=\ell+1}^{d} \log c_{a_{\ell j}j,j;a_{1j}\cdots a_{\ell-1,j}}(s_j, v_j; \theta_{\ell,j})$,

36:     update $\nabla \leftarrow \nabla + \sum_{j=\ell+1}^{d} \partial \log c_{a_{\ell j}j,j;a_{1j}\cdots a_{\ell-1,j}}/\partial u \cdot$ sder$[\cdot, j] + \partial \log c_{a_{\ell j}j,j;a_{1j}\cdots a_{\ell-1,j}}/\partial v \cdot$ fder$[\cdot, j]$, with the arguments $s_j, v_j, \theta_{\ell,j}$ in the jth term;

37:     update ∇ at position $q_{\ell,j}$ by adding $\partial \log c_{a_{\ell j}j,j;a_{1j}\cdots a_{\ell-1,j}}(s_j, v_j; \theta_{\ell,j})/\partial\theta$;

38:     reset $w'_j \leftarrow v'_j$, $w_j \leftarrow v_j$, $j = \ell+1, \ldots, d$;

39:     reset fderprev $\leftarrow$ fder and bderprev $\leftarrow$ bder.

40: **end for**

41: Return logden and ∇.

Next we derive recursion equations for the Hessian. For notation, write:

$$C_{k|j\cup S}(u_k|\boldsymbol{u}_{j\cup S};\boldsymbol{\theta}_{\mathrm{prev}},\theta_{jk}) = C_{k|j;S}\big(C_{k|S}(\cdot\,;\boldsymbol{\theta}_{\mathrm{prev}})|C_{j|S}(\cdot\,;\boldsymbol{\theta}_{\mathrm{prev}});\theta_{jk}\big)$$

(forward) and $C_{j|k\cup S} = C_{j|k;S}(C_{j|S}|C_{k|S})$ (backward) where $j < k$ and the indices in the set S are all less than k and $\theta_{jk} = \theta_{jk;S}$. Recursions can be obtained for derivatives of $C_{k|j\cup S}$ in terms of $C_{j|S}$ and $C_{k|S}$ for derivatives of parameters in $\boldsymbol{\theta}_{\mathrm{prev}}$.

For θ_{jk}, first and second order partial derivatives are:

$$\frac{\partial C_{k|j;S}}{\partial\theta_{jk}},\quad \frac{\partial^2 C_{k|j;S}}{\partial\theta_{jk}\partial\theta_{jk}^\top}.$$

If θ is any parameter in $\boldsymbol{\theta}_{\mathrm{prev}}$, then

$$\frac{\partial C_{k|j\cup S}}{\partial\theta} = \frac{\partial C_{k|j;S}}{\partial x}\cdot\frac{\partial C_{j|S}}{\partial\theta} + \frac{\partial C_{k|j;S}}{\partial y}\cdot\frac{\partial C_{k|S}}{\partial\theta}.$$

The second order partial derivative with respect $\theta\in\boldsymbol{\theta}_{\mathrm{prev}}$ and θ_{jk} is:

$$\frac{\partial^2 C_{k|j\cup S}}{\partial\theta\partial\theta_{jk}^\top} = \frac{\partial C_{j|S}}{\partial\theta}\cdot\frac{\partial^2 C_{k|j;S}}{\partial x\partial\theta_{jk}^\top} + \frac{\partial C_{k|S}}{\partial\theta}\cdot\frac{\partial^2 C_{k|j;S}}{\partial y\partial\theta_{jk}^\top}.$$

The second order partial derivative with respect $\theta_r,\theta_s\in\boldsymbol{\theta}_{\mathrm{prev}}$, with 6 summands, is:

$$\begin{aligned}
\frac{\partial^2 C_{k|j\cup S}}{\partial\theta_r\partial\theta_s^\top} &= \frac{\partial^2 C_{k|j;S}}{\partial x^2}\cdot\frac{\partial C_{j|S}}{\partial\theta_r}\frac{\partial C_{j|S}}{\partial\theta_s^\top} + \frac{\partial^2 C_{k|j;S}}{\partial x\partial y}\cdot\frac{\partial C_{j|S}}{\partial\theta_r}\frac{\partial C_{k|S}}{\partial\theta_s^\top}\\
&\quad + \frac{\partial^2 C_{k|j;S}}{\partial y^2}\cdot\frac{\partial C_{k|S}}{\partial\theta_r}\frac{\partial C_{k|S}}{\partial\theta_s^\top} + \frac{\partial^2 C_{k|j;S}}{\partial x\partial y}\cdot\frac{\partial C_{k|S}}{\partial\theta_r}\frac{\partial C_{j|S}}{\partial\theta_s^\top}\\
&\quad + \frac{\partial C_{k|j;S}}{\partial x}\cdot\frac{\partial^2 C_{j|S}}{\partial\theta_r\partial\theta_s^\top} + \frac{\partial C_{k|j;S}}{\partial y}\cdot\frac{\partial^2 C_{k|S}}{\partial\theta_r\partial\theta_s^\top}.
\end{aligned}$$

The operations on the derivatives of $\boldsymbol{\theta}_{\mathrm{prev}}$ can be vectorized.

Next for the partial derivatives (gradient and Hessian) of a log density:

$$\ell(u_j,u_j,\boldsymbol{u}_S,\theta_{jk},\boldsymbol{\theta}_{\mathrm{prev}}) := \log c_{jk;S}\big(C_{j|S}(u_j|\boldsymbol{u}_S;\boldsymbol{\theta}_{\mathrm{prev}}),C_{k|S}(u_k|\boldsymbol{u}_S;\boldsymbol{\theta}_{\mathrm{prev}});\theta_{jk}\big).$$

The gradient vector has

$$\frac{\partial\ell}{\partial\theta_{jk}} = \frac{\partial\log c_{jk;S}}{\partial\theta_{jk}},$$

$$\frac{\partial\ell}{\partial\theta} = \frac{\partial\log c_{jk;S}}{\partial u}\frac{\partial C_{j|S}}{\partial\theta} + \frac{\partial\log c_{jk;S}}{\partial v}\frac{\partial C_{k|S}}{\partial\theta},\quad \theta\in\boldsymbol{\theta}_{\mathrm{prev}}.$$

The Hessian has

$$\frac{\partial^2\ell}{\partial\theta_{jk}\partial\theta_{jk}^\top} = \frac{\partial^2\log c_{jk;S}}{\partial\theta_{jk}\partial\theta_{jk}^\top},$$

$$\frac{\partial^2\ell}{\partial\theta\partial\theta_{jk}^\top} = \frac{\partial C_{j|S}}{\partial\theta}\frac{\partial^2\log c_{jk;S}}{\partial u\partial\theta_{jk}^\top} + \frac{\partial C_{k|S}}{\partial\theta}\frac{\partial^2\log c_{jk;S}}{\partial v\partial\theta_{jk}^\top},\quad \theta\in\boldsymbol{\theta}_{\mathrm{prev}},$$

$$\begin{aligned}
\frac{\partial^2\ell}{\partial\theta_r\partial\theta_s^\top} &= \frac{\partial^2\log c_{jk;S}}{\partial u^2}\frac{\partial C_{j|S}}{\partial\theta_r}\frac{\partial C_{j|S}}{\partial\theta_s^\top} + \frac{\partial^2\log c_{jk;S}}{\partial u\partial v}\frac{\partial C_{j|S}}{\partial\theta_r}\frac{\partial C_{k|S}}{\partial\theta_s^\top}\\
&\quad + \frac{\partial^2\log c_{jk;S}}{\partial u\partial v}\frac{\partial C_{k|S}}{\partial\theta_r}\frac{\partial C_{j|S}}{\partial\theta_s^\top} + \frac{\partial^2\log c_{jk;S}}{\partial v^2}\frac{\partial C_{k|S}}{\partial\theta_r}\frac{\partial C_{k|S}}{\partial\theta_s^\top}\\
&\quad + \frac{\partial\log c_{jk;S}}{\partial u}\frac{\partial^2 C_{j|S}}{\partial\theta_r\partial\theta_s^\top} + \frac{\partial\log c_{jk;S}}{\partial v}\frac{\partial^2 C_{k|S}}{\partial\theta_r\partial\theta_s^\top},\quad \theta_r,\theta_s\in\boldsymbol{\theta}_{\mathrm{prev}}.
\end{aligned}$$

6.13   Generation of vines

Morales-Nápoles et al. (2009) and Morales-Nápoles (2011) have shown that the number of d-dimensional regular vines (with d nodes or variables) is $N_d = \frac{1}{2}d!\,2^{(d-3)(d-2)/2}$ for $d \geq 3$, and the number of vine arrays in natural order is $N_d^o = 2^{(d-3)(d-2)/2}$; see also Section 8.11. The number of equivalence classes of d-dimensional vines is less than N_d^o because some vines have two representations of vine arrays in natural order and others have one representation.

A vine array $A = (a_{kj})$ in a $d \times d$ upper triangular matrix is in *natural order* if (i) $a_{ii} = i$ for $i = 1, \ldots, d$ and (ii) $a_{i-1,i} = i - 1$ for $i = 2, \ldots, d$. Let NO(d) is the class of d-dimensional vine arrays A coded in natural order. Vine arrays have other constraints, one of which is that $a_{1,i}, \ldots, a_{i-2,i}$ is a permutation of $\{1, \ldots, i - 2\}$. There is also another condition based on binary vectors that restricts the number of possible permutations to 2^{i-3} in column i for $i = 4, \ldots, d$.

In a vine, each edge has a conditioned set of cardinality two, and tree $\mathcal{T}_\ell$ has $d - \ell$ edges for $\ell = 1, \ldots, d - 1$. So $\mathcal{T}_1$ has $d - 1$ edges and $\mathcal{T}_{d-1}$ has one edge. The natural order is a method for assigning indices to the variables of a regular vine on d variables, such that one of the two conditioned variables in $\mathcal{T}_{d-1}$ gets index d, and its partner gets index $d - 1$. Further, if indices $d, \ldots, j + 1$ have been assigned, index j is assigned to the conditioned-set-partner of the variable indexed $j + 1$ in the node of tree j. Hence, a natural ordering for a regular vine is a permutation of the indices defined with respect to that vine. For any regular vine, there are two natural orderings, according to the variable in $\mathcal{T}_{d-1}$ which is assigned index d. There are two vine arrays in natural order corresponding to the original vine, and in some cases these two vine arrays coincide (such as for C-vines and D-vines).

An algorithm for converting a vine array to natural order is given in Algorithm 12. Details of the vine array in natural order are illustrated below in Example 6.1 for the six equivalence classes of 5-dimensional vines. Example 6.2 discusses the form of $A \in \text{NO}(d)$ for the D-vine.

Algorithm 12 Converting from a vine array to one in NO(d).

1: Input A^*, a $d \times d$ vine array.
2: Input irev=1 as indicator to start with $a^*_{d-1,d}$, otherwise start with a^*_{dd}.
3: Create a $d \times d$ matrix $T = (t_{xy})$ based on looping over $1 \leq k < j \leq d$ with $x \leftarrow a^*_{jj}$, $y \leftarrow a^*_{kj}$, $t_{xy} = t_{yx} \leftarrow k$ (the tree k where $\{x, y\}$ is a conditioned set).
4: If (irev=1) then $a_{dd} \leftarrow a^*_{d-1,d}$ else $a_{dd} \leftarrow a^*_{dd}$.
5: **for** $k = d, \ldots, 2$: **do**
6:     $x \leftarrow a_{kk}$;
7:     for $(\ell = 1, \ldots, k - 1)$ $a_{\ell k} \leftarrow \{y : t_{xy} = \ell\}$;
8:     set row and column x of T to 0s;
9:     $a_{k-1,k-1} \leftarrow a_{k-1,k}$;
10: **end for**
11: A is now an array with $a_{i,i+1} = a_{ii}$ for $i = 1, \ldots, d - 1$.
12: Let P be a permutation of $1 : d$ with $P(a_{ii}) = i$ for $i = 1, \ldots, d$; replace a_{kj} by $P(a_{kj})$ for $1 \leq k \leq j \leq d$.
13: Return $A \in \text{NO}(d)$.

The representation of $A \in \text{NO}(d)$ in terms of binary vectors is given in Algorithm 13. This algorithm is based on the theory in Morales-Nápoles et al. (2009), and can be used to generate all vine arrays $A \in \text{NO}(d)$; see Section 8.11 for some explanations of the theory.

Consider vine copulas with a common bivariate parametric family $C(\cdot; \boldsymbol{\theta})$ for the edges. For bivariate Gaussian on edges, all vines lead to the same joint distribution, and the different vines just represent different parametrizations. If $C(\cdot; \boldsymbol{\theta})$ is any other family, then the vines are generally distinct and enumeration through them may be feasible only in dimension $3 \leq d \leq 6$. For d-dimensional data that have been converted to $U(0,1)$ margins, the best fitting vine might be one for which the simplifying assumption is closest to being met for the "true" multivariate distribution, or it might be one for which the edges in $\mathcal{T}_1$ have the strongest dependence. Enumeration of all or a random subset of vines might be a way to check for patterns in the best fitting vine to data with a specified bivariate family $C(\cdot; \boldsymbol{\theta})$.

Algorithm 13 Converting from a binary array to a vine array in $\mathrm{NO}(d)$ ($d \geq 4$); Joe et al. (2011).

1: Input $b_4, \ldots, b_d$ where $b_j = b_j(\cdot)$ is a binary vector of length j or a mapping from $(1, \ldots, j)$ to $\{0,1\}^j$. Assume $b_j(1) = b_j(j-1) = b_j(j) = 1$ for $j = 4, \ldots, d$. [b_1, b_2, b_3 with these properties can also be added but they would be fixed over all $A \in \mathrm{NO}(d)$.]
2: Initialize a $d \times d$ array A, with $A_{j,j} = j$ for $j = 1, \ldots, d$, $A_{j-1,j} = j-1$ for $j = 2, \ldots, d$, and $A_{1,3} = 1$.
3: **for** $j = 4, \ldots, d$: **do**
4:     ac = active column $\leftarrow j - 2$;
5:     **for** $k = j-2, \ldots, 1$: **do**
6:         if $b_j(k) = 1$ then $A_{k,j} = A_{\mathrm{ac,ac}}$ and ac$\leftarrow$ largest value among $1, \ldots, j-2$ not yet assigned in column k;
7:         else if $b_j(k) = 0$ then $A_{k,j} = A_{k-1,\mathrm{ac}}$.
8:     **end for**
9: **end for**
10: Return A.

Algorithm 14 Validating an array to be a vine array in $\mathrm{NO}(d)$; Joe et al. (2011).

1: Input A^*: assume $A^*_{j,j} = j$ for $j = 1, \ldots, d$ with $d \geq 4$, $A^*_{j-1,j} = j-1$ for $j = 2, \ldots, d$, $A^*_{1,3} = 1$ and $(A^*_{1,j}, \cdots, A^*_{j-2,j})$ is a permutation of $(1, \ldots, j-2)$ for $j = 4, \ldots, d$. Otherwise return -1 to indicate that $A^* \notin \mathrm{NO}(d)$.
2: For column 4: $b^*_4(2) = 1$ if $A^*_{2,4} = 2$ and $b^*_4(2) = 0$ if $A^*_{2,4} = 1$.
3: **for** $j = 5, \ldots, d$: **do**
4:     $b^*_j(1) = b^*_j(j-1) = b^*_j(j) = 1$.
5:     ac = active column $\leftarrow j - 2$;
6:     **for** $k = j-2, \ldots, 1$: **do**
7:         if $A^*_{k,j} = A^*_{\mathrm{ac,ac}}$ then $b^*_j(k) = 1$ and ac$\leftarrow$ the largest value among $A^*_{1,j}, \ldots, A^*_{k-1,j}$;
8:         else if $A^*_{k,j} = A^*_{k-1,\mathrm{ac}}$ then $b^*_j(k) = 0$;
9:         else A^* is not in $\mathrm{NO}(d)$ and return -1.
10:     **end for**
11: **end for**
12: Return $b^*_4, \ldots, b^*_d$ with b^*_j mapping $(1, \ldots, j)$ to $\{0,1\}^j$.

Example 6.1 (NO(5) with 6 equivalence classes). Let $d = 5$. The number of objects in $\mathrm{NO}(5)$ is $2^{(d-3)(d-2)/2} = 2^3 = 8$. We start with vine arrays in the form of edges given in Joe (2011a), then convert to two arrays that satisfy $a_{ii} = a_{i,i+1}$ for $i = 1, \ldots, d-1$ and finally convert to the two arrays to $\mathrm{NO}(d)$. In some cases, these two arrays in $\mathrm{NO}(d)$ are the same, and in some cases they are different. If the two arrays are different, the vine maps to an equivalence class of two objects in $\mathrm{NO}(d)$; otherwise the equivalence class consists of a singleton. We also show the corresponding b_4, b_5 vectors that appear in Algorithm 13.

In Table 6.1, the four equivalence classes of 5-dimensional vines between the C-vine and

D-vine are denoted as B0, B1, B2, B3; see Section 3.9.8 for their derivation. The B1 and B3 classes have less symmetry and are the ones that have two vine arrays in NO(5).

ec	original	$a_{i,i+1} = a_{ii}$		NO(5)		$b_4 b_5$	
	11111	11111	11111	11111		11	
	2222	2222	2222	2222		11	
C	333	333	333	333		11	
	44	44	55	44		11	
	5	5	4	5		1	
	11234	33324	33342	11123		11	
	2123	2233	4433	2211		00	
D	312	442	224	332		10	
	41	11	55	44		11	
	5	5	1	5		1	
	11111	11111	11111	11111		11	
	2223	2223	3332	2223		11	
B0	332	332	223	332		10	
	44	44	55	44		11	
	5	5	4	5		1	
	11112	11112	11121	11112	11121	11	11
	2221	2221	2212	2221	2212	10	01
B1	333	333	333	333	333	11	11
	44	44	55	44	44	11	11
	5	5	4	5	5	1	1
	11211	22211	22211	11122		11	
	2122	1122	1122	2211		00	
B2	333	333	333	333		11	
	44	44	55	44		11	
	5	5	4	5		1	
	11112	11112	11121	11113	11121	11	11
	2231	3331	2213	2221	2213	10	01
B3	323	223	332	332	332	10	10
	44	44	55	44	44	11	11
	5	5	4	5	5	1	1

Table 6.1 *Vine arrays of different forms for the 6 equivalence classes (ec) of 5-dimensional vines. For B1 and B3, there are 2 vine arrays in NO(5) and two sets of b_4, b_5 that lead to the equivalence class. For C,D,B0,B2, there is one vine array in NO(5).*

Example 6.2 (D-vine array in natural order). The D-vine in standard order has array $A^* = (a_{kj}^*)$ with $a_{jj}^* = j$ for $1 \leq j \leq d$ and $a_{kj}^* = j - k$ for $1 < k < j < d$. Let $M^* = (m_{kj}^*)$ in the upper triangle, where $m_{kj}^* = \max\{a_{1j}^*, \ldots, a_{kj}^*\}$ for $1 < k < j < d$. Then $m_{kj}^* = a_{kj}^*$ only if $k = 1$.

For the D-vine with array $A \in NO(d)$, the following holds in the jth column, (a) $j - 1$ is in row $j - 1$ and there is a decrement of 2 (until 2 or 1 is reached) as the row number decreases from $j - 1$; (b) $j - 2$ is in row 1 and there is a decrement of 2 (until 2 or 1 is reached) as the row number increases from 1. That is, $a_{kj} = j - 2k$ for $1 \leq k \leq [(j - 1)/2]$ and $a_{kj} = 2k + 1 - j$ for $[(j + 1)/2] \leq k \leq j - 1$. The first tree $\mathcal{T}_1$ of the vine has the form

$$\cdots - 6 - 4 - 2 - 1 - 3 - 5 - 7 - \cdots.$$

Let $M = (m_{kj})$ in the upper triangle, where $m_{kj} = \max\{a_{1j}, \ldots, a_{kj}\}$ for $1 < k < j < d$. Then $m_{kj} = a_{kj}$ in column j if $k = 1$ and $k = j - 1$.

In comparing M^* versus M, this might mean the some algorithms such as Algorithms 4 and 17 might be numerically faster for the D-vine in the natural order.   □

Algorithm 13 can be used to generate a random array $A \in \mathrm{NO}(d)$ by choosing random $b_4, \ldots, b_d$. As shown in Example 6.1, the mapping of the b vectors for the $\mathrm{NO}(d)$ form to a vine equivalence class is two-to-one in some cases. The mapping is many-to-one without the natural order form. Given an array $A \in \mathrm{NO}(d)$, one of $d!/2$ permutations of variables assigned to $1 : d$ can be made to get a random vine. Note that with the natural order, there are $\binom{d}{2}$ ways to assign variables to the last two indices and $(d-2)!$ ways to assign the remaining variables to indices $1, \ldots, d-2$ to get a product of $d!/2$. So the count of d-dimensional vines $N_d = \frac{1}{2} d! \, 2^{(d-3)(d-2)/2}$ comes from $d!/2$ permutations multiplied by $2^{(d-3)(d-2)/2}$ vine arrays in $\mathrm{NO}(d)$ — in the binary vectors $b_4, \ldots, b_d$, there are $\sum_{j=4}^{d} (j-3) = (d-3)(d-2)/2$ binary positions that are not fixed.

Algorithm 14 checks if an array A^* is a valid vine array in $\mathrm{NO}(d)$; it is the inversion of Algorithm 13. If valid, an array with the binary vectors is obtained. This can be useful when trying to generate vine arrays through random permutations of $1, \ldots, j$ in column j, instead of through the binary representation.

To enumerate all vines of dimension d, one can (i) generate the vine arrays in natural order through the non-fixed positions of the binary vectors $b_4, \ldots, b_d$; (ii) for each array, sequentially go through the $d!/2$ permutations $(i_1, \ldots, i_d)$, with $i_{d-1} < i_d$. Enumeration on the computer is possible for $d \leq 8$, with the values of N_d as shown in Table 8.2.

6.14   Simulation from vines and truncated vine models

In this section, we provide algorithms for simulation from R-vine copulas. In high dimensions, truncated vines may be adequate as models, and the general algorithm for a truncated R-vine can be obtained as a small modification. The truncated R-vine algorithm can be adapted for latent variable models, such as factor copulas and bi-factor copulas, which are truncated R-vines with a combination of latent and observed variables.

Consider a vine copula with vine array $A = (a_{\ell j})_{1 \leq \ell \leq j \leq d}$ and a set of bivariate copulas $\{C_{a_{\ell j} j; S(j, a_{\ell j})} : 1 \leq \ell < j \leq d\}$.

For the continuous case, simulation from the vine copula can be done via the conditional method or Rosenblatt's method by recursively applying (3.38), which is repeated here:

$$
\begin{aligned}
C_{j|S(j,a_{\ell j}) \cup a_{\ell j}} & (u_j | \boldsymbol{u}_{S(j,a_{\ell j}) \cup a_{\ell j}}) \\
&= C_{j|a_{\ell j}; S(j,a_{\ell j})}\big(C_{j|S(j,a_{\ell j})}(u_j | \boldsymbol{u}_{S(j,a_{\ell j})}) \mid C_{a_{\ell j}|S(j,a_{\ell j})}(u_{a_{\ell j}} | \boldsymbol{u}_{S(j,a_{\ell j})}) \big),
\end{aligned}
\tag{6.1}
$$

with $S(j, a_{\ell j}) = \{a_{1j}, \ldots, a_{\ell-1,j}\}$. If $\ell = j - 1$, and

$$
p = C_{j|S(j,a_{j-1,j}) \cup a_{j-1,j}}(u_j | \boldsymbol{u}_{S(j,a_{j-1,j}) \cup a_{j-1,j}}) = C_{j|1:(j-1)}(u_j | \boldsymbol{u}_{1:(j-1)}),
$$

then

$$
u_j = C^{-1}_{j|1:(j-1)}(p | \boldsymbol{u}_{1:(j-1)}).
\tag{6.2}
$$

Letting $q_{jj} = p$ to start the recursion, then for $1 \leq \ell < j$, (6.1) can be written as:

$$
q_{\ell j} = C^{-1}_{j|a_{\ell j}; a_{1j}, \ldots, a_{\ell-1,j}}\big(q_{\ell+1,j} | C_{a_{\ell j}|a_{1j}, \ldots, a_{\ell-1,j}}(u_{a_{\ell j}} | u_{a_{1j}}, \ldots, u_{a_{\ell-1,j}}) \big),
\tag{6.3}
$$

with $u_j = q_{1j} = C^{-1}_{j|a_{1j}}(q_{2j} | u_{a_{1j}})$ being the solution to (6.2).

To apply the conditional method to simulate from the vine, start with $P_1, \ldots, P_d$ as random independent $\mathrm{U}(0,1)$ variables. Then $(U_1, \ldots, U_d)$ is generated as a random vector from the vine copula through the sequence

$$
\begin{aligned}
U_1 &\leftarrow P_1, \; U_2 \leftarrow C^{-1}_{2|1}(P_2 | U_1), \ldots, U_j \leftarrow C^{-1}_{j|1:(j-1)}(P_j | \boldsymbol{U}_{1:(j-1)}), \\
&\ldots, U_d \leftarrow C^{-1}_{d|1:(d-1)}(P_d | \boldsymbol{U}_{1:(d-1)}),
\end{aligned}
\tag{6.4}
$$

where for $j \geq 3$, the jth step involves a recursion in (6.3) with the values of $\{C_{a_{\ell j}|a_{1j},\ldots,a_{\ell-1,j}}(U_{a_{\ell j}}|U_{a_{1j}},\ldots,U_{a_{\ell-1,j}})\}$ determined in a preceding step. So an algorithm for simulating from a vine copula has steps, somewhat like the algorithms in Section 6.10 for the log-likelihood, to keep track of conditional cdf values needed for future steps.

For vine models applied to discrete variables, such as in Section 3.9.5, the conditional method can be used but the conditional probabilities are computed using Algorithm 7.

6.14.1 Simulation from vine copulas

Algorithms are presented for the continuous case, and an outline is given for the discrete case. For continuous variables, details are first shown for the C-vine and D-vine, as understanding these helps to develop the general algorithm for the R-vine. The algorithms for simulating from the C-vine and D-vine are given in Aas et al. (2009) with simplifications in Joe (2011a). The first complete algorithm for simulating from a general R-vine is in Dissmann (2010) and Dissmann et al. (2013). but we will have some simplifications here. See also Chapter 5 of Mai and Scherer (2012b).

For the C-vine, $C_{a_{\ell j}|a_{1j},\ldots,a_{\ell-1,j}} = C_{\ell|1:(\ell-1)}$ and from (6.4), $C_{\ell|1:(\ell-1)}(u_\ell|\boldsymbol{u}_{1:(\ell-1)}) = p_\ell$, so the algorithm for the C-vine is that in Algorithm 15. Similar to other algorithms involving vines, the simplest occurs for the C-vine.

Algorithm 15 Simulating from a C-vine in dimension d; Joe (2011a).

1: Generate $p_1,\ldots,p_d$ to be independent $U(0,1)$ random variables.
2: Let $u_1 \leftarrow p_1$, $u_2 \leftarrow C_{2|1}^{-1}(p_2|p_1)$.
3: **for** $j = 3,\ldots,d$: **do**
4:      let $q \leftarrow p_j$;
5:      for $\ell = j-1, j-2,\ldots,1$: $q \leftarrow C_{j|\ell;1:(\ell-1)}^{-1}(q|p_\ell)$;
6:      let $u_j \leftarrow q$.
7: **end for**
8: Return $(u_1,\ldots,u_d)$.

For the D-vine, a notational simplification is $C_{a_{\ell j}|a_{1j},\ldots,a_{\ell-1,j}} = C_{j-\ell|(j-\ell+1):(j-1)}$ and $C_{j-\ell|(j-\ell+1):(j-1)}(u_{j-\ell}|\boldsymbol{u}_{(j-\ell+1):(j-1)})$ is obtained recursively, similar to the v_j' step in Algorithm 3. In general, after the $q_{\ell j}$ are obtained, a (backward or indirect) step of computing conditional cdf values is needed for possible use in the second argument of conditional quantiles in subsequent steps. With the u arguments of the conditional distributions matching the subscripts, define

$$v_{\ell j} = C_{a_{\ell j}|j,a_{1j}\cdots a_{\ell-1,j}} = C_{a_{\ell j},j;a_{1j}\cdots a_{\ell-1,j}}(C_{a_{\ell j}|a_{1j}\cdots a_{\ell-1,j}} \mid C_{j|a_{1j}\cdots a_{\ell-1,j}}) \qquad (6.5)$$

where the latter equality comes from (6.1). In (6.5), the first subscript indicates the number of elements in the conditioning set and the second subscript indicates the maximum of $\{j, a_{1j},\ldots,a_{\ell j}\}$.

In the second argument for $q_{\ell j}$ in (6.3), the value $C_{a_{\ell j}|a_{1i},\ldots,a_{\ell-1,j}}$ is equivalent to a previous q or v depending on whether $a_{\ell j} = m_{\ell j} = \max_{1 \leq i \leq \ell} a_{ij}$ or not. More precisely,

$$C_{a_{\ell j}|a_{1j},\ldots,a_{\ell-1,j}} = \begin{cases} q_{\ell,a_{\ell j}} & \text{if } a_{\ell j} = m_{\ell j}, \\ v_{\ell-1,m_{\ell j}} & \text{if } a_{\ell j} < m_{\ell j}, \end{cases}$$

where the first case above comes from inverting (6.3) to get

$$q_{\ell+1,j} = C_{j|a_{\ell j};a_{1j},\ldots,a_{\ell-1,j}}\big(q_{\ell j} \mid C_{a_{\ell j}|a_{1j},\ldots,a_{\ell-1,j}}(u_{a_{\ell j}}|u_{a_{1j}},\ldots,u_{a_{\ell-1,j}})\big).$$

Algorithm 16 has details for simulating from the D-vine, and Algorithm 17 from the general R-vine.

Algorithm 16 Simulating from a D-vine in dimension d; notation modification of Joe (2011a).

1: Generate $p_1, \ldots, p_d$ to be independent $U(0, 1)$ random variables.
2: Allocate $d \times d$ arrays $(q_{\ell j})$, $(v_{\ell j})$, but only the upper triangles are used.
3: Let $u_1 \leftarrow p_1$, $u_2 \leftarrow C_{2|1}^{-1}(p_2|p_1)$, $v_{12} \leftarrow C_{1|2}(u_1|u_2)$; $q_{22} \leftarrow p_2$;
4: **for** $j = 3, \ldots, d$: **do**
5:     $q_{jj} \leftarrow p_j$;
6:     **for** $\ell = j - 1, \ldots, 2$: $q_{\ell j} \leftarrow C_{j|j-\ell;j-\ell+1:j-1}^{-1}(q_{\ell+1,j}|v_{\ell-1,j-1})$;
7:     $u_j \leftarrow q_{1j} = C_{j|j-1}^{-1}(q_{2j}|u_{j-1})$;
8:     $v_{1j} \leftarrow C_{j-1|j}(u_{j-1}|u_j)$;
9:     **for** $\ell = 2, \ldots, j - 1$: $v_{\ell j} \leftarrow C_{j-\ell|j;j-\ell+1\cdots j-1}(v_{\ell-1,j-1}|q_{\ell,j})$.
10: **end for**
11: Return $(u_1, \ldots, u_d)$.

For the R-vine algorithm, sometimes the second argument of conditional quantile is a q and sometimes it is a v; for the D-vine, it is always a v.

Algorithm 17 Simulating from an R-vine in dimension d; modification of Dissmann et al. (2013).

1: Input vine array $A = (a_{kj})$.
2: Compute $M = (m_{kj})$ in the upper triangle, where $m_{kj} = \max\{a_{1j}, \ldots, a_{kj}\}$ for $k = 1, \ldots, j-1$, $j = 2, \ldots, d$.
3: Compute the $I = (I_{kj})$ indicator array as in Algorithm 5.
4: Generate $p_1, \ldots, p_d$ to be independent $U(0, 1)$ random variables.
5: Allocate $d \times d$ arrays $(q_{\ell j})$, $(v_{\ell j})$, $(z_{\ell j})$ (initialized to 0), but only the upper triangles are used.
6: $u_1 \leftarrow p_1$, $u_2 \leftarrow C_{2|1}^{-1}(p_2|p_1)$, $q_{22} \leftarrow p_2$; if $(I_{12} = 1)$ then $v_{12} \leftarrow C_{1|2}(u_1|u_2)$.
7: **for** $j = 3, \ldots, d$: **do**
8:     $q_{jj} \leftarrow p_j$;
9:     **for** $\ell = j - 1, \ldots, 2$: **do**
10:         if $(a_{\ell j} = m_{\ell j})$ then $s \leftarrow q_{\ell, a_{\ell j}}$ else $s \leftarrow v_{\ell-1, m_{\ell j}}$;
11:         let $z_{\ell j} \leftarrow s$ (for use in next loop);
12:         $q_{\ell j} \leftarrow C_{j|a_{\ell j};a_{1j}\cdots a_{\ell-1,j}}^{-1}(q_{\ell+1,j}|s)$;
13:     **end for**
14:     $u_j \leftarrow q_{1j} = C_{j|a_{1j}}^{-1}(q_{2j}|u_{a_{1j}})$;
15:     $v_{1j} \leftarrow C_{a_{1j}|j}(u_{a_{1j}}|u_j)$;
16:     **for** $\ell = 2, \ldots, j - 1$: **do**
17:         if $(I_{\ell j} = 1)$ then $v_{\ell j} \leftarrow C_{a_{\ell j}|j;a_{1j}\cdots a_{\ell-1,j}}(z_{\ell j}|q_{\ell j})$.
18:     **end for**
19: **end for**
20: Return $(u_1, \ldots, u_d)$.

To avoid unnecessary calculations of terms in the $(v_{\ell j})$ matrix, Algorithm 5 is used to get a matrix of indicators, and in the second for-loop of Algorithm 17 for the jth variable, $v_{\ell j}$ is evaluated only if $I_{\ell j} = 1$.

For an R-vine for d discrete variables, we assume the variables have been indexed in a vine array A with diagonal $a_{ii} = i$ for $i = 1, \ldots, d$. Let the joint pmf be $f_{1:d}$. The conditional method can be used by adapting Algorithm 7, which computes intermediate probabilities of $f_{1:k}(y_1, \ldots, y_k)$ for $k = 1, \ldots, d$ for a given input vector $y_1, \ldots, y_d$.

There are two approaches to the simulation depending on the number of possible values of $(y_1, \ldots, y_d)$ and on the desired simulation sample size. If the number of distinct d-vectors is finite and small relative to the simulation sample size, then Algorithm 7 can be used to compute the entire probability distribution from which the cdfs $F_1, F_{2|1}, \ldots, F_{k|1:(k-1)}, \ldots, F_{d|1:(d-1)}$ can be obtained. Then with $p_1, \ldots, p_d$ independent $U(0, 1)$, the conditional method in Section 6.9.1 can be applied even if the $F_{k|1:(k-1)}$ are cdfs of discrete random variables.

Alternatively the following is an approach for generating one d-vector.

- Generate $p_1, \ldots, p_d$ to be independent $U(0, 1)$ random variables.
- Let $y_1 \leftarrow F_1^{-1}(p_1)$, and let $y_2 \leftarrow F_{2|1}^{-1}(p_2|y_1)$ where $F_{2|1}$ is obtained from the bivariate distribution $C_{12}(F_1, F_2)$.
- Via steps in Algorithm 7, compute the probability masses $f_{3|12}(\cdot|y_1, y_2)$ to get y_3 such that $\sum_{y < y_3} f_{3|12}(y|y_1, y_2) < p_3 \leq \sum_{y \leq y_3} f_{3|12}(y|y_1, y_2)$.
- For $k = 4, \ldots, d$, continue in this way to get y_k such that $\sum_{y < y_k} f_{k|1:(k-1)}(y|\boldsymbol{y}_{1:(k-1)}) < p_k \leq \sum_{y \leq y_k} f_{k|1:(k-1)}(y|\boldsymbol{y}_{1:(k-1)})$.
- Return $(y_1, \ldots, y_d)$.

6.14.2 Simulation from truncated vines and factor copulas

For simulation from a truncated R-vine, Algorithm 17 can be modified slightly into Algorithm 18. Simplifications of Algorithm 18 can be made for truncated C-vines, truncated D-vines and 2-truncated R-vines; these are shown respectively in Algorithms 19, 20 and 21.

Algorithm 18 Simulating from a truncated R-vine in dimension d.

1: Input truncated level t (so that $C_{j|a_{\ell j};a_{1j}\cdots a_{\ell-1,j}}(u|v) \equiv u$ for $\ell > t$).
2: Input vine array $A = (a_{kj})$ (up to and including row t).
3: Compute $M = (m_{kj})$ in the upper triangle (up to row t), where $m_{kj} = \max\{a_{1j}, \ldots, a_{kj}\}$ for $k = 1, \ldots, \min\{j-1, t\}$, $j = 2, \ldots, d$.
4: Compute the $I = (I_{kj})$ indicator array as in Algorithm 5, up to row t.
5: Generate $p_1, \ldots, p_d$ to be independent $U(0, 1)$ random variables.
6: Allocate $d \times d$ arrays $(q_{\ell j}), (v_{\ell j}), (z_{\ell j})$ (initialized to 0), but only the upper triangles are used. (Only t rows are needed).
7: $u_1 \leftarrow p_1$, $u_2 \leftarrow C_{2|1}^{-1}(p_2|p_1)$, $q_{22} \leftarrow p_2$; if $(I_{12} = 1)$ then $v_{12} \leftarrow C_{1|2}(u_1|u_2)$.
8: **for** $j = 3, \ldots, d$: **do**
9:    let $t' \leftarrow \min\{j-1, t\}$, set $q_{t'+1,j} \leftarrow p_j$;
10:    **for** $\ell = \min\{j-1, t\}, \ldots, 2$: **do**
11:       if $(a_{\ell j} = m_{\ell j})$ then $s \leftarrow q_{\ell, a_{\ell j}}$ else $s \leftarrow v_{\ell-1, m_{\ell j}}$;
12:       let $z_{\ell j} \leftarrow s$ (for use in next loop);
13:       $q_{\ell j} \leftarrow C_{j|a_{\ell j};a_{1j}\cdots a_{\ell-1,j}}^{-1}(q_{\ell+1,j}|s)$;
14:    **end for**
15:    $u_j \leftarrow q_{1j} = C_{j|a_{1j}}^{-1}(q_{2j}|u_{a_{1j}})$;
16:    $v_{1j} \leftarrow C_{a_{1j}|j}(u_{a_{1j}}|u_j)$;
17:    **for** $\ell = 2, \ldots, \min\{j-1, t\}$: **do**
18:       if $(I_{\ell j} = 1)$ then $v_{\ell j} \leftarrow C_{a_{\ell j}|j;a_{1j}\cdots a_{\ell-1,j}}(z_{\ell j}|q_{\ell j})$.
19:    **end for**
20: **end for**
21: Return $(u_1, \ldots, u_d)$.

The algorithms for the 1-factor copula, 2-factor copula, bi-factor copula, and nested factor copula models are special cases of 1-truncated and 2-truncated C-vines/R-vines with latent variables.

Algorithm 19 Simulating from a truncated C-vine in dimension d.

1: Input truncated level t.
2: Generate $p_1, \ldots, p_d$ to be U$(0, 1)$ random variables.
3: Let $u_1 \leftarrow p_1$, $u_2 \leftarrow C_{2|1}^{-1}(p_2|p_1)$.
4: **for** $j = 3, \ldots, d$: **do**
5:     let $q \leftarrow p_j$;
6:     for $\ell = \min\{j-1, t\}, j-2, \ldots, 1$: $q \leftarrow C_{j|\ell;1:(\ell-1)}^{-1}(q|p_\ell)$;
7:     let $u_j \leftarrow q$.
8: **end for**
9: Return $(u_1, \ldots, u_d)$.

Algorithm 20 Simulating from a truncated D-vine in dimension d.

1: Input truncated level t.
2: Generate $p_1, \ldots, p_d$ to be U$(0, 1)$ random variables.
3: Allocate $d \times d$ arrays $q_{\ell j}, v_{\ell j}$ (initialized to 0),
4: $u_1 \leftarrow p_1$, $u_2 \leftarrow C_{2|1}^{-1}(p_2|p_1)$, $q_{22} \leftarrow p_2$; $v_{12} \leftarrow C_{1|2}(u_1|u_2)$.
5: **for** $j = 3, \ldots, d$: **do**
6:     let $t' \leftarrow \min\{j-1, t\}$, set $q_{t'+1,j} \leftarrow p_j$;
7:     for $\ell = \min\{j-1, t\}, \ldots, 2$: $q_{\ell j} \leftarrow C_{j|j-\ell;j-\ell+1:j-1}^{-1}(q_{\ell+1,j}|v_{\ell-1,j-1})$;
8:     $u_j \leftarrow q_{1j} = C_{j|j-1}^{-1}(q_{2j}|u_{j-1})$;
9:     $v_{1j} \leftarrow C_{j-1|j}(u_{j-1}|u_j)$;
10:     for $\ell = 2, \ldots, \min\{j-1, t\}$: $v_{\ell j} \leftarrow C_{j-\ell|j;j-\ell+1:j-1}(v_{\ell-1,j-1}|q_{\ell,j})$.
11: **end for**
12: Return $(u_1, \ldots, u_d)$.

Algorithm 21 Simulating from a 2-truncated R-vine in dimension d.

1: Input vine array $A = (a_{kj})$ (up to row $t = 2$).
2: Compute $M = (m_{kj})$ in the upper triangle (up to row t), where $m_{kj} = \max\{a_{1j}, \ldots, a_{kj}\}$ for $k = 1, \ldots, \min\{j-1, t\}$, $j = 2, \ldots, d$.
3: Compute the $I = (I_{kj})$ indicator array as in Algorithm 5, up to row t.
4: Generate $p_1, \ldots, p_d$ to be independent U$(0, 1)$ random variables.
5: $u_1 \leftarrow p_1$, $u_2 \leftarrow C_{2|1}^{-1}(p_2|p_1)$, $q_{22} \leftarrow p_2$; if $(I_{12} = 1)$ then $v_{12} \leftarrow C_{1|2}(u_1|u_2)$.
6: **for** $j = 3, \ldots, d$: **do**
7:     if $(a_{2j} = m_{2j})$ then $s \leftarrow q_{2,a_{2j}}$ else $s \leftarrow v_{1,m_{2j}}$;
8:     $q_{2j} \leftarrow C_{j|a_{2j};a_{1j}}^{-1}(p_j|s)$;
9:     $u_j \leftarrow C_{j|a_{1j}}^{-1}(q_{2j}|u_{a_{1j}})$;
10:     $v_{1j} \leftarrow C_{a_{1j}|j}(u_{a_{1j}}|u_j)$.
11: **end for**
12: Return $(u_1, \ldots, u_d)$.

For the 1-factor copula model, let V be the latent U$(0, 1)$ variable and let $U_1, \ldots, U_d$ be the observed variables. This can be represented as a 1-truncated C-vine with the edges of the tree denoted as $[V, U_1], \ldots, [V, U_d]$.

For the 2-factor copula model, let V_1, V_2 be the independent latent $U(0,1)$ variables and let $U_1, \ldots, U_d$ be the observed variables. This can be represented as a 2-truncated C-vine with the edges of tree $\mathcal{T}_1$ denoted as $[V_1, V_2], [V_1, U_1], \ldots, [V_1, U_d]$ and the edges of $\mathcal{T}_2$ denoted as $[V_2, U_1|V_1], \ldots, [V_2, U_d|V_1]$. A vine representation of the 2-factor model is shown in Figure 6.1.

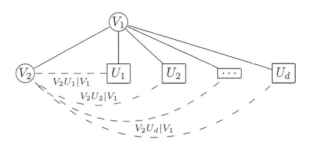

Figure 6.1 *2-factor model with d variables: latent variables are shown in circles and observed variables are shown in rectangles; solid lines are used for edges of $\mathcal{T}_1$ and dashed lines are used for edges of $\mathcal{T}_2$.*

By incorporating the latent variables into Algorithm 19, the algorithms for the 1-factor and 2-factor copulas are given respectively in Algorithms 22 and 23,

---

Algorithm 22 Simulating from a 1-factor copula model.

1: Generate $v, p_1, \ldots, p_d$ to be independent $U(0,1)$ random variables.
2: $u_j \leftarrow C_{j|V}^{-1}(p_j|v)$, for $j = 1, \ldots, d$.
3: Return $(u_1, \ldots, u_d)$.

Algorithm 23 Simulating from a 2-factor copula model.

1: Generate $v_1, v_2, p_1, \ldots, p_d$ to be independent $U(0,1)$ random variables.
2: **for** $j = 1, \ldots, d$: **do**
3: $\quad q_{2j} \leftarrow C_{j|V_2;V_1}^{-1}(p_j|v_2)$;
4: $\quad u_j \leftarrow C_{j|V_1}^{-1}(q_{2j}|v_1)$.
5: **end for**
6: Return $(u_1, \ldots, u_d)$.

For the bi-factor copula model with m groups of variables, let $V_0, V_1, \ldots, V_m$ be the independent latent $U(0,1)$ variables and let $U_1, \ldots, U_d$ be the observed variables. This can be represented as a 2-truncated R-vine with the edges of tree $\mathcal{T}_1$ denoted as $[V_0, V_1], \ldots, [V_0, V_m], [V_0, U_1], \ldots, [V_0, U_d]$, and the $d+m-1$ edges of $\mathcal{T}_2$ denoted as $[V_1, V_g|V_0]$ for $g = 2, \ldots, m$ and $[V_g, U_j|V_0]$ for all U_j belonging to group g with $g = 1, \ldots, m$. A vine representation of the bi-factor model is shown in Figure 6.2. By incorporating the latent variables into Algorithm 21, the algorithm for the bi-factor copula is given in Algorithm 24.

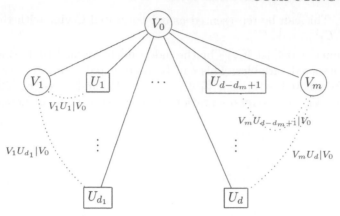

Figure 6.2 *Bi-factor model with m groups, d_j variables in the jth group with $U_1, \ldots, U_{d_1}$ in group 1, up to $U_{d-d_m+1}, \ldots, U_d$ in group m; latent variables are shown in circles and observed variables are shown in rectangles.*

Algorithm 24 Simulating from a bi-factor copula model in dimension d with m groups.

1: Generate $v_0, v_1, \ldots, v_m, p_1, \ldots, p_d$ to be independent $U(0,1)$ random variables.
2: **for** $j = 1, \ldots, d$: **do**
3:     determine the group g for variable j;
4:         $q_{2j} \leftarrow C_{j|V_g;V_0}^{-1}(p_j|v_g)$;
5:         $u_j \leftarrow C_{j|V_0}^{-1}(q_{2j}|v_0)$.
6: **end for**
7: Return $(u_1, \ldots, u_d)$.

For the nested factor copula model with m groups of variables, let $V_0, V_1, \ldots, V_m$ be the dependent latent $U(0,1)$ variables and let $U_1, \ldots, U_d$ be the observed variables. This can be represented as a 1-truncated R-vine with the edges of tree denoted as $[V_0, V_1], \ldots, [V_0, V_m]$ and $[V_g, U_j]$ for all variables U_j in group g, $g = 1, \ldots, m$. By incorporating the latent variables into Algorithm 18 or by considering a 1-truncated vine as a Markov tree, the algorithm for the nested factor copula is given in Algorithm 25.

Algorithm 25 Simulating from a nested factor copula in dimension d with m groups.

1: Generate $v_0, p_1', \ldots, p_m', p_1, \ldots, p_d$ to be independent $U(0,1)$ random variables.
2: **for** $g = 1, \ldots, m$: **do**
3:     $v_g \leftarrow C_{V_g|V_0}^{-1}(p_g'|v_0)$;
4: **end for**
5: **for** $j = 1, \ldots, d$: **do**
6:     determine the group g for variable j;
7:         $u_j \leftarrow C_{j|V_g}^{-1}(p_j|v_g)$;
8: **end for**
9: Return $(u_1, \ldots, u_d)$.

For these factor models applied to discrete or ordinal variables, one can convert from a vector of dependent $U(0,1)$ random variables to the discrete random variables via the univariate inverse cdfs.

6.15 Partial correlations and vines

In this section, we indicate how to compute all partial correlations from a correlation matrix. Also we provide algorithms and equations to map from one of the three equivalent forms to another with a specified vine array $A = (a_{\ell j})_{1 \leq \ell < j \leq d}$ that has $1:d$ on the diagonal; the three forms are (a) correlation matrix $\boldsymbol{R} = (\rho_{jk})_{1 \leq j,k \leq d}$, (b) partial correlation vine $\{\rho_{ja_{\ell j};a_{1j}\cdots a_{\ell-1,j}}\}$, (c) regression coefficients $\{\phi_{j\ell} : 1 \leq \ell < j \leq d\}$.

To calculate all possible partial correlations (conditioned on 1 to $d-2$ variables) from a $d \times d$ correlation matrix $\boldsymbol{R} = (\rho_{jk})$, Algorithm 26 makes use of the recursion:

$$\rho_{jk;m\cup T} = \frac{\rho_{jk;T} - \rho_{jm;T}\rho_{km;T}}{\sqrt{(1-\rho_{jm;T}^2)(1-\rho_{km;T}^2)}}, \tag{6.6}$$

where j,k,m are distinct and T is a subset of $\{1,\ldots,d\}\backslash\{j,k,m\}$. See Anderson (1958) for a derivation of this recursion. This algorithm is useful if one wants to enumerate through all partial correlation vines for comparisons.

Algorithm 26 Recursive computations of all partial correlations.

1: Input $\boldsymbol{R} = (\rho_{jk})_{1 \leq j < k \leq d}$.

2: Initialize P as a $d \times d \times (2^d - 4)$ array of 0s, where the third dimension has an indexing for the conditioning variables; the order is $1, 2, 12, 3, 13, 23, 123, 4, \ldots, 12\cdots d$ (third dimension has value m for subset S which is the decimal for the binary vector $(I(d \in S), \ldots, I(2 \in S), I(1 \in S))$).

3: **for** $k = 1, \ldots, d$: (conditioning on single variable k) **do**

4:     set $m \leftarrow 2^k$;

5:     **for** $i, j \neq k$: **do**

6:         $p_{ijm} \leftarrow \rho_{ij;k} = (\rho_{ij} - \rho_{ik}\rho_{jk})/[(1-\rho_{ik}^2)(1-\rho_{jk}^2)]^{1/2}$;

7:     **end for**

8: **end for**

9: **for** $m \in \{1, \ldots, 2^d - 4\}$ that is not a power of 2: (conditioning on two or more variables) **do**

10:     let the subset of conditioning variables be S and let $k \leftarrow \max S$; let q be the decimal index for the conditioning set $T = S\backslash\{k\}$;

11:     **for** $i, j \notin S$: **do**

12:         $p_{ijm} \leftarrow \rho_{ij;S} = (\rho_{ij;T} - \rho_{ik;T}\rho_{jk;T})/[(1-\rho_{ik;T}^2)(1-\rho_{jk;T}^2)]^{1/2}$

            $= (p_{ijq} - p_{ikq}p_{jkq})/[(1-p_{ikq}^2)(1-p_{jkq}^2)]^{1/2}$.

13:     **end for**

14: **end for**

15: Return $P = (\rho_{ij;S})_{1 \leq i,j \leq d, S \subset 1:d}$.

If $(Z_1, \ldots, Z_d) \sim N_d(\boldsymbol{0}, \boldsymbol{R})$ and the partial correlation vine is based on vine array $A = (a_{\ell j}; 1 \leq \ell \leq j)$, then there are d i.i.d. $N(0,1)$ random variables $\epsilon_1, \ldots, \epsilon_d$ and constants $(\phi_{j\ell}, 1 \leq \ell < j \leq d)$ and constants $\psi_1, \ldots, \psi_d$ such that a stochastic representation is

$$Z_j = \sum_{\ell=1}^{j-1} \phi_{j\ell} Z_{a_{\ell j}} + \psi_j \epsilon_j, \tag{6.7}$$

with $\epsilon_j \perp\!\!\!\perp (Z_1, \ldots, Z_{j-1})$. Note that $\psi_1 = 1$ and ψ_j satisfies

$$\psi_j^2 = 1 - \sum_{\ell=1}^{j-1}\sum_{k=1}^{j-1} \phi_{j\ell}\phi_{jk}\rho_{a_{\ell j}a_{kj}}, \quad j = 2, \ldots, d. \tag{6.8}$$

This stochastic representation with the regression coefficients $\phi_{j\ell}$ is useful for simulation from a multivariate Gaussian distribution with a structured correlation matrix that is a *truncated* partial correlation vine (see Example 6.3).

To map from the correlation matrix to the set $(\phi_{j\ell})$, equations for $1 \le k < j \le d$ are

$$\rho_{jk} = \mathrm{Cov}(Z_j, Z_k) = \sum_{\ell=1}^{j-1} \phi_{j\ell} \, \mathrm{Cov}(Z_{a_{\ell j}}, Z_k) = \sum_{\ell=1}^{j-1} \phi_{j\ell} \rho_{a_{\ell j}, k}. \tag{6.9}$$

If the set of $\binom{d}{2}$ $\phi_{j\ell}$'s is given, the above is a mapping to the $\binom{d}{2}$ ρ_{jk}'s. If $\boldsymbol{R} = (\rho_{jk})$ is given, then with j fixed sequentially from $2, \ldots, d$, the equations in (6.9) for $1 \le k < j$ form a linear system of $j - 1$ equations in $\phi_{j1}, \ldots, \phi_{j,j-1}$, so that the regression coefficients in the equation for Z_j can be solved for. For the C-vine, (6.9) is

$$\rho_{jk} = \sum_{\ell=1}^{j-1} \phi_{j\ell} \rho_{\ell k},$$

with $\phi_{j\ell}$ being the coefficient of Z_ℓ, and for the D-vine, (6.9) is

$$\rho_{jk} = \sum_{\ell=1}^{j-1} \phi_{j\ell} \rho_{j-\ell, k},$$

with $\phi_{j\ell}$ being the coefficient of $Z_{j-\ell}$.

For a C-vine, the conversion from correlations to partial correlations makes use of the recursion (6.6) and is given in Algorithm 27. This is the only vine where matrix inversion can be avoided. As is usual for algorithms with vines, it is simpler for the boundary case of the canonical C-vine. The algorithm for general R-vines is given in Algorithm 28; it makes use of the properties of conditional distributions for multivariate Gaussian (Section 2.6).

Algorithm 27 C-vine: correlations to partial correlations, and vice versa.

1: Input $\boldsymbol{R} = (\rho_{jk})_{1 \le j, k \le d}$.
2: Initialize $d \times d$ array $\boldsymbol{P} = (p_{jk})$ with the upper triangle of $\boldsymbol{R}$.
3: **for** $t = 1, \ldots, d-2$: (t is tree level minus 1) **do**
4:     **for** $j = t+1, \ldots, d-1$: **do**
5:         **for** $k = j+1, \ldots, d$: **do**
6:             $p_{jk} \leftarrow (p_{jk} - p_{tj} p_{tk})/[(1 - p_{tj}^2)(1 - p_{tk}^2)]^{1/2}$.
7:         **end for**
8:     **end for**
9: **end for**
10: Return the partial correlations in an array $\boldsymbol{P}$ with $p_{jk} = \rho_{jk;1:(j-1)}$.

11:
12: Input $d \times d$ array $\boldsymbol{P} = (p_{jk})$, where $p_{jk} = \rho_{jk;1:(j-1)}$ for $j < k$.
13: $\rho_{1k} = \rho_{k1} \leftarrow p_{1k}$ for $k = 2, \ldots, d$.
14: **for** $j = 2, \ldots, d-1$: **do**
15:     **for** $k = j+1, \ldots, d$: **do**
16:         $q \leftarrow p_{j,k}$;
17:         **for** $t = j-1, \ldots, 1$: **do**
18:             $q \leftarrow p_{tj} p_{tk} + q[(1 - p_{tj}^2)(1 - p_{tk}^2)]^{1/2}$;
19:         **end for**
20:         $\rho_{jk} = \rho_{kj} \leftarrow q$.
21:     **end for**
22: **end for**
23: Return matrix $\boldsymbol{R} = (\rho_{jk})$ with diagonal of 1s.

Algorithm 28 R-vine: correlations to partial correlations, and vice versa.

1: Input $\boldsymbol{R} = (\rho_{jk})_{1 \le j,k \le d}$ and vine array $A = (a_{kj})_{1 \le k < j \le d}$.

2: Initialize $d \times d$ array $\boldsymbol{P} = (p_{\ell j})$ with first row $p_{1j} = \rho_{a_{1j}j}$ ($j = 2, \ldots, d$) and second row
$p_{2j} = (\rho_{a_{2j}j} - \rho_{a_{1j}j}\rho_{a_{1j}a_{2j}})/[1 - \rho_{a_{1j}j}^2)(1 - \rho_{a_{1j}a_{2j}})^2]^{1/2}$ ($j = 3, \ldots, d$).

3: **for** $\ell = 3, \ldots, d - 1$: (ℓ is tree level) **do**

4:    **for** $j = \ell + 1, \ldots, d$: **do**

5:       let $\begin{pmatrix} \boldsymbol{\Sigma}_{11} & \boldsymbol{\Sigma}_{12} \\ \boldsymbol{\Sigma}_{21} & \boldsymbol{\Sigma}_{22} \end{pmatrix}$ be the submatrix of $\boldsymbol{R}$ with indices $(a_{1j}, \ldots, a_{\ell-1,j}), (a_{\ell j}, j)$;

6:       let $\boldsymbol{\Omega} = \boldsymbol{\Sigma}_{21}\boldsymbol{\Sigma}_{11}^{-1}\boldsymbol{\Sigma}_{12}$ be a 2×2 matrix with entries $\omega_{11}, \omega_{12} = \omega_{21}, \omega_{22}$;

7:       set $p_{\ell j} \leftarrow (\rho_{a_{\ell j},j} - \omega_{12})/[(1 - \omega_{11})(1 - \omega_{22})]^{1/2}$.

8:    **end for**

9: **end for**

10: Return the partial correlations in an array $\boldsymbol{P}$ with $p_{\ell j} = \rho_{a_{\ell j}j;a_{1j},\ldots,a_{\ell-1,j}}$, where row ℓ
has the partial correlations for tree ℓ for $j = \ell + 1, \ldots, d$ and $\ell = 1, \ldots, d - 1$.

11:

12: Input vine array $A = (a_{kj})_{1 \le k < j \le d}$ and $d \times d$ array $\boldsymbol{P} = (p_{\ell j})$, where $p_{\ell j} = \rho_{a_{\ell j}j;a_{1j},\ldots,a_{\ell-1,j}}$ for $j = \ell + 1, \ldots, d$ and $\ell = 1, \ldots, d - 1$.

13: Set $\rho_{a_{1j}j} = \rho_{ja_{1j}} = p_{1j}$ ($j = 2, \ldots, d$).

14: Set $\rho_{a_{2j}j} = \rho_{ja_{2j}} = \rho_{a_{1j}j}\rho_{a_{1j}a_{2j}} + p_{2j}[1 - \rho_{a_{1j}j}^2)(1 - \rho_{a_{1j}a_{2j}})^2]^{1/2}$ ($j = 3, \ldots, d$).

15: **for** $\ell = 3, \ldots, d - 1$: (ℓ is tree level) **do**

16:    **for** $j = \ell + 1, \ldots, d$: **do**

17:       let $\begin{pmatrix} \boldsymbol{\Sigma}_{11} & \boldsymbol{\Sigma}_{12} \\ \boldsymbol{\Sigma}_{21} & \boldsymbol{\Sigma}_{22} \end{pmatrix}$ be the submatrix of $\boldsymbol{R}$ with indices $(a_{1j}, \ldots, a_{\ell-1,j}), (a_{\ell,j}, j)$;

18:       let $\boldsymbol{\Omega} = \boldsymbol{\Sigma}_{21}\boldsymbol{\Sigma}_{11}^{-1}\boldsymbol{\Sigma}_{12}$ be a 2×2 matrix with entries $\omega_{11}, \omega_{12} = \omega_{21}, \omega_{22}$;

19:       let $q \leftarrow p_{\ell j}[(1 - \omega_{11})(1 - \omega_{22})]^{1/2}$;

20:       set $\rho_{a_{\ell j},j} = \rho_{j,a_{\ell j}} \leftarrow q + \omega_{12}$.

21:    **end for**

22: **end for**

23: Return matrix $\boldsymbol{R} = (\rho_{ij})$ with diagonal of 1s.

Example 6.3 (Transform from partial correlations to correlations). The results of the above algorithms are shown for a truncated C-vine and a truncated D-vine. Suppose $d = 6$ and correlations are $(0.3, 0.4, 0.5, 0.6, 0.7)$ for the first tree, and partial correlations $(0.1, 0.2, 0.3, 0.4)$ for the second tree and partial correlations of 0s for the third to fifth trees.

(a) For the (truncated) C-vine in standard order, the correlation matrix is

$$\boldsymbol{R} = \begin{pmatrix} 1.0 & 0.300 & 0.400 & 0.500 & 0.600 & 0.700 \\ 0.3 & 1.000 & 0.207 & 0.315 & 0.409 & 0.482 \\ 0.4 & 0.207 & 1.000 & 0.216 & 0.262 & 0.306 \\ 0.5 & 0.315 & 0.216 & 1.000 & 0.342 & 0.399 \\ 0.6 & 0.409 & 0.262 & 0.342 & 1.000 & 0.489 \\ 0.7 & 0.482 & 0.306 & 0.399 & 0.489 & 1.000 \end{pmatrix}$$

and the stochastic representation from (6.7) is:

$$Z_1 = \epsilon_1$$
$$Z_2 = 0.3Z_1 + \psi_2\epsilon_2$$
$$Z_3 = 0.371Z_1 + 0.096Z_2 + \psi_3\epsilon_3$$
$$Z_4 = 0.446Z_1 + 0.182Z_2 + \psi_4\epsilon_4$$
$$Z_5 = 0.525Z_1 + 0.252Z_2 + \psi_5\epsilon_5$$
$$Z_6 = 0.610Z_1 + 0.299Z_2 + \psi_6\epsilon_6,$$

where ψ_j for $j = 2, \ldots, 6$ can be obtained as (6.8).

(b) For the (truncated) D-vine in standard order, the correlation matrix with the above partial correlations is

$$
R = \begin{pmatrix}
1.000 & 0.300 & 0.207 & 0.145 & 0.124 & 0.112 \\
0.300 & 1.000 & 0.400 & 0.359 & 0.276 & 0.262 \\
0.207 & 0.400 & 1.000 & 0.500 & 0.508 & 0.425 \\
0.145 & 0.359 & 0.500 & 1.000 & 0.600 & 0.649 \\
0.124 & 0.276 & 0.508 & 0.600 & 1.000 & 0.700 \\
0.112 & 0.262 & 0.425 & 0.649 & 0.700 & 1.000
\end{pmatrix}
$$

and the stochastic representation from (6.7) is:

$$
\begin{aligned}
Z_1 &= \epsilon_1 \\
Z_2 &= 0.3 Z_1 + \psi_2 \epsilon_2 \\
Z_3 &= 0.371 Z_2 + 0.096 Z_1 + \psi_3 \epsilon_3 \\
Z_4 &= 0.424 Z_3 + 0.189 Z_2 + \psi_4 \epsilon_4 \\
Z_5 &= 0.461 Z_4 + 0.277 Z_3 + \psi_5 \epsilon_5 \\
Z_6 &= 0.486 Z_5 + 0.357 Z_4 + \psi_6 \epsilon_6.
\end{aligned}
$$

$\square$

Next, we indicate how to modify Algorithm 28 to convert partial correlations to correlations for a t-truncated vine; that is, the simplification when $p_{\ell j} = 0$ for $\ell > t$. This is relevant for latent variable models with a truncated vine.

Let $A = (a_{\ell j})$ be a vine array. For a t-truncated (partial correlation) vine, A just needs to be specified for rows $\ell = 1, \ldots, t$. The completion of the vine array in rows $t+1, \ldots, d-1$ does not matter if the partial correlations are 0 in trees $\mathcal{T}_{t+1}, \ldots, \mathcal{T}_{d-1}$.

For the 1-truncated partial correlation vine, there are $d-1$ correlations in tree $\mathcal{T}_1$: $\rho_{a_{1j},j}$, $j = 2, \ldots, d$. Then there is a stochastic representation

$$
Z_j = \phi_j Z_{a_{1j}} + \psi_j \epsilon_j, \quad \phi_j = \rho_{a_{1j},j}, \ \psi_j = \sqrt{1 - \phi_j^2}, \tag{6.10}
$$

for $j = 2, \ldots, d$, where ϵ_j are i.i.d. $N(0,1)$ random variables and $\epsilon_j \perp\!\!\!\perp (Z_1, \ldots, Z_{j-1})$. By recursion,

$$
\operatorname{Cor}(Z_i, Z_j) = \phi_j \operatorname{Cor}(Z_i, Z_{a_{1j}}), \quad 1 \le i < j \le d,
$$

for any correlation not in tree $\mathcal{T}_1$.

For the 2-truncated partial correlation vine, there are also $d-2$ partial correlations in tree $\mathcal{T}_2$: $\rho_{a_{2j},j;a_{1j}}$, $j = 3, \ldots, d$. There is a stochastic representation with (6.10) for $j = 2$ and

$$
Z_j = \phi_{j1} Z_{a_{1j}} + \phi_{j2} Z_{a_{2j}} + \psi_j \epsilon_j, \tag{6.11}
$$

for $j = 3, \ldots, d$, where ϵ_j are i.i.d. $N(0,1)$ random variables and $\epsilon_j \perp\!\!\!\perp (Z_1, \ldots, Z_{j-1})$. For the correlation matrix, $d-1$ correlations are completed from tree $\mathcal{T}_1$, and then from tree $\mathcal{T}_2$, one can obtain:

$$
\operatorname{Cor}(Z_{a_{2j}}, Z_j) = \rho_{a_{1j}j} \rho_{a_{1j}a_{2j}} + \rho_{a_{2j}j;a_{1j}} \sqrt{(1 - \rho_{a_{1j}j}^2)(1 - \rho_{a_{1j}a_{2j}}^2)}.
$$

For $j \ge 3$, the equations for ϕ_{j1}, ϕ_{j2} come from solving

$$
\begin{aligned}
\operatorname{Cov}(Z_{a_{2j}}, Z_j \mid Z_{a_{1j}}) &= \phi_{j2} \operatorname{Var}(Z_{a_{2j}} \mid Z_{a_{1j}}), \\
\operatorname{Cov}(Z_{a_{1j}}, Z_j) &= \phi_{j1} + \phi_{j2} \operatorname{Cov}(Z_{a_{1j}}, Z_{a_{2j}}),
\end{aligned}
$$

leading to

$$\phi_{j2} = \rho_{ja_{2j};a_{1j}} \sqrt{(1 - \rho_{a_{1j}j}^2)/(1 - \rho_{a_{1j}a_{2j}}^2)}, \quad \phi_{j1} = \rho_{a_{1j}j} - \phi_{j2}\rho_{a_{1j}a_{2j}}.$$

Then by recursion,

$$\mathrm{Cor}(Z_i, Z_j) = \phi_{j1}\, \mathrm{Cor}(Z_i, Z_{a_{1j}}) + \phi_{j2}\, \mathrm{Cor}(Z_i, Z_{a_{2j}}), \quad 1 \le i < j \le d,$$

for any correlation that doesn't come directly from trees $\mathcal{T}_1, \mathcal{T}_2$.

For the 3-truncated partial correlation vine, there are also $d - 3$ partial correlations in tree $\mathcal{T}_3$: $\rho_{a_{3j},j;a_{1j}a_{2j}}$, $j = 4, \ldots, d$. There is a stochastic representation with (6.10) for $j = 2$, (6.11) for $j = 3$, and

$$Z_j = \phi_{j1}Z_{a_{1j}} + \phi_{j2}Z_{a_{2j}} + \phi_{j3}Z_{a_{3j}} + \psi_j \epsilon_j,$$

for $j = 4, \ldots, d$, where ϵ_j are i.i.d. N$(0,1)$ random variables and $\epsilon_j \perp\!\!\!\perp (Z_1, \ldots, Z_{j-1})$. For the correlation matrix, $d - 1$ correlations are completed from tree $\mathcal{T}_1, \mathcal{T}_2$ as given above, and then from tree $\mathcal{T}_3$, one can obtain: $\mathrm{Cor}(Z_{a_{3j}}, Z_j)$ from $\rho_{a_{3j},j;a_{1j}a_{2j}}$ and the remaining five correlations from pairs among variables $Z_{a_{1j}}, Z_{a_{2j}}, Z_{a_{3j}}, Z_j$ (see Algorithm 28). For $j \ge 4$, the equations for $\phi_{j1}, \phi_{j2}, \phi_{j3}$ come from solving

$$\mathrm{Cov}(Z_{a_{3j}}, Z_j | Z_{a_{1j}}, Z_{a_{2j}}) = \phi_{j3}\, \mathrm{Var}(Z_{a_{3j}} | Z_{a_{1j}}, Z_{a_{2j}}),$$
$$\mathrm{Cov}(Z_{a_{2j}}, Z_j | Z_{a_{1j}}) = \phi_{j2}\, \mathrm{Var}(Z_{a_{2j}} | Z_{a_{1j}}) + \phi_{j3}\, \mathrm{Cov}(Z_{a_{2j}}, Z_{a_{3j}} | Z_{a_{1j}}),$$
$$\mathrm{Cov}(Z_{a_{1j}}, Z_j) = \phi_{j1} + \phi_{j2}\, \mathrm{Cov}(Z_{a_{1j}}, Z_{a_{2j}}) + \phi_{j3}\, \mathrm{Cov}(Z_{a_{1j}}, Z_{a_{3j}}),$$

leading to

$$\phi_{j3} = \left\{ \frac{1 - \rho_{a_{1j}a_{2j}}^2 - \rho_{a_{1j}j}^2 - \rho_{a_{2j}j}^2 + 2\rho_{a_{1j}a_{2j}}\rho_{a_{1j}j}\rho_{a_{2j}j}}{1 - \rho_{a_{1j}a_{2j}}^2 - \rho_{a_{1j}a_{3j}}^2 - \rho_{a_{2j}a_{3j}}^2 + 2\rho_{a_{1j}a_{2j}}\rho_{a_{1j}a_{3j}}\rho_{a_{2j}a_{3j}}} \right\}^{1/2},$$

$$\phi_{j2} = \frac{\rho_{ja_{2j};a_{1j}} \sqrt{1 - \rho_{a_{1j}j}^2} - \phi_{j3}\rho_{a_{3j}a_{2j};a_{1j}} \sqrt{1 - \rho_{a_{1j}a_{3j}}^2}}{\sqrt{1 - \rho_{a_{1j}a_{2j}}^2}},$$

$$\phi_{j1} = \rho_{a_{1j}j} - \phi_{j2}\rho_{a_{1j}a_{2j}} - \phi_{j3}\rho_{a_{1j}a_{3j}}.$$

Then by recursion,

$$\mathrm{Cor}(Z_i, Z_j) = \phi_{j1}\, \mathrm{Cor}(Z_i, Z_{a_{1j}}) + \phi_{j2}\, \mathrm{Cor}(Z_i, Z_{a_{2j}}), \quad 1 \le i < j \le d,$$

for any correlation that doesn't come directly from trees $\mathcal{T}_1, \mathcal{T}_2, \mathcal{T}_3$.

For t-truncation with $t \ge 3$, the pattern extends, except in general, it is simpler to use (6.9) to solve for the $\phi_{j\ell}$, instead of using equations based on partial covariances. Pseudocode for getting the correlation matrix for the t-truncated partial correlation vines with $t \ge 3$ is given in Algorithm 29.

Algorithm 29 Truncated R-vine: partial correlations to correlations and regression coefficients.

1: Input t=truncation level, vine array $A = (a_{kj})_{1 \le k < j \le d}$ and $d \times d$ array $P = (p_{\ell j})$, where $p_{\ell j} = \rho_{a_{\ell j} j; a_{1j}, \dots, a_{\ell-1,j}}$ for $j = \ell+1, \dots, d$. [Only the first t rows of A, P are used.]

2: The main steps, based on the second part of Algorithm 28 together with solving equations in (6.9), are the following: (i) obtain all possible correlations from the first t trees; (ii) obtain the regression coefficients $\phi_{j\ell}$, $j = 1, \dots, d$, $\ell = 1, \dots, t$; (iii) obtain the remaining correlations based on the $\phi_{j\ell}$.

3: Initialize $R = (\rho_{ij})_{1 \le i, j \le d}$ to a matrix of -2; set the diagonal to 1.

4: (i) Set $\rho_{a_1 j} = \rho_{j a_1} = p_{1j}$ $(j = 2, \dots, d)$;

5: set $\rho_{a_2 j} = \rho_{j a_2} = \rho_{a_1 j} \rho_{a_1 a_2} + p_{2j} [1 - \rho_{a_1 j}^2)(1 - \rho_{a_1 a_2})^2]^{1/2}$ $(j = 3, \dots, d)$;

6: **for** $\ell = 3, \dots, t$: (ℓ is tree level) **do**

7:     **for** $j = \ell+1, \dots, d$: **do**

8:         let $\begin{pmatrix} \Sigma_{11} & \Sigma_{12} \\ \Sigma_{21} & \Sigma_{22} \end{pmatrix}$ be the submatrix of R with indices $(a_{1j}, \dots, a_{\ell-1,j}), (a_{\ell j}, j)$;

9:         let $\Omega = \Sigma_{21} \Sigma_{11}^{-1} \Sigma_{12}$ be a 2×2 matrix with entries $\omega_{11}, \omega_{12} = \omega_{21}, \omega_{22}$;

10:         let $q \leftarrow p_{\ell j} [(1 - \omega_{11})(1 - \omega_{22})]^{1/2}$;

11:         set $\rho_{a_{\ell j}, j} = \rho_{j, a_{\ell j}} \leftarrow q + \omega_{12}$.

12:     **end for**

13: **end for**

14: (ii) Set the $d \times t$ matrix $(\phi_{j\ell})$ to 0; set $\phi_{21} = \rho_{12}$;

15: **for** $j = 3, \dots, d$: **do**

16:     let $n \leftarrow \min\{t, j-1\}$;

17:     $q = (\phi_{j1}, \dots, \phi_{jn})^\top$ solves $(\rho_{a_{jk} a_{j\ell}})_{1 \le k, \ell \le n} q = (\rho_{a_{jk} j})_{1 \le k \le n}$.

18: **end for**

19: (iii)

20: **for** $1 \le i < j \le d$: **do**

21:     if $(\rho_{ij} = -2)$ then $\rho_{ij} = \rho_{ji} \leftarrow \sum_{\ell=1}^{t} \phi_{j\ell} \rho_{i a_{j\ell}}$.

22: **end for**

23: Return matrix $R = (\rho_{ij})$ and return $d \times t$ matrix of $(\phi_{j\ell})$.

6.16 Partial correlations and factor structure

For multivariate t or Gaussian copula models with factor structures (Section 3.10.4), the following are useful for reparametrizations: (i) transform $d \times p$ matrix $A = (\alpha_{jk})$ of loadings to $d \times p$ matrix P of partial correlations; (ii) transform matrix of partial correlations to matrix of loadings; (iii) transform via a sequence of Givens rotations to set $p(p-1)/2$ zeros in a matrix of rotated loadings. It is assumed that $dp - p(p-1)/2 > 0$.

The details are these three reparametrizations are given below.

(i) (Loadings to partial correlations). $A = (\alpha_{jk})$ and $P = (\rho_{jk;1:(k-1)})$ are $d \times p$ matrices. The transform is the following for each $j \in \{1, \dots, d\}$: $\rho_{j1} = \alpha_{j1}$ and

$$\rho_{jk;1:(k-1)} = \alpha_{jk} \Big/ \sqrt{1 - \alpha_{j1}^2 - \cdots - \alpha_{j,k-1}^2}, \quad 2 \le k \le p.$$

Note that this requires $\alpha_{j1}^2 + \cdots + \alpha_{jp}^2 \le 1$.

(ii) (Partial correlations to loadings). $A = (\alpha_{jk})$ and $P = (\rho_{jk;1:(k-1)})$ are $d \times p$ matrices. The transform is the following for each $j \in \{1, \dots, d\}$: $\alpha_{j1} = \rho_{j1}$ and

$$\alpha_{jk} = \rho_{jk} \Big[(1 - \rho_{j1}^2) \prod_{i=2}^{k-1} (1 - \rho_{ji;1:(i-1)}^2) \Big]^{1/2}, \quad 2 \le k \le p.$$

The proof of this can be obtained by induction.

(iii) (Givens rotations to set $p(p-1)/2$ zeros in a $d \times p$ matrix of loadings). The algorithm is based on

$$(x, y) \begin{pmatrix} g_{11} & g_{12} \\ -g_{12} & g_{11} \end{pmatrix} = (x', 0), \quad x \neq 0, \ g_{11}^2 + g_{12}^2 = 1,$$

where $g_{11} = x/\sqrt{x^2 + y^2}$ and $g_{12} = -y/\sqrt{x^2 + y^2}$ leading to $x' = \sqrt{x^2 + y^2}$. If $y = 0$, then $g_{11} = \pm 1$, $g_{12} = 0$. Hence if $\alpha_{jk_1}, \alpha_{jk_2}$ are both non-zero, there is an orthogonal Givens rotation matrix $\boldsymbol{G} = (g_{ik})_{1 \leq i,k \leq p}$ which is zero everywhere except

$$g_{k_1 k_1} = g_{k_2 k_2} = \alpha_{jk_1} \Big/ \sqrt{\alpha_{jk_1}^2 + \alpha_{jk_2}^2}, \quad g_{k_2 k_1} = -g_{k_1 k_2} = \alpha_{jk_2} \Big/ \sqrt{\alpha_{jk_1}^2 + \alpha_{jk_2}^2}.$$

The sequence is the following.

- Pick a row j_1 such that $\alpha_{j_1 1} \neq 0$. Apply a product of $p-1$ Givens rotations, leading to $\boldsymbol{G}^{(1)}$, to get a new matrix $\boldsymbol{A}^{(1)} = \boldsymbol{AG}^{(1)}$ such that $\alpha_{j_1 2}^{(1)} = \cdots = \alpha_{j_1 p}^{(1)} = 0$. In general, other rows of $\boldsymbol{A}^{(1)}$ are different from the corresponding rows of $\boldsymbol{A}$ because of the rotation.

- If $p \geq 2$, pick a row j_2 such that $\alpha_{j_2 2}^{(1)} \neq 0$. (Note that a non-trivial loading matrix will not have all zeros in any column at this point.) Apply a product of $p-2$ Givens rotations, leading to $\boldsymbol{G}^{(2)}$, to get a new matrix $\boldsymbol{A}^{(2)} = \boldsymbol{A}^{(1)} \boldsymbol{G}^{(2)}$ such that $\alpha_{j_2 3}^{(2)} = \cdots = \alpha_{j_2 p}^{(2)} = 0$.

- Continue in this way for columns $3, \ldots, p-1$ ($p \geq 4$).

- For the last column, pick a row j_p such that $\alpha_{j_p p}^{(p-1)} \neq 0$. Apply a Givens rotation $\boldsymbol{G}^{(p)}$ to get a new matrix $\boldsymbol{A}^{(p)} = \boldsymbol{A}^{(p-1)} \boldsymbol{G}^{(p)}$ such that $\alpha_{j_p p}^{(p)} = 0$.

The bi-factor correlation structure with m groups is a special case of a $(m+1)$-factor correlation matrix with structured zeros. For the bi-factor Gaussian model with loading matrix in (3.64), for group g, let $\boldsymbol{\phi}_g$ be the loading vector for the global factor W_0 and $\boldsymbol{\alpha}_g$ be the loading vector for the gth group factor W_g. If $j \subset J_g$ is an index in group g and $\rho_{jW_g;W_0}$ is a partial correlation, then $\alpha_{gj} = (1 - \phi_{gj}^2)^{1/2} \rho_{jW_g;W_0}$. When the $\boldsymbol{\alpha}_g$ vectors are determined from all of the partial correlations and the $\boldsymbol{\phi}_g$ vectors, then the correlation for two variables j_1, j_2 in group J_g is $\phi_{gj_1} \phi_{gj_2} + \alpha_{gj_1} \alpha_{gj_2}$ and the correlation for two variables i_1, i_2 in separate groups J_{g_1}, J_{g_2} is $\phi_{g_1 i_1} \phi_{g_2 i_2}$.

6.17 Searching for good truncated R-vine approximations

In this section, we discuss methods for finding good truncated vines for high-dimensional data. Some general ideas are taken from Kurowicka (2011) and Dissmann et al. (2013). One principle is that the most important (and strongest) dependencies among variables can typically be captured best by the pair-copulas of tree $\mathcal{T}_1$. This is underlined by the property that, for all variables to be upper (lower) tail dependent, it is sufficient that all bivariate copulas in $\mathcal{T}_1$ are upper (lower) tail dependent (see Theorem 8.68). Also, this more likely leads to a parsimonious truncated R-vine, and the influence of rounding errors in the tree-by-tree calculation of the conditional distribution functions as arguments of the pair-copula densities is minimized.

The sequential algorithm of Dissmann et al. (2013) for R-vines uses pairwise empirical Kendall's τ values (or another dependence measure) as edge weights for a greedy tree selection; see Algorithm 30.

Algorithm 30 Sequential selection of R-vine and pair-copulas; Dissmann et al. (2013).

1: Input data that have been converted d-vectors $\boldsymbol{u}_1, \ldots \boldsymbol{u}_n$ that are marginally U(0, 1).

2: Compute empirical Kendall's tau $\hat{\tau}_{jk}$ for all pairs of variables:

3: Tree 1. Set tree level $\ell \leftarrow 1$.

    a. Find the spanning tree $\mathcal{T}_1$ with $d-1$ edges to maximize the sum of absolute Kendall taus (or another bivariate dependence measure): $\sum_{e=(j,k)\in\mathcal{T}_1} |\hat{\tau}_{jk}|$.

    b. Select parametric pair-copulas (from within a candidate set) for the edges $e \in \mathcal{T}_1$ and estimate the corresponding copula parameters $\hat{\theta}_e$; the pair-copulas are chosen to maximize the log-likelihood at each edge.

    c. If $e = (j, k)$ is an edge in the optimal spanning tree, transform to $C_{j|k}(u_{ij}|u_{ik}; \hat{\theta}_{jk})$ and $C_{k|j}(u_{ik}|u_{ij}; \hat{\theta}_{jk})$ for $i = 1, \ldots, n$.

4: Next tree. Increment the tree level $\ell \leftarrow \ell + 1$.

    a. Compute empirical conditional Kendall taus $\hat{\tau}_{pq;S}$ (or another bivariate dependence measure) from $C_{p|S}(u_{ip}|\boldsymbol{u}_{i,S}; \hat{\theta}_{p\cup S})$, $C_{q|S}(u_{iq}|\boldsymbol{u}_{i,S}; \hat{\theta}_{q\cup S})$ over all edges $[pq|S]$ that can be part of a tree $\mathcal{T}_\ell$ with nodes from $\mathcal{T}_{\ell-1}$.

    b. Select the spanning tree $\mathcal{T}_\ell$ with $d-\ell$ edges to maximize $\sum |\hat{\tau}_{pq;S}|$, where $\{p\} \cup S$ and $\{q\} \cup S$ are sets in tree $\mathcal{T}_{\ell-1}$.

    c. Select pair-copulas for the conditional distributions for the edges $e \in \mathcal{T}_\ell$ and estimate the copula parameters $\hat{\theta}_e$.

    d. If $[pq|S]$ is an edge in the optimal $\mathcal{T}_\ell$, transform to $C_{p|q;S}(u_{ip}|\boldsymbol{u}_{i,q\cup S}; \hat{\theta}_{\{p,q\}\cup S})$ and $C_{q|p;S}(u_{iq}|\boldsymbol{u}_{i,p\cup S}; \hat{\theta}_{\{p,q\}\cup S})$ for $i = 1, \ldots, n$.

5: If $\ell < d-1$, go to the above for the next tree.

For truncated R-vines, the greedy algorithm of Brechmann et al. (2012) augments Algorithm 30 by iteratively constructing R-vine copulas, which are 1-truncated, 2-truncated, and so on. In each step, the Vuong procedure in Section 5.13 for model comparison is used to assess the gain of extending the truncated R-vine copula by an additional tree. If the gain is determined to be negligible, the procedure is stopped.

Because greedy algorithms with local optimality cannot always be globally optimal, heuristics are desirable for alternative sequential approaches.

For Gaussian partial correlation vines, Kurowicka (2011) considered two heuristic methods for truncation.

1. One method starts with trying to get large absolute correlations and partial correlations in edges of low-indexed trees $\mathcal{T}_1, \mathcal{T}_2, \ldots$ so that small absolute partial correlations should occur in high-indexed trees $\mathcal{T}_{d-1}, \mathcal{T}_{d-2}, \ldots$, if in fact the correlation matrix has a truncated partial correlation vine representation. This is the idea that is elaborated on below.

2. The second method tries to assign edges to high-indexed trees $\mathcal{T}_{d-1}, \mathcal{T}_{d-2}, \ldots$ with smaller absolute partial correlations, by sequentially looking at some inverse correlation matrices for small non-diagonal elements (see Example 3.9). This method is harder to implement for large d.

The objective function for deciding on the edges for trees of the vines is based on the determinant of the correlation matrix $\boldsymbol{R}$. Kurowicka and Cooke (2006) and Lewandowski et al. (2009) showed that for any Gaussian partial correlation vine $\mathcal{V}$ representation of $\boldsymbol{R}$,

$$\det(\boldsymbol{R}) = \sum_{e\in\mathcal{E}(\mathcal{V})} \log(1 - \rho_e^2).$$

If the partial correlations of the vine are zero for trees $\mathcal{T}_{m+1}, \ldots, \mathcal{T}_{d-1}$, then the above simplifies to:

$$\det(\boldsymbol{R}) = \sum_{e \in \{\mathcal{T}_1, \dots, \mathcal{T}_m\}} \log(1 - \rho_e^2).$$

In particular, if $m = 1$ and the vine array is A, then $\det(\boldsymbol{R}) = \sum_{j=2}^{d} \log(1 - \rho_{a_{1j},j}^2)$.

If one is fitting a Gaussian copula, one could check if the dependence structure is close to a truncated vine for some $\mathcal{V}$ and truncated level m. Consider some maxima over all vines in d variables to get:

$$D_\ell := \max_{\mathcal{V} = \{\mathcal{T}_1, \dots, \mathcal{T}_{d-1}\}} \left\{ - \sum_{e \in \{\mathcal{T}_1, \dots, \mathcal{T}_\ell\}} \log(1 - \rho_e^2) \right\}.$$

D_ℓ is increasing in ℓ because there are extra non-negative terms of the form $-\log(1 - \rho^2)$ as ℓ increases. An m-truncated vine is a good approximation if

$$D_{m-1} < D_m, \quad D_m \approx D_{m+1} \approx \dots \approx D_{d-1}.$$

Note that D_{d-1} is always $\det(\boldsymbol{R})$.

The above idea depends on the evaluation of D_ℓ. If $d \le 8$, it is possible to enumerate through all $2^{(d-3)(d-2)}(d!/2)$ vines and find $D_1, \dots, D_{d-2}$. If one is only trying to evaluate D_1, D_2, D_3, say, in order to compare with $\det(\boldsymbol{R})$, then it becomes an issue of finding algorithms to enumerate through 1-truncated, 2-truncated and 3-truncated vines; the number of such ℓ-truncated vines is less than $2^{(d-3)(d-2)}(d!/2)$. For example, for 1-truncated vines or Markov trees, by Cayley's theorem, the total number is d^{d-2} and the Prüfer code is one way to systematically enumerate them.

In general, one could consider suboptimal search algorithms, such as forms of genetic algorithms, that might cover most of the relevant space of truncated vines. Then one would have suboptimal

$$\widetilde{D}_1 < \widetilde{D}_2 < \dots; \quad \text{with } \widetilde{D}_\ell \le D_\ell.$$

One could decide on a criterion to accept $\widetilde{D}_m$ and truncated level m if $\widetilde{D}_m$ is sufficiently close to $\det(\boldsymbol{R})$. With such a suboptimal algorithm, the value of $\tilde{m}$ to achieve a desired closeness might be one or two larger than the value m if one could evaluate D_ℓ exactly. Nevertheless, if $\tilde{m} < d - 1$, there is still a simplification of the dependence structure.

The next step relates to Remark 5.1 if the truncated R-vine is considered as a reasonable dependence structure. For the "suboptimal" procedure, let the "best" N truncated R-vines be $\mathcal{V}_1^*, \dots, \mathcal{V}_N^*$ with truncation level all equal m^*. For truncated R-vine $\mathcal{V}_k^* = \{\mathcal{T}_{k1}^*, \dots, \mathcal{T}_{km}^*\}$, replace a Gaussian copula with parameter ρ_e by a pair-copula $C_e(\cdot; \theta_e)$ (from candidate families that have appropriate tail behavior) for $e \in \{\mathcal{T}_{k1}^*, \dots, \mathcal{T}_{km}^*\}$, and carry out the numerical maximum likelihood. Based on AIC or BIC, there will be a best model starting from the N truncated R-vines.

If the algorithm for searching for truncated vines is good, and the candidate pair-copula families have a sufficient variety of tail asymmetry and strength of dependence in the tails, then the "best" model found should be adequate for inferences. We emphasize that parametric vine copulas should provide good approximations to many multivariate copulas, and if d is large, parametric truncated vine copulas should be more parsimonious approximations.

In subsections below, we outline greedy and non-greedy approaches to find suboptimal truncated partial correlation vines.

6.17.1 *Greedy sequential approach using minimum spanning trees*

Consider trying to find a m-truncated regular vine $\mathcal{V}_m = (\mathcal{T}_1, \dots, \mathcal{T}_m)$ to minimize $\sum_{e \in \mathcal{E}(\mathcal{V}_m)} \log(1 - \rho_e^2)$ or maximize $-\sum_{e \in \mathcal{E}(\mathcal{V}_m)} \log(1 - \rho_e^2)$. A greedy algorithm first finds a tree T_1^* that minimizes $h_1(T_1) = \sum_{e \in T_1} \log(1 - \rho_e^2)$. In the second stage, with $m \geq 2$ and $T_1 = T_1^*$, the algorithm finds a tree T_2^* minimizes $h_2(T_2) = \sum_{e \in T_2} \log(1 - \rho_e^2)$ over T_2 that are compatible with T_1^*. In the mth stage with $m \geq 3$ and $T_\ell = T_\ell^*$ for $\ell = 1, \ldots, m-1$, the algorithm finds a tree T_m^* minimizes $h_m(T_m) = \sum_{e \in T_m} \log(1 - \rho_e^2)$ over T_m that are compatible with $T_1^*, \ldots, T_{m-1}^*$. At stage m, a minimum spanning tree algorithm can be applied to find the tree T_m^*.

A numerical example of how trees are formed in a sequence with the greedy algorithm is in Example 6.4 below, followed by pseudo-code.

Example 6.4 (Specific details of greedy or sequential minimum spanning tree algorithm).

- Suppose there $d = 7$ variables with nodes labeled as $1, 2, 3, 4, 5, 6, 7$ and a complete graph is constructed with weight $\log(1 - \rho_{jk}^2)$ for the edge $e_{jk} = [j, k]$ that connects nodes j, k.

- Suppose the minimum spanning tree algorithm finds that the optimal tree $T_1^* = \{[13], [24], [34], [45], [46], [37]\}$ with $d - 1 = 6$ edges.

- Table 6.2 gives in indication for how the next few subsequent trees are found. The edges of tree T_1^* become nodes for tree T_2; using different symbols, we label them as a, b, c, d, e, f (in the order they appear above). After the best tree T_2^* is found, the edges become nodes of T_3 and we label them as A, B, C, D, E. After the next best tree T_3^* is found, the edges become nodes of T_4 and we label them as I, II, III, IV. At each stage in trees $T_2, \ldots, T_m$. an edge forms between two nodes only if the proximity condition in Definition 3.2 is satisfied. The vine structure guarantees that there are at least $d - \ell$ edges in tree T_ℓ to use for the minimum spanning tree algorithm.

T_2				T_3				T_4			
node1	node2	edge	T_2^*	node1	node2	edge	T_3^*	node1	node2	edge	T_4^*
a=13	b=24	⌢		A=bc	B=bd	35|24	I	I=AB	II=AE	57|234	i
a=13	c=34	14|3		A=bc	C=ce	26|34		I=AB	III=CE	⌢	
b=24	c=34	23|4	A	B=bd	C=ce	⌢		II=AE	III=CE	26|347	ii
a=13	d=45	⌢		A=bc	D=af	⌢		I=AB	IV=DE	⌢	
b=24	d=45	25|4	B	B=bd	D=af	⌢		II=AE	IV=DE	12|347	iii
c=34	d=45	35|4		C=ce	D=af	⌢		III=CE	IV=DE	16|347	
a=13	e=46	⌢		A=bc	E=cf	27|34	II				
b=24	e=46	26|4		B=bd	E=cf	⌢					
c=34	e=46	36|4	C	C=ce	E=cf	67|34	III				
d=45	e=46	56|4		D=af	E=cf	14|37	IV				
a=13	f=37	17|3	D								
b=24	f=37	⌢									
c=34	f=37	47|3	E								
d=45	f=37	⌢									
e=46	f=37	⌢									

Table 6.2 *Sequential minimum spanning trees where we suppose the edges of the optimal T_2^* are labeled as A,B,C,D,E; the edges of the optimal T_3^* are labeled as I,II,III,IV; the edges of the optimal T_4^* are labeled as i,ii,iii. The symbol ⌢ indicates that the proximity condition is not satisfied, or there are four (not three) distinct symbols in the union of the two nodes.*

Next is some pseudo-code for the sequential minimum spanning trees with d variables. The implementation would typically make use of some software for manipulating and creating graphs.

- Create a complete graph G_1 with d nodes with weight $\log(1 - \rho_{jk}^2)$ for edge $[jk]$.

- Find the spanning tree $\mathcal{T}_1^*$ with $d-1$ edges that minimizes $h_1(\mathcal{T}_1) = \sum_{e \in \mathcal{T}_1} \log(1 - \rho_e^2)$.
- Label the edges of $\mathcal{T}_1^*$ as nodes of a new graph G_2.
- For $\ell = 2, \ldots, m$ (trees $\mathcal{T}_2, \ldots, \mathcal{T}_m$):
- For the graph G_ℓ, enumerate with pairs of vertices node1, node2:
  - If node1=(a_1, b_1) and node2=(a_2, b_2) where a_1, b_1, a_2, b_2 are nodes in tree $G_{\ell-1}$, then the proximity condition is satisfied if there are three distinct elements in a_1, b_1, a_2, b_2 and the proximity condition is not satisfied if there are four distinct elements. Skip to the next pair of nodes in the latter case.
  - If the proximity condition is satisfied, determine the label $[T|S]$ for the edge of the form [conditioned|conditioning]: the conditioning set S is the intersection of the variables in node1 and node2, and the conditioned set T consists of the the symmetric difference of the variables in node1 and node2. The union $T \cup S$ is the constraint set of variables associated with $[T|S]$.
  - The weight assigned to this edge is $\log(1 - \rho_{T;S}^2)$ where $\rho_{T;S}$ is a partial correlation.
  - End of `for` loop for pairs of nodes for G_ℓ.
- There are at least $d - \ell$ edges with an assigned weight. Find the spanning tree $\mathcal{T}_\ell^*$ with $d - \ell$ edges to minimize $h_\ell(\mathcal{T}_\ell) = \sum_{e \in \mathcal{T}_\ell} \log(1 - \rho_e^2)$.
- Label the edges of $\mathcal{T}_\ell^*$ as nodes of a new graph $G_{\ell+1}$.
- End of `for` loop for trees.

6.17.2 Non-greedy algorithm

Brechmann and Joe (2014a) propose to consider the following tree specifications in addition to the greedy algorithm.

- *Neighbors of trees*: As the locally optimal solution, the minimum spanning tree (MST) is a reasonable starting point for the search for a better solution. Consider neighbors of the MST, where a 1-neighbor of a tree $\mathcal{T}$ is obtained as mentioned in Section 6.6.
- *Best spanning trees*: Take into account a certain number of best spanning trees.

However, even if only a small number of best spanning trees and neighbors is considered as candidate solutions, the number of potential ℓ-truncated partial correlation R-vines quickly increases. If s candidate solutions are considered at each level, there are up to $s^{\ell-1}$ different ℓ-truncated partial correlation R-vines to choose from (depending on the proximity condition). To reduce this number but still reasonably explore the search space of ℓ-truncated partial correlation R-vines, Brechmann and Joe (2014b) propose the following approach, making use of genetic algorithms (see, for example, Goldberg and Deb (1991) and Blickle and Thiele (1996)) and a fit index such as

$$\gamma(\widehat{\boldsymbol{R}}|\boldsymbol{R}) := \log |\widehat{\boldsymbol{R}}| \, / \, \log |\boldsymbol{R}| \in (0, 1],$$

where $\boldsymbol{R}$ is the empirical correlation matrix and $\widehat{\boldsymbol{R}}$ is an estimate based on a parsimonious model.

1. Calculate the index value γ_1 of the MST for $\mathcal{T}_1$, which is the best 1 truncated partial correlation R-vine.
2. Truncate at level 1 if γ_1 is sufficiently close to 1. Return the truncation level $\ell^* = 1$ and return the MST.
3. Otherwise, generate a range of s candidate solutions for $\mathcal{T}_1$ based on neighbors and best spanning trees.
4. For each of the s 1-truncated candidates, choose the MST as $\mathcal{T}_2$ and calculate the index value γ_2.

5. Truncate at level 2 if γ_2 is sufficiently close to 1 for one of the candidates. Return the truncation level $\ell^* = 2$ and the model with the largest fit index value γ_2 (best 2-truncated partial correlation R-vine).

6. Otherwise, for each of the s 1-truncated candidates, select a range of s candidates for $\mathcal{T}_2$ (neighbors, best spanning trees). The actual number of candidates may be less than s when the proximity condition only admits a smaller number of spanning trees.

7. Now there are up to s^2 2-truncated vines as candidates. Since these are too many candidates to continue with, choose an appropriate number $s_{(2)} < s^2$ of candidates according to their fit index value.

8. For each of the $s_{(2)}$ 2-truncated candidates, choose the MST as $\mathcal{T}_3$ and calculate the index γ_3.

9. Truncate at level 3 if γ_3 is sufficiently close to 1 for one of the candidates. Return the truncation level $\ell^* = 3$ and the model with the largest index γ_3 (best 3-truncated partial correlation R-vine).

10. Otherwise, for each of the $s_{(2)}$ 2-truncated candidates, select a range of s candidates for $\mathcal{T}_3$ (neighbors, best spanning trees), and so on, until truncation is possible.

That is, in order to prevent the number of candidate solutions from growing excessively, in each step a certain number of the candidates are selected and the rest are discarded.

6.18   Summary for algorithms

The flexible copula models that can be used for high-dimensional data (dimensions of 50–100 and beyond) are based on vines or pair-copula constructions, and factor copula models are truncated vines rooted at latent variables. The pair-copula construction with truncated vines should be computationally feasible for dimensions up to 50 or more because parameter estimation can proceed sequentially with numerical optimizations for subsets of parameters, There is more structure for factor copula models because they are closed under margins. With only one-dimensional numerical integrations, 1-factor copula models can probably be implemented for any dimension for which there is sufficient computer memory. Structural factor models, such as bi-factor, and 2-factor models, which require two-dimensional numerical integrations, are estimable for dimensions well over 100, when combined with an appropriate numerical optimization method. In general, copula and multivariate models that require one- or two-dimensional numerical integration are easily computationally feasible for likelihood inference.

This chapter has presented many numerical methods and algorithms for copula models that range from bivariate to arbitrary dimensions. The algorithms have been presented as pseudo-codes. When writing actual code, it is best to write code that can easily be ported to other programming languages. For algorithms that are defined for arbitrary dimensions, the implementation would have to be in a high-level programming language, such as Fortran90 or C/C++, for efficiency of memory (total storage space for temporary variables) as well as speed with multiple loops.

Many numerical methods are available in the public domain with code written in Fortran or C/C++, with older code in Fortran77. Fortran90 (or later version such as Fortran95 or Fortran2003) is especially useful because of its built-in vector and matrix operations. For statistical computations and data analysis, the open-source statistical software R (R-Core-Team (2013)) is convenient for providing a front end that makes it straightforward to link codes that are written in a mixture of Fortran77, Fortran90 and C. Provided that arguments of functions are defined appropriately, code in one of these programming languages can call functions written in another programming language. Also the random number generators used by R are accessible from C or Fortran.

Chapter 7

Applications and data examples

This chapter demonstrates a variety of applications of dependence modeling with copulas. Because in practice, all copula models are to some extent misspecified, there are some simulation-based examples to show what can be expected in terms of tail inference with misspecified copula models, and to illustrate some points concerning non-tail and tail inference with copulas. Some examples with multivariate data sets of different types are the following.

1. Insurance claims with two types of losses.
2. Count (longitudinal) response.
3. Count times series.
4. Extreme values with monthly sea levels at several sites.
5. Financial returns: market returns and stock returns.
6. Item response (ordinal) variables for several items in a survey.
7. Correlated sociological variables with latent variable dependence.

The first three sections compare copula models for continuous and discrete responses with simulated data. The remaining sections have data examples in several application areas and show a variety of uses of copulas for dependence modeling.

Section 7.1 compares misspecified copula models for adequacy of inferences on strength of dependence and tail probabilities. Section 7.2 goes into more detail on estimates of tail-weighted dependence measures with misspecified copula models. Section 7.3 shows how the discrete R-vine can be an excellent approximation to the multivariate probit model for ordinal response data. Section 7.4 compares tail asymmetric copula models for two loss variables in insurance claims. Section 7.5 compares copula models for multivariate/longitudinal ordinal response data with covariates. Section 7.6 compares copula-based Markov transition probabilities for modeling count time series data. Section 7.7 has analysis of monthly extreme sea levels with a multivariate extreme value copula model. Section 7.8 shows the use of high-dimensional factor copula models that can be applied to a very large number of financial returns. Section 7.9 shows how to use tail-comonotonic copula models to get conservative estimates of joint tail probabilities of several risks. Section 7.10 compares factor copula and discrete R-vine models for ordinal item response data. Sections 7.11 and 7.12 demonstrate the fitting of structural equation models with truncated vines.

Section 7.13 highlights some applications of copulas that have appeared in publications; especially some high-dimensional copula models.

7.1 Data analysis with misspecified copula models

In this section, simulated bivariate data sets are used to illustrate the quality of inference on some dependence measures and tail probabilities with misspecified models.

In the first subsection, simulated data are generated from one copula model which is reflection symmetric with tail dependence and another which is reflection asymmetric with intermediate tail dependence. In the second subsection, simulated data are generated from a copula model that is reflection asymmetric with lower and upper tail dependence. Comparisons are made with inferences based on copula models that might be selected based on some preliminary data analysis.

To reinforce some of the theory in Chapter 5, it will be shown that inference on dependence measures or quantities that are not much influenced by the tails can be adequately estimated from the copula if the fitted model is a good approximation. However, inferences on tail quantities can be much more sensitive to the fitted model.

These simulation studies are an imitation of real practice because generally there is no physical mechanism that leads to an obvious copula model for an application. Preliminary bivariate analyses of tail asymmetry and strength of dependence in the tails are used to come up with candidate models. That is, in practice, all copula models are "misspecified" but it is hoped that some of the candidate models will be good approximations to the "true" model for the purpose of making inferences.

For the simulation studies, we can compare the candidate models against the true model. But for practical applications, we need methodology to determine if any of the candidate models is an adequate fit to the data for relevant inferences; note that a model with a smaller value of AIC or BIC indicates a better fit but doesn't guarantee an adequate fit. The methodology we illustrate for the context of copula models is the comparison of empirical versus model-based tail-weighted quantities; the idea is that tail-weighted quantities such as lower and upper semi-correlations can differentiate copula models more than non-tail-weighted measures of monotone association. If tail inference is important, then adequate fitting parametric copula models are those for which model-based tail-weighted quantities match the empirical counterparts.

Kullback-Leibler divergence of a misspecified model to a true distribution depends on the log-likelihood ratio of the corresponding densities. For copula models with data converted to $U(0,1)$ margins, this log-likelihood ratio can approach $\pm\infty$ in the tails if the two models are quite different in tail behavior. That is, better fitting copula models will better match the tails of the copula of the true distribution.

7.1.1 *Inference for dependence measures*

For the first example in this subsection, simulated bivariate data sets were generated from the copula $\frac{1}{2}[C(u,v;\theta)+1-u-v+C(1-u,1-v;\theta)]$, where $C(\cdot;\theta)$ is the Gumbel copula, and θ is chosen to achieve a given value of Spearman's ρ_S. This was replicated 1000 times. With sample sizes of $n = 600$, the normal scores plots and semi-correlations suggest heavier tails than bivariate Gaussian and similar strength of dependence in the joint lower and upper tails. So we fitted several tail dependent models (see Table 4.1) via maximum likelihood; and also some models with tail quadrant independence, such as the Frank and Plackett copulas. The latter are used for "baseline" comparisons, even if they are not suggested from preliminary analysis. Note that the BB1 and BB7 copulas are close to reflection symmetric for some parameters.

For each simulated data set, empirical estimates were obtained for: (i) lower tail probability $C(0.2,0.2)$ and upper tail probability $\overline{C}(0.8,0.8)$; (ii) dependence measures β,τ,ρ_S. Also obtained for each simulated data set were model-based estimates for the preceding 5 quantities. Note that Patton (2009, 2012) make use of $C(p,p)/p$ and $\overline{C}(1-p,1-p)/p$ for $0 < p < \frac{1}{2}$ and refer to these as quantile dependence.

Summaries are given in Table 7.1 with true value $\rho_S = 0.5$. For this specific simulation and other values of ρ_S in the medium range, the pattern is similar. Some general conclusions are: (a) better fitting models lead to better match of empirical and model-based estimates of dependence measures; (b) of the 3 dependence measures, the most sensitive to model misspecification is Blomqvist's β and the least sensitive is Kendall's τ; (c) for estimation of $C(0.2,0.2)$ and $\overline{C}(0.8,0.8)$, a reflection asymmetric copula is worse, but otherwise all of the other reflection symmetric copulas were acceptable even with varying strength of dependence in the tails.

The symmetrized Gumbel copula has lower and upper tail dependence parameters equal to $\frac{1}{2}(2-2^{1/\theta})$. Note that the best three fitting misspecified models of the bivariate t, BB1 and BB7 copulas all have lower and upper tail dependence. The worst two fitting misspecified models, based on average log-likelihoods, are the Plackett and Frank copulas that have tail quadrant independence. Also note that the Gaussian copula does as well as the best fitting copulas in model-based estimation of the five quantities in Table 7.1.

When data have tail dependence, typically the best fitting parametric copula families in terms of log-likelihood have tail dependence, but this does not mean that these "best" parametric models lead to good model-based estimates of the tail dependence parameters (Section 7.1.2).

		$C(0.2, 0.2)$	$\overline{C}(0.8, 0.8)$	β	τ	ρ_S
true val.		0.092	0.092	0.349	0.351	0.500
model	loglik.			Means and SDs		
true	101.9 (14.1)	0.093 (0.003)	0.093 (0.003)	0.350 (0.022)	0.351 (0.021)	0.501 (0.028)
biv. t	101.5 (14.1)	0.093 (0.004)	0.091 (0.003)	0.353 (0.022)	0.352 (0.022)	0.504 (0.029)
BB1	101.4 (14.3)	0.092 (0.005)	0.092 (0.005)	0.337 (0.021)	0.345 (0.021)	0.493 (0.028)
BB7	99.7 (14.3)	0.092 (0.005)	0.090 (0.005)	0.311 (0.019)	0.331 (0.021)	0.472 (0.027)
Gaussian	97.2 (13.9)	0.090 (0.003)	0.090 (0.003)	0.352 (0.022)	0.352 (0.022)	0.508 (0.029)
Plackett	92.2 (12.8)	0.090 (0.003)	0.090 (0.003)	0.391 (0.023)	0.354 (0.021)	0.504 (0.027)
Frank	88.9 (12.6)	0.087 (0.003)	0.087 (0.003)	0.397 (0.024)	0.355 (0.022)	0.513 (0.029)
Gumbel	92.4 (13.8)	0.077 (0.003)	0.101 (0.004)	0.328 (0.022)	0.330 (0.022)	0.473 (0.029)
model				Mean absolute difference from empirical		
true		0.009	0.009	0.024	0.009	0.013
biv. t		0.009	0.009	0.024	0.009	0.012
BB1		0.009	0.008	0.028	0.012	0.017
BB7		0.009	0.009	0.044	0.023	0.031
Gaussian		0.009	0.009	0.025	0.010	0.015
Plackett		0.009	0.009	0.042	0.009	0.014
Frank		0.010	0.009	0.047	0.007	0.014
Gumbel		0.016	0.012	0.032	0.022	0.029

Table 7.1 *Symmetrized Gumbel copula: Simulation results based on 1000 samples of size $n = 600$ with parameter such that $\rho_S = 0.5$. Based on the average log-likelihood values at the MLEs, bivariate t is the best model (with shape parameter typically in the range of 6 to 10) and BB1 is the second best. For the bottom half, a column has the mean absolute difference of the empirical estimate and the model-based estimate for each of the 8 copula models.*

For the second example in this subsection, simulated bivariate data sets were generated from the copula of the bivariate skew-normal of Azzalini and Dalla Valle (1996). The bivariate skew-normal distribution is the limit as $\nu \to \infty$ of the bivariate skew-t_ν distribution in Section 3.17.2. Because we are using the copula only, we ignore the location and scale parameters and take the bivariate distribution from $(Y_1, Y_2) \stackrel{d}{=} [(Z_1, Z_2)|Z_0 > 0]$ where (Z_1, Z_2, Z_0) is trivariate Gaussian with correlation matrix $\boldsymbol{R} = \begin{pmatrix} 1 & \xi_1 & \xi_2 \\ \xi_1 & 1 & \rho \\ \xi_2 & \rho & 1 \end{pmatrix}$. Parameters of the skew-normal copula were chosen to have skewness to the joint upper tail.

Several bivariate copula families were selected from Table 4.1, with reflection if needed, to match the direction of tail asymmetry towards the upper tail. Because of Table 5.6, the reflected BB1 copula is considered but not the reflected BB7 copula. Summaries are given in Table 7.2, which lists the (misspecified) copula models that were fitted. The BB6 copula was also fitted but generally the MLE was at the boundary Gumbel copula, so results of this copula are not included.

		$C(0.2, 0.2)$	$\overline{C}(0.8, 0.8)$	β	τ	ρ_S
true val.		0.090	0.099	0.377	0.378	0.542
model	loglik.	\multicolumn{5}{c}{Means and SDs}				
true	116.4 (14.5)	0.090 (0.004)	0.099 (0.004)	0.377 (0.019)	0.378 (0.019)	0.542 (0.024)
Gauss.	114.5 (14.4)	0.095 (0.003)	0.095 (0.003)	0.380 (0.019)	0.380 (0.019)	0.544 (0.024)
iMLAr	114.2 (14.3)	0.091 (0.004)	0.100 (0.004)	0.381 (0.020)	0.374 (0.020)	0.534 (0.025)
ipsAr	112.6 (14.3)	0.095 (0.003)	0.096 (0.003)	0.386 (0.019)	0.377 (0.019)	0.537 (0.025)
BB1	111.8 (14.2)	0.090 (0.004)	0.099 (0.004)	0.353 (0.019)	0.360 (0.019)	0.513 (0.025)
BB1r	111.6 (14.1)	0.089 (0.005)	0.099 (0.004)	0.348 (0.019)	0.356 (0.019)	0.509 (0.025)
BB7	108.1 (13.9)	0.089 (0.005)	0.097 (0.005)	0.321 (0.017)	0.341 (0.019)	0.485 (0.024)
Gumbel	107.6 (14.1)	0.080 (0.003)	0.105 (0.003)	0.350 (0.020)	0.352 (0.020)	0.501 (0.026)
Frank	103.5 (13.4)	0.091 (0.003)	0.091 (0.003)	0.422 (0.023)	0.377 (0.021)	0.543 (0.027)
MTCJr	95.2 (13.0)	0.064 (0.002)	0.104 (0.004)	0.300 (0.020)	0.299 (0.020)	0.435 (0.026)
model		\multicolumn{5}{c}{Mean absolute difference from empirical}				
true		0.009	0.009	0.026	0.011	0.015
Gauss.		0.010	0.010	0.026	0.011	0.015
iMLAr		0.009	0.009	0.026	0.011	0.015
ipsAr		0.010	0.010	0.027	0.010	0.014
BB1		0.009	0.010	0.033	0.020	0.030
BB1r		0.009	0.010	0.036	0.024	0.034
BB7		0.009	0.010	0.058	0.038	0.056
Gumbel		0.012	0.011	0.035	0.028	0.041
Frank		0.009	0.011	0.047	0.007	0.011
MTCJr		0.027	0.011	0.077	0.080	0.105

Table 7.2 *Bivariate skew-normal copula: Simulation results based on 1000 samples of size $n = 600$ from the with parameters $\rho = \xi_1 = \xi_2 = 0.7$. The fitted copula models are skew-normal, Gaussian, Frank, reflected MTCJ, Gumbel, reflected Archimedean with positive stable LT (ipsAr), reflected Archimedean with integrated Mittag-Leffler LT (iMLAr), BB1, reflected BB1, BB7. Based on the average log-likelihood values at the MLEs, the Gaussian copula is the best misspecified model and iMLAr is the second best for the (ρ, ξ_1, ξ_2) used here, If the parametric vector changes so that there is more tail asymmetry, iMLAr becomes the best misspecified model, and if there is less dependence and tail asymmetry, the second best model is ipsAr as a one-parameter subfamily of iMLAr. For the bottom half, a column has the mean absolute difference of the empirical estimate and the model-based estimate for each of the 10 copula models.*

The general conclusions are similar to the preceding example: (a) better fitting models lead to better match of empirical and model-based estimates of dependence measures; (b) of the 3 dependence measures, the least sensitive to model misspecification is Kendall's τ; (c) for estimation of $C(0.2, 0.2)$ and $\overline{C}(0.8, 0.8)$, the most extremely tail asymmetric copula does the worse, but otherwise all of the other copulas were acceptable even with varying strength of dependence in the tails.

The bivariate skew-normal copula has lower and upper intermediate tail dependence with slight tail asymmetry. Note that the best two fitting misspecified models are the reflection symmetric Gaussian copula and the reflected Archimedean copula with integrated Mittag-Leffler LT; the latter copula has upper tail dependence but is not as tail asymmetric as the worse-fitting Gumbel copula. The worst two fitting misspecified models, based on average log-likelihoods, are the reflected MTCJ copula with lower tail quadrant independence and upper tail dependence and the Frank copula with tail quadrant independence.

To summarize this subsection, as expected based on Section 5.7, the examples here show that misspecified copula models tend to be better approximations if they better match tail asymmetry and tail orders. The additional results obtained from the simulation studies show (i) the sensitivity of some dependence measures to misspecified models and (ii) that comparison of empirical and model-based estimates of some dependence measures and tail probabilities can be used to assess the adequacy of fit of copula models. Other tail-weighted dependence measures are used for comparison in this next subsection.

7.1.2   Inference for tail-weighted dependence measures

The example with simulated data from the symmetrized Gumbel copula and related examples suggest that if data have both lower and upper tail dependence, then copula models with tail dependence tend have fit better in terms of AIC/BIC. But this does not mean that such models lead to good estimates of model-based tail dependence parameters, This subsection has an example with simulated data to illustrate that tail inference with copula models is not as accurate as estimation of the strength of dependence.

Simulated bivariate data sets were generated from the copula (4.87)

$$[1 - u^\eta - v^\eta + C(1 - u^\eta, 1 - v^\eta; \theta, \delta)]^{1/\eta}$$

in Section 4.31.4 where $C(\cdot; \theta, \delta)$ is the BB1 copula cdf, $\eta = 1.2$ and (θ, δ) were then chosen to achieve tail dependence parameters such as $\lambda_L = 0.6$, $\lambda_U = 0.4$. With sample sizes of $n = 600$, the normal scores plots and semi-correlations suggest the asymmetric lower and upper tail dependence, so we fitted three misspecified models, BB1, reflected BB1 and BB7 by maximum likelihood, as well as the 3-parameter model used to simulate the data. This was replicated 1000 times. For each simulated data set, empirical estimates were obtained for:

(i) lower and upper tail probabilities $C(0.2, 0.2), \overline{C}(0.8, 0.8)$;

(ii) dependence measures β, τ, ρ_S;

(iii) lower and upper tail-weighted dependence measures

$$\begin{aligned}
&\text{Cor}((1 - 2U)^r, (1 - 2V)^r | U < \tfrac{1}{2}, V < \tfrac{1}{2}) \\
&\text{Cor}((2U - 1)^r, (2V - 1)^r | U > \tfrac{1}{2}, V > \tfrac{1}{2}),
\end{aligned} \tag{7.1}$$

with powers $r = 2$ and 6.

The tail-weighted dependence measures are studied in Krupskii (2014). Also obtained for each simulated data set were model-based estimates for the preceding 9 quantities for models BB1, reflected BB1 and BB7 using the corresponding MLEs. In addition, model-based estimates of λ_L, λ_U were obtained without any empirical counterparts.

Summaries are given in Table 7.3. For this specific simulation and with other values of (λ_L, λ_U), the pattern is similar. Some general conclusions are: (a) better fitting models lead to better match of empirical and model-based estimates of dependence measures, (b) quantities that are based heavily on the joint tails are hard to estimate and can be quite biased with misspecified models; (c) quantities that are based mainly in the center of the distribution are less sensitive to misspecified models. For example, estimates of $C(0.2, 0.2)$ and $\overline{C}(0.8, 0.8)$ are quite good even with the BB7 model which is worse based on log-likelihood, but estimates of the tail dependence parameter based on a model can be quite biased (e.g., BB7) or quite variable (e.g., reflected BB1).

quantity	true val.	emp.	BB1	BB1r	BB7
$C(0.2, 0.2)$	0.132	0.133 (0.013)	0.132 (0.003)	0.132 (0.003)	0.132 (0.004)
$\overline{C}(0.8, 0.8)$	0.124	0.124 (0.014)	0.123 (0.004)	0.124 (0.004)	0.121 (0.004)
β	0.558	0.557 (0.034)	0.555 (0.014)	0.561 (0.014)	*0.512* (0.014)
τ	0.568	0.568 (0.019)	0.565 (0.014)	0.569 (0.014)	0.542 (0.014)
ρ_S	0.754	0.753 (0.020)	0.752 (0.015)	0.756 (0.015)	0.724 (0.015)
ltw2	0.636	0.633 (0.047)	0.632 (0.027)	0.629 (0.024)	0.645 (0.027)
utw2	0.556	0.554 (0.054)	0.550 (0.031)	0.549 (0.032)	0.564 (0.035)
ltw6	0.712	0.705 (0.055)	0.703 (0.031)	0.699 (0.025)	0.736 (0.028)
utw6	0.605	0.598 (0.063)	0.611 (0.034)	0.593 (0.041)	*0.654* (0.037)
λ_L	0.600		0.57 (0.04)	0.58 (0.03)	*0.64* (0.03)
λ_U	0.400		*0.49* (0.03)	*0.35* (0.09)	*0.56* (0.03)
		Mean absolute difference model-based vs. empirical			
β			0.025	0.025	0.048
τ			0.011	0.010	0.027
ρ_S			0.012	0.012	0.030
ltw2			0.031	0.032	0.033
utw2			0.037	0.036	0.038
ltw6			0.036	0.039	0.044
utw6			0.044	0.041	0.063
		Frequency for lower > upper			
λ		−	881	992	934
tw2		861	961	962	951
tw6		887	951	972	942

Table 7.3 *Three-parameter copula with asymmetric tail dependence: Simulation results based on 1000 samples of size $n = 600$ from the copula with parameters such that $\lambda_L = 0.6, \lambda_U = 0.4, \eta = 1.2$. Part 1: Mean and SDs of estimated quantities are shown for empirical and model-based with the BB1, reflected BB1 and BB7 copulas. ltw2 and utw2 (ltw6 and utw6) are lower and upper tail-weighted dependence measures in (7.1) with a power of 2 (respectively 6). The estimates with more bias are shown in italics. Part 2: mean absolute difference of empirical versus model-based estimates of dependence measures. Part 3: frequency for lower tail-weighted dependence measure larger than corresponding upper tail-weighted dependence measure. The average log-likelihoods are 296.6, 297.5, 292.6 for BB1, reflected BB1 and BB7 respectively, all with SDs of 22.3.*

In Table 7.3, generally the reflected BB1 copula is the best of the three and the BB7 copula is the worst. For the 1000 simulated data sets, based on log-likelihood, BB1 was best 242 times and worst 16 times, reflected BB1 was best 719 times and worst 64 times, BB7 was best 39 times and worst 920 times. The 3-parameter model with extra parameter η was only marginally better for maximum likelihood with an average log-likelihood value of 298.0, but model-based estimates of tail quantities can be be worse than for reflected BB1 because for a sample size of $n = 600$, sometimes one of the estimated parameter values goes to a boundary. Note that all misspecified models lead to reasonable parametric estimates of non-tail quantities and the worst-fitting BB7 has more bias. Model-based estimates of λ_L, λ_U are biased or more variable even with the better fitting models. Model-based estimates of the lower and upper tail-weighted dependence measures are generally close to their empirical counterparts; the difference is smaller for better fitting models based on the log-likelihood.

To conclude this section, we note that tail-weighted dependence measures are needed to assess adequacy of fit of models, because strength of dependence in the tails need not be relevant to whether copula models lead to good model-based estimates of β, τ, ρ_S. The tail-weighted dependence measures will be used again with financial return data in Section 7.8.

7.2 Inferences on tail quantities

In this section, we compare copula-based inferences on some tail quantities such as extreme quantiles of the mean and corresponding conditional tail expectation of the mean to show which quantities might be distinguishable among copulas for a sample size of the order of 1000.

For simplicity, the copulas that are compared are exchangeable multivariate t_5, Gaussian, Gumbel and Frank copulas for different dimensions $d \geq 2$ with univariate margins that are t_ν.

This is meant as a simplification of financial applications with the mean of the observations being a substitute for the equally-weighted portfolio return when the d variables represent (scaled) log returns of individual financial assets. The extreme quantiles of the portfolio return are Value-at-Risk values for porfolio losses (when negative) and for losses due to "shorting" (when positive). Conditional tail expectation (CTE) values are the expected values below a small quantile (e.g., 0.01, 0.025, 0.05, 0.10) or expected values above a large quantile (e.g., 0.99, 0.975, 0.95, 0.90).

From Chapter 4, the multivariate t_5 copula is reflection symmetric and has lower and upper tail dependence, the multivariate Gumbel copula has upper tail dependence and lower intermediate tail dependence, the Gaussian copula has lower and upper intermediate tail dependence, and the multivariate Frank copula has lower and upper orthant tail independence with slight skewness to the joint upper tail for dimension $d > 2$. Hence for the lower tail, the extreme quantiles and CTEs of the portfolio return should be most negative for the multivariate t_5 copula and least negative for the Frank copula; similarly for the upper tail, the most extreme quantiles and CTEs of the portfolio return should be most positive for the Gumbel and multivariate t_5 copulas and least positive for the Frank copula.

Tables 7.4 and 7.5 have results for 2-dimensional and 10-dimensional exchangeable Frank, Gumbel, Gaussian and t_5 copulas with t_ν margins ($\nu = 3, 5, 20$). The extreme quantiles of the mean are shown together with CTEs. Also shown are the average of $d(d-1)/2$ lower and upper semi-correlations; with a sample size of order $n = 1000$, these are more discriminating for the copulas than the CTEs values. The extreme quantiles of the mean and corresponding CTEs can distinguish strong tail asymmetry (such as in the Gumbel copula), or tail orthant independence versus strong tail dependence (such as Frank versus t_5 copulas) but they cannot distinguish the Gaussian and t_5 copulas. To distinguish CTEs better, the sample size would have to be larger and the quantiles more extreme.

These simulation results are a preview to results for real data in Section 7.8.

To summarize this subsection, for financial applications, model-based quantiles and conditional tail expectations of portfolio return might not be well differentiated among some copula models where strengths of dependence in the tails are not substantially different. This is partly because the distribution of portfolio returns is mostly influenced by the middle of the multivariate distribution unless the quantiles are much more extreme than 0.01 or 0.99. However, there are other tail-based measures that can better discriminate copula models.

quantity	Frank		Gumbel		Gaussian		t_5	
	value	SD	value	SD	value	SD	value	SD
ρ_N^-	0.316	0.048	0.340	0.050	0.463	0.045	0.550	0.044
ρ_N^+	0.316	0.048	0.662	0.035	0.463	0.045	0.550	0.044
$d = 2$, t_5 univariate margins								
$q_{0.01}$	-2.82	0.19	-2.91	0.22	-3.06	0.24	-3.10	0.26
$\text{CTE}_{0.01}$	-3.41	0.33	-3.61	0.39	-3.85	0.45	-4.10	0.50
$q_{0.025}$	-2.27	0.12	-2.28	0.13	-2.37	0.14	-2.37	0.15
$\text{CTE}_{0.025}$	-2.86	0.19	-2.96	0.22	-3.13	0.25	-3.25	0.28
$q_{0.05}$	-1.85	0.08	-1.82	0.09	-1.87	0.10	-1.86	0.10
$\text{CTE}_{0.05}$	-2.44	0.13	-2.49	0.15	-2.61	0.16	-2.66	0.18
$q_{0.10}$	-1.41	0.06	-1.36	0.06	-1.38	0.07	-1.36	0.07
$\text{CTE}_{0.10}$	-2.03	0.09	-2.03	0.10	-2.10	0.11	-2.12	0.11
$q_{0.99}$	2.82	0.19	3.23	0.28	3.06	0.24	3.10	0.26
$\text{CTE}_{0.99}$	3.41	0.33	4.12	0.54	3.85	0.45	4.10	0.50
$q_{0.975}$	2.27	0.12	2.46	0.16	2.37	0.14	2.37	0.15
$\text{CTE}_{0.975}$	2.86	0.19	3.31	0.29	3.13	0.25	3.25	0.28
$q_{0.95}$	1.85	0.08	1.91	0.11	1.87	0.10	1.86	0.10
$\text{CTE}_{0.95}$	2.44	0.13	2.73	0.19	2.61	0.16	2.66	0.18
$q_{0.90}$	1.41	0.06	1.38	0.07	1.38	0.07	1.36	0.07
$\text{CTE}_{0.90}$	2.03	0.09	2.17	0.12	2.10	0.11	2.12	0.11
$d = 10$, t_5 univariate margins								
$\text{ave}(\rho_N^-)$	0.32	0.02	0.34	0.02	0.46	0.04	0.55	0.04
$\text{ave}(\rho_N^+)$	0.32	0.02	0.66	0.04	0.46	0.04	0.55	0.04
$q_{0.01}$	-2.05	0.07	-2.39	0.15	-2.77	0.21	-2.85	0.24
$\text{CTE}_{0.01}$	-2.28	0.10	-2.90	0.23	-3.54	0.37	-3.77	0.46
$q_{0.025}$	-1.84	0.05	-1.95	0.10	-2.18	0.13	-2.19	0.14
$\text{CTE}_{0.025}$	-2.08	0.07	-2.45	0.14	-2.88	0.22	-3.00	0.25
$q_{0.05}$	-1.64	0.04	-1.61	0.07	-1.74	0.09	-1.72	0.09
$\text{CTE}_{0.05}$	-1.91	0.05	-2.11	0.10	-2.41	0.14	-2.46	0.16
$q_{0.10}$	-1.38	0.04	-1.25	0.05	-1.29	0.06	-1.26	0.06
$\text{CTE}_{0.10}$	-1.71	0.04	-1.76	0.07	-1.95	0.09	-1.96	0.11
$q_{0.99}$	2.30	0.10	3.11	0.27	2.77	0.21	2.85	0.24
$\text{CTE}_{0.99}$	2.61	0.13	4.15	0.52	3.54	0.37	3.77	0.46
$q_{0.975}$	1.98	0.08	2.37	0.15	2.18	0.13	2.19	0.14
$\text{CTE}_{0.975}$	2.32	0.09	3.28	0.28	2.88	0.22	3.00	0.25
$q_{0.95}$	1.69	0.06	1.83	0.10	1.74	0.09	1.72	0.09
$\text{CTE}_{0.95}$	2.07	0.07	2.68	0.18	2.41	0.14	2.46	0.16
$q_{0.90}$	1.34	0.05	1.31	0.07	1.29	0.06	1.26	0.06
$\text{CTE}_{0.90}$	1.79	0.06	2.11	0.12	1.95	0.09	1.96	0.11

Table 7.4 *Tail comparisons of extreme quantiles of the mean of the variables and corresponding conditional tail expectations, with $\rho = 0.7$ for exchangeable Gaussian and t copulas, and Kendall's tau=0.494 for all 4 copulas. SDs are based on 10000 simulations for sample size $n = 1000$; values of the tail quantities are determined by numerical integration for $d = 2$ and simulation for $d = 10$.*

quantity	Frank value	Frank SD	Gumbel value	Gumbel SD	Gaussian value	Gaussian SD	t_5 value	t_5 SD
				$d = 10$, t_3 univariate margins				
$q_{0.01}$	-2.74	0.14	-3.24	0.28	-3.72	0.39	-3.81	0.43
$CTE_{0.01}$	-3.32	0.33	-4.35	0.58	-5.36	0.91	-5.80	1.23
$q_{0.025}$	-2.34	0.09	-2.48	0.16	-2.72	0.20	-2.72	0.21
$CTE_{0.025}$	-2.84	0.17	-3.43	0.31	-4.04	0.46	-4.24	0.59
$q_{0.05}$	-2.02	0.07	-1.95	0.10	-2.07	0.12	-2.03	0.13
$CTE_{0.05}$	-2.50	0.11	-2.82	0.19	-3.21	0.27	-3.29	0.34
$q_{0.10}$	-1.64	0.06	-1.45	0.07	-1.47	0.08	-1.42	0.08
$CTE_{0.10}$	-2.16	0.07	-2.25	0.12	-2.47	0.17	-2.49	0.19
$q_{0.99}$	3.01	0.18	4.16	0.49	3.72	0.39	3.81	0.43
$CTE_{0.99}$	3.67	0.34	6.44	1.39	5.36	0.91	5.80	1.24
$q_{0.975}$	2.48	0.12	2.91	0.24	2.72	0.20	2.72	0.21
$CTE_{0.975}$	3.10	0.19	4.65	0.66	4.04	0.46	4.24	0.59
$q_{0.95}$	2.04	0.09	2.14	0.14	2.07	0.12	2.03	0.13
$CTE_{0.95}$	2.67	0.13	3.57	0.38	3.21	0.27	3.29	0.34
$q_{0.90}$	1.56	0.07	1.47	0.09	1.47	0.08	1.42	0.08
$CTE_{0.90}$	2.23	0.09	2.67	0.22	2.47	0.16	2.49	0.19
				$d = 10$, t_{20} univariate margins				
$q_{0.01}$	-1.62	0.04	-1.83	0.08	-2.12	0.12	-2.18	0.13
$CTE_{0.01}$	-1.74	0.05	-2.08	0.11	-2.49	0.16	-2.60	0.19
$q_{0.025}$	-1.49	0.03	-1.57	0.06	-1.77	0.08	-1.78	0.09
$CTE_{0.025}$	-1.63	0.04	-1.85	0.07	-2.16	0.11	-2.22	0.12
$q_{0.05}$	-1.36	0.03	-1.35	0.05	-1.47	0.06	-1.46	0.07
$CTE_{0.05}$	-1.53	0.03	-1.65	0.06	-1.88	0.08	-1.92	0.09
$q_{0.10}$	-1.18	0.03	-1.08	0.04	-1.14	0.05	-1.11	0.05
$CTE_{0.10}$	-1.40	0.03	-1.43	0.04	-1.59	0.06	-1.60	0.06
$q_{0.99}$	1.82	0.06	2.37	0.14	2.12	0.12	2.18	0.13
$CTE_{0.99}$	2.00	0.07	2.83	0.20	2.49	0.16	2.61	0.19
$q_{0.975}$	1.62	0.05	1.94	0.10	1.77	0.08	1.78	0.09
$CTE_{0.975}$	1.83	0.05	2.41	0.13	2.16	0.11	2.22	0.12
$q_{0.95}$	1.43	0.04	1.57	0.08	1.47	0.06	1.46	0.07
$CTE_{0.95}$	1.68	0.04	2.08	0.10	1.88	0.08	1.92	0.09
$q_{0.90}$	1.17	0.04	1.17	0.06	1.14	0.05	1.11	0.05
$CTE_{0.90}$	1.48	0.04	1.72	0.07	1.59	0.06	1.60	0.06

Table 7.5 *Tail comparisons of extreme quantiles of the mean of the variables and corresponding conditional tail expections, with $\rho = 0.7$ for exchangeable Gaussian and t copulas, and Kendall's tau=0.494 for all 4 copulas. All values are based on 10000 simulations for sample size $n = 1000$.*

7.3   Discretized multivariate Gaussian and R-vine approximation

In this section, we present an example with simulated multivariate ordinal data to show that the commonly used discretized multivariate Gaussian (or ordinal probit) model can be well approximated by discrete R-vines; the latter have the advantage of faster computation speed as the dimension increases. This expands on the results on Kullback-Leibler divergence in Section 5.7.4.

The setup for the simulated data set is as follows. There are $n = 600$ subjects and $d = 5$ repeated measures per subject; for $d = 5$, it is still possible to run within a reasonabe time the full likelihood for the discretized multivariate Gaussian. There is one predictor variable with values uniform in the interval $(-1, 1)$; its regression parameter is $\beta = 0.4$. There are $K = 3$ ordinal categories with cutpoints $-0.5 < \gamma_1 < \gamma_2 = 0.5$ in the $N(0,1)$ scale. The regression parameter and cutpoints are assumed to be constant over the $d = 5$ repeated measures; just the predictor variable changes. Data are $(y_{i1}, \ldots, y_{id}, x_{i1}, \ldots, x_{id})$, $i = 1, \ldots, n$, where $y_{ij} \in \{1, 2, 3\}$ and x_{ij} are real numbers. The unstructured correlation matrix with 10 correlation parameters is: $\boldsymbol{R} = \begin{pmatrix} 1 & 0.60 & 0.70 & 0.70 & 0.60 \\ 0.60 & 1 & 0.60 & 0.60 & 0.50 \\ 0.70 & 0.60 & 1 & 0.60 & 0.55 \\ 0.70 & 0.60 & 0.60 & 1 & 0.45 \\ 0.60 & 0.50 & 0.55 & 0.45 & 1 \end{pmatrix}$.

To get a good starting point for the full maximum likelihood of 13 parameters, a two-stage estimation is implemented first.

- Stage 1: estimate $\gamma_1, \gamma_2, \beta$ with the one-wise composite log-likelihood (which is essentially assuming independent repeated measures for each subject). This leads to $\tilde{\gamma}_1 = -0.550, \tilde{\gamma}_2 = 0.475, \tilde{\beta} = 0.388$,
- Stage 2: estimate the latent or polychoric correlation matrix with bivariate log-likelihoods for each pair of variables with $\tilde{\beta}, \tilde{\gamma}_1, \tilde{\gamma}_2$ fixed from stage 1. This leads to a positive definite matrix:

$$\widetilde{\boldsymbol{R}} = \begin{pmatrix} 1 & 0.635 & 0.675 & 0.663 & 0.587 \\ 0.635 & 1 & 0.625 & 0.608 & 0.532 \\ 0.675 & 0.625 & 1 & 0.642 & 0.553 \\ 0.663 & 0.608 & 0.642 & 1 & 0.496 \\ 0.587 & 0.532 & 0.553 & 0.496 & 1 \end{pmatrix}$$

$(\tilde{\beta}, \tilde{\gamma}_1, \tilde{\gamma}_2, \widetilde{\boldsymbol{R}})$ can be used as the starting point for the maximum likelihood with the multivariate ordinal probit model, and the maximum likelihood estimates are summarized in Table 7.6.

For this simulated data set, we look for whether $\widetilde{\boldsymbol{R}}$ matches a truncated vine structure such that high-order partial correlations are close to 0, leading to a possibly simpler dependence structure. Based on the computations with Kullback-Leibler sample sizes in Section 5.7.4, we would expect that many different vines would lead to roughly the same approximation. From enumerating through the 480 partial correlation vines for $\widetilde{\boldsymbol{R}}$, it looked like 3-truncated partial correlation vines could provide good approximations (this means that the partial correlation is tree $\mathcal{T}_4$ is close to 0).

We chose three vines where the correlation matrix based on the 3-truncated partial correlation vine has a log determinant within 0.001 of the optimal value over 3-truncated vines. For each of the three vines, the correlation parameters were estimated with $\tilde{\gamma}_1, \tilde{\gamma}_2, \tilde{\beta}$ fixed, and then a full likelihood with 13 parameters is optimized with the vector of two-stage estimates as the starting points. The three vines led to optimal negative log-likelihood values that were within one of each other and that from the ordinal probit model. Table 7.6 summarizes the parameter estimates only for the best of the three.

The 2-stage estimation followed by fitting of discrete R-vines is computationally fast at each stage.

$\hat{\gamma}_1, \hat{\gamma}_2$	$\hat{\beta}$	$\hat{R}$	-loglik.
		model = ordinal probit	
$-0.549, 0.474$	0.396 matrix	$\begin{pmatrix} 1 & 0.644 & 0.675 & 0.669 & 0.589 \\ 0.644 & 1 & 0.622 & 0.609 & 0.531 \\ 0.675 & 0.622 & 1 & 0.636 & 0.540 \\ 0.669 & 0.609 & 0.636 & 1 & 0.486 \\ 0.589 & 0.531 & 0.540 & 0.486 & 1 \end{pmatrix}$	2728.4
		model = discrete R-vine with Gaussian copulas	
$-0.546, 0.474$	0.396	$\begin{pmatrix} 1 & 0.680 & 0.676 & 0.721 & 0.591 \\ 0.680 & 1 & 0.624 & 0.647 & 0.589 \\ 0.676 & 0.624 & 1 & 0.639 & 0.586 \\ 0.721 & 0.647 & 0.639 & 1 & 0.558 \\ 0.591 & 0.589 & 0.586 & 0.558 & 1 \end{pmatrix}$	2728.9
$A = \begin{bmatrix} 3 & 3 & 3 & 3 & 1 \\ & 2 & 2 & 2 & 3 \\ & & 1 & 1 & 2 \\ & & & 4 & 4 \\ & & & & 5 \end{bmatrix}$	$P =$	$\begin{bmatrix} - & 0.624 & 0.676 & 0.639 & 0.591 \\ & - & 0.448 & 0.414 & 0.312 \\ & & - & 0.399 & 0.244 \\ & & & - & 0.104 \\ & & & & - \end{bmatrix}$	

Table 7.6 *A is the vine array for the fitted vine: tree T_1 has bivariate margins $32, 31, 34, 15$; T_2 has conditional distributions $21|3, 24|3, 35|1$; T_3 has conditional distributions $14|32, 25|13$; T_3 has conditional distribution $45|132$. P is the array of partial correlations: parameters in the first row are $\rho_{32}, \rho_{31}, \rho_{34}, \rho_{15}$ for the bivariate pairs in T_1; parameters in the second row for $\rho_{21;3}, \rho_{24;3}, \rho_{35;1}$ for the linking copulas in T_2. These are the dependence parameters of the vine. If these are assumed to come from a multivariate Gaussian distribution, they can be converted to a correlation matrix. But note that they do not match closely to the latent correlation matrix of the probit model, except in T_1 of the vine with pairs $(2,3), (1,3), (3,4), (1,5)$.*

To conclude this section, this example shows that for high-dimensional multivariate models with discrete responses having just a few categories, there can be many good approximating vines, and the advantage of the vine with discrete variables is fast computation of joint probabilities. As shown in Example 3.11, it is easier for the simplifying assumption to hold for multivariate discrete distributions with a few support points for each variable. Table 5.5 shows that the choice of bivariate linking copulas in vines has less influence for discrete responses with few categories. The comparisons here help in understanding examples of real data sets with multivariate discrete responses that are presented in later sections of this chapter.

7.4   Insurance losses: bivariate continuous

In this section, we compare bivariate copula models with tail asymmetry for two variables used in insurance. The modeling of insurance loss and ALAE (allocated loss adjustment expense) data is discussed in Frees and Valdez (1998) and Klugman and Parsa (1999); the example here is from the former reference. Sometimes the recorded loss (claim amount) is right-censored by a policy limit.

For notation, let y_1 denote ALAE and let y_2 denote loss. The variable y_2 can be right-censored. The copula model is $C(F_1(y_1; \boldsymbol{\eta}_1), F_2(y_2; \boldsymbol{\eta}_2); \boldsymbol{\delta})$, where the univariate cdf F_j has density f_j for $j = 1, 2$. Let $F_{12} = C(F_1, F_2)$ is the joint cdf of the two variables, and let $F_{2|1} = C_{2|1}(F_2|F_1)$ be the conditional distribution of loss given ALAE.

To summarize the data, the sample size is $n = 1500$, with 34 loss values being right-censored. Ignoring the censoring, the loss values range from 10 to 2.2×10^6 with quartiles of 4000, 12000 and 35000. The ALAE values range from 15 to 5.0×10^5 with quartiles of 2300,

5500 and 12600. In order that univariate parameter values are not too large, the variables y_1, y_2 for further model fits are loss and ALAE divided by 10000.

The data vectors are (y_{i1}, y_{i2}, I_i) for $i = 1, \ldots, n$, where $I_i = I(y_{i2}$ is right-censored). If $I_i = 1$, then the contribution to the log-likelihood for the ith observation is

$$\log \overline{C}_{2|1}(F_2(y_{i2}; \eta_2)|F_1(y_{i1}; \eta_1); \delta) + \log f_1(y_{i1}; \eta_1).$$

Hence the joint log-likelihood (compare Section 5.3) is:

$$L(\eta_1, \eta_2, \delta) = \sum_{i:I_i=0} \{\log c(F_1(y_{i1}; \eta_1), F_2(y_{i2}; \eta_2); \delta) + \log f_1(y_{i1}; \eta_1) + \log f_2(y_{i2}; \eta_2)\}$$

$$+ \sum_{i:I_i=1} \{\log \overline{C}_{2|1}(F_2(y_{i2}; \eta_2)|F_1(y_{i1}; \eta_1); \delta) + \log f_1(y_{i1}; \eta_1)\}. \tag{7.2}$$

For the IFM method, with univariate parameters held fixed, the relevant log-likelihood is:

$$L_{\text{IFM}}(\delta) = \sum_{i:I_i=0} \log c(F_1(y_{i1}; \eta_1), F_2(y_{i2}; \eta_2); \delta) + \sum_{i:I_i=1} \log \overline{C}_{2|1}(F_2(y_{i2}; \eta_2)|F_1(y_{i1}; \eta_1); \delta).$$

Similar to the analysis of the subset in Section 1.5.3, we consider the 2-parameter Pareto and the 3-parameter Burr distributions for $F_1(\cdot; \eta_1)$ and $F_2(\cdot; \eta_2)$. From the likelihood analysis, (i) for ALAE, the Burr distribution is preferred over the Pareto distribution based on AIC and the Pareto distribution is marginally preferred over the Burr distribution based on BIC; (ii) for loss, the Pareto distribution is preferred over the Burr distribution based on AIC and BIC. So we continue with a comparison of copula models with the Burr distribution for ALAE and the Pareto distribution for loss.

copula	-loglik.	AIC	BIC	parameter(s)	tau
bivariate Gaussian	4449.6	8911.1	8943.0	0.47	0.31
reflected ipsA	4440.4	8892.7	8924.6	2.47	0.32
Galambos	4426.4	8864.8	8896.7	0.72	0.31
Gumbel	4426.3	8864.7	8896.5	1.44	0.31
BB6	4426.3	8866.6	8903.8	1.02,1.42	0.31
reflected iMitLefA	4424.3	8862.6	8899.8	0.41,1.38	0.31
reflected MTCJ	4430.6	8873.1	8905.0	0.81	0.29

	MLE of the iMitLefA copula model			
	parameter	MLE	SE	
	α_{ALAE}	1.65	0.19	
	ζ_{ALAE}	1.11	0.04	
	σ_{ALAE}	0.99	0.15	
	α_{loss}	1.14	0.07	
	σ_{loss}	1.45	0.14	
	$\vartheta_{\text{copula}}$	0.41	0.10	
	δ_{copula}	1.38	0.13	

Table 7.7 *Loss-ALAE data: sample size $n = 1500$, with 34 loss values being right-censored. Log-likelihood, AIC and BIC values are for copula models with the Burr distribution for ALAE/10000 and the Pareto distribution for loss/10000; the variables are scaled so that the σ parameters are not large. Also shown are the maximum likelihood estimates of the copula parameters based on (7.2), and the corresponding Kendall tau values. The maximum likelihood estimates and their standard errors are shown for all univariate and dependence parameters only for the reflected iMitLefA copula model.*

The initial data analysis (see Figure 1.6) suggested skewness to the joint upper tail and no lower tail dependence. Hence we select some appropriate parametric copula families from Chapter 4 (see Table 4.1) with tail asymmetry and take the reflected form if needed to match the direction of tail asymmetry seen in the data. The suitable parametric copula families are the reflected MTCJ, Gumbel, Galambos, BB6 copulas, and also ipsA and iMitLefA (the reflected Archimedean copulas based on integrated positive stable LT and integrated Mittag-Leffler LT). The reflection symmetric bivariate Gaussian copula is used as a baseline comparison. Table 7.7 has comparisons of fit based on the joint log-likelihood (7.2) for these 1-parameter and 2-parameter bivariate copula families.

Of the copula families considered, the 1-parameter Galambos copula is best based on BIC, and the 2-parameter iMitLefA copula is best based on AIC. The full maximum likelihood estimates are given only for the latter. However, the fit of the Galambos, Gumbel and iMitLefA copulas are very similar.

An alternative to AIC/BIC for model comparisons is Vuong's log-likelihood ratio procedure; see Section 5.13. For two fitted models $\hat{f}^{(1)}, \hat{f}^{(2)}$, this essentially checks if enough of the ratios $\hat{f}^{(2)}(y_i)/\hat{f}^{(1)}(y_i)$ are sufficiently larger than 1 to declare $f^{(2)}$ to be a better model than $f^{(1)}$. Because some values of the loss variable are right-censored, we replace a right-censored value $t = y_2^+$ by $\widehat{\mathbb{E}}(Y_2|Y_2 > t) = t + (t + \tilde{\sigma}_{loss})/(\tilde{\alpha}_{loss} - 1)$ based on the estimated Pareto parameters $\tilde{\alpha}_{loss}, \tilde{\sigma}_{loss}$ from the first stage of the IFM procedure. A summary of comparisons of models is given in Table 7.8. The best model based on AIC, that is, the reflected iMitLefA copula, is significantly better than the bivariate Gaussian and reflected MTCJ copulas, but not the Gumbel, Galambos, BB6 and ipsA copulas.

copula 1	copula 2	95% interval
bivariate Gaussian	reflected iMitLefA	$(0.002, 0.024)$
reflected MTCJ	reflected iMitLefA	$(0.003, 0.014)$
reflected ipsA	reflected iMitLefA	$(-0.003, 0.014)$
Gumbel	reflected iMitLefA	$(-0.001, 0.005)$

Table 7.8 *Loss-ALAE data: Summary of Vuong's procedure, with sample size $n = 1500$, and 34 right-censored loss values imputed before computing 95% confidence intervals. (5.21) is used for the average log-likelihood ratio; an interval completely above 0 indicates that the copula 2 is significantly better than copula 1. The copula parameters are based on the second stage of IFM.*

Next we compare the copula models for inference on joint tail probabilities. Table 7.9 has some estimated exceedances in upper joint tail together with standard errors from the copula models in Table 7.7. The univariate quantiles are close to the 80th and 90th in the two cases. For a particular model, the estimated asymptotic covariance matrix V of the MLE from the numerical minimization. An upper tail probability has form

$$P(\boldsymbol{\theta}; q_1, q_2) = \overline{F}(q_1, q_2; \boldsymbol{\theta}) = 1 - F_1(q_1; \boldsymbol{\eta}_1) - F_2(q_2; \boldsymbol{\eta}_2) + C(F_1(q_1; \boldsymbol{\eta}_1), F_2(q_2; \boldsymbol{\eta}_2); \boldsymbol{\delta}),$$

where $\boldsymbol{\theta} = (\boldsymbol{\eta}_1, \boldsymbol{\eta}_2, \boldsymbol{\delta})$ and q_1, q_2 are upper quantiles. The gradient $\nabla = \partial P/\partial \boldsymbol{\theta}$ at the MLE can be computed using the Richardson extrapolation. Then the estimated standard error is $[\nabla^\top V \nabla]^{1/2}$.

Note that for the best fitting models based on AIC/BIC, the estimated upper tail probabilities are similar. For the bivariate Gaussian copula which does not have upper tail dependence, the estimated probability is smaller, and for the reflected MTCJ copula which is much more tail asymmetric, the estimated probability is larger. That is, the behavior of the tail of the copula has a big influence on tail inferences.

copula	$\overline{F}(1.5, 4.4)$	$\overline{F}(2.6, 9.4)$
bivariate Gaussian	0.088 (0.005)	0.031 (0.003)
reflected ipsA	0.090 (0.006)	0.032 (0.003)
Galambos	0.100 (0.006)	0.044 (0.004)
Gumbel	0.101 (0.006)	0.045 (0.004)
BB6	0.101 (0.007)	0.045 (0.004)
reflected imitlefA	0.101 (0.006)	0.044 (0.004)
reflected MTCJ	0.105 (0.007)	0.049 (0.004)
non-parametric	0.112	0.044

Table 7.9 *Loss-ALAE data: joint upper tail exceedances and their standard errors based on the estimated copula models. In column 2, $q_1 = q_{ALAE}/10000 = 1.5$ and $q_2 = q_{loss}/10000 = 4.4$. In column 3, $q_1 = q_{ALAE}/10000 = 2.6$ and $q_2 = q_{loss}/10000 = 9.4$. The univariate quantiles are close to 0.8 and 0.9 levels in the two cases respectively. The non-parametric estimates are obtained using the bivariate Kaplan-Meier estimation procedure in Campbell (1981).*

To conclude this section, we have shown (i) identification of potential copula models based on bivariate tail asymmetry seen in initial data analysis, (ii) likelihood inference including model comparisons with AIC, BIC, Vuong's procedure, and joint tail probabilities.

7.5 Longitudinal count: multivariate discrete

In this section, we show details in fitting a copula model to multivariate discrete data, such as longitudinal counts, with covariates. The steps include the following.

(i) Find a suitable univariate count regression model, accounting for possible overdispersion relative to Poisson.

(ii) Assess the dependence structure by estimating latent correlations, based on a preliminary working model of a multivariate Gaussian copula. Check if the dependence structure appears close to something simple like AR(1) or exchangeable.

(iii) For each bivariate margin, compute the multivariate Gaussian model-based expected frequencies and compare them to the observed frequencies (approach of Section 5.12.2). Check if there is pattern of overestimation or underestimation of bivariate joint tail probabilities; patterns can suggest a copula with different tail behavior to provide a better fit.

Before getting to the data set, we summarize suitable count regression models for count data that are overdispersed relative to Poisson.

Let $\text{NB}(\vartheta, \xi)$ be a parametrization with pmf

$$f_{\text{NB}}(y; \vartheta, \xi) = \frac{\Gamma(\vartheta + y)\, \xi^y}{\Gamma(\vartheta)\, y!\, (1 + \xi)^{\vartheta + y}}, \quad y = 0, 1, 2, \ldots, \ \vartheta > 0, \ \xi > 0, \tag{7.3}$$

with mean $\mu = \vartheta \xi$ and variance $\sigma^2 = \vartheta \xi (1 + \xi)$. The overdispersion index is $\sigma^2/\mu = 1 + \xi$. The parameter ϑ is a convolution parameter in that if ξ is fixed, the convolution of $\text{NB}(\vartheta_1, \xi)$ and $\text{NB}(\vartheta_2, \xi)$ random variables is $\text{NB}(\vartheta_1 + \vartheta_2, \xi)$. The Poisson distribution obtains as $\xi \to 0$, $\vartheta \to \infty$ with $\vartheta \xi = \mu$.

Two versions of negative binomial regression are considered, when x is a vector of covariates; see Cameron and Trivedi (1998) for more details and the interpolation of the two versions. With covariates, the mean function $\mu(x) = \vartheta \xi = \exp(\beta^\top x)$ is considered linear in x on the log scale. One of ϑ, ξ can be assumed to be a function of x with the other fixed over x. For NB1, the convolution parameter ϑ is a function of the covariate x and ξ is constant; that is, $\vartheta(x) = \xi^{-1}\mu(x)$. For NB2, the convolution parameter ϑ is constant and ξ is a function of the covariate $\boldsymbol{x}$; that is, $\xi(\boldsymbol{x}) = \mu(\boldsymbol{x})/\vartheta$. NB2 is the form of negative binomial regression in Lawless (1987).

Another model for overdispersed counts relative to Poisson is the generalized Poisson distribution. There are some differences in tail behavior for generalized Poisson compared with negative binomial — see Joe and Zhu (2005) and Nikoloulopoulos and Karlis (2008c). Let $\mathrm{GP}(\vartheta, \varrho)$ be a parametrization with pmf

$$f_{\mathrm{GP}}(y; \vartheta, \varrho) = \frac{\vartheta(\vartheta + \varrho y)^{y-1}}{y!} \exp\{-\vartheta - \varrho y\}, \quad y = 0, 1, 2, \ldots, \ \vartheta > 0, \ 0 \le \varrho < 1, \quad (7.4)$$

with mean $\mu = \vartheta/(1 - \varrho)$ and variance $\sigma^2 = \vartheta/(1 - \varrho)^3$. The overdispersion index is $\sigma^2/\mu = (1 - \varrho)^{-2} = 1 + \xi$ with $\xi = (1 - \varrho)^{-2} - 1 \ge 0$. The parameter ϑ is a convolution parameter in that if ϱ is fixed, the convolution of $\mathrm{GP}(\vartheta_1, \varrho)$ and $\mathrm{GP}(\vartheta_2, \varrho)$ random variables is $\mathrm{GP}(\vartheta_1 + \vartheta_2, \varrho)$. The Poisson distribution obtains as $\varrho \to 0$.

With covariates, one can define two versions of generalized Poisson regression with mean function $\mu(\boldsymbol{x}) = \vartheta/(1-\varrho) = \exp(\boldsymbol{\beta}^\top \boldsymbol{x})$; one of ϑ, ϱ can be assumed to be a function of $\boldsymbol{x}$ with the other fixed over $\boldsymbol{x}$. Similar to negative binomial, we call the two versions GP1 and GP2. For GP1, the convolution parameter ϑ is a function of the covariate $\boldsymbol{x}$ and ϱ is constant; that is, $\vartheta(\boldsymbol{x}) = (1 - \varrho)\mu(\boldsymbol{x})$, $1 - \varrho = (1 + \xi)^{-1/2}$. For GP2, the convolution parameter ϑ is constant and ϱ or ξ is a function of the covariate $\boldsymbol{x}$; that is, $1 - \varrho(\boldsymbol{x}) = \vartheta/\mu(\boldsymbol{x}) \in (0, 1]$. The constraint on GP2 implies that it may be difficult to implement.

Next, we describe the multivariate count response data set that is analyzed in this section. The data set comes from a study reported in Riphahn et al. (2003) and analyzed in Greene (2008), Nikoloulopoulos et al. (2011) with NB regression models. A count response variable is measured at five consecutive years 1984–1988 for each subject. The count is the number of doctor visits in last quarter prior to the survey. Here we also make comparisons with generalized Poisson regression models, and then compare copula models for the dependence.

When fitting different sets of covariates and looking at the diagnostics of comparing univariate model-based expected frequencies versus observed frequencies (Section 5.12.2), the GP1 and NB1 models are much better fits than NB2. Based on the maximum composite log-likelihood, GP1 fits a little better than NB1. The GP2 model had problems with the constraint of $1 - \varrho(\boldsymbol{x}) = \vartheta/\mu(\boldsymbol{x}) \in (0, 1]$ for some subsets of covariates so we do not include it in further comparisons. In the tables to summarize our analyses, we list a subset of covariates that have larger ratios of regression coefficient to standard error; that is, potential predictor variables that were included in some preliminary analyses and appear to be statistically insignificant are not shown.

Ignoring the repeated counts over subjects, 33% of the counts ($n = 295$ subjects, 5 repeated measurements each) were 0, the maximum is 60, the median is 2. Ignoring the covariates, the sample means, variances and dispersion indices for the five time points are in Table 7.10. Since the dispersion indices are so large, it should be safe to assume overdispersion counts relative to Poisson even after adjusting for covariates.

year	$\bar{y}$	s^2	$s^2/\bar{y}$
1984	3.28	25.0	7.6
1985	3.41	33.3	9.8
1986	3.84	42.1	11.0
1987	3.69	33.0	8.9
1988	3.56	33.1	9.3

Table 7.10: *Doctor visits: sample means, variances and dispersion indices by year.*

Table 7.11 has a summary of estimates of regression coefficients for NB1, NB2 and GP1 regression, with estimated standard errors based on one-wise composite log-likelihood (or misspecified log-likelihood assuming independent observations over the five years). The covariates are ifemale=I(female), agec=(age−50)/10, ageq=agec^2, healthsat=health satisfaction coded from 0 (low) to 10 (high), handicap=degree of handicap in 0 to 1, iuniversity=I(highest schooling degree is university degree). Other covariates were considered less important in the previous analyses; for these covariates, iuniversity might be the least important but is included to compare CLAIC/CLBIC with and without it.

Details on composite likelihood estimation is given in Section 5.6. Let $\tilde{\eta}$ be the max composite likelihood estimate for a model. The penalty term $\mathrm{tr}(\boldsymbol{JH}^{-1})$ can be written as $\mathrm{tr}(\boldsymbol{HV})$ where $\boldsymbol{H}$ is the Hessian of the one-wise log-likelihood (normalized by n^{-1}), $\boldsymbol{J}$ is the covariance matrix of the one-wise log-likelihood score function (normalized by n^{-1}), and $\boldsymbol{V} = \boldsymbol{H}^{-1}\boldsymbol{JH}^{-1}$ is the asymptotic covariance matrix of $n^{1/2}(\tilde{\boldsymbol{\eta}} - \boldsymbol{\eta})$, $n^{-1}\boldsymbol{V}$ can be computed by jackknife, and $\boldsymbol{H}$ can be output from the numerical minimization of the one-wise composite log-likelihood.

parameter	NB1	NB2	GP1	GP1
intercept	1.73 (0.16)	2.11 (0.17)	1.69 (0.16)	1.66 (0.16)
ifemale	0.34 (0.09)	0.27 (0.12)	0.36 (0.09)	0.40 (0.09)
agec	0.18 (0.05)	0.17 (0.07)	0.19 (0.05)	0.19 (0.05)
agec^2	0.11 (0.05)	0.07 (0.06)	0.12 (0.05)	0.11 (0.05)
healthsat	-0.15 (0.02)	-0.21 (0.02)	-0.15 (0.02)	-0.15 (0.02)
handicap	0.67 (0.15)	0.78 (0.20)	0.69 (0.14)	0.70 (0.14)
iuniversity	-0.52 (0.26)	-0.82 (0.39)	-0.53 (0.27)	
ξ	4.80 (0.50)		5.78 (0.70)	5.84 (0.71)
ϑ		0.77 (0.06)		
neg. comp. loglik.	3252	3267	3240	3245
$\mathrm{tr}(\boldsymbol{HV})$	18.2	21.3	17.7	15.3
CLAIC	6541	6577	6515	6521
CLBIC	6608	6656	6580	6577

Table 7.11 *Doctor visits: comparison of univariate count regression models; estimates (standard errors) based on one-wise composite likelihood, minimum negative composite log-likelihood, penalty* $\mathrm{tr}(\boldsymbol{HV})$, *CLAIC and CLBIC;* $\boldsymbol{V}$ *was estimated with the delete-one jackknife. The best univariate regression model with all criteria is GP1; the number of covariates is 5 for CLBIC and 6 for CLAIC.*

The GP1 univariate regression model in Table 7.11 is better than NB1 and NB2 based on CLAIC and CLBIC. There is one less covariate (iuniversity) for the best model based on CLBIC compared with that based on CLAIC. For the multivariate model, we will proceed with the GP1 regression model that includes iuniversity. For the next summary, the matrix of latent (polychoric) correlations from fitting a multivariate Gaussian copula is given below.

$$\widehat{\boldsymbol{R}} = \begin{pmatrix} 1.00 & 0.54 & 0.46 & 0.34 & 0.38 \\ 0.54 & 1.00 & 0.35 & 0.36 & 0.34 \\ 0.46 & 0.35 & 1.00 & 0.34 & 0.28 \\ 0.34 & 0.36 & 0.34 & 1.00 & 0.35 \\ 0.38 & 0.34 & 0.28 & 0.35 & 1.00 \end{pmatrix}$$

This matrix has some small deviations from an exchangeable correlation matrix.

As suggested in Section 5.12.2, in order to assess the adequacy of the fit of a multivariate Gaussian copula, Table 7.12 has the (aggregated) model-based expected versus observed counts for margin 1, margin 2 and the bivariate (1,2) margin for GP1 model with 6 covariates. Based on the univariate expected versus observed counts for margins 1–5, GP1 and

NB1 are much better fits than NB2; hence this diagnostic matches the values of the one-wise composite log-likelihood.

margin		count							
		0	1	2	3	4	5	6	≥ 7
1	E_1	105	50	32	22	16	13	10	48
	O_1	105	36	40	25	27	8	8	46
2	E_2	101	51	33	23	17	13	10	48
	O_2	105	36	39	24	19	12	15	45
(1,2)	E_{12}								
	0	62.6	18.8	9.0	5.0	3.0	1.9	1.3	3.2
	1	17.1	10.7	6.6	4.3	2.9	2.0	1.5	4.5
	2	8.0	6.4	4.4	3.1	2.3	1.7	1.3	4.5
	3	4.4	4.1	3.1	2.3	1.8	1.3	1.0	4.1
	4	2.7	2.8	2.2	1.7	1.4	1.1	0.9	3.7
	5	1.7	1.9	1.6	1.3	1.1	0.9	0.7	3.3
	6	1.1	1.4	1.2	1.0	0.9	0.7	0.6	2.9
	≥ 7	3.0	4.5	4.5	4.2	3.8	3.4	3.0	21.9
	O_{12}								
	0	75	12	8	4	2	1	0	3
	1	8	9	6	2	4	9	4	3
	2	10	4	6	4	5	3	2	6
	3	6	4	4	4	2	1	1	3
	4	2	3	4	5	3	3	2	5
	5	0	0	3	1	0	0	1	3
	6	2	1	3	2	0	0	0	0
	≥ 7	2	3	5	2	3	4	5	22

Table 7.12 *Doctor visits: bivariate Gaussian model-based expected counts compared with observed counts for margins 1, 2 and (1,2);* **E** *is the symbol for expected counts and* **O** *is the symbol for observed counts. The univariate model is GP1 with 6 covariates. The pattern is similar for other univariate margins; the fits are better in other bivariate margins with either the 4th or 5th year in a pair.*

Table 7.13 has comparisons for second stage estimation of dependence parameters; and AIC/BIC values for the multivariate Gaussian copula with exchangeable correlation structure, the multivariate Frank copula, the multivariate reflected Gumbel copula, and D-vines with pair-copulas that are bivariate Gaussian or reflected Gumbel. The reflected Gumbel copula is chosen because, based on the multivariate Gaussian copula, the expected versus observed counts (such as in Table 7.12) suggest that there might be more probability in the joint lower tail than would be expected with a Gaussian copula. The D-vine built from bivariate Gaussian copulas is faster computationally than the discretized multivariate Gaussian model and can approximate the latter well (see Section 7.3); rectangle probabilities for discretized exchangeable Gaussian can always be computed as 1-dimensional integrals. To check if a simpler model fits as well, a truncated D-vine with reflected Gumbel pair-copulas is also fitted because the conditional dependence in higher level trees is weaker.

For the best copula model based on AIC considered in Table 7.13, that is, the D-vine with reflected Gumbel pair-copulas, the bivariate tables of expected versus observed counts were a little better than with the multivariate Gaussian copula with the polychoric correlations.

copula	estimated parameter(s)	-loglik.	AIC	BIC
independence		3240	6480	6480
exchangeable Gaussian	0.37	3129	6260	6264
exchangeable Frank	2.19	3133	6267	6271
exchangeable Gumbel	1.27	3138	6279	6283
exchangeable r.Gumbel	1.30	3150	6301	6305
D-vine with Gaussian	0.54,0.35,0.35,0.35 0.35,0.28,0.19 0.13,0.21,0.18	3118	6255	6292
D-vine with Frank	3.54,2.20,2.14,2.25 2.35,1.71,1.29 0.76,1.00,1.03	3119	6258	6295
D-vine with r.Gumbel	1.61,1.33,1.32,1.31 1.28,1.23,1.15 1.08,1.15,1.14	3117	6254	6291
2-truncated vine, r.Gumbel	1.60,1.32,1.32,1.29 1.29,1.23,1.15	3128	6271	6297

Table 7.13 *Doctor visits: copula models and second stage negative log-likelihoods with univariate parameters held fixed. r.Gumbel is an abbreviation for the reflected Gumbel copula. The negative log-likelihood from the independence copula is the same as the one-wise negative composite log-likelihood in Table 7.11; it is included in this table as a baseline for comparisons. The dependence parameters for D-vines are $\theta_{12}, \theta_{23}, \theta_{34}, \theta_{45}, \theta_{13;2}, \theta_{24;3}, \theta_{35;4}, \theta_{14;23}, \theta_{25;34}, \theta_{15;234}$. AIC is $-2 \times \text{loglik} + 2 \times (\#\text{dependence parameters})$, and BIC is $-2 \times \text{loglik} + (\log n) \times (\#\text{dependence parameters})$ The best dependence model (among the above) based on AIC is D-vine with r.Gumbel and the best model based on BIC is exchangeable Gaussian.*

The two-stage estimation approach is a convenient way to quickly compare candidate copula models. If the total number of univariate and dependence parameters is not too large, then full likelihood should be applied for the best fitting dependence model. This is shown in Table 7.14. The parameter estimates are similar to the two-stage estimates.

univariate	estimate (SE)	dependence	estimate (SE)
β_0	1.54 (0.12)	θ_{12}	1.67 (0.10)
β_{ifemale}	0.33 (0.09)	θ_{23}	1.40 (0.08)
β_{agec}	0.18 (0.04)	θ_{34}	1.37 (0.08)
β_{agec^2}	0.08 (0.04)	θ_{45}	1.36 (0.08)
$\beta_{\text{healthsat}}$	-0.11 (0.01)	$\theta_{13;2}$	1.31 (0.08)
β_{handicap}	0.50 (0.13)	$\theta_{24;3}$	1.25 (0.07)
$\beta_{\text{iuniversity}}$	-0.45 (0.25)	$\theta_{35;4}$	1.18 (0.06)
ξ	5.83 (0.54)	$\theta_{14;23}$	1.08 (0.05)
		$\theta_{25;34}$	1.16 (0.06)
		$\theta_{15;234}$	1.15 (0.06)

Table 7.14 *Doctor visits: maximum likelihood estimates and standard errors from full likelihood with GP1 univariate regression and D-vine with reflected Gumbel for pair-copulas.*

In looking at the the standard errors for the full log-likelihood, it seems the subset of covariates based on CLBIC is better because the coefficient for iuniversity is no longer statistically significant (it was borderline significant based on the one-wise composite likelihood). The estimates of the dependence parameters do not change much if iuniversity is excluded as a covariate.

For comparison of the dependence models, we next show in Table 7.15 some applications of Vuong's log-likelihood ratio procedure in Section 5.13. Only a few of the models in Table 7.13 are included. Using Vuong's procedure, the copula models based on multivariate Frank, exchangeable Gaussian and D-vine with reflected Gumbel are all significantly better than the independence copula, and the D-vine with reflected Gumbel is significantly better than multivariate Frank. The discretized exchangeable Gaussian copula model is not significantly different from multivariate Frank or the D-vine. Vuong's procedure can provide an idea if differences based on AIC are large enough to affect model-based predictions.

copula 1	copula 2	95% interval
independence	mult. Frank	$(0.24, 0.48)$
independence	exch. Gaussian	$(0.24, 0.51)$
independence	D-vine/r.Gumbel	$(0.29, 0.55)$
mult. Frank	exch. Gaussian	$(-0.04, 0.06)$
mult. Frank	D-vine/r.Gumbel	$(0.01, 0.10)$
exch. Gaussian	D-vine/r.Gumbel	$(-0.01, 0.09)$

Table 7.15 *Doctor visits: 95% confidence intervals (5.21) for the average log-likelihood ratio; an interval completely above 0 indicates that the copula 2 is significantly better than copula 1. The univariate parameters are based on the one-wise composite likelihood. The copula parameters are based on the second stage with univariate parameters held fixed from the first stage.*

In summary, this example has illustrated the use of composite likelihood information criteria for choosing a univariate regression model for longitudinal count data, and also the use of exchangeable copulas and D-vine copulas with two-stage estimation. It is typical for multivariate discrete data that copula models with similar dependence structure provide approximately the same level of fit because the effects of the tail behavior of the copulas is attenuated.

7.6 Count time series

In this section, we use copula-based Markov models for count time series with covariates, where a parametric copula family is used for the joint distribution of two or three consecutive observations and then the corresponding transition probabilities are obtained. We assume that counts are small with values at or near 0, otherwise continuous time series models can be used if the counts are large and far away from 0. If the data were stationary, time series models with a Poisson stationary margin are possibilities but more general would be time series models with negative binomial or generalized Poisson margins; the latter two accommodate overdispersed count data relative to Poisson and include Poisson time series models at the boundary of the parameter space.

The literature for discrete time series has many models based on thinning operators; see Weiss (2008) for a survey. These models have nice properties including linear conditional expectation, and autocorrelation functions that are similar to autoregressive Gaussian time series except that only positive dependence is possible. The stationary margins of these models are typically infinitely divisible (such as negative binomial and generalized Poisson). However, these models based on thinning operators do not always extend nicely to nonstationary time series with covariates, and the higher order models might not be as interpretable.

Copula-based Markov time series models are more flexible, in that any univariate count regression model and negative serial dependence can be accommodated; but nonlinear conditional expectations result. The Gaussian copula is a special case, and this is called autoregressive-to-anything in Biller and Nelson (2005). Non-Gaussian copulas with upper tail dependence would be useful for the transition probability if there is more clustering of consecutive large values that would be expected with Gaussian.

We next introduce and briefly explain two models with thinning operators and also write the probabilities for the copula-based Markov time series models.

Let $\{Y_t : t = 1, 2, \ldots, n\}$ be a count time series with values in the set of non-negative integers. Negative binomial (NB) and generalized Poisson (GP) probability distributions are given in Section 7.5.

(a) (Binomial thinning): $Y_t = \alpha \circ Y_{t-1} + \epsilon_t(\alpha)$, where $\{\epsilon_t(\alpha)\}$ are innovations and $\alpha \circ Y$ refers to a random variable that is Binomial(y, α) conditioned on $Y = y$. This model is most plausible when a random proportion α of "counted individuals" continues from one time point to the next, and innovations are added at each time point. When the stationary margin is NB or GP, the distribution of $\epsilon_t(\alpha)$, the conditional distribution $\mathbb{P}(Y_t = w_2 | Y_{t-1} = w_1)$ and the joint distribution F_{12} of Y_{t-1}, Y_t can be computed with the method in Zhu and Joe (2006). For a higher order Markov model of order p, the INAR(p) extension to higher order terms is less interpretable. When there the covariates, the margin cannot be specified as NB or GP.

(b) (Thinning based on a joint distribution that uses the convolution closure property, Joe (1996b)): Let $Y_t \sim F(\cdot; \vartheta, \eta)$ where $\{F(\cdot; \vartheta, \eta)\}$ is infinitely divisible with convolution parameter ϑ and η is a non-convolution parameter. For Markov order 1, the stochastic representation for two consecutive observations is

$$Y_{t-1} = Z_{12} + Z_1, \quad Y_t = Z_{12} + Z_2, \tag{7.5}$$

where Z_1, Z_2, Z_{12} are independent random variables, $Z_j \sim F(\cdot; (1 - \alpha)\vartheta, \eta)$ for $j = 1, 2$, and $Z_{12} \sim F(\cdot; \alpha\vartheta, \eta)$. For NB and GP margins, the details of the joint distribution F_{12} of Y_{t-1}, Y_t and conditional distribution $\mathbb{P}(Y_t = w_2 | Y_{t-1} = w_1)$ are given in Section 8.4 of Joe (1997); for NB, the thinning operation involves a beta binomial distribution and for GP, the thinning operation involves a quasi-binomial distribution. This extends to higher order Markov but the computations are only managable up to Markov order 2 because the probabilities involve high-order summations. For Markov order 2, the stochastic representation for three consecutive observations is

$$Y_{t-2} = Z_{123} + Z_{12} + Z_{13} + Z_1, \quad Y_{t-1} = Z_{123} + Z_{12} + Z_{23} + Z_2, \quad Y_t = Z_{123} + Z_{13} + Z_{23} + Z_3,$$

where $Z_{123}, Z_{12}, Z_{23}, Z_{13}, Z_1, Z_3, Z_2$ are independent random variables with respective convolution parameters $\alpha_0 \vartheta$, $\alpha_1 \vartheta$, $\alpha_1 \vartheta$, $\alpha_2 \vartheta$, $(1 - \alpha_0 - \alpha_1 - \alpha_2)\vartheta$, $(1 - \alpha_0 - \alpha_1 - \alpha_2)\vartheta$, $(1 - \alpha_0 - 2\alpha_1)\vartheta$ and $\alpha_0, \alpha_1, \alpha_2 \geq 0$ with $0 < \alpha_0 + \alpha_1 + \alpha_2 < 1$, $0 < \alpha_0 + 2\alpha_1 < 1$. In order to have just one more parameter than Markov order 1, α_2 can be set to 0, and the special case of $\alpha_0 = \alpha^2$ and $\alpha_1 = \alpha(1 - \alpha)$ leads to the equivalent of model (7.5). This model can be adapted to include covariates with margin $F(\cdot; \vartheta_t, \eta)$ at time t.

(c) (Copula-based): For stationary Markov order 1, the joint distribution of Y_{t-1}, Y_t is $F_{12} = C(F(\cdot; \vartheta, \eta), F(\cdot; \vartheta, \eta); \boldsymbol{\delta})$ for a bivariate copula family $C(\cdot; \boldsymbol{\delta})$. This leads to the transition probability

$$\mathbb{P}(Y_t = w_2 | Y_{t-1} = w_1) = \frac{F_{12}(w_1, w_2) - F_{12}(w_1^-, w_2) - F_{12}(w_1, w_2^-) + F_{12}(w_1^-, w_2^-)}{f(w_1; \vartheta, \eta)},$$

where w_j^- shorthand for $w_j - 1$ and $f(\cdot; \vartheta, \eta)$ is the univariate pmf. For stationary Markov order 2, the joint distribution of Y_{t-2}, Y_{t-1}, Y_t is $F_{123} = C(F(\cdot; \vartheta, \eta), F(\cdot; \vartheta, \eta), F(\cdot; \vartheta, \eta); \boldsymbol{\delta})$ for a trivariate copula family $C(\cdot; \boldsymbol{\delta})$ such that the C_{12}, C_{23} bivariate margins are the same. This class of models can be adapted to include covariates with margin $F(\cdot; \vartheta_t, \eta)$ or $F(\cdot; \vartheta, \eta_t)$ at time t.

Note that the thinning operations in (b) bridge between (a) and (c).

For the models based on thinning operations in (a) and (b) above, we computed Jeffreys' divergence and Kullback-Leibler sample sizes (see Section 5.7) to assess the closeness of some copula models for the pmfs of two or three consecutive observations. With stationary NB margins as given in (7.3) with non-convolution parameter $\eta = \xi$ the models with thinning operators in (a) and (b) have behavior of intermediate tail dependence based on $\{\log F_{12}(y, y; \vartheta, \xi; \delta)\}/\{\log F_{\mathrm{NB}}(y; \vartheta, \xi)\}$ for $y = 4, \ldots, 1, 0$ and $\{\log \overline{F}_{12}(y, y; \vartheta, \xi; \delta)\}/\{\log \overline{F}_{\mathrm{NB}}(y; \vartheta, \xi)\}$ for larger values of y; also they have slight skewness in the direction of the joint upper tail, this being less so for the binomial thinning operator in (a). However, with moderate dependence (lag1 correlation in the range 0.4 to 0.7), the Kullback-Leibler sample sizes are typically over 500 when approximated with bivariate and trivariate Gaussian copulas for C; for (b), a copula with slight skewness to the joint upper tail leads to even larger Kullback-Leibler sample sizes. Similar patterns exists for stationary GP margins as given in (7.4) with non-convolution parameter $\xi = (1 - \varrho)^{-2} - 1$.

The copula-based models with Gaussian copulas are numerically fast for likelihood inference so we use this as a baseline to compare with other copula models in the data set that is described next.

The data set consists of monthly number of claims of short-term disability benefits made by injured workers to the B.C. Workers' Compensation Board (WCB). These data were reported for the period of 10 years from 1985 to 1994. There are multiple monthly time series for different classification of industry and injury, and we use one corresponding to cut injury in the logging industry; based on preliminary analysis, this series appears to be overdispersed relative to Poisson. The data are shown in Figure 7.1. Because of seasonality, we used seasonal covariates as $(x_{t1}, x_{t2}) = (\sin(2\pi t/12), \cos(2\pi t/12))$. The data set was analyzed in Zhu and Joe (2006) with a model that combines the operations in (a) and (b); the original source is Freeland (1998).

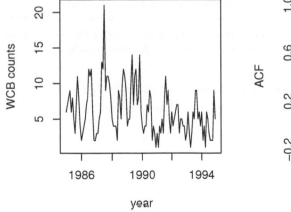
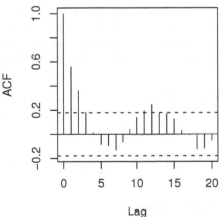

Figure 7.1 *WCB: monthly counts of compensation for cut injury in logging industry from January 1985 to December 1994; time series plot on left, autocorrelation function plot on right.*

For NB regression with convolution parameter ϑ and mean $\mu = \vartheta\xi$, time-varying covariates $\boldsymbol{x}_t$ lead to a mean $\mu_t = \exp\{\beta_0 + \boldsymbol{\beta}^\top \boldsymbol{x}_t\}$ at time t; for NB1, the dispersion index ξ

is fixed and ϑ_t varies with time, and for NB2, ϑ is fixed and ξ_t varies with time,. For GP regression, similarly there are the GP1 and GP2 versions with $\xi = (1 - \varrho)^{-2} - 1$ where the (ϑ, ϱ) parametrization is given in (7.4).

Summaries are given in Table 7.16 for Markov order 1 and 2 time series models with NB1, NB2 and GP1 regressions; GP2 is not included because the MLEs converge to a boundary where one of the ϱ_t values is 0. For Markov order 1 models with Gaussian, Frank, Gumbel and reflected Gumbel copulas, the model results are very similar. Based on log-likelihoods and root mean square prediction errors, the best model (by a small margin) is NB2 regression with the Frank copula. For each fixed regression model, the Frank copula is best, and then Gaussian, reflected Gumbel and Gumbel (the latter is not shown in Table 7.16). The best copula-based model is a little better in terms of AIC/BIC and root mean square prediction error compared with the best fitting model based on thinning operators in (a) and (b).

regr.	copula	p	β_0	β_1	β_2	ξ or ϑ	δ	-loglik.	rmspe
NB2	Frank	1	1.81(0.07)	-0.24(0.09)	-0.22(0.08)	12.3(5.2)	3.71(0.68)	276.8	2.64
NB2	r.Gumbel	1	1.76(0.08)	-0.31(0.09)	-0.18(0.09)	10.3(4.9)	1.49(0.15)	279.9	2.72
NB2	Gaussian	1	1.79(0.08)	-0.29(0.09)	-0.20(0.09)	11.6(5.6)	0.53(0.08)	277.2	2.71
NB1	Frank	1	1.82(0.07)	-0.21(0.08)	-0.21(0.08)	0.49(0.22)	3.69(0.67)	277.7	2.65
NB1	r.Gumbel	1	1.77(0.08)	-0.28(0.09)	-0.16(0.09)	0.55(0.27)	1.47(0.15)	280.9	2.73
NB1	Gaussian	1	1.79(0.08)	-0.26(0.09)	-0.18(0.09)	0.48(0.25)	0.51(0.08)	278.3	2.71
GP1	Frank	1	1.82(0.07)	-0.20(0.08)	-0.21(0.08)	0.50(0.22)	3.69(0.67)	277.6	2.65
GP1	r.Gumbel	1	1.77(0.08)	-0.28(0.09)	-0.16(0.09)	0.54(0.26)	1.47(0.14)	280.9	2.73
GP1	Gaussian	1	1.79(0.08)	-0.26(0.09)	-0.18(0.09)	0.49(0.26)	0.51(0.08)	278.2	2.71
NB1	Gaussian	2	1.79(0.09)	-0.26(0.09)	-0.18(0.09)	0.48(0.28)	0.51(0.09) 0.34(0.11)	277.7	2.70
NB2	Gaussian	2	1.79(0.09)	-0.29(0.09)	-0.19(0.09)	11.0(5.6)	0.54(0.09) 0.36(0.11)	276.6	2.70

Table 7.16 *WCB monthly count time series: maximum likelihood estimates and standard errors for univariate regression models combined with copula-based transition. The seasonal covariates are* $(x_{t1}, x_{t2}) = (\sin(2\pi t/12), \cos(2\pi t/12))$. *The root mean square prediction error (rmspe) is defined as* $\{(n-p)^{-1} \sum_{t=1+p}^{n} [y_t - \hat{\mathbb{E}}(Y_t | Y_{t-1} = y_{t-1}, \ldots, Y_{t-p} = y_{t-p})]^2\}^{1/2}$, *where* p *is the Markov order and* $\hat{\mathbb{E}}$ *means the conditional expectation with parameter vector equal to the MLE of the model. With no adjustment for seasonal effects, the rmpse is between 2.78 and 2.86 for these models.*

If $H_{12} = H_{32} = H_{23} = H$ are max-id bivariate copulas, and $H_{13}(u_1, u_3) = u_1 u_3$, then $K(u_1, u_2, u_3) = H(u_1^{1/2}, u_2^{1/2}) H(u_3^{1/2}, u_2^{1/2}) (u_1 u_3)^{1/2}$ is a trivariate max-id copula, and (3.19) leads to (3.22), that is,

$$C_{\psi, K}(\boldsymbol{u}) = \psi\Big(\sum_{j \in \{1, 3\}} \Big[-\log H\big(e^{-0.5\psi^{-1}(u_j)}, e^{-0.5\psi^{-1}(u_2)}\big) + \tfrac{1}{2}\psi^{-1}(u_j) \Big]\Big);$$

it has bivariate margins:

$$C_{j2} = \psi\Big(-\log H(e^{-0.5\psi^{-1}(u_j)}, e^{-0.5\psi^{-1}(u_2)}) + \tfrac{1}{2}\psi^{-1}(u_j) + \tfrac{1}{2}\psi^{-1}(u_2)\Big), \quad j = 1, 3,$$

and $C_{13}(u_1, u_3) = \psi(\psi^{-1}(u_1) + \psi^{-1}(u_3))$. This $C_{\psi, K}$ is a suitable copula, with closed form cdf, for the transition of a stationary time series of Markov order 2, when there is more dependence for measurements at nearer time points.

For Markov order 2, trivariate copulas of the above form $C_{\psi, K}$ or a reflection of $C_{\psi, K}$ were considered with ψ being of positive stable and logarithmic series LTs, and H being bivariate Frank. Also fitted were trivariate Gaussian copulas. The log-likelihoods were relatively flat and the convergences of the numerical optimizations were not always clean. The conclusion is that Markov order 2 models are not an improvement for this data set. Table 7.16 only reports the trivariate Gaussian copula with NB1/NB2 to show that these models are not improvements on the best Markov 1 models. Note that none of these mixture of max-id has $(1, 2)$ margin that is the Frank (or Gumbel) copula; there are no known trivariate Frank (Gumbel) copulas where all bivariate margins are Frank (Gumbel) copulas and the $(1, 3)$ margin has less dependence.

Much more flexibility in univariate regression models is an advantage of copula-based models over those based on thinning operators; with the latter, only the NB1 or GP1 type of regression margin is possible.

For the best fitting Markov order 1 model with the Frank copula, to obtain predictive distributions, we obtained a plug-in sampling distribution that should be closer to multivariate normal, with transformed $\log \vartheta$ and $\log \delta$ as parameters. This led to an MLE vector of $\hat{\boldsymbol{\theta}} = (1.815, -0.238, -0.219, 2.509, 1.310)^\top$ with asymptotic covariance matrix

$$\hat{V} = \begin{pmatrix} 0.00505 & 0.00112 & 0.00069 & 0.00161 & 0.00217 \\ 0.00112 & 0.00732 & 0.00078 & 0.00199 & 0.00056 \\ 0.00069 & 0.00078 & 0.00647 & 0.00157 & -0.00132 \\ 0.00161 & 0.00199 & 0.00157 & 0.17808 & -0.02458 \\ 0.00217 & 0.00056 & -0.00132 & -0.02458 & 0.03325 \end{pmatrix}.$$

This matrix is based on the inverse Hessian of the negative log-likelihood at the MLE; by results in Billingsley (1961), this is valid for Markov-dependent data.

For a parametric bootstrap sample size of 10000 from $N_5(\hat{\boldsymbol{\theta}}, \hat{V})$, for each replication the shortest 50% prediction intervals was obtained. Over the bootstrap samples with slightly varying parameter vectors, the most common 50% predictive intervals are $(2, 5)$, $(3, 6)$, and $(2, 4)$ for January, and $(5, 9)$, $(5, 10)$, $(6, 10)$, $(6, 11)$ for July. Similarly, if it is also known that the preceding month's value is 5, the most common 50% predictive intervals for July are $(4, 7)$ and $(5, 8)$; if the preceding month's value is 10, the most common 50% predictive intervals for July are $(8, 12)$ and $(7, 11)$.

An advantage of a model where univariate and conditional probabilities can be easily obtained is that predictive intervals with and without the previous observation can be obtained. A predictive interval without the previous time point is a regression inference and a predictive interval with the previous time point is an inference that combines regression and forecasting. The first type of predictive interval would not be easy to do with other classes of count time series models which are based on conditional specifications.

In summary, this example shows how bivariate and trivariate copula models can be used for the Markov transition probabilities for dependent count times series with a univariate margin that is a count regression model. Parametric likelihood inference methods, including assessment of prediction error and calculation of predictive intervals, are straightforward.

7.7 Multivariate extreme values

This section illustrates the application of fitting copula models to multivariate extremes. The data set consists of monthly extreme sea level maxima at a few sites along the British eastern coast, 23 years 1990–2012 or 276 months. A source is https://www.bodc.ac.uk/

We use six sites as shown in Figure 7.2: from south to north, they are Dover, Lowestoft, Cromer, Immingham, Whitby, N. Shields. The analysis will show that the shape of the coastline has an effect on dependence between locations and the dependence is not simply a function of distance between sites.

Figure 7.2: *UK sea levels: 6 sites on the east coast.*

For the monthly sea level maxima, there is an annual pattern, consistent over different sites, with lowest values in May to July and increasing to larger values from September to March except December. See Table 7.17 for the site of Dover; the pattern is somewhat similar for the other sites.

month	mean	SD	month	mean	SD
1	7.18	0.26	7	6.89	0.13
2	7.26	0.25	8	7.12	0.11
3	7.23	0.18	9	7.22	0.13
4	7.13	0.14	10	7.20	0.21
5	6.85	0.16	11	7.21	0.24
6	6.77	0.13	12	7.01	0.13

Table 7.17 *Sea level data for Dover: trend over months within a year based on averages and SDs for 23 years.*

Based on averages and SDs over years for each of the 12 months for each site, it was decided to fit univariate generalized extreme value (GEV) for three more homogeneous subsets of three months: (i) January to March, (ii) May to July and (iii) September to November. The transitional months of April, August and December are not used. A more sophisticated model approach that uses more information on hourly tides would follow ideas in Tawn (1992).

There were some missing values for all but one site, and in addition, also a few low outliers relative to GEV were removed for better fit when assessed with quantile plots. For further likelihood analysis, we assume that these values are missing at random. That is, univariate analysis for a single site proceeded with data on all available months at that site, and bivariate analysis for a pair of sites proceeded with data on all months where observations are available for both sites.

With three subsets, there are three GEV vectors (ξ, σ, μ), consisting of tail, scale and location parameters, to be estimated for each of six sites (total of 54 univariate parameters). This GEV analysis suggested that the the scale and location parameters were significantly different over the three periods, while the tail parameters were different but not always significantly different over the three periods.

The estimated tail parameters ξ are less than 0 except for the Lowestoft site which has the smallest maximum monthly sea levels among the six sites and also more variability in monthly values over 23 years. Hence the distributions of sea levels mostly have lighter tail than exponential.

For fitting a multivariate extreme value distribution/copula, the univariate maxima were converted via the GEV parameters to marginally uniform vectors $(u_{i1}, \ldots, u_{id})$ and to exponential survival vectors $(e_{i1}, \ldots, e_{id})$, separately for the three 3-month periods. The assumption of min-stable multivariate exponential distribution was assessed as acceptable based on quantile plots from weighted minima $e_i(\boldsymbol{w}) = \min_j \{w_j e_{ij}\}$ for some random $\boldsymbol{w} \in (0, \infty)^d$.

For a multivariate extreme value model, we use the multivariate Hüsler-Reiss copula, which has general dependence structure; its parameters can be estimated from separate bivariate log-likelihoods; this avoids the need to derive and code the 6-dimensional density of the Hüsler-Reiss distribution. This model has an advantage over the Gumbel copula because of its more flexible multivariate extension. An alternative is to fit multivariate t-EV (Section 4.16) with the parameter ν fixed, also based on bivariate log-likelihoods. The adequacy of the Hüsler-Reiss model for different bivariate margins can be checked by comparing the parametric exponent function B in (4.22) versus non-parametric estimators; see Naveau et al. (2009) for a discussion of several non-parametric estimators.

With multivariate Hüsler-Reiss copulas fitted separately to the three 3-month periods, the dependence structure seemed to be similar when considering the larger estimated standard errors (compared with those for the univariate parameters). Hence for better estimates of the dependence structure and the second stage of IFM estimation, the vectors $(u_{i1}, \ldots, u_{id})$ over the three 3-month periods were pooled together. The standard errors of parameters were obtained via jackknife (delete-one-year) of the 2-stage IFM estimation. A summary of estimated parameters and their standard errors is given in Table 7.18.

It has been checked that the estimated Hüsler-Reiss parameters satisfy the constraints of a d-variate Hüsler-Reiss distribution with $d = 6$. If $\delta_{jk} > 0$ is the parameter for the (j, k) margin with $j \neq k$, then the consistency condition for the $\binom{d}{2}$ parameters is that d correlation matrices of dimension $(d - 1) \times (d - 1)$ are positive definite. Defining $\delta_{ii} = \infty$ and $\delta_{ii}^{-1} = 0$ for all i, the jth matrix is $\boldsymbol{R}_j = (\rho_{ik;j})_{i,k \neq j}$ with

$$\rho_{ik;j} = \frac{\delta_{ij}^{-2} + \delta_{kj}^{-2} - \delta_{ik}^{-2}}{2\delta_{ij}^{-1}\delta_{kj}^{-1}}.$$

From the copula parameter estimates, dependence is not simply decreasing with distance. Even though sites Lowestoft and Cromer are close neighbors relative to the other four sites, the dependence of Lowestoft with other sites is much weaker than the dependence of Cromer with other sites. The three sites of Immingham, Whitby, and N. Shields have dependence that is close to an exchangeable cluster for dependence, so maybe the dependence structure can be simplified. Tawn (1992) has some analyses of annual maximum hourly sea levels at three sites including Lowestoft and Immingham, and mentions that the characteristics of Lowestoft are different because this site is surge dominant while Immingham is tidally dominant.

site	GEV for months 1–3			GEV for months 5–7			GEV for months 9–11		
	$\hat\xi$	$\hat\sigma$	$\hat\mu$	$\hat\xi$	$\hat\sigma$	$\hat\mu$	$\hat\xi$	$\hat\sigma$	$\hat\mu$
D	-.08 (.07)	.20 (.02)	7.13(.03)	-.36 (.11)	.15 (.01)	6.79(.03)	-.05 (.08)	.16 (.02)	7.13(.02)
L	.10 (.09)	.23 (.03)	3.02(.04)	-.11 (.07)	.10 (.01)	2.70(.01)	.17 (.10)	.18 (.02)	2.99(.02)
C	-.09 (.06)	.21 (.02)	5.45(.04)	-.26 (.05)	.14 (.01)	5.11(.02)	-.08 (.09)	.17 (.02)	5.44(.03)
I	-.18 (.06)	.19 (.02)	7.71(.03)	-.29 (.09)	.16 (.02)	7.31(.03)	-.37 (.10)	.25 (.02)	7.70(.04)
W	-.49 (.10)	.27 (.04)	5.98(.03)	-.55 (.33)	.20 (.03)	5.68(.04)	-.36 (.10)	.18 (.02)	6.06(.03)
N	-.12 (.07)	.17 (.01)	5.49(.03)	-.50 (.06)	.15 (.01)	5.21(.02)	-.24 (.20)	.15 (.02)	5.55(.02)

Hüsler-Reiss dependence parameters and Kendall taus			
D	L	1.21 (.11)	.34 (.04)
D	C	1.57 (.21)	.45 (.05)
L	C	1.38 (.14)	.39 (.04)
D	I	1.60 (.19)	.45 (.05)
L	I	1.04 (.06)	.27 (.02)
C	I	1.38 (.16)	.39 (.05)
D	W	1.21 (.15)	.34 (.05)
L	W	0.98 (.08)	.25 (.03)
C	W	1.35 (.25)	.38 (.08)
I	W	1.41 (.17)	.40 (.05)
D	N	1.27 (.13)	.36 (.04)
L	N	0.97 (.06)	.24 (.03)
C	N	1.55 (.23)	.44 (.06)
I	N	1.64 (.26)	.46 (.06)
W	N	1.44 (.27)	.41 (.07)

Table 7.18 *UK sea level data with two-stage estimation. Univariate GEV parameters with (ξ, σ, μ) estimated separate for three 3-month periods. The bivariate parameters are for the Hüsler-Reiss model, and the corresponding Kendall tau values are based on the inverse transform of the copula parameter. The standard errors were obtained by jackknife of the two-stage estimation procedure. Sites are Dover, Lowestoft, Cromer, Immingham, Whitby, N. Shields.*

The model parameters were estimated from univariate and bivariate margins, but inference can be performed with calculations from the d-variate Hüsler-Reiss distribution. As an example, we obtain the model-based estimates with jackknife-based SEs for the probability of at least one of the five sites Dover, Cromer, Immingham, Whitby, N. Shields exceeding a high quantile in January (July). The quantiles are $7.00, 5.35, 7.55, 5.95, 5.40$ respectively for the five sites; these are close to 90th percentile of monthly maxima for July, based on empirical values of 23 years and the fitted GEV distributions by site.

Estimates of the exceedance probabilities

month	point estimate	SE
January	0.984	0.008
July	0.254	0.062

The probability is much higher in January because the sea levels tend to be higher than in July. For comparison for July, if the sea levels at the 5 sites were independent, the probability would be about $1 - (0.9)^5 = 0.41$ and if the sea levels were perfectly dependent, the probability would be about 0.1.

In summary, this example shows the combination of fitting a multivariate extreme value copula with general dependence structure and GEV margins for multivariate data that are maxima. Because the multivariate density has high-dimensional integrals and the dependence parameters are identifiable from bivariate margins, likelihood inference proceeded with two-stage estimation of univariate and bivariate parameters.

7.8 Multivariate financial returns

This section has examples of fitting copula models to financial returns. Section 7.8.2 compares truncated R-vine and factor copula models for market returns, and Section 7.8.3 compares high-dimensional structural factor copula models for stock returns over different sectors. The former has 7 market returns and the latter has 29 stock returns in several sectors. A source for the data is http://quote.yahoo.com. Longin and Solnik (2001), Ang and Chen (2002), Hong et al. (2007), Okimoto (2008) and others have reported on tail asymmetries in the dependence in the extreme values of returns, so part of the data analysis involves assessing tail dependence and asymmetries.

Stylized facts for financial returns data (see Cont (2001) and Section 2.1.3 of Jondeau et al. (2007)) include the following; (i) fat tails (heavier tails than Gaussian); (ii) volatility clustering with typically significant serial correlations in absolute returns and squared returns. Item (i) motivates that use of the t_ν distribution and other scale mixture of Gaussian as univariate models for financial returns. Item (ii) is the basis of generalized autoregressive conditional heteroscedastic (GARCH) models for stochastic volatility; see Engel (2009) for details and multivariate versions of these models. Section 7.8.1 summarizes the copula model with univariate GARCH margins for individual financial time series.

7.8.1 Copula-GARCH

The multivariate GARCH model does not have univariate GARCH time series as margins, but the copula-GARCH model in Jondeau and Rockinger (2006) builds on univariate GARCH models for individual financial assets. Vine copulas and other copula models have been fitted to GARCH-filtered returns in, for example, Aas et al. (2009), Fantazzini (2009), Liu and Luger (2009), Nikoloulopoulos et al. (2012).

The stochastic volatility in financial return time series that are daily, weekly or monthly shows up in the autocorrelation functions of the squared or absolute returns (but not the actual returns). The GARCH-filtered times series have autocorrelation functions that do not have serial dependence in the squared or absolute values.

A summary of the copula-GARCH models is as follows. Let P_t $(t = 0, 1, \ldots, n)$ be the price time series of a financial asset such as a market index or stock; the time index could be day, week, or month. The (log) return R_t is defined as $\log(P_t/P_{t-1})$. For d assets, we denote the returns as $R_{t1}, \ldots, R_{td}$ at time t. Copula modeling can proceed in two steps. For a single financial return variable, a common choice is the GARCH(1,1) time series filter with innovation distribution being the symmetric Student t_ν distribution with variance 1 and $\nu > 2$; see Section 4.3.6 of Jondeau et al. (2007).

The univariate marginal model is:

$$R_{tj} = \mu_j + \sigma_{tj} Z_{tj}, \quad \sigma_{tj}^2 = \omega_j + \alpha_j R_{t-1,j}^2 + \beta_j \sigma_{t-1,j}^2, \quad j = 1, \ldots, d, \ t = 1, \ldots, n, \quad (7.6)$$

where for each j, $\omega_j > 0$, $\alpha_j > 0$, $\beta_j > 0$, the Z_{tj} are assumed to be innovations that are i.i.d. over t, and the random σ_{tj}^2 depends on $R_{t-1,j}^2$ and $\sigma_{t-1,j}^2$. For stationarity, $\alpha_j + \beta_j < 1$ for all j. The vectors $(Z_{t1}, \ldots, Z_{td})$ for $t = 1, \ldots, n$ are assumed to be i.i.d. with distribution:

$$F_Z(z; \nu_1, \ldots, \nu_d, \delta) = C\big(F_1(z_1; \nu_1), \ldots, F_d(z_d; \nu_d); \delta\big),$$

where δ is a dependence parameter of a d-dimensional copula C and F_j is the scaled Student t distribution with parameter $\nu_j > 2$. For reference below, we define $\eta_j = (\nu_j, \mu_j, \omega_j, \alpha_j, \beta_j)$. Sometimes an autoregressive coefficient ϕ_j is also added to η_j, when μ_j in (7.6) is replaced by $\mu_j + \phi_j R_{t-1,j}$.

The joint log-likelihood of the model is:

$$L(\boldsymbol{\eta}_1, \ldots, \boldsymbol{\eta}_d, \boldsymbol{\delta}) = \sum_{t=1}^{n} \log f_{\boldsymbol{R}_t}(r_{t1}, \ldots, r_{td}; \boldsymbol{\eta}_1, \ldots, \boldsymbol{\eta}_d, \boldsymbol{\delta}) = \sum_{j=1}^{d} \log \Big[s_{tj}^{-1} f_j([r_{tj} - \mu_j]/s_{tj}; \nu_j)$$

$$+ \sum_{t=1}^{n} \Big\{ \log c\big(F_1([r_{t1} - \mu_1]/s_{t1}; \nu_1), \ldots, F_d([r_{td} - \mu_d]/s_{td}; \nu_d); \boldsymbol{\delta})\big) \Big] \Big\}, \qquad (7.7)$$

with $s_{tj} = [\omega_j + \alpha_j r_{t-1,j}^2 + \beta_j s_{t-1,j}^2]^{1/2}$, and $c(\cdot; \boldsymbol{\delta})$ is the copula density for $C(\cdot; \boldsymbol{\delta})$.

When the dependence is not too strong, the IFM method (Section 5.5) can efficiently estimate the model parameters. In the first step, the GARCH(1,1) or AR(1)-GARCH(1,1) filter is applied to the return data to get the GARCH-filtered data $(z_{tj})_{1 \le t \le n, 1 \le j \le d}$ from the univariate parameter estimates $\hat{\nu}_j, \hat{\mu}_j, \hat{\omega}_j, \hat{\alpha}_j, \hat{\beta}_j$ (and also $\hat{\phi}_j$ for AR(1)) and estimated volatilities $\{\hat{s}_{tj}\}$. In the second step, the joint log-likelihood (7.7) is maximized over the copula parameter vector $\boldsymbol{\delta}$ with the ν_j and other univariate parameters fixed at the estimated values from the first step.

The estimated GARCH parameters $\hat{\omega}_j, \hat{\alpha}_j, \hat{\beta}_j$ are not sensitive to the innovation distribution being normal, Student t, or skewed t, etc., but following Section 5 of Aas et al. (2009), the dependence parameter $\boldsymbol{\delta}$ is usually estimated using the pseudo-likelihood method in Section 5.9. The idea is that to be less sensitive to the assumption that innovations have Student t distributions, the GARCH-filtered data $(z_{t1}, \ldots, z_{td})$ are converted to vectors of uniform scores for fitting and comparing copula models in the second step.

A resampling method can be used to get the asymptotic covariance matrix of the parameters of the copula-GARCH model based on two-stage parameter estimation. Standard errors obtained from maximizing the copula likelihood (7.7) do not include the variability of GARCH parameter estimates. One way to get standard errors for the two-stage estimation procedure is to use appropriate bootstrap methods. We use the following steps to get a bootstrap distribution for the two-stage estimates.

1. Compute GARCH parameter estimates $\hat{\boldsymbol{\eta}}_1, \ldots, \hat{\boldsymbol{\eta}}_d$ using the original data, separately for the d returns.

2. For the jth return, convert GARCH-filtered residuals $\boldsymbol{z}_j = (z_{j1}, \ldots, z_{jn})^\top$ to uniform scores $\boldsymbol{u}_j = (u_{j1}, \ldots, u_{jn})^\top$.

3. Compute copula parameter estimates $\hat{\boldsymbol{\delta}}$ from the d-dimensional data set from (7.7).

4. For the bth bootstrap sample, resample the filtered residuals as d-vectors at different time points (see Pascual et al. (2006) for more details on bootstrapping the GARCH model).

5. Use the resampled filtered data and estimated GARCH parameters $\hat{\boldsymbol{\eta}}_1, \ldots, \hat{\boldsymbol{\eta}}_d$, to get a bootstrap sample of log-returns $\mathbf{r}^{(b)} = (\mathbf{r}_1^{(b)}, \ldots, \mathbf{r}_d^{(b)})$, where $\mathbf{r}_j^{(b)} = (r_{j1}^{(b)}, \ldots, r_{jn}^{(b)})^\top$.

6. From the bootstrap sample $\mathbf{r}^{(b)}$, compute GARCH parameter estimates $\hat{\boldsymbol{\eta}}_1^{(b)}, \ldots, \hat{\boldsymbol{\eta}}_d^{(b)}$ and copula parameter $\hat{\boldsymbol{\delta}}^{(b)}$.

7. Repeat steps 4 to 6 for $b = 1, \ldots, B$, where B is the number of bootstrap samples. For example, B can be chosen to be between 1000 and 5000. Then one has a $B \times n_p$ matrix where n_p is the total number of parameters in the vectors $\boldsymbol{\eta}_1, \ldots, \boldsymbol{\eta}_d, \boldsymbol{\delta}$.

From a bootstrap distribution of the two-stage estimates one can compute standard errors and confidence intervals for $\hat{\boldsymbol{\eta}}_1, \ldots, \hat{\boldsymbol{\eta}}_d$, and $\hat{\boldsymbol{\delta}}$ as well as for the model-based estimates of different quantities which are functions of these parameter vectors. For example, to compute a confidence interval for the model-based Value-at-Risk (VaR) estimate (which is a complicated function of the model parameters), for each bootstrap estimate $\hat{\boldsymbol{\eta}}_1^{(b)}, \ldots, \hat{\boldsymbol{\eta}}_d^{(b)}, \hat{\boldsymbol{\delta}}^{(b)}$, a large data set of log-returns can be simulated to compute portfolio VaR and hence get a bootstrap distribution.

7.8.2 Market returns

We consider seven European market indices for the "recession" years of 2008–2009. Some initial data analyses are first presented followed by some comparison of copula-GARCH models with factor and truncated vine copulas for the GARCH-filtered returns. The $d = 7$ European market indices are (1) OSE=Oslo Exchange (Norway); (2) FTSE=Financial Times and the London Stock Exchange (UK); (3) AEX=Amsterdam Exchange (Netherlands); (4) CAC=Cotation Assistée en Continu (Paris, France); (5) SMI=Swiss Market Index (Switzerland); (6) DAX=Deutscher Aktien Index (Germany); (7) ATX=Austrian Traded Index (Austria). For the years 2008–2009. there are $n = 484$ daily log returns based on trading days available on all seven markets.

Index	μ_j	ϕ_j	ω_j	α_j	β_j	ν_j
OSE	7.58×10^{-4}	-0.044	6.88×10^{-6}	0.103	0.885	20.0
FTSE	6.44×10^{-4}	-0.103	4.55×10^{-6}	0.075	0.910	9.4
AEX	6.09×10^{-4}	-0.058	5.70×10^{-6}	0.081	0.905	11.3
CAC	2.14×10^{-4}	-0.101	5.54×10^{-6}	0.070	0.916	10.2
SMI	4.38×10^{-4}	-0.043	3.62×10^{-6}	0.096	0.890	15.3
DAX	1.28×10^{-4}	-0.062	5.68×10^{-6}	0.068	0.918	7.4
ATX	1.98×10^{-4}	-0.012	0.10×10^{-6}	0.089	0.896	12.4

Table 7.19 *European market index returns 2008–2009: AR(1)-GARCH(1,1) parameters with standardized t_{ν_j}-distributed innovations.*

Table 7.19 has the parameters when the AR(1)-GARCH(1,1) univariate model is fitted to each of the 7 log return time series. The GARCH-filtered data were converted to normal scores for some initial data analysis. Correlations ρ_N and lower/upper semi-correlations are summarized in Table 7.20.

	OSE	FTSE	AEX	CAC	SMI	DAX	ATX
OSE		0.550	0.551	0.516	0.437	0.491	0.523
FTSE	0.768		0.804	0.850	0.705	0.769	0.606
AEX	0.769	0.917		0.871	0.677	0.789	0.632
CAC	0.528	0.858	0.879		0.700	0.838	0.605
SMI	0.677	0.866	0.850	0.881		0.676	0.549
DAX	0.723	0.900	0.910	0.950	0.849		0.615
ATX	0.748	0.806	0.823	0.822	0.767	0.812	
	OSE	FTSE	AEX	CAC	SMI	DAX	ATX
OSE		0.487	0.614	0.483	0.362	0.493	0.527
FTSE	0.562		0.796	0.856	0.667	0.733	0.578
AEX	0.572	0.844		0.873	0.656	0.757	0.642
CAC	0.528	0.858	0.879		0.700	0.838	0.606
SMI	0.467	0.777	0.700	0.770		0.594	0.422
DAX	0.573	0.819	0.843	0.908	0.742		0.599
ATX	0.557	0.621	0.692	0.599	0.585	0.617	

Table 7.20 *European market GARCH-filtered index returns 2008–2009 with sample size $n = 484$: top: correlations of normal scores below the diagonal and semi-correlations assuming bivariate Gaussian above the diagonal; bottom: lower and upper semi-correlations of normal scores below and above the diagonal respectively. The pattern is slight skewness to the joint lower tail for all but two bivariate margins.*

The dependence is generally quite strong and based on the semi-correlations, there is slight tail asymmetry toward the joint lower tail for most pairs, The tail asymmetry is only noticeable in a few of the bivariate normal scores plots of the GARCH-filtered returns; however, sharper corners than ellipses can be seen in many of the plots suggesting stronger dependence in the tails than Gaussian (compare Figure 1.5).

Using the correlation matrix $\widehat{R}$ of normal scores of GARCH-filtered data, the "best" $R(\widehat{\theta})$ was obtained for (i) 1-factor, 2-factor and 3-factor based on fitting Gaussian factor models (ii) 1-truncated, 2-truncated and 3-truncated partial correlation vines based on minimizing the log determinant criterion in Section 6.17. Based on the log determinant criterion, the best m-truncated vine was found for $m = 1, 2, 3, 4$ by enumerating through all 7-dimensional vine structures. Also the best sequential minimum spanning tree (MST) m-truncated partial correlation vines for $m = 2, 3, 4$ were founding based on applying the algorithm after Example 6.4. A summary is presented in Table 7.21.

With variable 1=OSE, variable 2=FTSE, variable 3=AEX, variable 4=CAC, variable 5=SMI, variable 6=DAX, variable 7=ATX.

- A best (possibly non-unique) 2-truncated vine has edges 41, 31, 24, 74, 52, 67; $34|1, 21|4, 71|4, 54|2, 64|7$.
- A best (possibly non-unique) 3-truncated vine has edges 27, 47, 32, 57, 12, 67; $42|7, 37|2, 54|7, 17|2, 64|7; 34|27, 52|74, 13|27, 65|74$.
- The best sequential MST up to tree T_3 has edges 43, 73, 24, 64, 13, 54; $74|3, 23|4, 63|4, 17|3, 52|4; 27|34, 67|34, 14|37, 53|24$. The first six edges also correspond to those for the best 1-truncated vine.

Note that the edges for T_1 for the best 1-truncated, 2-truncated and 3-truncated are different. Also there are other 2-truncated and 3-truncated vines that can have discrepancy values close to those in Table 7.21.

The various measures of discrepancy of the empirical and model-based correlation matrices of normal scores suggest the 2-factor and the 3-truncated partial correlation vine structures as the best fits. The next step is to replace bivariate Gaussian copulas in the truncated vine/factor dependence structures with other parametric bivariate copula families.

structure	max. abs. dif.	Dfit	AIC	BIC
1-factor	0.095	0.254	4584	4613
2-factor	0.019	0.054	4499	4554
3-factor	0.018	0.020	4493	4568
best 1-truncated	0.115	0.407	4656	4681
best 2-truncated	0.036	0.080	4508	4554
best 3-truncated	0.022	0.022	4488	4551
best 4-truncated	0.004	0.001	4484	4559
2-truncated, seq. MST	0.053	0.116	4525	4571
3-truncated, seq. MST	0.045	0.072	4512	4575
4-truncated, seq. MST	0.013	0.004	4485	4561

Table 7.21 *European market GARCH-filtered index returns 2008–2009 with $d = 7$ and sample size $n = 484$: measures of closeness of the empirical correlation matrix $\widehat{R}$ of normal scores and $R(\widehat{\theta})$ for different fitted dependence structures. The second column has $\max \|R(\widehat{\theta}) - \widehat{R}\|$ and Dfit in the third column refers to: $D_{model} = \log(|R(\widehat{\theta})|) - \log(|R_{obs}|) + \mathrm{tr}[R^{-1}(\widehat{\theta})R_{obs}] - d$. The dimension of $\widehat{\theta}$ is $pd - (p-1)p/2$ for p-factor, and $md - m(m+1)/2$ for m-truncated partial correlation vine. For sequential minimum spanning trees (MST), the best next tree is found given the previous trees; these are suboptimal truncated vines. The AIC/BIC values are based on the Gaussian log-likelihood.*

dep.structure	biv.copula	#dep.param	-2 loglik.	AIC	BIC
1-factor	Frank	7	-4728	-4714	-4685
1-factor	BB1r	14	-5045	-5017	-4958
1-factor	Gaussian	7	-5039	-5025	-4996
1-factor	t_{15}	8	-5101	-5085	-5051
2-factor	Frank	14	-5039	-5011	-4952
2-factor	Gaussian	13	-5126	-5100	-5045
2-factor	BB1/Frank	21	-5210	-5168	-5080
2-factor	t_{15}/t_{10}	16	-5203	-5171	-5104
1-truncated	Frank	6	-4622	-4610	-4585
1-truncated	Gaussian	6	-4979	-4967	-4941
1-truncated	BB1	12	-5015	-4991	-4941
1-truncated	t_{10}	7	-5055	-5041	-5012
2-truncated	Frank	11	-4873	-4851	-4805
2-truncated	BB1/Frank	17	-5075	-5041	-4970
2-truncated	Gaussian	11	-5138	-5116	-5070
MST 2-truncated	t_{10}/t_{10}	13	-5213	-5187	-5133
2-truncated	t_{10}/t_{10}	13	-5214	-5188	-5133
3-truncated	Frank	15	-4913	-4883	-4820
3-truncated	BB1/Frank/Frank	21	-5040	-4998	-4910
3-truncated	Gaussian	15	-5166	-5136	-5073
MST 3-truncated	$t_{15}/t_{10}/t_{15}$	18	-5226	-5190	-5115
3-truncated	$t_{15}/t_{10}/t_{15}$	18	-5246	-5210	-5135

Table 7.22 *European market GARCH-filtered index returns 2008–2009: negative log-likelihood, AIC and BIC values. For simpler interpretation and summary, and to see the effect of dependence structure and tail behavior, there is a common copula family for a factor or vine tree. There could be small improvements for any of the dependence structures with a mix of several bivariate parametric copula families.*

Table 7.22 has log-likelihood and AIC/BIC values for some parametric copula models, evaluated at the maximum likelihood estimates with univariate GARCH parameters held fixed. For the 1-factor and 2-factor copulas, 25 quadrature points were used in the likelihood calculations. For simpler interpretation and to see the effect of dependence structure and tail behavior, a common parametric copula family is used for each factor or vine tree. For bivariate t copulas, because the log-likelihood is quite flat over the shape parameters ν, this parameter was fixed at several multiples of 5 and then optimized over the (partial) correlation parameters. The inclusion of factor and vine models with bivariate Frank copulas on each edge is to show the differences when tail dependent and tail quadrant independent bivariate copulas are used. The improvement of t over Gaussian (with intermediate tail dependence) is generally less than the improvement of the Gaussian versus the Frank copula.

Based on the AIC/BIC values, the best model all have tail dependence in the first factor or vine tree: 2-truncated or 3-truncated vine models with bivariate t, 2-factor models with bivariate t, and 2-factor models with BB1 for factor 1 and Frank for factor 2. The tail asymmetry was not sufficiently significant enough for tail asymmetric models to have the smallest values of AIC/BIC. The 2-factor BB1/Frank model has smaller negative log-likelihood value because of a few more parameters. On average, the lower and upper semi-correlation scores only suggested slightly more dependence in the tails than with Gaussian, and the tail asymmetry seems insignificant when compared with Table 2.4; the sample size $n = 484$ is smaller than 600 used in Table 2.4 and the dependence is stronger than $\rho_N = 0.6$ so difference of mostly less than 0.1 for the lower and upper semi-correlations would not be significant.

Table 7.22 also shows some comparisons of the best m-truncated and the sequential MST m-truncated for $m = 2, 3$. The small improvement of best m-truncated over sequential MST m-truncated carries over from Gaussian to non-Gaussian for the bivariate copulas.

Note that the heuristics in Remark 5.1 hold here; the ordering is t with the best fit, then Gaussian or BB1 in the middle, and finally Frank for all of the dependence structures of 1-factor, 2-factor, 1-truncated vine, 2-truncated vine, and 3-truncated vine. The 2-truncated vine and 2-factor copulas may be adequate (see further tables below to compare model-based and empirical dependence measures), as the improvement of 3-truncated vines over 2-truncated vines seems very little compared with improvement of 2-factor over 1-factor or 2-truncated over 1-truncated. Even though the 2-truncated vine copula with t is better based on AIC/BIC, the 2-factor copula with t might have a more interpretable dependence structure.

A summary of comparisons of models via Vuong's procedure with the Schwarz or BIC correction is given in Table 7.23. The several models that are the best based on AIC/BIC are considered equivalent based on Vuong's procedure (with or without the Schwarz correction).

copula 1	copula 2	95% interval
1-truncated t_{10}	3-truncated $t_{15}/t_{10}/t_{15}$	$(0.07, 0.18)$
2-truncated BB1/Frank	3-truncated $t_{15}/t_{10}/t_{15}$	$(0.13, 0.26)$
2-truncated t_{10}/t_{10}	3-truncated $t_{15}/t_{10}/t_{15}$	$(-0.04, 0.04)$
2-factor BB1/Frank	3-truncated $t_{15}/t_{10}/t_{15}$	$(-0.00, 0.12)$
2-factor t_{15}/t_{10}	3-truncated $t_{15}/t_{10}/t_{15}$	$(-0.01, 0.08)$
2-factor t_{15}/t_{10}	2-truncated t_{10}/t_{10}	$(-0.02, 0.08)$
2-factor t_{15}/t_{10}	2-factor BB1/Frank	$(-0.08, 0.03)$
1-factor t_{15}	2-factor t_{15}/t_{10}	$(0.00, 0.10)$

Table 7.23 *European market GARCH-filtered index returns 2008–2009 with sample size $n = 484$: a summary of Vuong's procedure with the Schwarz or BIC correction. (5.22) is used for the average log-likelihood ratio; an interval completely above 0 indicates that the copula 2 is significantly better than copula 1. The copula parameters are based on the second stage of IFM.*

Generally, the improvement of the log-likelihood with a non-Gaussian bivariate copulas in vine/factor models implies a better match in the tails relative to Gaussian. Table 7.20 has an indication that there is slightly more dependence in the joint lower tails than Gaussian. The copula models with bivariate t copulas are best in terms of AIC and BIC for different dependence structures, and those with Frank copulas are the worst.

Bivariate copulas with tail dependence have densities that quickly increase to infinity in the corners of the unit square. Hence there can be a big difference in the model log-likelihood when the data exhibits tail dependence and tail dependent pair-copulas are used in tree $\mathcal{T}_1$ of a vine copula or factor 1 of a factor copula. Theorems 8.68, 8.76 and 8.77 imply that lower (upper) tail dependence in all observed pairs will be achieved the copulas in tree $\mathcal{T}_1$ or factor 1 are lower (upper) tail dependent.

To check if the log-likelihood is large enough for any of the fitted copula models to match the tails, empirical and model-based Kendall taus and tail-weighted dependence measures will be compared for the factor and truncated vine copulas. For a multivariate vector $\boldsymbol{U} = (U_1, \ldots, U_d)$, the tail-weighted dependence measures are

$$\varrho_{L,jk} = \mathrm{Cor}\big((1 - 2U_j)^6, (1 - 2U_k)^6 | U_j < \tfrac{1}{2}, U_k < \tfrac{1}{2}\big),$$
$$\varrho_{U,jk} = \mathrm{Cor}\big((2U_j - 1)^6, (2U_k - 1)^6 | U_j > \tfrac{1}{2}, U_k > \tfrac{1}{2}\big) \tag{7.8}$$

for $j \neq k$. We use these rather than the semi-correlations for model-based comparisons because they have good properties (Krupskii (2014)) and are computationally faster than semi-correlations for both the empirical and model versions. The empirical versions are the sample correlations of two transformed variables over pairs in the lower and upper quadrants respectively, and they are summarized in Table 7.24. The model-based versions are calculated as integrals for truncated distributions of form $C_{jk}(u, v)/C_{jk}(\frac{1}{2}, \frac{1}{2})$ on $[0, \frac{1}{2}]^2$ and $\overline{C}_{jk}(u, v)/\overline{C}_{jk}(\frac{1}{2}, \frac{1}{2})$ on $[\frac{1}{2}, 1]^2$. For dimension d, there are $d(d-1)/2$ lower and upper tail-weighted dependence coefficients to compute, empirically and for the bivariate margins of each parametric copula candidate model $C_{1:d}(\cdot; \boldsymbol{\delta})$.

	OSE	FTSE	AEX	CAC	SMI	DAX	ATX
OSE	1.000	0.583	0.575	0.555	0.498	0.543	0.554
FTSE	0.583	1.000	0.740	0.776	0.668	0.731	0.608
AEX	0.575	0.740	1.000	0.799	0.658	0.741	0.619
CAC	0.555	0.776	0.799	1.000	0.693	0.811	0.628
SMI	0.498	0.668	0.658	0.693	1.000	0.662	0.564
DAX	0.543	0.731	0.741	0.811	0.662	1.000	0.616
ATX	0.554	0.608	0.619	0.628	0.564	0.616	1.000
OSE		0.436	0.553	0.465	0.296	0.474	0.467
FTSE	0.539		0.810	0.867	0.690	0.717	0.552
AEX	0.537	0.858		0.876	0.667	0.797	0.629
CAC	0.522	0.861	0.867		0.720	0.861	0.606
SMI	0.505	0.818	0.752	0.813		0.606	0.450
DAX	0.549	0.808	0.837	0.928	0.754		0.594
ATX	0.500	0.626	0.685	0.592	0.597	0.596	

Table 7.24 *European market GARCH-filtered index returns 2008–2009: the top table has empirical Kendall taus and the bottom table has lower and upper tail-weighted dependence ϱ_L, ϱ_U as in (7.8), below and above the diagonal respectively.*

model	avg. abs. dif.: model vs. emp			max. abs. dif.: model vs. emp		
	τ	ϱ_L	ϱ_U	τ	ϱ_L	ϱ_U
1-factor Frank	0.01	0.30	0.23	0.08	0.43^-	0.30^-
1-factor BB1	0.03	0.04	0.09	0.11	0.10^+	0.23^+
1-factor t_{15}	0.01	0.04	0.06	0.10	0.12^-	0.19^+
2-factor Frank	0.01	0.18	0.12	0.05	0.28^-	0.20^-
2-factor BB1/Frank	0.02	0.03	0.05	0.04	0.10^-	0.17^+
2-factor t_{15}/t_{10}	0.01	0.04	0.06	0.03	0.12^-	0.20^+
2-truncated BB1/Frank	0.06	0.06	0.07	0.10	0.15^-	0.18^+
MST 2-truncated t_{10}/t_{10}	0.02	0.04	0.05	0.07	0.08^-	0.15^+
2-truncated t_{10}/t_{10}	0.02	0.05	0.06	0.05	0.09^-	0.18^+
3-truncated Frank	0.04	0.24	0.18	0.07	0.32^-	0.27^-
MST 3-trunc. $t_{15}/t_{10}/t_{15}$	0.01	0.04	0.05	0.06	0.08^-	0.15^+
3-truncated $t_{15}/t_{10}/t_{15}$	0.01	0.03	0.06	0.02	0.08^-	0.19^+

Table 7.25 *European market GARCH-filtered index returns 2008–2009: deviations of model-based versus empirical for three dependence measures. Values for the factor models were computed via numerical integration and values for the truncated vine models were computed via Monte Carlo simulation with 10^5 replications (and checked against numerical integration for trees T_1, T_2). The superscript $+$ or $-$ in last two columns gives an indication of whether the maximum deviation is positive or negative respectively; for example, a superscript of $-$ for the lower tail indicates the model underestimates the dependence in one bivariate lower tail in the extreme case.*

Table 7.25 has average and maximum absolute difference for the model-based versus empirical $\tau, \varrho_L, \varrho_U$ over the 21 bivariate margins. All 2-factor, 2-truncated and 3-truncated models match quite well on Kendall's tau, but the better models based on bivariate t copulas better match the empirical tail-weighted dependence measures. Because of the slight tail asymmetry, the models based on t underestimate some tail dependence in the joint lower tail and overestimate some tail dependence in the joint upper tail. The 2-factor BB1/Frank model does marginally better on average in matching the empirical tail-weighted ϱ_L, ϱ_U values and marginally worse on average in matching the empirical τ values. If the dependence structure has a sufficient number of parameters, the match to the empirical Kendall's tau is quite good even if the fitted copula does not match the empirical tails, such as the models with all Frank copulas that severely underestimate dependence in the tails. For a sample size of $n = 484$, the sampling variability of around 0.04–0.05 is about the best possible (compare Table 2.4 for semi-correlations). The empirical tail-weighted dependence measures have higher sampling variability than Kendall's tau because they use only between a quarter to a half of the sample size n.

To summarize this subsection, we showed some details on how to decide on factor and vine dependence structures, followed by choice of bivariate copulas that take into account some preliminary analysis on tail asymmetry and strength of dependence in bivariate tails. Because there are many potential choices of dependence structures and then bivariate copula families within factor and vine models, an assessment is needed to determine if the models with the most negative log-likelihood or AIC/BIC values are adequate. The comparison of model-based versus empirical tail-weighted dependence measures can help in this regard, since model-based estimates of tail dependence parameters λ_L, λ_U are not reliable.

7.8.3 Stock returns over several sectors

In this subsection, some analysis and fitting of copula-GARCH models are applied to the daily log returns of $d = 29$ stocks listed in Deutscher Aktien Index (DAX). Table 7.26 contains a summary of the stocks and their sectors.

group	stock	company	group	stock	company
1	ALV	Allianz	5	ADS	Adidas
1	CBK	Commerzbank	5	BEI	Beiersdorf
1	DB1	Deutsche Börse	5	HEN3	Henkel
1	DBK	Deutsche Bank	5	MEO	Metro
1	MUV2	Munich Re			
2	BAS	BASF	6	DPW	Deutsche Post
2	BAYN	Bayer	6	LHA	Lufthansa
2	LIN	Linde			
2	SDF	K+S	7	EOAN	E.ON
2	MRK	Merck	7	RWE	RWE
3	SIE	Siemens			
3	TKA	ThyssenKrupp	8	FME	Fresenius Medical Care
3	HEI	HeidelbergCement	8	FRE	Fresenius
4	BMW	BMW	9	DTE	Deutsche Telekom
4	DAI	Damiler	9	IFX	Infioneon
4	VOW3	Volkswagen	9	SAP	SAP

Table 7.26 *DAX with $d = 29$ stocks in 2011–2012, grouped by sectors. Group 1 is financial (insurance/banking), group 2 is pharmaceutical/chemical, group 3 is industrial, group 4 is automobile, group 5 is consumer goods, group 6 is logistics and transportation, group 7 is energy, group 8 is health care, group 9 is telecommunications and information technology.*

AR(1)-GARCH(1,1) univariate models were fitted to each of the 29 log return time series. For the GARCH-filtered data, Table 7.27 has a summary of lower and upper semi-correlations. Figure 7.3 shows some normal score plots GARCH-filtered returns for some pairs of stocks in group 2. Sharper corners than ellipses and tail asymmetry can be seen only in some of the plots. Hence the summary statistics for semi-correlations are useful additional diagnostics.

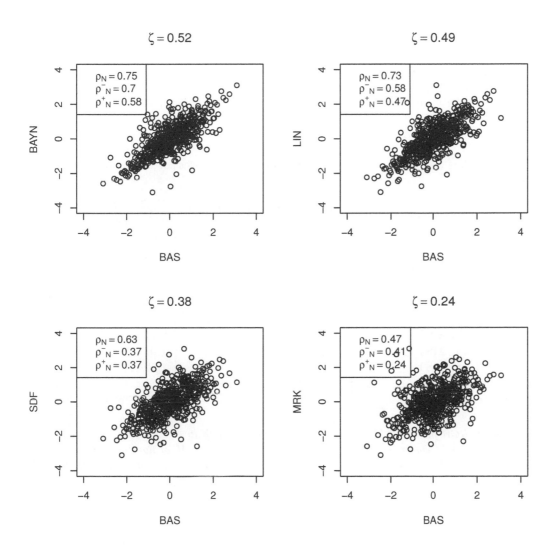

Figure 7.3 *GARCH-filtered DAX stock returns 2011–2012: some normal scores plots for group 2, sample size is $n = 511$. $\zeta(\rho_N)$ is the semi-correlation for bivariate Gaussian with correlation ρ_N.*

Table 7.27 shows that there is slight bivariate tail asymmetry towards the lower tail as the overall pattern over 406 pairs. The median $\hat{\rho}_N$ is 0.51, the median lower semi-correlation is 0.40, the median upper semi-correlation is 0.31. For a bivariate Gaussian distribution with correlation 0.51, the lower and upper semi-correlations are 0.28. Also, the last two columns of Table 7.27 suggest that the dependence in the joint lower tails is generally slightly stronger than bivariate Gaussian tails.

	ρ_N	ρ_N^-	ρ_N^+	$\rho_N^- - \rho_N^+$	$\rho_N^- - \zeta(\rho_N)$	$\rho_N^+ - \zeta(\rho_N)$
min	0.13	0.15	- 0.01	-0.20	-0.10	-0.21
Q1	0.43	0.33	0.23	0.03	0.07	-0.02
median	0.51	0.40	0.31	0.09	0.12	0.03
Q3	0.59	0.48	0.39	0.15	0.17	0.07
max	0.85	0.70	0.69	0.35	0.33	0.21

Table 7.27 *GARCH-filtered DAX stock returns 2011–2012 with d = 29 stocks and sample size n = 511: summaries of quartiles for the d(d−1)/2 = 406 correlations of normal scores, lower/upper semi-correlations, difference of the lower and upper semi-correlations of normal scores, and difference of the empirical semi-correlations from the semi-correlations assuming Gaussian. $\zeta(\rho_N)$ is the semi-correlation for bivariate Gaussian with correlation ρ_N; see Section 2.17. The pattern is slight overall skewness to the joint lower tail, and a little more dependence than Gaussian in the joint tails (especially lower).*

Table 7.28 has a summary of different dependence structures for the correlation matrix of normal scores.

structure	#par.	max. abs. dif.	avg. abs. dif.	Dfit	AIC/n	BIC/n
1-factor	29	0.39	0.044	4.67	63.6	63.9
nested-factor	38	0.19	0.041	2.29	61.3	61.6
2-factor	57	0.27	0.030	3.14	62.2	62.7
3-factor	84	0.23	0.019	1.95	61.1	61.8
bi-factor	55	0.18	0.035	1.98	61.0	61.5
tri-factor	84	0.13	0.022	1.20	60.4	61.1
1-truncated, seq. MST	28	0.37	0.180	4.78	63.7	64.0
2-truncated, seq. MST	55	0.25	0.090	2.71	61.8	62.2
3-truncated, seq. MST	81	0.20	0.053	1.91	61.1	61.7+
4-truncated, seq. MST	106	0.19	0.037	1.52	60.8	61.7−
5-truncated, seq. MST	130	0.19	0.030	1.31	60.7	61.7+
13-truncated, seq. MST	286	0.17	0.011	0.47	60.4	62.8

Table 7.28 *GARCH-filtered DAX stock returns 2011–2012 with d = 29 stocks and sample size n = 511: measures of closeness of the empirical correlation matrix $\widehat{R}$ of normal scores and $R(\hat{\theta})$ for different fitted dependence structures. The third and fourth columns have $\max \|R(\hat{\theta}) - \widehat{R}\|$ and $average\|R(\hat{\theta}) - \widehat{R}\|$. Dfit in the fifth column refers to: $D_{\mathrm{model}} = \log(|R(\hat{\theta})|) - \log(|R_{\mathrm{obs}}|) + \mathrm{tr}[R^{-1}(\hat{\theta})R_{\mathrm{obs}}] - d$. The dimension of $\hat{\theta}$ is $pd − (p−1)p/2$ for p-factor, and $md − m(m+1)/2$ for m-truncated partial correlation vine. The are 9 groups for the bi-factor and nested factor models as shown in Table 7.26. For bi-factor, the dimension of $\hat{\theta}$ is $2d−$#(group size of 2)=55; for nested factor, the dimension of $\hat{\theta}$ is $d+$#groups=38. For sequential minimum spanning trees (MST), the best next tree is found given the previous trees; these are suboptimal truncated vines. The scaled AIC/BIC values are based on the Gaussian log-likelihood*

Using the correlation matrix $\widehat{R}$ of normal scores of GARCH-filter data, the "best" $R(\hat{\theta})$ was obtained for (i) 1-factor, nested factor, 2-factor, 3-factor and bi-factor structures based on fitting Gaussian factor models (ii) 1-truncated up to 13-truncated partial correlation vines based on the log determinant criterion in Section 6.17. The suboptimal truncated vines were found using the sequential minimum spanning tree (MST) algorithm after Example 6.4. For the best sequential MST truncated R-vines with 2 or 3 trees, the stocks are clustered similar to the groups for the the bi-factor structure (except for two or three of the 29 stocks).

When considering all of the measures of fit in Table 7.28, the bi-factor structure looks like a good model. One could check further on the large deviations of the empirical and model-based correlations. The largest absolute deviations are cross-correlations of FME and FRE (health-care sector) with some stocks in other sectors, namely CBK and DBK (finance), LIN and MRK (pharmaceutical/chemical), BEI and HEN3 (consumer goods). Because the pattern suggests something that may not random, one could add an extra factor that loads on some stocks over several (but not all) sectors.

The empirical correlations of the normal scores of GARCH-filtered data are in the range 0.13–0.85. The empirical correlation of the normal scores of FME and FRE is 0.61, and the within group empirical correlations are in the 0.51–0.82 range for group 1, 0.43–0.75 range for group 2, 0.67–0.73 range for group 3, 0.73–0.82 range for group 4, 0.36–0.63 range for group 5, 0.61–0.85 range for group 6–8, 0.39–0.52 range for group 9. The bi-factor model fits well the within-group correlations but less so for some between-group correlations.

To improve on the bi-factor model, accounting for some residual dependence, a model with the groups of health-care, pharmaceutical/chemical and consumer goods combined in a larger supergroup 1 and the remaining groups combined in a larger second supergroup 2. The resulting "tri-factor" model (see Section 3.11.1) has 12 factors in total: one global factor, one factor loading only on stocks in supergroup 1, one factor loading only on stocks in supergroup 2, and nine sector factors that load only on stocks in a given sector. This led to an improved fit for all measures in Table 7.28, and reduced the larger correlation discrepancies within supergroup 1.

Table 7.29 has AIC and BIC values of copula models fitted to the GARCH-filtered stock returns. Factor copula (1-factor, 2-factor, nested factor and bi-factor) models are included as well as 1-truncated to 4-truncated vines based on the best sequential MST found with the partial correlation criterion. The multivariate t copula with the tri-factor structure was also added.

The choice of models in Table 7.29 is to show the separate effect of the dependence structure and the strength of dependence in the tails of the pair-copulas. The fits of truncated vine copulas can be improved with the sequential procedure of Brechmann et al. (2012) and the structured copula models could also be improved by a mix of bivariate t linking copulas and others linking copulas like BB1 that are tail dependent and asymmetric.

For structured factor models based on t copulas, it is more convenient to fit multivariate t_ν copula with nested factor, bi-factor and tri-factor correlation structure, as mentioned in Section 3.10.4. Results are similar with (3.66) but the former eliminates the numerical integration so is computationally faster.

Regarding Table 7.29, for simplicity of interpretation, the bi-factor BB1r/Frank and multivariate t_{20} with nested factor, bi-factor or tri-factor structure are best, but the 4-truncated vines with the corresponding choices of pair-copulas can match on AIC and BIC values. For improvement of the dependence structure, one possibility is a bi-factor copula combined with a truncated vine for the residual dependence. For improvements of modeling of tails for any of the dependence structures, one could mix BB1, reflected BB1 and bivariate t copula families in the first factor or tree. One could continue with structured factor models with more factors and tail asymmetric bivariate linking copulas. But with three- or higher-dimensional numerical integrals, the computing effort would be substantially increased.

dep.structure	biv.copula	#dep.param	-2loglik/n	AIC/n	BIC/n
1-factor	Frank	29	-18.3	-18.2	-18.0
1-factor	Gaussian	29	-18.8	-18.7	-18.5
1-factor	BB1	58	-19.5	-19.3	-18.8
1-factor	t_5	30	-19.7	-19.6	-19.4
nested-factor	Frank	38	-20.3	-20.2	-19.8
nested-factor	Gaussian	38	-21.2	-21.0	-20.7
nested-factor	refl. Gumbel/BB1	67	-21.8	-21.5	-20.9
nested-factor	mvt$_{20}$	39	-22.0	-21.8	-21.5*
2-factor	Frank	58	-21.0	-20.8	-20.3
2-factor	Gaussian	57	-20.4	-20.2	-19.7
2-factor	BB1r/Frank	87	-21.5	-21.2	-20.5
2-factor	t_5/t_5	60	-21.4	-21.2	-20.7
bi-factor	Frank	55	-21.4	-21.1	-20.7
bi-factor	Gaussian	55	-21.5	-21.3	-20.8
bi-factor	BB1r/Frank	87	-22.3	-22.0	-21.2*
bi-factor	mvt$_{20}$	56	-22.3	-22.0	-21.6*
tri-factor	mvt$_{20}$	85	-23.0	-22.7	-22.0*
1-truncated	Frank	28	-18.2	-18.0	-17.8
1-truncated	Gaussian	28	-18.7	-18.6	-18.3
1-truncated	BB1r	56	-19.4	-19.1	-18.7
1-truncated	t_5	29	-19.5	-19.4	-19.1
MST 2-truncated	Frank	55	-20.9	-20.7	-20.2
MST 2-truncated	Gaussian	55	-20.7	-20.5	-20.1
MST 2-truncated	BB1r/Frank	83	-21.6	-21.3	-20.6
MST 2-truncated	$t_{10}t_{15}$	57	-21.6	-21.4	-20.9
MST 3-truncated	Frank	81	-21.9	-21.6	-20.9
MST 3-truncated	Gaussian	81	-21.5	-21.2	-20.6
MST 3-truncated	BB1r/Frank/Frank	109	-22.5	-22.0	-21.1
MST 3-truncated	$t_{10}t_{15}t_{25}$	84	-22.5	-22.1	-21.4*
MST 4-truncated	Frank	106	-22.4	-22.0	-21.1
MST 4-truncated	Gaussian	106	-21.9	-21.5	-20.6
MST 4-truncated	BB1r/Frank[3]	134	-22.9	-22.4	-21.3*
MST 4-truncated	$t_{10}t_{15}t_{25}t_{25}$	110	-22.9	-22.5	-21.6*

Table 7.29 *GARCH-filtered DAX stock returns 2011–2012 with $d = 29$ stocks and sample size $n = 511$: negative log-likelihood, AIC and BIC values. For simpler interpretation and summary, there is a common copula family for a factor or vine tree. Something like BB1r/Frank means reflected BB1 for the first factor or tree and Frank for the remaining factors/trees; and something like $t_{\nu_1} t_{\nu_2} \cdots$ means shape parameter ν_1 for the first tree of the truncated vine, ν_2 for the second tree etc. For comparsion, there are $d(d-1)/2 = 406$ correlation parameters in the Gaussian model with an unstructured correlation matrix. The purpose of this table is to show the various tail behaviors for each of several dependence structures. The better of BB1 and reflected BB1 (BB1r) is given each dependence structure and the best ν_ℓ parameters was chosen for t among 5,10,15,20,25 because the log-likelihoods are relatively flat near the best ν_ℓ values. The truncated vines, based on sequential MST, are those used in Table 7.28 with different pair-copulas assigned to the edges. The best models in this table based on BIC (value ≤ -21.2) are indicated with an asterisk.*

Tables 7.28 and 7.29 illustrate the comments in Remark 5.1. Maximum likelihood estimation with copulas can suggest that models with tail dependence are better than those without tail dependence when dependence is strong enough in the tails. For the log return data set considered here, the models with t or BB1 copulas in the first tree of the vine models or the first latent variable of factor models have the smaller AIC/BIC values for any dependence structure. The use of bivariate copulas with intermediate tail dependence or tail quadrant independence have larger AIC/BIC values for any dependence structure.

A summary of comparisons of models via Vuong's procedure with the Schwarz or BIC correction is given in Table 7.30. For this example, unlike in Section 7.8.2, there is a difference in the conclusions with and without the BIC correction. Without this correction, the model with more parameters and a larger log-likelihood tends to be favored. With the correction, more parsimonious models compare more favorably; after the "best" model of tri-factor multivariate t, the 3-truncated vine with bivariate t, 4-truncated vine with bivariate t, 4-truncated vine with BB1/Frank, nested factor multivariate t are considered are about the same as bi-factor multivariate t. The bi-factor BB1/Frank model did not compare as favorably because it has more parameters in achieving the approximately same log-likelihood. The comparisons are roughly the same as with the BIC values.

copula 1	copula 2	95% interval
2-truncated $t_{10}t_{15}$	bi-factor mvt_{20}	$(0.12, 0.50)$
3-truncated $t_{10}t_{15}t_{25}$	bi-factor mvt_{20}	$(-0.09, 0.23)$
4-truncated $t_{10}t_{15}t_{25}t_{25}$	bi-factor mvt_{20}	$(-0.15, 0.15)$
4-truncated BB1r/Frank^3	bi-factor mvt_{20}	$(-0.04, 0.32)$
1-factor t_5	bi-factor mvt_{20}	$(0.88, 1.32)$
nested-factor mvt_{20}	bi-factor mvt_{20}	$(-0.02, 0.08)$
2-factor BB1r/Frank	bi-factor mvt_{20}	$(0.35, 0.76)$
2-factor t_5/t_5	bi-factor mvt_{20}	$(0.26, 0.64)$
bi-factor BB1r/Frank	bi-factor mvt_{20}	$(0.03, 0.31)$
tri-factor mvt_{20}	bi-factor mvt_{20}	$(-0.29, -0.11)$

Table 7.30 *GARCH-filtered DAX stock returns 2011–2012 with $d = 29$ stocks and sample size $n = 511$: a summary of Vuong's procedure with the Schwarz or BIC correction. (5.22) is used for the average log-likelihood ratio; an interval completely positive (negative) indicates that the copula 2 is significantly better (worse) than copula 1. The copula parameters are based on the second stage of IFM.*

To decide if it is worthwhile to continue to look for copula models with parsimonious dependence structures and smaller values of AIC or BIC, we compare some of the above fitted models for portfolio Value-at-Risk (VaR) and conditional tail expectation (CTE), and model versus empirical tail-weighted dependence measures. Because this further analysis is much more computationally demanding, the comparisons are only made for some of the structured factor copula models.

VaR is defined as a quantile of the distribution of a portfolio return. Let $(\bar{R}_1, \ldots, \bar{R}_n)$ be the portfolio returns for n consecutive time units (such as trading days) and let $\hat{F}_{\bar{R}}$ be the corresponding empirical cdf. The empirical $100\alpha\%$ VaR values of the portfolio are:

$$\widehat{\text{VaR}}_\alpha = \{\inf \bar{r} : \hat{F}_{\bar{R}}(\bar{r}) \geq \alpha\}.$$

Common α values are 0.01, 0.05, 0.95 and 0.99. With a small α near 0, VaR_α represents the quantile of possible loss for the portfolio. Similarly, with a large α near 1, VaR_α represents a quantile of possible loss for investors who short sell the portfolio.

The second measure CTE is defined as a conditional mean of a portfolio return given that the return falls below or exceeds some threshold. The empirical lower (upper) CTE at level r^* is:

$$\widehat{\text{CTE}}^-(r^*) = \frac{\sum_{i^*: \bar{R}_{i^*} \leq r^*} \bar{R}_{i^*}}{\sum_i I\{\bar{R}_i \leq r^*\}}, \quad \widehat{\text{CTE}}^+(r^*) = \frac{\sum_{i^*: \bar{R}_{i^*} \geq r^*} \bar{R}_{i^*}}{\sum_i I\{\bar{R}_i \geq r^*\}}.$$

The lower CTE^- is used for small quantiles near 0.01, and the upper CTE^+ is used for large quantiles near 0.99.

For each value of α, a value of VaR_α for the data set is obtained assuming the stationarity of the log-returns. Similarly, some values of CTE^- and CTE^+ are obtained.

The bootstrapping procedure in Section 7.8.1 can be used to get model-based interval estimates of portfolio VaR and CTE. For each bootstrap $\hat{\boldsymbol{\theta}}^{(b)}$ replication (a vector of univariate GARCH parameters from d returns and copula parameters), the copula-GARCH model is simulated for a large sample size (say 20000). Each value of $\hat{\boldsymbol{\theta}}^{(b)}$ leads to a model-based estimate of VaR and CTE at different levels. Interval estimates for these quantities can be obtained from a sufficiently large bootstrap sample size.

model	$VaR_{0.01}$	$VaR_{0.05}$	$VaR_{0.95}$	$VaR_{0.99}$
empirical estimate	-0.043	-0.024	0.025	0.039
nested-factor Gaussian	(-.053, -.026)	(-.031, -.017)	(.017, .031)	(.027, .053)
bi-factor Gaussian	(-.053, -.026)	(-.031, -.017)	(.017, .031)	(.027, .054)
bi-factor t(20)	(-.054, -.027)	(-.031, -.017)	(.017, .031)	(.027, .054)
nested-factor t(20)	(-.055, -.027)	(-.031, -.017)	(.017, .031)	(.027, .055)
nested-factor Frank	**(-.042, -.019)**	(-.029, -.016)	(.016, .029)	(.020, .042)
bi-factor Frank	(-.043, -.020)	(-.029, -.016)	(.016, .029)	(.020, .042)
nested refl. Gumbel/BB1	(-.061, -.031)	(-.032, -.018)	(.016, .029)	(.025, .050)
bi-factor BB1r/Frank	(-.061, -.030)	(-.032, -.017)	(.017, .030)	(.027, .055)
model	$CTE^-(-0.04)$	$CTE^-(-0.025)$	$CTE^+(0.025)$	$CTE^+(0.04)$
empirical estimate	-0.051	-0.037	0.033	0.046
nested-factor Gaussian	(-.062, -.045)	(-.040,-.031)	(.031, .040)	(.045, .062)
bi-factor Gaussian	(-.062, -.045)	(-.040,-.031)	(.031, .040)	(.045, .063)
bi-factor t(20)	(-.061, -.046)	(-.040, -.031)	(.031, .040)	(.046, .061)
nested-factor t(20)	(-.061, -.046)	(-.040, -.031)	(.031, .040)	(.046, .061)
nested-factor Frank	(-.069, -.041)	(-.037, -.027)	(.027, .037)	(.041, .069)
bi-factor Frank	(-.072, -.041)	(-.037, -.027)	(.027, .037)	(.041, .071)
nested refl. Gumbel/BB1	(-.065, -.050)	(-.044, -.034)	(.030, .041)	(.045, .065)
bi-factor BB1r/Frank	(-.064, -.049)	(-.043, -.034)	(.032, .042)	**(.047, .064)**

Table 7.31 *GARCH-filtered DAX stock returns 2011–2012 with $d = 29$ stocks and sample size $n = 511$. Empirical estimates of VaR_α for $\alpha = 0.01, 0.05, 0.95, 0.99$ and the model-based 95% confidence intervals. Also, empirical estimates of $CTE^-(r^*)$, for $r^* = -0.04, -0.025$, and $CTE^+(r^*)$, for $r^* = 0.025, 0.04$, and the model-based 95% confidence intervals. The model-based confidence intervals are based on the parametric bootstrap with 1000–2000 replications, depending on the speed of the computations. Intervals that don't contain the empirical value are shown in bold font.*

Table 7.31 has empirical estimates as well as model-based 95% confidence intervals for VaR_α with $\alpha = 0.01, 0.05, 0.95, 0.99$ and for $CTE^-(-0.04)$, $CTE^-(-0.025)$, $CTE^+(0.025)$, $CTE^+(0.04)$. The values $\pm0.025, \pm0.04$ for CTE approximately correspond to the 5% and 1% empirical lower and upper quantiles of the portfolio returns. In this case, VaR and CTE are not very sensitive to the copula model choice; intervals based on factor models with bivariate Frank linking copulas have interval estimates that are shifted towards 0 compared with those based on BB1 and t copulas. But the intervals still almost all contain the empirical values (one is slightly low in absolute value for the Frank copula, and one is slightly high for the BB1/Frank bi-factor model). A possible reason is that inferences on VaR and CTE are mainly dominated by the fit in the middle and more extreme quantiles of the portfolios are needed to differentiate the models.

For each copula model, using the bootstrap procedure in Section 7.8.1, we also compute model-based 95% confidence intervals for overall and sector group averages for Spearman ρ_S and lower/upper tail-weighted dependence measures ϱ_L, ϱ_U. There are too many intervals if we consider all pairs of stocks. Also corresponding values for specific pairs have more variability, but the sector group averages do give an indication of average strength of dependence and the direction of tail asymmetry. The GARCH-filtered log-returns are used to compute the corresponding empirical values. The results are presented in Table 7.32. There is a slight tail asymmetry toward the joint lower tail, so the tail symmetric models underestimate dependence in the lower tail. In addition, the multivariate t model overestimates dependence in the upper tail and the bi-factor Frank copula model underestimate dependence in both tails. Based on the tail-weighted dependence measures, the best fit is for the BB1/Frank bi-factor model.

model	ρ_S(all)	ρ_S(group1)	ρ_S(group2)	ρ_S(group3)
empirical estimate	0.50	0.63	0.58	0.68
nested-factor Gaussian	(0.46, 0.53)	(0.58, 0.66)	(0.53, 0.61)	(0.63, 0.71)
bi-factor Gaussian	(0.45, 0.52)	(0.57, 0.65)	(0.52, 0.61)	(0.62, 0.71)
bi-factor t(20)	(0.47, 0.54)	(0.58, 0.66)	(0.53, 0.61)	(0.63, 0.72)
nested-factor t(20)	(0.47, 0.55)	(0.58, 0.67)	(0.53, 0.62)	(0.63, 0.72)
nested-factor Frank	(0.50, 0.57)	(0.61, 0.68)	(0.56, 0.64)	(0.66, 0.74)
bi-factor Frank	(0.48, 0.55)	(0.60, 0.67)	(0.56, 0.64)	(0.67, 0.75)
nested refl. Gumbel/BB1	(0.44, 0.51)	(0.56, 0.64)	(0.51, 0.60)	(0.63, 0.71)
bi-factor BB1r/Frank	(0.44, 0.51)	(0.56, 0.64)	(0.52, 0.60)	(0.63, 0.71)

model	ϱ_L(all)	ϱ_L(group1)	ϱ_L(group2)	ϱ_L(group3)
empirical estimate	0.41	0.47	0.49	0.53
nested-factor Gaussian	**(0.24, 0.30)**	**(0.36, 0.44)**	**(0.30, 0.38)**	**(0.39, 0.49)**
bi-factor Gaussian	**(0.24, 0.30)**	**(0.35, 0.43)**	**(0.29, 0.37)**	**(0.39, 0.49)**
bi-factor t(20)	**(0.29, 0.35)**	**(0.39, 0.46)**	**(0.34, 0.42)**	**(0.42, 0.52)**
nested-factor t(20)	**(0.29, 0.35)**	(0.39, 0.47)	**(0.34, 0.42)**	**(0.42, 0.52)**
nested-factor Frank	**(0.10, 0.12)**	**(0.17, 0.22)**	**(0.14, 0.18)**	**(0.18, 0.24)**
bi-factor Frank	**(0.11, 0.14)**	**(0.23, 0.29)**	**(0.18, 0.24)**	**(0.23, 0.32)**
nested refl. Gumbel/BB1	**(0.42, 0.50)**	**(0.48, 0.59)**	(0.45, 0.57)	(0.52, 0.66)
bi-factor BB1r/Frank	(0.38, 0.49)	(0.46, 0.57)	(0.42, 0.55)	(0.52, 0.66)

model	ϱ_U(all)	ϱ_U(group1)	ϱ_U(group2)	ϱ_U(group3)
empirical estimate	0.29	0.40	0.31	0.48
nested-factor Gaussian	(0.24, 0.30)	(0.36, 0.44)	(0.30, 0.38)	(0.39, 0.49)
bi-factor Gaussian	(0.24, 0.30)	(0.35, 0.43)	(0.29, 0.37)	(0.39, 0.49)
bi-factor t(20)	(0.29, 0.35)	(0.39, 0.46)	**(0.34, 0.42)**	(0.42, 0.52)
nested-factor t(20)	(0.29, 0.35)	(0.39, 0.47)	**(0.34, 0.42)**	(0.42, 0.52)
nested-factor Frank	**(0.10, 0.12)**	**(0.17, 0.22)**	**(0.14, 0.18)**	**(0.18, 0.24)**
bi-factor Frank	**(0.11, 0.14)**	**(0.23, 0.29)**	**(0.18, 0.24)**	**(0.23, 0.32)**
nested refl. Gumbel/BB1	**(0.19, 0.25)**	(0.33, 0.50)	(0.27, 0.42)	(0.40, 0.59)
bi-factor BB1r/Frank	(0.24, 0.34)	(0.32, 0.49)	(0.29, 0.44)	(0.39, 0.57)

Table 7.32 *GARCH-filtered DAX stock returns 2011–2012 with $d = 29$ stocks and sample size $n = 511$: Overall and sector group 1,2,3 estimated averages of ρ_S, ϱ_L, ϱ_U, and the model-based 95% confidence intervals. Patterns for the other 6 sector groups are similar except there are wider interval estimates for the smaller groups. The model-based confidence intervals are based on the parametric bootstrap with 1000–2000 replications, depending on the speed of the computations. Intervals that don't contain the empirical value are shown in bold font.*

Conclusions for this subsection are the following.

1. For multivariate financial asset returns, structural factor copula models are amongst the best fitting copula models. If there are simple explanatory latent variables, then they tend to fit a little better than truncated R-vine copulas with the same number of parameters, and also they have the property of closure under margins (so that the copula model is interpreted in the same way when assets are added or deleted). Because truncated R-vine copulas can approximate many multivariate distributions well, they are also good fits to financial asset returns but maybe not with as parsimonious a dependence structure.

2. A typical experience with financial return data is that a vine or factor copula based on t is best based on AIC/BIC if the lower and upper semi-correlations do not suggest much tail asymmetry, and a vine or factor copula based on BB1 in the first tree/factor is best if the semi-correlations suggest significant tail asymmetry.

3. The example in this subsection shows that a model that is better in AIC or BIC need not be better for some tail inferences. The tail-weighted dependence measures are more sensitive to the tails of different copula models than portfolio VaR/CTE values. The choice of which copula models to use or the search for improved copula models would depend on considerations of parsimony, interpretability and adequacy for (tail) inferences.

4. Further research for asymmetric version of multivariate t copulas could be useful, perhaps based on some of the constructions in Section 3.17. McNeil et al. (2005) mention the importance of having copulas with tail asymmetry for finance applications, and Longin and Solnik (2001) and Ang and Chen (2002) have empirical studies of tail asymmetry for financial returns.

7.9 Conservative tail inference

As hinted in Section 7.1, accurate tail inference with copula models is difficult. In this section, we demonstrate a copula modeling approach to get conservative estimate of the probability of joint exceedance of a tail threshold. By conservative estimation, we mean the estimated probability should be larger than the "true" probability. By using copula models with differing amount of tail asymmetry and tail dependence, different degree of conservativeness can be achieved.

We use weekly log return data (sample size $n = 465$) in the period of July 1997 to May 2006 for four Asian financial markets: Hong Kong Hang Seng (HSI), Singapore Straights (ST)), Seoul Kospi (KS11), and Taiwan Taiex (TWII). A source is http://quote.yahoo.com . We will compare estimation of the probability that all of these four indexes will have log weekly returns below a threshold of q, where $q = -0.03$, -0.04 and -0.05 representing about 3%, 4% and 5% decrease respectively in a week. GARCH filtering will not be applied because it is easier to estimate the joint tail probabilities without the GARCH model; to account for the serial dependence, a stationary bootstrap (Section 5.5.1) is used for interval estimates.

Some summaries for the log returns are given in Table 7.33. As is usual for financial returns, there are significant autocorrelations and the absolute log returns but not the log returns.

For univariate models, we fitted t_ν with location/scale, and normal inverse Gaussian, generalized hyperbolic, and sinh-arcsinh-t (Rosco et al. (2011)); the latter three have skewness parameters. The sinh-arcsinh-t distribution was marginally the best fit, and for copula modeling, we use it as a simple skewing of the t_ν density which doesn't change the tail indices.

index	$q_{0.01}$	$q_{0.05}$	$q_{0.10}$	$q_{0.25}$	$q_{0.5}$	$q_{0.75}$	skew
HSI	-0.097	-0.055	-0.044	-0.020	0.001	0.024	-0.50
STI	-0.092	-0.050	-0.037	-0.015	0.001	0.016	-0.46
KS11	-0.143	-0.075	-0.054	-0.028	0.004	0.031	-0.44
TWII	-0.097	-0.065	-0.048	-0.021	0.002	0.021	0.03

indices	τ_{emp}	
HSI,STI	0.427	
HSI,KS11	0.361	
STI,KS11	0.359	
HSI,TWII	0.279	
STI,TWII	0.270	
KS11,TWII	0.289	

Table 7.33 *Weekly log returns: descriptive statistics with some quantiles and skewness coefficients (scaled third central moment), and bivariate empirical Kendall's taus.*

For bivariate skewness, lower and upper semi-correlations suggest tail dependence and only slight tail asymmetry to the joint upper tail. In this situation, a multivariate t copula is generally a good fit, so we use it for comparisons. The dependence pattern of the empirical Kendall's tau values suggest a model with hierarchical dependence with HSI, STI in cluster 1 having the most dependence, addition of KS11 for the second level of dependence and then finally addition of TWII at the third level.

To get a conservative estimates of joint lower tail probabilities, we use models with extreme tail asymmetry and lower tail dependence: (i) the hierarchical MTCJ copula with a gamma LT has lower tail dependence parameter in $(0, 1)$ and upper orthant tail independence; and (ii) the hierarchical BB2 copula, with fixed θ parameter and a gamma stopped gamma LT, has lower tail dependence parameter of 1 and upper orthant tail independence. This approach is then a multivariate version of an analysis in Hua and Joe (2012a) with tail comonotonicity and conservative risk measures. If the pattern of dependence were farther from hierarchical dependence, conservative models with copulas in the mixture of max-id family based on the gamma and gamma stopped gamma LTs could be used. Note that we are not looking for the best fitting model, but are intentionally fitting models which should have stronger lower tail dependence than the "true" joint distribution.

For maximum pseudo likelihood with uniform scores, the analysis is similar for all bivariate margins and we explain the pattern with the (HSI, STI) pair. The empirical Kendall tau value for (HSI, STI) is 0.427. From fitting different bivariate parametric copula families, the following are obtained.

- Bivariate t-copula: loglik=119.9 with parameter estimates $\tilde{\rho} = 0.62$, $\tilde{\nu} = 3.9$, and model-based $\tilde{\tau} = 0.426$.
- Bivariate MTCJ copula: loglik=103.6 with parameter estimate $\tilde{\theta} = 1.13$, and model-based $\tilde{\tau} = 0.361$; if estimation of θ matches with the empirical tau, then $\hat{\theta} = 1.49$ with more dependence and a larger probability of $C(p, p)$ for small $p > 0$.
- BB2 copula: loglik$\approx$ 102.1 with $\tilde{\theta} \to 0$, $\tilde{\delta} \to \infty$; and if $\theta = 0.1$ is fixed, loglik=99.2 with $\tilde{\delta} = 8.4$ and model-based $\tilde{\tau} = 0.337$; if estimation of δ matches with the empirical tau, then $\hat{\delta} = 12.8$ when $\theta = 0.1$, and this also represents more dependence.

By matching model-based taus and the empirical Kendall's tau, the copula parameter estimates are larger than those with maximum likelihood; because the MTCJ and BB2 copulas are increasing in concordance as the copula parameter increases, there are resulting larger values for estimated lower tail probabilities.

Hence to get conservative estimates from misspecified copula models, we estimate parameters that match the model-based Kendall tau to the empirical Kendall tau for two hierarchical Archimedean copula models (Section 3.4) that are mentioned below. The estimation steps for the original data set, and 100 resampled data series with the stationary bootstrap (Section 5.5.1) are summarized next.

1. The univariate sinh-arcsinh-t distribution (sast) with parameter $\boldsymbol{\eta}_j$ for variable j is fitted via maximum likelihood, and transform to $U(0,1)$ with $F_{\text{sast}}(\cdot; \hat{\boldsymbol{\eta}}_j)$, for $j = 1, 2, 3, 4$.

2. For second stage IFM, the 4-variate t_ν distribution is fitted via maximum likelihood.

3. For second stage estimation, the three parameters of hierarchical Archimedean MTCJ and BB2 are estimated by matching with (appropriate averages of) empirical Kendall tau values. The 4-variate hierarchical Archimedean copulas here have the form:

$$C_{1234}(\boldsymbol{u}) = \psi_1 \Big[\omega_1 \big(\omega_2 [\psi_3^{-1}(u_1) + \psi_3^{-1}(u_2)] + \psi_2^{-1}(u_3) \big) + \psi_1^{-1}(u_4) \Big].$$

For MTCJ, $\psi_i(s) = (1 + s)^{-1/\theta_i}$, $\omega_i(s) = (1 + s)^{\gamma_i} - 1$ with $\gamma_i = \theta_i/\theta_{i+1} \leq 1$ and $\theta_1 > 0$; and for BB2, $\psi_i(s) = [1 + \delta_i^{-1} \log(1+s)]^{-1/\theta}$, $\omega_i(s) = (1 + s)^{\gamma_i} - 1$ with $\gamma_i = \delta_i/\delta_{i+1} \leq 1$, $\delta_1 > 0$ and fixed $\theta > 0$.

4. With three resulting models $F_{1234}^{(m)}$ with two-stage parameter estimates, model-based probability estimates of $F_{1234}^{(m)}(q, q, q, q) = \mathbb{P}(\text{all indices} \leq q; \text{model } m)$ are obtained, where the lower threshold is one of $q = -0.03, -0.04, -0.05$. For comparison, the joint probabilities of form $\min_j \{ F_{\text{sast}}(q; \hat{\boldsymbol{\eta}}_j) \}$ based on the comonotonic copula are also obtained.

Summaries are given in Table 7.34; included are standard error estimates based on the 100 stationary bootstrap samples. As the fixed θ value becomes larger for the hierarchical BB2 copula, then the joint lower boundary may increase a bit and then eventually decrease for large θ values. Note the the assumption of comonotonicity is too conservative here, with estimated probabilities that are more than double those from hierarchical BB2; the conservativeness from the hierarchical MTCJ or BB2 copulas might be acceptable.

model	parameter estimates
HSI, sast	$(\nu, \mu, \sigma, \text{skew}) = (5.6, 0.004, 0.029, -0.08)$
STI, sast	$(\nu, \mu, \sigma, \text{skew}) = (3.1, 0.001, 0.022, -0.02)$
KS11, sast	$(\nu, \mu, \sigma, \text{skew}) = (5.0, 0.010, 0.038, -0.15)$
TWII, sast	$(\nu, \mu, \sigma, \text{skew}) = (5.0, 0.004, 0.029, -0.12)$
4-variate t	$(\nu, \rho_{12}, \rho_{13}, \rho_{23}, \rho_{14}, \rho_{14}, \rho_{24} = (5.1, 0.63, 0.53, 0.54, 0.42, 0.41, 0.44)$
hier. MTCJ	$(\theta_1, \theta_2, \theta_3) = (0.77, 1.12, 1.49)$
hier. BB2	$(\theta, \delta_1, \delta_2, \delta_3) = (0.1, 6.18, 9.42, 12.8)$
model	$F_{1234}(t, t, t, t)$ and bootstrap SEs for $t = -0.03, -0.04, -0.05$
4-variate t	0.034 (0.005), 0.021 (0.004), 0.013 (0.003)
hier. MTCJ	0.049 (0.008), 0.032 (0.006), 0.020 (0.004)
hier. BB2	0.053 (0.008), 0.035 (0.006), 0.024 (0.004)
C^+	0.127 (0.019), 0.080 (0.014), 0.051 (0.011)

Table 7.34 *Weekly log returns: Parameter estimates from 2-stage estimation procedures. The empirical values for $F_{1234}(q, q, q, q)$ are 0.030, 0.017, 0.011 for $q = -0.03, -0.04, -0.05$. For another comparison, the joint probabilities assuming independence are respectively 0.0009, 0.0002, 0.00004.*

To conclude this section, we have illustrated a copula modeling method to get conservative estimate of joint tail probabilities. This is important because likelihood methods are based mainly on data in the middle and might not lead to reliable estimates of tail quantities.

7.10 Item response: multivariate ordinal

One purpose of this section is to show that if multivariate ordinal data have a factor structure based on latent variables, then a factor copula model can provide a better fit than a truncated vine model; also the factor model is more interpretable.

An item response data set comes from the Consumer Protection and Perceptions of Science and Technology section of the 1992 Euro-Barometer Survey based on a sample of $n = 392$ from Great Britain; see Karlheinz and Melich (1992) and Rizopoulos (2006). The questions (items) are the following.

1. Y_1: Science and technology are making our lives healthier, easier and more comfortable;
2. Y_2: Scientific and technological research cannot play an important role in protecting the environment and repairing it;
3. Y_3: The application of science and new technology will make work more interesting;
4. Y_4: Thanks to science and technology, there will be more opportunities for the future generations;
5. Y_5: New technology does not depend on basic scientific research;
6. Y_6: Scientific and technological research do not play an important role in industrial development;
7. Y_7: The benefits of science are greater than any harmful effect it may have.

All of the items were measured on a four-point scale with response categories "0=strongly disagree," "1=disagree to some extent," "2=agree to some extent" and "3=strongly agree."

The summary of findings from the preliminary bivariate analysis based on Section 5.12.2 shows that there is more probability in the joint upper tail compared with discretized multivariate Gaussian, and also sometimes in the joint lower tail. Nikoloulopoulos and Joe (2014) have an intuitive explanation for latent maxima (or latent high quantiles). For some items such as the first and third, it is plausible that a respondent might be thinking about the maximum benefit (or a high quantile) of many past events. Nikoloulopoulos and Joe (2014) found that a 2-factor model that combines Gumbel copulas for the first factor with t_ν copulas for the second factor, to add some additional dependence in both tails, provided a better fit.

For comparison with the factor copula models, we also fit discrete R-vine models. From the polychoric correlation matrix, if $4 \le d \le 8$, it is possible to find (via enumeration with Algorithm 13 in Chapter 6) the best approximations with truncated R-vine correlation matrices with truncation level $1, 2, \ldots, d - 1$.

The polychoric correlation matrix is

$$
\boldsymbol{R}_{\text{poly}} = \begin{pmatrix}
1.000 & 0.099 & 0.201 & 0.346 & 0.090 & 0.182 & 0.408 \\
0.099 & 1.000 & -0.083 & -0.028 & 0.464 & 0.411 & -0.037 \\
0.201 & -0.083 & 1.000 & 0.479 & -0.104 & -0.008 & 0.209 \\
0.346 & -0.028 & 0.479 & 1.000 & -0.036 & 0.103 & 0.377 \\
0.090 & 0.464 & -0.104 & -0.036 & 1.000 & 0.435 & -0.014 \\
0.182 & 0.411 & -0.008 & 0.103 & 0.435 & 1.000 & 0.118 \\
0.408 & -0.037 & 0.209 & 0.377 & -0.014 & 0.118 & 1.000
\end{pmatrix}.
$$

Through enumeration of all 7-dimensional vines, the minimum log determinants of best truncated vines at levels 1 to 6 are: $-1.081, -1.222, -1.237, -1.247, -1.248, -1.248$, with the last number being the log determinant of the above polychoric correlation matrix. This suggests that there is not much improvement after the 2-truncated vine; the correlation matrix discrepancy measures in (5.12) are 0.167, 0.024 and 0.011 respectively for the best 1-truncated, 2-truncated and 3-truncated vines. For comparison, the discrepancy measure is 0.075 for the best fitting 2-factor structure correlation matrix. The vine array A_2 of best

2-truncated vine and the first three rows of its partial correlation array P_2 are:

$$A_2 = \begin{bmatrix} 6 & 6 & 6 & 4 & 6 & 4 & 5 \\ & 4 & 4 & 6 & 3 & 1 & 6 \\ & & 3 & & & & \\ & & & 1 & & & \\ & & & & 5 & & \\ & & & & & 7 & \\ & & & & & & 2 \end{bmatrix}, \quad P_2 = \begin{bmatrix} 0.103 & -0.008 & 0.346 & 0.435 & 0.377 & 0.464 \\ & 0.482 & 0.157 & -0.112 & 0.319 & 0.262 \\ & & 0.054 & -0.042 & 0.038 & -0.052 \end{bmatrix}.$$

The vine array A_3 of best 3-truncated vine and the first three rows of its partial correlation array P_3 are:

$$A_3 = \begin{bmatrix} 4 & 4 & 4 & 2 & 2 & 5 & 2 \\ & 2 & 2 & 4 & 4 & 2 & 4 \\ & & 6 & 6 & 6 & 4 & 1 \\ & & & 5 & & & \\ & & & & 1 & & \\ & & & & & 3 & \\ & & & & & & 7 \end{bmatrix}, \quad P_3 = \begin{bmatrix} -0.028 & 0.103 & 0.464 & 0.099 & -0.104 & -0.037 \\ & 0.416 & -0.026 & 0.351 & -0.039 & 0.376 \\ & & 0.308 & 0.121 & 0.478 & 0.325 \end{bmatrix}.$$

Note that the trees of the best 3-truncated vine are quite different than the trees of the best 2-truncated vine.

For another measure of discrepancy, consider the maximum absolute difference between the model-based and polychoric correlation matrices. For the best 2-truncated and 3-truncated vines, this leads to $\max|R_{2\text{truncated}} - R_{\text{poly}}| = 0.065$ and $\max|R_{3\text{truncated}} - R_{\text{poly}}| = 0.057$; for comparison with the 2-factor structure, $\max|R_{2\text{factor}} - R_{\text{poly}}| = 0.144$.

Next we improve on the truncated R-vines based on multivariate Gaussian by allowing the pair-copulas to be Gumbel or t_ν. Several combinations were tried with a fixed copula family for each tree of the truncated vines. Based on AIC, the best 2-truncated vine found was with Gumbel in tree T_1 and t_5 in tree T_2, and the best 3-truncated vine found was with t_5 in trees T_1, T_2 and Gumbel in tree T_3.

model	-loglik.	#parameters	AIC	BIC	95% interval
2-factor Gumbel/t_2	2864.6	15	5759	5819	
3-truncated t_5/t_5/Gumbel	2882.2	16	5796	5860	$(-0.004, 0.09)$
2-truncated Gumbel/t_5	2889.2	12	5802	5850	$(0.01, 0.12)$
2-factor Gaussian	2921.4	13	5869	5920	$(0.08, 0.21)$
1-factor t_2	2957.0	8	5930	5962	$(0.16, 0.31)$
1-factor Gumbel	2992.6	7	5999	6027	$(0.22, 0.43)$

Table 7.35 *Science item response data with $n = 392$, 7 items, 4 ordinal categories: negative log-likelihood, AIC, BIC values and the interval in (5.21) for Vuong's procedure relative to the 2-factor Gumbel/t_2 copula model. Except for the 3-truncated vine model, all of the compared models are significantly worse since the 95% confidence intervals are all above 0. The univariate cutpoints are estimated from univariate margins and are held fixed for estimation of dependence parameters in all of the models considered here. For the models with t_ν, several integer values were used, and the best value in terms of log-likelihood is reported; because of the relative flatness of the log-likelihood in ν there would not be improvement in the AIC/BIC with ν varying over different bivariate copulas.*

Table 7.35 has AIC and BIC values of different fitted models, including discretized 2-factor Gaussian, 2-factor Gumbel/t_2, and the best 2-truncated and 3-truncated R-vine

models; it also has the 95% confidence intervals from Vuong's log-likelihood ratio procedure in comparison with the best-fitting 2-factor copula model, and shows that all of the
compared models are significantly worse except for the 3-truncated vine,

Note that no effort was made to optimize over 3-truncated R-vine copula models; experience suggests that there will be different truncated R-vines that can fit approximately
the same. The truncated R-vine is not as interpretable for this item response data set as
the 2-factor model. The idea is to show that without much effort, we can find a 3-truncated
R-vine that is not significantly different (based on Vuong's procedure) from the best 2-factor
model. In other simple examples, we have found that often 2-truncated R-vines can be good
approximations to 1-factor models and 3-truncated R-vines can be good approximations to
2-factor models.

To conclude the analysis, Table 7.36 has parameter estimates and standard errors based
on the jackknife for the 2-factor model with Gumbel and t_2 linking copulas for factors 1
and 2 respectively.

item	cutpoint1 a_{j1}	cutpoint2 a_{j2}	cutpoint3 a_{j3}	Gumbel(θ_{j1})	$t_2(\theta_{j2})$
1	0.01 (0.01)	0.09 (0.01)	0.77 (0.02)	1.36 (0.11)	0.31 (0.12)
2	0.07 (0.01)	0.30 (0.02)	0.67 (0.02)	1.56 (0.14)	-0.47 (0.11)
3	0.08 (0.01)	0.33 (0.02)	0.86 (0.02)	1.18 (0.09)	0.54 (0.07)
4	0.04 (0.01)	0.22 (0.02)	0.76 (0.02)	1.38 (0.14)	0.70 (0.08)
5	0.05 (0.01)	0.28 (0.02)	0.68 (0.02)	1.57 (0.17)	-0.55 (0.10)
6	0.03 (0.01)	0.15 (0.02)	0.59 (0.02)	1.80 (0.19)	-0.33 (0.16)
7	0.05 (0.01)	0.31 (0.02)	0.80 (0.02)	1.26 (0.09)	0.45 (0.10)

Table 7.36 *Science item response data with $n = 392$, 7 items, 4 ordinal categories. Parameter estimates of 2-factor copula model based on 2-stage IFM and standard errors (in parentheses) based on
jackknife. There are three cutpoints on $U(0, 1)$ scale for each item, seven Gumbel dependence parameters linking items to factor 1, and seven t_2 dependence parameters linking items to factor 2. The ρ_N
values corresponding to the Gumbel parameters are respectively $0.41, 0.53, 0.24, 0.42, 0.54, 0.64, 0.32$
for easier comparison with the parameters of t_2.*

In summary, this example illustrates the use of a factor copula model for multivariate
ordinal data when the discretized multivariate Gaussian model underestimates some joint
tail probabilities.

7.11 SEM model as vine: alienation data

The purpose is of this section is to show how a structural equation model (with a structured
correlation matrix for multivariate Gaussian) can be estimated if the dependence structure
can be represented as a truncated vine rooted at some latent variables. With a two-stage
estimation approach, variances would be estimated in step 1 followed by the correlation
structure in step 2.

We use an example that appears in the structural equation modeling (SEM) literature,
originally in Wheaton et al. (1977). A brief description is given below.

In order to study the effects of industrial development on rural regions, a longitudinal
study was set up in Illinois. A set of sociological variables was measured from subjects
in a panel study. The characteristic of alienation (toward minority groups) is considered
as a latent variable expressed through the feeling of powerlessness and anomia. Attitude
scales measuring Anomia and Powerlessness were administered to a sample of 932 people
in 1967 and 1971; denote $Y_1 = $ anomia1967, $Y_2 = $ powerlessness1967, $Y_3 = $ anomia1971,
$Y_4 = $ powerlessness1971. Additional variables are the educational attainment measured
as $Y_5 = $ years of education completed as of 1967, and socioeconomic position expressed in

terms of the $Y_6 = $ SEI (Duncan socioeconomic index). The manifest variables are regarded as indicators of the latent variables $W_1 = $ alienation1967 and $W_2 = $ alienation1971; another latent variable affecting all manifest variables is $W_0 = $ SES (socioeconomic status).

The sample correlation matrix of the 6 observed (or manifest) variables is in the following table.

	anomia67	powerls67	anomia71	powerls71	education	SEI
anomia67	1					
powerls67	0.66	1				
anomia71	0.56	0.47	1			
powerls71	0.44	0.52	0.67	1		
education	-0.36	-0.41	-0.35	-0.37	1	
SEI	-0.30	-0.29	-0.29	-0.28	0.54	1

In the path diagram for the structural equation model in Browne and Arminger (1995), the direction of flows are:

1. SES $\rightarrow$ education,
2. SES $\rightarrow$ SEI,
3. SES $\rightarrow$ alienation1967,
4. SES $\rightarrow$ alienation1971,
5. alienation1967 $\rightarrow$ alienation1971,
6. alienation1967 $\rightarrow$ anomia1967,
7. alienation1967 $\rightarrow$ powerlessness1967,
8. alienation1971 $\rightarrow$ anomia1971,
9. alienation1971 $\rightarrow$ powerlessness1971.

In addition, the regressions of anomia1967 on alienation1967 and anomia1971 on alienation1971 have dependent residuals, and similarly the regressions of powerlessness1967 on alienation1967 and powerlessness1971 on alienation1971 have dependent residuals. Hence there are a total of 11 parameters in the correlation structure; since this is less than $\binom{6}{2} = 15$, the model should be identifiable.

We can convert the path diagram to a partial correlation vine on nine variables with 3 latent and 6 observed. Note that all observed variables are correlated with $W_0 = $ SES.

To get a 3-level truncated vine, there are 8 correlations for tree T_1, 6 edges with conditional independence and 1 edge with conditional dependence for tree T_2; and 4 edges with conditional independence and 2 edges with conditional dependence for tree T_3. Specifically, the edges of the trees are summarized below with the pairs of variables for the nodes of the edges.

1. Edges for tree T_1: $W_1 W_0$, $W_2 W_1$, $Y_1 W_1$, $Y_2 W_1$, $Y_3 W_2$, $Y_4 W_2$, $Y_5 W_0$, $Y_6 W_0$.
2. Edges for tree T_2: $W_2 W_0 | W_1$ with dependence; $Y_1 W_2 | W_1$, $Y_2 W_2 | W_1$, $Y_3 W_1 | W_2$, $Y_4 W_1 | W_2$, $Y_5 W_1 | W_0$, $Y_6 W_1 | W_0$ with conditional independence.
3. Edges for tree T_3: $Y_3 Y_1 | W_1 W_2$ and $Y_4 Y_2 | W_1 W_2$ with conditional dependence. Four additional edges are needed with conditional independence to verify the truncated vine structure — simple choices are $Y_1 W_0 | W_1 W_2$, $Y_2 W_0 | W_1 W_2$, $Y_5 W_2 | W_1 W_0$, $Y_6 W_2 | W_1 W_0$.

The combined path diagram and truncated vine showing the non-zero partial correlations is in Figure 7.4.

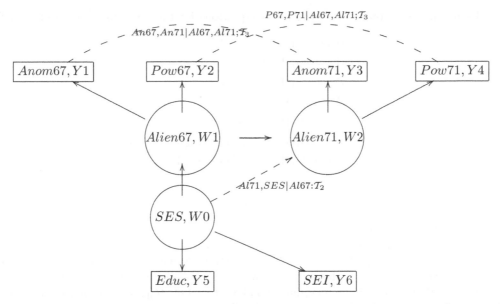

Figure 7.4 *Wheaton alienation data, with 3 latent variables; a combined path diagram and truncated vine.*

With the variable order of $W_0, W_1, W_2, Y_1, Y_2, Y_3, Y_4, Y_5, Y_6$ for variables 1 to 9, the vine array for the first three trees has:

$$A = \begin{bmatrix} 1 & 1 & 2 & 2 & 2 & 3 & 3 & 1 & 1 \\ & 2 & 1 & 3 & 3 & 2 & 2 & 2 & 2 \\ & & 3 & 1 & 1 & 4 & 5 & 3 & 3 \\ & & & 4 & & & & & \\ & & & & 5 & & & & \\ & & & & & 6 & & & \\ & & & & & & 7 & & \\ & & & & & & & 8 & \\ & & & & & & & & 9 \end{bmatrix} = \begin{bmatrix} W_0 & W_0 & W_1 & W_1 & W_1 & W_2 & W_2 & W_0 & W_0 \\ & W_1 & W_0 & W_2 & W_2 & W_1 & W_1 & W_1 & W_1 \\ & & W_2 & W_0 & W_0 & Y_1 & Y_2 & W_2 & W_2 \\ & & & Y_1 & & & & & \\ & & & & Y_2 & & & & \\ & & & & & Y_3 & & & \\ & & & & & & Y_4 & & \\ & & & & & & & Y_5 & \\ & & & & & & & & Y_6 \end{bmatrix}.$$

The vine array can be completed non-uniquely in rows 4 to 8, but this would have no effect on estimation of the model parameters. The non-zero (partial) correlations are in the following array (indexed in same way as the vine array):

$$\begin{bmatrix} - & \rho_{W_1 W_0} & \rho_{W_2 W_1} & \rho_{Y_1 W_1} & \rho_{Y_2 W_1} & \rho_{Y_3 W_2} & \rho_{Y_4 W_2} & \rho_{Y_5 W_0} & \rho_{Y_6 W_0} \\ - & - & \rho_{W_2 W_0; W_1} & 0 & 0 & 0 & 0 & 0 & 0 \\ - & - & - & 0 & 0 & \rho_{Y_3 Y_1; W_1 W_2} & \rho_{Y_4 Y_2; W_1 W_2} & 0 & 0 \end{bmatrix}$$

There are 11 parameters, which are algebraically independent in $(-1, 1)$.

Using the notation of (6.7), the linear equations for the structural equation model are:

$$W_0 = \psi_1 \epsilon_1$$
$$W_1 = \phi_{21} W_0 + \psi_2 \epsilon_2$$
$$W_2 = \phi_{31} W_1 + \phi_{32} W_0 + \psi_3 \epsilon_3$$
$$Y_1 = \phi_{41} W_1 + \psi_4 \epsilon_4$$
$$Y_2 = \phi_{51} W_1 + \psi_5 \epsilon_5 \qquad (7.9)$$
$$Y_3 = \phi_{61} W_2 + \phi_{62} W_1 + \phi_{63} Y_1 + \psi_6 \epsilon_6$$
$$Y_4 = \phi_{71} W_2 + \phi_{72} W_1 + \phi_{73} Y_2 + \psi_7 \epsilon_7$$
$$Y_5 = \phi_{81} W_0 + \psi_8 \epsilon_8$$
$$Y_6 = \phi_{91} W_0 + \psi_9 \epsilon_9$$

where the ψ's satisfy (6.8).

To get the model-based correlation matrix of the six observed variables, Algorithm 29 in Chapter 6 can be applied to get a 9×9 correlation matrix from the partial correlations and the vine array A, and the last six rows/columns can be extracted. Let the submatrix be $R_{\text{model}}(\theta)$ where θ is an 11-dimensional vector. Let R_{obs} be the observed correlation matrix. Then n^{-1} times the negative log-likelihood is

$$\tfrac{1}{2} \log |R_{\text{model}}(\theta)| + \tfrac{1}{2} \operatorname{tr}[R_{\text{model}}^{-1}(\theta) R_{\text{obs}}] + 3 \log(2\pi),$$

where $|R| = \det(R)$. This function can be input to a numerical minimizer.

The resulting maximum likelihood estimates for the 11 parameters are:

$$\hat{\rho}_{W_1 W_0} = -0.563, \ \hat{\rho}_{W_2 W_1} = 0.684,$$
$$\hat{\rho}_{Y_1 W_1} = 0.775, \ \hat{\rho}_{Y_2 W_1} = 0.852, \ \hat{\rho}_{Y_3 W_2} = 0.806,$$
$$\hat{\rho}_{Y_4 W_2} = 0.832, \ \hat{\rho}_{Y_5 W_0} = 0.841, \ \hat{\rho}_{Y_6 W_0} = 0.642,$$
$$\hat{\rho}_{W_2 W_0; W_1} = -0.235, \ \hat{\rho}_{Y_3 Y_1; W_1 W_2} = 0.356, \ \hat{\rho}_{Y_4 Y_2; W_1 W_2} = 0.121.$$

The maximum absolute value $\|R_{\text{model}} - R_{\text{obs}}\|$ is 0.020, and the fit value in (5.12) is

$$D_{\text{model}} = \log(|R_{\text{model}}|) - \log(|R_{\text{obs}}|) + \operatorname{tr}(R_{\text{model}}^{-1} R_{\text{obs}}) - 6 = 0.005.$$

Using Algorithm 29, the linear representation (7.9) is:

$$W_0 = \psi_1 \epsilon_1, \qquad Y_1 = 0.775 W_1 + \psi_4 \epsilon_4,$$
$$W_1 = -0.563 W_0 + \psi_2 \epsilon_2, \qquad Y_2 = 0.852 W_1 + \psi_5 \epsilon_5,$$
$$W_2 = 0.567 W_1 - 0.208 W_0 + \psi_3 \epsilon_3, \quad Y_3 = 0.806 W_2 - 0.258 W_1 + 0.333 Y_1 + \psi_6 \epsilon_6,$$
$$Y_4 = 0.832 W_2 - 0.109 W_1 + 0.128 Y_2 + \psi_7 \epsilon_7,$$
$$Y_5 = 0.841 W_0 + \psi_8 \epsilon_8,$$
$$Y_6 = 0.642 W_0 + \psi_9 \epsilon_9,$$

where

$$(\psi_1, \ldots, \psi_9) = (1.00, 0.826, 0.709, 0.633, 0.523, 0.554, 0.551, 0.540, 0.767),$$
$$(\psi_1^2, \ldots, \psi_9^2) = (1.00, 0.683, 0.503, 0.400, 0.274, 0.307, 0.303, 0.292, 0.588).$$

To conclude, we have shown an approach different from the recticular action model formulation in McArdle and McDonald (1984) for estimating parameters of a multivariate Gaussian structural equation model that can be represented as a truncated partial correlation vine. With the theory and algorithms in Chapter 3 and Section 6.15, the first step of this approach is to convert a path diagram to a truncated vine.

7.12 SEM model as vine: attitude-behavior data

In this section, there is an example of a structured factor model with residual dependence. There are 11 observed variables and 5 latent variables. The background for this data set is given in Fredricks and Dossett (1983) and Breckler (1990). The path diagram leads to a 3-truncated vine with a total of $11+4=15$ correlations and $3+1=4$ partial correlations for a total of 19 parameters to be estimated. The 11 correlations of observed variables with latent variables can be considered as loadings. Then there are four correlations, three partial correlations of order 1 and one partial correlation of order 2 for the five latent variables. Two remaining partial correlations among the latent variables are set to 0, as well as partial correlations of observed and latent given other latent variables.

Latent variables are: $W_1 = \mathtt{PB} = \mathtt{prior\ behaviors}$, $W_2 = \mathtt{SN} = \mathtt{subjective\ norms}$, $W_3 = \mathtt{attitudes}$, $W_4 = \mathtt{behavioral\ intentions}$, and $W_5 = \mathtt{target\ behaviors}$. Observed variables are: X_1, X_2 as two measures of *attitudes towards the behaviors* (loading on W_3), X_3, X_4 as two measures of *subjective norms* (loading on W_2), X_5, X_6, X_7 as three (longitudinal) measures of *prior behaviors* (loading on W_1), Y_1, Y_2 as two measures of *behavioral intentions* (loading on W_4), Y_3, Y_4 as two (longitudinal) measures of *target behaviors* (loading on W_5). $X_1, \ldots, X_7$ are considered as exogenous and $Y_1, \ldots, Y_4$ are considered as endogenous.

The sample correlation matrix of the 11 observed variables is in the following table.

	Y_1	Y_2	Y_3	Y_4	X_1	X_2	X_3	X_4	X_5	X_6	X_7
Y_1	1										
Y_2	0.660	1									
Y_3	0.285	0.332	1								
Y_4	0.292	0.363	0.432	1							
X_1	0.364	0.334	0.244	0.142	1						
X_2	0.407	0.329	0.260	0.211	0.534	1					
X_3	0.251	0.154	0.164	0.037	0.357	0.334	1				
X_4	0.281	0.184	0.149	−0.010	0.307	0.352	0.722	1			
X_5	0.219	0.414	0.310	0.467	0.142	0.209	−0.065	−0.032	1		
X_6	0.258	0.300	0.423	0.419	0.200	0.225	−0.003	0.052	0.455	1	
X_7	0.276	0.420	0.365	0.424	0.284	0.249	0.073	0.096	0.462	0.441	1

The path diagram in Breckler (1990) shows that W_1, W_2, W_3 are mutually dependent (double arrows), directional arrows from each of W_1, W_2, W_3 to W_4 and directional arrows from W_1, W_4 to W_5.

A truncated vine structure that matches this description is given below. Note that this is only one example of the vine structure that can be used; there are other 3-truncated vines that would work as well.

1. Edges for tree $\mathcal{T}_1$: W_2W_1, W_3W_1, W_4W_1, W_5W_1, Y_1W_4, Y_2W_4, Y_3W_5, Y_4W_5, X_1W_3, X_2W_3, X_3W_2, X_4W_2, X_5W_1, X_6W_1, X_7W_1.
2. Edges for tree $\mathcal{T}_2$: $W_3W_2|W_1$, $W_4W_2|W_1$, $W_5W_4|W_1$ and other edges with zero partial correlations could be: $Y_1W_1|W_4$, $Y_2W_1|W_4$, $Y_3W_1|W_5$, $Y_4W_1|W_5$, $X_1W_1|W_3$, $X_2W_1|W_3$, $X_3W_1|W_2$, $X_4W_1|W_2$, $X_5W_2|W_1$, $X_6W_2|W_1$, $X_7W_2|W_1$.
3. Edges for tree $\mathcal{T}_3$: $W_4W_3|W_1W_2$ and other edges with zero partial correlations could be: $W_5W_2|W_1W_4$ $Y_1W_2|W_1W_4$, $Y_2W_2|W_1W_4$, $Y_3W_4|W_1W_5$, $Y_4W_4|W_1W_5$, $X_1W_2|W_1W_3$, $X_2W_2|W_1W_3$, $X_3W_3|W_1W_2$, $X_4W_3|W_1W_2$, $X_5W_3|W_1W_2$, $X_6W_3|W_1W_2$, $X_7W_3|W_1W_2$.

The combined path diagram and truncated vine showing the non-zero partial correlations is in Figure 7.5.

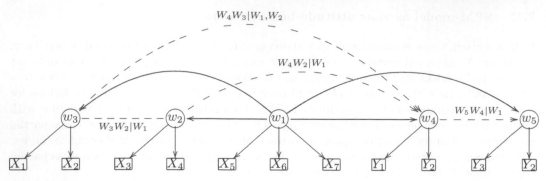

Figure 7.5 *Attitude-behavior with 5 latent variables and 11 observed variables; a combined path diagram and truncated vine.*

The vine array (with diagonal flattened because of truncation) is:

$$A = \begin{bmatrix} W_1 & W_1 & W_1 & W_1 & W_1 & W_4 & W_4 & W_5 & W_5 & W_3 & W_3 & W_2 & W_2 & W_1 & W_1 & W_1 \\ & W_2 & W_2 & W_2 & W_4 & w_1 & w_1 & w_1 & w_1 & w_1 & w_1 & w_1 & w_1 & w_2 & w_2 & w_2 \\ & & W_3 & W_3 & w_2 & w_2 & w_2 & w_4 & w_4 & w_2 & w_2 & w_3 & w_3 & w_3 & w_3 & w_3 \\ & & & W_4 & & & & & & & & & & & & \\ & & & & W_5 & Y_1 & Y_2 & Y_3 & Y_4 & X_1 & X_2 & X_3 & X_4 & X_5 & X_6 & X_7 \end{bmatrix}.$$

In this array, the lower-case letters show the zero partial correlations.

For the above SEM, the maximum absolute value $||\boldsymbol{R}_{\text{SEM}} - \boldsymbol{R}_{\text{obs}}||$ is 0.12, and the fit value in (5.12) is

$$D_{\text{SEM}} = \log(|\boldsymbol{R}_{\text{SEM}}|) - \log(|\boldsymbol{R}_{\text{obs}}|) + \text{tr}(\boldsymbol{R}_{\text{SEM}}^{-1}\boldsymbol{R}_{\text{obs}}) - 11 = 0.24.$$

The correlation matrix of the SEM matches well with the sample correlation matrix when the (absolute) correlations are larger; the larger deviations occur for smaller correlations. For the 1-factor correlation structure, the corresponding values are $||\boldsymbol{R}_{\text{1factor}} - \boldsymbol{R}_{\text{obs}}|| = 0.63$ and $D_{\text{1factor}} = 1.65$.

The sample size is $n = 236$ and the saturated correlation matrix has 55 parameters; this SEM model has 19 parameters. The log-likelihood ratio statistic is $n \cdot D_{\text{SEM}} = 56.14$ for the null SEM model with $55 - 19 = 36$ degrees of freedom. Hence the P-value is around 0.017.

The resulting maximum likelihood estimates for the correlation and partial correlation parameters are in Table 7.37.

edge	estimate	edge	estimate	edge	estimate	edge	estimate
W_2W_1	0.053	Y_1W_4	0.780	X_1W_3	0.712	$W_3W_2|W_1$	0.580
W_3W_1	0.441	Y_2W_4	0.846	X_2W_3	0.749	$W_4W_2|W_1$	0.331
W_4W_1	0.592	Y_3W_5	0.609	X_3W_2	0.862	$W_5W_4|W_1$	0.187
W_5W_1	0.897	Y_4W_5	0.709	X_4W_2	0.838	$W_4W_3|W_1W_2$	0.340
				X_5W_1	0.682		
				X_6W_1	0.664		
				X_7W_1	0.679		

Table 7.37 *Attitude-behavior data: maximum likelihood estimates of the correlation and partial correlation parameters in the vine representation.*

Using Algorithm 29 in Chapter 6, the linear representation is:

$$
\begin{aligned}
W_1 &= \psi_1\epsilon_1, & X_1 &= 0.712W_3 + \psi_{10}\epsilon_{10}, \\
W_2 &= 0.053W_1 + \psi_2\epsilon_2, & X_2 &= 0.749W_3 + \psi_{11}\epsilon_{11}, \\
W_3 &= 0.413W_1 + 0.521W_2 + \psi_3\epsilon_3, & X_3 &= 0.862W_3 + \psi_{12}\epsilon_{12}, \\
W_4 &= 0.431W_1 + 0.083W_2 + 0.035W_3 + \psi_4\epsilon_4, & X_4 &= 0.838W_3 + \psi_{13}\epsilon_{13}, \\
W_5 &= 0.836W_1 + 0.103W_4 + \psi_5\epsilon_5, & X_5 &= 0.682W_1 + \psi_{14}\epsilon_{14}, \\
Y_1 &= 0.780W_4 + \psi_6\epsilon_6, & X_6 &= 0.664W_1 + \psi_{15}\epsilon_{15}, \\
Y_2 &= 0.846W_4 + \psi_7\epsilon_7, & X_7 &= 0.679W_1 + \psi_{16}\epsilon_{16}, \\
Y_3 &= 0.609W_5 + \psi_8\epsilon_8, & & \\
Y_4 &= 0.709W_5 + \psi_9\epsilon_9, & &
\end{aligned}
$$

where

$$
(\psi_1, \ldots, \psi_{16}) = (1.00, 0.997, 0.535, 0.512, 0.188,
$$
$$
0.392, 0.284, 0.629, 0.497, 0.492, 0.438, 0.257, 0.298, 0.534, 0.559).
$$

To conclude, we have shown another example of using a truncated partial correlation vine to estimate parameters of a multivariate Gaussian structural equation model. The algorithms in Section 6.15 are useful for conversion among the different parametrizations.

7.13   Overview of applications

In this chapter, we have shown a variety of applications of dependence models with copulas, for discrete and continuous response data.

In the scientific literature, there are many applications of copula models, far too many to summarize here. The list below just gives an indication of different types of applications.

- Li (2001) uses the Gaussian copula for default correlations of credit risks. Did Li's article have the "formula that killed Wall Street" in 2008? See Salmon (2009). See also Embrechts (2009) for caveats on use of copulas and extreme value theory for quantitative risk management and the financial crisis of 2008.
- Patton (2006) uses a copula time series model with time-varying dependence to explain the dependence of two exchange rates. Rodriguez (2007) models dependence with switching-parameter copulas to study financial contagion. Chollete et al. (2009) uses a multivariate regime-switching copula for international financial returns. Other applications of time-varying copulas are in Hafner and Manner (2012) and Manner and Reznikova (2012). Patton (2009, 2012) have extensive reviews of copula models for financial and economic time series.
- Dissmann et al. (2013) and Brechmann et al. (2012) have applications of copula-GARCH models to financial return data with high-dimensional vines and truncated vines.
- Wang et al. (2009) use a 1-factor heavy-tailed copula model for pricing credit default index swap tranches,
- Mendes and Kolev (2008) fit copulas to co-exceedances of pairs of financial returns to assess volatility in international equities.
- Karlis and Pedeli (2013) propose a bivariate copula model with univariate margins that integer autoregressive count time series models, and apply it to numbers of quotes of intraday bid and intraday ask for a specific stock.
- In Escarela et al. (2006), copulas are used for the transition probability for discrete response time series with univariate marginal regression models.
- Smith et al. (2010) use the pair-copula construction for modeling longitudinal data, with application to forecasting electricity load at an intraday resolution.

- Bernard and Czado (2013) use copula models for multivariate option pricing.
- Brechmann et al. (2014) use copula models for operational risk losses.
- Danaher and Smith (2011) use elliptical copulas for several applications in marketing with large databases.
- There have been many applications in insurance. Frees et al. (1996) and Carriere (2000) use bivariate copulas as dependent mortality models with application to life insurance. Sun et al. (2008) use copula models for heavy-tailed longitudinal data, with an application to nursing home care utilization. Erhardt and Czado (2012) use copulas to model dependent yearly claim totals in private health insurance. Czado et al. (2012) study a copula model for insurance claim amounts and claim sizes.
- Zimmer and Trivedi (2006) use trivariate copula families with univariate discrete response regression models with applications to family health care demand.
- Bárdossy and Li (2008) implement geostatistical interpolation with copulas that have a correlation matrix as a parameter. Kazianka and Pilz (2010) use these copulas for geostatistics modeling with spatial isotropic dependence and univariate regression models for continuous and discrete response. Serinaldi (2008) use bivariate copulas for spatial dependence analysis of rainfall fields.
- Some hydrology applications are the following. Favre et al. (2004) use copulas for multivariate hydrological frequency analysis. Grimaldi and Serinaldi (2006) use copulas for multivariate flood frequency analysis. De Michele et al. (2007) use copulas for a multivariate model of sea storms. Salvadori and De Michele (2010) use multivariate extreme value copulas to estimate multivariate return periods from maximum annual flood data.
- Dupuis (2005) uses the MM1 multivariate extreme value copula to fit data on weekly maximum ozone concentrations at four monitoring sites. For spatial extremes of annual maximum rainfall data, Smith and Stephenson (2009) use an extended Gaussian max-stable process model which has the bivariate Hüesler-Reiss model for pairs of sites; composite pairwise likelihood is used for estimation of the spatial dependence parameters.
- Some application in medical statistics and biostatistics are the following. Bogaerts and Lesaffre (2008) use a copula model for bivariate interval-censored data with application in a dental study of children. Nikoloulopoulos and Karlis (2008b) use a multivariate logit copula model for application to dental data with binary response (indicator of tooth decay); the model is a mixture of max-id copula and covariate information is incorporated into the copula parameters as well as univariate logistic regressions. Braeken and Tuerlinckx (2009) use copula models for multivariate binary anomaly data in teratology studies. de Leon and Wu (2011) use a bivariate copula model on mixed discrete-continuous response variables with application to burn injury data.
- In psychometrics, in modeling item response data, Braeken et al. (2007) use a copula to model exchangeable subgroup of items with residual dependency.
- Bauer et al. (2012) have a pair-copula construction for non-Gaussian directed acyclic graph models, with application to financial return data. Morales-Nápoles et al. (2008) use vines and continuous Bayesian belief nets to elicit conditional and unconditional rank correlations for an application in aviation safety. Hanea et al. (2010) use Bayesian belief nets for multivariate ordinal data of pollutants emissions; a criterion for edges is based on the partial correlation vine.
- Elidan (2013) surveys copula-based constructions in machine learning.

Applications with copulas should continue to grow as multivariate Gaussian dependence can be improved on.

Chapter 8

Theorems for properties of copulas

This chapter has theorems and more advanced examples on dependence properties and tail behavior of copula families, such as multivariate Gaussian, Archimedean, elliptical, extreme value, mixture of max-id, vine, and factor. Also there are general results on copulas and dependence properties: absolutely and singular components of continuous distributions, tail heaviness, regular variation, dependence orderings, relations of dependence concepts, tail dependence and tail order functions, and Laplace transforms. This chapter is intended as a reference for results to determine dependence and tail properties of multivariate distributions which are constructed from a combination of methods in Chapter 3. For a theorem, a proof is given if (i) the proof is simpler than the original proof, (b) if the theorem is a variation of previously published results, or (c) the proof is short and provides some insight such as illustrating a useful technique.

The chapter starts with a few sections with preliminary results on multivariate distributions and dependence concepts. Section 8.1 has results on absolutely continuous and singular multivariate distributions. Section 8.2 has results on continuity of copulas. Section 8.3 has results on dependence concepts and orderings. Section 8.4 has results on Fréchet classes and compatibility of marginal distributions.

Then there are sections with results on dependence and tail behavior for different multivariate families: Section 8.5 for Archimedean copulas, Section 8.6 for multivariate extreme value distributions, Section 8.7 for mixture of max-infinitely divisible copulas and Section 8.8 for elliptical distributions.

For concepts for copula tail behavior in general, Section 8.9 has results on tail dependence and tail dependence functions that are especially useful for copula models built from mixing conditional distributions, and Section 8.10 has results on the tail order. Next, Section 8.11 has results on the combinatorics of vines, Section 8.12 has results for mixture of conditional distributions and vine copulas, and Section 8.13 has results on dependence and tail behavior for factor copula models.

For auxiliary results, Section 8.14 is about Kendall functions (a multivariate probability integral transform), Section 8.15 is about Laplace transforms and Section 8.16 is about regular variation.

Section 8.17 contains a chapter summary.

8.1 Absolutely continuous and singular components of multivariate distributions

A d-dimensional copula cdf is continuous and increasing, so right derivatives of first order, i.e., $\partial C(\boldsymbol{u})/\partial u_j$, $j = 1, \ldots, d$, exist. Hence if $\boldsymbol{U} \sim C$, conditional distributions of the form

$$C_{1,\ldots,j-1,j+1,\ldots,d|j}(u_1, \ldots, u_{j-1}, u_{j+1}, \ldots, u_d | u_j)$$

exist, and C has a singular component if one or more of these conditional distributions has a jump discontinuity. A similar result holds if C has mixed derivatives of kth order everywhere, $2 \leq k < d$. A result for identifying the total mass of the singular component is stated below for multivariate distributions including copulas. It is a technical result, but needed because copulas in the Archimedean and extreme value families can have singular components.

Theorem 8.1 (Mass of singular component). *Let $F(x_1, \ldots, x_d)$ be a continuous d-variate distribution with univariate margins $F_1, \ldots, F_d$.*

(a) Let $d = 2$. Suppose that F_1 is absolutely continuous, and suppose that the conditional distribution $F_{2|1}$ exists using the right-hand partial derivative of F with respect to x_1. Suppose that $F_{2|1}(\cdot|x_1)$ has jump discontinuities totaling a mass of $m(x_1)$, and $m(\cdot)$ is continuous and positive on an interval. Then F has a singular component and the mass of the singular component is $\int m(x_1) \, dF_1(x_1)$.

(b) Suppose that the margin $F_{1:(d-1)}$ is absolutely continuous, and suppose that the conditional distribution $F_{d|1:(d-1)}$ exists using the right-hand partial derivative of F with respect to $x_1, \ldots, x_{d-1}$. Suppose that $F_{d|1:(d-1)}(\cdot|x_1, \ldots, x_{d-1})$ has jump discontinuities totaling a mass of $m(x_1, \ldots, x_{d-1})$, and $m(\cdot)$ is continuous and positive on an open set. Then F has a singular component and the mass of the singular component is $\int m(x_1, \ldots, x_{d-1}) \, dF_{1:(d-1)}(x_1, \ldots, x_{d-1})$.

(c) Let $1 \leq k \leq d-1$ be an integer. Suppose that the margin $F_{1\cdots k}$ is absolutely continuous, and suppose that the conditional distribution $F_{(k+1):d|1\cdots k}$ exists using the right-hand partial derivative of F with respect to $x_1, \ldots, x_k$. Suppose that $F_{(k+1):d|1\cdots k}(\cdot|x_1, \ldots, x_k)$ has jump discontinuities totaling a mass of $m(x_1, \ldots, x_k)$, and $m(\cdot)$ is continuous and positive on an open set. Then F has a singular component and the mass of the singular component is $\int m(x_1, \ldots, x_k) \, dF_{1:k}(x_1, \ldots, x_k)$.

Proof: Because we assume F is continuous, there is not a discrete component, so there remain the absolutely continuous and singular components. Then the integral of $\frac{\partial^d F}{\partial x_1 \cdots \partial x_d}$ (treated as right derivatives everywhere) is the total probability of the absolutely continuous component.

(a) is a special case of (b), so we prove (b) next. Let $f_{d|1:(d-1)}(\cdot|x_1, \ldots, x_{d-1})$ be the right derivative of $F_{d|1:(d-1)}(\cdot|x_1, \ldots, x_{d-1})$. Because of the jump discontinuities, $\int f_{d|1:(d-1)}(x_d|x_1, \ldots, x_{d-1}) \, dx_d = 1 - m(x_1, \ldots, x_{d-1})$. The integral of the absolutely continuous part is:

$$\int f_{d|1:(d-1)}(x_d|x_1, \ldots, x_{d-1}) f_{1:(d-1)}(x_1, \ldots, x_{d-1}) \, dx_d dx_1 \cdots dx_{d-1}$$

$$= \int [1 - m(x_1, \ldots, x_{d-1})] f_{1:(d-1)}(x_1, \ldots, x_{d-1}) \, dx_1 \cdots dx_{d-1}.$$

The mass of the singular component is 1 minus the above.

(c) The proof is similar to that of (b). Let $f_{(k+1):d|1\cdots k}(\cdot|x_1, \ldots, x_k)$ be the right derivative of $F_{(k+1):d|1\cdots k}(\cdot|x_1, \ldots, x_k)$. Because of the jump discontinuities, $\int f_{(k+1):d|1\cdots k}(x_{k+1}, \ldots, x_d|x_1, \ldots, x_k) \, dx_{k+1} \cdots dx_d = 1 - m(x_1, \ldots, x_k)$. The integral of the absolutely continuous part is:

$$\int f_{(k+1):d|1\cdots k}(x_{k+1}, \ldots, x_d|x_1, \ldots, x_k) f_{1:k}(x_1, \ldots, x_k) \, dx_{k+1} \cdots dx_d dx_1 \cdots dx_k$$

$$= \int [1 - m(x_1, \ldots, x_k)] f_{1:k}(x_1, \ldots, x_k) \, dx_1 \cdots dx_k.$$

□

Example 8.1 (Illustration of singular components of multivariate distributions). We include two examples here to illustrate the above theorem. Other examples are in Sections 2.1.2 and 3.3.

1. Trivariate comonotonicity copula.

$$C^+(\boldsymbol{u}) = \min\{u_1, u_2, u_3\} = \begin{cases} u_1, & u_1 \le u_2 \text{ and } u_1 \le u_3, \\ u_2, & u_2 \le u_1 \text{ and } u_2 \le u_3, \\ u_3, & u_3 \le u_1 \text{ and } u_3 \le u_2. \end{cases}$$

Take $k = 1$ in part (c) of Theorem 8.1. F_1 is the $U(0,1)$ cdf and it is absolutely continuous. Next the conditional distribution is:

$$F_{23|1}(u_2, u_3|u_1) = C_{23|1}(u_2, u_3|u_1) = \begin{cases} 1, & u_1 \le u_2 \text{ and } u_1 \le u_3, \\ 0, & \text{otherwise.} \end{cases}$$

Hence the total mass of jump discontinuities is $m(u_1) = 1$ for $0 < u_1 < 1$. Theorem 8.1 just confirms that the mass of the singular component is $\int_0^1 m(u_1)\, du_1 = 1$, which we know from the stochastic representation $U_1 = U_2 = U_3$.

2. Trivariate Archimedean copula family based on $\psi(s) = (1 + \theta s)_+^{-1/\theta}$, $\theta \ge -1$; the boundary for $\mathcal{L}_3$ has $\theta = -\frac{1}{2}$ or $\psi(s) = (1 - \frac{1}{2}s)_+^2$ and $\psi^{-1}(t) = 2(1 - t^{1/2})$ for $0 \le t \le 1$. The trivariate copula is

$$C(\boldsymbol{u}) = (u_1^{1/2} + u_2^{1/2} + u_3^{1/2} - 2)_+^2$$

In the above, notationally $(x)_+ = \max\{0, x\}$. Take $k = 2$ in part (c) of Theorem 8.1. $F_{12}(u_1, u_2) = C_{12}(u_1, u_2) = (u_1^{1/2} + u_2^{1/2} - 1)_+^2$ is absolutely continuous with density $c_{12}(u_1, u_2) = I(u_1^{1/2} + u_2^{1/2} - 1 > 0) \cdot \frac{1}{2} u_1^{-1/2} u_2^{-1/2}$. Next the conditional distribution is:

$$C_{3|12}(u_3|u_1, u_2) = \begin{cases} \dfrac{I(u_1^{1/2} + u_2^{1/2} + u_3^{1/2} - 2 > 0) \cdot \frac{1}{2} u_1^{-1/2} u_2^{-1/2}}{\frac{1}{2} u_1^{-1/2} u_2^{-1/2}}, & u_1^{1/2} + u_2^{1/2} > 1, \\ \text{doesn't exist,} & \text{otherwise.} \end{cases}$$

Hence this conditional distribution is 1 if $u_3^{1/2} > 2 - u_1^{1/2} - u_2^{1/2}$ and 0 if $u_3^{1/2} < 2 - u_1^{1/2} - u_2^{1/2}$, and the total mass of the jump discontinuities is $m(u_1, u_2) = 1$ when $u_1^{1/2} + u_2^{1/2} > 1$ ($m(u_1, u_2)$ can be defined as 0 for $u_1^{1/2} + u_2^{1/2} \le 1$). Theorem 8.1 implies that the mass of the singular component is $\int_0^1 m(u_1, u_2)\, c_{12}(u_1, u_2)\, du_1 du_2 = 1$. So this copula has a singularity with mass 1 on the set $\{\boldsymbol{u} : u_1^{1/2} + u_2^{1/2} + u_3^{1/2} - 2 = 0\}$.

8.2 Continuity properties of copulas

In this section, we state some results on bounds for multivariate distribution functions and uniform continuous of copula cdfs.

The lemma below is based on Lemma 2.10.4 of Nelsen (1999), but given in the form of multivariate distributions.

Lemma 8.2 (Bound of difference of multivariate cdf at two points). *Let $d \ge 2$ and suppose $F \in \mathcal{F}(F_1, \ldots, F_d)$ where $F_1, \ldots, F_d$ are univariate cdfs (that could be continuous or discrete or mixed). Then*

$$\left| F(x_1, \ldots, x_d) - F(y_1, \ldots, y_d) \right| \le \sum_{j=1}^{d} \left| F_j(x_j) - F_j(y_j) \right|.$$

Proof: Let $(X_1, \ldots, X_d) \sim F$, then

$$
\begin{aligned}
\left|F(x_1, \ldots, x_d) - F(y_1, \ldots, y_d)\right| &= \left|\mathbb{P}(X_j \leq x_j, j = 1, \ldots, d) - \mathbb{P}(X_j \leq y_j, j = 1, \ldots, d)\right| \\
&\leq \mathbb{P}(X_j \leq x_j \vee y_j, j = 1, \ldots, d) - \mathbb{P}(X_j \leq x_j \wedge y_j, j = 1, \ldots, d) \\
&= \mathbb{P}(x_j \wedge y_j < X_j \leq x_j \vee y_j, \text{ for at least one } j) \\
&\leq \sum_{j=1}^{d} \{F_j(x_j \vee y_j) - F_j(x_j \wedge y_j)\} = \sum_{j=1}^{d} |F_j(x_j) - F_j(y_j)|
\end{aligned}
$$

$\square$

The above lemma can be used to prove uniform continuity of copula cdfs (Theorem 2.10.7 of Nelsen (1999) is stated for subcopulas).

Theorem 8.3 (Uniform continuity of copulas, Nelsen (1999)). *Let C be a d-dimensional copula cdf and let $\boldsymbol{u}, \boldsymbol{v} \in [0,1]^d$. Then*

$$
|C(\boldsymbol{u}) - C(\boldsymbol{v})| \leq \sum_{j=1}^{d} |u_j - v_j|.
$$

Hence C is uniformly continuous on $[0,1]^d$.

8.3   Dependence concepts

Several dependence concepts are given in Section 2.8. Some relations among them are summarized here. Other forms of the definitions of the dependence concepts are in the Appendix of Salvadori et al. (2007). Additional dependence concepts and results are presented in Tong (1980).

The abbreviations from Section 2.8 are summarized below.

- PQD/POD: positive quadrant/orthant dependent;
- SI: stochastically increasing;
- RTI: right-tail increasing;
- LTD: left-tail decreasing;
- TP$_2$: totally positive of order 2;
- PUOD/PLOD: positive upper/lower orthant dependent;
- CIS: conditional increasing in sequence;
- max-id, min-id: max/min infinitely divisible.

The next theorem is a statement of an invariance property that applies to many dependence concepts.

Theorem 8.4 (Invariance of dependence properties to increasing transformations). *The dependence properties of PQD/POD, SI, RTI, LTD, association, TP$_2$ are invariant with respect to strictly increasing transformations on the components of the random vector. For example, if (X_1, X_2) is PQD then so is $(a_1(X_1), a_2(X_2))$ for strictly increasing functions a_1, a_2.*

The results in the next two theorems, as well as additional related results, are in Barlow and Proshan (1981).

Theorem 8.5 *Relations for bivariate dependence concepts are:*

(a) TP$_2$ density $\Rightarrow$ SI $\Rightarrow$ LTD, RTI;

(b) LTD or RTI $\Rightarrow$ association $\Rightarrow$ PQD;

(c) TP$_2$ density $\Rightarrow$ TP$_2$ cdf and TP$_2$ survival function;

(d) TP$_2$ cdf $\Rightarrow$ LTD, and TP$_2$ survival function $\Rightarrow$ RTI.

Theorem 8.6 *Relations in multivariate dependence concepts are:*

(a) a random subvector of an associated random vector is associated;

(b) association $\Rightarrow$ PUOD and PLOD;

(c) positively dependent through conditional stochastic ordering $\Rightarrow$ PUOD and PLOD;

(d) CIS $\Rightarrow$ association.

Theorem 8.7 (max-id and TP$_2$, association).
1. (Marshall and Olkin (1990)). Let F be a bivariate cdf.

(a) F is max-id if and only if F is TP$_2$.

(b) F is min-id if and only if $\overline{F}$ is TP$_2$.

2. (Resnick (1987)). Let F be a multivariate max-id cdf and let $\boldsymbol{X} \sim F$. Then $\boldsymbol{X}$ is associated. Also F is PUOD and PLOD.

The conclusion of association in part 2 of the above theorem is given as Proposition 5.29 and depends on characterizations of max-id distributions in Resnick (1987). Hence, by Theorem 8.6, max-id distributions satisfy the positive dependence conditions of PUOD and PLOD.

Theorem 8.8 (Condition for max-id, Joe and Hu (1996)). *Let $d \geq 2$. Suppose $F(\boldsymbol{x})$ is an absolutely continuous d-variate distribution, and let $R = \log F$. For a subset S of $\{1, \ldots, d\}$, let R_S denote the partial derivative of R with respect to $\{x_i : i \in S\}$. A necessary and sufficient condition for F to be max-id is that $R_S \geq 0$ for all (non-empty) subsets S of $\{1, \ldots, d\}$.*

Proof: We look at the derivatives of $H = F^q = e^{qR}$ with respect to $x_1, \ldots, x_d$, $i = 1, \ldots, d$, and then permute indices. All of the derivatives must be non-negative for all $q > 0$ if F is max-id. The derivatives are:

- $\partial H / \partial x_1 = qHR_1$,
- $\partial^2 H / \partial x_1 \partial x_2 = q^2 H R_1 R_2 + qHR_{12}$,
- $\partial^3 H / \partial x_1 \partial x_2 \partial x_3 = q^3 R R_1 R_2 R_3 + q^2 H[R_1 R_{23} + R_2 R_{13} + R_3 R_{12}] + qHR_{123}$, etc.

Let $|S|$ be the cardinality of the set S. For the non-negativity of $\partial^{|S|} H / \prod_{i \in S} \partial x_i$ for $q > 0$ arbitrarily small, a necessary condition is that $R_S \geq 0$. From the form of the derivatives above, it is clear that $R_S \geq 0$ for all S is a sufficient condition. □

The next theorem is useful to establish dependence properties of distributions obtained as limits, such as an extreme value limit. For example, it implies that the bivariate Galambos copula family has TP$_2$ survival functions and satisfies a concordance ordering.

Theorem 8.9 (Preservation of dependence property under limit in distribution). *Let $\{F_n\}$ be a sequence of d-variate cdfs that converge in distribution F, and let $\{G_n\}$ be a sequence of d-variate cdfs that converge in distribution G.*

(a) If each of $\{F_n\}$ satisfies the property of (i) PLOD; (ii) PUOD; (iii) max-id; (iv) min-id; then so does the limit F.

(b) If the F_n are bivariate and each of $\{F_n\}$ satisfies the property of (i) LTD; (ii) RTI; (iii) TP$_2$ cdf; (iv) TP$_2$ survival function; (v) association; then so does the limit F.

(c) If the F_n are bivariate with $F_n \to_d F$, and also (i) $\{F_{n,2|1}(\cdot|x)\}$ (conditional distribution of second given first variable for F_n) is SI for all n; (ii) $F_{n,2|1}(\cdot|x) \to_d F_{2|1}(\cdot|x)$ for all x; (iii) the support of F_n does not change; then $\{F_{2|1}(\cdot|x)\}$ is SI.

(d) Suppose the F_n are bivariate with $F_n \to_d F$, F_n has density $\{f_n\}$ and F has density f. If also $f_n \to f$ pointwise, then F_n with TP_2 density for all n implies that F has TP_2 density.

(e) If $F_n \prec_c G_n$ for each n, and $F_n \to_d F$, $G_n \to_d G$, then $F \prec_c G$.

(f) If (i) $F_n \prec_{SI} G_n$ for each n; (ii) $F_n \to_d F$, $G_n \to_d G$; (iii) $F_{n,2|1}(\cdot|x)$, $G_{n,2|1}(\cdot|x)$ are all continuous and strictly increasing with x fixed; and (iv) $F_{n,2|1}(\cdot|x) \to F_{2|1}(\cdot|x)$ and $G_{n,2|1}^{-1}(\cdot|x) \to G_{2|1}^{-1}(\cdot|x)$ as $n \to \infty$ for all x; then $F \prec_{SI} G$.

The next two results are stated as theorems because of the operations on copula mentioned in Section 3.18.

Theorem 8.10 (Effect of reflected copula for dependence properties).
(a) Let C be a bivariate copula and let $\widehat{C}(u,v) = u + v - 1 + C(1-u, 1-v)$. If C is PQD, RTI, LTD, SI or has TP_2 cdf, then respectively $\widehat{C}$ is PQD, LTD, RTI, SI or has TP_2 survival function.
(b) Let C be a multivariate copula and let $\widehat{C}$ be the reflected copula. If C is PLOD, PUOD, or CIS, then respectively $\widehat{C}$ is PUOD, PLOD, or CIS.

Proof: (a) The only non-immediate result is for SI. Note that

$$1 - \widehat{C}_{2|1}(v|u) = \mathbb{P}(1 - V \geq v | 1 - U = u) = \mathbb{P}(V \leq 1 - v | U = 1 - u)$$

is decreasing as $1 - u$ increases or increasing as u increases.
(b) The proof for CIS is similar.   □

Theorem 8.11 (Effect of power parameter in bivariate copula for dependence properties).
Let C be a bivariate copula and suppose $C_\gamma(u,v) = C^{1/\gamma}(u^\gamma, v^\gamma)$ is a copula with parameter $\gamma > 0$. If C is PQD or LTD, then C_γ retains the dependence condition. If $C_{2|1}$ is SI, then the corresponding conditional cdf of C_γ is SI for $0 < \gamma < 1$.

Proof: If C is PQD, then $C_\gamma(u,v) \geq (u^\gamma v^\gamma)^{1/\gamma} = uv$.
   If C is LTD, $C_\gamma(u,v)/u = [C(u^\gamma, v^\gamma)/u^\gamma]^{1/\gamma}$ is decreasing in u.
   If $\{C_{2|1}(\cdot|u)\}$ is SI (and hence LTD), then

$$(D_1 C_\gamma)(u,v) = C^{1/\gamma-1}(u^\gamma, v^\gamma)\, C_{2|1}(v^\gamma|u^\gamma)\, u^{\gamma-1} = \left[\frac{C(u^\gamma, v^\gamma)}{u^\gamma}\right]^{1/\gamma-1} \cdot C_{2|1}(v^\gamma|u^\gamma).$$

If $0 < \gamma < 1$, this is the product of two positive terms that are decreasing in $u \in (0,1)$.   □

In Sections 2.19 and 2.20, boundary cdfs and conditional tail expectations are introduced as alternative ways to assess dependence in the joint tails.

Theorem 8.12 (Tail dependence with SI, Hua and Joe (2014)). *Suppose C is a bivariate copula that is SI for both conditionings. If C has lower tail dependence with $\lambda_L > 0$, then the boundary conditional cdfs $C_{1|2}(\cdot|0)$ and $C_{2|1}(\cdot|0)$ have strictly positive mass p_0 at 0, and $p_0 \geq \lambda_L$; if C has upper tail dependence with $\lambda_U > 0$, then the boundary conditional cdfs $C_{1|2}(\cdot|1)$ and $C_{2|1}(\cdot|1)$ have strictly positive mass p_1 at 1, and $p_1 \geq \lambda_U$.*

Proof: Suppose $(U, V) \sim C$. From the SI assumption, for any $0 \leq w \leq 1$, $\mathbb{P}(V \leq w|U = u)$ is decreasing in u. Then

$$\mathbb{P}(V \leq w|U \leq w) = w^{-1} \int_0^w \mathbb{P}(V \leq w|U = u)\, du \leq \mathbb{P}(V \leq w|U = 0). \qquad (8.1)$$

If C has lower tail dependence, then letting $w \to 0^+$ on both sides of (8.1) leads to:

$$0 < \lambda_L \leq \mathbb{P}(V \leq 0|U = 0).$$

So the conditional cdf $C_{2|1}(\cdot|0)$ has strictly positive mass at 0. Similarly, the results for $C_{1|2}(\cdot|0)$, $C_{2|1}(\cdot|1)$ and $C_{1|2}(\cdot|1)$ hold. $\qquad \square$

Theorem 8.13 (Conditional tail expectation with SI, Hua and Joe (2014)). *Let X_1, X_2 be identically distributed with cdf F, density function f, support on $[0, \infty)$ and finite mean. Suppose X_1, X_2 satisfy the SI condition of X_1 given X_2. If $\limsup_{t \to \infty} \int_t^\infty \overline{F}(x)\, dx/[t\overline{F}(t)] < \infty$, then $\mathbb{E}[X_1|X_2 = t]$ increases at a fastest rate of $O(t)$ as $t \to \infty$. With the limsup condition but not the SI condition, the same conclusion holds for $\mathbb{E}[X_1|X_2 > t]$.*

Proof: Let C be the copula of X_1, X_2, and $\widehat{C}$ be the survival copula. Write

$$\mathbb{E}[X_1|X_2 = t] = \int_0^\infty \widehat{C}_{1|2}(\overline{F}(x)|\overline{F}(t))\, dx = \frac{1}{\overline{F}(t)} \int_0^\infty \int_0^{\overline{F}(t)} \widehat{C}_{1|2}(\overline{F}(x)|\overline{F}(t))\, dz\, dx$$

$$\leq \frac{1}{\overline{F}(t)} \int_0^\infty \int_0^{\overline{F}(t)} \widehat{C}_{1|2}(\overline{F}(x)|z)\, dz\, dx$$

$$= \frac{1}{\overline{F}(t)} \int_0^\infty \widehat{C}(\overline{F}(x), \overline{F}(t))\, dx = \mathbb{E}[X_1|X_2 > t] \qquad (8.2)$$

$$\leq \frac{1}{\overline{F}(t)} \int_0^\infty \min\{\overline{F}(x), \overline{F}(t)\}\, dx = \frac{1}{\overline{F}(t)} \left[\int_0^t \overline{F}(t)\, dx + \int_t^\infty \overline{F}(x)\, dx \right]$$

$$= t + \int_t^\infty \frac{\overline{F}(x)}{\overline{F}(t)}\, dx,$$

where the first inequality follows from SI and the second inequality follows from the comonotonic copula as an upper bound. If $\limsup \int_t^\infty \overline{F}(x)\, dx/[t\overline{F}(t)] < \infty$, then $\mathbb{E}[X_1|X_2 = t]$ is bounded by $O(t)$.

For $\mathbb{E}[X_1|X_2 > t]$, the same conclusion holds without the SI condition, because the integral for $\mathbb{E}[X_1|X_2 > t]$ appears in (8.2) above. $\qquad \square$

8.4 Fréchet classes and compatibility

In this section, some results on Fréchet classes and compatibility of marginal distributions from Chapter 3 of Joe (1997) are summarized, some with generalizations. An example at the end of the section illustrates how to use the results concerning compatibility of bivariate margins.

The basic results on bounds on a multivariate probability given marginal probabilities follow from the inequalities for sets in the next lemma.

Lemma 8.14 *Let $A_1, \ldots, A_d$ be events such that $\mathbb{P}(A_i) = a_i$, $i = 1, \ldots, d$. Then*

$$\max\left\{0, \sum_j a_j - (d - 1)\right\} \leq \mathbb{P}(A_1 \cap \cdots \cap A_d) \leq \min_j a_j$$

and the bounds are sharp.

Let A_i^0 *be the complement of* A_i *and let* $A_i^1 = A_i$, $i = 1, \ldots, d$. *Then it is possible to assign probabilities to*

$$A_1^{\epsilon_1} \cap \cdots \cap A_d^{\epsilon_d}, \quad \epsilon_i = 0 \text{ or } 1, \quad i = 1, \ldots, d,$$

(in a continuous way over a_i*) such that* $\mathbb{P}(A_1 \cap \cdots \cap A_d) = \max\{0, \sum_j a_j - (d-1)\}$ *and* $a_i = \sum_{\epsilon:\epsilon_i=1} \mathbb{P}(A_1^{\epsilon_1} \cap \cdots \cap A_d^{\epsilon_d})$.

In the next theorem, parts (b) and (c) are due to Fréchet (1951) and Dall'Aglio (1972), covering both discrete and continuous margins. Hoeffding (1940) covers just the continuous case. The proof of part (d) in Joe (1997) makes use of the second part of the above lemma, due to Fréchet (1935).

Theorem 8.15 (Pointwise bounds for multivariate cdf). *Consider* $\mathcal{F}(F_1, \ldots, F_d)$, *where each* F_j *could be a discrete or continuous univariate cdf.*

(a) *Let* $F \in \mathcal{F}(F_1, \ldots, F_d)$. *For all* $\boldsymbol{x} \in \mathbb{R}^d$,

$$\max\{0, F_1(x_1) + \cdots + F_d(x_d) - (d-1)\} \leq F(\boldsymbol{x}) \leq \min_{1 \leq j \leq d} F_j(x_j).$$

(b) $F^+(\boldsymbol{x}) := \min_{1 \leq j \leq d} F_j(x_j)$ *is a cdf, called the Fréchet upper bound cdf.*
(c) *For* $d = 2$, $F^-(x_1, x_2) := \max\{0, F_1(x_1) + F_2(x_2) - 1\}$ *is a cdf, called the bivariate Fréchet lower bound cdf.*
(d) *If the cdfs* F_j *are continuous, then* $F^-(\boldsymbol{x}) := \max\{0, F_1(x_1) + \cdots + F_d(x_d) - (d-1)\}$ *is sharp for* $\mathcal{F}(F_1, \ldots, F_d)$, *that is, for any* $\boldsymbol{x} \in \mathbb{R}^d$, *there is at least one* $F \in \mathcal{F}(F_1, \ldots, F_d)$ *for which* $F(\boldsymbol{x}) = F^-(\boldsymbol{x})$.

The results in the next theorem are harder to obtain than those in the preceding theorem.

Theorem 8.16 (Condition for lower bound to be distribution; Dall'Aglio (1972).) *Let* $F^-(\boldsymbol{x}) := \max\{0, F_1(x_1) + \cdots + F_d(x_d) - (d-1)\}$ *be defined pointwise for* $\boldsymbol{x} \in \mathbb{R}^d$. *A necessary and sufficient condition for* F^- *to be a cdf in* $\mathcal{F}(F_1, \ldots, F_d)$ *to be a cdf is that either*

(a) $\sum_j F_j(x_j) \leq 1$ *whenever* $0 < F_j(x_j) < 1$, $j = 1, \ldots, d$; *or*
(b) $\sum_j F_j(x_j) \geq d - 1$ *whenever* $0 < F_j(x_j) < 1$, $j = 1, \ldots, d$.

Bounds can also be obtained for the class $\mathcal{F}(F_{iS}, F_{jS})$ where $i \notin S$ and $j \notin S$, based on for mixtures of conditional distributions.

Theorem 8.17 (Bounds for trivariate cdf based on bivariate margins). *The upper bound of* $\mathcal{F}(F_{12}, F_{13})$ *is given by* $F_{123;1}^+(x_1, x_2, x_3) := \int_{-\infty}^{x_1} \min\{F_{2|1}(x_2|z), F_{3|1}(x_3|z)\} \, dF_1(z)$ *and the lower bound is given by* $F_{123;1}^-(x_1, x_2, x_3) := \int_{-\infty}^{x_1} \max\{0, F_{2|1}(x_2|z) + F_{3|1}(x_3|z) - 1\} \, dF_1(z)$, *and both of these bounds are proper cdfs.*

For $F \in \mathcal{F}(F_{12}, F_{13}, F_{23})$, *and* $\boldsymbol{x} \in \mathbb{R}^3$,

$$\max\{0, b_1(\boldsymbol{x}), b_2(\boldsymbol{x}), b_3(\boldsymbol{x})\} \leq F(\boldsymbol{x}) \leq \min\{a_{12}(\boldsymbol{x}), a_{13}(\boldsymbol{x}), a_{23}(\boldsymbol{x}), a_{123}(\boldsymbol{x})\},$$

where

$a_{12}(\boldsymbol{x}) = F_{12}(x_1, x_2)$, $a_{13}(\boldsymbol{x}) = F_{13}(x_1, x_3)$, $a_{23}(\boldsymbol{x}) = F_{23}(x_2, x_3)$,
$a_{123}(\boldsymbol{x}) = 1 - F_1(x_1) - F_2(x_2) - F_3(x_3) + F_{12}(x_1, x_2) + F_{13}(x_1, x_3) + F_{23}(x_2, x_3)$,
$b_1(\boldsymbol{x}) = F_{12}(x_1, x_2) + F_{13}(x_1, x_3) - F_1(x_1)$, $b_2(\boldsymbol{x}) = F_{12}(x_1, x_2) + F_{23}(x_2, x_3) - F_2(x_2)$,
$b_3(\boldsymbol{x}) = F_{13}(x_1, x_3) + F_{23}(x_2, x_3) - F_3(x_3)$.

For F_{12}, F_{13}, F_{23} *to be compatible bivariate margins, the condition:*

$$\max\{0, b_1(\boldsymbol{x}), b_2(\boldsymbol{x}), b_3(\boldsymbol{x})\} \le \min\{a_{12}(\boldsymbol{x}), a_{13}(\boldsymbol{x}), a_{23}(\boldsymbol{x}), a_{123}(\boldsymbol{x})\} \quad \forall \boldsymbol{x},$$

must hold as well as

$$\int_{-\infty}^{\infty} \max\{0, F_{i|j}(x_i|z) + F_{k|j}(x_k|z) - 1\}\, \mathrm{d}F_j(z) \le F_{ik}(x_i, x_k)$$

$$\le \int_{-\infty}^{\infty} \min\{F_{i|j}(x_i|z), F_{k|j}(x_k|z)\}\, \mathrm{d}F_j(z) \quad \forall x_i, x_k,$$

with (i, j, k) *being any permutation of* $(1, 2, 3)$.

In the next result, which is a variation of the preceding, if F is the cdf of a random variable or vector $\boldsymbol{X}$, then $\tilde{F}$ is the measure associated with F and $\tilde{F}(A) = \mathbb{P}(\boldsymbol{X} \in A)$ for a measurable set A.

Theorem 8.18 (Bounds for trivariate probability based on bivariate margins). *Let* $F \in \mathcal{F}(F_{12}, F_{13}, F_{23})$ *and* $(X_1, X_2, X_3) \sim F$. *Suppose* $A_j \subset \mathbb{R}$ *for* $j = 1, 2, 3$ *are measurable sets. Then* $\max\{0, b_1, b_2, b_3\} \le \tilde{F}(A_1 \cap A_2 \cap A_3) \le \min\{a_{12}, a_{13}, a_{23}, a_{123}\}$, *where*

$a_{12} = \tilde{F}_{12}(A_1 \cap A_2)$, $a_{13} = \tilde{F}_{13}(A_1 \cap A_3)$, $a_{23} = \tilde{F}_{23}(A_2 \cap A_3)$,
$a_{123} = 1 - \tilde{F}_1(A_1) - \tilde{F}_2(A_2) - \tilde{F}_3(A_3) + \tilde{F}_{12}(A_1 \cap A_2) + \tilde{F}_{13}(A_1 \cap A_3) + \tilde{F}_{23}(A_2 \cap A_3)$,
$b_1 = \tilde{F}_{12}(A_1 \cap A_2) + \tilde{F}_{13}(A_1 \cap A_3) - \tilde{F}_1(A_1)$, $b_2 = \tilde{F}_{12}(A_1 \cap A_2) + \tilde{F}_{23}(A_2 \cap A_3) - \tilde{F}_2(A_2)$,
$b_3 = \tilde{F}_{13}(A_1 \cap A_3) + \tilde{F}_{23}(A_2 \cap A_3) - \tilde{F}_3(A_3)$.

For F_{12}, F_{13}, F_{23} *to be compatible bivariate margins,*

$$\max\{0, b_1, b_2, b_3\} \le \min\{a_{12}, a_{13}, a_{23}, a_{123}\} \quad \forall A_1, A_2, A_3$$

must hold as well as

$$\int_{-\infty}^{\infty} \max\{0, \tilde{F}_{i|j}(A_i|z) + \tilde{F}_{k|j}(A_k|z) - 1\}\, \mathrm{d}F_j(z) \le \tilde{F}_{ik}(A_i \cap A_k)$$

$$\le \int_{-\infty}^{\infty} \min\{\tilde{F}_{i|j}(A_i|z), \tilde{F}_{k|j}(A_k|z)\}\, \mathrm{d}F_j(z) \quad \forall A_i, A_k,$$

with (i, j, k) *being any permutation of* $(1, 2, 3)$.

Next are results with inequality constraints from dependence measures; some of these results are from Joe (1996a, 1997). Consider the range of $(\zeta_{jk})_{1 \le j,k \le d}$, where ζ_{jk} is a measure of dependence for the (j, k) bivariate margin such as Kendall's τ, Blomqvist's β, Spearman's ρ_S or the correlation of normal scores. Note that $\zeta_{jj} = 1$ for $1 \le j \le d$.

The next result can be used to show that some bivariate marginal copulas are not compatible with multivariate distributions.

Theorem 8.19 (Constraints on sets of dependence measures).
Let $C \in \mathcal{F}(C_{jk}, 1 \le j < k \le d)$, *where* C_{jk} *are bivariate copulas. For* C_{jk} *with* $j \ne k$, *let* $\tau_{jk}, \beta_{jk}, \rho_{S,jk}, \rho_{N,jk}$ *be the value of Kendall's tau, Blomqvist's beta, Spearman's rho and the correlation of normal scores respectively.*

(a) *For distinct* $h, j, k \in \{1, \ldots, d\}$,

$$-1 + |\tau_{hj} + \tau_{jk}| \le \tau_{hk} \le 1 - |\tau_{hj} - \tau_{jk}|. \tag{8.3}$$

For $d = 3$, *the inequalities are sharp.*

(b) For distinct $h, j, k \in \{1, \ldots, d\}$,

$$-1 + |\beta_{hj} + \beta_{jk}| \leq \beta_{hk} \leq 1 - |\beta_{hj} - \beta_{jk}|. \qquad (8.4)$$

For $d = 3$, the inequalities are sharp.

(c) The matrix $(\rho_{N,jk})_{1 \leq j,k \leq d}$ is non-negative definite.

(d) Consider $\boldsymbol{R}_S = (\rho_{S,jk})_{1 \leq j,k \leq d}$. If $d = 2, 3, 4$, the only constraint on $\boldsymbol{R}_S$ is non-negative definiteness. For $d \geq 5$, the set of possible $\boldsymbol{R}_S$ is a strict subset of non-negative definite correlation matrices.

Proof: In Joe (1997), the proof (a) is given as Theorem 3.12, and the inequality in (b) is given as Exercise 3.9. We prove (b) here using the technique of the proof for (a).

Let $(U_1, \ldots, U_d) \sim C$. Take $(h, j, k) = (1, 2, 3)$ without loss of generality. The other inequalities follow by permuting indices. Let $\eta_{jk} = \mathbb{P}((U_j - \frac{1}{2})(U_k - \frac{1}{2}) > 0)$ so that $\beta_{jk} = 2\eta_{jk} - 1$ and $\eta_{jk} = (1 + \beta_{jk})/2$. Then

$$\begin{aligned}
\eta_{13} &= \mathbb{P}((U_{11} - \tfrac{1}{2})(U_{12} - \tfrac{1}{2})^2(U_{13} - \tfrac{1}{2}) > 0) \\
&= \mathbb{P}((U_{11} - \tfrac{1}{2})(U_{12} - \tfrac{1}{2}) > 0, (U_{12} - \tfrac{1}{2})(U_{13} - \tfrac{1}{2}) > 0) \\
&\quad + \mathbb{P}((U_{11} - \tfrac{1}{2})(U_{12} - \tfrac{1}{2}) < 0, (U_{12} - \tfrac{1}{2})(U_{13} - \tfrac{1}{2}) < 0).
\end{aligned}$$

Hence, from Lemma 8.14, an upper bound for η_{13} is $\min\{\eta_{12}, \eta_{23}\} + \min\{1 - \eta_{12}, 1 - \eta_{23}\}$ and a lower bound is $\max\{0, \eta_{12} + \eta_{23} - 1\} + \max\{0, (1 - \eta_{12}) + (1 - \eta_{23}) - 1\}$. After substituting for β_{jk} and simplifying, inequality in (8.4) results. The sharpness follows from the trivariate Gaussian (or t_ν) case as given next.

For the multivariate Gaussian distribution with correlation matrix (ρ_{jk}), from Theorem 8.53, $\beta_{jk} = (2/\pi) \arcsin(\rho_{jk})$. For the trivariate Gaussian distributions, the constraint $-1 \leq \rho_{13;2} \leq 1$ (for the partial correlation) is the same as

$$\rho_{12}\rho_{23} - [(1 - \rho_{12}^2)(1 - \rho_{23}^2)]^{1/2} \leq \rho_{13} \leq \rho_{12}\rho_{23} + [(1 - \rho_{12}^2)(1 - \rho_{23}^2)]^{1/2}$$

or

$$-\cos[\tfrac{1}{2}\pi(\beta_{12} + \beta_{23})] \leq \sin(\tfrac{1}{2}\pi\beta_{13}) \leq \cos[\tfrac{1}{2}\pi(\beta_{12} - \beta_{23})]$$

or equivalently (8.4) with $(i, j, k) = (1, 2, 3)$. Equality holds when the $\rho_{13;2}$ is ± 1.

The result in (c) follows because ρ_N is the same as the correlation parameter for multivariate Gaussian distributions.

The result in (d) is non-trivial; it is a special case of results in Joe (2006).[1] $\square$

It is a research problem to find improved inequalities for the 4-dimensional and higher dimensional cases.

The next result generalizes the inequality in Theorem 3.14 of Joe (1997) for the tail dependence parameter. For dependent U$(0,1)$ random variables U_1, U_2 and $A \subset [0, 1]$, let $\lambda^A = \mathbb{P}(U_1 \in A | U_2 \in A) = \mathbb{P}(U_2 \in A | U_1 \in A)$. With $A = [0, u]$ or $[1 - u, 1]$ such that $u \to 0^+$, the lower and upper tail dependence parameters are obtained.

Theorem 8.20 (Constraints for three bivariate copula margins). *Let $C \in \mathcal{F}(C_{12}, C_{23}, C_{13})$,*

---

[1] Exercise 4.17 of Joe (1997) stated a result that all positive correlation matrices are possible for $\mathcal{F}(F, \ldots, F)$ in dimension d iff the univariate margin F is a location-scale transform of a univariate margin of a d-dimensional spherical distribution. In fact, only the "if" part is simple to prove and the converse should say "$(d - 1)$-dimensional spherical distribution." Thanks to Dorota Kurowicka and Soumyadip Ghosh for pointing out in the year 2003 that the converse should not have been in the statement of the exercise. Eventually, I was able to prove the converse in 8 pages,

where C_{jk} are bivariate copulas. Let $\lambda_{jk}^A = \mathbb{P}(U_j \in A | U_k \in A)$, where $A \subset [0,1]$ satisfies $\pi^A = \mathbb{P}(U_j \in A) > 0$. Then

$$\max\{0, \lambda_{ij}^A + \lambda_{jk}^A - 1\} \le \lambda_{ik}^A \le 1 - |\lambda_{ij}^A - \lambda_{jk}^A| \qquad (8.5)$$

holds for all permutations (i,j,k) of $(1,2,3)$. If $\eta_{jk}^A = \mathbb{P}(U_j \in A, U_k \in A)$, then this inequality can be written as

$$\max\{0, \eta_{ij}^A + \eta_{jk}^A - \pi^A\} \le \eta_{ik}^A \le \pi^A - |\eta_{ij}^A - \eta_{jk}^A|.$$

Proof: The proof of (8.5) will be given in the case of $(i,j,k) = (1,2,3)$. The other inequalities follow by permuting indices. Let $(U_1, U_2, U_3) \sim C$ and let A^c be the complement of A. Then

$$
\begin{aligned}
\mathbb{P}(U_3 &\in A \mid U_1 \in A) \\
&= \mathbb{P}(U_3 \in A, U_2 \in A \mid U_1 \in A) + \mathbb{P}(U_3 \in A, U_2 \in A^c \mid U_1 \in A) \\
&= \mathbb{P}(U_3 \in A, U_1 \in A \mid U_2 \in A) + \mathbb{P}(U_3 \in A, U_2 \in A^c \mid U_1 \in A) \\
&\le \min\{\mathbb{P}(U_1 \in A \mid U_2 \in A), \mathbb{P}(U_3 \in A \mid U_2 \in A)\} + 1 - \mathbb{P}(U_2 \in A \mid U_1 \in A) \\
&= \min\{\lambda_{12}^A, \lambda_{23}^A\} + 1 - \lambda_{12}^A.
\end{aligned}
$$

Similarly, by interchanging the subscripts 1 and 3, $\lambda_{13}^A \le \min\{\lambda_{12}^A, \lambda_{23}^A\} + 1 - \lambda_{23}^A$. From combining these two upper bounds, $\lambda_{13}^A \le 1 - |\lambda_{12}^A - \lambda_{23}^A|$. For the lower bound,

$$
\begin{aligned}
\lambda_{13}^A &= \mathbb{P}(U_3 \in A \mid U_1 \in A) = \mathbb{P}(U_3 \in A, U_1 \in A)/\mathbb{P}(U_2 \in A) \\
&\ge \mathbb{P}(U_3 \in A, U_1 \in A \mid U_2 \in A) \\
&\ge \max\{0, \mathbb{P}(U_3 \in A \mid U_2 \in A) + \mathbb{P}(U_1 \in A \mid U_2 \in A) - 1\} \\
&= \max\{0, \lambda_{ij}^A + \lambda_{jk}^A - 1\}.
\end{aligned}
$$

$\square$

For the special case of tail dependence parameters, it is shown in Nikoloulopoulos et al. (2009) that (8.5) is sharp.

Some examples are given below to illustrate the theorems of this section and to check which conditions in the preceding theorems are most stringent.

Example 8.2 (Compatibility based on trivariate equalities of dependence measures). We do the following analysis for a 1-parameter copula family $C(\cdot; \delta)$ to check if it is impossible for a trivariate copula with bivariate margins $C(\cdot; \delta_{12}), C(\cdot; \delta_{23}), C(\cdot; \delta_{13})$. If $\delta_{12} = \delta_{23}$, then Theorems 8.19 and 8.20 imply for the Blomqvist betas, Kendall taus, Spearman's rhos, correlations of normal scores and upper/lower tail dependence parameters that

(a) $-1 + 2|\beta_{12}| \le \beta_{13} \le 1$;
(b) $-1 + 2|\tau_{12}| \le \tau_{13} \le 1$;
(c) $-1 + 2\rho_{S,12}^2 \le \rho_{S,13} \le 1$;
(d) $-1 + 2\rho_{N,12}^2 \le \rho_{N,13} \le 1$;
(e) $\max\{0, -1 + 2\lambda_{12}\} \le \lambda_{13} \le 1$.

If the family only has positive dependence, then the above lower bound of a dependence measure is replaced by 0 if it is less than 0. Given $\delta_{12} = \delta_{23} = \delta$, the value ζ of each of the five dependence measures can be obtained for $C(\cdot; \delta)$. Then a lower bound ζ_{13} can be obtained and the value δ_{13} such that the dependence measure of $C(\cdot; \delta_{13})$ is ζ_{13} can be solved for. The lower bound for δ_{13} can be increased using the conditions in Theorem 8.17.

Some representative results of this form are shown in Table 8.1 for the Plackett, Frank, MTCJ and Gumbel families.

copula	$\beta_{12} = \beta_{23}$	δ_{12}	Lower bound on δ_{13} from measure:					Lower bound from Theorem 8.17
			β	τ	ρ_S	ρ_N	λ	
Plackett	0.7	32.11	5.44	4.38	3.43	2.95	-	5.44
Frank	0.7	9.10	3.60	2.69	2.65	1.77	-	3.60
MTCJ	0.7	4.08	1.31	1.04	0.90	0.65	1.85	1.88
Gumbel	0.7	3.29	1.67	1.65	1.59	1.59	1.80	1.85

Table 8.1 *Compatibilty of bivariate margins based on inequalities on dependence measures when the (1,2) and (2,3) margins have the same copula. The pattern is similar for other β_{12} values in column 2. For the MTCJ and Gumbel copulas, the inequality based on the tail dependence parameter λ is most stringent and then the inequality based on β. For the Plackett and Frank copulas that do not have tail dependence, the inequality based on β is most stringent. Even more stringent are the inequalities in Theorem 8.17; these can be checked over a grid of values in $(0,1)^3$. An example of interpretation is as follows. There is no trivariate copula that has bivariate margins of Gumbel(3.29), Gumbel(3.29) and Gumbel(1.79); the inequalities do not establish that there is a trivariate copula with bivariate margins of Gumbel(3.29), Gumbel(3.29) and Gumbel(1.86). From the hierarchical Archimedean construction, there exists a trivariate copula with bivariate margins of Gumbel(3.29), Gumbel(3.29) and Gumbel(δ) with $\delta \geq 3.29$.*

Other results on compatibility of multivariate margins are given in Kellerer (1964) and Rüschendorf (1985).

8.5 Archimedean copulas

In this section, there are results on dependence and concordance properties of Archimedean copulas, as well as results on tail properties. Some results are restatements of theorems in Joe (1997); corresponding bivariate results in Nelsen (1999) are stated in terms of the inverse Laplace or Williamson transform. For statements of the results, the family $\mathcal{L}_d$ is defined in (3.10), and

$$\mathcal{L}_n^* = \{\omega : [0,\infty) \to [0,\infty) \mid \omega(0) = 0,\ \omega(\infty) = \infty,\ (-1)^{j-1}\omega^{(j)} \geq 0,\ j = 1, 2, \ldots, n\};$$
(8.6)

the family $\mathcal{L}_\infty^*$, defined in (2.10), is the same as the above with $n \to \infty$.

The first few results are used to establish positive or negative dependence of Archimedean copulas, and to show that a parametric family of Archimedean copulas is increasing in concordance.

Theorem 8.21 (Invariance of copula to scale changes of ψ). *Let $\psi \in \mathcal{L}_d$ and let $\psi_a(s) = \psi(as)$ for a positive constant a. The d-variate Archimedean copula based on ψ_a is the same as that based on ψ.*

Proof: It is straightforward to check that $\psi_a \in \mathcal{L}_d$. Since $\psi_a^{-1}(t) = a^{-1}\psi^{-1}(t)$. then

$$\psi_a\Big(\sum_{j=1}^{d} \psi_a^{-1}(u_j)\Big) = \psi_a\Big(a^{-1}\sum_{j=1}^{d} \psi^{-1}(u_j)\Big) = \psi\Big(\sum_{j=1}^{d} \psi^{-1}(u_j)\Big).$$

$\square$

Theorem 8.22 (Concordance and PLOD ordering for multivariate Archimedean copulas). *Let $\psi_1, \psi_2 \in \mathcal{L}_d$. Let $\omega = \psi_1^{-1} \circ \psi_2$ and $\omega^{-1} = \psi_2^{-1} \circ \psi_1$.*

(a) *For $d = 2$, let $C_i(u_1, u_2) = \psi_i(\psi_i^{-1}(u_1) + \psi_i^{-1}(u_2))$ for $i = 1, 2$. Then $C_1 \prec_c C_2$ if and only if ω^{-1} is superadditive ($\omega^{-1}(x + y) \geq \omega^{-1}(x) + \omega^{-1}(y)$ for all $x, y \geq 0$).*

(b) For the multivariate extensions, $C_{id}(\boldsymbol{u}) = C_{\psi_i}(\boldsymbol{u}) = \psi_i(\sum_{j=1}^{d} \psi_i^{-1}(u_j))$, $i = 1, 2$, $C_{1d} \prec_{cL} C_{2d}$ if and only if ω^{-1} is superadditive.

Note that since $\omega(0) = 0$, sufficient conditions for ω^{-1} to be superadditive are (i) ω^{-1} convex (ω concave) and (ii) ω^{-1} star-shaped with respect to the origin ($\omega^{-1}(s)/s$ increasing in s, or ω anti-star-shaped with respect to the origin).

Proof: (a) Let $\eta = \omega^{-1}$. Let $u_j = \psi_1(x_j)$, $j = 1, 2$. Because ψ_2 is decreasing, then

$$C_1 \prec_c C_2$$
$$\iff \psi_1(\psi_1^{-1}(u_1) + \psi_1^{-1}(u_2)) \leq \psi_2(\psi_2^{-1}(u_1) + \psi_2^{-1}(u_2)) \; \forall 0 \leq u_1, u_2 \leq 1$$
$$\iff \eta(\psi_1^{-1}(u_1) + \psi_1^{-1}(u_2)) \geq \psi_2^{-1}(u_1) + \psi_2^{-1}(u_2) \; \forall 0 \leq u_1, u_2 \leq 1$$
$$\iff \eta(x_1 + x_2) \geq \eta(x_1) + \eta(x_2) \; \forall x_1, x_2 \geq 0.$$

The sufficient conditions for superadditivity are given in Result 5.4 and Exercise 5 of Chapter 4 of Barlow and Proshan (1981). The convexity assumption is the strongest. We give brief explanations here for convexity $\Rightarrow$ star-shapedness $\Rightarrow$ superadditivity. Because η is increasing and $\eta(0) = 0$, then for $0 < x_1 < x_2$, the convexity of η and $x_1 = [(x_2 - x_1) \cdot 0 + x_1 \cdot x_2]/x_2$ implies $x_2 \eta(x_1) \leq (x_2 - x_1)\eta(0) + x_1\eta(x_2) = x_1\eta(x_2)$ or $\eta(x_1)/x_1 \leq \eta(x_2)/x_2$. If η is star-shaped, then $(x_1 + x_2)^{-1} x_j \eta(x_1 + x_2) \geq \eta(x_j)$ for $j = 1, 2$ with $x_1, x_2 > 0$; add these two inequalities to get $\eta(x_1 + x_2) \geq \eta(x_1) + \eta(x_2)$. $\square$

Corollary 8.23 *Consider the Archimedean copula C_ψ based on $\psi \in \mathcal{L}_d$.*

(a) *For $d = 2$, if $-\log\psi$ is concave, then C_ψ is positive quadrant dependent (PQD), and if $-\log\psi$ is convex, then C_ψ is negative quadrant dependent (NQD).*
(b) *For $d > 2$, if $-\log\psi$ is concave, then C_ψ is positive lower orthant dependent (PLOD), and if if $-\log\psi$ is convex, then C_ψ is negative lower orthant dependent (NLOD).*

Proof: The independence copula $C^\perp$ is an Archimedean copula with $\psi_\perp(s) = e^{-s}$.

(a) PQD is the same $C^\perp \prec_c C_\psi$ and NQD is the same as $C_\psi \prec_c C^\perp$. In (a), Theorem 8.22, take $\psi_2 = \psi$, $\psi_1 = \psi_\perp$, then $\eta(s) = \psi_2^{-1}(\psi_1(s)) = \psi^{-1}(e^{-s})$ is convex iff $-\log\psi$ is concave. Hence this condition is sufficient for C_ψ to be PQD. Next take $\psi_2 = \psi_\perp$, $\psi_1 = \psi$, then $\omega(s) = \psi_2^{-1}(\psi_1(s)) = -\log\psi(s)$ convex implies that $-\log\psi(s)$ is superadditive and C_ψ is NQD.

(b) The proof is similar to that of part (a). Note that PLOD implied PQD for $d = 2$ but not necessarily positive upper orthant dependence for $d > 2$. $\square$

Note that if $\psi \in \mathcal{L}_\infty$ is a LT, then ψ is a survival function on $[0, \infty)$ as a rate mixture of exponential survival functions. Then from Barlow and Proshan (1981), $-\log\psi$ concave (convex) is the same as ψ being a survival function with decreasing (increasing) failure/hazard rate.

Corollary 8.24 *Let $C_i(u_1, u_2) = \psi_i(\psi_i^{-1}(u_1) + \psi_i^{-1}(u_2))$, where $\psi_i \in \mathcal{L}_\infty$ is a LT, $i = 1, 2$. Suppose $\omega = \psi_1^{-1} \circ \psi_2 \in \mathcal{L}_\infty^*$, then $C_1 \prec_c C_2$.*

Proof: ω has non-negative first derivative and non-positive second derivative, and satisfies $\omega(0) = 0$. Therefore ω is concave and $\omega^{-1} = \psi_2^{-1} \circ \psi_1$ is convex, and Theorem 8.22 applies. $\square$

Theorem 8.25 (Concordance ordering of two d-variate Archimedean copulas). *Suppose $\psi_1, \psi_2 \in \mathcal{L}_\infty$ are such that $\psi_1^{-1} \circ \psi_2 \in \mathcal{L}_\infty^*$. Let C_{ψ_1}, C_{ψ_2} be two d-variate Archimedean copulas. Then $C_{\psi_1} \prec_c C_{\psi_2}$.*

Proof: For $d \geq 3$, this is not as simple as the bivariate case. See the proof of Theorem 4.7 in Joe (1997). □

Next are results on Kendall's tau for Archimedean copulas. Note that there is no analogous reduction to one-dimensional integrals for Spearman's rho for Archimedean copulas.

Theorem 8.26 (Kendall's tau for bivariate Archimedean, Genest and MacKay (1986)). *For the bivariate Archimedean copula C_ψ, Kendall's tau can be written as the one-dimensional integral:*

$$\tau = 1 - 4 \int_0^\infty s[\psi'(s)]^2 ds = 4 \int_0^1 \frac{\psi^{-1}(t)}{(\psi^{-1})'(t)} dt + 1. \tag{8.7}$$

Proof: This proof is an improvement of that in Genest and MacKay (1986). Using the formula in (2.42),

$$I := \int_{[0,1]^2} C_{2|1}(v|u) C_{1|2}(u|v) \, dudv = \int_{[0,1]^2} \frac{[\psi'(\psi^{-1}(u) + \psi^{-1}(v))]^2}{\psi'(\psi^{-1}(u)) \, \psi'(\psi^{-1}(v))} dudv$$

$$= \int_0^\infty \int_0^\infty [\psi'(x+y)]^2 dxdy = \int_{s=0}^\infty \int_{r=0}^s [\psi'(s)]^2 drds = \int_0^\infty s[\psi'(s)]^2 ds,$$

so that $\tau = 1 - 4I$ becomes (8.7). □

For numerical computations, the first form in (8.7) from Joe (1997) should be better than the second form from Genest and MacKay (1986).

Theorem 8.27 (Lower bound cdf for d-variate Archimedean, McNeil and Nešlehová (2009)). *Consider the d-variate Archimedean copula $C_\psi(\boldsymbol{u}_{1:d})$ based on $\psi \in \mathcal{L}_d$.*
(a) The lower bound in the PLOD ordering $\prec_{\mathrm{cL}}$ occurs for $\psi(s) = [\max\{0, 1-s\}]^{d-1} = (1-s)_+^{d-1}$ or $\psi(s) = (1 - s/[d-1])_+^{d-1}$.
(b) For bivariate margins of $C_\psi(\boldsymbol{u}_{1:d})$, the lower bound on Kendall's tau is $-1/(2d - 3)$.

Proof: The proof is based on the Williamson transform. See Proposition 4.6 and Corollary 4.1 of McNeil and Nešlehová (2009). Part (b) solves a conjecture in Chapter 5 of Joe (1997). □

Next are results on total positivity and max-infinite divisibility. The property of TP_2 density implies max-infinite divisibility in the bivariate case, and if an Archimedean copula is max-id, it can be used in the construction in Section 3.5.

Theorem 8.28 (Archimedean copula and TP_2, Marshall and Olkin (1988)). *The bivariate Archimedean copula C_ψ based on a LT $\psi \in \mathcal{L}_\infty$ has TP_2 density. Hence it is also SI and PQD.*

Theorem 8.29 (Archimedean copula and stochastically increasing, Müller and Scarsini (2005)). *If the d-variate Archimedean copula C_ψ based on $\psi \in \mathcal{L}_d$ is such that $(-1)^{d-1}\psi^{(d-1)}$ is log-convex, then C_ψ is conditionally increasing in sequence.*

Theorem 8.30 (Archimedean copula and multivariate TP_2, Müller and Scarsini (2005)).
(a) Consider the d-variate Archimedean copula C_ψ based on $\psi \in \mathcal{L}_d$. Then $(-1)^d\psi^{(d)}$ is log-convex iff C_ψ satisfies the multivariate TP_2 condition: $g := \log \frac{\partial^d}{\partial u_1 \cdots \partial u_d} C$ is supermodular, or

$$g(\boldsymbol{u} \vee \boldsymbol{v}) + g(\boldsymbol{u} \wedge \boldsymbol{v}) \geq g(\boldsymbol{u}) + g(\boldsymbol{v}), \quad \forall \boldsymbol{u}, \boldsymbol{v} \in (0,1)^d.$$

[The lattice operators $\vee, \wedge$ are defined as componentwise maxima and componentwise minima respectively.]

(b) If the Archimedean copula C_ψ is based on $\psi \in \mathcal{L}_\infty$, then C_ψ is multivariate TP$_2$ for any $d \geq 2$.

Theorem 8.31 (Condition for d-variate Archimedean to be max-id, Joe and Hu (1996)). *For $\psi \in \mathcal{L}_d$, $C(\boldsymbol{u}) = C_\psi(\boldsymbol{u}) = \psi(\sum_{j=1}^{d} \psi^{-1}(u_j))$ is max-id if $\Psi = -\log\psi \in \mathcal{L}_d^*$ where $\mathcal{L}_d^*$ is defined in (8.6). For $\psi \in \mathcal{L}_\infty$, C_ψ is max-id for all $d \geq 2$ if $\Psi = -\log\psi \in \mathcal{L}_\infty^*$. (The condition $\Psi \in \mathcal{L}_\infty^*$ is the same as ψ being a LT of an infinitely divisible random variable.)*

Proof: Write $C^r(\boldsymbol{u}) = \exp\{-r\Psi(\sum_{j=1}^{d} \xi_j(u_j))\}$, where $\xi_j(u_j) = \psi^{-1}(u_j)$ and $r > 0$. The ith-order mixed derivatives of $H = C^r$ for $i = 2, 3, \ldots$ are:

- $h_{12} = \partial^2 H / \partial u_1 \partial u_2 = C^r \xi_1' \xi_2' [r^2 \Psi'^2 - r\Psi'']$,
- $h_{123} = \partial^3 H / \partial u_1 \partial u_2 \partial u_3 = C^r \xi_1' \xi_2' \xi_3' [-r^3 (\Psi')^3 + 3r^2 \Psi' \Psi'' - r\Psi''']$, etc.

If $\Psi \in \mathcal{L}_2^*$, each of the two summands (distribute the leading product term) in h_{12} is non-negative, so that $h_{12} \geq 0$. If $\Psi \in \mathcal{L}_3^*$, each of the three summands in h_{123} is non-negative, so that $h_{123} \geq 0$. Continue to take derivatives and assume derivatives of Ψ continue to alternative in sign. The partial derivative of $C^r = e^{-r\Psi}$ in a summand with respect to the next u_j leads to the summand multiplied by $-r\xi_j' \Psi'$; since $\xi_j' \leq 0$ and $\Psi' \geq 0$, the sign is unchanged. The partial derivative of a positive power $[\Psi^{(k)}]^n$ in a summand with respect to the next u_j leads to the summand multiplied by $n\xi_j' \Psi^{(k+1)} / \Psi^{(k)}$, and the sign is unchanged if $\Psi^{(k+1)}$ and $\Psi^{(k)}$ have opposite signs. Hence from the pattern of the derivatives, C is max-id for up to dimension d if $\Psi \in \mathcal{L}_d^*$, and C is max-id for all d if $\Psi \in \mathcal{L}_\infty^*$. $\quad\square$

Remark 8.1 From Corollary 8.23, $-\log\psi \in \mathcal{L}_2^*$ is the same as $-\log\psi$ concave since ψ is decreasing. This is a weaker condition than $\psi \in \mathcal{L}_\infty$ for a bivariate Archimedean copula $C_\psi(u_1, u_2)$ to be max-id. If $\psi \in \mathcal{L}_\infty$, then $C_\psi(u_1, u_2)$ is max-id, following from the combination of Theorem 8.28, part (c) of Theorem 8.5 and Theorem 8.7.

Theorem 8.32 (Condition for hierarchical Archimedean to be max-id, Joe (1997)).
(a) For the 1-level hierarchical Archimedean copula, let

$$C(\boldsymbol{u}) = \psi\big[\psi^{-1}(C_{S_1}(u_1, \ldots, u_m)) + \psi^{-1}(C_{S_2}(u_{m+1}, \ldots, u_d))\big],$$

where $2 \leq m < d$, $S_1 = \{1, \ldots, m\}$, $S_2 = \{m+1, \ldots, d\}$, $C_{S_1}(u_1, \ldots, u_m) = \varphi_1(\sum_{j=1}^{m} \varphi_1^{-1}(u_j))$, $C_{S_2}(u_{m+1}, \ldots, u_d) = \varphi_2(\sum_{j=m+1}^{d} \varphi_2^{-1}(u_j))$. Assume $\Psi = -\log\psi \in \mathcal{L}_\infty^*$ and $\omega_i = \psi^{-1} \circ \varphi_i \in \mathcal{L}_\infty^*$.
(b) For the multi-level hierarchical Archimedean copula, let C be a defined in (3.16). Assume $\Psi = -\log\psi \in \mathcal{L}_\infty^*$ and $\varphi_{l,a}^{-1} \circ \varphi_{l+1,a'} \in \mathcal{L}_\infty^*$ for all nestings of LTs.
Then C is max-id.

Proof: (a) Let $\eta_j(u_j) = \varphi_i^{-1}(u_j)$ if $j \in S_i$. Each η_j is decreasing so that $\eta_j' \leq 0$. Write

$$H = C^r = \exp\big\{-r\Psi\big[\omega_1(\eta_1 + \cdots + \eta_m) + \omega_2(\eta_{m+1} + \cdots + \eta_d)\big]\big\}$$

Take partial derivatives sequentially in an appropriate order. We show that all of them are non-negative. The partial derivatives are:

- $\partial H / \partial u_1 = H[-r\Psi' \omega_1' \eta_1']$,
- $\partial^2 H / \partial u_1 \partial u_2 = H[r^2 \Psi' \omega_1' \eta_1' \Psi' \omega_1' \eta_2'] + H[-r\Psi'' \omega_1' \eta_1' \omega_1' \eta_2'] + H[-r\Psi' \omega_1'' \eta_1' \eta_2']$,
- $\partial^3 H / \partial u_1 \partial u_2 \partial u_{m+1} = H[-r^3 (\Psi')^3 (\omega_1')^2 \omega_2' \eta_1' \eta_2' \eta_{m+1}'] + H[2r^2 \Psi' \Psi'' (\omega_1')^2 \omega_2' \eta_1' \eta_2' \eta_{m+1}'] +$
- $\cdots$,
- etc.

Similar to the proof of Theorem 8.31, the derivative of each summand leads to a few new summands, each of which is overall non-negative because for the jth derivative in the sequence,

- $-r\eta'_j \Psi' \omega'_i \geq 0$ if $j \in S_i$,
- $\eta'_j \omega'_i \Psi^{(k+1)}/\Psi^{(k)} \geq 0$ if $j \in S_i$ and $k \geq 1$,
- $\eta'_j \omega_i^{(k+1)}/\omega_i^{(k)} \geq 0$ if $j \in S_i$ and $k \geq 1$.

(b) The proof is the same but with more complex notation and more functions of the form of ω_i. $\qquad\qquad\qquad\qquad\qquad\qquad\qquad\qquad\qquad\qquad\qquad\qquad\square$

Remark 8.2 In a careful look at the proof, for part (a), the condition on the ω functions can be weakened from $\mathcal{L}^*_\infty$ to $\mathcal{L}^*_{|S_i|}$ where $|S_i|$ is the cardinality of subset S_i. For part (b), it is $\mathcal{L}^*_{|S|}$ where S is the cardinality of the subset associated with $\varphi_{l+1,a'}$. Also $\Psi \in \mathcal{L}^*_d$ is sufficient for dimension d.

The next results summarize upper and lower tail behavior of Archimedean copulas C_ψ. These properties are important to know when a parametric Archimedean is suitable as a model.

If ψ is the LT of a resilience random variable Q (see (3.2)), then the behavior of $F_Q(q)$ as $q \to \infty$ affects $\psi(s)$ as $s \to 0^+$ and the behavior of $F_Q(q)$ as $q \to 0^+$ affects $\psi(s)$ as $s \to \infty$. In the extreme cases, if Q has a mass at 0, then $\psi(\infty) = \mathbb{P}(Q = 0) > 0$ and if Q is a defective random variable with mass at ∞, then $\psi(0) < 1$. For an Archimedean copula, ψ must correspond to a LT with no mass at 0 or ∞.

From the resilience representation

$$C_\psi(\boldsymbol{u}) = \int_0^\infty \prod_{j=1}^d G^q(u_j)\, dF_Q(q).$$

where $G(u) = \exp\{-\psi^{-1}(u)\}$ for $0 \leq u \leq 1$. As explained in Section 2.1.5, as $q \to \infty$, G^q converges in distribution to a degenerate distribution at 1, and as $q \to 0^+$, G^q converges in distribution to a degenerate distribution at 0. Hence if F_Q has more probability for larger values, then C_ψ has more probability near $\boldsymbol{1}_d$, and if F_Q has more probability for smaller values, then C_ψ has more probability near $\boldsymbol{0}_d$.

Therefore the joint upper tail behavior of C_ψ is influenced by the behavior of $\psi(s)$ as $s \to 0^+$, and the joint lower tail behavior of C_ψ is influenced by the behavior of $\psi(s)$ as $s \to \infty$. The theorems below show conditions on ψ for upper/lower tail dependence and tail order. Results on tail order are from Hua and Joe (2011). Bivariate tail dependence results from Joe (1997) are included as parts of the theorems.

Theorem 8.33 (Upper tail dependence of Archimedean).

$$\lambda_{U,d} = \lim_{u \to 0^+} \frac{\overline{C}_\psi((1-u)\boldsymbol{1}_d)}{u} = d + d \lim_{s \to 0^+} \sum_{i=2}^d \frac{(-1)^{i-1}\binom{d-1}{i-1}\psi'(is)}{\psi'(s)}$$

and for $d = 2$, this becomes

$$\lambda_U = \lambda_{U,2} = 2 - 2 \lim_{s \to 0^+} [\psi'(2s)/\psi'(s)].$$

If $\psi'(0)$ is finite, then $\lambda_U = 0$ and the Archimedean copula C_ψ does not have upper tail dependence. If C_ψ has upper tail dependence, then $\psi'(0) = -\infty$.

If $1 - \psi(s)$ is $\mathrm{RV}_\alpha(0^+)$ for some $0 \leq \alpha < 1$, then $0 < \lambda_{U,d} \leq 1$. In this case, the limiting extreme value copula is Gumbel.

Proof: Consider:

$$
\lim_{u\to 0^+}\frac{\overline{C}_\psi((1-u)\mathbf{1}_d)}{u} = \lim_{u\to 0^+}\frac{1-d(1-u)+\sum_{i=2}^d(-1)^i\binom{d}{i}\psi(i\psi^{-1}(1-u))}{u}
$$

$$
= \lim_{s\to 0^+}\frac{1-d\psi(s)+\sum_{i=2}^d(-1)^i\binom{d}{i}\psi(is)}{1-\psi(s)}
$$

$$
= d\lim_{s\to 0^+}\frac{\sum_{i=1}^d(-1)^{i-1}\binom{d-1}{i-1}\psi'(is)}{\psi'(s)}.
$$

If $\psi'(0)\in(-\infty,0)$, then the above limit is zero because $\sum_{j=0}^{d-1}(-1)^j\binom{d-1}{j}=0$, otherwise the first term with $i=1$ is d. Note that $\psi'(0)$ cannot equal 0 because it is the negative of the expectation of a positive random variable.

The result on the limiting extreme value copula is in Genest and Rivest (1989) and Joe et al. (2010). $\qquad\square$

Theorem 8.34 (Lower tail dependence of Archimedean). *The d-variate Archimedean copula C_ψ has lower tail dependence parameter equal to*

$$
\lambda_{L,d} = \lim_{s\to\infty}[\psi(ds)/\psi(s)] = d\lim_{s\to\infty}[\psi'(ds)/\psi'(s)].
$$

For the bivariate case, this is

$$
\lambda_L = \lambda_{L,2} = \lim_{s\to\infty}[\psi(2s)/\psi(s)] = 2\lim_{s\to\infty}[\psi'(2s)/\psi'(s)].
$$

If $\psi\in\mathrm{RV}_{-\alpha}$ as $s\to\infty$ where $\alpha>0$, then $\lambda_{L,d}=d^{-\alpha}$. In this case, the limiting lower extreme value copula is Galambos.

Proof: Straightforward. The result on the limiting extreme value copula is in Charpentier and Segers (2009) and Joe et al. (2010). $\qquad\square$

Theorem 8.35 (Upper tail order of Archimedean, Hua and Joe (2011)). *Let ψ be the LT of a positive random variable and assume that ψ satisfies the condition of Theorem 8.88. Assume that $k<M_\psi<k+1$ with some $k\in\{1,\dots,d-1\}$, then the Archimedean copula C_ψ has upper intermediate tail dependence. The corresponding tail order is $\kappa_U=M_\psi$. If $\psi^{(i)}(0)$ is finite for all $i=1,\dots,d$, then the upper tail order $\kappa_U=d$. If $\psi'(0)$ is infinite and $0<M_\psi<1$, then the upper tail order is $\kappa_U=1$, and particularly for the bivariate case, $\lambda_U=2-2^{M_\psi}$.*

Proof: See details in Hua and Joe (2011). $\qquad\square$

Theorem 8.36 (Upper tail order function of Archimedean, Hua and Joe (2011)). *Let C_ψ be a multivariate Archimedean copula with $\kappa_U=M_\psi$ being a non-integer in the interval $(1,d)$, and suppose ψ satisfies the condition of Theorem 8.88. With notation $I_d=\{1,\dots,d\}$, $M=M_\psi-[M_\psi]$ and $k=[M_\psi]$, the upper tail order parameter is*

$$
\Upsilon_U(C_\psi) = \frac{Mh}{[-\psi'(0)]^{M_\psi}\prod_{j=0}^k(M_\psi-j)}\sum_{\emptyset\neq I\subset I_d}(-1)^{|I|+k+1}|I|^{M_\psi},
$$

and the upper tail order function is

$$
b^*(\boldsymbol{w}) = \frac{\sum_{\emptyset\neq I\subset I_d}(-1)^{|I|}\left(\sum_{i\in I}w_i\right)^{M_\psi}}{\sum_{\emptyset\neq I\subset I_d}(-1)^{|I|}|I|^{M_\psi}},
$$

where $h=\lim_{s\to 0^+}\ell(s)$ with $|\psi^{(k)}(s)-\psi^{(k)}(0)|=s^M\ell(s)$ as $s\to 0^+$.

Proof: See details in Hua and Joe (2011).   □

Remark 8.3 For a d-variate Archimedean copula, the pattern of the upper tail order function also depends on the upper tail order κ. For example, in $d = 3$, the homogeneous function b^* is positively proportional to

$$-w_1^\kappa - w_2^\kappa - w_3^\kappa + (w_1 + w_2)^\kappa + (w_1 + w_3)^\kappa + (w_2 + w_3)^\kappa - (w_1 + w_2 + w_3)^\kappa, \quad 1 < \kappa < 2;$$

$$w_1^\kappa + w_2^\kappa + w_3^\kappa - (w_1 + w_2)^\kappa - (w_1 + w_3)^\kappa - (w_2 + w_3)^\kappa + (w_1 + w_2 + w_3)^\kappa, \quad 2 < \kappa < 3.$$

The signs of all terms depend on whether $1 < \kappa < 2$ or $2 < \kappa < 3$. The pattern of alternating signs extends to $d > 3$. This pattern, together with Lemma 2 of Hua and Joe (2011), also shows why we don't have a general form of the tail order function when M_ψ is a positive integer.

Theorem 8.37 (Lower tail order of Archimedean; Charpentier and Segers (2009), Hua and Joe (2011)). *Suppose a LT ψ satisfies the condition in (8.42) with $0 \le p \le 1$. If $p = 0$, then C_ψ has lower tail dependence or its lower tail order is 1. If $p = 1$, then $\kappa_L(C_\psi) = d$. If $0 < p < 1$, then C_ψ has intermediate lower tail dependence with $1 < \kappa_L(C_\psi) = d^p < d$, $\ell(u) = d^q a_1^{1-\kappa} a_2^{-\zeta}(-\log u)^\zeta$ with $\zeta = (q/p)(1 - d^p)$, and the tail order function is $b(\boldsymbol{w}) = \prod_{i=1}^d w_i^{d^{p-1}}$.*

Proof: See details in Hua and Joe (2011).   □

Remark 8.4 Charpentier and Segers (2009) obtain the lower tail order function of Archimedean copulas with the condition $-\psi(s)/\psi'(s) \in \mathrm{RV}_{1-p}$. Condition (8.42) does not cover all possibilities. It is possible that as $s \to \infty$, $\psi(s)$ goes to 0 slower than anything of form (8.42). Examples are given by the gamma stopped gamma and positive stable stopped gamma LT families (see the Appendix), leading to Archimedean families such that $\lim_{u\to 0^+} C_\psi(u\mathbf{1}_d)/u = 1$ (for the bivariate case, these are copula families BB2 and BB3). Note that, for the gamma stopped gamma LT family, $\psi(s) = [1 + \delta^{-1}\log(1 + s)]^{-1/\theta}$ with $\delta > 0$ and $\theta > 0$ and as $s \to \infty$, $\psi(s) \sim \delta^{1/\theta}(\log s)^{-1/\theta}$; for the positive stable stopped gamma LT family, $\psi(s) = \exp\{-[\delta^{-1}\log(1 + s)]^{1/\theta}\}$ with $\delta > 0$, $\theta > 1$ and as $s \to \infty$, $\psi(s) \sim \exp\{-\delta^{-1/\theta}(\log s)^{1/\theta}\}$.

Larsson and Nešlehová (2011) studied the tail behavior and tail dependence of Archimedean copulas via Williamson transform in Section 3.3. The results depend on the maximum domain of attraction of F_R or $F_{1/R}$ where R is the radial random variable in the stochastic representation (3.12) of an Archimedean copula C_ψ with Williamson transform $\psi = \mathcal{W}_d F_R$. With related conditions for R, additional tail order results for intermediate tail dependence are proved in Hua and Joe (2013).

Theorem 8.38 (Tail order in northwest and southeast corners). *For the bivariate Archimedean copula C_ψ, the tail order in the northwest and southeast corners can be obtained from the rate of*

$$-\psi'(\psi^{-1}(u)) \cdot \psi^{-1}(1 - u)$$

as $u \to 0^+$.

Proof: Let $(U, V) \sim C_\psi$. The copula for $(U, 1 - V)$ is $u - C_\psi(u, 1 - v) = u - \psi(\psi^{-1}(u) + \psi^{-1}(1 - v))$. The tail order of C_ψ in the northwest and southeast corners and be obtained from $u - \psi(\psi^{-1}(u) + \psi^{-1}(1 - u))$ as $u \to 0^+$. Since $\psi^{-1}(u) \to \infty$ and $\psi^{-1}(1 - u) \to 0$, the first term dominates and a Taylor expansion leads to:

$$u - \psi\big(\psi^{-1}(u) + \psi^{-1}(1 - u)\big) \sim -\psi'(\psi^{-1}(u)) \cdot \psi^{-1}(1 - u), \quad u \to 0^+.$$

□

The next result says that if the LT ψ is slowly varying at one of the tails, then C_ψ must have the tail dependence function $\min(w_1, \ldots, w_d)$; this implies the strong possible tail dependence. These types of copulas can be used to get conservative bounds on tail risks.

Theorem 8.39 (Tail comonotonicity for Archimedean, Hua and Joe (2012b)). *Let the Archimedean copula C_ψ be based on the LT $\psi \in \mathcal{L}_\infty$. If $\psi(s) \in \mathrm{RV}_0$ as $s \to \infty$ then the lower tail dependence function exists and C_ψ is lower tail comonotonic; if $1 - \psi(s) \in \mathrm{RV}_0(0^+)$ then the upper tail dependence function exists and C_ψ is upper tail comonotonic.*

Proof: We only need to prove the tail dependence parameter is 1. For the lower tail, letting $s := \psi^{-1}(u)$, then because $\psi(s) \in \mathrm{RV}_0$,

$$\lambda_L = \lim_{u \to 0^+} \frac{C_\psi(u \mathbf{1}_d)}{u} = \lim_{u \to 0^+} \frac{\psi(d\psi^{-1}(u))}{u} = \lim_{s \to +\infty} \frac{\psi(ds)}{\psi(s)} = 1.$$

For the upper tail, it suffices to prove the bivariate case. Letting $s := \psi^{-1}(1 - u)$, then

$$\lambda_U = \lim_{u \to 0^+} \frac{\widehat{C}_\psi(u, u)}{u} = \lim_{u \to 0^+} \left[2 + \frac{\psi(2\psi^{-1}(1 - u)) - 1}{u} \right] = \lim_{s \to +} \left[2 - \frac{1 - \psi(2s)}{1 - \psi(s)} \right] = 1.$$

□

Next are results on another form of tail behavior. The properties of boundary conditional cdfs are relevant for conditional tail expectations of the form $\mathbb{E}(X_1 | X_2 = t)$ or $\mathbb{E}(X_1 | X_2 > t)$ as $t \to \infty$, where $(X_1, X_2) \sim C(F, F)$.

Theorem 8.40 (Lower boundary conditional cdf for Archimedean, Hua and Joe (2014)). *Consider a bivariate Archimedean copula $C = C_\psi$.*
(a) If $\psi \in \mathrm{RV}_{-\alpha}$ with $\alpha \geq 0$, then the conditional distribution $C_{1|2}(\cdot|0)$ is a cdf of a degenerate random variable at 0; that is, $C_{1|2}(u|0) = 1$ for any $0 \leq u \leq 1$.
(b) Suppose ψ satisfies (8.42). If $0 < p < 1$, then $C_{1|2}(0|0) = 1$. If $p = 1$, then $C_{1|2}(u|0) < 1$ for any $0 \leq u < 1$ and $C_{1|2}(0|0) = 0$.

Proof: Write

$$C_{1|2}(u|0) = \lim_{v \to 0^+} \frac{\psi'(\psi^{-1}(u) + \psi^{-1}(v))}{\psi'(\psi^{-1}(v))} = \lim_{s \to \infty} \frac{\psi'(s + \psi^{-1}(u))}{\psi'(s)}. \tag{8.8}$$

(a) Since $\psi'(s)$ is increasing in s, by the Monotone Density Theorem (see Theorem 8.90), $\psi'(s) \sim -\alpha s^{-\alpha-1} \ell(s)$ as $s \to \infty$ for a slowly varying $\ell(s)$. So $-\psi' \in \mathrm{RV}_{-\alpha-1}$, and thus for any $0 < u \leq 1$, $\lim_{s \to \infty} \psi'(s + \psi^{-1}(u))/\psi'(s) = 1$, since for any small $\epsilon > 0$, as s is sufficiently large, $-\psi'((1 - \epsilon)s) \leq -\psi'(s + \psi^{-1}(u)) \leq -\psi'((1 + \epsilon)s)$. That is, $C_{1|2}(u|0) = 1$ for $0 < u < 1$.

(b) For $0 < p \leq 1$, with assumption (8.42) and keeping the dominating term in

$$\psi'(s) \sim T'(s) \sim -u_1 u_2 s^{q+p-1} \exp[\ a_2 s^p\},$$

and for large s, the term on the right-hand side of (8.8) becomes

$$[s^{-1}\psi^{-1}(u) + 1]^{q+p-1} \exp\{-a_2[\psi^{-1}(u) + s]^p + a_2 s^p\} \sim \exp\{-a_2[\psi^{-1}(u) + s]^p + a_2 s^p\}$$
$$\sim \exp\{-a_2 s^p[1 + ps^{-1}\psi^{-1}(u) + o(s^{-2})] + a_2 s^p\} = \exp\{-a_2[ps^{p-1}\psi^{-1}(u) + o(s^{p-2})]\}.$$

The limit as $s \to \infty$ is 1 if $0 < p < 1$. If $p = 1$, the limit is $\exp\{-a_2\psi^{-1}(u)\} < 1$ and $\lim_{u \to 0^+} \exp\{-a_2\psi^{-1}(u)\} = 0$.

□

Theorem 8.41 (Upper boundary conditional cdf for Archimedean. Hua and Joe (2014)). *Consider a bivariate Archimedean copula $C = C_\psi$.*
(a) If $\psi'(0) = -\infty$, then the conditional distribution $C_{1|2}(\cdot|1)$ is a cdf of a degenerate random variable at 1; that is, $C_{1|2}(u|1) = 0$ for any $0 \le u < 1$.
(b) If the upper tail order is κ_U and $1 < \kappa_U$, then $C_{1|2}(u|1) > 0$ for $0 < u < 1$. Furthermore $C_{1|2}(1^-|1) = 1$. If the expansion of ψ at 0 is $\psi(s) \sim 1 - a_1 s + a_2 s^m$, $s \to 0$, where $m = \min\{2, \kappa_U\} \in (1, 2]$, then $1 - C_{1|2}(1 - y|1) = O(y^{m-1})$ as $y \to 0^+$.

Proof: (a) If $\psi'(0) = -\infty$, then for any $0 \le u < 1$,

$$C_{1|2}(u|1) = \lim_{v \to 1^-} \frac{\psi'(\psi^{-1}(u) + \psi^{-1}(v))}{\psi'(\psi^{-1}(v))} = \lim_{s \to 0^+} \frac{\psi'(s + \psi^{-1}(u))}{\psi'(s)} = 0,$$

since $\psi'(\psi^{-1}(u)) < \infty$ for $0 \le u < 1$.

(b) From results in Hua and Joe (2011), if $1 < \kappa_U$ for a bivariate Archimedean copula, then $-\infty < \psi'(0) < 0$. Furthermore,

$$\lim_{v \to 1} C_{1|2}(u|v) = \lim_{s \to 0^+} \frac{\psi'(\psi^{-1}(u) + s)}{\psi'(s)} = \frac{\psi'(\psi^{-1}(u))}{\psi'(0)}. \tag{8.9}$$

Therefore, $0 < C_{1|2}(u|1) < 1$ for $0 < u < 1$ and $C_{1|2}(1|1) = 1$.

From Theorem 8.88, if $1 < m = \kappa_U < 2$ then $\psi(s) \sim 1 - a_1 s + a_2 s^m$, and if the tail order is greater than 2, then $\psi(s) \sim 1 - a_1 s + a_2 s^2$ is the expansion to order 2. Based on this expansion, $\psi^{-1}(1 - \epsilon) = a_1^{-1}\epsilon + a_1^{-1-m} a_2 \epsilon^m + o(\epsilon^m)$ as $\epsilon \to 0^+$ and

$$\frac{\psi'(\psi^{-1}(1-y))}{\psi'(0)} = 1 - a_2 a_1^{-m} m y^{m-1} + o(y^{m-1}).$$

Therefore $1 - C_{1|2}(1 - y|1) = O(y^{m-1})$. $\square$

The above two theorems show that bivariate Archimedean copulas cannot approximate bivariate Gaussian copulas with correlation parameter $\rho > 0$, or more generally cannot approximate all bivariate permutation symmetric copulas that have positive dependence properties such as SI. The bivariate Gaussian copula with $\rho > 0$ is reflection symmetric and has tail order $1 < 2/(1 + \rho) < 2$, and the lower and upper boundary conditional cdfs are degenerate at 0 and 1 respectively. The above two theorems, and also the theorems on tail order functions of Archimedean copulas with intermediate tail dependence, show that Archimedean copulas with lower and upper intermediate tail dependence have quite different behavior in the lower and upper tails.

However, with tail order 2, the Frank copula is reflection symmetric, and with tail order 1, there are members of the BB1 and BB7 copula families that are close to reflection symmetric (see Table 5.6).

8.6   Multivariate extreme value distributions

This section has some results on dependence and tail behavior for bivariate and multivariate extreme value distributions.

A bivariate extreme value copula C can be written as

$$C(u, v) = C_A(u, v) = \exp\{-A(-\log u, -\log v)\}, \tag{8.10}$$

where $A : [0, \infty)^2 \to [0, \infty)$ is convex, homogeneous of order 1 and satisfies $\max(x_1, x_2) \le A(x_1, x_2) \le x_1 + x_2$ (Pickands, 1981). Let $B(w) := A(w, 1 - w)$ for $0 \le w \le 1$, then B is convex and $\max\{w, 1 - w\} \le B(w) \le 1$. The conditions on B imply that $B(0) = B(1) = 1$, $-1 \le B'(0) \le 0$, $0 \le B'(1) \le 1$. and $B(w) \ge \frac{1}{2}$. The next result follows from Pickands (1981), with a proof in Joe (1997).

Theorem 8.42 (Characterization of bivariate extreme value copula). *Suppose B is a continuous non-negative function on $[0,1]$ with $B(0) = B(1) = 1$. Suppose that B has right and left derivatives up to second order except for at most a countable number of points. Then $\overline{G}(x,y) = \exp\{-(x+y)B(x/(x+y))\}$ is a bivariate exponential survival function if and only if B is convex and $\max\{w, 1-w\} \le B(w) \le 1$ for $0 \le w \le 1$.*

Based on Theorem 8.7, a bivariate extreme value copula has TP_2 cdf since it is max-id. Another form of positive dependence is stochastically increasing positive dependence; it is proved in Garralda Guillem (2000) as another property of bivariate extreme value copulas. Stochastically increasing positive dependence in general does not imply nor is it implied by the positive dependence condition of TP_2 cdf.

Theorem 8.43 (Bivariate extreme value copula is SI; Garralda Guillem (2000)). *Consider a bivariate extreme value copula C with $C(u,v) = C_A(u,v) = \exp\{-A(-\log u, -\log v)\}$, where $A : [0,\infty)^2 \to [0,\infty)$ is convex, homogeneous of order 1 and satisfies $\max(x,y) \le A(x,y) \le x+y$. Then the conditional distribution $C_{2|1}(\cdot|u)$ is stochastically increasing in u and $C_{1|2}(\cdot|v)$ is stochastically increasing in v.*

Proof: As defined after (8.10), let $B(w) := A(w, 1-w)$ for $0 \le w \le 1$, so that B is convex and $\max\{w, 1-w\} \le B(w) \le 1$. The conditions on B imply that $B(0) = B(1) = 1$, $-1 \le B'(0) \le 0$ and $0 \le B'(1) \le 1$; from the convexity $B'(w) \le [1 - B(w)]/(1 - w)$ or $(1-w)B'(w) + B(w) \le 1$ for $0 < w < 1$. The proof of SI is given only for $C_{2|1}$ and then, analogously, SI also holds for $C_{1|2}$.

Since B is convex, one-sided derivatives of B exist in $(0,1)$ and B'' exists almost everywhere (except at corner points of B). The first derivative of C with respect to u is

$$C_{2|1}(v|u) = \frac{\partial C(u,v)}{\partial u} = u^{-1}C(u,v)\{B(w) + (1-w)B'(w)\},$$

where $w = (\log u)/(\log u + \log v)$. It suffices to show that $C_{2|1}(v|u)$ is decreasing in u for all $0 < v < 1$ or that $\frac{\partial C_{2|1}}{\partial u}$ is non-positive almost everywhere. When B'' exists at w, the second partial derivative of $C(u,v)$ with respect to u is

$$\frac{\partial C_{2|1}(v|u)}{\partial u} = \frac{\partial^2 C(u,v)}{\partial u^2} = u^{-2}C(u,v)\Big\{[B(w) + (1-w)B'(w)]^2 - [B(w) + (1-w)B'(w)]$$
$$+ (1-w)B''(w) \cdot (1-w)[\log(uv)]^{-1}\Big\}$$
$$\le -u^{-2}C(u,v)\,t(1-t) \le 0,$$

since $\log(uv) \le 0$, $t := B(w) + (1-w)B'(w) \le 1$, and $t \ge \max\{w, 1-w\} - (1-w) \ge 0$. $\square$

Next is a result to show some dependence measures for bivariate extreme value copulas that can be reduced to one-dimensional integrals. This is useful for numerical calculations.

Theorem 8.44 (τ, ρ_S, ρ_E for bivariate extreme value copulas). *For a bivariate extreme value copula $C(u,v) = \exp\{-A(x,y)\}$ with $x = -\log u$, $y = -\log v$ and $B(w) = A(w, 1-w)$,*

$$\tau = \int_0^1 \frac{(2w-1)B'(w)B(w) + w(1-w)[B'(w)]^2}{[B(w)]^2}\,dw, \tag{8.11}$$

$$\rho_S = 12\int_0^1 [1 + B(w)]^{-2}dw - 3, \tag{8.12}$$

$$\rho_E = \int_0^1 [B(w)]^{-2}dw - 1,$$

where $\overline{G}(x,y) = \exp\{-(x+y)B(x/[x+y])\}$ *is the survival function of a bivariate min-stable exponential distribution and ρ_E is the correlation of $\overline{G}$.*

Proof: Since B is convex, B' exists as a right derivative everywhere in $(0, 1)$. The calculations for ρ_S and τ make use of (2.47) and (2.42). Transform to $w = x/(x+y)$, $z = x+y$, with Jacobian $|\frac{\partial(x,y)}{\partial(w,z)}| = z$. Let $(X, Y) \sim \overline{G}$. Then

$$\mathbb{E}(XY) = \int_0^\infty \int_0^\infty \overline{G}(x,y)\,\mathrm{d}x\mathrm{d}y = \int_0^1 \int_0^\infty e^{-zB(w)}z\mathrm{d}z\mathrm{d}w = \int_0^1 [B(w)]^{-2}\mathrm{d}w,$$

leading to ρ_E. Next,

$$\int_{[0,1]^2} C(u,v)\,\mathrm{d}u\mathrm{d}v = \int_0^\infty \int_0^\infty \overline{G}(x,y)e^{-x-y}\,\mathrm{d}x\mathrm{d}y$$

$$= \int_0^1 \int_0^\infty e^{-z[B(w)+1]}z\,\mathrm{d}z\mathrm{d}w = \int_0^1 [1+B(w)]^{-2}\mathrm{d}w,$$

leading to ρ_S. The conditional distributions of C are

$$C_{2|1}(v|u) = C(u,v) \cdot [B(w)+(1-w)B'(w)] \cdot u^{-1}, \quad C_{1|2}(u|v) = C(u,v) \cdot [B(w)-wB'(w)] \cdot v^{-1},$$
$$(8.13)$$

with $w = (\log u)/(\log u + \log v)$, so that

$$I := \int_{[0,1]^2} C_{2|1}(v|u)\,C_{1|2}(u|v)\,\mathrm{d}u\mathrm{d}v$$

$$= \int_{[0,1]^2} e^{-2zB(w)}[B(w)+(1-w)B'(w)]\,[B(w)-wB'(w)]\,(uv)^{-1}\mathrm{d}u\mathrm{d}v$$

$$= \int_0^1 \int_0^\infty e^{-2zB(w)}[B(w)+(1-w)B'(w)]\,[B(w)-wB'(w)]\,z\,\mathrm{d}z\mathrm{d}w$$

$$= \tfrac{1}{4}\int_0^1 [B(w)+(1-w)B'(w)]\,[B(w)-wB'(w)][B(w)]^{-2}\mathrm{d}w$$

$$= \tfrac{1}{4}\Big[1 + \int_0^1 \big\{(1-2w)B'(w)B(w) - w(1-w)[B'(w)]^2\big\}[B(w)]^{-2}\mathrm{d}w\Big],$$

so that $\tau = 1 - 4I$ becomes (8.11). $\qquad\square$

For bivariate extreme value copulas, Capéraà et al. (1997a) has the above formula (8.12) for ρ_S but a different result for Kendall's tau:

$$\tau = \int_0^1 \frac{w(1-w)}{B(w)}\mathrm{d}B'(w).$$

Hürlimann (2003) has formula (8.11) for Kendall's tau. For bivariate extreme value copula families in Chapter 4, $B''(w)$ is analytically more cumbersome to obtain, so (8.11) is easier for numerical calculations.

Next are results on properties of boundary conditional cdfs; these are relevant for conditional tail expectations of the form $\mathbb{E}(X_1|X_2 = t)$ or $\mathbb{E}(X_1|X_2 > t)$ as $t \to \infty$.

Theorem 8.45 (Boundary conditional cdfs for bivariate extreme value; Hua and Joe (2014)). *Suppose C is a bivariate extreme value copula and $B(w) := A(w, 1-w)$ for $0 \le w \le 1$. Then*

$$C_{1|2}(u|1) = [1 - B'(1)]\,u \quad and \quad C_{1|2}(u|0) = u^{1+B'(0)}.$$

That is, $C_{1|2}(u|1)$ has mass $B'(1)$ at $u = 1$ if $B'(1) > 0$, and is degenerate at 1 if $B'(1) = 1$. $C_{1|2}(u|0)$ degenerates at $u = 0$ with mass 1 if $B'(0) = -1$, and otherwise it has no mass at 0. Similarly,

$$C_{2|1}(v|1) = [1 + B'(0)]\, v \quad and \quad C_{2|1}(v|0) = v^{1-B'(1)}.$$

Therefore $C_{2|1}(\cdot|1)$ has a mass of 1 at 1 if $B'(0) = -1$ and $C_{2|1}(\cdot|0)$ has a mass of 1 at 0 if $B'(1) = 1$.

Proof: For $j = 1, 2$, let A_j be the right-hand partial derivative of $A(x_1, x_2)$ with respect to x_j. Since A is convex, A_j always exists and A_j is homogeneous of order 0.

(Upper tail) Write $C_{1|2}(u|v) = C(u, v)\, A_2(-\log u, -\log v)\, v^{-1}$. Then $C_{1|2}(u|1) = u\, A_2(\log u, 0) = u\, A_2(1, 0)$. Since $B(w) := A(w, 1 - w)$, and thus letting $y := (1 - w)/w$, we have $A(1, y) = (1+y) B(1/(1+y))$. Therefore, $A_2(1, y) = B(1/(1+y)) - (1+y)^{-1} B'(1/(1+y))$, and $A_2(1, 0) = B(1) - B'(1) = 1 - B'(1)$. So $C_{1|2}(u|1) = [1 - B'(1)]\, u$.

(Lower tail) write

$$C_{1|2}(u|v) = \exp\{-A(-\log u, -\log v)\}\, A_2(-\log u, -\log v)\, v^{-1}$$
$$= \exp\{\log v \cdot A(\log u/\log v, 1)\}\, A_2(1, \log v/\log u)\, v^{-1}.$$

Moreover, $A(z, 1) = (1+z)\, B(z/(1+z))$, and $A_1(z, 1) = B(z/(1+z)) + (1+z)^{-1} B'(z/(1+z))$. So, $A_1(0, 1) = B(0) + B'(0) = 1 + B'(0)$. Note that $\lim_{v \to 0^+} A_2(1, \log v/\log u) = A_2(1, \infty) = B(0) = 1$. Therefore,

$$
\begin{aligned}
C_{1|2}(u|0) &= \lim_{v \to 0^+} \exp\{\log v \cdot A(\log u/\log v, 1)\}\, v^{-1} \\
&= \lim_{v \to 0^+} \exp\{\log v \cdot [1 + A_1(\log u/\log v, 1)\, (\log u/\log v)]\}\, v^{-1} \\
&= \lim_{v \to 0^+} \exp\{\log v + A_1(\log u/\log v, 1)\, \log u\}\, v^{-1} \\
&= \lim_{v \to 0^+} \exp\{A_1(\log u/\log v, 1)\, \log u\} = \lim_{v \to 0^+} u^{A_1(\log u/\log v, 1)} \\
&= u^{1+B'(0)}.
\end{aligned}
$$

For the opposite conditioning, $C_{2|1}(v|1) = v\, A_1(0, 1) = [1 + B'(0)]\, v$ and $C_{2|1}(v|0) = \lim_{u \to 0^+} v^{A_2(1, \log v/\log u)} = v^{1-B'(1)}$. □

The class of bivariate extreme value copulas is illuminating to show what is happening in the tails. For any bivariate extreme value copula, the lower tail order is $\kappa_L = A(1, 1) = 2B(\frac{1}{2}) \in [1, 2]$; if the bivariate copula is not the independence copula, then $\kappa_U = 1$ and $\lambda_U = 2 - A(1, 1) = 2 - 2B(\frac{1}{2})$. However, the boundary conditional cdfs $C_{2|1}(\cdot|0)$, $C_{2|1}(\cdot|1)$, $C_{1|2}(\cdot|0)$, $C_{1|2}(\cdot|1)$ depend on $B'(0)$ and $B'(1)$. That is, the strength of lower tail order and upper tail dependence depend on $B(\frac{1}{2})$, and the strength of dependence in the edges of the unit square depend on $B'(0)$ or $B'(1)$. The conditions of $B'(1) = 1$ and $B'(0) = -1$ hold for the Gumbel, Galambos and Hüsler-Reiss copula families (except the boundary case of independence). But the conditions do not hold for the t-EV bivariate extreme value copula.

The density of multivariate extreme value copula or multivariate min-stable exponential is given next. This is needed for likelihood inference.

Theorem 8.46 (Density of multivariate extreme value copula). *Let $\overline{G}(x_1, \ldots, x_d) = \exp\{-A(\boldsymbol{x})\}$ be a min-stable exponential survival function, where A is homogeneous of order 1 and has mixed derivatives are up to order d. Let the corresponding extreme value copula be $C(\boldsymbol{u}) = \exp\{-A(-\log u_1, \ldots, -\log u_d)\}$. Let $A_S = \frac{\partial^{|S|} A}{\prod_{j \in S} \partial x_j}$. Then for $1 \le m \le d$,*

$$(-1)^m \frac{\partial^m \overline{G}}{\partial x_1 \cdots \partial x_m} = e^{-A}\Big\{ \sum_{P=(S_1,\ldots,S_{|P|})} (-1)^{m-|P|} \prod_i A_{S_i}\Big\}, \tag{8.14}$$

where the summation is over all of the partitions P of $\{1, \ldots, m\}$. Furthermore,

$$\frac{\partial^m C(\boldsymbol{u})}{\partial u_1 \cdots \partial u_m} = (-1)^m \frac{\partial^m \overline{G}(-\log u_1, \ldots, -\log u_d)}{\partial x_1 \cdots \partial x_m} \cdot \prod_{j=1}^{m} u_j^{-1}.$$

When $m = d$, these are the densities of the min-stable exponential distribution and extreme value copula.

For the special case of bivariate, if A has mixed derivatives up to order 2, and $B(w) = A(w, 1 - w)$, then with $w = x/(x + y)$,

$$A_1(x, y) = B(w) + (1 - w)B'(w) \geq 0, \quad A_2(x, y) = B(w) - wB'(w) \geq 0,$$
$$A_{12}(x, y) = -(x + y)^{-1} w(1 - w)B''(w) \leq 0.$$

Proof: For $m = 1, 2, 3$,

$$-\frac{\partial \overline{G}}{\partial x_1} = e^{-A} \cdot A_1,$$

$$\frac{\partial^2 \overline{G}}{\partial x_1 \partial x_2} = e^{-A} \cdot \{A_1 A_2 - A_{12}\},$$

$$-\frac{\partial^2 \overline{G}}{\partial x_1 \partial x_2 \partial x_3} = e^{-A} \cdot \{A_1 A_2 A_3 - A_{12}A_3 - A_{13}A_2 - A_{23}A_1 + A_{123}\}.$$

We obtain the general result by induction by assuming $m < d$, differentiating (8.14) with respect to x_{m+1} and showing that the result has the same form. The additional partial derivative with respect to x_{m+1} with a factor of -1 is:

$$(-1)^{m+1} \frac{\partial^{m+1}\overline{G}}{\partial x_1 \cdots \partial x_m \partial x_{m+1}} = e^{-A} \Bigg\{ \sum_{P=(S_1, \ldots, S_{|P|})} (-1)^{m+1-|P|} A_{m+1} \prod_{i=1}^{|P|} A_{S_i}$$

$$+ \sum_{P=(S_1, \ldots, S_{|P|})} (-1)^{m+1-|P|} \sum_{k=1}^{|P|} \frac{A_{S_k \cup \{m+1\}}}{A_{S_k}} \prod_{i=1}^{|P|} A_{S_i} \Bigg\},$$

and this is the same as (8.14) with m replaced by $m + 1$.

For the bivariate case, the derivatives of A are in the calculations of preceding proofs. $\square$

8.7 Mixtures of max-id distributions

This section has dependence and tail properties for copulas constructed as mixtures of max-id distributions. The first result is on a condition on when the mixture distribution is also max-id.

Theorem 8.47 (Condition for max-id for mixture of max-id; Joe and Hu (1996)). *Consider the d-variate distribution,*

$$F(\boldsymbol{x}) = \psi(-\log K(\boldsymbol{x})), \quad \boldsymbol{x} \in [0, 1]^d,$$

where K is max-id and $-\log \psi \in \mathcal{L}_d^*$, with $\mathcal{L}_d^*$ defined in (8.6). Then F is max-id.

Proof: We use Theorem 8.8. Let $\Psi = -\log \psi$ and $R = \log K$, so that $F = \exp\{-\Psi(-R)\}$. Let $H = \exp\{-q\Psi(-R)\}$, and let R_S denote the partial derivative of R with respect to $\{x_j : j \in S\}$. Then

- $\partial H/\partial x_1 = qH\Psi'R_1 \geq 0$,
- $\partial^2 H/\partial x_1\partial x_2 = q^2H\Psi'^2R_1R_2 - qH\Psi''R_1R_2 + qH\Psi'R_{12} \geq 0$,

since each summand is non-negative. The pattern of derivatives of each term being non-negative continues for higher-order partial derivatives of H. For example, differentiation of H in a term with respect to x_i leads to an extra factor $q\Psi'R_i \geq 0$, differentiation of $[\Psi^{(j)}(-R)]^m$ ($1 \leq j \leq d-1$, $m \geq 1$), in a term leads to an extra factor $-m\Psi^{(j+1)}R_i/\Psi^{(j)} \geq 0$, and differentiation of R_S in a term leads to an extra factor $R_{S\cup\{i\}}/R_S \geq 0$. $\square$

The next theorem is on concordance properties. It extends a result in Joe and Hu (1996) and is mostly a restatement of Theorem 4.11 of Joe (1997). The extra multivariate result makes use of part 2 of Theorem 8.7.

Theorem 8.48 (Concordance for mixture of max-id $C_{\psi,K}$). *Consider the functional form in* (3.19), *namely*

$$C_{\psi,K}(\boldsymbol{u}) = \psi\big(-\log K(\exp\{-\psi^{-1}(u_1)\},\ldots,\exp\{-\psi^{-1}(u_d)\})\big).$$

Then (a) *the multivariate $C_{\psi,K}$ is increasing in $\prec_c^{pw}$ and the PLOD ordering as K increases in $\prec_c$, and* (b) *$C_{\psi,K}$ is more PLOD than the d-variate Archimedean copula C_ψ if K is max-id. Also if $C_{\psi,K}$ is a valid copula with K that is negative lower orthant dependent, then $C_{\psi,K}$ is smaller in the PLOD ordering than the d-variate Archimedean copula C_ψ.*

The next result shows that if the max-id distribution used for mixing is a bivariate extreme value distribution, then Kendall's tau involve terms that are one-dimensional integrals appearing in Kendall's tau for bivariate Archimedean copulas and extreme value copulas.

Theorem 8.49 (Kendall's τ for mixture of bivariate extreme value copulas or Archimax; Capéraà et al. (2000)). *Consider the bivariate copula*

$$C_{\psi,K} = \psi\big(-\log K[\exp\{-\psi^{-1}(u)\}, \exp\{-\psi^{-1}(v)\}]\big) = \psi\big(A[\psi^{-1}(u), \psi^{-1}(v)]\big),$$

where $\psi \in \mathcal{L}_2$ and $K(u,v) = \exp\{-A(-\log u, -\log v)\}$ is a bivariate extreme value copula with exponent function A. The Kendall tau value for $C_{\psi,K}$ is

$$\tau(C_{\psi,K}) = 1 - [1 - \tau(C_\psi)][1 - \tau_A],$$

where $\tau(C_\psi) = 1 - 4\int_0^\infty s[\psi'(s)]^2 ds$ is the Kendall tau value for the Archimedean copula $C_\psi(u,v) = \psi(\psi^{-1}(u) + \psi^{-1}(v))$, and $\tau(K) = \tau_A$ is the Kendall tau value for K.

Proof: This is a different proof from the original. From (2.42), with $C = C_{\psi,K}$, $\tau(C_{\psi,K}) = 1 - 4\int_{[0,1]^2} C_{2|1}(v|u)\,C_{1|2}(u|v)\,dudv$. Let A_1, A_2 be the partial derivatives of $A(x,y)$ with respect to the first and second arguments respectively, and let $B(w) = A(w, 1-w)$. With $w = x/(x+y)$, by Theorem 8.46, $A_1(x,y) = B(w) + (1-w)B'(w)$ and $A_2(x,y) = B(w) - wB'(w)$. With $x = \psi^{-1}(u)$ and $y = \psi^{-1}(v)$,

$$C_{2|1}(v|u) = \psi'\big(A(x,y)\big)\,A_1(x,y)/\psi'(x), \quad C_{1|2}(u|v) = \psi'\big(A(x,y)\big)\,A_2(x,y)/\psi'(y).$$

Then, with $g(w) = [B(w) + (1-w)B'(w)]\,[B(w) - wB'(w)]$,

$$I := \int_{[0,1]^2} C_{2|1}(v|u)\,C_{1|2}(u|v)\,dudv = \int_0^\infty\int_0^\infty [\psi'(A(x,y))]^2 A_1(x,y)\,A_2(x,y)\,dxdy$$

$$= \int_0^1\int_0^\infty [\psi'(zB(w))]^2 g(w)\,z\,dzdw = \int_0^1\int_0^\infty [\psi'(s)]^2 s ds \cdot g(w)\,[B(w)]^{-2}dw$$

$$= \int_0^\infty s[\psi'(s)]^2 ds \cdot \int_0^1 g(w)\,[B(w)]^{-2}dw = \tfrac{1}{4}[1 - \tau(C_\psi)] \cdot [1 - \tau_A].$$

The last equality follows from the proofs of Theorems 8.26 and 8.44. The conclusion follows.
□

The next two theorems concern tail properties and combine results in Hua and Joe (2011) and Chapter 4 of Joe (1997). The tail properties are important in understanding the tail behavior of this class and in constructing copulas in this class with desired tail behavior.

Theorem 8.50 (Upper tail of bivariate mixture of max-id $C_{\psi,K}$). *Let*

$$C_{\psi,K}(u_1,u_2) = \psi\Big(-\log K\big(e^{-\psi^{-1}(u_1)}, e^{-\psi^{-1}(u_2)}\big)\Big),$$

where $\psi \in \mathcal{L}_\infty$ and K is a bivariate max-id copula. Let M_ψ be the maximal non-negative moment degree in (2.11) for a random variable with the LT ψ.

(a) *The copula $C_{\psi,K}$ has upper tail dependence if $0 < M_\psi < 1$ (so that $\psi'(0) = -\infty$) or K has upper tail dependence or both. In this case, if $\lambda_U(K)$ is the upper tail dependence parameter of K, then the upper tail dependence parameter of $C_{\psi,K}$ is*

$$\lambda_U(C_{\psi.K}) = 2 - [2 - \lambda_U(K)] \lim_{s \to 0^+} \frac{\psi'([2 - \lambda_U(K)]s)}{\psi'(s)}.$$

If $\lambda_U(K) = 0$, this is the same as the upper tail dependence parameter of the bivariate Archimedean copula C_ψ. If $\lambda_U(K) > 0$, then $\lambda_U(C_{\psi.K}) \geq \lambda_U(C_\psi)$.

(b) *If $M_\psi > 1$ and K has upper tail order $\kappa_U(K) > 1$, then $\kappa_U(C_{\psi,K}) = \kappa_U(C_\psi)$, that is, the upper tail order of $C_{\psi,K}$ is the same as the upper tail order of the bivariate Archimedean copula C_ψ.*

Proof: For simpler notation, write $\lambda(K)$ for $\lambda_U(K)$ and $\kappa(K)$ for $\kappa_U(K)$.

Let $\widehat{K}(u,v) = u + v - 1 + K(1-u, 1-v)$ be the reflected copula of K, and suppose $\widehat{K}(u,u) \sim u^{\kappa(K)}\ell(u)$ as $u \to 0^+$. If $\kappa(K) = 1$, let $\ell(u) = \lambda(K)$, and if $\kappa(K) > 1$, let $\lambda(K) = 0$.

For (a), the upper tail dependence parameter is:

$$\lambda_U(C_{\psi,K}) = \lim_{u \to 0^+} \frac{\overline{C}_{\psi,K}(1-u, 1-u)}{u} = \lim_{u \to 0^+} \frac{2u - 1 + C_{\psi,K}(1-u, 1-u)}{u}$$

$$= \lim_{s \to 0^+} \frac{1 - 2\psi(s) + \psi\big(-\log K(e^{-s}, e^{-s})\big)}{1 - \psi(s)}$$

$$= \lim_{s \to 0^+} \frac{1 - 2\psi(s) + \psi\big(-\log[\widehat{K}(1 - e^{-s}, 1 - e^{-s}) + 2e^{-s} - 1]\big)}{1 - \psi(s)}$$

$$= \lim_{s \to 0^+} \frac{1 - 2\psi(s) + \psi\big(-\log[s^{\kappa(K)}\ell(s) + 1 - 2s]\big)}{1 - \psi(s)}$$

$$= \lim_{s \to 0^+} \frac{1 - 2\psi(s) + \psi\big(2s - s^{\kappa(K)}\ell(s)\big)}{1 - \psi(s)}$$

$$= \frac{\lim_{s \to 0^+} 1 - 2\psi(s) + \psi([2 - \lambda(K)]s)}{1 - \psi(s),}$$

and the conclusion in (a) follows by applying l'Hopital's rule. If $\lambda(K) = 0$, this limit is the same as that in Theorem 8.33 with $d = 2$.

(b) Let $u = 1 - \psi(s)$. Then $M_\psi > 1$ implies that $u \sim -\psi'(0)s$ as $s \to 0^+$. From the above,

$$\widehat{C}_{\psi,K}(u,u) \sim 1 - 2\psi(s) + \psi\big(2s - s^{\kappa(K)}\ell(s)\big) \sim 2u - 1 + \psi(2\psi^{-1}(1 - u)) = \widehat{C}_\psi(u,u).$$

The conclusion follows.
□

Theorem 8.51 (Lower tail of bivariate mixture of max-id $C_{\psi,K}$). *Let*

$$C_{\psi,K}(u_1, u_2) = \psi\Big(-\log K\big(e^{-\psi^{-1}(u_1)}, e^{-\psi^{-1}(u_2)}\big)\Big),$$

where $\psi \in \mathcal{L}_\infty$ and K is a bivariate max-id copula. Suppose $\psi(s) \sim a_1 s^q \exp\{-a_2 s^p\}$ as $s \to \infty$ with $0 \le p \le 1$, $a_1 > 0$ and $a_2 > 0$.

(a) *If K has lower tail dependence, then $C_{\psi,K}$ has lower tail dependence. Let $\lambda_L(K)$ and $\lambda_L(C_{\psi,K})$ be the tail dependence parameters in this case. Then $\lambda_L(C_{\psi,K}) = 1$ (lower tail comonotonicity) if $0 \le p < 1$ and $\lambda_L(C_{\psi,K}) = [\lambda_L(K)]^{a_2}$ if $p = 1$.*

(b) *Suppose K has lower tail order $\kappa_L(K) > 1$ and $K(u,u) \sim u^{\kappa_L}(K)\ell(u)$ as $u \to 0^+$ where $\ell(u)$ is slowly varying. For ψ, if $0 < p \le 1$, then $\kappa_L(C_{\psi,K}) = [\kappa_L(K)]^p$.*

(c) *Suppose K has lower tail order $\kappa_L(K) > 1$ and $K(u,u) \sim u^{\kappa_L}(K)\ell(u)$ as $u \to 0^+$ with $\ell(u)$ being an ultimately monotone slowly varying function. For ψ, if $p = 0$ and $q < 0$, then $\kappa_L(C_{\psi,K}) = 1$ and $\lambda_L(C_{\psi,K}) = [\kappa_L(K)]^q$.*

Proof: For simpler notation, write $\kappa(K)$ for $\kappa_L(K)$. If $\kappa(K) = 1$, let $\lambda_L(K) \equiv \ell(u)$.
(a) If $K(u,u) \sim \lambda_L(K)u$ as $u \to 0^+$, then

$$\lambda_L(C_{\psi,K}) = \frac{\lim_{u\to 0} \frac{C_{\psi,K}(u,u)}{u} = \lim_{u\to 0} \psi(-\log \lambda_L(K) + \psi^{-1}(u))}{u}$$

$$= \lim_{s\to\infty} \frac{\psi(-\log \lambda_L(K) + s)}{\psi(s)}.$$

The conclusion for the different cases follows from substitution and taking the limit.
(b) As $u \to 0^+$, $e^{-\psi^{-1}(u)} \to 0$, and

$$\psi^{-1}(u) \sim (-a_2^{-1} \log u)^{1/p} + \ell^*(-\log u),$$

where ℓ^* is slowly varying. Then

$$\begin{aligned} C_{\psi,K}(u,u) &\sim \psi\big[-\log\{(e^{-\psi^{-1}(u)})^{\kappa(K)}\ell(e^{-\psi^{-1}(u)})\}\big] \\ &\sim \psi\big[\kappa(K)\psi^{-1}(u) - \log\ell(e^{-\psi^{-1}(u)})\big] \\ &\sim \psi\big[\kappa(K)(-a_2^{-1}\log u)^{1/p} + \kappa(K)\ell^*(-\log u) - \log\ell(e^{-\psi^{-1}(u)})\big] \\ &\sim \ell^{**}(u)u^{[\kappa(K)]^p}, \end{aligned}$$

where ℓ^{**} is slowly varying. Therefore, if $\kappa(K) = \kappa_L(K) > 1$ and $0 < p \le 1$, then $\kappa_L(C_{\psi,K}) = [\kappa_L(K)]^p$.
(c) As $u \to 0^+$ and $\psi^{-1}(u) \sim (u/a_1)^{1/q}$, so $t := e^{-\psi^{-1}(u)} \to 0^+$ and $(-\log t)^q \sim u/a_1$. Then

$$\begin{aligned} C_{\psi,K}(u,u) &\sim \psi\big(-\log\ell(t)t^{\kappa}(K)\big) = \psi\big(-\log\ell(t) - \kappa(K)\log t\big) \\ &\sim a_1[\kappa(K)]^q(-\log t)^q\Big[1 + \frac{-\log\ell(t)}{-\kappa\log t}\Big]^q \sim u[\kappa(K)]^q\Big[1 + \frac{-\log\ell(t)}{\kappa\log t}\Big]^q. \end{aligned}$$

If $\ell(t)$ converges to a positive constant as $t \to 0^+$, then $\frac{-\log\ell(t)}{-\kappa\log t}$ converges to 0. If $\ell(t)$ converges to 0 or ∞ as $t \to 0^+$, then $|\log\ell(t)| \to \infty$, and by l'Hopital's rule and Theorem 8.91,

$$\lim_{t\to 0} \frac{-\log\ell(t)}{-\kappa\log t} = \lim_{t\to 0} \frac{t\ell'(t)}{\kappa\ell(t)} = 0.$$

The conclusion follows. $\qquad\square$

In Hua and Joe (2011), the condition about the LT is stated as $-\psi/\psi' \in \mathrm{RV}_{1-p}$ where $0 \leq p < 1$, but then the proof has more technical details. Note that part (a) of the above theorem explains the lower tail comonotonicity of the BB2 and BB3 copulas, where K is the MTCJ copula and ψ is respectively the gamma LT and the positive stable LT. Part (c) of the above theorem explains the lower tail dependence parameter of the BB1 copula, where K is the Gumbel copula with parameter $\delta > 1$, ψ is the gamma LT with parameter $\theta > 0$ and $q = -\theta^{-1}$, and $\lambda_L(C_{\psi,K}) = (2^{1/\delta})^{-1/\theta}$.

The above two theorems can suggest constructions of bivariate copulas with tail dependence in one joint tail and intermediate tail dependence in the other. Further explanations of possibilities are mentioned in Example 8.3 when ψ is one of the first four one-parameter families of LTs in the Appendix, and in Example 8.4 when K is bivariate Gaussian with positive dependence.

Example 8.3 (Tail orders for mixture of max-id copulas). This example summarizes how the above two theorems can be applied to four LT families.

(a) Positive stable with $\theta > 1$, $M_\psi = \theta^{-1}$: For the upper tail, $\kappa_U(C_{\psi,K}) = 1$ and $\lambda_U(C_{\psi,K}) = 2 - [2 - \lambda_U(K)]^{1/\theta}$. For the lower tail, since $p = \theta^{-1} \in (0,1)$, if K has lower tail dependence then $\lambda_L(C_{\psi,K}) = 1$, and otherwise there is intermediate tail dependence with $\kappa_L(C_{\psi,K}) = [\kappa_L(K)]^{1/\theta}$ if $\kappa_L(K) > 1$.

(b) Gamma with $\theta > 0$, $M_\psi = \infty$: For the upper tail, if K has upper tail dependence then $\kappa_U(C_{\psi,K}) = 1$ with $\lambda_U(C_{\psi,K}) = \lambda_U(K)$ and otherwise $\kappa_U(C_{\psi,K}) = 2$. For the lower tail, since $p = 0$, $\lambda_L(C_{\psi,K}) = 1$ if K has lower tail dependence, and otherwise $\lambda_L(C_{\psi,K}) = [\kappa_L(K)]^{-1/\theta}$ if $\kappa_L(K) > 1$.

(c) Sibuya with $\theta > 1$, $M_\psi = \theta^{-1}$: For the upper tail, $\kappa_U(C_{\psi,K}) = 1$ and $\lambda_U(C_{\psi,K}) = 2 - [2 - \lambda_U(K)]^{1/\theta}$. For the lower tail, since $p = 1$, $\lambda_L(C_{\psi,K}) = \lambda_L(K)$ if K has lower tail dependence, and $\kappa_L(C_{\psi,K}) = \kappa_L(K)$ if $\kappa_L(K) > 1$.

(d) Logarithmic series with $\theta > 0$, $M_\psi = \infty$: For the upper tail, if K has upper tail dependence then $\kappa_U(C_{\psi,K}) = 1$ with $\lambda_U(C_{\psi,K}) = \lambda_U(K)$ and otherwise $\kappa_U(C_{\psi,K}) = 2$. For the lower tail, since $p = 1$, $\lambda_L(C_{\psi,K}) = \lambda_L(K)$ if K has lower tail dependence, and $\kappa_L(C_{\psi,K}) = \kappa_L(K)$ if $\kappa_L(K) > 1$. So there is no interesting new tail behavior in this case, and there were no 2-parameter copula families of this form in Joe and Hu (1996) and Joe (1997).

Example 8.4 (Mixture of power of bivariate Gaussian). We introduce an example of $C_{\psi,K}$ that interpolates an Archimedean copula and a Gaussian copula with positive dependence. Also, we show that the invariance to scale changes of the LT for Archimedean copulas (i.e., Theorem 8.21) does not necessarily extend to mixture of max-id copulas.

Let $\psi_\alpha(s) = \psi(\alpha s)$ and $\psi_\alpha^{-1}(t) = \alpha^{-1}\psi^{-1}(t)$ with $\alpha > 0$. Then

$$C_{\psi_\alpha,K} = \psi\left(-\alpha \log K\left(e^{-\alpha^{-1}\psi^{-1}(u_1)}, e^{-\alpha^{-1}\psi^{-1}(u_2)}\right)\right). \tag{8.15}$$

For some choices of one-parameter families for ψ and one-parameter non-Archimedean max-id copula family $K(\cdot; \delta)$, this can be a 3-parameter family. For example, if ψ is the Sibuya LT with parameter $\theta > 1$ and K is the Gaussian copula with parameter $\rho > 0$, then (8.15) is:

$$1 - \left\{1 - \Phi_2^\alpha\left(\Phi\{[1 - (1-u)^\theta]^{1/\alpha}\}, \Phi\{[1 - (1-v)^\theta]^{1/\alpha}\}; \rho\right)\right\}^{1/\theta}.$$

From part (a) of Theorem 8.50 and part (b) of Theorem 8.51, this family has upper tail dependence and lower intermediate tail dependence.

8.8 Elliptical distributions

This section has results for elliptical distributions concerning the concordance ordering, tail dependence and tail order, and the boundary cdfs of the copulas.

Theorem 8.52 (Concordance ordering for elliptical, Das Gupta et al. (1972), Joe (1990b)). *Consider the elliptical family based on radial random variable R with random d-vector having stochastic representation $\mathbf{X} = R\mathbf{A}\mathbf{U}$, where $\mathbf{\Sigma} = (\rho_{jk})$ is a $d \times d$ correlation matrix. Let $F_{\mathbf{X}}(\mathbf{x}; \mathbf{\Sigma})$ be the cdf. Then $F_{\mathbf{X}}$ is increasing in concordance as ρ_{jk} increases while staying within the family of positive definite matrices). That is, for all $\mathbf{x} \in \mathbb{R}^d$,*

$$\mathbb{P}(X_1 \leq x_1, \ldots, X_d \leq x_d)$$

is increasing in ρ_{jk} for all $j \neq k$.

The next result also implied the formula for $\tau = \beta = (2/\pi)\arcsin(\rho)$ for bivariate elliptical in Section 2.12.5.

Theorem 8.53 (Bivariate Gaussian quadrant probability). *Let (Z_1, Z_2) be bivariate Gaussian with means of 1, variances of 1 and correlation ρ. Then*

$$\beta_0(\rho) := \mathbb{P}(Z_1 > 0, Z_2 > 0; \rho) = \mathbb{P}(Z_1 < 0, Z_2 < 0; \rho) = \tfrac{1}{4} + (2\pi)^{-1}\arcsin(\rho). \qquad (8.16)$$

It follows that Kendall's τ and Blomqvist's β are both $(2/\pi)\arcsin(\rho)$ for bivariate Gaussian.

Proof: This proof follows that of Kepner et al. (1989). Since $[Z_2|Z_1 = z] \sim N(\rho z, 1 - \rho^2)$,

$$\beta_0(\rho) = \int_{-\infty}^0 \phi(z_1)\Phi\left(\frac{-\rho z_1}{\sqrt{1-\rho^2}}\right) dz_1.$$

Then $\frac{d}{d\rho}\rho(1-\rho^2)^{-1/2} - (1-\rho^2)^{-3/2}$, so that

$$\begin{aligned}
\beta_0'(\rho) &= \int_{-\infty}^0 \phi(z_1)(-z_1)(1-\rho^2)^{-3/2}\phi\left(\frac{-\rho z_1}{\sqrt{1-\rho^2}}\right) dz_1 \\
&= (1-\rho^2)^{-3/2}(2\pi)^{-1}\int_0^\infty z\exp\{-\tfrac{1}{2}z^2 - \tfrac{1}{2}\rho^2 z^2/(1-\rho^2)\} dz \\
&= (1-\rho^2)^{-3/2}(2\pi)^{-1}\int_0^\infty z\exp\{-\tfrac{1}{2}z^2/(1-\rho^2)\} dz \\
&= (1-\rho^2)^{-1/2}(2\pi)^{-1}\int_0^\infty x\exp\{-\tfrac{1}{2}x^2\} dx = (2\pi)^{-1}(1-\rho^2)^{-1/2}.
\end{aligned}$$

By taking the antiderivative and matching at $\rho = 0$ or ± 1, (8.16) obtains.

Hence Blomqvist's beta is $\beta(\rho) = 4\beta_0(\rho) - 1 = (2/\pi)\arcsin(\rho)$. For Kendall's tau, let (Z_1, Z_2) and (Z_1', Z_2') be i.i.d. bivariate Gaussian with zero means, unit variances and correlation ρ). Then $(Z_1 - Z_1', Z_2 - Z_2')$ is bivariate Gaussian with zero means, variances of 2 and correlation $2\rho/2 = \rho$. Therefore, $\mathbb{P}(Z_1 - Z_1' > 0, Z_2 - Z_2' > 0) = \beta_0(\rho)$ and $\tau(\rho) = 4\mathbb{P}(Z_1 - Z_1' > 0, Z_2 - Z_2' > 0) - 1 = (2/\pi)\arcsin(\rho)$. $\qquad\square$

Next is a result on tail dependence in elliptical distributions. There are additional results on this topic in Hult and Lindskog (2002) and Schmidt (2002).

Theorem 8.54 (Tail dependence of elliptical). *Suppose $\boldsymbol{X}$ is an elliptical random d-vector, with stochastic representation $\boldsymbol{X} = R\boldsymbol{A}\boldsymbol{U}$, where $\boldsymbol{\Sigma} = (\rho_{jk}) = \boldsymbol{A}\boldsymbol{A}^{\top}$ is a $d \times d$ correlation matrix, $\boldsymbol{U}$ is a uniform random vector on the surface of the d-dimensional unit hypersphere, and $R \geq 0$ is a radial variable.*
(a) (Hult and Lindskog (2002), Schmidt (2002)). If $\mathbb{P}(R > s) \in \mathrm{RV}_{-\alpha}$ as $s \to \infty$, then for the (j, k) bivariate margin, the upper and lower tail dependence coefficients are both

$$\lambda_{jk} = \frac{\int_{(\pi/2-\arcsin \rho_{jk})/2}^{\pi/2}(\cos t)^{\alpha}\mathrm{d}t}{\int_{0}^{\pi/2}(\cos t)^{\alpha}\mathrm{d}t} = \frac{\int_{0}^{\sqrt{(1+\rho_{jk})/2}} u^{\alpha}(1 - u^2)^{-1/2}\mathrm{d}u}{\int_{0}^{1} u^{\alpha}(1 - u^2)^{-1/2}\mathrm{d}u} \tag{8.17}$$

(b) (Hua and Joe (2012b)). If $\mathbb{P}(R > s) \in \mathrm{RV}_0$ as $s \to \infty$, then

$$\lambda_{jk} = \tfrac{1}{2} + \pi^{-1} \arcsin \rho_{jk},$$

which is the limit of (8.17) as $\alpha \to 0^{+}$.

Proof: See the cited references. Note that $(\pi/2 - \arcsin \rho)/2 = \tfrac{1}{2}\arccos \rho$ is also equal to $\arcsin(\sqrt{(1 - \rho)/2}) = \arccos(\sqrt{(1 + \rho)/2})$. $\qquad\square$

An extension of Theorem 8.54 is that elliptical distributions have multivariate tail dependence in the joint upper and lower tails. It also has tail dependence in all other $2^d - 2$ orthants because the tail dependence is due to the radial random variable which affects all directions.

If $(U_1, U_2) \sim C$ for a bivariate copula C, define the tail dependence parameters $\lambda_{NW}, \lambda_{SE}$ in the "Northwest (NW)" and "Southeast (SE)" as:

$$\lambda_{\mathrm{NW}} := \lim_{u \to 0^+} \mathbb{P}(U_2 > 1 - u | U_1 \leq u) = \lim_{u \to 0^+} \mathbb{P}(U_1 \leq u | U_2 > 1 - u) = \lim_{u \to 0^+} \frac{u - C(u, 1 - u)}{u},$$

$$\lambda_{\mathrm{SE}} := \lim_{u \to 0^+} \mathbb{P}(U_1 > 1 - u | U_2 \leq u) = \lim_{u \to 0^+} \mathbb{P}(U_2 \leq u | U_1 > 1 - u) = \lim_{u \to 0^+} \frac{u - C(1 - u, u)}{u}.$$

Then we can also say that $\lambda_U = \lambda_{NE}$ (Northeast) and $\lambda_L = \lambda_{SW}$ (Southwest).

For a bivariate elliptical density with dependence parameter $-1 < \rho < 1$, a rotation of the density by $\pi/2$ radians leads to bivariate elliptical density with dependence parameter $-\rho$. Hence under the conditions of Theorem 8.54, the opposite corner tail dependence parameters $\lambda_{NW}, \lambda_{SE}$ are both equal to (8.17) with ρ replaced by $-\rho$. This is indicated in Embrechts et al. (2009) for the bivariate t_ν copula.

Note that $\lambda_{\mathrm{NE}} + \lambda_{\mathrm{SE}}$ is still a limiting conditional probability and thus $0 \leq \lambda_{\mathrm{NE}} + \lambda_{\mathrm{SE}} \leq 1$. The upper bound of 1 is attained for part (b) of Theorem 8.54, since the RV_0 condition implies that for the bivariate elliptical distribution,

$$\lambda_{\mathrm{NE}} = \tfrac{1}{2} + \pi^{-1} \arcsin \rho, \quad \lambda_{\mathrm{SE}} = \tfrac{1}{2} - \pi^{-1} \arcsin \rho,$$

with a sum of 1. For part (a) of Theorem 8.54 with the $\mathrm{RV}_{-\alpha}$ condition with $\alpha > 0$,

$$\lambda_{\mathrm{NE}} + \lambda_{\mathrm{SE}} = \frac{\int_{(\pi/2-\arcsin \rho)/2}^{\pi/2}(\cos t)^{\alpha}\mathrm{d}t + \int_{(\pi/2+\arcsin \rho)/2}^{\pi/2}(\cos t)^{\alpha}\mathrm{d}t}{\int_{0}^{\pi/2}(\cos t)^{\alpha}\mathrm{d}t} < 1,$$

and this sum is decreasing in α (Hua and Joe (2012b)). So the maximum tail dependence possible when all tails are considered is when the radial random variable R has survival function that is RV_0. See below for related comments on the boundary conditional cdfs of the copula.

For elliptical distributions, for the case where R has lighter tails than any regularly varying tails, some asymptotic study has been conducted for R in the maximum domain of attraction of Gumbel (see Hashorva (2007), Hashorva (2008) and Hashorva (2010)). Below is a result that is useful to find the tail order of a bivariate elliptical copula where the radial random variable $R \in$ MDA(Gumbel).

Theorem 8.55 (Tail order of bivariate elliptical, Hua and Joe (2013)). *Let C be the copula for an elliptical random vector $\boldsymbol{X} := (X_1, X_2)$ with radial random variable $R \geq 0$. Let $b_\rho = \sqrt{2/(1+\rho)}$. If $R \in$ MDA(Gumbel), then the upper and lower tail orders of C are*

$$\kappa = \lim_{r \to \infty} \frac{\log[1 - F_R(b_\rho r)]}{\log[1 - F_R(r)]} \geq 1, \tag{8.18}$$

provided that the limit exists.

It is convenient to apply the above result to derive the tail order if we know the tail behavior of R, and R belongs to MDA of Gumbel. A specific application is given in the Example 8.5 below.

By Theorem 3.1 of Hashorva (2007), this result can also be extended to multivariate case. For a d-dimensional exchangeable elliptical copula, of which the off-diagonal entries of $\boldsymbol{\Sigma}$ are all ρ and the diagonals are all 1's, the tail order is similar to (8.18) where b_ρ is replaced by $b_{\rho,d} = \sqrt{d/[1 + (d-1)\rho]}$.

Example 8.5 (Bivariate symmetric Kotz type distribution). The density generator

$$g(x) = Kx^{N-1} \exp\{-\beta x^\xi\}, \quad x > 0, \ \beta, \xi, N > 0,$$

where K is a normalizing constant. By Theorem 2.9 of Fang et al. (1990), the density of R is $f_R(x) = 2\pi x g(x^2) = 2K\pi x^{2N-1} \exp\{-\beta x^{2\xi}\}$ for $x > 0$. So, the survival function is

$$\overline{F}_R(x) = \int_x^\infty 2K\pi t^{2N-1} \exp\{-\beta t^{2\xi}\} \, \mathrm{d}t = \int_{\beta x^{2\xi}}^\infty \frac{K\pi}{\xi} \beta^{-N/\xi} w^{N/\xi - 1} \exp\{-w\} \, \mathrm{d}w$$

$$= \frac{K\pi}{\xi} \beta^{-N/\xi} \Gamma(N/\xi) \cdot \left[1 - F_\Gamma(\beta x^{2\xi}; N/\xi)\right] \sim \frac{K\pi}{\xi} \beta^{-1} x^{2N-2\xi} \exp\{-\beta x^{2\xi}\}, \ x \to \infty,$$

where F_Γ is the gamma cdf and the asymptotic relation is in Section 6.5 of Abramowitz and Stegun (1964). Then by (8.18), we can get that $\kappa = b_\rho^{2\xi} = [2/(1+\rho)]^\xi$. Therefore, the tail order for the symmetric Kotz type copula is $\kappa = [2/(1+\rho)]^\xi$. The Gaussian copula belongs to this class with $\xi = 1$, so its tail order is $2/(1+\rho)$. □

Next, we comment on the boundary conditional cdfs for bivariate elliptical distributions. Let (X_1, X_2) have an elliptical distribution with dependence parameter $0 < \rho < 1$ and let C be its copula. From Fang et al. (1990), we can write a stochastic representation for the conditional distributions:

$$[X_1 | X_2 = y] \stackrel{d}{=} \rho y + \sqrt{1 - \rho^2} \, \sigma(y) \, Z(y),$$

where $Z(y) \sim F_Z(\cdot; y)$ has a scale of 1 (on some measure of spread) and $\sigma(y)$ is the (conditional) scale parameter. Then

$$F_{1|2}(x|y) = \mathbb{P}(X_1 \leq x | X_2 = y) = F_Z\left(\frac{x - \rho y}{\sigma(y)\sqrt{1 - \rho^2}}; y\right).$$

If the scale parameter $\sigma(y)$ is (asymptotically) constant or increasing in $|y|$ at a sublinear rate, then $C_{1|2}(\cdot|1)$ has mass of 1 at 1 and $C_{1|2}(\cdot|0)$ has mass of 1 at 0. If the scale $\sigma(y)$ is (asymptotically) increasing linearly in $|y|$, then $C_{1|2}(\cdot|0)$ and $C_{1|2}(\cdot|1)$ are 2-point distributions with masses at 0 and 1. For the bivariate t_ν distribution, $Z(y)$ has the univariate $t_{\nu+1}$ distribution (for any y) and $\sigma(y) = [(\nu + y^2)/(\nu + 1)]^{1/2}$ is asymptotically linear.

For bivariate elliptical distributions, tail dependence occurs in all four corners if the survival function of the radial random variable has a regularly varying tail and this corresponds to the boundary conditional cdfs having masses and support on the set $\{0, 1\}$.

8.9   Tail dependence

This section has some results on tail dependence parameters and tail dependence functions. These are not specific to special classes of copulas or multivariate distributions as in the preceding sections. Some of the results are applied in the sections on vine and factor copulas.

The first result establishes that if a bivariate copula has lower (upper) tail dependence, then the tail dependence parameter in the copula is the same as the extreme value limit.

Theorem 8.56 (Tail dependence parameter/function for bivariate copula). *The results here are given only for the bivariate case, but there are multivariate counterparts.*

(i) *Let C be a bivariate copula with non-zero lower tail dependence function $b(w_1, w_2)$ and related function $a(w_1, w_2) = w_1 + w_2 - b(w_1, w_2)$ as given in (2.63). Then the lower tail dependence parameter is $\lambda_L = b(1, 1) = 2 - a(1, 1)$; λ_L is also the upper tail dependence parameter of the lower extreme value limit*

$$\lim_{n \to \infty} \left[u_1^{1/n} + u_2^{1/n} - 1 + C(1 - u_1^{1/n}, 1 - u_2^{1/n}) \right]^n$$

as given in (2.67).

(ii) *Let C be a bivariate copula with non-zero upper tail dependence function $b^*(w_1, w_2)$ and related function $a^*(w_1, w_2) = w_1 + w_2 - b^*(w_1, w_2)$. Then the upper tail dependence parameter is $\lambda_U = b^*(1, 1) = 2 - a^*(1, 1)$; λ_U is also the upper tail dependence parameter of the upper extreme value limit $\lim_{n \to \infty} C^n(u_1^{1/n}, u_2^{1/n})$.*

(iii) *If $C(u_1, u_2) = \exp\{-A(-\log u_1, -\log u_2)\}$ is a bivariate extreme value copula, the its upper tail dependence function is $a^* = A$.*

Proof: (i) By the definition of the tail dependence function, $C(uw_1, uw_2) \sim ub(w_1, w_2)$ as $u \to 0^+$. Hence $\lambda_L = \lim_{u \to 0^+} u^{-1} C(u, u) = b(1, 1)$. Next, with $u_j^{1/n} \sim 1 - n^{-1}\tilde{u}_j$ and $\tilde{u}_j = -\log u_j$ for $j = 1, 2$, as $n \to \infty$,

$$\left[u_1^{1/n} + u_2^{1/n} - 1 + C(1 - u_1^{1/n}, 1 - u_2^{1/n}) \right]^n$$
$$\sim \left[1 - n^{-1}\tilde{u}_1 - n^{-1}\tilde{u}_2 + C(n^{-1}\tilde{u}_1, n^{-1}\tilde{u}_2) \right]^n$$
$$\sim \left[1 - n^{-1}\tilde{u}_1 - n^{-1}\tilde{u}_2 + n^{-1}b(\tilde{u}_1, \tilde{u}_2) \right]^n$$
$$\to \exp\{-\tilde{u}_1 - \tilde{u}_2 + b(\tilde{u}_1, \tilde{u}_2)\} = \exp\{-a(\tilde{u}_1, \tilde{u}_2)\} =: C_{\mathrm{EV}}(u_1, u_2).$$

The upper tail dependence parameter of C_{EV} is

$$\lambda = \lim_{u \to 0^+} u^{-1} \overline{C}_{\mathrm{EV}}(1 - u, 1 - u) = \lim_{u \to 0^+} u^{-1}\{2u - 1 + e^{-a(-\log[1-u], -\log[1-u])}\}$$
$$= \lim_{u \to 0^+} u^{-1}\{2u - 1 + e^{\log[1-u] \cdot a(1,1)}\} = \lim_{u \to 0^+} u^{-1}\{2u - 1 + (1 - u)^{a(1,1)}\}$$
$$= 2 - a(1, 1) = b(1, 1) = \lambda_L.$$

This result is also Theorem 3.3 of Joe (1996a).

(ii) Follows from (i) since $b^*(w_1, w_2; C) = b(w_1, w_2; \widehat{C})$, where $\widehat{C}(u, v) = u + v - 1 + C(1 - u, 1 - v)$.

(iii) We make use of $C(1 - uw_1, 1 - uw_2) \sim 1 - ua^*(w_1, w_2)$ in (2.66). Since A is homogeneous of order 1, as $u \to 0^+$,

$$C(1 - uw_1, 1 - uw_2) \sim e^{-A(uw_1, uw_2)} = e^{-uA(w_1, w_2)} \sim 1 - uA(w_1, w_2),$$

so that $a^* = A$. $\hspace{1cm}$ $\square$

The next result shows a method to calculate the tail dependence parameter when the conditional cdfs have a simple form but not the copula cdf.

Theorem 8.57 (Tail dependence parameter from conditional distributions; Demarta and McNeil (2005)). *Let $(Y_1, Y_2) \sim F_{12} \in \mathcal{F}(F_1, F_2)$, where F_{12} has first order partial derivatives, and the univariate densities are f_1, f_2. Suppose $F_1 = F_2$. Then the tail dependence parameters are*

$$\lambda_L = \lim_{y \to -\infty} \{F_{2|1}(y|y) + F_{1|2}(y|y)\},$$

$$\lambda_U = \lim_{y \to \infty} \{\overline{F}_{2|1}(y|y) + \overline{F}_{1|2}(y|y)\}.$$

Proof: We prove only one of these as the technique of l'Hopital's rule is the same for both tails. Because the two univariate margins are the same, let $F = F_1 = F_2$ and $f = f_1 = f_2$. By the definition of λ_L,

$$\lambda_L = \lim_{y \to -\infty} \frac{\mathbb{P}(Y_1 \le y, Y_2 \le y)}{\mathbb{P}(Y_1 \le y)} = \lim_{y \to -\infty} \frac{F_{12}(y, y)}{F(y)}$$

$$= \lim_{y \to -\infty} \frac{(D_1 F_{12})(y, y) + (D_2 F_{12})(y, y)}{f(y)}$$

$$- \lim_{y \to -\infty} \frac{(D_1 F_{12})(y, y)}{f(y)} + \lim_{y \to -\infty} \frac{(D_2 F_{12})(y, y)}{f(y)}$$

$$= \lim_{y \to -\infty} \mathbb{P}(Y_2 \le y | Y_1 = y) + \lim_{y \to -\infty} \mathbb{P}(Y_1 \le y | Y_2 = y).$$

$\hspace{1cm}$ $\square$

Next, some results in Joe et al. (2010) are summarized. These additional results to those in Section 2.18 are useful to compute tail dependence functions and parameters for vine and factor copulas; see Sections 8.12 and 8.13.

Let C be a d-variate copula with continuous second order partial derivatives. Let S_1, S_2 be non-empty, non-overlapping subsets of $\{1, \ldots, d\}$, define the *lower and upper conditional tail dependence functions* as

$$b_{S_1|S_2}(\boldsymbol{w}_{S_1} \mid \boldsymbol{w}_{S_2}) = \lim_{u \to 0^+} C_{S_1|S_2}(u\boldsymbol{w}_{S_1} \mid u\boldsymbol{w}_{S_2}) \hspace{2cm} (8.19)$$

$$b^*_{S_1|S_2}(\boldsymbol{w}_{S_1} \mid \boldsymbol{w}_{S_2}) = \lim_{u \to 0^+} C_{S_1|S_2}(1 - u\boldsymbol{w}_{S_1} \mid 1 - u\boldsymbol{w}_{S_2})$$

Note that $b_{S_1|S_2}(\cdot \mid \boldsymbol{w}_{S_2})$ and $b^*_{S_1|S_2}(\cdot \mid \boldsymbol{w}_{S_2})$ are sub-distribution functions with possible mass at ∞. Let b and b^* be respectively the lower and upper tail dependence function of C. Then b, b^* are differentiable almost surely and homogeneous of order 1, and by Euler's formula on homogeneous functions, for $\boldsymbol{w} \in \mathbb{R}^d_+$,

$$b(\boldsymbol{w}) = \sum_{j=1}^d w_j \frac{\partial b(\boldsymbol{w})}{\partial w_j}, \quad b^*(\boldsymbol{w}) = \sum_{j=1}^d w_j \frac{\partial b^*(\boldsymbol{w})}{\partial w_j},$$

where the partial derivatives $\partial b / \partial w_j$ and $\partial b^* / \partial w_j$ are homogeneous of order zero and bounded.

Theorem 8.58 (Margins and derivatives of tail dependence functions; Joe et al. (2010)). *Let C be a d-variate copula with continuous partial derivatives to order d. Let $I_j = \{1, \ldots, d\} \backslash \{j\}$ for $j = 1, \ldots, d$. Let S be a non-empty subset of $\{1, \ldots, d\}$. Below are results for lower tail dependence functions, and there are parallel results for upper tail dependence functions.*

(a) *Let S_1, S_2 be non-empty, non-overlapping subsets of $\{1, \ldots, d\}$. If $C_{S_1 \cup S_2}$ has multivariate lower tail dependence, then for $b_{S_1|S_2}$ in (8.19) is not the zero function, and*

$$b_{S_1|S_2}(\boldsymbol{w}_{S_1}|\boldsymbol{w}_{S_2}) = \frac{\partial^{|S_2|} b_{S_1 \cup S_2}(\boldsymbol{w}_{S_1}, \boldsymbol{w}_{S_2})}{\prod_{j \in S_2} \partial w_j} \Big/ \frac{\partial^{|S_2|} b_{S_2}(\boldsymbol{w}_{S_2})}{\prod_{j \in S_2} \partial w_j}. \tag{8.20}$$

As a special case,

$$b_{I_j|\{j\}}(w_i, i \in I_j | w_j) = \frac{\partial b(\boldsymbol{w})}{\partial w_j}.$$

(b) *As a special case of (b), suppose $|S| = m \geq 2$ and let $S_{-k} = S \backslash \{k\}$ for $k \in S$. If C_S has multivariate lower tail dependence, then for any $k \in S$, $b_{k|S_{-k}}$ in (8.19) is not the zero function, and*

$$b_{k|S_{-k}}(w_k|\boldsymbol{w}_{S_{-k}}) = \frac{\partial^{m-1} b_S}{\prod_{j \in S_{-k}} \partial w_j} \Big/ \frac{\partial^{m-1} b_{S_{-k}}}{\prod_{j \in S_{-k}} \partial w_j}.$$

(c) *Suppose C has a non-zero tail lower dependence function. Then $\frac{\partial b}{\partial w_j} = b_{I_j|\{j\}}(\boldsymbol{w}_{I_j}|w_j)$ is a bounded function that is homogeneous of order zero, increasing in w_i ($i \in I_j$) and decreasing in w_j;*

$$b(\boldsymbol{w}) = \int_0^{w_j} b_{I_j|\{j\}}(\boldsymbol{w}_{I_j} \mid w) \, \mathrm{d}w;$$

$$b_S(\boldsymbol{w}_S) \geq \lim_{w_i \to \infty, i \notin S} b(\boldsymbol{w}), \quad \text{for } S \text{ with } 1 \leq |S| < d;$$

if $b_{I_j|\{j\}}(\boldsymbol{w}_{I_j}|w_j)$ are proper distribution functions, then

$$b_S(\boldsymbol{w}_S) = \lim_{w_i \to \infty, i \notin S} b(\boldsymbol{w}), \quad 1 \leq |S| < d.$$

Proof: (a) With the existence and continuity of all relevant derivatives,

$$b_{S_1|S_2}(\boldsymbol{w}_{S_1}|\boldsymbol{w}_{S_2}) = \lim_{u \to 0^+} C_{S_1|S_2}(u\boldsymbol{w}_{S_1} \mid u\boldsymbol{w}_{S_2}) = \lim_{u \to 0^+} \frac{\partial^{|S_2|} C(u\boldsymbol{w}_{S_1}, u\boldsymbol{w}_{S_2}) / \prod_{j \in S_2} \partial u_j}{\partial^{|S_2|} C_{S_2}(u\boldsymbol{w}_{S_2}) / \prod_{j \in S_2} \partial u_j}$$

$$= \frac{\partial^{|S_2|} b(\boldsymbol{w}_{S_1}, \boldsymbol{w}_{S_2}) / \prod_{j \in S_2} \partial w_j}{\partial^{|S_2|} b_{S_2}(\boldsymbol{w}_{S_2}) / \prod_{j \in S_2} \partial w_j}.$$

For the special case where $S_2 = \{j\}$ has cardinality of 1, then $b_{\{j\}}(w_j) = w_j$ and the denominator of (8.20) is 1.

(c) Since $b_{I_j|\{j\}}(w_i, i \in I_j \mid w_j)$ is a limiting probability, and it must be bounded and increasing in $w_i, i \in I_j$. Since $b_{I_j|\{j\}}(w_i, i \in I_j \mid w_j) = \partial b / \partial w_j$ is homogeneous of order zero, we have from Euler's formula on homogeneous functions that $b = \sum_{k=1}^d w_k b_{I_k|\{k\}}$. Differentiate with respect to w_j to get

$$0 = w_j \frac{\partial b_{I_j|\{j\}}}{\partial w_j} + \sum_{k \neq j} w_k \frac{\partial b_{I_k|\{k\}}}{\partial w_j}.$$

Since $\partial b_{I_k|\{k\}} / \partial w_j \geq 0$ for any $k \neq j$, then $\partial b_{I_j|\{j\}} / \partial w_j \leq 0$ and $b_{I_j|\{j\}}(w_i, i \in I_j \mid w_j)$ is decreasing in w_j.

The identity $b(\boldsymbol{w}) = \int_0^{w_j} b_{I_j|\{j\}}(\boldsymbol{w}_{I_j} \mid w)\, dw$ is just an antiderivative result.
Next, consider

$$b_{I_j}(w_i, i \in I_j) = \lim_{u \to 0+} u^{-1} \int_0^1 C_{I_j|\{j\}}(uw_i, i \in I_j|v)\, dv = \lim_{u \to 0+} \int_0^{1/u} C_{I_j|\{j\}}(uw_i, i \in I_j|uw)\, dw.$$

For fixed $w_i, i \in I_j$, let $g_u(w_j) = C_{I_j|\{j\}}(uw_i, i \in I_j \mid uw_j) I_{[0,1/u]}(w_j) \ge 0$, where $I_{[0,1/u]}(w_j)$ is the indicator function of $[0, 1/u]$. Since $b_{I_j|\{j\}}(w_i, i \in I_j \mid w_j) = \lim_{u \to 0+} g_u(w_j)$ pointwise, then Fatou's lemma implies that

$$\lim_{w_j \to \infty} b(\boldsymbol{w}) = \int_0^\infty b_{I_j|\{j\}}(w_i, i \in I_j \mid w)\, dw$$

$$= \int_0^\infty \lim_{u \to 0+} g_u(w)\, dw \le \lim_{u \to 0+} \int_0^\infty g_u(w)\, dw = b_{I_j}(w_i, i \in I_j). \quad (8.21)$$

For any $\emptyset \ne S \subset I$, $b_S(w_j, j \in S) \ge \lim_{w_i \to \infty, i \notin S} b(\boldsymbol{w})$ follows from the repeated use of (8.21).

If $b_{I_j|\{j\}}(\cdot \mid w_j)$ is a proper distribution function, we have for any subset S with $j \in S$,

$$\lim_{w_i \to \infty, i \notin S} \frac{\partial b}{\partial w_j} = \lim_{w_i \to \infty, i \notin S} b_{I_j|\{j\}}(w_i, i \in I_j \mid w_j)$$

$$= \lim_{u \to 0+} \mathbb{P}\{U_i \le uw_i, \forall i \in S \backslash \{j\} \mid U_j = uw_j\} = \frac{\partial b_S}{\partial w_j}.$$

The assumption of the proper distribution function is used in the second equality in the above equation. The bounded convergence theorem implies that $\lim_{w_i \to \infty, i \notin S} b(\boldsymbol{w}) = \int_0^{w_j} \lim_{w_i \to \infty, i \notin S} \frac{\partial b}{\partial v_j}\, dv_j = \int_0^{w_j} \frac{\partial b_S}{\partial v_j}\, dv_j = b_S(w_i, i \in S)$. □

Example 8.6 (Conditional tail dependence functions). We show two examples where the conditional tail dependence functions are not proper so that one cannot get marginal tail dependence functions by letting some arguments increase to ∞.

(1) This distribution comes from the geometric mixture of $C^\perp$ and the bivariate Galambos copula. This is an extreme value copula $C(u_1, u_2) = \exp\{-A(-\log u_1, -\log u_2)\}$, where $A(w_1, w_2) = w_1 + w_2 - \gamma(w_1^{-\delta} + w_2^{-\delta})^{-1/\delta}$, $\delta > 0$, $0 < \gamma < 1$. From Theorem 8.56, the upper tail dependence function is $b^*(w_1, w_2) = w_1 + w_2 - A(w_1, w_2) = \gamma(w_1^{-\delta} + w_2^{-\delta})^{-1/\delta}$ and $b^*(1, \infty) = \gamma < 1$. However the univariate marginal tail dependence functions are $b_{\{j\}}^*(w) = w$ for $j = 1, 2$. The marginal limiting result in part (c) of Theorem 8.58 does not hold because $b_{1|2}(w|z) = b_{2|1}(w|z) = \gamma(1 + [w/z]^{-\delta})^{-1/\delta - 1}$ is not a proper distribution function with $b_{1|2}(\infty|z) = \gamma$.

(2) By Theorem 8.56 and the reflection symmetry of the bivariate t distribution, the lower/upper tail dependence function of the bivariate t_ν copula is the same as the upper tail dependence function of its t-EV extreme value limit. Let T_η, t_η be the univariate t cdf and pdf with parameter $\eta > 0$. From Section 4.16, the exponent of bivariate t-EV is:

$$A(w_1, w_2) = w_1 T_{\nu+1}\left(\frac{\sqrt{\nu+1}}{\sqrt{1-\rho^2}}\left[\left(\frac{w_1}{w_2}\right)^{1/\nu} - \rho\right]\right) + w_2 T_{\nu+1}\left(\frac{\sqrt{\nu+1}}{\sqrt{1-\rho^2}}\left[\left(\frac{w_2}{w_1}\right)^{1/\nu} - \rho\right]\right)$$

and the tail dependence function of the bivariate t_ν copula

$$b(w_1, w_2) = w_1 + w_2 - A(w_1, w_2)$$

$$= w_1 T_{\nu+1}\left(-\frac{\sqrt{\nu+1}}{\sqrt{1-\rho^2}}\left[\left(\frac{w_1}{w_2}\right)^{1/\nu} - \rho\right]\right) + w_2 T_{\nu+1}\left(-\frac{\sqrt{\nu+1}}{\sqrt{1-\rho^2}}\left[\left(\frac{w_2}{w_1}\right)^{1/\nu} - \rho\right]\right).$$

Note that $b(1, \infty) = T_{\nu+1}(\rho[(\nu+1)/(1-\rho^2)]^{1/2}) < 1$ and

$$b_{1|2}(w|z) = b_{2|1}(w|z) = T_{\nu+1}\left(-\frac{\sqrt{\nu+1}}{\sqrt{1-\rho^2}}\left[\left(\frac{z}{w}\right)^{1/\nu} - \rho\right]\right)$$

$$-zt_{\nu+1}\left(-\frac{\sqrt{\nu+1}}{\sqrt{1-\rho^2}}\left[\left(\frac{z}{w}\right)^{1/\nu} - \rho\right]\right) \cdot \nu^{-1}\frac{\sqrt{\nu+1}}{\sqrt{1-\rho^2}} z^{1/\nu-1}w^{-1/\nu}$$

$$+wt_{\nu+1}\left(-\frac{\sqrt{\nu+1}}{\sqrt{1-\rho^2}}\left[\left(\frac{w}{z}\right)^{1/\nu} - \rho\right]\right) \cdot \nu^{-1}\frac{\sqrt{\nu+1}}{\sqrt{1-\rho^2}} z^{-1/\nu-1}w^{1/\nu},$$

with $b_{1|2}(\infty|z) = T_{\nu+1}(\rho[(\nu+1)/(1-\rho^2)]^{1/2}) < 1$, so that $b_{1|2}(\cdot|z)$ is not a proper distribution function.

8.10   Tail order

The tail orders can be used to assess strength of dependence in the tails of a copula as well as tail asymmetry, so would be considered an important property to obtain for any new copula family. This section has results on alternative ways to determine the lower or upper tail order of a copula, or a bound on the tail order. Only the details for the lower tail are shown as the results for the upper tail can then be obtained from the reflected copula.

One approach is through the limit of a ratio of logarithms, as shown in Hua and Joe (2013). Suppose $C(u\mathbf{1}_d) \sim u^\kappa \ell(u)$ as $u \to 0^+$. Then $\kappa_L = \kappa = \lim_{u \to 0^+} \frac{\log C(u\mathbf{1}_d)}{\log u}$. In the case of a multivariate cdf $F_{1:d}$ with univariate margins $F_1 = \cdots = F_d$, then the tail equivalence becomes $F_{1:d}(y\mathbf{1}_d) \sim [F_1(y)]^\kappa \ell(F_1(y))$ as $y \to F_1^{-1}(0)$, and

$$\kappa_L = \kappa = \lim_{y \to F_1^{-1}(0)} \frac{\log F_{1:d}(y\mathbf{1}_d)}{\log F_1(y)} \tag{8.22}$$

The proof combines l'Hopital's rule and the Monotone Density Theorem (Theorem 8.90):

$$\lim_{u \to 0^+} \frac{\log C(u\mathbf{1}_d)}{\log u} = \lim_{u \to 0^+} \frac{\frac{\mathrm{d}}{\mathrm{d}u}C(u\mathbf{1}_d)}{[u^\kappa \ell(u)]'} \times \frac{u[u^\kappa \ell(u)]'}{C(u\mathbf{1}_d)} = \lim_{u \to 0^+} \frac{\kappa u^\kappa \ell(u)}{C(u\mathbf{1}_d)} = \kappa.$$

The next result computes the tail order with conditional distributions in a similar way to Theorem 8.57 for computing tail dependence with conditional distributions.

Theorem 8.59 (Tail order from conditional distributions). *Let* $(Y_1, Y_2) \sim F_{12} \in \mathcal{F}(F_1, F_2)$, *where* F_{12} *has first order partial derivatives, and the univariate densities are* f_1, f_2, *and* $F_1 = F_2$. *Suppose* $F_{12}(y, y) \sim [F_1(y)]^\kappa \ell(F_1(y))$ *as* $y \to F_1^{-1}(0)$ *and* $\ell(u)$ *is slowly varying as* $u \to 0^+$. *Then the lower tail order* κ_L *can be obtained from:*

$$\{F_{2|1}(y|y) + F_{1|2}(y|y)\} \sim \kappa_L[F_1(y)]^{\kappa_L - 1}\ell(F_1(y)), \quad y \to F_1^{-1}(0).$$

Similarly, the upper tail order κ_U *can be obtained from:*

$$\{\overline{F}_{2|1}(y|y) + \overline{F}_{1|2}(y|y)\} \sim \kappa_U[\overline{F}_1(y)]^{\kappa_U - 1}\ell(\overline{F}_1(y)), \quad y \to F_1^{-1}(1).$$

Proof: [2] By l'Hopital's rule and (8.22),

$$\kappa_L = \lim_{y \to F_1^{-1}(0)} \frac{\log F_{12}(y,y)}{\log F_1(y)} = \lim_{y \to F_1^{-1}(0)} \frac{\{F_{2|1}(y|y) + F_{1|2}(y|y)\} f_1(y)/F_{12}(y,y)}{f_1(y)/F_1(y)}$$

$$= \lim_{y \to F_1^{-1}(0)} \frac{[F_1(y)]^{\kappa_L} \ell(F_1(y))}{F_{12}(y,y)} \cdot \frac{\{F_{2|1}(y|y) + F_{1|2}(y|y)\}}{[F_1(y)]^{\kappa_L - 1} \ell(F_1(y))}$$

$$= \lim_{y \to F_1^{-1}(0)} \frac{\{F_{2|1}(y|y) + F_{1|2}(y|y)\}}{[F_1(y)]^{\kappa_L - 1} \ell(F_1(y))}.$$

$\square$

Theorem 8.60 (Bounds of tail order). *Let C be a d-variate copula with constants $1 \le \kappa_1 \le \kappa_2$ such that*

$$\liminf_{u \to 0^+} C(u \mathbf{1}_d)/u^{\kappa_1} = 0, \tag{8.23}$$

$$\liminf_{u \to 0^+} u^{\kappa_2}/C(u \mathbf{1}_d) = 0. \tag{8.24}$$

If $\frac{C(u\mathbf{1}_d)}{u^\kappa \ell(u)} \to \zeta > 0$ as $u \to 0^+$ for a slowly varying function, then $\kappa_1 \le \kappa \le \kappa_2$ and $u^{\kappa_2} \le u^\kappa \ell(u) \le u^{\kappa_1}$.

Proof: The inequality in (8.23) implies that

$$0 < \lim_{u \to 0^+} \frac{C(u\mathbf{1}_d)}{u^\kappa \ell(u)} \le \liminf_{u \to 0^+} \frac{C(u\mathbf{1}_d)}{u^{\kappa_1}} \frac{u^{\kappa_1}}{u^\kappa \ell(u)}$$

so that $\frac{u^{\kappa_1}}{u^\kappa \ell(u)} \to \infty$ and $\kappa \ge \kappa_1$. The inequality in (8.24) implies that

$$\lim_{u \to 0^+} \frac{C(u\mathbf{1}_d)}{u^\kappa \ell(u)} \ge \limsup_{u \to 0^+} \frac{C(u\mathbf{1}_d)}{u^{\kappa_2}} \frac{u^{\kappa_2}}{u^\kappa \ell(u)}$$

so that $\frac{u^{\kappa_2}}{u^\kappa \ell(u)} \to 0$ and $\kappa \le \kappa_2$. $\square$

Example 8.7 (Tail order of gamma convolution model). We consider only the bivariate exchangeable as more details are needed to get the slowly varying function in the non-exchangeable case.

Let $X_1 = Z_0 + Z_1$, $X_2 = Z_0 + Z_2$, where Z_0, Z_1, Z_2 are independent Gamma random variables with respective shape parameters $\theta_0, \theta_1, \theta_2$ and scale parameters all equal to 1. Then X_1, X_2 are dependent Gamma random variables with shape parameter $\eta_1 = \theta_0 + \theta_1$ and $\eta_2 = \theta_0 + \theta_2$ respectively. Given $X_j = x$ ($j = 1, 2$), then $Z_0 = xB$, where $B \sim \text{Beta}(\theta_0, \theta_j)$ independently of Z_{3-j}.

As an aside, we note that copula of this density is unusual if $\theta_1 + \theta_2 \le 1$ because the density f_{X_1, X_2} is infinite along the line $x_1 = x_2$ which means the copula density is infinite along a curve; see Section 4.28.

For the exchangeable case with $\theta_1 = \theta_2$ and $\eta_1 = \eta_2$, Theorem 8.59 is used to get the tail orders, in combination with tail expansions (Abramowitz and Stegun (1964)) of the cdf of gamma random variables.

---

[2]Thanks to Lei Hua for this improved proof.

- (Lower) As $x \to 0^+$, with $1 - B \sim \text{Beta}(\theta_1, \theta_0)$,

$$F_{1|2}(x|x) = \mathbb{P}(X_1 \le x | X_2 = x) = \mathbb{P}(xB + Z_1 \le x) = \mathbb{P}(Z_1 \le (1 - B)x)$$

$$= \int_0^1 \mathbb{P}(Z_1 \le bx)[B(\theta_1, \theta_0)]^{-1} b^{\theta_1 - 1}(1 - b)^{\theta_0 - 1} db$$

$$\sim \int_0^1 \frac{(bx)^{\theta_1}}{\Gamma(\theta_1 + 1)} \frac{b^{\theta_1 - 1}(1 - b)^{\theta_0 - 1}}{B(\theta_1, \theta_0)} db = O(x^{\theta_1}).$$

Also $F_1(x) \sim x^{\theta_1 + \theta_0}/\Gamma(\theta_1 + \theta_0 + 1)$ as $x \to 0^+$. Hence $\kappa = \kappa_L$ satisfies $\theta_1 = (\kappa_L - 1)(\theta_1 + \theta_0)$ or $\kappa_L = 1 + \theta_1/(\theta_1 + \theta_0)$.

- (Upper)

$$\lim_{x \to \infty} \mathbb{P}(X_1 > x | X_2 = x) = \lim_{x \to \infty} \mathbb{P}(xB + Z_1 > x) = \lim_{x \to \infty} \mathbb{P}(Z_1/(1 - B) > x) = 0,$$

so that from Theorem 8.57, the upper tail dependence parameter is $\lambda_U = 0$. For the upper tail order, as $x \to \infty$, write

$$\mathbb{P}(Z_1 > (1 - B)x) = [B(\theta_1, \theta_0]^{-1} \int_0^1 \mathbb{P}(Z_1 > bx) b^{\theta_1 - 1}(1 - b)^{\theta_0 - 1} db$$

$$\sim B(\theta_1, \theta_0]^{-1} \int_0^1 [\Gamma(\theta_1)]^{-1} [bx]^{\theta_1 - 1} e^{-bx} \cdot b^{\theta_1 - 1}(1 - b)^{\theta_0 - 1} db = O(x^{-\theta_1}).$$

Also $\overline{F}_1(x) \sim x^{\theta_1 + \theta_0 - 1} e^{-x}/\Gamma(\theta_1 + \theta_0)$ as $x \to \infty$. With $\kappa = \kappa_U$ and $\ell(u)$ slowly varying at 0, matching

$$\mathbb{P}(X_1 > x | X_2 = x) = \mathbb{P}(Z_1 > (1 - B)x) \sim \kappa_U [\overline{F}_1(x)]^{\kappa_U - 1} \ell(\overline{F}_1(x)), \quad x \to \infty,$$

leads to $\kappa_U = 1$ and $\ell(u) = (-\log u)^{-\theta_1}$.

$\square$

8.11   Combinatorics of vines

This section has results on the number of vines in d variables and the number of equivalence classes of d-dimensional vines, and also some background details on the binary representation for vine arrays in the natural order. The binary representation is used in algorithms in Section 6.13.

The property of a d-variable vine array $A = (a_{\ell j})$ in natural order is that A has $a_{ii} = i$ for $i = 1, \ldots, d$ and also $a_{i-1,i} = i - 1$ for $i = 2, \ldots, d$. The set of vine arrays in natural order is denoted as $\text{NO}(d)$. In particular, for the edge in last tree $\mathcal{T}_{d-1}$, one conditioned variable is denoted with index d and the other conditioned variable is denoted with index $d - 1$.

Lemma 8.61 (Extension a vine array by one column; Morales-Nápoles (2011)). *Let $j \ge 4$. Consider a vine in $j-1$ variables with vine array $A^* \in \text{NO}(j-1)$. The number of extensions of A^* to $A \in \text{NO}(j)$ is 2^{j-3} by keeping the first $j - 1$ columns and adding a jth column with elements $a_{1j} = i_1, a_{2j} = i_2, \ldots, a_{j-2,j} = i_{j-2}, a_{j-1,j} = j - 1, a_{jj} = j$.*

Proof: We outline a proof that is different from the original proof. First note that $(i_1, \ldots, i_{j-2})$ is a permutation of $(1, \ldots, j - 2)$ so the number of extensions is bounded by $(j - 2)!$ but most of these are not possible because of the definition of a vine. The enumeration depends on the following properties P1 and P2.

P1. If the added edges in the jth column of A are $[i_1 j], [i_2 j | i_1], \ldots, [i_{j-2} j | i_1, \ldots, i_{j-3}]$ and $[j-1, j | 1 : (j-2)]$, then $\{i_1, i_2\}, \{i_1, i_2, i_3\}, \ldots, \{i_1, \ldots, i_{j-2}\}$ must be constraint sets from the first $j-2$ columns of A^*. A constraint set, as defined in Kurowicka and Cooke (2006), is the union of the two conditioned variables (appearing before | in the edges) and the conditioning variables (appearing after | in the edges).

P2. If $[k_1, k_2 | S]$ is a node in a tree $\mathcal{T}_\ell$ with $2 \le \ell \le j-2$ and $|S| = \ell - 1$, then there is a sub-vine with indices in $\{k_1, k_2\} \cup S$ leading to $[k_1, k_2 | S]$ as the edge in the final tree.

The proof of the enumeration comes from showing sequentially that there are two choices of indices for $j-3$ positions $a_{j-2,j}, a_{j-3,j}, \ldots, a_{2j}$.

- Suppose $j \ge 4$. Consider position $a_{j-2,j}$. The edge represented in this position is $[i_{j-2} j | i_1, \ldots, i_{j-3}]$. One node of this edge is $[i_{j-3} j | i_1, \ldots, i_{j-4}]$ with constraint set $\{i_1, \ldots, i_{j-3}, j\}$ and the other edge must have constraint set $\{i_1, \ldots, i_{j-2}\} = \{1, \ldots, j-2\}$ so that the symmetric difference of the two constraint sets is $\{i_{j-2}, j\}$. The set $\{1, \ldots, j-2\}$ must be the constraint set of an edge based on two nodes n_1, n_2 with constraint sets of cardinality $j-3$, but there are only two nodes in variables $1, \ldots, j-2$ with cardinality $j-3$. The constraint sets of these two nodes have a symmetric difference of two elements and these are the two choices for i_{j-2}. Because $A^* \in \mathrm{NO}(j-1)$, one choice is $i_{j-2} = j-2$ since $[j-2, j-1 | 1 \ldots, j-3]$ is represented by A^*.
- Suppose $j \ge 5$ and consider position $a_{j-3,j}$ with i_{j-2} fixed. The edge represented in this position is $[i_{j-3} j | i_1, \ldots, i_{j-4}]$. One node of this edge is $[i_{j-4} j | i_1, \ldots, i_{j-5}]$ with constraint set $\{i_1, \ldots, i_{j-4}, j\}$ and the other edge must have constraint set $\{i_1, \ldots, i_{j-3}\}$. The set $\{i_1, \ldots, i_{j-3}\}$ must be the constraint set of an edge and by Properties P1 and P2, there is a sub-vine with the indices in this set. In this sub-vine, there are exactly two nodes n_1, n_2 with constraint sets of cardinality $j-4$. The constraint sets of these two nodes have a symmetric difference of two elements and these are the two choices for i_{j-3}.
- For $j \ge 6$, continue in this way to position $a_{3,j}$ with $i_4, \ldots, i_{j-2}$ fixed, Let α, β, γ be the three remaining indices for the first three values i_1, i_2, i_3 in column j. By Property P2, there is a sub-vine with α, β, γ. If $i_3 = \gamma$, then by Property P1, $[\alpha, \beta]$ must be an edge represented by A^*. If $i_3 = \beta$, then $[\alpha, \gamma]$ must be an edge represented by A^*. If $i_3 = \alpha$, then $[\beta, \gamma]$ must be an edge represented by A^*. Exactly two of the edges out of $[\alpha, \beta]$, $[\alpha, \gamma]$, $[\beta, \gamma]$ exist in the sub-vine with α, β, γ. Hence there are two choices for i_3.
- This leaves two indices α', β' for the first two positions, and any of these can be assigned to i_2.

Hence, sequentially there are two choices for each of $i_{j-2}, \ldots, i_2$ for the extension of A^* to a vine array $A \in \mathrm{NO}(j)$, leading to a total of 2^{j-3} extensions. $\qquad \square$

Theorem 8.62 (Number of vine arrays in natural order; Morales-Nápoles (2011)). *The cardinality of* $\mathrm{NO}(d)$ *is* $2^{(d-3)(d-2)/2}$ *for* $d \ge 3$.

Proof: With the conditions imposed for vine arrays in natural order, there is one member of $\mathrm{NO}(3)$ with third column $1, 2, 3$. By Lemma 8.61, there are $2^{4-3} = 2$ extensions to $\mathrm{NO}(4)$ with $d = 4$. By extending one column at the time, the cardinality of $\mathrm{NO}(d)$ is $2^{1+\cdots+(d-3)}$ for $d \ge 4$. $\qquad \square$

Example 8.8 Consider the following vine array in $\mathrm{NO}(6)$.

$$
\begin{bmatrix}
- & 12 & 13 & 24 & 25 & 26 \\
  & - & 23|1 & 14|2 & 15|2 & 16|2 \\
  &   & - & 34|12 & 35|12 & 46|12 \\
  &   &   & - & 45|123 & 36|124 \\
  &   &   &   & - & 56|1234 \\
  &   &   &   &   & -
\end{bmatrix}
,
\begin{bmatrix}
1 & 1 & 1 & 2 & 2 & 2 \\
  & 2 & 2 & 1 & 1 & 1 \\
  &   & 3 & 3 & 3 & 4 \\
  &   &   & 4 & 4 & 3 \\
  &   &   &   & 5 & 5 \\
  &   &   &   &   & 6
\end{bmatrix}
= A.
$$

- Temporarily ignore the first four elements of column 6 and consider possibilities $1, 2, 3, 4$ for a_{46}. If $a_{46} = 4$, then the constraint set $\{1, 2, 3\}$ must exist. If $a_{46} = 3$, then the constraint set $\{1, 2, 4\}$ must exist. If $a_{46} = 2$, then the constraint set $\{1, 3, 4\}$ must exist. If $a_{46} = 1$, then the constraint set $\{2, 3, 4\}$ must exist. In checking the second row of the first four columns, $[23|1]$ leads to constraint set $\{1, 2, 3\}$ and $[14|2]$ leads to constraint set $\{1, 2, 4\}$. The constraint sets $\{1, 3, 4\}$ and $\{2, 3, 4\}$ are not available. Hence the choices for a_{46} are 3 and 4.
- Suppose we take $a_{46} = 3$. Temporarily ignore the first three elements of column 6 and consider possibilities $1, 2, 4$ for a_{36}. If $a_{36} = 4$, then the constraint set $\{1, 2\}$ must exist. If $a_{36} = 2$, then the constraint set $\{1, 4\}$ must exist. If $a_{36} = 1$, then the constraint set $\{2, 4\}$ must exist. In check the first row of the first four columns, $[12]$ and $[24]$ exist, but not $[14]$. Hence the choices for a_{36} are 1 and 4.
- Suppose we take $a_{36} = 4$. Then $a_{26} = 2$ and $a_{16} = 1$ lead to the edge $[26|1]$ which is acceptable because the edge $[12]$ exists. Similarly, $a_{26} = 1$ and $a_{16} = 2$ lead to the edge $[16|2]$ which is also acceptable.

Remark 8.5 Algorithm 13 in Chapter 6 for the binary representation of $A \in \mathrm{NO}(d)$ follows the order in Lemma 8.61, that is, decreasing positions $k = j - 2, \ldots, 2$ in column j. Observe that for the two choices in the proof of Lemma 8.61, (i) one is the largest index (denoted as ac for active column) that remains (ac starts as $j - 2$ when $k = j - 2$ for column j because of the natural order); (ii) the other is the index in row $k - 1$ of column ac.

For row k of column j with ac $= m$, the algorithm selects a value i_k equal to m or $A_{k-1,m}$ and this leads to the edge $[i_k, j|S]$, where $S = \{A_{1,m}, \ldots, A_{k-1,m}, m\} \backslash i_k$, so that the corresponding constraint set has cardinality $k + 1$. The edge is formed from $[A_{k-1,m}, m | A_{1m}, \ldots, A_{k-2,m}]$ with constraint set $\{A_{1,m}, \ldots, A_{k-1,m}, m\}$ and a node with constraint set $S \cup \{j\}$. The symmetric difference of these two sets consists of i_k and j, whether $i_k = m$ or $i_k = A_{k-1,m}$.

Theorem 8.63 (Number of vines and equivalence classes).
(a) (Morales-Nápoles (2011)). *The number of vines in d variables is $2^{(d-3)(d-2)/2} \cdot d!/2$.*
(b) (Joe et al. (2011)). *The number of equivalence classes of vines in d variables is $E_d = (N_d + E_{1d})/2$, where*

$$E_{1d} = \sum_{k=1}^{[d/2]-1} N_d \, \ell_k \cdot 2^{-k} \cdot 2^{-\sum_{i=0}^{k-1}(d-4-2i)}, \quad N_d = 2^{(d-3)(d-2)/2},$$

and $\ell_k = 1$ for all k except $\ell_{[d/2]-1} = 2$.

Proof: The proof is mentioned here for (a) only. There are $2^{(d-3)(d-2)/2}$ vine arrays in natural order according to Theorem 8.62. There are $\binom{d}{2}$ ways to assign variables to the last two indices and $(d-2)!$ ways to assign the remaining variables to indices $1, \ldots, d-2$ to get a product of $d!/2$. $\square$

The number of vines and their equivalence classes increases quickly with d. Table 8.2 shows the count of equivalence classes and vines up to $d = 9$.

d	#vines	#equiv classes
4	24	2
5	480	6
6	23 040	40
7	2 580 480	560
8	660 602 880	17 024
9	380 507 258 880	1 066 496

Table 8.2: *Number of vines and their equivalence classes in dimension d based on Theorem 8.63.*

8.12 Vines and mixtures of conditional distributions

This section has results on mixtures of conditional distributions and the vine pair-copula construction. The first result is used in the likelihood of the vine for continuous variables.

Theorem 8.64 (Density of absolutely continuous vine copula; Bedford and Cooke (2001)). *Let $\mathcal{V} = (\mathcal{T}_1, \ldots, \mathcal{T}_{d-1})$ be a regular vine on d elements. For an edge $e \in E(\mathcal{V})$ with conditioned elements e_1, e_2 and conditioning set $S(e)$, let the bivariate copula and its density be $C_{e_1,e_2;S(e)}$ and $c_{e_1,e_2;S(e)}$, respectively. Let the resulting vine copula be the copula for $F \in \mathcal{F}(F_1, \ldots, F_d)$ where the univariate marginal distributions F_j has density f_j for $j = 1, \ldots, d$. Then the density of F is given by*

$$f_{1\cdots d} = f_1 \cdots f_d \prod_{e \in E(\mathcal{V})} c_{e_1,e_2;S(e)}(F_{e_1|S(e)}, F_{e_2|S(e)}),$$

where $F_{e_1|S(e)}$ and $F_{e_2|S(e)}$ are conditional distributions of the margin $F_{\{e_1,e_2\}\cup S(e)}$.

Proof: Some details of the proof are given in Section 3.9.3. □

The next result is concerned with the concordance ordering when a bivariate copula in the vine is increasing in dependence.

Theorem 8.65 (Concordance ordering for parametric vine copula; Joe (1996a)). *Consider a d-dimensional vine copula where $[jk|S]$ is an edge in one of the trees $\mathcal{T}_2, \ldots, \mathcal{T}_{d-1}$. As $C_{jk;S}$ increases in concordance, with other bivariate copulas held fixed, then $C_{\{j,k\}\cup S}$ increases in the $\prec_{cL}$ ordering and hence C_{jk} increases in concordance (and hence Kendall's τ, Spearman's ρ_S and the tail dependence parameters increase).*

Proof: Suppose $C_{jk;S}(\cdot; \theta)$ is increasing in θ. The result follows from (3.34) by writing

$$C_{\{j,k\}\cup S}(u_j, u_k, \boldsymbol{u}_S) = \int_{(\boldsymbol{0}, \boldsymbol{u}_S)} C_{jk;S}(C_{j|S}(u_j|\boldsymbol{v}_S), C_{k|S}(u_k|\boldsymbol{v}_S); \theta) \, d\boldsymbol{v}_S.$$

□

A wide range of dependence can result, by allowing the copulas $C_{jk;S}$ to range from the C^- to C^+. For the trivariate case, the bounds for the Kendall tau values $\tau_{12}, \tau_{13}, \tau_{23}$ in part (a) of Theorem 8.19 can be achieved when the conditional Fréchet bound copulas are used for T_2 of the vine.

Theorem 8.66 (Kendall's taus for 3-dimensional vine; Joe (1996a)). *Let C_{123} be the trivariate distribution of the vine copula based on $C_{12}, C_{23}, C_{13;2}$. Let τ_{jk} be the Kendall tau value for the (j, k) bivariate margin, $j < k$. If $C_{13;2} = C^+$ and $C_{12} \prec_{SI} C_{32}$ (that is, $C_{3|2}^{-1}(C_{1|2}(y_1|y_2)|y_2)$ is (strictly) increasing in y_2), then $\tau_{13} = 1 - |\tau_{12} - \tau_{23}|$. Similarly, if $C_{13;2} = C^-$ and $C_{3|2}^{-1}(1 - C_{1|2}(y_1|y_2)|y_2)$ is (strictly) increasing in y_2, then $\tau_{13} = -1 + |\tau_{12} + \tau_{23}|$.*

A sufficient condition for $C_{3|2}^{-1}(1 - C_{1|2}(y_1|y_2)|y_2)$ to be strictly increasing is that both $C_{1|2}(\cdot|y)$ and $C_{3|2}(\cdot|y)$ are SI. More generally, the condition is equivalent to the $\prec_{\mathrm{SI}}$ ordering on C_{12}^* and C_{32}, where $C_{12}^*(u,v) = v - C_{12}(1-u,v)$. Regarding the assumption $C_{12} \prec_{\mathrm{SI}} C_{32}$, the 1-parameter bivariate copula families in Sections 4.3 to 4.10 have been shown to satisfy the SI ordering as the dependence parameter increases. The proof of this property for the 1-parameter bivariate extreme value copula families makes use of Theorem 2.14 of Joe (1997), adapted to the form

$$\frac{\partial^2 h}{\partial y \partial x} \cdot \frac{\partial h}{\partial \delta} - \frac{\partial^2 h}{\partial \delta \partial x} \cdot \frac{\partial h}{\partial y} \geq 0,$$

where $A(x,y;\delta) = -\log C(e^{-x}, e^{-y}; \delta)$ and

$$h(x,y;\delta) = \log C_{2|1}(e^{-y}|e^{-x};\delta) = -A(x,y;\delta) + \log \frac{\partial A(x,y;\delta)}{\partial x} + x.$$

Next we indicate how the above theorems can be used to consider for measures of conditional monotone association.

Example 8.9 (Conditional monotone association). For a measure of conditional monotone association, some desirable properties might be that it increases in value with stronger conditional dependence in the sense of concordance, and having values of $-1, 0, 1$ respectively for the conditional Fréchet lower bound, conditional independence and the conditional Fréchet upper bound. Here, we discuss the partial tau, due to Kendall (1942) and presented in Kendall and Gibbons (1990), and its variations.

Suppose there are n observations of three variables leading to (y_{ih}, y_{ij}, y_{ik}) for $i = 1, \ldots, n$. Suppose there are no ties on any variable. Without loss of generality, suppose $y_{1h} < \cdots < y_{nh}$ have been indexed to be in increasing order. For variables j, k given variable h, produce the following two-way classification: among $\binom{n}{2}$ pairs indexed as $i < i'$, dichotomy 1 is $y_{ij} < y_{i'j}$ versus $y_{ij} > y_{i'j}$, and dichotomy 2 is $y_{ik} < y_{i'k}$ versus $y_{ik} > y_{i'k}$. This can be summarized into the following table.

	$y_{ik} < y_{i'k}$	$y_{ik} > y_{i'k}$	margin
$y_{ij} < y_{i'j}$	$n_{<<}$	$n_{<>}$	$\#\{y_{ij} < y_{i'j}\}$
$y_{ij} > y_{i'j}$	$n_{><}$	$n_{>>}$	$\#\{y_{ij} > y_{i'j}\}$
margin	$\#\{y_{ik} < y_{i'k}\}$	$\#\{y_{ik} > y_{i'k}\}$	$n(n-1)/2$

Kendall (1942) defines a partial tau as

$$\tau_{jk;h}^K = \frac{n_{<<}n_{>>} - n_{><}n_{<>}}{\sqrt{(n_{<<} + n_{<>})(n_{><} + n_{>>})(n_{<<} + n_{><})(n_{<>} + n_{<>})}}$$

and shows that this is equal to

$$\tau_{jk;h}^K = \frac{\tau_{jk} - \tau_{hj}\tau_{hk}}{\sqrt{(1 - \tau_{hj}^2)(1 - \tau_{hk}^2)}}, \qquad (8.25)$$

where $\tau_{hj}, \tau_{hk}, \tau_{jk}$ are the usual pairwise Kendall's taus. This is bounded between -1 and 1, but it is shown below in specific cases that the bounds are not attained for the conditional Fréchet bounds (extremes of $\mathcal{F}(C_{hj}, C_{hk})$ as defined in Section 8.4).

Let $\zeta_{ab} = \sin(\pi\tau_{ab}/2)$ for a, b being any two choices among h, j, k. Based on Theorems 8.66 and 8.19, a better definition which is a function of $\tau_{hj}, \tau_{hk}, \tau_{jk}$, might be

$$\tau_{jk;h}^{\mathrm{pc}} = \frac{2}{\pi} \arcsin\left(\frac{\zeta_{jk} - \zeta_{hj}\zeta_{hk}}{\sqrt{(1 - \zeta_{hj}^2)(1 - \zeta_{hk}^2)}}\right). \qquad (8.26)$$

This is based on the identities (achievements of bounds for triplets of Kendall taus in (8.3)):

$$\tau_{jk} = 1 - |\tau_{hj} - \tau_{hk}| \iff \zeta_{jk} = \zeta_{hj}\zeta_{hk} + \sqrt{(1 - \zeta_{hj}^2)(1 - \zeta_{hk}^2)}$$

$$\iff \frac{\zeta_{jk} - \zeta_{hj}\zeta_{hk}}{\sqrt{(1 - \zeta_{hj}^2)(1 - \zeta_{hk}^2)}} = 1,$$

$$\tau_{jk} = -1 + |\tau_{hj} + \tau_{hk}| \iff \zeta_{jk} = \zeta_{hj}\zeta_{hk} - \sqrt{(1 - \zeta_{hj}^2)(1 - \zeta_{hk}^2)}$$

$$\iff \frac{\zeta_{jk} - \zeta_{hj}\zeta_{hk}}{\sqrt{(1 - \zeta_{hj}^2)(1 - \zeta_{hk}^2)}} = -1.$$

Because ζ is the partial correlation for trivariate Gaussian distributions, then in this case, $\tau_{jk;h}^{\mathrm{pc}}$ is $-1, 0, 1$ for $\rho_{jk;h}$ equalling $-1, 0, 1$ respectively. With the definition in (8.26), the bounds of ± 1 can be achieved with conditional Fréchet lower and upper bounds provided the bivariate copulas C_{hj}, C_{hk} satisfy the conditions of Theorem 8.66 (when matching $h = 2, j = 1, k = 3$).

$C_{j|h}, C_{k|h}, C_{jk;h}$	$\tau_{hj} = 0.5, \tau_{hk} = 0.5$			$\tau_{hj} = 0.5, \tau_{hk} = 0.6$		
	τ_{jk}	$\tau_{jk;h}^{K}$	$\tau_{jk;h}^{\mathrm{pc}}$	τ_{jk}	$\tau_{jk;h}^{K}$	$\tau_{jk;h}^{\mathrm{pc}}$
$\prec_{\mathrm{SI}}$ ordered,CFUB	1.000	1.000	1.000	0.900	0.866	1.000
Gumbel,Frank,CFUB	0.916	0.888	0.881	0.860	0.809	0.849
Frank,Gumbel,CFUB	0.916	0.888	0.881	> 0.899	0.866	0.997
Gumbel,Gumbel,CI	0.328	0.104	-0.009	0.385	0.123	-0.005
Frank,Frank,CI	0.331	0.107	-0.005	0.386	0.124	-0.004
Gumbel,Frank,CI	0.325	0.100	-0.015	0.379	0.114	-0.017
Frank,Gumbel,CI	0.325	0.100	-0.015	0.379	0.114	-0.017
triv. Gaussian,CI	0.333	0.111	0.000	0.388	0.127	0.000
SI,SI,CFLB	0.000	-0.333	-1.000	0.100	-0.289	-1.000
$C_{j|h}, C_{k|h}, C_{jk;h}$	$\tau_{hj} = 0.5, \tau_{hk} = 0.7$			$\tau_{hj} = 0.2, \tau_{hk} = 0.3$		
	τ_{jk}	$\tau_{jk;h}^{K}$	$\tau_{jk;h}^{\mathrm{pc}}$	τ_{jk}	$\tau_{jk;h}^{K}$	$\tau_{jk;h}^{\mathrm{pc}}$
$\prec_{\mathrm{SI}}$ ordered,CFUB	0.800	0.728	1.000	0.900	0.899	1.000
Gumbel,Frank,CFUB	0.782	0.699	0.848	0.876	0.873	0.921
Frank,Gumbel,CFUB	0.800	0.728	1.000	0.900	0.899	1.000
Gumbel,Gumbel,CI	0.433	0.134	-0.003	0.091	0.033	0.002
Frank,Frank,CI	0.433	0.134	-0.003	0.087	0.029	-0.003
Gumbel,Frank,CI	0.426	0.123	-0.019	0.086	0.028	-0.004
Frank,Gumbel,CI	0.427	0.124	-0.017	0.086	0.028	-0.005
triv. Gaussian,CI	0.434	0.136	0.000	0.090	0.032	0.000
SI,SI,CFLB	0.200	-0.243	-1.000	-0.500	-0.599	-1.000

Table 8.3 *Values of two versions of partial tau under different scenarios for the conditional Fréchet upper bound (CFUB), conditional independence (CI) and the conditional Fréchet lower bound (CFLB); SI in column 1 refers to the "stochastically increasing" positive dependence condition or ordering. An "ideal" conditional dependence measure would take values $1, 0, -1$ respectively in these three cases. $\tau_{jk;h}^{\mathrm{pc}}$ can attain ± 1 under some conditions for conditional Fréchet bounds. $\tau_{jk;h}^{K}$ can attain 1 for the conditional Fréchet upper bound when $\tau_{hj} = \tau_{hk}$ and -1 for the conditional Fréchet lower bound when $\tau_{hj} = -\tau_{hk}$. The values of τ_{jk} were obtained via numerical integration for conditional independence, and via simulation with sample size 10^6 for the conditional Fréchet upper bound.*

Table 8.3 has some examples to show the behavior of $\tau_{jk;h}^{K}$ and $\tau_{jk;h}^{\mathrm{pc}}$ for different scenarios for the conditional Fréchet upper bound, conditional independence and the conditional Fréchet lower bound. $\tau_{jk;h}^{\mathrm{pc}}$ does not always yield $1, 0, -1$ for perfect conditional positive dependence, conditional independence and perfect conditional negative dependence; in cases where it does not achieve these values in Table 8.3, it is quite close. Note that $\tau_{jk;h}^K$ can be quite far from the desirable values in the three cases.

The value of $\tau_{jk;h}^K$ can be optimized under the assumptions of (i) $\tau_{jk} = 1 - |\tau_{hj} - \tau_{hk}|$ (conditional Fréchet upper bound) or (ii) $\tau_{jk} = -1 + |\tau_{hj} + \tau_{hk}|$ (conditional Fréchet lower bound). For (i), if $\tau_{hj} = \tau_{hk}$, then $\tau_{jk} = 1$ and $\tau_{jk;h}^K = 1$; if $\tau_{hj} = 0$, then $\tau_{jk} = 1 - |\tau_{hk}|$ and $\tau_{jk;h}^K = (1 - |\tau_{hk}|)/(1 - \tau_{hk}^2)^{1/2} \to 0$ as $\tau_{hk} \to 1$. For (ii), if $\tau_{hj} = -\tau_{hk}$, then $\tau_{jk} = -1$ and $\tau_{jk;h}^K = -1$; if $\tau_{hj} = 0$, then $\tau_{jk} = -1 + |\tau_{hk}|$ and $\tau_{jk;h}^K = (-1 + |\tau_{hk}|)/(1 - \tau_{hk}^2)^{1/2} \to 0$ as $\tau_{hk} \to 1$. Hence $\tau_{jk;h}^K$ can take the range of $[0, 1]$ and $[-1, 0]$ for the conditional Fréchet upper and lower bound respectively.

The value of $\tau_{jk;h}^K$ can also be optimized under the assumption of (iii) $\zeta_{jk} = \zeta_{hj}\zeta_{hk}$ for trivariate Gaussian with conditional independence of variables j, k given variable h. Under (iii), the range of $\tau_{jk;h}^K$ is $-(1 - 1/\sqrt{2})$ to $1 - 1/\sqrt{2} = 0.2929$. The bounds come from $\zeta = \sin(\pi\tau/2) = \zeta_{hj} = \pm\zeta_{hk}$ with $\tau \to 1^-$; the limits can be obtained with l'Hopital's rule.

As a variation of Theorem 8.65, we show that both versions of partial tau are increasing as the bivariate conditional distributions increase in the concordance ordering. Suppose the (h, j) and (h, k) bivariate margins are held fixed. In terms of copulas, write

$$C_{jk}(u_j, u_k) = \int_0^1 C_{jk;h}\big(C_{j|h}(u_j|u_h), C_{k|h}(u_k|u_h); u_h\big)\,\mathrm{d}u_h.$$

As the $C_{jk;h}(\cdot; u_h)$ increase in concordance for different u_h, then the bivariate margin $C_{jk}(u_j, u_k)$ increases for all u_h, u_k, that is, C_{jk} increases in concordance and hence τ_{jk} increases. Hence $\tau_{jk;h}^K$ and $\tau_{jk;h}^{\mathrm{pc}}$ also increase since τ_{hj} and τ_{hk} are fixed when their respective bivariate margins are fixed; that is, both versions of partial tau have the desirable property of being larger with stronger conditional dependence in the sense of concordance.

There are analogies of (8.25) with τ_{ab} replaced by Spearman's rho $\rho_{S,ab}$ or the correlation of normal scores $\rho_{N,ab}$, leading to $\rho_{S,jk;h}$ and $\rho_{N,jk;h}$. These will also have the property of being larger with more concordance in the copulas $C_{jk;h}$, and the behavior for the Fréchet lower and upper bounds is similar to that of (8.25) in Table 8.3. Note that $\rho_{N,jk;h}$ is the same as the partial correlation for the trivariate Gaussian copula. For $\rho_{S,jk;h}$, the upper bound of 1 can be achieved with the conditional Fréchet upper bound if $C_{k|h}^{-1}(C_{j|h}(v|u_h)|u_h)$ is linear in v for all u_h. Similarly, the lower bound of -1 can be achieved with the conditional Fréchet lower bound if $C_{k|h}^{-1}(1 - C_{j|h}(v|u_h)|u_h)$ is linear in v for all u_h. From the proof of part (d) of Theorem 8.19 in the trivariate case, these are possible from elliptical distributions with partial correlations $\rho_{jk;h} = \pm 1$ that are copulas arising from uniform distributions on the surface of 3-dimensional ellipsoids. The bounds of ± 1 for $\rho_{S,jk;h}$ can also be obtained when $C_{hj} = C_{hk}$ (so that $C_{jk} = C^+$) and $C_{hk}(u, v) = u - C_{hj}(u, 1 - v)$ respectively.

In summary, (8.26) is the best measure of conditional concordance in the sense of being able to attain the bounds of ± 1 under more cases of the conditional Fréchet lower and upper bounds. $\qquad\square$

The next result indicates that reflection symmetry in all pair-copulas imply reflection symmetry in the vine copula.

Theorem 8.67 (Reflection symmetry for vine copula; Joe (2011b)).
Let $\{C_{i_1,i_2;S(i_1,i_2)} : 1 \le i_1 < i_2 \le d\}$ be the set of pair-copulas associated with a regular vine $\mathcal{V}$. If all of these bivariate copulas are reflection symmetric with densities, then the resulting d-dimensional vine copula density $c_{1:d}$ is reflection symmetric.

Proof: The proof is straightforward, but requires two auxiliary results.

(a) If C is a m-dimensional reflection symmetric copula with a density, then

$$C_{m|1...m-1}(1 - u_m \mid 1 - u_1, \ldots, 1 - u_{m-1}) = 1 - C_{m|1...m-1}(u_m \mid u_1, \ldots, u_{m-1}),$$

with similar identities for any conditional distribution of one variable given the other $m - 1$ variables.

(b) Let $2 \leq \ell < d$ be an integer for level ℓ of the vine. Suppose $C_{i_1, j_1, \ldots j_{\ell-1}}$ and $C_{i_2, j_1 \cdots j_{\ell-1}}$ are reflection symmetric copulas with densities, and $C_{i_1 i_2; j_1, \ldots, j_{\ell-1}}$ is a reflection symmetric bivariate copula with density $c_{i_1 i_2; j_1, \ldots, j_{\ell-1}}$, then the function

$$c_{i_1 i_2; j_1, \ldots, j_{\ell-1}}\left(C_{i_1|j_1, \ldots j_{\ell-1}}(u_{i_1} \mid u_{j_1}, \ldots, u_{j_{\ell-1}}), C_{i_2|j_1, \ldots j_{\ell-1}}(u_{i_2} \mid u_{j_1}, \ldots, u_{j_{\ell-1}})\right)$$

is reflection symmetric.

These two results can be combined with the vine copula density representation in Theorem 8.64. □

The next result shows how the tail dependence of a regular vine depends on its pair-copulas in tree $\mathcal{T}_1$. For notation, a conditional distribution of variables indexed by i_1, i_2, is in tree $\mathcal{T}_\ell$ of the vine if the cardinality of conditioning set $S(i_1, i_2) = \{j_1, \ldots, j_{\ell-1}\}$ is $\ell - 1$ for $\ell = 1, \ldots, d - 1$. The notation of conditional tail dependence functions given in Section 8.9 is used.

Theorem 8.68 (Tail dependence in tree $\mathcal{T}_1$ of vine copula implies tail dependence of all margins; Joe et al. (2010), Joe (2011b)). *Consider a vine copula $C_{1:d}$ constructed from the bivariate copulas $\{C_{i_1, i_2; S(i_1, i_2)} : 1 \leq i_1 < i_2 \leq d, \text{ with } d - \ell \text{ pairs in } \mathcal{T}_\ell, \ell = 1, \ldots, d - 1\}$. If all the bivariate copulas have continuous second-order partial derivatives, then the lower and upper tail dependence functions are given respectively by recursions. For $\ell > 1$, with $\boldsymbol{v} = (v_{j_1}, \ldots, v_{j_{\ell-1}})$ and $S(i_1, i_2) = \{j_1, \ldots, j_{\ell-1}\}$ as the conditioning set for $\{i_1, i_2\}$, the recursion for the lower tail is:*

$$b_{i_1, i_2, j_1, \ldots, j_{\ell-1}}(w_{i_1}, w_{i_2}, w_{j_1}, \ldots, w_{j_{\ell-1}})$$
$$= \int_0^{w_{j_1}} \cdots \int_0^{w_{j_{\ell-1}}} C_{i_1, i_2; j_1, \ldots, j_{\ell-1}}\left(b_{i_1|j_1, \ldots, j_{\ell-1}}(w_{i_1}|\boldsymbol{v}), b_{i_2|j_1, \ldots, j_{\ell-1}}(w_{i_2}|\boldsymbol{v})\right)$$
$$\times \frac{\partial^{l-1} b_{j_1, \ldots, j_{\ell-1}}(\boldsymbol{v})}{\partial v_{j_1} \cdots \partial v_{j_{\ell-1}}} \, d\boldsymbol{v},$$

Analogous expressions can be obtained for the upper tail with the upper tail functions b^* replacing the lower tail b functions. If the supports of the bivariate copulas are the entire $(0, 1)^2$ and the copulas $\{C_{i_1, i_2} : (i_1, i_2) \text{ in tree } \mathcal{T}_1\}$ are all lower (upper) tail dependent, then $C_{1:d}$ is lower (upper) tail dependent.

Proof: Theorem 8.58 is used. □

Note that the $(0, 1)^2$ support of the bivariate copulas is a sufficient but not necessary condition. It is possible for the conclusion not to hold if (a) some bivariate copulas for $\mathcal{T}_2$ have negative dependence such that there is no probability near $(0, 0)$ and the conditional tail dependence functions are sub-distribution functions bounded by $\frac{1}{2}$.

Example 8.10 (Illustrations for Theorem 8.68). For $d = 3$ and 4, special cases of the theorem for the C-vine are given as follows, together with an outline of the derivations. For $d = 3$, with copulas $C_{12}, C_{13}, C_{23;1}$ and conditional tail dependence functions $b_{2|1}, b_{3|1}$,

$$C_{123}(uw_1, uw_2, uw_3) = \int_0^{uw_1} C_{23;1}\big(C_{2|1}(uw_2|v_1),\, C_{3|1}(uw_3|v_1)\big)\, dv_1$$

$$= u \int_0^{w_1} C_{23;1}\big(C_{2|1}(uw_1|uz),\, C_{3|1}(uw_3|uz)\big)\, dz$$

$$\sim u \int_0^{w_1} C_{23;1}\big(b_{2|1}(w_2|z),\, b_{3|1}(w_3|z)\big)\, dz, \quad \text{as } u \to 0^+.$$

By Theorem 8.68,

$$b_{123}(w_1, w_2, w_3) = \int_0^{w_1} C_{23;1}\big(b_{2|1}(w_2|z),\, b_{3|1}(w_3|z)\big)\, dz.$$

C_{123} has trivariate lower tail dependence if $b_{12}, b_{13}, b_{2|1}, b_{3|1}$ are positive and $C_{23;1}$ is positive in $(0,1)^2$. With these conditions, $b_{3|12}$ is also positive.

For $d = 4$, with additional copulas $C_{14}, C_{24;1}, C_{34;12}$, the above construction means that b_{124} is positive if also $b_{14}, b_{4|1}$ are positive and $C_{24;1}$ is positive in $(0,1)^2$. With these conditions, $b_{4|12}$ is also positive. If $C_{3|12}$ is a conditional distribution of C_{123} and $C_{4|12}$ is a conditional distribution of C_{124}, then

$$C_{1234}(uw_1, \ldots, uw_4)$$

$$= \int_0^{uw_1} \int_0^{uw_2} C_{34;12}\big(C_{3|12}(uw_3|v_1, v_2),\, C_{4|12}(uw_4|v_1, v_2)\big)\, c_{12}(v_1, v_2)\, dv_1 dv_2$$

$$= u^2 \int_0^{w_1} \int_0^{w_2} C_{34;12}\big(C_{3|12}(uw_3|uz_1, uz_2),\, C_{4|12}(uw_4|uz_1, uz_2)\big)\, c_{12}(uz_1, uz_2)\, dz_1 dz_2$$

$$\sim u^1 \int_0^{w_1} \int_0^{w_2} C_{34;12}\big(b_{3|12}(w_3|z_1, z_2),\, b_{4|12}(w_4|z_1, z_2)\big)\, \frac{\partial^2 b_{12}(z_1, z_2)}{\partial z_1 \partial z_2}\, dz_1 dz_2.$$

The last approximation follows from (2.68). C_{1234} has multivariate lower tail dependence if $b_{123}, b_{124}, b_{3|12}, b_{4|12}$ are positive (and hence b_{12} is positive), and $C_{34;12}$ is positive in $(0,1)^2$.

Example 8.11 (Numerical evaluation of tail dependence parameters for vines). This example goes further than the preceding one, and gives expressions for lower tail dependence parameters $\lambda_{23}, \lambda_{24}, \lambda_{34}$ when C_{12}, C_{13}, C_{14} are MTCJ copulas (with lower tail dependence). In this case, we show that $b_{12|3}, b_{12|4}, b_{123|4}$ etc. are proper distribution functions, so that by part (a) of Theorem 8.58, $b_{23}(w_2, w_3) = b_{123}(\infty, w_2, w_3)$, $b_{24}(w_2, w_4) = b_{124}(\infty, w_2, w_4)$, $b_{34}(w_3, w_4) = b_{1234}(\infty, \infty, w_3, w_4)$. Then $\lambda_{23} = b_{23}(1,1) = 2 - A_{23}(1,1)$, $\lambda_{24} = b_{24}(1,1) = 2 - A_{24}(1,1)$ and $\lambda_{34} = b_{34}(1,1) = 2 - A_{34}(1,1)$, where A_{23}, A_{24}, A_{34} are the exponent functions of the bivariate marginal limiting extreme value copulas.

Suppose C_{12}, C_{13}, C_{14} are MTCJ copulas with respective parameters $\theta_{12}, \theta_{13}, \theta_{14} > 0$. Then their lower tail dependence functions are $b_{1j}(w_1, w_j) = (w_1^{-\theta_{1j}} + w_j^{-\theta_{1j}})^{-1/\theta_{1j}}$ and $b_{j|1}(w_j|w_1) = [(w_j/w_1)^{-\theta_{1j}} + 1]^{-1/\theta_{1j}-1}$ for $j = 2, 3, 4$. Note that $b_{j|1}(w|w_1)$ is increasing from 0 to 1 as w increases from 0 to ∞. From Example 8.10,

$$b_{123}(\boldsymbol{w}) = \int_0^{w_1} C_{23;1}\big([(w_2/z)^{-\theta_{12}} + 1]^{-1/\theta_{12}-1}, [(w_3/z)^{-\theta_{13}} + 1]^{-1/\theta_{13}-1}\big)\, dz,$$

and a similar expression holds for b_{124} and b_{134}. Let's next check some derivatives of b_{123}:

$$b_{23|1}(w_2, w_3|w_1) = \frac{\partial b_{123}}{\partial w_1} = C_{23;1}\big([(w_2/w_1)^{-\theta_{12}} + 1]^{-1/\theta_{12}-1}, [(w_3/w_1)^{-\theta_{13}} + 1]^{-1/\theta_{13}-1}\big),$$

so that $b_{23|1}(\infty, \infty|w_1) = C_{23;1}(1,1) = 1$;

$$b_{12|3}(w_1, w_2|w_3) = \int_0^{w_1} C_{2|3;1}\big(b_{2|1}(w_2|z)|b_{3|1}(w_3|z)\big) \cdot \frac{\partial^2 b_{13}(z, w_3)}{\partial z \partial w_3}\, dz,$$

so that $b_{12|3}(\infty, \infty|w_3) = \int_0^\infty \frac{\partial^2 b_{13}}{\partial z \partial w_3}\, dz = b_{1|3}(z|w_3)|_0^\infty = 1$. Similarly, $b_{13|2}(\infty, \infty|w_2) = 1$. Note also that with one more derivative,

$$b_{3|12} = \frac{\partial^2 b_{123}}{\partial w_1 \partial w_2} \Big/ \frac{\partial^2 b_{12}}{\partial w_1 \partial w_2} = C_{3|2;1}(b_{3|1}|b_{2|1}) \frac{\partial b_{2|1}}{\partial w_2} \Big/ \frac{\partial^2 b_{12}}{\partial w_1 \partial w_2} = C_{3|2;1}(b_{3|1}|b_{2|1}),$$

and $b_{3|12}(\infty|w_1, w_2) = C_{3|2;1}(1|b_{2|1}(w_2|w_1)) = 1$. Similarly, $b_{4|12}(\infty|w_1, w_2) = 1$.

Because $b_{j|1}(\cdot|w_1)$ are proper distributions, there are no extra support conditions on $C_{2j;1}$ in order that $b_{123}, b_{124}, b_{134}$ are non-zero.

Then

$$\lambda_{23} = b_{123}(\infty, 1, 1) = \int_0^\infty C_{23;1}\big([z^{\theta_{12}} + 1]^{-1/\theta_{12}-1}, [z^{\theta_{13}} + 1]^{-1/\theta_{13}-1}\big)\, dz,$$

with similar expressions for $\lambda_{24}, \lambda_{34}$.

Next for b_{1234} from Example 8.10,

$$b_{1234}(\boldsymbol{w}) = \int_0^{w_1} \int_0^{w_2} C_{34;12}\big(b_{3|12}(w_3|z_1, z_2), b_{4|12}(w_4|z_1, z_2)\big) \frac{\partial^2 b_{12}(z_1, z_2)}{\partial z_1 \partial z_2}\, dz_2 dz_1,$$

with $\partial^2 b_{12}(z_1, z_2)/\partial z_1 \partial z_2 = (1 + \theta_{12})(z_1^{-\theta_{12}} + z_2^{-\theta_{12}})^{-1/\theta_{12}-2}(z_1 z_2)^{-\theta_{12}-1}$. Let's next check some derivatives of b_{1234}:

$$b_{234|1} = \frac{\partial b_{1234}}{\partial w_1} = \int_0^{w_2} C_{34;12}\big(b_{3|12}(w_3|w_1, z_2), b_{4|12}(w_4|w_1, z_2)\big) \frac{\partial^2 b_{12}(w_1, z_2)}{\partial z_1 \partial z_2}\, dz_2,$$

so that $b_{234|1}(\infty, \infty, \infty|w_1) = b_{2|1}(z_1|w_1)|_0^\infty = 1$;

$$b_{123|4} = \int_0^{w_1} \int_0^{w_2} C_{3|4;12}\big(b_{3|12}(w_3|z_1, z_2)|b_{4|12}(w_4|z_1, z_2)\big) \frac{\partial b_{4|12}(w_4|z_1, z_2)}{\partial w_4} \frac{\partial^2 b_{12}(z_1, z_2)}{\partial z_1 \partial z_2}\, dz_2 dz_1,$$

$$= \int_0^{w_1} \int_0^{w_2} C_{3|4;12}\big(b_{3|12}(w_3|z_1, z_2)|b_{4|12}(w_4|z_1, z_2)\big) \frac{\partial^3 b_{124}(z_1, z_2, w_4)}{\partial z_1 \partial z_2 \partial w_4}\, dz_2 dz_1,$$

(using (8.20)), so that $b_{123|4}(\infty, \infty, \infty|w_4) = \int_0^\infty \int_0^\infty \frac{\partial^3 b_{124}}{\partial z_1 \partial z_2 \partial w_4}\, dz_2 dz_1 = b_{12|4}(z_1, z_2|w_4)|_0^\infty|_0^\infty = 1$. Similarly, $b_{134|2}$ and $b_{124|3}$ are proper distributions.

Then from all of the above,

$$\lambda_{34} = b_{1234}(\infty, \infty, 1, 1) = \int_0^\infty \int_0^\infty C_{34;12}\big(C_{3|2;1}[b_{3|1}(1|z_1)|b_{2|1}(z_2|z_1)],$$

$$C_{4|2;1}[b_{4|1}(1|z_1)|b_{2|1}(z_2|z_1)]\big) \frac{\partial^2 b_{12}(z_1, z_2)}{\partial z_1 \partial z_2}\, dz_2 dz_1$$

can be numerically evaluated.

Some numerical values of $\lambda_{23}, \lambda_{24}, \lambda_{34}$ are given in the Table 8.4. $\qquad\square$

T_2, T_3	β_{12}	β_{13}	β_{14}	β_{23}	β_{24}	β_{34}	λ_{23}	λ_{24}	λ_{34}
Frank	0.3	0.4	0.5	0.0	0.0	0.0	0.346	0.389	0.333
Gumbel	0.3	0.4	0.5	0.0	0.0	0.0	0.346	0.389	0.333
MTCJ	0.3	0.4	0.5	0.0	0.0	0.0	0.346	0.389	0.333
Frank	0.3	0.4	0.5	0.3	0.4	0.1	0.440	0.496	0.457
Gumbel	0.3	0.4	0.5	0.3	0.4	0.1	0.450	0.509	0.471
MTCJ	0.3	0.4	0.5	0.3	0.4	0.1	0.576	0.604	0.544
Frank	0.3	0.3	0.3	0.3	0.3	0.1	0.376	0.376	0.285
Gumbel	0.3	0.3	0.3	0.3	0.3	0.1	0.388	0.388	0.294
MTCJ	0.3	0.3	0.3	0.3	0.3	0.1	0.567	0.567	0.446
Frank	0.3	0.3	0.3	0.8	0.8	0.1	0.612	0.612	0.549
Gumbel	0.3	0.3	0.3	0.8	0.8	0.1	0.731	0.731	0.654
MTCJ	0.3	0.3	0.3	0.8	0.8	0.1	0.903	0.903	0.869

Table 8.4 *Lower tail dependence parameters for $(2,3),(2,4),(3,4)$ margins when T_1 of C-vine for $d = 4$ has MTCJ copulas with given values of Blomqvist's betas, and T_2, T_3 has Frank, Gumbel or MTCJ copulas. For the first and second set of betas, $\lambda_{12} = 0.448$. $\lambda_{13} = 0.590$, $\lambda_{14} = 0.696$. For the third and fourth set of betas, $\lambda_{12} = \lambda_{13} = \lambda_{14} = 0.448$.*

Something like Example 8.11 does not hold for any regular vine when some bivariate copulas are t copulas. The conditional tail dependence functions for bivariate t copulas are not proper distributions (see Example 8.6 or Nikoloulopoulos et al. (2009)).

The 2-factor models built from a truncated C-vine with latent variables (Section 8.13) are examples where relevant tail dependence can be obtained without tail dependence for all bivariate copulas in tree T_1.

8.13   Factor copulas

The results in this section on dependence and tail properties are mostly given in Krupskii and Joe (2013) and Krupskii (2014), but expressions for the tail dependence parameters are obtained under some conditions and a few more details are given for the p-factor copula.

The first few results are on dependence properties of bivariate margin $C_{1,2}$ (or $C_{j,k}$) in 1-factor copula model. The dependence results on positive quadrant dependence and stochastically increasing positive dependence are presented for factor models with continuous response variables. But they also apply to the factor models with ordinal responses because the continuous latent variables have the same factor structure.

Theorem 8.69 (PQD for bivariate margins of 1-factor copula; Krupskii and Joe (2013)). *For $j = 1, \ldots, d$, suppose $C_{j|V_1} = C_{U_j|V_1}$ is SI, that is, $\mathbb{P}(U_j > u|V_1 = v) = 1 - C_{j|V_1}(u|v)$ is increasing in $v \in (0,1)$ for any $0 < u < 1$. Let $(U_j, U_{j'}) \sim C_{jj'}$, where $C_{jj'}$ is a bivariate margin of the 1-factor copula (3.52) for $j \neq j'$. Then $\mathrm{Cov}(U_j, U_{j'}) \geq 0$ and $C_{jj'}$ is PQD, that is, $C_{jj'}(u_j, u_{j'}) \geq u_j u_{j'}$ for any $0 < u_j < 1$ and $0 < u_{j'} < 1$.*

Proof: Without loss of generality take $j = 1$ and $j' = 2$. We have

$$\mathrm{Cov}(U_1, U_2) = \mathbb{E}[\mathrm{Cov}(U_1, U_2|V_1)] + \mathrm{Cov}(\mathbb{E}(U_1|V_1), \mathbb{E}(U_2|V_1)) = \mathrm{Cov}(\mathbb{E}(U_1|V_1), \mathbb{E}(U_2|V_1)) \geq 0$$

since $\mathbb{E}(U_1|V_1 = v)$ and $\mathbb{E}(U_2|V_1 = v)$ are increasing in v from the SI assumption, and the covariance of two increasing functions is non-negative from Chebyshev's inequality for similarly ordered functions (see Hardy et al. (1952)). Similarly, for any $0 < u_1 < 1$ and $0 < u_2 < 1$,

$$\mathbb{P}\{U_1 \geq u_1, U_2 \geq u_2\} - (1 - u_1)(1 - u_2) = \mathrm{Cov}(I\{U_1 \geq u_1\}, I\{U_2 \geq u_2\})$$
$$= \mathrm{Cov}(\mathbb{E}[I\{U_1 \geq u_1\}|V_1], \mathbb{E}[I\{U_2 \geq u_2\}|V_1]) \geq 0,$$

because $\mathbb{E}[I\{U_i \geq u_i\}|V_1 = v] = \mathbb{P}\{U_i \geq u_i|V_1 = v\}$ for $i = 1, 2$ are increasing functions of v from the SI assumption. Then $\overline{C}_{12}(u_1, u_2) \geq (1 - u_1)(1 - u_2)$ which is the same thing as $C_{12}(u_1, u_2) \geq u_1 u_2$. $\qquad\square$

Theorem 8.70 (TP$_2$ dependence for 1-factor copula model). *If $c_{V_11}(v, u_1)$ and $c_{V_12}(v, u_2)$ have TP$_2$ densities, then $c_{12}(u_1, u_2)$ is TP$_2$. If $C_{1|V_1}(u_1|v)$ and $C_{2|V_1}(u_2|v)$ are TP$_2$ functions, then $C_{12}(u_1, u_2)$ is TP$_2$.*

Proof: Take (3.53) and (3.52) with $d = 2$ to get expressions for c_{12} and C_{12}. Applying the TP$_2$ Basic Composition Theorem (Karlin (1968) or Theorem 18.A.4.a in Marshall et al. (2011)) to these two equations leads to the conclusion.

With a proof similar to that of part (c) of Theorem 8.5, note that $c_{V_1j}(v, u_j)$ being TP$_2$ implies that $C_{j|V_1}(u_j|v)$ is TP$_2$. $\qquad\square$

Theorem 8.71 (Concordance ordering and SI for bivariate margins of 1-factor copula; Krupskii and Joe (2013)). *Consider the bivariate margin C_{12} of (3.52). Assume all bivariate linking copulas are twice continuously differentiable. Assume that C_{V_12} is fixed and that $C_{2|V_1}$ is stochastically increasing (respectively decreasing). (a) Assume that $C_{V_11}(\cdot; \theta)$ increases in the concordance ordering as θ increases. Then C_{12} is increasing (respectively decreasing) in concordance as θ increases. (b) Assume that $C_{V_1|1}$ is stochastically increasing. Then $C_{2|1}$ is stochastically increasing (respectively decreasing).*

Proof: (a) Suppose C_{V_11} has a parameter θ and C_{V_12} is fixed. The increasing in concordance assumption implies that $C_{V_11}(\cdot; \theta') - C_{V_11}(\cdot; \theta) \geq 0$ for $\theta < \theta'$. Using the integration by parts formula we get:

$$C_{12}(u_1, u_2; \theta) = \int_0^1 C_{1|V_1}(u_1|v; \theta)\, C_{2|V_1}(u_2|v)\, dv$$
$$= u_1 C_{2|V_1}(u_2|1) - \int_0^1 C_{V_11}(v, u_1; \theta) \frac{\partial C_{2|V_1}(u_2|v)}{\partial v}\, dv. \qquad (8.27)$$

With the assumption of twice continuous differentiability of C_{2V_1}, then $\partial C_{2|V_1}(u_2|v)/\partial v$ is a continuous function of v for $v \in (0, 1)$ but can be unbounded at 0 or 1. Nevertheless, the integrand is an integrable function since

$$\int_0^1 \left| C_{V_11}(v, u_1; \theta) \frac{\partial C_{2|V_1}(u_2|v)}{\partial v} \right| dv \leq \left| \int_0^1 \frac{\partial C_{2|V_1}(u_2|v)}{\partial v}\, dv \right| = \left| C_{2|V_1}(u_2|0) - C_{2|V_1}(u_2|1) \right|.$$

Therefore the formula (8.27) is valid. For $\theta' > \theta$ we have:

$$C_{12}(u_1, u_2; \theta') - C_{12}(u_1, u_2; \theta) = \int_0^1 \left[C_{V_11}(v, u_1; \theta) - C_{V_11}(v, u_1; \theta') \right] \cdot \frac{\partial C_{2|V_1}(u_2|v)}{\partial v}\, dv.$$

Since $C_{V_1}(v, u_1; \theta') \geq C_{V_11}(v, u_1; \theta)$ and $\partial C_{2|V_1}(u_2|v)/\partial v \leq (\geq) 0$ by the assumption of stochastically increasing (decreasing), we get $C_{12}(u_1, u_2; \theta') \geq (\leq) C_{12}(u_1, u_2; \theta)$ respectively, that is, C_{12} is increasing (decreasing) in concordance as θ increases.

(b) The parameter θ for C_{V_11} is not relevant for this result, For $u_1 \in (0, 1)$, (8.27) can be differentiated with respect to u_1 twice to get

$$\frac{\partial^2 C_{12}(u_1, u_2)}{\partial u_1^2} = \frac{\partial C_{2|1}(u_2|u_1)}{\partial u_1} = -\int_0^1 \frac{\partial C_{V_1|1}(v|u_1)}{\partial u_1} \cdot \frac{\partial C_{2|V_1}(u_2|v)}{\partial v}\, dv.$$

Assuming $C_{V_1|1}$ is SI we get $\partial C_{V_1|1}(v|u_1)/\partial u_1 \leq 0$; since also $\partial C_{2|V_1}(u_2|v)/\partial v \leq (\geq) 0$

by the assumption of stochastically increasing (decreasing), then $\partial C_{2|1}(u_2|u_1)/\partial u_1 \leq 0$ (respectively, ≥ 0), that is, $C_{2|1}$ is stochastically increasing (decreasing). $\qquad\square$

Next are theorems for dependence properties of bivariate margins in the 2-factor and p-factor ($p \geq 3$) models.

While the above result with a conclusion of SI can not be readily extended to the 2-factor and p-factor models, the results for PQD and increasing in the concordance ordering hold in under similar assumptions. The next three results extend those Krupskii and Joe (2013) from 2-factor to p-factor copulas.

Lemma 8.72 (SI in latent variables for p-factor ($p \geq 2$) copula). *Consider the p-factor model (3.51) with (3.58) where $V_1, \ldots, V_p$ are independent U$(0,1)$ random variables. Let U be an observed U$(0,1)$ random variable such that $V_1, \ldots, V_p, U$ are ordered as a C-vine. Suppose the bivariate copulas $C_{U,V_1}, C_{U,V_2;V_1}, \ldots, C_{UV_p;V_1\cdots V_{p-1}}$ have conditional distributions $C_{U|V_1}(\cdot|v_1)$, $C_{U|V_2;V_1}(\cdot|v_2)$, $\ldots, C_{U|V_p;V_1\cdots V_{p-1}}(\cdot|v_p)$ that are stochastically increasing. Then U is stochastically increasing in $v_1, \ldots, v_p$, that is, the conditional distribution $C_{U|V_1,\ldots,V_p}(u|v_1,\ldots,v_p)$ is decreasing in $v_1, \ldots, v_p$ for every $u \in (0,1)$.*

Proof: From the recursion equation (3.38) for conditional distributions in vines, making use of the independence of $V_1, \ldots, V_p$, one has for $m = 2, \ldots, p$:

$$C_{U|V_1\cdots V_m}(u|v_1,\ldots,v_m)$$
$$= C_{U|V_m;V_1\cdots V_{m-1}}\big(C_{U|V_1\cdots V_{m-1}}(u|v_1,\ldots,v_{m-1})|C_{V_m|V_1\cdots V_{m-1}}(v_m|v_1,\ldots,v_{m-1})\big)$$
$$= C_{U|V_m;V_1\cdots V_{m-1}}\big(C_{U|V_1\cdots V_{m-1}}(u|v_1,\ldots,v_{m-1})|v_m\big). \qquad (8.28)$$

With the assumption of stochastically increasing, the left-hand side of (8.28) is decreasing in v_1 for $m = 1$. Then, with $m = 2$, the right-hand side of (8.28) is decreasing in v_1 with fixed v_2 and decreasing in v_2 with fixed v_1, so that $C_{U|V_1,V_2}$ is decreasing in v_1, v_2. By induction, $C_{U|V_1\cdots V_m}(u|v_1,\ldots,v_m)$ is decreasing in $v_1, \ldots, v_m$ for $m = 2, \ldots, p$. $\qquad\square$

Theorem 8.73 (PQD for bivariate margins of p-factor copula with C-vine structure).
    (a) *For the 2-factor copula model in (3.54), suppose that $C_{j|V_1}(\cdot|v_1)$ and $C_{j|V_2;V_1}(\cdot|v_2)$ are SI for $j = 1, \ldots, d$. Then the margin $C_{jj'}$ of (3.54) is PQD for $j \neq j'$.*
    (b) *For the p-factor ($p \geq 3$) model (3.51) and (3.58) based on a C-vine, suppose that $C_{j|V_1}(\cdot|v_1)$ and $C_{j|V_m;V_1\ldots V_{m-1}}(\cdot|v_m)$ ($m = 2, \ldots, p$) are SI for $j = 1, \ldots, d$. Then the margin $C_{jj'}$ of (3.51) is PQD for $j \neq j'$.*

Proof: Without loss of generality take $j = 1$ and $j' = 2$.
    (a) From Lemma 8.72, $1 - C_{i|V_1,V_2}(u|v_1,v_2) = 1 - C_{i|V_2;V_1}(C_{i|V_1}(u|v_1)|v_2)$ is an increasing function of v_1 and v_2, for $i = 1, 2$. Fix $0 < u_1, u_2 < 1$. Let $(U_1, U_2) \sim C_{12}$. Define

$$g_i(v_1,v_2) = \mathbb{P}\{U_i > u_i|V_1 = v_1, V_2 = v_2\} = \mathbb{E}[I\{U_i > u_i\}|V_1 = v_1, V_2 = v_2], \ i = 1,2.$$

From the SI assumption, $g_i(v_1,v_2)$ is an increasing function of v_1 and v_2. By Chebyshev's inequality for similarly ordered functions, for every v_1 we have

$$h_{12}(v_1) := \mathbb{E}[g_1(v_1,V_2)\, g_2(v_1,V_2)|V_1 = v_1]$$
$$\geq \mathbb{E}[g_1(v_1,V_2)|V_1 = v_1]\, \mathbb{E}[g_2(v_1,V_2)|V_1 = v_1] =: h_1(v_1)\, h_2(v_1).$$

Because V_2 is independent of V_1 and g_1, g_2 are increasing in v_1, then h_1, h_2 are increasing in v_1. Next, with another application of Chebyshev's inequality,

$$\mathbb{E}[h_{12}(V_1)] = \mathbb{E}[g_1(V_1,V_2)\, g_2(V_1,V_2)] \geq \mathbb{E}[h_1(V_1)\, h_2(V_1)]$$
$$\geq \mathbb{E}[h_1(V_1)]\, \mathbb{E}[h_2(V_1)] = \mathbb{E}[g_1(V_1,V_2)]\, \mathbb{E}[g_2(V_1,V_2)],$$

and this implies

$$\mathbb{P}\{U_1 \geq u_1, U_2 \geq u_2\} - (1 - u_1)(1 - u_2) = \text{Cov}(I\{U_1 \geq u_1\}, I\{U_2 \geq u_2\})$$
$$= \text{Cov}(\mathbb{E}[I\{U_1 \geq u_1\}|V_1, V_2], \mathbb{E}[I\{U_2 \geq u_2\}|V_1, V_2]) \geq 0.$$

(b) From Lemma 8.72, $1 - C_{i|V_1,\ldots,V_p}(u|v_1,\ldots,v_p)$ is an increasing function of $v_1,\ldots,v_p$ for $i = 1, 2$. The technique in part (a) above can be applied inductively since

$$\mathbb{P}\{U_1 \geq u_1, U_2 \geq u_2\} - (1 - u_1)(1 - u_2) = \text{Cov}(g_1(V_1,\ldots,V_p), g_2(V_1,\ldots,V_p)),$$

where $g_i(v_1,\ldots,v_p) = \mathbb{P}\{U_i > u_i|V_1 = v_1,\ldots,V_p = v_p\}$. $\qquad\square$

The PQD result of Theorem 8.69 and the above theorem overlap with a multivariate result in Lemma 2.2.1 and Theorem 5.3.1 of Tong (1980) with a conclusion of positive orthant dependence.

Theorem 8.74 (Concordance ordering for bivariate margins of p-factor ($p \geq 2$) copula). *Consider the bivariate margin C_{12} of (3.51) with (3.58), where $V_1,\ldots,V_p$ are independent U(0,1) random variables and $V_1,\ldots,V_p, U_j$ are ordered as a C-vine for each j. Assume all bivariate linking copulas are twice continuously differentiable. Suppose that $C_{1V_p;V_1\cdots V_{p-1}}(\cdot;\theta)$ is increasing in the concordance ordering as θ increases and $C_{2|V_p;V_1\cdots V_{p-1}}$ is stochastically increasing. With other bivariate linking copulas held fixed, then C_{12} is increasing in concordance as θ increases.*

Proof: This result is stated for $p = 2$ in Krupskii and Joe (2013). This has some similarity to the proof of Theorem 8.71. Using the integration by parts formula over v_p we get:

$$C_{12}(u_1, u_2; \theta) = \int_{[0,1]^{p-1}} \int_0^1 C_{1|V_p;V_1\ldots V_{p-1}}(C_{1|V_1\ldots V_{p-1}}(u_1|\boldsymbol{v}_{1:(p-1)})|v_p; \theta)$$
$$\cdot C_{2|V_p;V_1\ldots V_{p-1}}(C_{2|V_1\ldots V_{p-1}}(u_2|\boldsymbol{v}_{1:(p-1)})|v_p)\,dv_p d\boldsymbol{v}_{1:(p-1)}$$
$$= I_0 - \int_{[0,1]^{p-1}} \int_0^1 C_{V_p1;V_1,\ldots,V_{p-1}}(v_p, C_{1|V_1,\ldots,V_{p-1}}(u_1|\boldsymbol{v}_{1:(p-1)}); \theta)$$
$$\cdot \frac{\partial C_{2|V_p;V_1,\ldots,V_{p-1}}(C_{2|V_p;V_1\ldots V_{p-1}}(u_2|\boldsymbol{v}_{1:(p-1)})|v_p)}{\partial v_p}\,dv_p d\boldsymbol{v}_{1:(p-1)} \qquad (8.29)$$

where

$$I_0 = \int_{[0,1]^{p-1}} C_{1|V_1,\ldots,V_{p-1}}(u_1|\boldsymbol{v}_{1:(p-1)})\, C_{2|V_p;V_1,\ldots,V_{p-1}}(C_{2|V_1\ldots V_{p-1}}(u_2|\boldsymbol{v}_{1:(p-1)})|1)\,d\boldsymbol{v}_{1:(p-1)}.$$
$$(8.30)$$

The assumption of twice continuous differentiability of the copula cdfs $C_{2V_p;V_1,\ldots,V_{p-1}}$ implies that the right-hand integral in (8.29) is well defined; see the proof of Theorem 8.71. In (8.30), I_0 does not depend on θ, and in (8.29), $C_{V_p1;V_1,\ldots,V_{p-1}}(v_p, \cdot; \theta)$ is increasing in θ, and $\partial C_{2|V_p,V_1,\ldots,V_{p-1}}/\partial v_p \leq 0$. Hence $C_{12}(u_1, u_2; \theta)$ in (8.29) is increasing in θ. $\qquad\square$

The next result is on reflection symmetry when all bivariate linking copulas are reflection symmetric.

Theorem 8.75 (Reflection symmetry for factor copula). *Consider the p-factor copula (3.51) with (3.58). If all of the bivariate linking copulas to the latent variables are reflection symmetric with densities, then the resulting d-dimensional factor copula density $c_{1:d}$ is reflection symmetric.*

Proof: The copula density $c_{V_1,\ldots,V_p,U_1,\ldots,U_d}$ with p latent variables and d observed variables is a regular vine copula density. Bivariate independence copulas are reflection symmetric. so by Theorem 8.67, it is reflection symmetric. By integration over the p independent $U(0,1)$ latent variables, the p-factor copula density satisfies $c_{U_1,\ldots,U_d}(u_1,\ldots,u_d) = c_{U_1,\ldots,U_d}(1 - u_1,\ldots,1 - u_d)$ and is reflection symmetric. □

Next are results on tail dependence and tail order of factor copulas. These are important results on how tail properties of bivariate linking copulas of observed and latent variables are transferred to tail properties of the observed variables. Hence they provide guidelines in choosing the bivariate linking copulas to match observed tail behavior in multivariate data with factor structure,

Theorem 8.76 (Tail dependence for 1-factor copula). *Consider the 1-factor copula model (3.52). Assume all bivariate linking copulas are twice continuously differentiable. Let $b_{V_1 j}$ be the tail dependence function for (V_1, U_j) and suppose this is non-zero for $j = 1,\ldots,d$. Suppose $C_{V_1 j}$ has lower tail dependence with positive conditional tail dependence functions $b_{j|V_1}$, for $j = 1,\ldots,d$, where*

$$b_{j|V_1}(w_j|w) = \lim_{u \to 0^+} C_{j|V_1}(uw_j|uw), \quad \text{for } w_j > 0, w > 0.$$

(a) Let $b_{jj'}$ be lower the tail dependence function for $(U_j, U_{j'})$, then

$$b_{jj'}(w_j, w_{j'}) \geq \int_0^\infty b_{j|V_1}(w_j|z)\, b_{j'|V_1}(w_{j'}|z)\, \mathrm{d}z.$$

(b) Let $\lambda_{jj'} = \lim_{u \to 0^+} u^{-1} C_{jj'}(u,u)$ be the lower tail dependence parameter of $(U_j, U_{j'})$. Then $\lambda_{jj'} \geq \int_0^\infty b_{j|V_1}(1|z)\, b_{j'|V_1}(1|z)\, \mathrm{d}z$. That is, $C_{jj'}$ has lower tail dependence.
(c) Let $b_{V_1|j}(w|w_j) = \lim_{u \to 0^+} C_{V_1|j}(uw|uw_j)$. If the conditional tail dependence functions $b_{j|V_1}$ and $b_{V_1|j}$ are all proper distributions on $[0,\infty)$, then equalities hold in the results of (a) and (b), and

$$\lambda_{jj'} = \int_0^\infty b_{j|V_1}(1|z)\, b_{j'|V_1}(1|z)\, \mathrm{d}z. \tag{8.31}$$

There are parallel results for upper tail dependence.

Proof: Without loss of generality take $j = 1$ and $j' = 2$.

Note that (V_1, U_1, U_2) is distributed as a C-vine copula with $C_{V_1 1}, C_{V_1 2}$ and $C_{12;V_1} = C^\perp$. Let $b_{V_1 12}$ be the tail dependence function of (V_1, U_1, U_2). Then using Theorem 8.68,

$$b_{V_1 12}(w_1', w_1, w_2) = \int_0^{w_1'} b_{1|V_1}(w_1|z)\, b_{2|V_1}(w_2|z)\, \mathrm{d}z.$$

From Theorem 8.58, $b_{12}(w_1, w_2) \geq b_{V_1 12}(\infty, w_1, w_2)$, so that (a) follows. The tail dependence parameter is $\lambda_{12} = b_{12}(1,1)$ by Theorem 8.56 so that (b) follows.

For the conclusion of part (c), using Theorem 8.58, we must show that $b_{12|V_1}, b_{V_1 1|2}, b_{V_1 2|1}$ are proper distributions. For the first of these conditional tail dependence functions,

$$b_{12|V_1}(w_1, w_2|w_1') = \frac{\partial b_{V_1 12}}{\partial w_1'} = b_{1|V_1}(w_1|w_1')\, b_{2|V_1}(w_2|w_1')$$

and

$$\lim_{w_1 \to \infty, w_2 \to \infty} b_{12|V_1}(w_1, w_2|w_1') = b_{1|V_1}(\infty|w_1')\, b_{2|V_1}(\infty|w_1') = 1$$

since $b_{1|V_1}, b_{2|V_1}$ are proper conditional distributions. Next,

$$b_{V_1 1|2}(w_1', w_1|w_2) = \int_0^{w_1'} b_{1|V_1}(w_1|z)\, \frac{\partial^2 b_{V_1 2}(z, w_2)}{\partial z \partial w_2}\, \mathrm{d}z$$

and

$$b_{V_1 1|2}(\infty, \infty | w_2) = \int_0^\infty \frac{\partial^2 b_{V_1 2}(z, w_2)}{\partial z \partial w_2}\, dz = b_{V_1|2}(z|w_2) \Big|_0^\infty = 1,$$

since $b_{1|V_1}(\infty|z) = 1$, $b_{V_1|2}(\infty|w_2) = 1$ and $b_{V_1|2}(0|w_2) = 0$ when $b_{1|V_1}, b_{V_1|2}$ are proper conditional distributions. The result for $b_{V_1 2|1}$ follows analogously.   □

These results show that for the 1-factor copula, there is upper (lower) tail dependence for all bivariate pairs if there is upper (lower) tail dependence for all $C_{V_1 j}$ ($j = 1, \ldots, d$) linking observed variables to the latent variable.

For the 2-factor copula (3.54), Theorem 8.68 does not apply because we are assuming V_1, V_2 to be independent so that one edge of T_1 of the C-vine (with variables order as $V_1, V_2, U_j, U_{j'}$ for any $j \neq j'$) does not have tail dependence. Instead, we have to do some direct calculations and apply Theorem 8.58. Krupskii and Joe (2013) have an alternative way of stating the conditions and giving a proof, but without getting an explicit bound on the tail dependence parameters.

Theorem 8.77 (Tail dependence for 2-factor copula). *Consider the 2-factor copula model (3.54). Assume all bivariate linking copulas are twice continuously differentiable.*

(a) *Suppose $C_{V_1 j}$ has lower tail dependence with positive conditional tail dependence functions $b_{j|V_1}$, for $j = 1, \ldots, d$, where*

$$b_{j|V_1}(w_j|w) = \lim_{u \to 0^+} C_{j|V_1}(uw_j|uw), \quad \text{for } w_j > 0, w > 0.$$

Then, $C_{jj'}$ has lower tail dependence for $j \neq j'$.

(b) *Supppose that the conditional lower tail dependence functions $b_{j|V_1}, b_{V_1|j}$ are all proper (conditional) distributions on $[0, \infty)$, then the lower tail dependence parameter of the bivariate margin $C_{jj'}$*

$$\lambda_{jj'} = \int_0^1 \int_0^\infty C_{j|V_2;V_1}\big(b_{j|V_1}(1|z_1)|v_2\big)\, C_{j'|V_2;V_1}\big(b_{j'|V_1}(1|z_1)|v_2\big)\, dz_1 dv_2. \tag{8.32}$$

(c) *There are parallel results for upper tail dependence.*

Proof: The proofs are given for the case of lower tail dependence; the proof for upper tail dependence follows from the reflection of the uniform random variables. Without loss of generality take $j = 1$ and $j' = 2$.

(a) Let $u_i = uw_i$ with $w_i > 0$ for $i = 1, 2$. Then for $u > 0$ small enough,

$$C_{12}(uw_1, uw_2) = \int_0^1 \int_0^1 C_{1|V_2;V_1}\big(C_{1|V_1}(uw_1|v_1)|v_2\big)\, C_{2|V_2;V_1}\big(C_{2|V_1}(uw_2|v_1)|v_2\big)\, dv_2 dv_1$$

$$= u \int_0^1 \int_0^{1/u} C_{1|V_2;V_1}\big(C_{1|V_1}(uw_1|uz_1)|v_2\big)\, C_{2|V_2;V_1}\big(C_{2|V_1}(uw_2|uz_1)|v_2\big)\, dz_1 dv_2.$$

Since $C_{i|V_1}(uw_i|uz_1) \to b_{i|V_1}(w_i|v_1)$ as $u \to 0^+$, then by Fatou's lemma,

$$b_{12}(w_1, w_2) = \lim_{u \to 0^+} \frac{C_{12}(uw_1, uw_2)}{u}$$

$$\geq \int_0^1 \int_0^\infty C_{1|V_2;V_1}\big(b_{1|V_1}(w_1|z_1)|v_2\big)\, C_{2|V_2;V_1}\big(b_{2|V_1}(w_2|z_1)|v_2\big)\, dz_1 dv_2,$$

and this limit is positive because the $b_{i|V_1}$ functions are continuous and positive. Hence C_{12} has lower tail dependence.

(b) We start with the trivariate tail dependence function $b_{V_1 12}$ of (V_1, U_1, U_2). Similar to the above, for $u > 0$ small enough,

$$u^{-1} C_{V_1 12}(uw_1', uw_1, uw_2)$$

$$= u^{-1} \int_0^{uw_1'} \int_0^1 C_{1|V_2;V_1} \left(C_{1|V_1}(uw_1|v_1)|v_2 \right) C_{2|V_2;V_1} \left(C_{2|V_1}(uw_2|v_1)|v_2 \right) dv_2 dv_1$$

$$= \int_{v_2=0}^1 \int_{z_1=0}^{w_1'} C_{1|V_2;V_1} \left(C_{1|V_1}(uw_1|uz_1)|v_2 \right) C_{2|V_2;V_1} \left(C_{2|V_1}(uw_2|uz_1)|v_2 \right) dz_1 dv_2.$$

The integrand is dominated by 1 and the integral is dominated by w_1'. By the Lebesgue dominated convergence theorem,

$$b_{V_1 12}(w_1', w_1, w_2) = \lim_{u \to 0+} u^{-1} C_{V_1 12}(uw_1', uw_1, uw_2)$$

$$= \int_0^1 \int_0^{w_1'} C_{1|V_2;V_1} \left(b_{1|V_1}(w_1|z_1)|v_2 \right) C_{2|V_2;V_1} \left(b_{2|V_1}(w_2|z_1)|v_2 \right) dz_1 dv_2.$$

By Theorem 8.58, we need to show that $b_{12|V_1}, b_{V_1 1|2}, b_{V_1 2|1}$ are proper distributions in order that $b_{12}(w_1, w_2) = b_{V_1 12}(\infty, w_1, w_2)$ and then the tail dependence parameter follows from Theorem 8.56.

Assume $b_{i|V_1}(\infty|z) = 1$ and $b_{V_1|i}(\infty|w_i) = 1$ for $i = 1, 2$. The required calculations are:

$$b_{12|V_1}(w_1, w_2|w_1') = \int_0^1 C_{1|V_2;V_1} \left(b_{1|V_1}(w_1|w_1')|v_2 \right) C_{2|V_2;V_1} \left(b_{2|V_1}(w_2|w_1')|v_2 \right) dv_2.$$

and

$$\lim_{w_1 \to \infty, w_2 \to \infty} b_{12|V_1}(w_1, w_2|w_1') = \int_0^1 C_{1|V_2;V_1}(1|v_2) C_{2|V_2;V_1}(1|v_2) dv_2 = 1;$$

$$b_{V_1 1|2}(w_1', w_1|w_2) = \int_0^{w_1'} \int_0^1 C_{1|V_2;V_1} \left(b_{1|V_1}(w_1|z_1)|v_2 \right) c_{V_2 2;V_1} \left(v_2, b_{2|V_1}(w_2|z_1) \right) \frac{\partial^2 b_{V_1 2}}{\partial z_1 \partial w_2} dv_2 dz_1$$

and

$$b_{V_1 1|2}(\infty, \infty|w_2) = \int_0^\infty \int_0^1 c_{V_2 2;V_1} \left(v_2, b_{2|V_1}(w_2|z_1) \right) dv_2 \frac{\partial^2 b_{V_1 2}(z_1, w_2)}{\partial z_1 \partial w_2} dz_1$$

$$= \int_0^\infty \frac{\partial^2 b_{V_1 2}(z_1, w_2)}{\partial z_1 \partial w_2} dz_1 = b_{V_1|2}(z|w_2) \Big|_0^\infty = 1,$$

since $\int_0^1 c(v, u) dv = 1$ for a bivariate copula density c. The result for $b_{V_1 2|1}$ follows analogously. $\qquad\square$

Next are some illustrations of the above theorems.

Example 8.12 (Numerical evaluation of tail dependence parameters for 1-factor and 2-factor copula models). We use Theorems 8.76 and 8.77 to compute tail dependence parameters for (i) 1-factor copula model with reflected Gumbel linking copulas; (ii) 2-factor copula model with reflected Gumbel linking copulas for factor 1 and Frank copula for factor 2. Because of Example 8.6, note that these theorems do not apply if the linking copulas are bivariate t.

(i) For the bivariate Gumbel copula, $A(w_1, w_2) = (w_1^\delta + w_2^\delta)^{1/\delta}$ for $\delta > 1$, so that the lower tail dependence function of the reflected Gumbel copula is $b(w_1, w_2) = w_1 + w_2 - (w_1^\delta + w_2^\delta)^{1/\delta}$. If (V_1, U_j) has the reflected Gumbel copula with parameter δ_j for $j = 1, 2$, then

$$b_{j|V_1}(w|z) = 1 - ([w/z]^{\delta_j} + 1)^{1/\delta_j - 1},$$

and from (8.31), the lower tail dependence parameter is

$$\lambda_{12} = \int_0^\infty \prod_{j=1}^2 \{1 - (z^{-\delta_j} + 1)^{1/\delta_j - 1}\}\, dz.$$

Some numerical examples with δ_j chosen to achieve different Blomqvist beta values are given in Table 8.5.

(ii) Assume that $C_{V_1 j}$ are the same as in part (i), and let $C_{V_2 j; V_1}$ be the linking copulas for the second latent variable. From (8.32), the lower tail dependence parameter is

$$\lambda_{12} = \int_0^\infty \int_0^1 \prod_{j=1}^2 C_{j|V_2; V_1}\left(1 - (z^{-\delta_j} + 1)^{1/\delta_j - 1}|v\right)\, dv\, dz.$$

Some numerical examples with Frank copulas for $C_{V_2 j; V_1}$ are given in Table 8.5.

$\beta_{V_1 1}$	$\beta_{V_1 2}$	$\beta_{V_2 1}$	$\beta_{V_2 2}$	λ_{12}
\multicolumn{5}{c}{1-factor copula}
0.5	0.5			0.428
0.5	0.8			0.559
0.8	0.8			0.782
\multicolumn{5}{c}{2-factor copula}
0.5	0.5	0.2	0.2	0.444
0.5	0.5	0.5	0.5	0.523
0.5	0.5	0.3	0.6	0.495
0.5	0.8	0.2	0.2	0.567
0.5	0.8	0.5	0.5	0.610
0.5	0.8	0.3	0.6	0.595
0.8	0.8	0.2	0.2	0.789
0.8	0.8	0.5	0.5	0.827
0.8	0.8	0.3	0.6	0.814

Table 8.5 *Lower tail dependence parameter λ_{12} in 1-factor and 2-factor copulas, given Blomqvist beta values of observed with latent variables, reflected Gumbel copulas for latent variable 1, and Frank copulas for latent variable 2. The tail dependence with latent variable 1 is 0.585 when $\beta = 0.5$ and 0.848 when $\beta = 0.8$.*

A result on tail order of the 1-factor model is given next. The qualitative conclusion should hold more generally; the lower tail behavior for extreme value copulas allows for more concrete calculations.

Theorem 8.78 (Lower tail order of 1-factor copula with bivariate extreme value copulas; Krupskii and Joe (2013)). *(1) Suppose $C_{V_1 j}$ is a bivariate extreme value copula*

$$C_{V_1 j}(v, u) = \exp\left\{-(x + y)B_j\left(\frac{y}{x + y}\right)\right\} = (vu)^{B_j(\log v / \log(uv))}, \quad x = -\log u, \ y = \log v,$$
(8.33)

where $B_j(\cdot) : [0, 1] \mapsto [\frac{1}{2}, 1]$ is a convex function such that $B_j(t) \geq \max\{t, 1 - t\}$, $j = 1, 2$. Assume that B_j is continuously differentiable, and $B_j'(t) > -1$ for $0 < t < \frac{1}{2}$, $j = 1, 2$. The lower tail order of the bivariate marginal copula C_{12} in (3.52) is equal to $\xi^* = \min_{0 < s < \infty}\{\xi(s)\} \in [1, 2]$, where

$$\xi(s) = (s + 1)\left[B_1\left(\frac{s}{s + 1}\right) + B_2\left(\frac{s}{s + 1}\right)\right] - s.$$
(8.34)

(2) If, in addition, copulas for the second factor are also extreme value copulas:

$$C_{V_2j;V_1}(v,u) = (vu)^{B_{j;1}(\log v/\log(uv))}, \tag{8.35}$$

where $B_{j;1}(\cdot) : [0,1] \mapsto [\frac{1}{2}, 1]$ is a convex function such that $B_{j;1}(t) \geq \max\{t, 1-t\}$, $j = 1, 2$. Assume $B_j(t), B_{j;1}(t)$ in (8.33) and (8.35) are continuously differentiable functions and $B'_j(t) > -1$, $B'_{j;1}(t) > -1$ for $0 < t < \frac{1}{2}$. The lower tail order of the bivariate marginal copula C_{12} in (3.54) is equal to $\xi_2^* = \min_{0 < s_1, s_2 < \infty} \tilde{\xi}_2(s_1, s_2) \in [1, 2]$, where

$$\tilde{\xi}_2(s_1, s_2) = (s_1 + \tilde{m}_1(s_2))B_{1;1}\left(\frac{s_1}{s_1 + \tilde{m}_1(s_2)}\right) + (s_1 + \tilde{m}_2(s_2))B_{2;1}\left(\frac{s_1}{s_1 + \tilde{m}_2(s_2)}\right) - s_1 + s_2, \tag{8.36}$$

$$\tilde{m}_j(s_2) = (s_2 + 1)B_j\left(\frac{s_2}{s_2 + 1}\right) - s_2, \quad j = 1, 2.$$

Proof: See Krupskii and Joe (2013). □

In the next example, we show some intuition for the above theorem using Laplace's method or approximation; this approach might also work for other copulas with lower intermediate tail dependence.

Example 8.13 (Laplace's method for analysis of tail order in factor models).

For the 1-factor model, let the exponent function of C_{V_1j} be A_j with $B_j(t) = A_j(t, 1-t)$ and $A_j(s, 1) = (1 + s)B_j(s/[s + 1])$. Then, using the conditional distributions of a bivariate extreme value copula in (8.13),

$$C_{12}(u, u) = \int_0^1 C_{1|V_1}(u|v)\, C_{2|V_1}(u|v)\, dv = (-\log u)\int_0^\infty C_{1|V_1}(u|u^s)\, C_{2|V_1}(u|u^s)\, u^s ds$$

$$= (-\log u)\int_0^\infty u^{A_1(s,1) + A_2(s,1) - s} h_1(s)\, h_2(s)\, ds$$

$$= z\int_0^\infty e^{-z\xi(s)} h_1(s)\, h_2(s)\, ds, \tag{8.37}$$

where $z = -\log u$,

$$h_j(s) = B_j\left(\frac{s}{s+1}\right) + \frac{1}{s+1}B'_j\left(\frac{s}{s+1}\right), \quad j = 1, 2,$$

and $\xi(s)$ in (8.34) is the same as $A_1(s, 1) + A_2(s, 1) - s$. If $\xi(s)$ is twice differentiable and has a unique minimum at $s_0 > 0$, then Laplace's method applied to (8.37) leads to the approximation:

$$ze^{-z\xi(s_0)} h_1(s_0)h_2(s_0)\frac{\sqrt{2\pi}}{\sqrt{\xi''(s_0)}} = (-\log u)u^{\xi(s_0)} h_1(s_0)h_2(s_0)\frac{\sqrt{2\pi}}{\sqrt{\xi''(s_0)}}, \quad z \to \infty, u \to 0^+,$$

so that the lower tail order is $\xi(s_0)$.

For the 2-factor model, in addition, let the exponent function of $C_{V_2j;V_1}$ be $A_{j;1}$ with

$B_{j;1}(t) = A_{j;1}(t, 1 - t)$. Then similar to the above,

$$C_{12}(u, u) = \int_0^1 \int_0^1 C_{1|V_2;V_1}\big[C_{1|V_1}(u|v_1)|v_2\big] C_{2|V_2;V_1}\big[C_{2|V_1}(u|v_1)|v_2\big] \, dv_1 dv_2$$

$$= (-\log u)^2 \int_0^\infty \int_0^\infty C_{1|V_2;V_1}\big[C_{1|V_1}(u|u^{s_1})|u^{s_2}\big] C_{2|V_2;V_1}\big[C_{2|V_1}(u|u^{s_1})|u^{s_2}\big] \, u^{s_1+s_2} ds_1 ds_2$$

$$= (-\log u)^2 \int_0^\infty \int_0^\infty \prod_{j=1}^2 \Big\{ u^{A_{j;1}[s_2, A_j(s_1,1)-1+\log_u h_j(s_1)]-s_1} h_{j;1}(s_2) \Big\} \, u^{s_1+s_2} ds_1 ds_2$$

$$= z^2 \int_0^\infty \int_0^\infty e^{-z\xi_2(s_1,s_2)} h_{1;1}(s_2) h_{2;1}(s_2) \, ds_1 ds_2, \tag{8.38}$$

where
$$h_{j;1}(s) = B_{j;1}\Big(\frac{s}{s+1}\Big) + \frac{1}{s+1} B'_{j;1}\Big(\frac{s}{s+1}\Big),$$

and $\xi_2(s_1, s_2) = s_2 - s_1 + \sum_{j=1}^2 A_{j;1}[s_2, A_j(s_1, 1) - 1 + \log_u h_j(s_1)]$. Laplace's method can be applied to the double integral in (8.38).

More generally, for intermediate tail dependence, the transform $v_j \to u^{s_j}$ followed by Laplace's method might be applicable to determine the tail order coefficient. ◻

The next example has some interpretation of tail orders for 1-factor and 2-factor copulas.

Example 8.14 (1-factor and 2-factor models with Gumbel bivariate linking copulas).
Suppose $A_j(t) = [t^{\theta_j} + (1 - t)^{\theta_j}]^{1/\theta_j}$, $\theta_j > 1$, for $j = 1, 2$. Then in (8.34),

$$\xi(s) = (1 + s^{\theta_1})^{1/\theta_1} + (1 + s^{\theta_2})^{1/\theta_2} - s.$$

If $\theta_1 = \theta_2 = \theta_0$, $s^* = (2^{\frac{\theta_0}{\theta_0-1}} - 1)^{-1/\theta_0}$ and $\xi^* = (2^{\frac{\theta_0}{\theta_0-1}} - 1)^{(\theta_0-1)/\theta_0}$. In general, $1 < \xi^* < 2$ for any $\theta_1 > 1$, $\theta_2 > 1$. Hence, C_{12} has intermediate lower tail independence. The Gumbel copula has upper tail dependence and therefore C_{12} also has upper tail dependence. As a result, C_{12} has tail asymmetry skewed to the upper tail.

If $\theta_1 = \theta_2 = \theta_0 > 1$, the tail orders of Gumbel linking copulas are $2^{1/\theta_0}$ and this is less than ξ^*. This demonstrates the general pattern that the conditional independence in the 1-factor model "dampens" the strength of dependence in the tail (or the tail order increases).

For the 2-factor copula, note that $\tilde{\xi}_2(0, s_2) = \xi(s_2)$ as given in (8.34) and (8.36), and therefore $\xi_2^* \le \xi^*$. That is, the lower tail order of C_{12} in 2-factor model is lower than that in 1-factor model when the same linking copulas $C_{V_{1j}}$ are used for the first factor. It means the intermediate tail dependence is stronger if the second factor is added to the model. For example, if all linking copulas are Gumbel copulas, the lower tail order of C_{12} is always smaller than 2.

8.14 Kendall functions

Kendall functions have appeared in the copula literature as a multivariate probability integral transform. It has origins and uses for the family of Archimedean copulas (see Genest and Rivest (1993)). Let C be a d-variate copula, and suppose $\boldsymbol{U} = (U_1, \ldots, U_d) \sim C$; then the Kendall function is the distribution of $C(\boldsymbol{U})$. Denote this as:

$$F_K(z; C) = \mathbb{P}(C(\boldsymbol{U}) \le z; \boldsymbol{U} \sim C).$$

Some special cases are the following.

- C^+: $F_K(z; C^+) = z$ for $0 < z < 1$, for any dimension $d \geq 2$. Let $(U_1, \ldots, U_d) \sim C^+$, so that $U_1 = \cdots = U_d$ and then $\mathbb{P}(C^+(U_1, \ldots, U_d) \leq z) = \mathbb{P}(U_1 \leq z) = z$.
- C^- with $d = 2$: $F_K(z; C^-) = 1$ for $0 < z < 1$ (degenerate distribution at 0).
- $C^\perp$ for $d = 2$: $F_K(z; C^\perp) = z - z \log z$ for $0 < z < 1$.
- $F_K(z; C) \geq z$ for $0 < z < 1$ for any copula C. A proof follows from the recursion of Imlahi et al. (1999).
- bivariate Archimedean (Genest and Rivest (1993)): $F_K(z; C_\psi) = z - \psi^{-1}(z) \cdot \psi'[\psi^{-1}(z)]$ for $0 < z < 1$.
- multivariate Archimedean with $\psi \in \mathcal{L}_\infty$ (Barbe et al. (1996)):

$$F_K(z; C_{\psi,d}) = z + \sum_{k=1}^{d-1} [-\psi^{-1}(z)]^k \psi^{(k)}[\psi^{-1}(z)]/k!.$$

- $C^\perp$ for $d \geq 3$:

$$F_K(z; C^\perp_{1:d}) = z + z \sum_{k=1}^{d-1} (-\log z)^k/k!,$$

as the special case of the preceding with $\psi(s) = e^{-s}$.

Remark 8.6 Capéraà et al. (1997b) show that for bivariate Archimedean, larger in concordance ordering implies $F_K(\cdot; C_\psi)$ is stochastically larger, that is, the cdf F_K is smaller. *Proof:* Let ψ_1, ψ_2 be two LTs or functions in $\mathcal{L}_2$. Then from Theorem 8.22, $C_{\psi_1} \prec_c C_{\psi_2}$ if $\omega = \psi_2^{-1} \circ \psi_1$ is superadditive and a sufficient condition for this is that $\psi_2^{-1} \circ \psi_1$ is starshaped with respect the origin or $\psi_2^{-1} \circ \psi_1(s)/s$ increasing in $s > 0$.

The derivative of $\omega(s)/s$ is

$$\frac{\psi_1'(s)}{s\psi_2'[\psi_2^{-1}(\psi_1(s))]} - \frac{\psi_2^{-1}(\psi_1(s))}{s^2}.$$

Since this is non-negative, then the non-positivity of ψ_2' implies

$$s\psi_1'(s) \leq \psi_2^{-1}(\psi_1(s)) \cdot \psi_2'[\psi_2^{-1}(\psi_1(s))],$$

and with $s = \psi_1^{-1}(t)$ for $0 < t < 1$, this becomes

$$\psi_1^{-1}(t) \cdot \psi_1'[\psi_1^{-1}(t))] \leq \psi_2^{-1}(t) \cdot \psi_2'[\psi_2^{-1}(t)].$$

This implies $F_K(z; C_{\psi_1}) \geq F_K(z; C_{\psi_2})$, that is, $F_K(z; C_{\psi_1})$ is stochastically smaller than $F_K(z; C_{\psi_1})$. $\square$

Suppose the density $f_K = F_K'$ exists. Then, using Theorems 8.37 and 8.35 it can be shown that for bivariate Archimedeam copulas, (a) intermediate lower tail dependence or lower tail quadrant independence implies f_K is infinite at 0, and strong lower tail dependence implies f_K is finite at 0; (b) intermediate upper tail dependence or upper tail quadrant independence implies f_K is 0 at 1, and strong upper tail dependence implies f_K is positive at 1.

In general, patterns might be similar; the strength of dependence in the lower tail of C affects $F_K(z)$ as $z \to 0^+$, and the strength of dependence in the upper tail of C affects $F_K(z)$ as $z \to 1^-$.

Although Capéraà et al. (1997b), Genest and Rivest (2001), Nelsen et al. (2001), Nelsen et al. (2003) have derived some properties of Kendall functions, not much is known about how the dependence of C transfers to inequalities of the Kendall function. The aforementioned authors have shown for the bivariate case that the bivariate concordance ordering of

C_1, C_2 need not imply a stochastic ordering of $F_K(\cdot; C_1), F_K(\cdot; C_2)$. More informative results can be obtained based on the stochastically increasing dependence concept and dependence ordering. These are stated in the following theorems.

Theorem 8.79 (Stochastic ordering of Kendall functions with copula satisfying SI).
(a) *Let C be a bivariate copula such that either $C_{2|1}(\cdot|u)$ is stochastically increasing in $0 < u < 1$ or $C_{1|2}(\cdot|v)$ is stochastically increasing in $0 < v < 1$ or both. Then $F_K(z; C) \leq z - z \log z$ for $0 < z < 1$. If at least one of $C_{2|1}(\cdot|u)$ and $C_{1|2}(\cdot|v)$ is stochastically decreasing, then $F_K(z; C) \geq z - z \log z$ for $0 < z < 1$.*
(b) *If C is a d-variate copula that is conditionally increasing in sequence, that is, after a permutation of indices, $C_{j|1:(j-1)}(\cdot|\boldsymbol{u}_{1:(j-1)})$ is stochastically increasing in $\boldsymbol{u}_{1:(j-1)}$ for $j = 2, \ldots, d$, then $F_K(z; C) \leq F_K(z; C_{1:d}^{\perp})$.*

Proof: (a) Without loss of generality, suppose the indexing of the variables is such that $C_{2|1}(\cdot|u)$ is stochastically increasing, that is, $C_{2|1}(v|u)$ is decreasing in u for all v. Let $(U_1, U_2) \sim C$ and $F_{U_2|U_1\leq}(v|u) = \mathbb{P}(U_2 \leq v|U_1 \leq u) = C(u, v)/u$. Then for $0 < v < 1$,

$$F_{U_2|U_1\leq}(v|u) = u^{-1} \int_0^u C_{2|1}(v|w)\, dw \geq C_{2|1}(v|u) \cdot u^{-1} \int_0^u dw = C_{2|1}(v|u).$$

Hence $F_{U_2|U_1\leq}^{-1}(p|u) \leq C_{2|1}^{-1}(p|u)$ for $0 < p < 1$. This leads to

$$C\big[u, C_{2|1}^{-1}(v|u)\big] \geq C\big[u, F_{U_2|U_1\leq}^{-1}(v|u)\big] = uF_{U_2|U_1\leq}\big[F_{U_2|U_1\leq}^{-1}(v|u)\big] = uv$$

for any $0 < u, v < 1$. Let V be independent of U_1, and then we can assume that $U_2 = C_{2|1}^{-1}(V|U_1)$ since $(U_1, C_{2|1}^{-1}(V|U_1)) \sim C$ in this case. Hence,

$$F_K(z; C) = \mathbb{P}\big(C\big[U_1, C_{2|1}^{-1}(V|U_1)\big] \leq z\big) \leq \mathbb{P}(U_1 V \leq z) = z - z \log z.$$

For stochastically decreasing, all of the preceding inequalities are reversed.

(b) Let $(U_1, \ldots, U_d) \sim C$ and let $F_{U_j|U_1\cdots U_{j-1}\leq}(\cdot|\boldsymbol{u}_{1:(j-1)})$ be the conditional cdf of U_j given $U_k \leq u_k$ for $k = 1, \ldots, j-1$. If $V_1, V_2, \ldots, V_d$ are independent $U(0,1)$ random variables, then a stochastic representation for $(U_1, \ldots, U_d)$ is

$$U_1 = V_1, U_2 = C_{2|1}^{-1}(V_2|U_1), U_3 = C_{3|12}^{-1}(V_3|U_1, U_2), \ldots, U_d = C_{d|1:(d-1)}^{-1}(V_d|U_1, \ldots, U_{d-1}).$$

Let $0 < v_j < 1$ for $j = 1, \ldots, d$ and let $u_j = C_{j|1:(j-1)}^{-1}(v_j|u_1, \ldots, u_{j-1})$ for $j = 2, \ldots, d$. Similar to part (a), it suffices to show that

$$C_{1:j}(u_1, \ldots, u_j) \geq u_1 v_2 \cdots v_j, \quad j = 2, \ldots, d.$$

The $d = 2$ case has been proved in part (a) and we proceed by induction. Since $C_{j|1:(j-1)}(\cdot|\boldsymbol{u}_{1:(j-1)})$ is stochastically increasing in $\boldsymbol{u}_{1:(j-1)}$, then

$$C_{1:j}\big(u_1, \ldots, u_{j-1}, C_{j|1:(j-1)}^{-1}(v_j|\boldsymbol{u}_{1:(j-1)})\big) \geq C_{1:j}\big(u_1, \ldots, u_{j-1}, F_{U_j|U_1\cdots U_{j-1}\leq}^{-1}(v_j|\boldsymbol{u}_{1:(j-1)})\big)$$
$$= C_{1:(j-1)}(u_1, \ldots, u_{j-1}) v_j \geq u_1 v_2 \cdots v_{j-1} v_j,$$

where the last inequality is the inductive step. $\qquad\square$

Theorem 8.80 (Stochastic ordering of Kendall functions with two bivariate copulas that are $\prec_{SI}$ ordered). *Let C, C' be bivariate copulas such that $C \prec_{SI} C'$ with continuity and increasingness of $h(x, y) = C'^{-1}_{2|1}[C_{2|1}(y|x)|x]$ in $x, y \in [0, 1]$. Then $K(z; C') \leq K(z; C)$, that is, $K(\cdot; C')$ is stochastically larger than $K(\cdot; C)$.*

Proof: If U, V are independent $\mathrm{U}(0, 1)$, then $(U, C_{2|1}^{-1}(V|U)) \sim C$ and $(U, C'^{-1}_{2|1}(V|U)) \sim C'$. To show the ordering of $K(\cdot; C')$ and $K(\cdot; C)$, it suffices to that

$$C[u, C_{2|1}^{-1}(v|u)] \leq C'[u, C'^{-1}_{2|1}(v|u)], \ \forall 0 < u, v < 1.$$

Fix $0 < u, v < 1$ and let $q = C_{2|1}^{-1}(v|u)$ and $q' = C'^{-1}_{2|1}(v|u)$, which are in the interval $(0, 1)$. Then the above inequality is equivalent to

$$\int_0^u C_{2|1}(q|w) \, \mathrm{d}w \leq \int_0^u C'_{2|1}(q'|w) \, \mathrm{d}w. \tag{8.39}$$

Let $g(w) = C_{2|1}(q|w)$ and $g'(w) = C'_{2|1}(q'|w)$ (this is not a derivative) for $0 \leq w \leq u$. Note that $g(u) = g'(u) = v$ so that $g(u) - g'(u) = 0$.

Next we compare $g(w)$ and $g'(w)$ for $0 \leq w < u$. The definition of q implies that $v = C_{2|1}(q|u)$. The assumption of the SI ordering means that $h(x, y) = C'^{-1}_{2|1}[C_{2|1}(y|x)|x]$ is increasing in $x, y \in (0, 1)$. In particular, with $u > 0$ and $0 \leq w < u$, $h(w, q) \leq h(u, q)$ or $C'^{-1}_{2|1}[C_{2|1}(q|w)|w] \leq C'^{-1}_{2|1}[C_{2|1}(q|u)|u]$ or $C_{2|1}(q|w) \leq C'_{2|1}(C'^{-1}_{2|1}[v|u]|w) = C'_{2|1}(q'|w)$. Hence $g'(w) - g(w) \geq 0$ for $0 < w < u$ and (8.39) holds. $\qquad \square$

Note that there is currently no definition of a multivariate version of the bivariate $\prec_{\mathrm{SI}}$ ordering.

8.15   Laplace transforms

This section has a few results on LTs that are used for construction of and determining properties of Archimedean and other copulas built from LTs. Some of the results make connections between the tails of the LT and tails of the density of the positive random variable.

In the theorems below, a Laplace transform (LT) ψ is said to be *infinitely divisible* if it is the LT of an infinitely divisible distribution or random variable; this means that ψ^a is a LT for all $a > 0$. Also the family $\mathcal{L}_\infty^*$ is defined in (2.10) and $\mathcal{L}_m^*$ is defined in (8.6).

A basic result is the form of the LT under scale transform. If Q is a positive random variable with LT ψ_Q and $\sigma > 0$, then the LT of σQ is $\psi_{\sigma Q}(s) = \mathbb{E}(e^{-s\sigma Q}) = \psi_Q(s\sigma)$.

Theorem 8.81 (Infinitely divisible LT; Feller (1971)). *Let ζ be a LT. Then ζ is infinitely divisible (or ζ^a is completely monotone for all $a > 0$) if and only if $\eta = -\log \zeta \in \mathcal{L}_\infty^*$, that is, $(-1)^{i-1}\eta^{(i)} \geq 0$ for $i = 1, 2, \ldots$.*

Theorem 8.82 (LT with composition; Feller (1971)). *If ζ is an infinitely divisible LT such that $-\log \zeta \in \mathcal{L}_\infty^*$ and ψ is another LT, then $\varphi(s) = \psi(-\log \zeta(s))$ is a LT.*

Corollary 8.83 (Infinitely divisible LT with composition). *If ζ is an infinitely divisible LT such that $-\log \zeta \in \mathcal{L}_\infty^*$ and ψ is another infinitely divisible LT, then $\varphi(s) = \psi(-\log \zeta(s))$ is an infinitely divisible LT.*

Proof: This follows from the preceding Theorem 8.82 since ψ^a is a LT for all $a > 0$ if ψ is infinitely divisible.

We also give a proof with a stochastic representation. Let $\{Z(t) : t \geq 0\}$ be a Lévy process (Lévy (1954), Bertoin (1996), Schoutens (2003)) with independent and stationary increments such that the LT of $Z(t)$ is ζ^t for $t > 0$. Let Q be a non-negative random variable with LT ψ, then $Z(Q)$ has LT φ since

$$\mathbb{E}[e^{-sZ(Q)}|Q = q] = \zeta^q(s), \quad \mathbb{E}[e^{-sZ(Q)}] = E[\zeta^Q(s)] = \psi(-\log \zeta(s)).$$

If Q_a has LT ψ^a, then the above implies that $Z(Q_a)$ has LT $\psi^a(-\log\zeta) = \varphi^a$. Hence φ is infinitely divisible. $\qquad\qquad\square$

Remark 8.7 In the above proof, the random variable Q with LT ψ can be considered as a stopping time of a Lévy process based on LT ζ, so we will say that $\varphi(s) = \psi(-\log\zeta(s))$ is the "ψ stopped ζ LT"; for example, if ψ is the gamma LT and ζ is the positive stable LT, then we say that $\varphi(s) = \psi(-\log\zeta(s))$ is the gamma stopped positive stable LT. The stochastic representation in the proof leads to a useful method of simulation of a random variable with LT φ.

Theorem 8.84 (Composition of functions without infinite divisibility). *If ζ and ψ are LTs in $\mathcal{L}_\infty$, then $\varphi(s) = \psi(-\log\zeta(s)) \in \mathcal{L}_2$ (that is, $\varphi(0) = 1$, $\varphi(\infty) = 0$, $\varphi' \leq 0$, $\varphi'' \geq 0$).*

Proof: The boundary conditions are satisfied: $\varphi(0) = \psi(0) = 1$ and $\varphi(\infty) = \psi(-\log 0) = 0$. For the first two derivatives,

$$\varphi' = -\psi'(-\log\zeta) \cdot \frac{\zeta'}{\zeta} \leq 0,$$

$$\varphi'' = \psi''(-\log\zeta) \cdot \frac{\zeta'^2}{\zeta^2} - \psi'(-\log\zeta) \cdot \left[\frac{\zeta''}{\zeta} - \frac{\zeta'^2}{\zeta^2}\right] \geq 0,$$

because $\psi' \leq 0$, $\zeta' \leq 0$, $\psi'' \geq 0$, $\zeta'' \geq 0$, and $\frac{\zeta''}{\zeta} - \frac{\zeta'^2}{\zeta^2} \geq 0$. If ζ is the LT of positive random variable X, then the latter function when evaluated at s is the variance of the random variable with cdf $F_Y(y) = \int_0^y e^{-sx} \mathrm{d}F_X(x)/\zeta(s)$. $\qquad\square$

Theorem 8.85 (Composition of functions in $\mathcal{L}_m^*$). *If ω_1, ω_2 are increasing function in $\mathcal{L}_m^*$, with $m \geq 2$, satisfying $\omega_1(0) = \omega_2(0) = 0$ and $\omega_1(\infty) = \omega_2(\infty) = \infty$, then $\omega_1 \circ \omega_2 \in \mathcal{L}_m^*$. For the case with $m = \infty$, if $e^{-\omega_1}$ and $e^{-\omega_2}$ are LTs of infinitely divisible positive random variables, then $e^{-\omega_1\circ\omega_2}$ is also the LT of an infinitely divisible positive random variable.*

Proof: Let $\eta = \omega_1 \circ \omega_2$; clearly $\eta(0) = 0$, $\eta(\infty) = \infty$ and η is increasing. The first two derivatives are: $\eta' = (\omega_1' \circ \omega_2) \cdot \omega_2' \geq 0$, $\eta'' = (\omega_1'' \circ \omega_2) \cdot (\omega_2')^2 + (\omega_1' \circ \omega_2) \cdot \omega_2'' \leq 0$, if ω_1, ω_2 have derivatives that alternate in sign. Continuing to higher order derivatives, each summand when differentiated leads to several terms opposite in sign. For example, differentiation of $(\omega_1^{(k)} \circ \omega_2)$ in a summand leads to the summand multiplied by $(\omega_1^{(k+1)} \circ \omega_2)\omega_2'/(\omega_1^{(k)} \circ \omega_2)$, and differentiation of $[\omega_2^{(k)}]^i$ for $i \geq 2$ in a summand leads to the summand multiplied by $i\omega_2^{(k+1)}/\omega_2^{(k)}$. The mth derivative of η involves up to the mth derivatives of ω_1, ω_2.

The second result with $m = \infty$ follows from Theorem 8.81. $\qquad\square$

Corollary 8.86 (Transitivity ordering of LTs with respect to $\mathcal{L}_m^*$). *Let ψ_1, ψ_2, ψ_3 be LTs in $\mathcal{L}_\infty$. If $\omega_1 = \psi_1^{-1} \circ \psi_2$ and $\omega_2 = \psi_2^{-1} \circ \psi_3$ are in $\mathcal{L}_m^*$ with $m \geq 2$, then $\omega_1 \circ \omega_2 = \psi_1^{-1} \circ \psi_3$ is in $\mathcal{L}_m^*$.*

The above corollary is useful for showing that a two-parameter Archimedean copula family $C(\cdot; \theta, \delta)$, based on a two-parameter LT family, is increasing in concordance in one of the parameters θ, δ when the other parameter held fixed. For the BB1, BB2 copula families etc., the corollary holds with $m = 2$.

Below are a few theorems that are useful for establishing the tail order or proving tail comonotonicity of some copulas based on a LT. The next result shows that maximal non-negative moment degree M_ψ in Definition 2.2 is related to the behavior of ψ at 0 when $0 < M_\psi < 1$. The result for a general non-integer M_ψ, such that $k < M_\psi < k + 1$ for a non-negative integer, is the second theorem given below.

Theorem 8.87 (LT at zero and maximal non-negative moment degree; Hua and Joe (2011)). *Suppose $\psi(s)$ is the LT of a positive random variable Y, with $0 < M_Y < 1$. If $1 - \psi(s)$ is regularly varying at 0^+, then $1 - \psi(s) \in \mathrm{RV}_{M_Y}(0^+)$.*

Proof: This proof is summarized from Hua and Joe (2011), to show the technique of proof.

Let Z be an exponential random variable, independent of Y, with mean 1. Choose any fixed m with $0 < m < 1$. Then $\mathbb{E}(Z^{-m}) = \Gamma(1 - m)$. If we define $W_m = (Y/Z)^m$, then for any $w > 0$,

$$\mathbb{P}(W_m \geq w) = \mathbb{P}(Z \leq Yw^{-1/m}) = \int_0^\infty \left(1 - \exp\{-yw^{-1/m}\}\right) \mathrm{d}F_Y(y) = 1 - \psi(w^{-1/m}).$$

Therefore, $\mathbb{E}(Y^m) < \infty$ implies $\mathbb{E}(W_m) < \infty$ and $\lim_{w \to \infty} w[1 - \psi(w^{-1/m})] = 0$, i.e.,

$$\lim_{s \to 0^+} [1 - \psi(s)]/s^m = 0. \tag{8.40}$$

If $1 - \psi(s)$ is regularly varying at 0^+, then we can write $1 - \psi(s) = s^\alpha \ell(s)$ with $\alpha \neq 0$, where $\ell(s) \in \mathrm{RV}_0(0^+)$. Then, (8.40) implies that $\lim_{s \to 0^+} s^{\alpha - m} \ell(s) = 0$. Let $\epsilon > 0$ be arbitrarily small. If $m = M_Y - \epsilon$, then we have $\mathbb{E}(Y^{M_Y - \epsilon}) < \infty$ and thus $\lim_{s \to 0^+} s^{\alpha - M_Y + \epsilon} \ell(s) = 0$. Therefore, $\alpha \geq M_Y - \epsilon$.

Also by a result on page 49 of Chung (1974), (8.40) implies that for any $0 < \delta < 1$, $\mathbb{E}(Y^{m(1-\delta)}) < \infty$. If we assume that there exists an $\epsilon > 0$ with $m = M_Y + \epsilon$ such that, $\lim_{s \to 0^+} s^{\alpha - M_Y - \epsilon} \ell(s) = 0$, then for any small $\delta > 0$, $\mathbb{E}(Y^{(M_Y + \epsilon)(1-\delta)}) < \infty$. Then we may choose some $\delta_\epsilon < \epsilon/(\epsilon + M_Y)$, and get $\mathbb{E}(Y^{(M_Y + \epsilon)(1-\delta_\epsilon)}) < \infty$ with $(M_Y + \epsilon)(1 - \delta_\epsilon) > M_Y$, which gives rise to a contradiction. Thus, for any $\epsilon > 0$, we must have $\lim_{s \to 0^+} s^{\alpha - M_Y - \epsilon} \ell(s) \neq 0$, and hence, $\alpha - M_Y - \epsilon \leq 0$. So,

$$M_Y - \epsilon \leq \alpha \leq M_Y + \epsilon.$$

Since $\epsilon > 0$ is arbitrary, this completes the proof. $\square$

Theorem 8.88 (LT at zero and maximal non-negative moment degree; Hua and Joe (2011)). *Suppose $\psi(s)$ is the LT of a positive random variable Y, with $k < M_Y < k+1$ where k is a non-negative integer. If $\left|\psi^{(k)}(0) - \psi^{(k)}(s)\right|$ is regularly varying at 0^+, then $\left|\psi^{(k)}(0) - \psi^{(k)}(s)\right| \in \mathrm{RV}_{M_Y - k}(0^+)$. In particular, if the slowly varying component is $\ell(s)$ and $\lim_{s \to 0^+} \ell(s) = h_{k+1}$ with $0 < h_{k+1} < \infty$, then*

$$\psi(s) = 1 - h_1 s + h_2 s^2 - \cdots + (-1)^k h_k s^k + (-1)^{k+1} h_{k+1} s^{M_Y} + o(s^{M_Y}), \quad s \to 0^+, \tag{8.41}$$

where $0 < h_i < \infty$ for $i = 1, \ldots, k+1$.

Proof: This proof in Hua and Joe (2011) extends that of Theorem 8.87, which corresponds to the case where $k = 0$. The steps are similar but are more technical. $\square$

Remark 8.8 Even if $\mathbb{E}(Y) = \infty$, we may have $M_Y = 1$. However, Theorem 8.87 does not hold in general for this case. The above two theorems can be summarized as follows. If $M_\psi = \infty$, the LT $\psi(s)$ has an infinite Taylor expansion about $s = 0$. If M_ψ is finite and non-integer-valued, then with some regularity conditions, $\psi(s)$ has a Taylor expansion about $s = 0$ up to order $[M_\psi]$, and the next term after this has order M_ψ.

When $M_Y = M_\psi = k$ is a positive integer, the above techniques of proof do not apply. In this case, it is conjectured that the expansion in (8.41) goes up to $s^k \ell(s)$ where ℓ is a slowly varying function.

Theorem 8.89 (LT at zero for very heavy-tailed random variable; Hua and Joe (2012b)). *The condition $1 - \psi \in \mathrm{RV}_0(0^+)$ means that if the random variable Y has LT ψ, then the density $f_Y(y)$ has a very heavy tail as $y \to \infty$ so that $\mathbb{E}[Y^m] = \infty$ for all $m > 0$. The condition $\psi \in \mathrm{RV}_0$ means that the density $f_Y(y)$ has a very heavy tail as $y \to 0^+$ so that $\mathbb{E}[Y^{-m}] = \infty$ for all $m > 0$.*

Proof: This proof is summarized from Hua and Joe (2012b). The implication of the condition of $1 - \psi$ at 0 follows by the proof Theorem 8.87. Similarly, we can show the implication of the condition of ψ at ∞ as follows. Suppose to the contrary that $\mathbb{E}[Y^{-m}] < \infty$ for some $m > 0$. Let $W_m = (Z/Y)^m$, where $Z \sim \mathrm{Exponential}(1)$, independent of Y, and Y has the LT ψ. Then $\mathbb{P}(W_m \geq w) = \mathbb{P}(Z \geq Yw^{1/m}) = \int_0^\infty \exp(-yw^{1/m})\mathrm{d}F_Y(y) = \psi(w^{1/m})$. Then $\mathbb{E}[Y^{-m}] < \infty$ and Z having all positive moments implies that $\mathbb{E}[W_m] < \infty$ and thus $w\mathbb{P}(W_m \geq w) \to 0$, as $w \to \infty$; that is, $w\psi(w^{1/m}) \to 0$, or equivalently $s^m\psi(s) \to 0$, as $s \to \infty$. If $\psi(s) \in \mathrm{RV}_0$ and $m > 0$, then we must have $s^m\psi(s) \to \infty$. The contradiction implies that $\mathbb{E}[Y^{-m}] = \infty$ for all $m > 0$. $\qquad\square$

The condition below on the LT $\psi(s)$ as $s \to \infty$ covers almost all of the LT families in the Appendix, as well as other LT families that can be obtained by integration or differentiation. Suppose

$$\psi(s) \sim T(s) = a_1 s^q \exp\{-a_2 s^p\} \text{ and } \psi'(s) \sim T'(s), \quad s \to \infty, \text{ with } a_1 > 0, a_2 \geq 0, \quad (8.42)$$

where $p = 0$ implies $a_2 = 0$ and $q < 0$, and $p > 0$ implies $p \leq 1$ and q can be 0, negative or positive. Note that $p > 1$ is not possible because of the complete monotonicity property of a LT.

The condition (8.42) can be interpreted as follows. As $\psi(s)$ decreases to 0 more slowly as $s \to \infty$, then the random variable Y with LT ψ has a heavier "tail" at 0. Let $f_Y(0) = \lim_{y \to 0} f_Y(y) \in [0, \infty)$, where f_Y is the density of Y and is assumed well-behaved near 0. As $f_Y(0)$ increases, then the "tail" at 0 is heavier. If $f_Y(0) = 0$, then the tail is lighter as the rate of decrease to 0 is faster. If $f_Y(0) = \infty$, then the tail is heavier as the rate of increase to ∞ is faster. In terms of the LT and the condition in (8.42), as p increases (with fixed q), the tail of Y at 0 gets lighter, and as q increases (with fixed p), the tail of Y at 0 gets heavier.

If $0 < p \leq 1$ in (8.42), then the LT is decreasing at an exponential or subexponential rate as $s \to \infty$. If $p = 0$ and $q < 0$, then $\psi \in \mathrm{RV}_q$ as $s \to \infty$. It is possible for the tail of the LT to go to 0 slower than this as $s \to \infty$. Examples in the Appendix are the gamma stopped gamma LT and positive stable stopped gamma LT. A result from Hua and Joe (2012b) is the following. Suppose the LT $\psi(t)$ satisfies condition (8.42), and let $\zeta(s) = (1 + s)^{-1/\theta}$ be a gamma LT. Then $\varphi(s) = \psi(-\log(\zeta(s)))$ is also a LT and $\varphi \in \mathrm{RV}_0$ as $s \to \infty$.

Example 8.15 (Relation of tail of LT and density at 0). This example provides some intuition for the LT tail condition in (8.42). Suppose the positive random variable Y has density

$$f_Y(y) = k_1 y^{k_2} \exp\{-k_3 y^{-b}\}, \quad 0 < y < \upsilon,$$

where $\upsilon > 0$ is the upper support point, and $k_1 > 0$, $-\infty < k_2 < \infty$, $k_3 > 0$, and $b > 0$. Then the LT of Y has the asymptotic form

$$\psi_Y(s) \sim a_1 s^q \exp\{-(1 + b^{-1})(k_3 b)^{1/(b+1)} s^{b/(b+1)}\}, \quad s \to \infty,$$

where

$$a_1 = k_1 (k_3 b)^{(k_2 + 1/2)/(b+1)} \sqrt{\frac{2\pi}{(b+1)}}, \quad q = -\frac{2k_2 + 2 + b}{2(b+1)}.$$

Hence $p = b/(b+1) \in (0,1)$ for (8.42). Note that $b > 0$ implies $f_Y(0^+) = 0$.

Only an outline of the result based on Laplace's method is given here.

Suppose $g(y,s)$ is twice continuously differentiable, is minimized at $y_0(s)$ and $0 < g''(y_0(s); s) \to \infty$ as $s \to \infty$. Then, from Laplace's method, as $s \to \infty$.

$$\int_0^\infty e^{-g(y;s)} dy \sim e^{-g(y_0(s);s)} \int_0^\infty \exp\{-\tfrac{1}{2}[y - y_0(s)]^2 g''(y_0(s); s)\} dy$$

$$\sim e^{-g(y_0(s);s)} \frac{\sqrt{2\pi}}{\sqrt{g''(y_0(s); s)}}.$$

(8.43)

For large $s > 0$,

$$\psi(s) = \int_0^v e^{-sy} k_1 y^{k_2} \exp\{-k_3 y^{-b}\} dy = \int_0^v e^{-g(y;s)} dy,$$

with

$$g(y; s) = sy - \log k_1 - k_2 \log y + k_3 y^{-b},$$
$$g'(y; s) = s - k_2 y^{-1} - k_3 b y^{-b-1},$$
$$g''(y; s) = k_2 y^{-2} + k_3 b(b+1) y^{-b-2},$$
$$y_0(s) \sim (k_3 b/s)^{1/(b+1)} \to 0, \text{ as } s \to \infty,$$
$$g(y_0(s); s) \sim (1 + b^{-1})(k_3 b)^{1/(b+1)} s^{b/(b+1)} + k_2(b+1)^{-1} \log s - \log k_1 - k_2(b+1)^{-1} \log(k_3 b),$$
$$g''(y_0(s); s) \sim (b+1)(k_3 b)^{-1/(b+1)} s^{(b+2)/(b+1)}.$$

Then (8.43) becomes

$$\frac{k_1 (k_3 b)^{k_2/(b+1)} s^{-k_2/(b+1)} \exp\{-(1 + b^{-1})(k_3 b)^{1/(b+1)} s^{b/(b+1)}\}}{\cdot \sqrt{\dfrac{2\pi}{(b+1)(k_3 b)^{-1/(b+1)} s^{(b+2)/(b+1)}}}}$$

$$= a_1 s^{-(k_2 + 1 + b/2)/(b+1)} \exp\{-(1 + b^{-1})(k_3 b)^{1/(b+1)} s^{b/(b+1)}\}.$$

The above approximation can be rigorized using Theorem 1 in Chapter II of Wong (1989).

Since the tail of the LT as $s \to \infty$ depends mainly on the density near 0, the assumption on f_Y can be weakened

$$f_Y(y) \sim k_1 y^{k_2} \exp\{-k_3 y^{-b}\}, \text{ as } y \to 0^+.$$

It can be shown that this result matches the tail of the LT for the inverse Gaussian and inverse gamma densities.   □

8.16 Regular variation

This section summarizes some basic results on integrals and derivatives of regularly varying functions that are used in proofs of theorems in this chapter. The definition of regular variation is given in Definition 2.3.

Karamata's theorem (page 17 of Resnick (1987)) says that

(i) if $\alpha \geq -1$ then $g \in \mathrm{RV}_\alpha(0^+)$ implies $G(x) := \int_0^x g(t)\, dt \in \mathrm{RV}_{\alpha+1}(0^+)$;

(ii) if $\alpha < -1$ then $g \in \mathrm{RV}_\alpha$ implies $G(x) := \int_x^\infty g(t)\, dt$ is finite and $G(x) \in \mathrm{RV}_{\alpha+1}$.

An interpretation is that for regularly varying functions such that the integral exists, the antiderivative is regularly varying with index incremented by 1 (similar to the integral of a power $h(x) = x^\alpha$). However, regular variation does not hold for derivatives unless additional conditions are satisfied.

For derivatives, the following versions of the Monotone Density Theorem (Bingham et al. (1987), Embrechts et al. (1997)) can be applied to regular varying and slowly varying functions.

Theorem 8.90 *(a) Let $g \in \mathrm{RV}_{-\alpha}$ with $\alpha \neq 0$ and suppose $g'(s)$ is ultimately monotone as $s \to \infty$. Then $g' \in \mathrm{RV}_{-\alpha-1}$.*
(b) Let $\ell \in \mathrm{RV}_0$ be slowly varying at ∞, and $\ell(s) = \int_0^s \ell'(t)\,\mathrm{d}t$ (or $\ell(s) = -\int_s^\infty \ell'(t)\,\mathrm{d}t$). If $\ell'(s)$ is ultimately monotone as $s \to \infty$, then $\ell'(s) = o(s^{-1}\ell(s))$, $s \to \infty$.

Theorem 8.91 *(a) Let $g \in \mathrm{RV}_\alpha(0^+)$ with $\alpha \neq 0$ and suppose $g'(s)$ is ultimately monotone as $s \to 0^+$. Then $g' \in \mathrm{RV}_{\alpha-1}(0^+)$.*
(b) Let $\ell \in \mathrm{RV}_0(0^+)$ be slowly varying at 0^+, and $\ell(s) = \int_0^s \ell'(t)\,\mathrm{d}t$. If $\ell'(s)$ is ultimately monotone as $s \to 0^+$, then $\ell'(s) = o(s^{-1}\ell(s))$, $s \to 0^+$.

Example 8.16 (Examples of regular variation and derivatives).

1. Let $\alpha > 0$. Then $g(s) = s^{-\alpha}(1 + s^{-1}\cos s)$ $(s > 0)$ is a regularly varying function at ∞ with index $-\alpha$. But $g'(s) = -s^{-\alpha+1}(\alpha + \sin s) - (\alpha+1)s^{-\alpha+2}\cos s$ is not ultimately monotone. Hence part (a) of Theorem 8.90 does not hold and it can be checked directly that g' is not regularly varying (but it has bounds that are $\mathrm{RV}_{-\alpha-1}$).

2. The function $\ell(s) = (\log s)(1 + s^{-1}\cos s)$ $(s > 0)$ is slowly varying at ∞. But $\ell'(s) = -s^{-1}(\sin s)(\log s) + s^{-1} + s^{-2}\cos s - s^{-2}(\cos s)(\log s)$ is not ultimately monotone. Part (b) of Theorem 8.90 does not hold since $s\ell'(s)/\ell(s) = O(1)$ as $s \to \infty$.

8.17 Summary for further reseach

This chapter has summarized dependence and tail properties for commonly used copula families such as Archimedean, elliptical and multivariate extreme value. For parametric copula models, knowledge about the properties helps in selecting bivariate members of these families for vine pair-copula constructions and factor copulas.

In summary, a few topics for further research are listed.

- Because multivariate (with dimension $d \geq 3$) parametric copula families that have simple form do not have flexible dependence, the vine/factor copula approach based on a sequence of parametric bivariate copulas is convenient, and the simplifying assumption (3.35) for copulas applied to conditional distributions is used. Without the simplifying assumption, any continuous multivariate distribution has many different vine representations. The use of vines in practice is based on the heuristic that vine models with the simplifying assumption can approximate many multivariate copulas. There remains the research problem of finding conditions for classes of multivariate distributions so that vine or factor copulas do not provide good approximations.

- For many applications of parametric copula models, the variables are monotonically related; in this case, parametric bivariate copulas with one to three parameters can be used within vine/factor copulas. However, in the non-parametric sense, none of the bivariate Archimedean, elliptical and extreme value families are "dense" for the space of (permutation symmetric) bivariate copulas with monotonic dependence. Therefore, for parametric modeling, there is scope for new approaches to derive convenient copula families.

- As alternatives to vines, further research can be done on constructions of multivariate distributions that build on having copulas for different non-overlapping subsets of variables. Section 3.11 has an outline of constructions in this direction.

- The tail order has been important to understand how different copula models behave in the tails, and it helps to explain Kullback-Leibler divergences. For multivariate models that are constructed to have simple form of density functions but not cdfs, such as skew-elliptical models, it might be useful to have a tail density approach analogous to Li and Wu (2013) to determine tail orders.

- Tail dependence functions can be characterized in terms of the representation in (3.78). When the tail order κ is greater than 1, discovering more properties of tail order functions could be useful for tail inference when there is intermediate tail dependence.

- Chapter 2 has introduced several concepts that can quantify the strength of dependence in the tails of multivariate distributions. Some results are given in Section 2.20 for conditional tail expectations on one variable given that another is large. Further results could be obtained for conditional tail expectations for a function of a subset of variables given that variables in a non-overlapping subset are large. Such results will help in understanding how tails of copula models affect tail inference.

- Copula modeling for variables that are not monotonically related has been less developed, partly because, in most applications, the monotonic dependence assumption and the use of measures of monotonic dependence (such as pairwise Kendall's τ and Spearman's ρ_S) are adequate. This is a direction for further research that can be driven by applications.

Appendix A

Laplace transforms and Archimedean generators

This appendix has a list of Laplace transform (LT) families that are used in this book. Section A.1 has a list of 1-parameter and 2-parameter families. Section A.2 explains how the 1-parameter families of Archimedean generators in Nelsen (1999) fit within these 1-parameter and 2-parameter LT families. For notation, $\mathcal{L}_\infty^*$ is defined in (2.10) and $\mathcal{L}_n^*$ is defined in (8.6).

A.1 Parametric Laplace transform families

Laplace transform families are used in several classes of copulas: exchangeable Archimedean copulas, hierarchical Archimedean copulas and mixture of max-id copulas. The LT families that are used in this book are listed below, together with some of their properties.

The LT families are divided into four groups. Families A–M were in the Appendix of Joe (1997), but LT family K is not included here as it wasn't used for copulas. LT families N, O, P, Q, R are additions and they are grouped into appropriate subsections. Some families are now reparametrized differently in order that a larger value means more dependence for the Archimedean copulas.

For the list of LT families: ψ is used for one-parameter families, and φ or ζ is used for two-parameter families. The first group consists of one-parameter LT families that are infinitely divisible; for them the function $\omega(s; \theta_1, \theta_2) = \psi^{-1}(\psi(s; \theta_2); \theta_1)$ with $\theta_1 < \theta_2$. is listed because ω is used in extension from exchangeable Archimedean to hierarchical Archimedean copulas; see Section 3.4. The parametrization of the LT families uses the reciprocal of the most commonly used parametrization of the corresponding distributions; the reason is to get concordance increasing as the parameter increases for the corresponding family of Archimedean copulas.

Properties are mentioned are each family of LTs, including: (i) functional inverse; (ii) infinitely divisibility (iii) maximal non-negative moment degree (see Definition 2.2); (iv) probability mass function in the case of discrete random variables; (v) simulation of random variables with the LT. An infinitely divisible LT is one that is the LT of an infinitely divisible random variable.

A.1.1 One-parameter LT families

A. (positive stable) $\psi(s; \theta) = \exp\{-s^{1/\theta}\}$, $\theta \geq 1$; functional inverse $\psi^{-1}(t; \theta) = (-\log t)^\theta$.

- Infinitely divisible with $\psi^a(s; \theta) = \psi(a^\theta s; \theta)$ (scale change a^θ from the a-fold convolution.
- For $1 < \theta_1 < \theta_2$, $\omega(s) = \psi_1^{-1}(\psi_2(s; \theta_2); \theta_1) \in \mathcal{L}_\infty^*$ and $e^{-\omega(s)} = \exp\{-s^\gamma\}$, with $\gamma = \theta_1/\theta_2$, is also positive stable.

- As $s \to 0^+$, $1 - \psi(s) = s^{1/\theta} + o(s^{1/\theta})$ so that the maximal non-negative moment degree is $M_\psi = \theta^{-1} < 1$, and the mean is infinite.

- Simulation of Postable(α) with $0 < \alpha = \theta^{-1} < 1$ (Chambers et al. (1976), Kanter (1975)). Generate as $[h(V)/W]^{(1-\alpha)/\alpha}$ where W is Exponential(1), $V \sim \mathrm{U}(0, \pi)$ independent of W and

$$h(v) = \frac{\sin[(1 - \alpha)v]\,(\sin \alpha v)^{\alpha/(1-\alpha)}}{(\sin v)^{1/(1-\alpha)}}.$$

B. (gamma) $\psi(s; \theta) = (1 + s)^{-1/\theta}$, $\theta \geq 0$; functional inverse $\psi^{-1}(t; \theta) = t^{-\theta} - 1$.

- Infinitely divisible: $\psi^a(s; \theta) = \psi(s; \theta/a)$ and θ^{-1} is the convolution parameter.

- For $0 < \theta_1 < \theta_2$, $\omega(s) = \psi_1^{-1}(\psi_2(s; \theta_2); \theta_1) \in \mathcal{L}_\infty^*$ and $e^{-\omega(s)} = \exp\{-(1 + s)^\gamma + 1\}$ is LT family L with $\vartheta = \gamma^{-1} = \theta_2/\theta_1$ and $\delta = 1$. If $\psi_1(s) = (1 + \delta s)^{-1/\theta_1}$ with scale parameter $\delta > 0$ and $\psi_2(s) = (1 + s)^{-1/\theta_2}$, then $\exp\{-\psi_1^{-1}(\psi_2(s))\}$ leads to LT family L with parameters $\vartheta = \theta_2/\theta_1$ and $\delta > 0$.

- $1 - \psi(s)$ has infinite Taylor expansion about 0 and the maximal non-negative moment degree is $M_\psi = \infty$.

- Simulation: see Devroye (1986) for efficient algorithms.

C. (Sibuya[1]) $\psi(s; \theta) = 1 - (1 - e^{-s})^{1/\theta}$, $\theta \geq 1$; functional inverse $\psi^{-1}(t; \theta) = -\log[1 - (1 - t)^\theta]$.

- Infinitely divisible (a proof is in Joe (1993)).

- With $\alpha = \theta^{-1} \in (0, 1)$, the (discrete) Sibuya($\alpha$) distribution has probability masses $p_1 = \alpha$ and $p_i = \alpha\,[\prod_{k=1}^{i-1}(k - \alpha)]/i!$ for $i = 2, 3, \ldots$. Hofert (2011) has cdf $F_{\mathrm{Sibuya}}(x; \alpha) = 1 - [x\,B(x, 1 - \alpha)]^{-1}$ for $x = 1, 2, \ldots$.

- For $1 < \theta_1 < \theta_2$, $\omega(s) = \psi_1^{-1}(\psi_2(s; \theta_2); \theta_1) \in \mathcal{L}_\infty^*$, and $e^{-\omega(s)} = 1 - (1 - e^{-s})^\gamma$, with $\gamma = \theta_1/\theta_2$, is also Sibuya.

- As $s \to 0^+$, $1 - \psi(s) = s^{1/\theta} + o(s^{1/\theta})$ so that the maximal non-negative moment degree is $M_\psi = \theta^{-1}$. The mean is infinite for $\theta > 1$.

- For simulation, the method in Devroye (1993) is as follows. Let Z_1, Z_2, Z_3 as independent random variables with Exponential(1), Gamma($1 - \alpha$) and Gamma(α) distributions respectively. Given $(Z_1, Z_2, Z_3) = (z_1, z_2, z_3)$, generate a Poisson random variable with mean $z_1 z_2/z_3$ and add 1 to get a Sibuya(α) random variable with $0 < \alpha < 1$. Proposition 3.2 of Hofert (2011) has an alternative approach based on the cdf.

D. (logarithmic series) $\psi(s; \theta) = -\theta^{-1}\log[1 - (1 - e^{-\theta})e^{-s}]$, $\theta > 0$; functional inverse $\psi^{-1}(t; \theta) = -\log[(1 - e^{-\theta t})/(1 - e^{-\theta})]$.

- Infinitely divisible (a proof is in Joe (1993)).

- The masses of the corresponding discrete distribution are $(1 - e^{-\theta})^i/(i\theta)$ for $i = 1, 2, \ldots$. The mean is $(e^\theta - 1)\theta^{-1}$ and the variance is $(e^\theta - 1)\theta^{-2}[(\theta - 1)e^\theta + 1]$.

- $1 - \psi(s)$ has infinite Taylor expansion about 0 and the maximal non-negative moment degree is $M_\psi = \infty$.

- For $1 < \theta_1 < \theta_2$, $\omega(s) = \psi_1^{-1}(\psi_2(s; \theta_2); \theta_1) \in \mathcal{L}_\infty^*$; with $\delta = 1 - e^{-\theta_1}$, $\vartheta = \gamma^{-1} = \theta_2/\theta_1$, and $1 - e^{-\theta_2} = 1 - (1 - \delta)^\vartheta$, then $e^{-\omega(s)} = \{1 - [1 - (1 - e^{-\theta_2})e^{-s}]^\gamma\}/(1 - e^{-\theta_1})$ is LT family J.

- Simulation of Logseries(θ) with $\alpha = 1 - e^{-\theta} \in (0, 1)$ (see Kemp (1981) or page 548 of Devroye (1993)): algorithms are based on the stochastic representation $\lfloor 1 + \log(V_2)/\log(1 - e^{-\theta V_1})\rfloor$, with V_1, V_2 independent U(0, 1) random variables for a Logseries(θ) random variable. The parametrization with θ is better than that with α because α becomes 1 in double precision computer arithmetic with θ exceeding 35.3

---

[1]See Sibuya (1979), Christoph and Schreiber (2000)

corresponding to a Kendall tau value of around 0.9. A Taylor approximation can be used when underflow occurs.

P. (inverse gamma) The LT of X^{-1} when $X \sim \text{Gamma}(\theta^{-1}, 1)$ is

$$\psi(s; \theta) = \frac{2}{\Gamma(a)} s^{\alpha/2} K_\alpha(2\sqrt{s}), \quad s > 0, \quad \alpha = \theta^{-1} > 0,$$

where K_α is the modified Bessel function of the second kind. The functional inverse does not have closed form.

- The maximal non-negative moment degree is $M_\psi = \theta^{-1}$.

A.1.2 Two-parameter LT families: group 1

The next group are two-parameter LT families that are of the form $\varphi(s; \theta, \delta) = \psi(-\log \zeta(s; \delta); \theta)$ with $\varphi^{-1}(t; \theta, \delta) = \zeta^{-1}(\exp\{-\psi^{-1}(s; \theta)\}; \delta)$, where both ψ, ζ are chosen from the positive stable, gamma, Sibuya, logarithmic series LT families. There are 16 possibilities. If ψ, ζ are both positive stable (respectively Sibuya), then $\varphi = \psi(-\log \zeta)$ is positive stable (respectively Sibuya). Also the cases of ψ or ζ being logarithmic series are less interesting from a tail dependence viewpoint for Archimedean copulas and their extensions. This leaves 7 possibilities of which 5 (families E to I) were used in Joe (1997). The other two possibilities are included below as family N and O; they are added for completeness here but are not useful for Archimedean copulas because of computational difficulties.

Because ψ, ζ are infinitely divisible, then so is φ by Corollary 8.83.

Let $\varphi_i(s) = \varphi(s; \theta_i, \delta_i)$, $\psi_i(s) = \psi(s; \theta_i)$ and $\zeta_i(s) = \zeta(s; \delta_i)$ for $i = 1, 2$. For $\theta_1 = \theta_2$ and $\delta_1 < \delta_2$, $\omega_\varphi(s) = \varphi_1^{-1} \circ \varphi_2(s) = \zeta_1^{-1} \circ \zeta_2(s) = \omega_\zeta(s)$ so that property of $\omega_\varphi \in \mathcal{L}_\infty^*$ is inherited from ω_ζ. Let $\omega_\psi(s) = \psi_1^{-1} \circ \psi_2(s)$. For $\theta_1 < \theta_2$, $\omega_\varphi(s) = \varphi_1^{-1} \circ \varphi_2 = \zeta_1^{-1}(e^{-\omega_\psi(-\log \zeta_2)})$, and this has to investigated by case to determine whether it is in $\mathcal{L}_2^*$ or $\mathcal{L}_d^*$ or $\mathcal{L}_\infty^*$. For $\delta_1 = \delta_2 = \delta$, the functions $\omega_\varphi(s; \theta_1, \theta_2, \delta)$ with $\theta_1 < \theta_2$ are included below. If $\omega_\varphi \in \mathcal{L}_2^*$ (that is, concave), it is the same as the functional inverse ω_φ^{-1} being convex.

See Section 8.15 for the terminology involving composition of two LTs in the form $\varphi = \psi(-\log \zeta)$. This form has the interpretation of a T-stopped infinitely divisible process $\{Z(t)\}$, where T has LT ψ and then conditional on $T = t$, $Z(T)$ has LT ζ^t. Simulation methods based on the stochastic representation are included below.

E. (gamma stopped positive stable, or Mittag-Leffler[2]) . $\varphi(s; \theta, \delta) = (1 + s^{1/\delta})^{-1/\theta}$, for $\theta > 0$ and $\delta \geq 1$; functional inverse $\varphi^{-1}(t; \theta, \delta) = (t^{-\theta} - 1)^\delta$.

- Infinitely divisible due to Corollary 8.83.
- $\varphi = \psi(-\log \zeta)$ where ψ is gamma and ζ is positive stable.
- For $\theta_1 < \theta_2$, $\omega_\varphi(s; \theta_1, \theta_2, \delta) = [(1 + s^{1/\delta})^\gamma - 1]^\delta$ with $\gamma = \theta_1/\theta_2$. $\omega_\varphi \in \mathcal{L}_2^*$ (increasing and concave) with $0 < \gamma < 1$.
- As $s \to 0^+$, $1 - \varphi(s) = \theta^{-1} s^{1/\delta} + o(s^{1/\delta})$ so that $M_\varphi = \delta^{-1}$. The mean is infinite for $\delta > 1$.
- Simulation: let $T \sim \text{Gamma}(\theta^{-1}, 1)$, and given $T = t$, generate $Z(t) \sim \text{Postable}(\alpha = \delta^{-1}, \text{scale} = t^\delta)$.

F. (gamma stopped gamma) $\psi(s; \theta, \delta) = [1 + \delta^{-1} \log(1 + s)]^{-1/\theta}$, for $\theta, \delta > 0$; functional inverse $\varphi^{-1}(t; \theta, \delta) = \exp\{\delta(t^{-\theta} - 1)\} - 1$.

- Infinitely divisible due to Corollary 8.83.
- $\varphi = \psi(-\log \zeta)$ where ψ is gamma and ζ is gamma.
- For $\theta_1 < \theta_2$, $\omega_\varphi(s; \theta_1, \theta_2, \delta) = \exp\{\delta([1 + \delta^{-1} \log(1 + s)]^\gamma - 1)\} - 1$ with $\gamma = \theta_1/\theta_2$. $\omega_\varphi \in \mathcal{L}_2^*$ (increasing and concave) with $0 < \gamma < 1$.

---

[2]See Pillai (1990), Lin (1998)

- φ has an infinite Taylor expansion about $s = 0$ so that the maximal non-negative moment degree is $M_\varphi = \infty$.
- Simulation: let $T \sim \text{Gamma}(\theta^{-1}, 1)$, and given $T = t$, generate $Z(t) \sim \text{Gamma}(t/\delta, 1)$.

G. (positive stable stopped gamma) $\varphi(s; \theta, \delta) = \exp\{-[\delta^{-1}\log(1 + s)]^{1/\theta}\}$, for $\theta \geq 1$ and $\delta > 0$; functional inverse $\varphi^{-1}(t; \theta, \delta) = \exp\{\delta(-\log t)^\theta\} - 1$.

- Infinitely divisible due to Corollary 8.83.
- $\varphi = \psi(-\log \zeta)$ where ψ is positive stable and ζ is gamma.
- For $\theta_1 < \theta_2$, $\omega_\varphi(s; \theta_1, \theta_2, \delta) = \exp\{\delta^{1-\gamma}[\log(1+s)]^\gamma\} - 1$ with $\gamma = \theta_1/\theta_2$. For $0 < \delta < e^2$, $\omega_\varphi \in \mathcal{L}_2^*$ (increasing and concave) with $0 < \gamma < 1$.
- As $s \to 0^+$, $1 - \varphi(s) = \delta^{-1/\theta}s^{1/\theta} + o(s^{1/\theta})$ so that $M_\varphi = \theta^{-1}$. The mean is infinite for $\theta > 1$.
- Simulation: let $T \sim \text{Postable}(\alpha = \theta^{-1}, \text{scale} = 1)$, and given $T = t$, generate $Z(t) \sim \text{Gamma}(t/\delta, 1)$.

H. (Sibuya stopped positive stable) $\varphi(s; \theta, \delta) = 1 - [1 - \exp\{-s^{1/\delta}\}]^{1/\theta}$, for $\theta, \delta \geq 1$; functional inverse $\varphi^{-1}(t; \theta, \delta) = \{-\log[1 - (1 - t)^\theta]\}^\delta$.

- Infinitely divisible due to Corollary 8.83.
- $\varphi = \psi(-\log \zeta)$ where ψ is Sibuya and ζ is positive stable.
- For $\theta_1 < \theta_2$, $\omega_\varphi(s; \theta_1, \theta_2, \delta) = \{-\log[1 - (1 - e^{-s^{1/\delta}})^\gamma]\}^\delta$ with $\gamma = \theta_1/\theta_2$. $\omega_\varphi \in \mathcal{L}_2^*$ (increasing and concave) with $0 < \gamma < 1$.
- As $s \to 0^+$, $1 - \varphi(s) = s^{1/(\theta\delta)} + o(s^{1/(\theta\delta)})$ so that the maximal non-negative moment degree is $M_\varphi = (\theta\delta)^{-1}$. The mean is infinite for $\theta\delta > 1$.
- Simulation: let $T \sim \text{Sibuya}(\theta^{-1})$, and given $T = t$, generate $Z(t) \sim \text{Postable}(\alpha = \delta^{-1}, \text{scale} = t^\delta)$.

I. (Sibuya stopped gamma) $\varphi(s; \theta, \delta) = 1 - [1 - (1 + s)^{-1/\delta}]^{1/\theta}$, for $\theta \geq 1$ and $\delta > 0$; functional inverse $\varphi^{-1}(t; \theta, \delta) = [1 - (1 - t)^\theta]^{-\delta} - 1$.

- Infinitely divisible due to Corollary 8.83.
- $\varphi = \psi(-\log \zeta)$ where ψ Sibuya gamma and ζ is gamma.
- For $\theta_1 < \theta_2$, $\omega_\varphi(s; \theta_1, \theta_2, \delta) = \{1 - [1 - (1 + s)^{-1/\delta}]^\gamma\}^{-\delta} - 1$ with $\gamma = \theta_1/\theta_2$. For $0 < \delta \leq 1$, $\omega_\varphi \in \mathcal{L}_2^*$ (increasing and concave) with $0 < \gamma < 1$.
- As $s \to 0^+$, $1 - \varphi(s) = \delta^{-1/\theta}s^{1/\theta} + o(s^{1/\theta})$ so that $M_\varphi = \theta^{-1}$. The mean is infinite for $\theta > 1$.
- Simulation: let $T \sim \text{Sibuya}(\theta^{-1})$, and given $T = t$, generate $Z(t) \sim \text{Gamma}(t/\delta, 1)$.

N. (positive stable stopped Sibuya) $\varphi(s; \theta, \delta) = \exp\{-(-\log[1 - (1 - e^{-s})^{1/\delta}])^{1/\theta}\}$ for $\theta > 1$ and $\delta \geq 1$; functional inverse $\varphi^{-1}(t; \theta, \delta) = -\log\{1 - [1 - e^{-(-\log t)^\theta}]^\delta\}$.

- Infinitely divisible due to Corollary 8.83.
- $\varphi = \psi(-\log \zeta)$ where ψ is positive stable and ζ is Sibuya.
- For $\theta_1 < \theta_2$, $\omega_\varphi(s; \theta_1, \theta_2, \delta) = -\log\{1 - [1 - e^{-(-\log[1-(1-e^{-s})^{1/\delta}])^\gamma}]^\delta\}$ with $\gamma = \theta_1/\theta_2$.
- As $s \to 0^+$, $1 - \varphi(s) = s^{1/(\theta\delta)} + o(s^{1/(\theta\delta)})$, so that the maximal non-negative moment degree is $M_\varphi = (\theta\delta)^{-1}$.

O. (gamma stopped Sibuya) $\varphi(s; \theta, \delta) = \{1 - \log[1 - (1 - e^{-s})^{1/\delta}]\}^{-1/\theta}$ for $\theta > 0$ and $\delta \geq 1$; functional inverse $\varphi^{-1}(t; \theta, \delta) = -\log[1 - (1 - e^{1-t^{-\theta}})^\delta]$.

- Infinitely divisible due to Corollary 8.83.
- $\varphi = \psi(-\log \zeta)$ where ψ is gamma and ζ is Sibuya.
- For $\theta_1 < \theta_2$, $\omega_\varphi(s; \theta_1, \theta_2, \delta) = -\log[1 - (1 - e^{1-\{1-\log[1-(1-e^{-s})^{1/\delta}]\}^\gamma})^\delta]$ with $\gamma = \theta_1/\theta_2$.
- As $s \to 0^+$, $1 - \varphi(s) \sim \theta^{-1}s^{1/\delta}$, so that the maximal non-negative moment degree is $M_\varphi = \delta^{-1}$.

A.1.3 Two-parameter LT families: group 2

The third group of LT includes two that come from generalizations or reparametrizations of $\zeta(s) = \exp\{-\psi^{-1}(\psi(s;\theta_2);\theta_1)\}$ with $\theta_1 < \theta_2$ and ψ in gamma and logarithmic series families. There is a LT family K in the Appendix of Joe (1997) that is not included here; it is a variation of LT family J with extended parameter range.

J. (generalized Sibuya) $\zeta(s;\vartheta,\delta) = \delta^{-1}\left[1 - \{1 - \eta e^{-s}\}^{1/\vartheta}\right]$, $\vartheta \geq 1$, $0 < \delta \leq 1$, and $\eta = 1 - (1-\delta)^\vartheta \in (0,1]$; functional inverse $\zeta^{-1}(t;\vartheta,\delta) = -\log\{[1 - (1-\delta t)^\vartheta]/\eta\}$.

- Infinitely divisible because $-\log\zeta \in \mathcal{L}_\infty^*$.

- With $\delta = \eta = 1$, this becomes the Sibuya family.

- This is the LT of a discrete distribution on positive integers with mass $p_i = \eta/(\vartheta\delta)$ for $i = 1$ and $\eta^i[\prod_{j=1}^{i-1}(j - \vartheta^{-1})]/[\delta\vartheta i!]$ for $i > 1$. The mean is $\eta(1-\delta)/[\vartheta\delta(1-\eta)]$ and the variance is $\eta(1-\delta)(\delta\vartheta - \eta)/[\vartheta\delta(1-\eta)]^2$. Since $(1-\eta)^{1/\vartheta} = 1 - \delta$, $\zeta''(0) = (\vartheta - 1)\eta^2(1-\eta)^{1/\vartheta-2}/[\vartheta^2\delta] + \eta(1-\eta)^{1/\vartheta-1}/[\vartheta\delta] = \eta(1-\delta)(1-\eta)^{-2}\vartheta^{-2}\delta^{-1}(\vartheta - \eta)$.

- The maximal non-negative moment degree is $M_\zeta = \infty$ if $0 < \delta < 1$.

- If $\zeta(\cdot;\vartheta,\delta)$ is as above and $\psi(\cdot;\theta_1)$ is logarithmic series such that $\delta = 1 - e^{-\theta_1} \in (0,1)$, $1 - e^{-\theta_2} = \eta = 1 - (1-\delta)^\vartheta$ and $\vartheta = \theta_2/\theta_1 > 1$, then $\varphi = \psi(-\log\zeta)$ is log series with parameter $\theta_2 > \theta_1$. The transform is also: $(\theta_1,\theta_2) = (-\log(1-\delta), -\vartheta\log(1-\delta))$.

- Simulation: Proposition 3.3 of Hofert (2011) has an algorithm for simulation from this LT. With the transformed parameters, the algorithm becomes the following.

  If $0 < \delta \leq 1 - e^{-1}$, simulate $U \sim U(0,1)$ and $X \sim$ Logseries($-\vartheta\log(1-\delta)$) until $U \leq 1/[(X - \vartheta^{-1})B(X, 1 - \vartheta^{-1})]$, and then return X as a variable with LT ζ.

  If $1 - e^{-1} < \delta < 1$, simulate $U \sim U(0,1)$ and $X \sim$ Sibuya(ϑ^{-1}) until $U \leq \eta^{X-1}$, and then return X as a variable with LT ζ.

L. (exponentially tilted positive stable) $\zeta(s;\vartheta,\delta) = \exp\{-(\delta^{-\vartheta} + s)^{1/\vartheta} + \delta^{-1}\}$, $\vartheta \geq 1$ and $\delta > 0$; functional inverse $\zeta^{-1}(t;\vartheta,\delta) = (\delta^{-1} - \log t)^\vartheta - \delta^{-\vartheta}$.

- Infinitely divisible because $-\log\zeta \in \mathcal{L}_\infty^*$.

- This two-parameter family that includes the positive stable LT when $\delta \to \infty$ and the inverse Gaussian[3] LT when $\vartheta = 2$.

- If ζ is as above with parameters $\vartheta = \theta_2/\theta_1 > 1$ and $\delta = 1$, and ψ is gamma with parameter $\theta_1 > 0$, then $\varphi = \psi(-\log\zeta)$ is gamma with parameter $\theta_2 > \theta_1$.

- The maximal non-negative moment degree is $M_\zeta = \infty$ if $0 < \delta < \infty$.

- Simulation: A method is based on noted that ζ is the LT of a positive stable random variable modified with a Laplace transform parameter (page 218 of Marshall and Olkin (2007) and Section 3.18). That is, $\zeta(s) = \eta(s+\varsigma)/\eta(\varsigma)$ where $\eta(s) = \exp\{-s^{-1/\vartheta}\}$ is the positive stable LT and $\varsigma = \delta^{-\vartheta}$. Hence a random variable with this LT can be simulated via the standard rejection algorithm as "sample $X \sim$ Postable(ϑ^{-1}) and $U \sim U(0,1)$ independently until $U \leq e^{-\varsigma X}$, and then return X" (see Hofert (2008)). The expected number of iterations is $e^\delta = 1/\mathbb{P}(U \leq e^{-\varsigma X})$. Since ζ is infinitely divisible, to reduce the computational complexity for large δ, one can simulate m independent random variables from LT $\zeta^{1/m}(s;\vartheta,\delta)$ and take the sum (see Algorithm 3.5 of Hofert (2011) for an optimal m). Note that $\zeta^{1/m}(s) = \eta^{1/m}(s+\varsigma)/\eta^{1/m}(\varsigma)$ and $\eta^{1/m}(s)$ is the LT of a scaled positive stable random variable with shape parameter ϑ^{-1} and scale parameter $m^{-\vartheta}$.

---

[3]With density $f_{\text{IG}}(x;\mu,\varsigma) = \varsigma^{1/2}[2\pi x^3]^{-1/2}\exp\{-\varsigma(x-\mu)^2/[2\mu^2 x]\}$, $x > 0$, $\mu,\varsigma > 0$, the inverse Gaussian LT is $\exp\{\varsigma\mu^{-1}[1 - (1 + 2\mu^2 s/\varsigma)^{1/2}]\}$, and this matches with $\delta^{-1} = \varsigma/\mu$, $\vartheta = 2$ and scale parameter 2ς (latter doesn't affect the Archimedean copula). In the frailty model literature, sometimes the inverse Gaussian frailty model is considered in addition to the gamma and positive stable frailty models.

M. (negative binomial) $\varphi(s; \theta, \pi) = [(1 - \pi)e^{-s}/(1 - \pi e^{-s})]^{1/\theta} = [(1 - \pi)/(e^s - \pi)]^{1/\theta}$, $\theta > 0$ and $0 \leq \pi < 1$; functional inverse $\varphi^{-1}(t; \theta, \pi) = \log[(1 - \pi)t^{-\theta} + \pi]$.

- Infinitely divisible because θ is a parameter that is a power.
- LT of a shifted negative binomial distribution with convolution parameter θ^{-1}, probability parameter $1 - \pi$ and masses on $\theta^{-1} + i$ for $i = 0, 1, 2, \ldots$;
- The maximal non-negative moment degree is $M_\varphi = \infty$.
- Simulation can be done as a gamma mixture of Poisson, or by the standard method for discrete random variables (for a value $p \in (0, 1)$, iterate from the mode. then increment/decrement by 1 at a time until the cumulative probability exceeds p).

A.1.4   LT families via integration

This is based on results on LTs in Section 3.18. Let $\psi \in \mathcal{L}_\infty$. Consider

$$v(s) = \int_s^\infty \psi(t)\, \mathrm{dt} \Big/ \int_0^\infty \psi(t)\, \mathrm{dt}$$

where the integral in the denominator is finite. Then $v(0) = 1$, $v(\infty) = 0$ and v is completely monotone, so that $v \in \mathcal{L}_\infty$.

     Examples of v that are useful for Archimedean copulas are normalized integrals of the positive stable and Mittag-Leffler LTs.

Q. (Integrated positive stable). For $\theta > 1$,

$$v(s; \theta) = 1 - F_\Gamma(s^{1/\theta}; \theta) = \frac{1}{\Gamma(\theta)} \int_{s^{1/\theta}}^\infty z^{\theta-1}e^{-z}\mathrm{dz} = \frac{1}{\Gamma(1+\theta)} \int_s^\infty \exp\{-y^{1/\theta}\}\mathrm{dy},$$

where $F_\Gamma(\cdot; \theta)$ is the cdf of a Gamma$(\theta, 1)$ random variable. The functional inverse is $v^{-1}(t; \theta) = [F_\Gamma^{-1}(1 - t; \theta)]^\theta$.

R. (Integrated Mittag-Leffler). The Mittag-Leffler LT family is integrable for part of its parameter range.

Let B be the Beta function, and let $F_B(\cdot; \alpha_1, \alpha_2)$ denote the cdf for the Beta(α_1, α_2) distribution with functional inverse F_B^{-1}. Let $\zeta = 1/\theta$. Then the integral of the Mittag-Leffler LT is:

$$\int_s^\infty (1 + t^{1/\delta})^{-\zeta}\mathrm{dt} = \int_{s^{1/\delta}}^\infty (1 + w)^{-\zeta}\delta w^{\delta-1}\mathrm{dw}$$

$$= \delta \int_{s^{1/\delta}/(1+s^{1/\delta})}^1 v^{\delta-1}(1 - v)^{\zeta-\delta-1}\mathrm{dv} \quad (v = w/(1 + w))$$

$$= \delta B(\delta, \zeta - \delta)\left[1 - F_B\big(s^{1/\delta}/(1 + s^{1/\delta}); \delta, \zeta - \delta\big)\right],$$

provided $\zeta > \delta > 1$. By normalizing to get $v(0) = 1$, one gets

$$v(s; \zeta, \delta) = 1 - F_B\big(s^{1/\delta}/(1 + s^{1/\delta}); \delta, \zeta - \delta\big), \quad s \geq 0,$$

with functional inverse

$$v^{-1}(t; \zeta, \delta) = \left\{\frac{1}{1 - F_B^{-1}(1 - t; \delta, \zeta - \delta)} - 1\right\}^\delta, \quad 0 < t < 1.$$

For use with an Archimedean copula, it is convenient to reparametrize to $\vartheta = (\zeta - \delta)^{-1}$ to get:

$$v(s; \vartheta, \delta) = 1 - F_B\big(s^{1/\delta}/(1 + s^{1/\delta}); \delta, \vartheta^{-1}\big).$$

A.2 Archimedean generators in Nelsen's book

Table 4.1 of Nelsen (1999) has a list of one-parameter generators of bivariate Archimedean copulas. Hence, these are inverses of Laplace transforms or Williamson 2-transforms. Because no link was made to the Laplace transform families in Joe (1997) (replicated in Section A.1), a summary is given in Table A.1 to show members that are special cases of our listed one-parameter and two-parameter LT families. It appears that those families not based on LTs are not statistically useful.

Nelsen's label	inverse of generator	resulting family
4.2.1	gamma LT (B) + extension to $\mathcal{L}_2$	MTCJ + neg. dependence
4.2.4	positive stable LT (A)	Gumbel
4.2.5	log. series LT (D) + extension to $\mathcal{L}_2$	Frank
4.2.6	Sibuya LT (C)	Joe/B5
4.2.12	Mittag-Leffler LT (E)	BB1 with $\theta = 1$, $\delta = \vartheta$
4.2.14	Mittag-Leffler LT (E)	BB1 with $\delta = \vartheta$, $\theta = \vartheta^{-1}$
4.2.13	tilted positive stable LT (L)	BB9 with $\delta = 1$, $\vartheta = \vartheta$
4.2.19	gamma stopped gamma LT (F)	BB2 with $\delta = \vartheta$, $\theta = 1$
4.2.20	gamma stopped gamma LT (F)	BB2 with $\delta = 1$, $\theta = \vartheta$
4.2.3	geometric LT + extension to $\mathcal{L}_2$	BB10 with $\theta = 1$, $\pi = \vartheta$
4.2.9	boundary of $\mathcal{L}_2$ extension of LT (L)	Barnett (1980), neg. dependence
4.2.10	$\mathcal{L}_2$ extension of NegBin LT (M)	BB10 with $0 < \theta \leq 1$, $\pi = -1$
4.2.16	in $\mathcal{L}_2$, not a LT	includes C^- but not $C^\perp, C^+$, in $\mathcal{L}_{d(\vartheta)}$ with $d(\vartheta)$ increasing
4.2.17	LT for $\vartheta > 0$	$C^\perp$ for $\vartheta = -1$, C^+ as $\vartheta \to \infty$, $\kappa_L = 2$, $\kappa_U = 2$
4.2.2	$\mathcal{L}_2$ with bounded support	copula support not all of $[0,1]^2$
4.2.7	$\mathcal{L}_2$ with bounded support	copula support not all of $[0,1]^2$
4.2.8	$\mathcal{L}_2$ with bounded support	copula support not all of $[0,1]^2$
4.2.11	$\mathcal{L}_2$ with bounded support	copula support not all of $[0,1]^2$
4.2.15	$\mathcal{L}_2$ with bounded support	copula support not all of $[0,1]^2$
4.2.18	$\mathcal{L}_2$ with bounded support	copula support not all of $[0,1]^2$
4.2.21	$\mathcal{L}_2$ with bounded support	copula support not all of $[0,1]^2$
4.2.22	$\mathcal{L}_2$ with bounded support	copula support not all of $[0,1]^2$

Table A.1 *Relation of Archimedean generators of Nelsen (1999) with LT families in Joe (1997); Nelsen's parameter is denoted here as ϑ and $\mathcal{L}_2$ refers to the class of Williamson 2-transforms.*

Note that the bivariate copula family (4.2.9) in Nelsen (1999) is a boundary case of an extension of copula family BB9 to negative dependence. Nelsen has generator $\psi^{-1}(t) = \log(1 - \delta \log t)$ for $0 < \delta \leq 1$ which leads to $\psi(s) = \exp\{\delta^{-1}(1 - e^s)\}$. Joe (1997) on page 161–162 has family MB9E extending BB9 to negative dependence with function $\zeta(s; \theta, \delta) = \exp\{-(\delta^{-\theta} + s)^{1/\theta} + \delta^{-1}\}$ that is in $\mathcal{L}_2$ for $0 < \theta < 1$ and $\delta^{-1} \geq 1 - \theta$. Since the Archimedean copula remains the same with a scale change on ζ, introduce a scale parameter as θ and get

$$\lim_{\theta \to 0^+} \exp\{-(\delta^{-\theta} + \theta s)^{1/\theta} + \delta^{-1}\} = \lim_{\theta \to 0^+} \exp\{-\delta^{-1}(1 + \delta^\theta \theta s)^{1/\theta} + \delta^{-1}\} = \exp\{\delta^{-1}(1 - e^s)\}.$$

Family (4.2.17) in Nelsen (1999) is a bit different from other Archimedean copula families. It is

$$C(u, v; \vartheta) = \left\{1 + \frac{[(1 + u)^{-\vartheta} - 1][(1 + v)^{-\vartheta} - 1]}{2^{-\vartheta} - 1}\right\}^{-1/\vartheta} - 1, \quad \vartheta \in (-\infty, \infty) \setminus \{0\},$$

with a limit as $\vartheta \to 0$. The generator is $\psi^{-1}(t; \vartheta) = -\log\{[(1 + t)^{-\vartheta} - 1]/[2^{-\vartheta} - 1]\}$, and the inverse generator is

$$\psi(s; \vartheta) = [1 + (2^{-\vartheta} - 1)e^{-s}]^{-1/\vartheta} - 1. \tag{A.1}$$

It can be checked that (A.1) is the LT of a random variable on the positive integers when $\vartheta > 0$ (the Taylor expansion with $z = e^{-s}$ leads to a valid probability generating function). When $\vartheta = 1$, it is the LT of a Geometric($\frac{1}{2}$) distribution with support on the positive integers. For $-1 < \vartheta < 0$, with $z = e^{-s}$, (A.1) has a Taylor series in z with negative coefficients, and therefore is not a LT. As $\vartheta \to 0$, one gets $\psi(s; 0) = \exp\{(\log 2)e^{-s}\} - 1$. Through a combination of numerical calculations and approximation for large s, it seems that $-\log \psi(s)$ is concave for $-1 < \vartheta < 0$, and this implies that the copula has positive quadrant dependence. Hence for part of the parameter space this appears to be a copula with positive dependence such that $\psi \in \mathcal{L}_2$ but $\psi \notin \mathcal{L}_\infty$.

Bibliography

Aas, K., Czado, C., Frigessi, A., and Bakken, H. (2009). Pair-copula constructions of multiple dependence. *Insurance: Mathematics and Economics*, 44(2):182–198.

Aas, K. and Hobæk Haff, I. (2006). The generalized hyperbolic skew Student's t-distribution. *Journal of Financial Econometrics*, 4(2):275–309.

Abramowitz, M. and Stegun, I. A. (1964). *Handbook of Mathematical Functions with Formulas, Graphs, and Mathematical Tables*. Dover Publications.

Acar, E. F., Craiu, R. V., and Yao, F. (2011). Dependence calibration in conditional copulas: A nonparametric approach. *Biometrics*, 67(2):445–453.

Acar, E. F., Genest, C., and Nešlehová, J. (2012). Beyond simplified pair-copula constructions. *Journal of Multivariate Analysis*, 110:74–90.

Agresti, A. (1984). *Analysis of Ordinal Categorical Data*. Wiley, New York.

Ali, H. M., Mikhail, N. N., and Haq, M. S. (1978). A class of bivariate distribution including the bivariate logistic. *Journal of Multivariate Analysis*, 8(3):405–412.

Alink, S., Löwe, M., and Wüthrich, M. V. (2007). Diversification for general copula dependence. *Statistica Neerlandica*, 61(4):446–465.

Anderson, T. W. (1958). *An Introduction to Multivariate Statistical Analysis*. Wiley, New York.

Andrews, D. W. K. (1988). Chi-square diagnostic tests for econometric models. *Journal of Econometrics*, 37(1):135–156.

Ang, A. and Chen, J. (2002). Asymmetric correlations of equity portfolios. *Journal of Financial Economics*, 63(3):443–494.

Arbenz, P., Hummel, C., and Mainik, G. (2012). Copula based hierarchical risk aggregation through sample reordering. *Insurance: Mathematics and Economics*, 51(1):122–133.

Azzalini, A. (1985). A class of distributions which includes the normal ones. *Scandinavian Journal of Statistics*, 12(2):171–178.

Azzalini, A. and Capitanio, A. (2003). Distributions generated by perturbation of symmetry with emphasis on a multivariate skew t-distribution. *Journal of the Royal Statistical Society B*, 65(2):367–389.

Azzalini, A. and Dalla Valle, A. (1996). The multivariate skew-normal distribution. *Biometrika*, 83(4):715–726.

Balakrishnan, N. and Lai, C. D. (2009). *Continuous Bivariate Distributions*. Springer, Dordrecht, second edition.

Barbe, P., Genest, C., Ghoudi, K., and Rémillard, B. (1996). On Kendall's process. *Journal of Multivariate Analysis*, 58(2):197–229.

Bárdossy, A. and Li, J. (2008). Geostatistical interpolation using copulas. *Water Resources Research*, 44(7).

Barlow, R. E. and Proshan, F. (1981). *Statistical Theory of Reliability and Life Testing*. To Begin With, Silver Spring, MD.

Barndorff-Nielsen, O. E. (1977). Exponentially decreasing distributions for the logarithm of particle size. *Proceedings of the Royal Society of London, Series A*, 353(1674):409–419.

Barndorff-Nielsen, O. E. (1997). Normal inverse Gaussian distributions and stochastic volatility modelling. *Scandinavian Journal of Statistics*, 24(1):1–13.

Barnett, V. (1980). Some bivariate uniform distributions. *Communications in Statistics — Theory and Methods*, 9(4):453–461.

Bauer, A., Czado, C., and Klein, T. (2012). Pair-copula constructions for non-Gaussian DAG models. *Canadian Journal of Statistics*, 40(1):86–109.

Beare, B. K. (2010). Copulas and temporal dependence. *Econometrica*, 78(1):395–410.

Bedford, T. and Cooke, R. M. (2001). Probability density decomposition for conditionally dependent random variables modeled by vines. *Annals of Mathematics and Artificial Intelligence*, 32(1-4):245–268.

Bedford, T. and Cooke, R. M. (2002). Vines — a new graphical model for dependent random variables. *Annals of Statistics*, 30(4):1031–1068.

Belgorodski, N. (2010). Selecting Pair-Copula Families for Regular Vines with Application to the Multivariate Analysis of European Stock Market Indices. Master's thesis, Technische Universität München.

Berg, D. (2009). Copula goodness-of-fit testing: an overview and power comparison. *European Journal of Finance*, 15(7-8):675–701.

Bernard, C. and Czado, C. (2013). Multivariate option pricing using copulae. *Applied Stochastic Models in Business and Industry*, 29(5):509–526.

Bernstein, S. (1928). Sur les fonctions absolument monotones. *Acta Mathematica*, 52:1–66.

Berntsen, J., Espelid, T., and Genz, A. (1991). An adaptive algorithm for the approximate calculation of multiple integrals. *ACM Transactions on Mathematical Software*, 17(4):437–451.

Bertoin, J. (1996). *Lévy Processes*. Cambridge University Press, Cambridge.

Bhuchongkul, S. (1964). A class of nonparametric tests for independence in bivariate populations. *Annals of Mathematical Statistics*, 35(1):138–149.

Biller, B. (2009). Copula-based multivariate input models for stochastic simulation. *Operations Research*, 57(4):878–892.

Biller, B. and Nelson, B. L. (2005). Fitting time-series input processes for simulation. *Operations Research*, 53(3):549–559.

Billingsley, P. (1961). *Statistical Inference for Markov Processes*. University of Chicago Press, Chicago.

Bingham, N. H., Goldie, C. M., and Teugels, J. L. (1987). *Regular Variation*. Cambridge University Press, Cambridge.

Blickle, T. and Thiele, L. (1996). A comparison of selection schemes used in evolutionary algorithms. *Evolutionary Computation*, 4(4):361–394.

Blomqvist, N. (1950). On a measure of dependence between two random variables. *Annals of Mathematical Statistics*, 21(4):593–600.

Bogaerts, K. and Lesaffre, E. (2008). Modeling the association of bivariate interval-censored data using the copula approach. *Statistics in Medicine*, 27(30):6379–6392.

Bollen, K. A. (1989). *Structural Equations with Latent Variables*. Wiley, New York.

Bouezmarni, T., Rombouts, J. V. K., and Taamouti, A. (2010). Asymptotic properties of the Bernstein density copula estimator for alpha-mixing data. *Journal of Multivariate Analysis*, 101(1):1–10.

Braeken, J. and Tuerlinckx, F. (2009). A mixed model framework for teratology studies. *Biostatistics*, 10(4):744–755.

Braeken, J., Tuerlinckx, F., and de Boeck, P. (2007). Copula functions for residual dependency. *Psychometrika*, 72(3):393–411.

Brechmann, E. C. (2013). *Hierarchical Kendall Copulas and the Modeling of Systemic and Operational Risk*. PhD thesis, Technische Universität München.

Brechmann, E. C. (2014). Hierarchical Kendall copulas: properties and inference. *Canadian Journal of Statistics*, 42(1):78–108.

Brechmann, E. C., Czado, C., and Aas, K. (2012). Truncated regular vines in high dimensions with application to financial data. *Canadian Journal of Statistics*, 40(1):68–85.

Brechmann, E. C., Czado, C., and Paterlini, S. (2014). Flexible dependence modeling of operational risk losses and its impact on total capital requirements. *Journal of Banking & Finance*, 40:271–285.

Brechmann, E. C. and Joe, H. (2014a). Parsimonious parameterization of correlation matrices using truncated vines and factor analysis. *Computational Statistics and Data Analysis*. in press.

Brechmann, E. C. and Joe, H. (2014b). Truncation of vine copulas using fit indices. *Submitted*.

Breckler, S. J. (1990). Applications of covariance structure modeling in psychology: a cause for concern? *Psychological Bulletin*, 107(2):260–273.

Browne, M. W. and Arminger, G. (1995). Specification and estimation of mean- and covariance-structure models. In Arminger, G., Clogg, C. C., and Sobel, M. F., editors, *Handbook of Statistical Modeling for the Social and Behavioral Science*, chapter 4, pages 185–249. Plenum Press, New York.

Cameron, A. C. and Trivedi, P. K. (1998). *Regression Analysis of Count Data*. Cambridge University Press, Cambridge.

Campbell, G. (1981). Nonparametric bivariate estimation with randomly censored data. *Biometrika*, 68(2):417–422.

Capéraà, P., Fougères, A.-L., and Genest, C. (1997a). A nonparametric estimation procedure for bivariate extreme value copulas. *Biometrika*, 84(3):567–577.

Capéraà, P., Fougères, A.-L., and Genest, C. (1997b). A stochastic ordering based on a decomposition of Kendall's tau. In Beneš, V. and Štěpán, J., editors, *Distributions with Given Marginals and Moment Problems*, pages 81–86, Dordrecht. Kluwer.

Capéraà, P., Fougères, A.-L., and Genest, C. (2000). Bivariate distributions with given extreme value attractor. *Journal of Multivariate Analysis*, 72(1):30–49.

Capéraà, P. and Genest, C. (1993). Spearman's ρ is larger than Kendall's τ for positively dependent random variables. *Nonparametric Statistics*, 2(2):183–194.

Carriere, J. F. (2000). Bivariate survival models for coupled lives. *Scandinavian Actuarial Journal*, (1):17–32.

Chaganty, N. R. and Joe, H. (2006). Range of correlation matrices for dependent Bernoulli random variables. *Biometrika*, 93(1):197–206.

Chambers, J. M., Mallows, C. L., and Stuck, B. W. (1976). A method for simulating stable random variables. *Journal of the American Statistical Association*, 71(354):340–344.

Chan, Y. and Li, H. (2008). Tail dependence for multivariate t-copulas and its monotonicity. *Insurance: Mathematics and Economics*, 42(2):763–770.

Charpentier, A. and Juri, A. (2006). Limiting dependence structures for tail events, with applications to credit derivatives. *Journal of Applied Probability*, 43(2):563–586.

Charpentier, A. and Segers, J. (2009). Tails of multivariate Archimedean copulas. *Journal of Multivariate Analysis*, 100(7):1521–1537.

Chen, X. and Fan, Y. (2006). Estimation of copula-based semiparametric time series models. *Journal of Econometrics*, 130(2):307–335.

Cherubini, U., Gobbi, F., Mulinacci, S., and Romagnoli, S. (2012). *Dynamic Copula Methods in Finance*. Wiley, Chichester.

Cherubini, U., Luciano, E., and Vecchaito, W. (2004). *Copula Methods in Finance*. Wiley, Chichester.

Cheung, K. C. (2009). Upper comonotonicity. *Insurance: Mathematics and Economics*, 45(1):35–40.

Chollete, L., Heinen, A., and Valdesogo, A. (2009). Modeling international financial returns with a multivariate regime-switching copula. *Journal of Financial Econometrics*, 7(4):437–480.

Christoph, G. and Schreiber, K. (2000). Scaled Sibuya distribution and discrete self-decomposability. *Statistics & Probability Letters*, 48(2):181–187.

Chung, K. L. (1974). *A Course in Probability Theory, second edition*. Academic, New York.

Clayton, D. G. (1978). A model for association in bivariate life tables and its application in epidemiological studies of familial tendency in chronic disease incidence. *Biometrika*, 65(1):141–151.

Colangelo, A. (2008). A study on LTD and RTI positive dependence orderings. *Statistics & Probability Letters*, 78(14):2222–2229.

Coles, S. (2001). *An Introduction to Statistical Modeling of Extreme Values*. Springer-Verlag, London.

Cont, R. (2001). Empirical properties of asset returns: stylized facts and statistical issues. *Quantitative Finance*, 1(2):223–236.

Cook, R. D. and Johnson, M. E. (1981). A family of distributions for modelling non-elliptically symmetric multivariate data. *Journal of the Royal Statistical Society B*, 43(2):210–218.

Cooke, R. M. (1997). Markov and entropy properties of tree and vines-dependent variables. In *Proceedings of the ASA Section of Bayesian Statistical Science*.

Cooke, R. M., Joe, H., and Aas, K. (2011a). Vines arise. In Kurowicka, D. and Joe, H., editors, *Dependence Modeling: Vine Copula Handbook*, chapter 3, pages 37–71. World Scientific Publishing Company, Singapore.

Cooke, R. M., Kousky, C., and Joe, H. (2011b). Micro correlations and tail dependence. In Kurowicka, D. and Joe, H., editors, *Dependence Modeling: Vine Copula Handbook*, chapter 5, pages 89–112. World Scientific, Singapore.

Cooley, D., Davis, R. A., and Naveau, P. (2010). The pairwise beta distribution: A flexible parametric multivariate model for extremes. *Journal of Multivariate Analysis*, 101(9):2103–2117.

Crowder, M. (1989). A multivariate distribution with Weibull connections. *Journal of the Royal Statistical Society B*, 51(1):93–107.

Csárdi, G. and Nepusz, T. (2012). *igraph Reference Manual*.

Cuadras, C. and Augé, J. (1981). A continuous general multivariate distribution and its properties. *Communications in Statistics — Theory and Methods*, 10(4):339–353.

Czado, C., Kastenmeier, R., Brechmann, E. C., and Min, A. (2012). A mixed copula model for insurance claims and claim sizes. *Scandinavian Actuarial Journal*, (4):278–305.

Dall'Aglio, G. (1972). Fréchet classes and compatibility of distribution functions. *Sympos. Math.*, IX:131–150.

Danaher, P. J. and Smith, M. S. (2011). Modeling multivariate distributions using copulas: applications in marketing. *Marketing Science*, 30(1):4–21.

Darsow, W. F., Nyugen, B., and Olsen, E. T. (1992). Copulas and Markov processes. *Illinois Journal of Mathematics*, 36(4):600–642.

Das Gupta, S., Eaton, M., Olkin, I., Perlman, M., Savage, L. J., and Sobel, M. (1972). Inequalities on the probability content of convex regions for elliptically contoured distributions. In *Proceedings, Sixth Berkeley Symposium on Mathematical Statistics and Probability*, volume 2, pages 241–264, Berkeley, CA. University of California Press.

Daul, S., De Giorgi, E., Lindskog, F., and McNeil, A. (2003). Using the grouped t-copula. *RISK Magazine*, 6(11):73–76.

de Haan, L. (1984). A spectral representation for max-stable processes. *Annals of Probability*, 12(4):1194–1204.

de Haan, L. and Ferreira, A. (2006). *Extreme Value Theory: An Introduction.* Springer Series in Operations Research and Financial Engineering. Springer, New York.

de Haan, L. and Resnick, S. I. (1977). Limit theory for multivariate sample extremes. *Zeitschrift für Wahrscheinlichkeitstheorie und Verwandte Gebiete*, 40(4):317–337.

de Haan, L. and Zhou, C. (2011). Extreme residual dependence for random vectors and processes. *Advances in Applied Probability*, 43(1):217–242.

de Leon, A. R. and Wu, B. (2011). Copula-based regression models for a bivariate mixed discrete and continuous outcome. *Statistics in Medicine*, 30(2):175–185.

De Michele, C., Salvadori, G., Passoni, G., and Vezzoli, R. (2007). A multivariate model of sea storms using copulas. *Coastal Engineering*, 54(10):734–751.

Deheuvels, P. (1978). Caractérisation complète des lois extrêmes multivariées et de la convergence des types extrêmes. *Publications de l'Institut de Statistique de l'Université de Paris*, XXIII(3):1–36.

Dehgani, A., Dolati, A., and Úbeda Flores, M. (2013). Measures of radial asymmetry for bivariate random vectors. *Statistical Papers*, 54(2):271–286.

Demarta, S. and McNeil, A. J. (2005). The t copula and related copulas. *International Statistical Review*, 73(1):111–129.

Denuit, M., Dhaene, J., Goovaerts, M., and Kaas, R. (2005). *Actuarial Theory for Dependent Risks.* Wiley, Chichester.

Devroye, L. (1986). *Non-Uniform Random Variate Generation.* Springer-Verlag, New York.

Devroye, L. (1993). A triptych of discrete distributions related to the stable law. *Statistics & Probability Letters*, 18(5):349–351.

Dharmadhikari, S. and Joag-dev, K. (1988). *Unimodality, Convexity, and Applications.* Academic, San Diego.

Dissmann, J., Brechmann, E. C., Czado, C., and Kurowicka, D. (2013). Selecting and estimating regular vine copulae and application to financial returns. *Computational Statistics and Data Analysis*, 59:52–69.

Dissmann, J. F. (2010). Statistical Inference for Regular Vines and Application. Master's thesis, Technische Universität München.

Dobrić, J. and Schmid, F. (2005). Nonparametric estimation of the lower tail dependence λ_l in bivariate copulas. *Journal of Applied Statistics*, 32(4):387–407.

Donnelly, T. G. (1973). Algorithm 462: bivariate normal distribution. *Communications of the Association for Computing Machinery*, 16(10):638.

du Sautoy, M. (2008). *Symmetry. A Journey into the Patterns of Nature*. Harper, New York.

Dupuis, D. J. (2005). Ozone concentrations: A robust analysis of multivariate extremes. *Technometrics*, 47(2):191–201.

Durante, F. and Sempi, C. (2010). Copula theory: an introduction. In Jaworski, P., Durante, F., Härdle, W. K., and Rychlik, T., editors, *Copula Theory and Its Applications, Lecture Notes in Statistics*, volume 198, pages 3–31, Heidelberg. Springer-Verlag.

El-Shaarawi, A. H., Zhu, R., and Joe, H. (2011). Modelling species abundance using the Poisson-Tweedie family. *Environmetrics*, 22(2):152–164.

Elidan, G. (2013). Copulas in machine learning. In Jaworski, P., Durante, F., and Härdle, W. K., editors, *Copulae in Mathematical and Quantitative Finance*, pages 39–60. Springer-Verlag, Berlin.

Embrechts, P. (2009). Linear correlation and EVT: Properties and caveats. *Journal of Financial Econometrics*, 7(1):30–39.

Embrechts, P., Klüppelberg, C., and Mikosch, T. (1997). *Modelling Extremal Events*, volume 33. Springer-Verlag, Berlin.

Embrechts, P., Lambrigger, D. D., and Wüthrich, M. V. (2009). Multivariate extremes and the aggregation of dependent risks: examples and counter-examples. *Extremes*, 12(2):107–127.

Embrechts, P., McNeil, A. J., and Straumann, D. (2002). Correlation and dependence in risk management: properties and pitfalls. In Dempster, M., editor, *Risk Management: Value at Risk and Beyond*, pages 176–223. Cambridge University Press, Cambridge.

Engel, R. F. (2009). *Anticipating Correlations: a New Paradigm for Risk Management*. Princeton University Press, Princeton.

Erhardt, V. and Czado, C. (2012). Modeling dependent yearly claim totals including zero claims in private health insurance. *Scandinavian Actuarial Journal*, (2):106–129.

Escarela, G., Mena, R. H., and Castillo-Morales, A. (2006). A flexible class of parametric transition regression models based on copulas: application to poliomyelitis incidence. *Statistical Methods in Medical Research*, 15(6):593–609.

Everitt, B. (1974). *Cluster Analysis*. Heinemann Educational Books, London.

Falk, M., Hüsler, J., and Reiss, R. D. (2010). *Laws of Small Numbers: Extremes and Rare Events*. Springer, Basel, second edition.

Fang, K. T., Kotz, S., and Ng, K. W. (1990). *Symmetric Multivariate and Related Distributions*. Chapman & Hall, London.

Fang, Z. and Joe, H. (1992). Further developments on some dependence orderings for continuous bivariate distributions. *Annals of the Institute of Statistical Mathematics*, 44(3):501–517.

Fantazzini, D. (2009). The effects of misspecified marginals and copulas on computing the value at risk: A Monte Carlo study. *Computational Statistics & Data Analysis*, 53(6):2168–2188.

Farlie, D. J. G. (1960). The performance of some correlation coefficients for a general bivariate distribution. *Biometrika*, 47(3-4):307–323.

Favre, A. C., El Adlouni, S., Perreault, L., Thiémonge, N., and Bobée, B. (2004). Multivariate hydrological frequency analysis using copulas. *Water Resources Research*, 40(1).

Feller, W. (1971). *An Introduction to Probability Theory and Its Applications. Vol. II*. Wiley, New York, second edition.

Fermanian, J.-D., Radulović, D., and Wegkamp, M. (2004). Weak convergence of empirical copula processes. *Bernoulli*, 10(5):847–860.

Fermanian, J.-D. and Scaillet, O. (2003). Nonparametric estimation of copulas for time series. *Journal of Risk*, 5(1):25–54.

Fienberg, S. E., Bromet, E. J., Follman, D., Lambert, D., and May, S. M. (1985). Longitudinal analysis of categorical epidemiological data: a study of Three Mile Island. *Environmental Health Perspectives*, 63:241–248.

Frahm, G., Junker, M., and Schmidt, R. (2005). Estimating the tail-dependence coefficient: properties and pitfalls. *Insurance: Mathematics and Economics*, 37(1):80–100.

Frahm, G., Junker, M., and Szimayer, A. (2003). Elliptical copulas: applicability and limitations. *Statistics & Probability Letters*, 63(3):275–286.

Frank, M. J. (1979). On the simultaneous associativity of $f(x,y)$ and $x + y - f(x,y)$. *Aequationes Mathematicae*, 19(1):194–226.

Fréchet, M. (1935). Généralisations du théorème des probabilités totales. *Fundamenta Mathematicae*, 25:379–387.

Fréchet, M. (1951). Sur les tableaux de corrélation dont les marges sont données. *Annales de l'Université de Lyon I, Sciences*, 14:53–77.

Fréchet, M. (1958). Remarques au sujet de la note précédente. *Comptes Rendus de l'Académie des Sciences, Paris*, 246:2719–2720.

Fredricks, A. J. and Dossett, D. L. (1983). Attitude-behavior relations: A comparison of the Fishbein-Ajzen and the Bentler-Speckart models. *Journal of Personality and Social Psychology*, 45(3):501–512.

Fredricks, G. A. and Nelsen, R. B. (2007). On the relationship between Spearman's rho and Kendall's tau for pairs of continuous random variables. *Journal of Statistical Planning and Inference*, 137(7):2143–2150.

Freeland, R. K. (1998). *Statistical Analysis of Discrete Time Series with Application to the Analysis of Workers' Compensation Claims Data*. PhD thesis, University of British Columbia.

Frees, E. W., Carriere, J., and Valdez, E. A. (1996). Annual valuation of dependent mortality. *Journal of Risk and Insurance*, 63(2):229–261.

Frees, E. W. and Valdez, E. A. (1998). Understanding relationships using copulas. *North American Actuarial Journal*, 2(1):1–25.

Fritsch, F. N. and Carlson, R. E. (1980). Monotone piecewise cubic interpolation. *SIAM Journal of Numerical Analysis*, 17(2):238–246.

Gabbi, G. (2005). Semi-correlations as a tool of geographical and sector asset allocation. *European Journal of Finance*, 11(3):271–281.

Gabow, H. N. (1977). Two algorithms for generating weighted spanning trees in order. *SIAM Journal on Computing*, 6(1):139–150.

Gaenssler, P. and Stute, W. (1987). *Seminar on Empirical Processes. DMV Seminar*, volume 9. Birkhäuser, Basel.

Galambos, J. (1975). Order statistics of samples from multivariate distributions. *Journal of the American Statistical Association*, 70(351):674–680.

Galambos, J. (1987). *The Asymptotic Theory of Extreme Order Statistics*. Kreiger Publishing Co., Malabar, FL, second edition.

Gao, X. and Song, P. X. K. (2010). Composite likelihood Bayesian information criteria for model selection in high-dimensional data. *Journal of the American Statistical Association*, 105(492):1531–1540.

Garralda Guillem, A. I. (2000). Structure de dépendance des lois de valeurs extrêmes bivariées. *Comptes Rendus de l'Académie des Sciences, Probabilités*, 330:593–596.

Genest, C. (1987). Frank's family of bivariate distributions. *Biometrika*, 74(3):549–555.

Genest, C., Ghoudi, K., and Rivest, L.-P. (1995a). A semiparametric estimation procedure of dependence parameters in multivariate families of distributions. *Biometrika*, 82(3):543–552.

Genest, C., Ghoudi, K., and Rivest, L.-P. (1998). Discussion of understanding relationships using copulas. *North American Actuarial Journal*, 2:143–149.

Genest, C. and MacKay, R. J. (1986). Copules archimédiennes et familles de lois bidimensionnelles dont les marges sont données. *Canadian Journal of Statistics*, 14(2):145–159.

Genest, C., Masiello, E., and Tribouley, K. (2009a). Estimating copula densities through wavelets. *Insurance: Mathematics and Economics*, 44(2):170–181.

Genest, C., Molina, J. J. Q., and Lallena, J. A. R. (1995b). De l'impossibilité de construire des lois à marges multidimensionnelles données à partir de copules. *Comptes Rendus de l'Académie des Sciences, Probabilités*, 320(6):723–726.

Genest, C. and Nešlehová, J. (2007). A primer on copulas for count data. *Astin Bulletin*, 37(2):475–515.

Genest, C., Nešlehová, J., and Quessy, J.-F. (2012). Tests of symmetry for bivariate copula. *Annals of the Institute of Statistical Mathematics*, 64(4):811–834.

Genest, C., Rémillard, B., and Beaudoin, D. (2009b). Goodness-of-fit tests for copulas: a review and a power study. *Insurance: Mathematics and Economics*, 44(2):199–213.

Genest, C. and Rivest, L.-P. (1989). A characterization of Gumbel's family of extreme value distributions. *Statistics & Probability Letters*, 8(3):207–211.

Genest, C. and Rivest, L.-P. (1993). Statistical inference procedures for bivariate Archimedean copulas. *Journal of the American Statistical Association*, 88(423):1034–1043.

Genest, C. and Rivest, L.-P. (2001). On the multivariate probability integral transformation. *Statistics & Probability Letters*, 53(4):391–399.

Genest, C. and Verret, F. (2002). The TP_2 ordering of Kimeldorf and Sampson has the normal-agreeing property. *Statistics & Probability Letters*, 57(4):387–391.

Genton, M. G., Ma, Y., and Sang, H. (2011). On the likelihood function of Gaussian max-stable processes. *Biometrika*, 98(2):481–488.

Genz, A. and Bretz, F. (2009). *Computation of Multivariate Normal and t Probabilities*, volume 195 of *Lecture Notes in Statistics*. Springer, Heidelberg.

Gibbons, R. and Hedeker, D. (1992). Full-information item bi-factor analysis. *Psychometrika*, 57(3):423–436.

Gijbels, I. and Mielniczuk, J. (1990). Estimation the density of a copula function. *Communications in Statistics — Theory and Methods*, 19(2):445–464.

Gijbels, I., Veraverbeke, N., and Omelka, M. (2011). Conditional copulas, association measures and their applications. *Computational Statistics & Data Analysis*, 55(5):1919–1932.

Goldberg, D. E. and Deb, K. (1991). A comparative analysis of selection schemes used in genetic algorithms. In *Foundations of Genetic Algorithms*, pages 69–93, San Francisco. Morgan Kaufmann.

Goldie, C. M. and Resnick, S. (1988). Distributions that are both subexponential and in the domain of attraction of an extreme-value distribution. *Advances in Applied Probability*, 20(4):706–718.

Greene, W. (2008). Functional forms for the negative binomial model for count data. *Economics Letters*, 99(3):585–590.

Grimaldi, S. and Serinaldi, F. (2006). Asymmetric copula in multivariate flood frequency analysis. *Advances in Water Resources*, 29(8):1155–1167.

Gumbel, E. J. (1958). Distributions à plusieures variables dont les marges sont donnée. *Comptes Rendus de l'Académie des Sciences, Paris*, 246(19):2717–2719.

Gumbel, E. J. (1960a). Bivariate exponential distributions. *Journal of the American Statistical Association*, 55(292):698–707.

Gumbel, E. J. (1960b). Distributions des valeurs extrêmes en plusieurs dimensions. *Publications de l'Institut de Statistique de l'Université de Paris*, 9:171–173.

Gumbel, E. J. (1961). Bivariate logistic distributions. *Journal of the American Statistical Association*, 56(294):335–349.

Hafner, C. M. and Manner, H. (2012). Dynamic stochastic copula models: estimation, inference and applications. *Journal of Applied Econometrics*, 27(2):269–295.

Hanea, A. M., Kurowicka, D., Cooke, R. M., and Ababei, D. A. (2010). Mining and visualising ordinal data with non-parametric continuous BBNs. *Computational Statistics and Data Analysis*, 54(3):668–687.

Hansen, B. (1994). Autoregressive conditional density estimation. *International Economic Review*, 35(3):705–730.

Hardy, G. H., Littlewood, J. E., and Pólya, G. (1952). *Inequalities*. Cambridge University Press, second edition.

Harville, D. A. (1997). *Matrix Algebra From a Statistician's Perspective*. Springer, New York.

Hashorva, E. (2006). On the regular variation of elliptical random vectors. *Statistics & Probability Letters*, 76(14):1427–1434.

Hashorva, E. (2007). Asymptotic properties of type I elliptical random vectors. *Extremes*, 10(4):175–206.

Hashorva, E. (2008). Tail asymptotic results for elliptical distributions. *Insurance: Mathematics and Economics*, 43(1):158–164.

Hashorva, E. (2010). On the residual dependence index of elliptical distributions. *Statistics & Probability Letters*, 80(13-14):1070–1078.

Heffernan, J. E. (2000). A directory of coefficients of tail dependence. *Extremes*, 3(3):279–290 (2001).

Hering, C., Hofert, M., Mai, J.-F., and Scherer, M. (2010). Constructing hierarchical Archimedean copulas with Levy subordinators. *Journal of Multivariate Analysis*, 101(6):1428–1433.

Hobæk Haff, I. (2012). Comparison of estimators for pair-copula constructions. *Journal of Multivariate Analysis*, 110:91–105.

Hobæk Haff, I. (2013). Parameter estimation for pair-copula constructions. *Bernoulli*, 19(2):462–491.

Hobæk Haff, I., Aas, K., and Frigessi, A. (2010). On the simplified pair-copula construction — simply useful or too simplistic? *Journal of Multivariate Analysis*, 101(5):1296–1310.

Hoeffding, W. (1940). Maßstabinvariante Korrelationstheorie. *Schriften des Mathematischen Instituts und des Instituts für Angewandte Mathematik der Universität Berlin*, 5:181–233.

Hoeffding, W. (1948). A class of statistics with asymptotically normal distribution. *Annals of Mathematical Statistics*, 19(3):293–325.

Hofert, M. (2008). Sampling Archimedean copulas. *Computational Statistics & Data Analysis*, 52(12):5163–5174.

Hofert, M. (2011). Efficiently sampling nested Archimedean copulas. *Computational Statistics & data Analysis*, 55(1):57–70.

Hofert, M. and Maechler, M. (2011). Nested Archimedean copulas meet R: The nacopula package. *Journal of Statistical Software*, 39(9):1–20.

Hofert, M. and Pham, D. (2013). Densities of nested Archimedean copulas. *Journal of Multivariate Analysis*, 118:37–52.

Holzinger, K. J. and Swineford, F. (1937). The bi-factor method. *Psychometrika*, 2(1):41–54.

Hong, Y., Tu, J., and Zhou, G. (2007). Asymmetries in stock returns: statistical tests and economic evaluation. *The Review of Financial Studies*, 20(5):1547–1581.

Hougaard, P. (1986). Survival models for heterogeneous populations derived from stable distributions. *Biometrika*, 73(2):387–396.

Hougaard, P. (2000). *Analysis of Multivariate Survival Data*. Springer-Verlag, New York.

Hua, L. and Joe, H. (2011). Tail order and intermediate tail dependence of multivariate copulas. *Journal of Multivariate Analysis*, 102(10):1454–1471.

Hua, L. and Joe, H. (2012a). Tail comonotonicity and conservative risk measures. *ASTIN Bulletin*, 42(2):602–629.

Hua, L. and Joe, H. (2012b). Tail comonotonicity: properties, constructions, and asymptotic additivity of risk measures. *Insurance: Mathematics and Economics*, 51(2):492–503.

Hua, L. and Joe, H. (2013). Intermediate tail dependence: a review and some new results. In Li, H. and Li, X., editors, *Stochastic Orders in Reliability and Risk*, volume 208 of *Lecture Notes in Statistics*, pages 291–311, New York. Springer.

Hua, L. and Joe, H. (2014). Strength of tail dependence based on conditional tail expectation. *Journal of Multivariate Analysis*, 123:143–159.

Huang, W. and Prokhorov, A. (2014). A goodness-of-fit test for copulas. *Econometric Reviews*, 33(7):751–771.

Hull, J. and White, A. (2004). Valuation of a CDO and an nth to default CDS without Monte Carlo simulation. *Journal of Derivatives*, 12(2):8–23.

Hult, H. and Lindskog, F. (2002). Multivariate extremes, aggregation and dependence in elliptical distributions. *Advances in Applied Probability*, 34(3):587–608.

Hürlimann, W. (2003). Hutchinson-Lai's conjecture for bivariate extreme value copulas. *Statistics & Probability Letters*, 61(2):191–198.

Hüsler, J. and Reiss, R.-D. (1989). Maxima of normal random vectors: between independence and complete dependence. *Statistics & Probability Letters*, 7(4):283–286.

Hutchinson, T. P. and Lai, C. D. (1990). *Continuous Bivariate Distributions, Emphasising Applications*. Rumsby, Sydney, Australia.

Ibragimov, R. (2009). Copula-based characterizations for higher-order Markov processes. *Econometric Theory*, 25(3):819–846.

Imlahi, L., Ezzerg, M., and Chakak, A. (1999). Estimación de la curva mediana de una cópula $c(x_1, \ldots, x_m)$. *Revista de la Real Academia de Ciencias Exactas, Físicas y Naturales. Serie A: Matemáticas*, 93(2):241–250.

Janssen, P., Swanepoel, J., and Veraverbeke, N. (2012). Large sample behavior of the Bernstein copula estimator. *Journal of Statistical Planning and Inference*, 142(5):1189–1197.

Jaworski, P. (2006). On uniform tail expansions of multivariate copulas and wide convergence of measures. *Applicationes Mathematicae (Warsaw)*, 33(2):159–184.

Joe, H. (1990a). Families of min-stable multivariate exponential and multivariate extreme value distributions. *Statistics & Probability Letters*, 9(1):75–81.

Joe, H. (1990b). Multivariate concordance. *Journal of Multivariate Analysis*, 35(1):12–30.

Joe, H. (1993). Parametric families of multivariate distributions with given margins. *Journal of Multivariate Analysis*, 46(2):262–282.

Joe, H. (1994). Multivariate extreme-value distributions with applications to environmental data. *Canadian Journal of Statistics*, 22(1):47–64.

Joe, H. (1996a). Families of m-variate distributions with given margins and $m(m - 1)/2$ bivariate dependence parameters. In Rüschendorf, L., Schweizer, B., and Taylor, M. D., editors, *Distributions with Fixed Marginals and Related Topics*, volume 28, pages 120–141, Hayward, CA. Institute of Mathematical Statistics.

Joe, H. (1996b). Time series models with univariate margins in the convolution-closed infinitely divisible class. *Journal of Applied Probability*, 33(3):664–677.

Joe, H. (1997). *Multivariate Models and Dependence Concepts*. Chapman & Hall, London.

Joe, H. (2005). Asymptotic efficiency of the two-stage estimation method for copula-based models. *Journal of Multivariate Analysis*, 94(2):401–419.

Joe, H. (2006). Range of correlation matrices for dependent random variables with given marginal distributions. In Balakrishnan, N., Castillo, E., and Sarabia, J. M., editors, *Advances in Distribution Theory, Order Statistics and Inference, in honor of Barry Arnold*, pages 125–142. Birkhäuser, Boston.

Joe, H. (2011a). Dependence comparisons of vine copulae in four or more variables. In Kurowicka, D. and Joe, H., editors, *Dependence Modeling: Vine Copula Handbook*, chapter 7, pages 139–164. World Scientific, Singapore.

Joe, H. (2011b). Tail dependence in vine copulae. In Kurowicka, D. and Joe, H., editors, *Dependence Modeling: Vine Copula Handbook*, chapter 8, pages 165–187. World Scientific, Singapore.

Joe, H., Cooke, R. M., and Kurowicka, D. (2011). Regular vines: generation algorithm and number of equivalence classes. In Kurowicka, D. and Joe, H., editors, *Dependence Modeling: Vine Copula Handbook*, chapter 10, pages 219–231. World Scientific, Singapore.

Joe, H. and Hu, T. (1996). Multivariate distributions from mixtures of max-infinitely divisible distributions. *Journal of Multivariate Analysis*, 57(2):240–265.

Joe, H., Li, H., and Nikoloulopoulos, A. K. (2010). Tail dependence functions and vine copulas. *Journal of Multivariate Analysis*, 101(1):252–270.

Joe, H. and Ma, C. (2000). Multivariate survival functions with a min-stable property. *Journal of Multivariate Analysis*, 75(1):13–35.

Joe, H., Smith, R. L., and Weissman, I. (1992). Bivariate threshold methods for extremes. *Journal of the Royal Statistical Society B*, 54(1):171–183.

Joe, H. and Xu, J. J. (1996). The estimation method of inference functions for margins for multivariate models. *UBC, Department of Statistics, Technical Report*, 166.

Joe, H. and Zhu, R. (2005). Generalized Poisson distribution: the property of mixture of Poisson and comparison with negative binomial distribution. *Biometrical Journal*, 47(2):219–229.

Johnson, M. E. (1987). *Multivariate Statistical Simulation*. Wiley, New York.

Johnson, N. L., Kemp, A. W., and Kotz, S. (2005). *Univariate Discrete Distributions*. Wiley, New York, third edition.

Johnson, N. L. and Kotz, S. (1972). *Continuous Multivariate Distributions*. Wiley, New York.

Johnson, N. L. and Kotz, S. (1975). On some generalized Farlie-Gumbel-Morgenstern distributions. *Communications in Statistics — Theory and Methods*, 4(5):415–427.

Jondeau, E., Poon, S.-H., and Rockinger, M. (2007). *Financial Modeling under non-Gaussian Distributions*. Springer, London.

Jondeau, E. and Rockinger, M. (2006). The copula-GARCH model of conditional dependencies: an international stock market application. *Journal of International Money and Finance*, 25(5):827–853.

Jones, M. C. and Faddy, M. J. (2003). A skew extension of the t distribution, with applications. *Journal of the Royal Statistical Society B*, 65(1):159–174.

Jones, M. C. and Pewsey, A. (2009). Sinh-arcsinh distributions. *Biometrika*, 96(4):761–780.

Juri, A. and Wüthrich, M. V. (2002). Copula convergence theorems for tail events. *Insurance: Mathematics and Economics*, 30(3):405–420.

Kahaner, D., Moler, C. B., and Nash, S. (1989). *Numerical Methods and Software*. Prentice Hall, Englewood Cliffs, NJ.

Kanter, M. (1975). Stable densities under change of scale and total variation inequalities. *Annals of Probability*, 3(4):697–707.

Karlheinz, R. and Melich, A. (1992). Euro-barometer 38.1: Consumer protection and perceptions of science and technology. *Inter-university Consortium for Political and Social Research*, ICPSR 6045.

Karlin, S. (1968). *Total Positivity*, volume I. Stanford University Press, Stanford, CA.

Karlis, D. and Pedeli, X. (2013). Flexible bivariate INAR(1) processes using copulas. *Communications in Statistics — Theory and Methods*, 42(4):723–740.

Kauermann, G., Schellhase, C., and Ruppert, D. (2013). Flexible copula density estimation with penalized hierarchical B-splines. *Scandinavian Journal of Statistics*, 40(4):685–705.

Kazianka, H. and Pilz, J. (2010). Copula-based geostatistical modeling of continuous and discrete data including covariates. *Stochastic Environmental Research and Risk Assessment*, 24(5):661–673.

Kelker, D. (1970). Distribution theory of spherical distributions and a location-scale parameter generalization. *Sankhyā A*, 32:419–430.

Kellerer, H. G. (1964). Verteilungsfunktionen mit gegebenen Marginalverteilungen. *Zeitschrift für Wahrscheinlichkeitstheorie verwandte Gebiete*, 3(3):247–270.

Kemp, A. W. (1981). Efficient generation of logarithmically distributed pseudo-random variables. *Applied Statistics*, 30(3):249–253.

Kendall, M. G. (1938). A new measure of rank correlation. *Biometrika*, 30(1-2):81–93.

Kendall, M. G. (1942). Partial rank correlation. *Biometrika*, 32(3-4):277–283.

Kendall, M. G. (1948). *Rank Correlation Methods*. Griffin, London.

Kendall, M. G. and Gibbons, J. D. (1990). *Rank Correlation Methods*. Edward Arnold, London, fifth edition.

Kepner, J. L., Harper, J. D., and Keith, S. Z. (1989). A note on evaluating a certain orthant probability. *American Statistician*, 43(1):48–49.

Kimberling, C. H. (1974). A probabilistic interpretation of complete monotonicity. *Aequationes Mathematicae*, 10(2):152–164.

Kimeldorf, G. and Sampson, A. R. (1975). Uniform representations of bivariate distributions. *Communications in Statistics — Theory and Methods*, 4(7):617–627.

Kimeldorf, G. and Sampson, A. R. (1978). Monotone dependence. *Annals of Statistics*, 6(4):895–903.

Kimeldorf, G. and Sampson, A. R. (1989). A framework for positive dependence. *Annals of the Institute of Statistical Mathematics*, 41(1):31–45.

Klement, E. P. and Mesiar, R. (2006). How non-symmetric can a copula be? *Commentationes Mathematicae Universitatis Carolinae*, 47:141–148.

Klugman, S. A., Panjer, H. H., and Willmot, G. E. (2010). *Loss Models: From Data to Decisions*. Wiley, New York, third edition.

Klugman, S. A. and Parsa, R. (1999). Fitting bivariate loss distributions with copulas. *Insurance: Mathematics and Economics*, 24(1-2):139–148.

Klüppelberg, C. and Kuhn, G. (2009). Copula structure analysis. *Journal of the Royal Statistical Society B*, 71(3):737–753.

Klüppelberg, C., Kuhn, G., and Peng, L. (2008). Semi-parametric models for the multivariate tail dependence function — the asymptotically dependent case. *Scandinavian Journal of Statistics*, 35(4):701–718.

Knight, W. R. (1966). A computer method for calculating Kendall's tau with ungrouped data. *Journal of the American Statistical Association*, 61(314):436–439.

Kojadinovic, I. and Yan, J. (2010). Comparison of three semiparametric methods for estimating dependence parameters in copula models. *Insurance Mathematics & Economics*, 47(1):52–63.

Kojadinovic, I. and Yan, J. (2011). A goodness-of-fit test for multivariate multiparameter copulas based on multiplier central limit theorems. *Statistics and Computing*, 21(1):17–30.

Kojadinovic, I., Yan, J., and Holmes, M. (2011). Fast large-sample goodness-of-fit tests for copulas. *Statistica Sinica*, 21(2):841–871.

Kokonendji, C. C., Dossou-Gbété, S., and Demétrio, C. G. B. (2004). Some discrete exponential dispersion models: Poisson-Tweedie and Hinde-Demétrio classes. *Statistics and Operations Research Transactions – SORT*, 28:201–214.

Krupskii, P. (2014). *Structured Factor Copulas and Tail Inference*. PhD thesis, University of British Columbia.

Krupskii, P. and Joe, H. (2013). Factor copula models for multivariate data. *Journal of Multivariate Analysis*, 120:85–101.

Krupskii, P. and Joe, H. (2014). Structured factor copula models: theory, inference and computation. *Submitted*.

Kurowicka, D. (2011). Optimal truncation of vines. In Kurowicka, D. and Joe, H., editors, *Dependence Modeling: Vine Copula Handbook*, chapter 11, pages 233–247. World Scientific, Singapore.

Kurowicka, D. and Cooke, R. (2006). *Uncertainty Analysis with High Dimensional Dependence Modelling*. Wiley, Chichester.

Kurowicka, D. and Joe, H. (2011). *Dependence Modeling: Vine Copula Handbook*. World Scientific, Singapore.

Larsson, M. and Neslehová, J. (2011). Extremal behavior of Archimedean copulas. *Advances in Applied Probability*, 43(1):195–216.

Lawless, J. F. (1987). Negative binomial and mixed Poisson regression. *Canadian Journal of Statistics*, 15(2):209–225.

Ledford, A. W. and Tawn, J. A. (1996). Statistics for near independence in multivariate extreme values. *Biometrika*, 83(1):169–187.

Lévy, P. (1954). *Théorie de l'Addition des Variables Aléatoires*. Gauthier-Villars, Paris, second edition.

Lewandowski, D., Kurowicka, D., and Joe, H. (2009). Generating random correlation matrices based on vines and extended onion method. *Journal of Multivariate Analysis*, 100(9):1989–2001.

Li, D. (2001). On default correlation: A copula function approach. *Journal of Fixed Income*, 9(4):43–54.

Li, H. (2008a). Duality of the multivariate distributions of Marshall-Olkin type and tail dependence. *Communications in Statistics — Theory and Methods*, 37(11):1721–1733.

Li, H. (2008b). Tail dependence comparison of survival Marshall-Olkin copulas. *Methodology and Computing in Applied Probability*, 10(1):39–54.

Li, H. (2013). Toward a copula theory for multivariate regular variation. In Jaworski, P., Durante, F., and Härdle, W. K., editors, *Copulae in Mathematical and Quantitative Finance*, pages 177–199. Springer-Verlag, Berlin.

Li, H. and Sun, Y. (2009). Tail dependence for heavy-tailed scale mixtures of multivariate distributions. *Journal of Applied Probability*, 46(4):925–937.

Li, H. and Wu, P. (2013). Extremal dependence of copulas: A tail density approach. *Journal of Multivariate Analysis*, 114(1):99–111.

Lin, G. W. (1998). On the Mittag-Leffler distributions. *Journal of Statistical Planning and Inference*, 74(1):1–9.

Lindsay, B. G. (1988). Composite likelihood methods. *Contemporary Mathematics*, 80:221–239.

Lindskog, F., McNeil, A., and Schmock, U. (2003). Kendall's tau for elliptical distributions. In Bol, G., Nakhaeizadeh, G., Rachev, S. T., Ridder, T., and Vollmer, K.-H., editors, *Credit Risk: Measurement, Evaluation and Management*, pages 149–156. Physica-Verlag, Heidelberg.

Liu, Y. and Luger, R. (2009). Efficient estimation of copula-GARCH models. *Computational Statistics & Data Analysis*, 53(6):2284–2297.

Longin, F. and Solnik, B. (2001). Extreme correlations in international equity markets. *Journal of Finance*, 56(2):649–676.

Luo, X. and Shevchenko, P. V. (2010). The t copula with multiple parameters of degrees of freedom: bivariate characteristics and application to risk management. *Quantitative Finance*, 10(9):1039–1054.

Mai, J.-F. and Scherer, M. (2010). The Pickands representation of survival Marshall-Olkin copulas. *Statistics & Probability Letters*, 80(5-6):357–360.

Mai, J.-F. and Scherer, M. (2012a). H-extendible copulas. *Journal of Multivariate Analysis*, 110:151–160.

Mai, J.-F. and Scherer, M. (2012b). *Simulating Copulas: Stochastic Models, Sampling Algorithms, and Applications*. World Scientific, Singapore.

Malevergne, Y. and Sornette, D. (2006). *Extreme Financial Risks From Dependence to Risk Management*. Springer, Berlin.

Malov, S. V. (2001). On finite-dimensional Archimedean copulas. In Balakrishnan, N., Ibragimov, I. A., and Nevzorov, V. B., editors, *Asymptotic Methods in Probability and Statistics with Applications*, pages 19–35. Birkhäuser, Boston.

Manner, H. and Reznikova, O. (2012). A survey on time-varying copulas: specification, simulations, and application. *Econometric Reviews*, 31(6):654–687.

Marco, J. M. and Ruiz-Rivas, C. (1992). On the construction of multivariate distributions with given nonoverlapping multivariate marginals. *Statistics & Probability Letters*, 15(4):259–265.

Mardia, K. V. (1962). Multivariate Pareto distributions. *Annals of Mathematical Statistics*, 33(3):1008–1015.

Mardia, K. V. (1967). Some contributions to contingency-type bivariate distributions. *Biometrika*, 54(1):235–249.

Mardia, K. V. (1970). *Families of Bivariate Distributions*. Griffin, London.

Marshall, A. W. and Olkin, I. (1967a). A generalized bivariate exponential distribution. *Journal of Applied Probability*, 4(2):291–302.

Marshall, A. W. and Olkin, I. (1967b). A multivariate exponential distribution. *Journal of the American Statistical Association*, 62(317):30–44.

Marshall, A. W. and Olkin, I. (1988). Families of multivariate distributions. *Journal of the American Statistical Association*, 83(403):834–841.

Marshall, A. W. and Olkin, I. (1990). Multivariate distributions generated from mixtures of convolutions and product families. In Block, H. W., Sampson, A. R., and Savits, T. H., editors, *Topics in Statistical Dependence*, volume 18, pages 371–393, Hayward, CA. Institute of Mathematical Statistics.

Marshall, A. W. and Olkin, I. (2007). *Life Distributions, Structure of Nonparametric, Semiparametric, and Parametric Families*. Springer Series in Statistics. Springer, New York.

Marshall, A. W., Olkin, I., and Arnold, B. C. (2011). *Inequalities: Theory of Majorization and Its Applications*. Springer, New York, second edition.

Maydeu-Olivares, A. and Joe, H. (2006). Limited information goodness-of-fit testing in multidimensional contingency tables. *Psychometrika*, 71(4):713–732.

McArdle, J. J. and McDonald, R. P. (1984). Some algebraic properties of the reticular action model. *British Journal of Mathematical and Statistical Psychology*, 37:234–251.

McFadden, D. (1974). Conditional logit analysis of qualitative choice behavior. In Zarembka, P., editor, *Frontiers in Econometrics*, pages 105–142. Academic Press, New York.

McKenzie, E. (1988). Some ARMA models for dependent sequences of Poisson counts. *Advances in Applied Probability*, 20(4):822–835.

McNeil, A. J. (2008). Sampling nested Archimedean copulas. *Journal of Statistical Computation and Simulation*, 78(6):567–581.

McNeil, A. J., Frey, R., and Embrechts, P. (2005). *Quantitative Risk Management*. Princeton Series in Finance. Princeton University Press, Princeton, NJ.

McNeil, A. J. and Nešlehová, J. (2009). Multivariate Archimedean copulas, d-monotone functions and l_1-norm symmetric distributions. *Annals of Statistics*, 37(5B):3059–3097.

McNeil, A. J. and Nešlehová, J. (2010). From Archimedean to Liouville copulas. *Journal of Multivariate Analysis*, 101(8):1772–1790.

Meester, S. and MacKay, J. (1994). A parametric model for cluster correlated categorical data. *Biometrics*, 50(4):954–963.

Mendes, B. V. M. and Kolev, N. (2008). How long memory in volatility affects true dependence structure. *International Review of Financial Analysis*, 17(5):1070–1086.

Mesfioui, M. and Quessy, J.-F. (2008). Dependence structure of conditional Archimedean copulas. *Journal of Multivariate Analysis*, 99(3):372–385.

Michiels, F. and De Schepper, A. (2012). How to improve the fit of Archimedean copulas by means of transforms. *Statistical Papers*, 53(2):345–355.

Min, A. and Czado, C. (2010). Bayesian inference for multivariate copulas using pair-copula constructions. *Journal of Financial Econometrics*, 8(4):511–546.

Min, A. and Czado, C. (2011). Bayesian model selection for D-vine pair-copula constructions. *Canadian Journal of Statistics*, 39(2):239–258.

Molenberghs, G. and Lesaffre, E. (1994). Marginal modeling of correlated ordinal data using a multivariate Plackett distribution. *Journal of the American Statistical Association*, 89(426):633–644.

Morales-Nápoles, O. (2011). Counting vines. In Kurowicka, D. and Joe, H., editors, *Dependence Modeling: Vine Copula Handbook*, chapter 9, pages 189–218. World Scientific, Singapore.

Morales-Nápoles, O., Cooke, R. M., and Kurowicka, D. (2009). The number of vines and regular vines in *n* nodes. *Technical report*.

Morales-Nápoles, O., Kurowicka, D., and Roelen, A. (2008). Eliciting conditional and unconditional rank correlations from conditional probabilities. *Reliability Engineering and System Safety*, 93(5):699–710.

Morettin, P. A., Toloi, C. M. C., Chiann, C., and de Miranda, J. C. (2010). Wavelet-smoothed empirical copula estimators. *Brazilian Review of Finance*, 8(3):263–281.

Morgenstern, D. (1956). Einfache Beispiele zweidimensionaler Verteilungen. *Mitteilungsblatt für Mathematische Statistik*, 8:234–235.

Mulaik, S. A. (2009). *Linear Causal Modeling with Structural Equations*. Chapman & Hall/CRC, Boca Raton.

Müller, A. and Scarsini, M. (2005). Archimedean copulae and positive dependence. *Journal of Multivariate Analysis*, 93(2):434–445.

Nadaraya, E. A. (1964). On estimating regression. *Theory of Probability and its Applications*, 9(1):141–142.

Nash, J. C. (1990). *Compact Numerical Methods for Computers: Linear Algebra and Function Minimisation*. Hilger, New York, second edition.

Naveau, P., Guillou, A., Cooley, D., and Diebolt, J. (2009). Modelling pairwise dependence of maxima in space. *Biometrika*, 96(1):1–17.

Nelsen, R. B. (1986). Properties of a one-parameter family of bivariate distributions with specified marginals. *Communications in Statistics — Theory and Methods*, 15(11):3277–3285.

Nelsen, R. B. (1999). *An Introduction to Copulas*. Lecture Notes in Statistics. Springer, New York.

Nelsen, R. B. (2006). *An Introduction to Copulas*. Springer, New York, second edition.

Nelsen, R. B. (2007). Extremes of nonexchangeability. *Statistical Papers*, 48(2):329–336.

Nelsen, R. B., Quesada-Molina, J. J., Rodriguez-Lallena, J. A., and Ubeda-Flores, M. (2001). Distribution functions of copulas: a class of bivariate probability integral transforms. *Statistics & Probability Letters*, 54(3):277–282.

Nelsen, R. B., Quesada-Molina, J. J., Rodriguez-Lallena, J. A., and Ubeda-Flores, M. (2003). Kendall distribution functions. *Statistics & Probability Letters*, 65(3):263–268.

Nikoloulopoulos, A. K. and Joe, H. (2014). Factor copula models for item response data. *Psychometrika*, 79. in press.

Nikoloulopoulos, A. K., Joe, H., and Chaganty, N. R. (2011). Weighted scores method for regression models with dependent data. *Biostatistics*, 12(4):653–665.

Nikoloulopoulos, A. K., Joe, H., and Li, H. (2009). Extreme value properties of multivariate t copulas. *Extremes*, 12(2):129–148.

Nikoloulopoulos, A. K., Joe, H., and Li, H. (2012). Vine copulas with asymmetric tail dependence and applications to financial return data. *Computational Statistics & Data Analysis*, 56(11):3659–3673.

Nikoloulopoulos, A. K. and Karlis, D. (2008a). Copula model evaluation based on parametric bootstrap. *Computational Statistics & Data Analysis*, 52(7):3342–3353.

Nikoloulopoulos, A. K. and Karlis, D. (2008b). Multivariate logit copula model with an application to dental data. *Statistics in Medicine*, 27(30):6393–6406.

Nikoloulopoulos, A. K. and Karlis, D. (2008c). On modeling count data: a comparison of some well-known discrete distributions. *Journal of Statistical Computation and Simulation*, 78(5):437–457.

Oakes, D. (1982). A model for association for bivariate survival data. *Journal of the Royal Statistical Society B*, 44(3):414–422.

Oakes, D. (1989). Bivariate survival models induced by frailties. *Journal of the American Statistical Association*, 84(406):487–493.

Oh, D. H. and Patton, A. J. (2012). Modeling dependence in high dimensions with factor copulas. *Submitted*.

Okhrin, O., Okhrin, Y., and Schmid, W. (2013). On the structure and estimation of hierarchical Archimedean copulas. *Journal of Econometrics*, 173(2):189–204.

Okimoto, T. (2008). New evidence of asymmetric dependence structures in international equity markets. *Journal of Financial and Quantitative Analysis*, 43(3):787–815.

Olsson, F. (1979). Maximum likelihood estimation of the polychoric correlation coefficient. *Psychometrika*, 44(4):443–460.

Omelka, M., Gijbels, I., and Veraverbeke, N. (2009). Improved kernel estimation of copulas: weak convergence and goodness-of-fit testing. *Annals of Statistics*, 37(5B):3023–3058.

Padoan, S. A. (2011). Multivariate extreme models based on underlying skew-t and skew-normal distributions. *Journal of Multivariate Analysis*, 102(5):977–991.

Padoan, S. A. (2013). Extreme dependence models based on event magnitude. *Journal of Multivariate Analysis*, 122:1–19.

Panagiotelis, A., Czado, C., and Joe, H. (2012). Pair copula constructions for multivariate discrete data. *Journal of the American Statistical Association*, 107(499):1063–1072.

Papadimitriou, C. H. and Steiglitz, K. (1982). *Combinatorial Optimization: Algorithms and Complexity*. Prentice-Hall, Englewood Cliffs, NJ.

Pascual, L., Romo, J., and Ruiz, E. (2006). Bootstrap prediction for returns and volatilities in GARCH models. *Computational Statistics & Data Analysis*, 50(9):2293–2312.

Patton, A. J. (2006). Modelling asymmetric exchange rate dependence. *International Economic Review*, 47(2):527–556.

Patton, A. J. (2009). Copula-based models for financial time series. In Andersen, T. G., Davis, R. A., Kreiß, J.-P., and Mikosch, T., editors, *Handbook of Financial Time Series*, pages 767–785. Springer, Berlin.

Patton, A. J. (2012). A review of copula models for economic time series. *Journal of Multivariate Analysis*, 110:4–18.

Pickands, J. (1975). Statistical inference using extreme order statistics. *Annals of Statistics*, 3(1):119–131.

Pickands, J. (1981). Multivariate extreme value distributions. In *Proceedings 43rd Session International Statistical Institute*, pages 859–878.

Pillai, R. N. (1990). On Mittag-Leffler functions and related distributions. *Annals of the Institute of Statistical Mathematics*, 42(1):157–161.

Plackett, R. L. (1965). A class of bivariate distributions. *Journal of the American Statistical Association*, 60(310):516–522.

Politis, D. N. and Romano, J. P. (1994). The stationary bootstrap. *Journal of the American Statistical Association*, 89(428):1303–1313.

Prim, R. C. (1957). Shortest connection networks and some generalizations. *Bell System Technical Journal*, 36(6):1389–1401.

Qu, L., Qian, Y., and Xie, H. (2009). Copula density estimation by total variation penalized likelihood. *Communications in Statistics — Simulation and Computation*, 38(9):1891–1908.

Qu, L. and Yin, W. (2012). Copula density estimation by total variation penalized likelihood with linear equality constraints. *Computational Statistics and Data Analysis*, 56(2):384–398.

R-Core-Team (2013). *R: A Language and Environment for Statistical Computing*. R Foundation for Statistical Computing, Vienna, Austria.

Ralston, A. and Rabinowitz, P. (1978). *A First Course in Numerical Analysis*. McGraw-Hill, New York, second edition.

Resnick, S. I. (1987). *Extreme Values, Regular Variation, and Point Processes*. Springer-Verlag, New York.

Resnick, S. I. (2007). *Heavy-tail Phenomena: Probabilistic and Statistical Modeling*. Springer Series in Operations Research and Financial Engineering. Springer, New York.

Riphahn, R. T., Wambach, A., and Million, A. (2003). Incentive effects in the demand for health care: a bivariate panel count data estimation. *Journal of Applied Econometrics*, 18(4):387–405.

Rizopoulos, D. (2006). ltm: An R package for latent variable modeling and item response theory analyses. *Journal of Statistical Software*, 17(5).

Rodriguez, J. C. (2007). Measuring financial contagion: A copula approach. *Journal of Empirical Finance*, 14(3):401–423.

Rosco, J. F. and Joe, H. (2013). Measures of tail asymmetry for bivariate copulas. *Statistical Papers*, 54(3):709–726.

Rosco, J. F., Jones, M. C., and Pewsey, A. (2011). Skew t distributions via the sinh-arcsinh transformation. *Test*, 20(3):630–652.

Rosenblatt, M. (1952). Remarks on a multivariate transformation. *Annals of Mathematical Statistics*, 23(3):470–472.

Ruiz-Rivas, C. (1981). Un nuevo sistema bivariante y sus propiedades. *Estadíst. Española*, 87:47–54.

Rüschendorf, L. (1976). Asymptotic distributions of multivariate rank order statistics. *Annals of Statistics*, 4(5):912–923.

Rüschendorf, L. (1985). Construction of multivariate distributions with given marginals. *Annals of the Institute of Statistical Mathematics*, 37(2):225–223.

Salmon, F. (2009). The formula that killed Wall Street. *Wired*. Reproduced in *Significance*, February 2012, pp 16-20.

Salvadori, G. and De Michele, C. (2010). Multivariate multiparameter extreme value models and return periods: A copula approach. *Water Resources Research*, 46.

Salvadori, G., De Michele, C., Kottegoda, N. T., and Rosso, R. (2007). *Extremes in Nature*.

An Approach using Copulas. Springer, Dordrecht.

Sancetta, A. and Satchell, S. (2004). The Bernstein copula and its applications to modeling and approximations of multivariate distributions. *Econometric Theory*, 20(3):535–562.

Savu, C. and Trede, M. (2010). Hierarchies of Archimedean copulas. *Quantitative Finance*, 10(3):295–304.

Scarsini, M. (1984). On measures of concordance. *Stochastica*, 8:201–218.

Schepsmeier, U. (2010). Maximum Likelihood Estimation of C-vine Pair-Copula Constructions Based on Bivariate Copulas from Different Families. Master's thesis, Technische Universität München.

Schepsmeier, U. (2014). *Estimating Standard Errors and Efficient Goodness-of-Fit Tests for Regular Vine Copula Models.* PhD thesis, Technische Universität München.

Schepsmeier, U. and Stöber, J. (2014). Derivatives and Fisher information of bivariate copulas. *Statistical Papers*, 55:525–542.

Schepsmeier, U., Stöber, J., Brechmann, E. J., and Graeler, B. (2013). *VineCopula: Statistical inference of vine copulas. R package version 1.2.*

Schervish, M. J. (1984). Multivariate normal probabilities with error bound. *Applied Statistics*, 33(1):81–94.

Schlather, M. (2002). Models for stationary max-stable random fields. *Extremes*, 5(1):33–44.

Schmid, F. and Schmidt, R. (2007a). Multivariate conditional versions of Spearman's rho and related measures of tail dependence. *Journal of Multivariate Analysis*, 98(6):1123–1140.

Schmid, F. and Schmidt, R. (2007b). Nonparametric inference on multivariate versions of Blomqvist's beta and related measures of tail dependence. *Metrika*, 66(3):323–354.

Schmidt, R. (2002). Tail dependence for elliptically contoured distributions. *Mathematical Methods of Operations Research*, 55(2):301–327.

Schoutens, W. (2003). *Lévy Processes in Finance.* Wiley, Chichester.

Schriever, B. F. (1987). An ordering of positive dependence. *Annals of Statistics*, 15(3):1208–1214.

Schweizer, B. and Sklar, A. (1983). *Probabilistic Metric Spaces.* North Holland, New York.

Segers, J. (2012). Asymptotics of empirical copula processes under non-restrictive smoothness assumptions. *Bernoulli*, 18(3):764–782.

Serfling, R. J. (1980). *Approximation Theorems of Mathematical Statistics.* Wiley, New York.

Serinaldi, F. (2008). Analysis of inter-gauge dependence by Kendall's τ_k, upper tail dependence coefficient, and 2-copulas with application to rainfall fields. *Stochastic Environmental Research and Risk Assessment*, 22(6):671–688.

Shah, S. M. and Parikh, N. T. (1964). Moments of singly and doubly truncated standard bivariate normal distribution. *Vidya*, 7(1):81–91.

Shaked, M. (1975). A note on the exchangeable generalized Farlie-Gumbel-Morgenstern distributions. *Communications in Statistics — Theory and Methods*, 4(8):711–721.

Shaked, M. and Shanthikumar, J. G. (2007). *Stochastic Orders.* Springer, New York.

Shao, J. and Tu, D. (1995). *The Jackknife and Bootstrap.* Springer, New York.

Shea, G. A. (1983). Hoeffding's lemma. In Kotz, S., Johnson, N. L., and Read, C. B., editors, *Encyclopedia of Statistical Science*, volume 3, pages 648–649. Wiley, New York.

Shen, X., Zhu, Y., and Song, L. (2008). Linear B-spline copulas with applications to nonparametric estimation of copulas. *Computational Statistics and Data Analysis*,

52(7):3806–3819.

Sibuya, M. (1960). Bivariate extreme statistics. *Annals of the Institute of Statistical Mathematics*, 11(3):195–210.

Sibuya, M. (1979). Generalized hypergeometric digamma and trigamma distributions. *Annals of the Institute of Statistical Mathematics*, 31(3):373–390.

Sklar, A. (1959). Fonctions de répartition à n dimensions et leurs marges. *Publications de l'Institut de Statistique de l'Université de Paris*, 8:229–231.

Smith, E. L. and Stephenson, A. G. (2009). An extended Gaussian max-stable process model for spatial extremes. *Journal of Statistical Planning and Inference*, 139(4):1266–1275.

Smith, M. and Khaled, M. (2012). Estimation of copula models with discrete margins. *Journal of the American Statistical Association*, 107(497):290–303.

Smith, M., Min, A., Almeida, C., and Czado, C. (2010). Modeling longitudinal data using a pair-copula decomposition of serial dependence. *Journal of the American Statistical Association*, 105(492):1467–1479.

Smith, R. L. (1990a). Extreme value theory. In Lederman, W., editor, *Handbook of Applicable Mathematics*, pages 437–472. Wiley, Chichester.

Smith, R. L. (1990b). Max-stable processes and spatial extremes. *Technical Report*.

Spearman, C. (1904). The proof and measurement of association between two things. *American Journal of Psychology*, 15:72–101.

Stephenson, A. (2003). Simulating multivariate extreme value distributions of logistic type. *Extremes*, 6(1):49–59.

Stewart, G. W. (1973). *Introduction to Matrix Computations*. Academic Press, Orlando, FL.

Stöber, J., Hong, H. G., Czado, C., and Ghosh, P. (2013a). Comorbidity of chronic diseases in the elderly: Longitudinal patterns identified by a copula design for mixed responses. *Preprint*.

Stöber, J., Joe, H., and Czado, C. (2013b). Simplified pair copula constructions — limitations and extensions. *Journal of Multivariate Analysis*, 119:101–118.

Stöber, J. and Schepsmeier, U. (2013). Estimating standard errors in regular vine copula models. *Computational Statistics*, 28(6):2679–2707.

Stroud, A. and Secrest, D. (1966). *Gaussian Quadrature Formulas*. Prentice-Hall, Englewood Cliffs, NJ.

Sun, J., Frees, E. W., and Rosenberg, M. A. (2008). Heavy-tailed longitudinal data modeling using copulas. *Insurance Mathematics & Economics*, 42(2):817–830.

Takahasi, K. (1965). Note on the multivariate Burr's distribution. *Annals of the Institute of Statistical Mathematics*, 17(2):257–260.

Tallis, G. M. (1961). The moment generating function of the truncated multi-normal distribution. *Journal of the Royal Statistical Society B*, 23(1):223–229.

Tawn, J. A. (1988). Bivariate extreme value theory: models and estimation. *Biometrika*, 75(3):397–415.

Tawn, J. A. (1990). Modelling multivariate extreme value distributions. *Biometrika*, 77(2):245–253.

Tawn, J. A. (1992). Estimating probabilities of extreme sea-levels. *Journal of the Royal Statistical Society C*, 41(1):77–93.

Taylor, M. D. (2007). Multivariate measures of concordance. *Annals of the Institute of Statistical Mathematics*, 59(4):789–806.

Tchen, A. H. (1980). Inequalities for distributions with given marginals. *Annals of Probability*, 8(4):814–827.

Tiago de Oliveira, J. (1980). Bivariate extremes: foundations and statistics. In Krishnaiah, P. R., editor, *Multivariate Analysis V*, pages 349–366. North-Holland, Amsterdam.

Tong, Y. L. (1980). *Probability Inequalities in Multivariate Distributions*. Academic Press, New York.

Tsukahara, H. (2005). Semiparametric estimation in copula models. *Canadian Journal of Statistics*, 33(3):357–375.

van der Vaart, A. W. and Wellner, J. A. (1996). *Weak Convergence and Empirical Processes: with Applications to Statistics*. Springer, New York.

Varin, C. (2008). On composite marginal likelihoods. *Advances in Statistical Analysis*, 92(1):1–28.

Varin, C., Reid, N., and Firth, D. (2011). An overview of composite likelihood methods. *Statistica Sinica*, 21(1):5–42.

Varin, C. and Vidoni, P. (2005). A note on composite likelihood inference and model selection. *Biometrika*, 92(3):519–528.

Vuong, Q. H. (1989). Likelihood ratio tests for model selection and non-nested hypotheses. *Econometrica*, 57(2):307–333.

Wang, D., Rachev, S. T., and Fabozzi, F. J. (2009). Pricing of credit default index swap tranches with one-factor heavy-tailed copula models. *Journal of Empirical Finance*, 16(2):201–215.

Wang, R., Peng, L., and Yang, J. (2013). Jackknife empirical likelihood for parametric copulas. *Scandinavian Actuarial Journal*, (5):325–339.

Watson, G. S. (1964). Smooth regression analysis. *Sankhyā A*, 26(4):359–372.

Weiss, C. H. (2008). Thinning operations for modeling time series of counts — a survey. *ASTA-Advances in Statistical Analysis*, 92(3):319–341.

Wheaton, B., Muthén, B., Alwin, D. F., and Summers, G. F. (1977). Assessing reliability and stability in panel models. *Sociological Methodology*, 8:84–136.

Whelan, N. (2004). Sampling from Archimedean copulas. *Quantitative Finance*, 4(3):339–352.

White, H. (2004). Maximum likelihood estimation of misspecified models. *Econometrica*, 50(1):1–25.

Whittaker, J. (1990). *Graphical Models in Applied Multivariate Statistics*. Wiley, Chichester.

Williamson, R. E. (1956). Multiply monotone functions and their laplace transforms. *Duke Mathematical Journal*, 23(2):189–207.

Wong, R. (1989). *Asymptotic Approximations of Integrals*. Academic Press, Boston.

Yanagimoto, T. and Okamoto, M. (1969). Partial orderings of permutations and monotonicity of a rank correlation statistic. *Annals of the Institute of Statistical Mathematics*, 21(3):489–506.

Yang, J., Qi, Y., and Wang, R. (2009). A class of multivariate copulas with bivariate Fréchet marginal copulas. *Insurance Mathematics & Economics*, 45(1):139–147.

Yang, X., Frees, E. W., and Zhang, Z. (2011). A generalized beta copula with applications in modeling multivariate long-tailed data. *Insurance: Mathematics and Economics*, 49(2):265–284.

Zhao, Y. and Joe, H. (2005). Composite likelihood estimation in multivariate data analysis. *Canadian Journal of Statistics*, 33(3):335–356.

Zhu, R. and Joe, H. (2006). Modelling count data time series with Markov processes based on binomial thinning. *Journal of Time Series Analysis*, 27(5):725–738.

Zimmer, D. M. and Trivedi, P. K. (2006). Using trivariate copulas to model sample selection and treatment effects: Application to family health care demand. *Journal of Business & Economic Statistics*, 24(1):63–76.

Index